# Linear algebra and geometry

# Linear algebra and geometry

DAVID M. BLOOM
Associate Professor of Mathematics
Brooklyn College of the City University of New York



CAMBRIDGE UNIVERSITY PRESS
CAMBRIDGE
LONDON · NEW YORK · MELBOURNE

Published by the Syndics of the Cambridge University Press
The Pitt Building, Trumpington Street, Cambridge CB2 1RP
Bentley House, 200 Euston Road, London NW1 2DB
32 East 57th Street, New York, NY 10022, USA
296 Beaconsfield Parade, Middle Park, Melbourne 3206, Australia



First published 1979

Printed in the United States of America
Typeset by Offset Composition Services, Inc., Washington, D.C.
Printed and bound by R. R. Donnelley & Sons Company,
Crawfordsville, Ind.

*Library of Congress Cataloging in Publication Data*

Bloom, David M., 1936–

Linear algebra and geometry.

Includes index.

1. Algebras, Linear. 2. Geometry. I. Title.
QA184.B59    512'.5    77-26666

ISBN 0 521 21959 0 hard covers
ISBN 0 521 29324 3 paperback

# Contents

# Preface

In the spring semester of 1966, I taught two undergraduate courses at Brooklyn College: an "advanced elective" course in linear algebra, which I had taught before; and a course entitled "Higher Analytic Geometry," which I had not taught before. I was aware from the beginning of the semester, of course, that connections existed between the subject matter of the first course and that of the second, having first been exposed to such connections in a major way as a student in the beautiful course taught by O. Zariski at Harvard in 1957–58; but I was not fully aware of the extent of these connections. Thus, it came as something of a revelation to me to discover, as the term progressed, that *every* topic which we covered in the first course (literally without exception) was *significantly* applicable to the second course.

At this point I asked myself: if linear algebra and geometry can be so well integrated mathematically, why not integrate them pedagogically? Specifically, instead of two one-term courses, why not teach one full-year course in which the relations between linear algebra and geometry could be explored to their fullest extent? Here, it seemed, might lie an opportunity to illustrate the unity of mathematics and to counteract the prevalent tendency toward compartmentalization of knowledge.

It is in the spirit of the preceding remarks that I decided to write this book. There are other books available whose object is similar (Jaeger's *Introduction to Analytic Geometry and Linear Algebra*, for example), but they are still sufficiently few in number so that I feel no need for an excuse to contribute another one to the collection. To confess the truth, I am really writing this book for myself; every teacher eventually develops his or her own pet approaches to this or that topic, and the urge to self-expression is strong. However, because self-expression requires an audience to be fully satisfying, let me describe the probable audience for this book: students at the advanced undergraduate level who have already had a one-term introduction to abstract ("modern") algebra, and *have understood the latter reasonably well.* The necessary algebraic background is fully summarized in Chapter 1, but this chapter is intended mainly for reference and review; the student who is not already familiar with the concepts of group, ring, field, homomorphism, equivalence relation, and so on, is urged not only to read Chapter 1 carefully before going further, but also to consult one of the good books on abstract algebra (for example, [10] or [11]) for further details.

Some of the material in Chapter 1 will probably be new even to those readers (hopefully the majority of readers) who have already taken a course in abstract algebra. One may, if desired, postpone such material until it is needed. (For example, Theorem 1.3.6 is not needed until Section 4.7, and in that section a specific reference to Theorem 1.3.6 is given so that one will know where to look.) However, it is recommended that the reader spend at least some time with Chapter 1 to refresh his or her memory of abstract concepts, before going on to Chapter 2, which is the "real" beginning of the book.

Mathematically, this book is self-contained with four exceptions. *First exception*: Some of the results in Sections 1.12 (the integers), 1.15 (properties of rational, real, and complex numbers), and 1.16 (properties of polynomials) are stated without proof. (For example, I felt that this was not the place to construct the real number system; thus, the existence of such a system is assumed.) However, references are given in each of the above-mentioned sections to books in which the missing proofs (including constructions) may be found. *Second exception*: Certain concepts and results external to algebra, whose full development would be impractical in this book, are nevertheless needed for the treatment of some specific topics that I wished to include, namely: applications to area and volume (Sections 5.7 and 5.14), a treatment of rigid motions and orientation using the concept of continuity (Sections 6.6 and 6.7), and the use of determinants in analysis (Section 9.8 and Appendix E). These applications of algebra to other fields are ends in themselves; their deletion would result in no harm to the self-contained structure of the book (though the *spirit* of the book would suffer; this sort of thing is what makes mathematics interesting). *Third exception* (perhaps this is a subcase of the second exception): The proof of the main result of Section 7.6 depends on the fact that a real polynomial function of $n$ variables is continuous. *Fourth exception*: Some of the proofs are left as exercises for the reader. Those exercises whose results are needed later are marked by asterisks.

With the exceptions just noted, any result that is used is proved. So as not to intimidate the reader, a few of the most formidable proofs are relegated to appendixes. There is both a pedagogical and a philosophical reason to include proofs of theorems in a textbook. The pedagogical reason is that after working one's way through a certain number of proofs, one begins to recognize patterns, methods, styles of proof and eventually develops skill in constructing proofs oneself. The philosophical reason is that in the presence of proof, one is freed from *dependence on authority*. It is not too good an idea to believe everything you're told, whether the teller be a politician, a newspaper, someone you love – or even a textbook on linear algebra and geometry. We, too, make mistakes; more than one statement made in a textbook (or even in a prestigious professional journal) has turned out to be false. The reader, coming across any given statement in this or another mathematics textbook, will perhaps assume that the statement has a high probability (say, 99.97%) of being true; but if he or she can verify the statement by following a printed proof, step-by-step, the probability increases. (One recalls the old advertisement for the multisection Sunday edition of the *New York Times*: "You don't have to read it all – but it's nice to know it's all there.")

A few words about my approach to geometry. In Chapter 2, all of the geometry is informal; we use it to illustrate and motivate the algebraic concepts of the chapter. Starting with Chapter 3, the approach changes: geometry is made formal, rigorous, algebraic. However, formal definitions are usually preceded by informal discussions for the purpose of motivation, and I have tried to formulate such definitions, whenever possible, so as to parallel the reader's previous geometric experience. For example, to define $\cos(\mathbf{u},\mathbf{v})$ as $(\mathbf{u}\cdot\mathbf{v})/|\mathbf{u}||\mathbf{v}|$ seems unnecessarily artificial when the right-triangle approach can be made rigorous. (Not *all* such artificialities are avoidable; e.g., distance is defined via the square-root formula. We have to start *somewhere*.)

The major geometric areas covered in this book are affine geometry, orthogonal (Euclidean) geometry, isometries, and quadric surfaces; other topics present but receiving less emphasis include area and volume and a very brief introduction to projective geometry (in an affine context). The *n*-dimensional case is considered throughout, both for the sake of generality and as a means of unifying the treatment.

For use in a two-semester undergraduate course, Chapters 2 through 7, excluding Sections 7.6 and 7.7, should suffice for all but the best students. Sections 5.14, 6.6, and 6.7 could be omitted without loss of continuity, but the material in these sections is likely to interest the student and should be covered if time permits. Sections 7.6 and 7.7 and Chapters 8 and 9 are more abstract than what precedes them and require a greater degree of mathematical maturity of the reader.

Even in mathematics there can be different paths to the same goal. Thus, I try now and then to present more than one proof of a theorem, more than one way of solving a problem, more than one way of defining something. Sometimes, these alternate approaches appear in the exercises.

Numerical examples appear throughout. Andrew Gleason, under whom I took two graduate courses, used to tell us, "You're no good if you can't compute." I agree.

Considerable use is made of the "equivalence relation" concept. For instance, parallelism is an equivalence relation among lines; an "angle" is an equivalence class of pairs of vectors (or pairs of rays); an "orientation" in real $n$-space is defined to be an equivalence class of orthonormal $n$-tuples.

Numbering (except in appendixes) is by chapter, section, and item, with the figures numbered separately from other items. Thus, for example, the first two numbered items in Chapter 3, Section 4 (other than figures) are Definition 3.4.1 and Theorem 3.4.2; the first two figures in this same section are Figure 3.4.1 and Figure 3.4.2. "1.11.1(d)" refers to the fourth part of item 1.11.1.

At the end of almost every section, exercises appear. The abbreviation SOL following an exercise means that a full or partial solution to the exercise appears in Appendix G at the end of the book. The abbreviation ANS means that an answer (but not the steps leading to the answer) appears in the appendix; the abbreviation SUG means that a mere hint or suggestion appears. As noted before, an asterisk beside an exercise means that the result of the exercise will be needed later.

The Halmos symbol ▌ is used throughout the book to indicate the end of a proof. The abbreviation R.A.A. (reductio ad absurdum) appears in several proofs and means "contradiction". (I do not understand why the latter abbreviation is so seldom used in mathematical writing; it seems no more abstruse than Q.E.D.)

Of the courses I have taken and the texts from which I have taught, so many have influenced this book that I could not possibly list them all. However, I wish to express special gratitude to Professor Robert Taylor, formerly of Columbia University, who first interested me in algebra; to Oscar Zariski, whose course in geometry was a revelation to me; to Richard Brauer, whose advice and encouragement during my struggle with a doctoral thesis (and subsequently) was invaluable; to the late Moses Richardson of Brooklyn College, who encouraged me to start this book; to Melvin Hausner, whose own book served as a model in some respects; to my students, who have stimulated and challenged me and have provided more than one of the ideas that I have used in the book; and to my family, who patiently did without me for too many days and nights.

September, 1978 David M. Bloom

# 1 Background in abstract algebra

## 1.0 Introduction

This chapter is for the purpose of reference and review. It summarizes those fundamentals of abstract algebra that we will need later, and largely parallels a standard first course in abstract algebra. If you have already taken such a course, you probably should start right in with Chapter 2, though you will find it necessary to refer back to this chapter from time to time. The proofs are all here (except in Sections 1.15 and 1.16, where references to the literature replace proofs); examples are fewer in number than in an abstract algebra text per se, but I have tried to explain concepts clearly. If supplementary reading is desired, either of the references [10] or [11] listed in the References should prove helpful.

## 1.1 Sets, elements, basic notations

You probably have some idea of what is meant by a "set" and by an "element" of a set. For example, we can speak of "the set of all integers from 1 to 10"; the elements of this set are the integers from 1 to 10. Similarly, if we speak of "the set of all Presidents of the United States from 1955 to 1965", this set has three elements, namely Eisenhower, Kennedy, Johnson. These examples do *not* constitute a formal definition; in fact, we cannot really give a formal definition at all. A concept can be defined only in terms of simpler concepts, and there are no concepts logically simpler than those of "set" and "element". Thus we must formally treat the words "set" and "element" as *undefined terms*. However, this should not prevent us from believing that we know what the words mean.

Sets are often denoted by capital letters, and elements of sets by small letters. (There are exceptions.) The set of all integers is commonly denoted **Z**; we shall do so throughout this book. The symbolic expression

$$x \in A$$

means "$x$ is an element of (the set) $A$" (or some grammatical modification thereof; for instance, the sentence, "Let $x \in A$" is of course to be read, "Let $x$ be an element of $A$", not "Let $x$ is an element of $A$"). Phrases like "$x$ belongs to $A$", "$x$ is in $A$", "$A$ contains $x$", etc., mean the same thing as "$x$ is an element of $A$."

When we wish to specify the elements of a set, we enclose them in brackets which look like { }; thus, for example, if $A$ is the set of all integers from 1 to 16, we may write

$$A = \{1, 2, 3, 4, 5, 6, 7, 8, 9, 10, 11, 12, 13, 14, 15, 16\}.$$

If we do not feel like doing so much writing, an alternative is

$$A = \{x \in \mathbf{Z} : 1 \leqslant x \leqslant 16\},$$

which can be read, "$A$ is the set consisting of all elements $x$ of **Z** (that is, all integers $x$) such that $1 \leqslant x \leqslant 16$". (The colon means "such that".) In short, the elements of a set may be specified either by listing them or by describing them in terms of one or more properties. As a further illustration, if $B$ is the set of all *even* integers from 1 to 10, then

$$B = \{2, 4, 6, 8, 10\} = \{x \in \mathbf{Z} : 1 \leqslant x \leqslant 10, x \text{ is even}\}.$$

The symbol $\varnothing$ denotes the *empty set*, a set having no elements. It is important not to confuse $\varnothing$ (which has no elements) with $\{\varnothing\}$ (which has one element, namely $\varnothing$).

Two sets are equal (i.e., the same) if they have the same elements; thus, to prove that the sets $A$ and $B$ are equal it suffices to show that (1) every element of $A$ belongs to $B$, and (2) every element of $B$ belongs to $A$. If (1) is true (whether (2) is true or not), $A$ is called a *subset* of $B$ (*notation*: $A \subseteq B$), and $B$ is called an *overset*, or *superset*, of $A$. The notation $A \subset B$ means that $A$ is a *proper* subset of $B$; that is, $A$ is a subset of $B$ but is not equal to $B$. (In this situation, we naturally call $B$ a *proper overset* of $A$.) For example, $\{0, 2\}$ is a proper subset of $\{0, 1, 2, 3\}$; $\{3, 5, 7, 9\}$ is a proper overset of $\{3, 7\}$; $\{3\} \subset \{3, 5\}$; $\{1\} \subseteq \{1\}$ but not $\{1\} \subset \{1\}$.

A diagonal line through one of the symbols $=, \in, \subset, \subseteq, <, >$ (etc.) means "is not"; thus, for example, $3 \neq 5$; $7 \notin \{1, 2, 3, 4, 5, 6\}$; $\{3, 4, 5\} \not\subseteq \{3, 4\}$; $10 \not< 8$.

If $A$ and $B$ are sets, then $A \cup B$ (the *union* of $A$ and $B$) is the set whose elements are the elements of $A$ together with the elements of $B$. More generally, if $A_1, A_2, \ldots, A_n$ are sets, then the union of these $n$ sets, denoted either $A_1 \cup \cdots \cup A_n$ or

$$\bigcup_{i=1}^{n} A_i,$$

is defined to be the set consisting of all elements which belong to at least one of the sets $A_i$. For example,

$$\{1, 2, 4, 6\} \cup \{2, 8\} \cup \{1, 7\} = \{1, 2, 4, 6, 7, 8\},$$

and if, say, $A_1 = \{1\}, A_2 = \{2\}, \ldots, A_n = \{n\}$, then

$$\bigcup_{i=1}^{5} A_i = \{1, 2, 3, 4, 5\}.$$

The *intersection* of two sets $A$ and $B$ (denoted $A \cap B$) is the set whose elements are the elements common to $A$ and $B$. Similar definitions apply to more than two sets. For example,

$$\{1, 3, 5\} \cap \{3, 5, 7\} = \{3, 5\}$$
$$\{2, 3, 5\} \cap \{2, 5, 6\} \cap \{2, 3, 6\} = \{2\}$$
$$\{1, 2\} \cap \{3, 4\} = \varnothing.$$

Two sets are *disjoint* if they have empty intersection; that is, if they have no elements in common.

The sets $\{2, 3\}$, $\{3, 2\}$, and $\{2, 3, 2\}$ are equal, since they have the same elements; the order in which we write the elements does not matter, nor does it matter if some element is listed more than once. In some situations, however, such distinctions matter. If, for example, we assign coordinates $(x, y)$ to points in the plane (as in elementary analytic geometry), the points $(2, 3)$ and $(3, 2)$ are different; similarly, the point $(2, 3)$ in the plane is not the same as the point $(2, 3, 2)$ in 3-space. Expressions like $(2, 3)$ and $(3, 2)$, written with ordinary parentheses instead of brackets, are called *ordered pairs*; expressions like $(2, 3, 2)$ are called *ordered triples*. More generally, if $n$ is any positive integer, then an expression of the form $(a_1, a_2, \ldots, a_n)$ is called an *ordered n-tuple*. (The word "ordered" is occasionally omitted to save space.) In the $n$-tuple $(a_1, a_2, \ldots, a_n)$, $a_1$ is called the *first coordinate*, $a_2$ is called the *second coordinate*, and so on. Two ordered $n$-tuples are equal only if they have the same first coordinate, the same second coordinate, ..., the same $n$th coordinate.

If $A$ and $B$ are sets, then $A \times B$ is the standard notation for the set of all ordered pairs $(a, b)$ such that $a \in A$ and $b \in B$. For example, if $A = \{1, 2\}$ and $B = \{a, b, c\}$, then

$$A \times B = \{(1, a), (1, b), (1, c), (2, a), (2, b), (2, c)\}.$$

Observe that in this example $A$ has two elements, $B$ has three elements, $A \times B$ has six elements. Does this suggest to you a general result concerning the number of elements in $A \times B$? Does it suggest a reason for the notation "$A \times B$"?

## 1.2 Mappings

Let $A$, $B$ be sets. A *mapping of $A$ into $B$* is (to be informal) a rule that assigns to each element $x \in A$ exactly one element $x' \in B$. A mapping of $A$ into $B$ is normally denoted by a single letter, such as $\mathbf{T}$; the notation

$$\mathbf{T} : A \to B$$

is read, "$\mathbf{T}$ is a mapping of $A$ into $B$". (The set $A$ is then called the *domain* of $\mathbf{T}$.) If the mapping $\mathbf{T}$ assigns to the element $a \in A$ the element $b \in B$, then $b$ is called the *image* of $a$ under $\mathbf{T}$, or (more concisely) the $\mathbf{T}$-*image* of $a$; and we say that "$a$ is mapped into $b$ by $\mathbf{T}$". In this situation, we write

(1.2.1) $$a\mathbf{T} = b$$

or

$$\mathbf{T} : a \to b.$$

If $b$ is the image of $a$ under $\mathbf{T}$, then $a$ is said to be a *pre-image* of $b$ under $\mathbf{T}$. Note that although an element of $A$ has exactly one image in $B$, an element $b \in B$ may have any number of pre-images in $A$; for that matter, it may have no pre-images in $A$.

The type of notation exhibited in 1.2.1 is called *right-hand notation*. Some authors prefer *left-hand notation*, in which one writes $\mathbf{T}(a) = b$ instead of $a\mathbf{T} = b$. (Left-hand notation is almost universal in elementary calculus textbooks, in which mappings are called "functions" and the sets $A$ and $B$ are usually intervals of real numbers.) Unfortunately, mathematicians as a body have never been able to agree on which hand to use. This is not entirely the fault of mathematicians; each notation is at times more convenient to use than the other, depending on the context. In this book, we shall usually use the right-hand notation 1.2.1, but exceptions will occur.

If $A_0$ is any subset of $A$ and if $\mathbf{T} : A \to B$, then we use the notation $A_0\mathbf{T}$ for the set of all images (under $\mathbf{T}$) of elements of $A_0$. That is,

$$A_0\mathbf{T} = \{x\mathbf{T} : x \in A_0\}.$$

For example, if $A = B = \mathbf{Z}$ and the mapping $\mathbf{T} : \mathbf{Z} \to \mathbf{Z}$ is defined by

$n\mathbf{T} = n^2$, and if $A_0 = \{2, 5, 7\}$, then $A_0\mathbf{T} = \{4, 25, 49\}$. By analogy with our terminology regarding single elements, the set $A_0\mathbf{T}$ is called the *image* of the set $A_0$ under $\mathbf{T}$.

If $\mathbf{T} : A \to B$, then the set $A\mathbf{T}$ is called the *range* of $\mathbf{T}$; it consists of all elements of $B$ which are images of elements of $A$. The range of $\mathbf{T}$ is a subset of $B$; it may or may not be the whole set $B$. Both possibilities are illustrated in Figure 1.2.1. In each part of Figure 1.2.1, dots denote distinct elements of $A$ and $B$, and arrows indicate the correspondence between each element of $A$ and its image in $B$. In part (a) of the figure, every element of $A$ has exactly one image in $B$ (as required), but two of the elements of $B$, namely, $b_3$ and $b_6$, have no pre-images in $A$; the range of $\mathbf{T}$ is, thus, $A\mathbf{T} = \{b_1, b_2, b_4, b_5\}$. If $\mathbf{T} : A \to B$ is a mapping such that *every* element of $B$ has a pre-image in $A$, then $\mathbf{T}$ is said to be *onto* and we say that $\mathbf{T}$ maps $A$ *onto* $B$. In other words, $\mathbf{T}$ is onto if and only if $A\mathbf{T} = B$. The mapping shown in Figure 1.2.1(b) is onto; the mapping shown in Figure 1.2.1(a) is not.

In Figure 1.2.1(b), every element of $A$ has exactly one image in $B$, as required; but the images are not all distinct, since $a_1$ and $a_2$ have the same image. A mapping of $A \to B$ is called *one-to-one* (1–1) if no two elements of $A$ have the same image (equivalently, if no element of $B$ has more than one pre-image). Thus in Figure 1.2.1(a), the mapping is one-to-one, but not in Figure 1.2.1(b). To show that a given mapping $\mathbf{T} : A \to B$ is one-to-one, we must show that if $a_1 \neq a_2$ (in $A$) then $a_1\mathbf{T} \neq a_2\mathbf{T}$; or, equivalently, that if $a_1\mathbf{T} = a_2\mathbf{T}$ then $a_1 = a_2$. The latter is actually the more common method of proof; in a typical proof that a mapping $\mathbf{T}$ is one-to-one, the first sentence reads, "Assume $a_1\mathbf{T} = a_2\mathbf{T}$", and the last sentence reads, "Therefore $a_1 = a_2$".

Figure 1.2.1

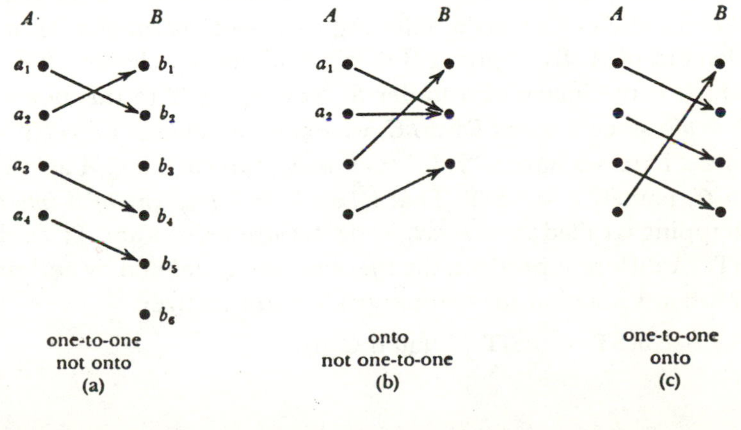

The mapping shown in Figure 1.2.1(c) is both one-to-one and onto. If $\mathbf{T} : A \to B$ is both one-to-one and onto, $\mathbf{T}$ is called a *one-to-one correspondence between A and B*. The word "correspondence" is not used if $\mathbf{T}$ is merely one-to-one but not onto.

In parts (a) and (c) of Figure 1.2.1, in which the mappings are one-to-one, observe that $B$ has at least as many elements as $A$. Will this always be true of one-to-one mappings? What is the situation with respect to "onto" mappings? What about mappings that are both one-to-one and onto?

Let $\mathbf{T}$ be any mapping of $A \to B$ and let $A_0$ be a subset of $A$. By considering what $\mathbf{T}$ does to the elements of $A_0$ (and forgetting about those elements of $A$ which are not in $A_0$), we obtain a mapping of $A_0 \to B$. This mapping is called the *restriction of* $\mathbf{T}$ *to* $A_0$ and will be denoted $\mathbf{T} \restriction A_0$. For example: if $A = B = \{3, 4, 5, 6\}$, $A_0 = \{3, 5\}$, and $\mathbf{T}$ maps

$$3 \to 3, \qquad 4 \to 5, \qquad 5 \to 6, \qquad 6 \to 5,$$

then $\mathbf{T} \restriction A_0$ maps $3 \to 3$ and $5 \to 6$.

The opposite of a restriction is an *extension*: if $\mathbf{S}$ is a restriction of $\mathbf{T}$, then $\mathbf{T}$ is an extension of $\mathbf{S}$. To say the same thing another way: if $\mathbf{S}$ is a mapping with domain $D$, then an *extension* of $\mathbf{S}$ is any mapping $\mathbf{T}$ such that (1) the domain of $\mathbf{T}$ is an overset of $D$, and (2) $\mathbf{T}$ and $\mathbf{S}$ do the same thing to the elements of $D$; that is, $x\mathbf{T} = x\mathbf{S}$ for all $x \in D$. Under these circumstances, we say that $\mathbf{T}$ *extends* $\mathbf{S}$, or that $\mathbf{S}$ is *extended* to $\mathbf{T}$.

## 1.3 Multiplication of mappings

Let $A$, $B$, $C$ be sets, and let $\mathbf{S} : A \to B$ and $\mathbf{T} : B \to C$ be mappings. We can then construct a mapping of $A \to C$ as follows: let $a$ be any element of $A$. By applying $\mathbf{S}$ to $a$, we obtain an element $b = a\mathbf{S}$ in $B$ which is the image of $a$ under $\mathbf{S}$. Now apply $\mathbf{T}$ to this new element $b = a\mathbf{S}$; since $\mathbf{T}$ maps $B$ into $C$, we obtain an element $c = b\mathbf{T} = (a\mathbf{S})\mathbf{T}$ in $C$. Thus we have a "rule" that assigns to each $a \in A$ an element $c$ in $C$, namely $c = (a\mathbf{S})\mathbf{T}$. That is, *we have a mapping of A into C*. This mapping is called the *product*, or *right composite*, of $\mathbf{S}$ and $\mathbf{T}$; it is denoted $\mathbf{ST}$. (A different product, the *left composite*, is defined by authors using left-hand notation for mappings.) To summarize:

**(1.3.1)** $a(\mathbf{ST}) = (a\mathbf{S})\mathbf{T} \qquad (\text{all } a \in A).$

To repeat: **S** maps $A \to B$, **T** maps $B \to C$, and **ST** maps $A \to C$. The whole situation can be represented conveniently by the diagram

$$\begin{array}{c} \quad\mathbf{S} \qquad \mathbf{T} \\ A \longrightarrow B \longrightarrow C \\ \xrightarrow{\qquad\qquad\quad} \\ \mathbf{ST} \end{array}$$

which pictorially suggests the same fact that equation 1.3.1 states formally: the effect of applying **ST** is the same as the effect of applying first **S** and then **T**. The domain of **ST** is the same as the domain of **S**, namely $A$.

*In general,* **ST** *does not equal* **TS**, not even if the sets $A$, $B$, $C$ are the same set. For example, if $A = B = C = \mathbf{Z}$ and if $n\mathbf{S} = n^2$, $n\mathbf{T} = n + 2$, then

$$n(\mathbf{ST}) = (n\mathbf{S})\mathbf{T} = (n^2)\mathbf{T} = n^2 + 2$$
$$n(\mathbf{TS}) = (n\mathbf{T})\mathbf{S} = (n + 2)\mathbf{S} = (n+ 2)^2 \neq n^2 + 2.$$

This example illustrates the general rule for deciding whether two mappings are equal: *two mappings with domain A are equal if and only if they do the same thing to each element of A.* To show that $\mathbf{T} = \mathbf{S}$ (where $\mathbf{T} : A \to A'$ and $\mathbf{S} : A \to A'$), we must show that $a\mathbf{T} = a\mathbf{S}$ for all $a \in A$. Of course, if two mappings have different domains, then they cannot be equal.

**1.3.2 Theorem.** The multiplication of mappings is always associative: that is, if $\mathbf{S} : A \to B$, $\mathbf{T} : B \to C$, and $\mathbf{U} : C \to D$, then $\mathbf{S}(\mathbf{TU}) = (\mathbf{ST})\mathbf{U}$.

*Proof.* First of all, the mapping **S(TU)** has the same domain as **S**, namely, $A$; the mapping **(ST)U** has the same domain as **ST**, which has the same domain as **S**, namely, $A$. Thus, **S(TU)** and **(ST)U** at least have the same domain $A$. To show that they do the same thing to each $a \in A$, we have

| | |
|---|---|
| $a[\mathbf{S}(\mathbf{TU})] = (a\mathbf{S})\,(\mathbf{TU})$ | (by 1.3.1 with **TU** in place of **T**) |
| $= [(a\mathbf{S})\mathbf{T}]\mathbf{U}$ | (by 1.3.1 with $a$**S**, **T**, **U** in place of $a$, **S**, **T**) |
| $= [a(\mathbf{ST})]\mathbf{U}$ | (by 1.3.1) |
| $= a[(\mathbf{ST})\mathbf{U}]$ | (by 1.3.1, with what in place of what?) |

so that the image of $a$ is the same under **S(TU)** as under **(ST)U**, as required. ▮

1.3.3 **Definition.** If $A$ is any set, then the mapping $\mathbf{I}_A : A \to A$ (called the *identity mapping on A*) is defined by

$$\mathbf{I}_A : a \to a \qquad \text{(all } a \in A);$$

that is, every element of $A$ is mapped into itself by $\mathbf{I}_A$. For simplicity, we often write just $\mathbf{I}$ instead of $\mathbf{I}_A$, especially if $A$ is the only set under discussion. In the notation of equation 1.2.1,

$$a\mathbf{I}_A = a \qquad \text{(all } a \in A). \tag{1.3.4}$$

1.3.5 **Theorem.** If $\mathbf{I}$ is the identity mapping on $A$, and if $B$ is any set, then

(a) $\mathbf{IS} = \mathbf{S}$ for every mapping $\mathbf{S} : A \to B$;

(b) $\mathbf{TI} = \mathbf{T}$ for every mapping $\mathbf{T} : B \to A$.

The proof of Theorem 1.3.5 is left as an exercise. In proving part (a), for example, you must first verify that **IS** and **S** have the same domain, and then that **IS** and **S** do the same thing to every element $x$ in that domain. *Question*: Why is it *not* true that $\mathbf{SI} = \mathbf{S}$ for every mapping $\mathbf{S} : A \to B$?

1.3.6 **Theorem.** Let $A$, $B$ be sets with $A$ nonempty, and let $\mathbf{S}$ be a mapping of $A \to B$. Then

(a) $\mathbf{S}$ is one-to-one if and only if there exists a mapping $\mathbf{T}_1 : B \to A$ such that $\mathbf{ST}_1 = \mathbf{I}_A$.

(b) $\mathbf{S}$ is onto if and only if there exists a mapping $\mathbf{T}_2 : B \to A$ such that $\mathbf{T}_2\mathbf{S} = \mathbf{I}_B$.

*Proof of (a).* Assume $\mathbf{S}$ is one-to-one and let $b \in B$. If $b$ has a pre-image in $A$ under $\mathbf{S}$, then it has only one such pre-image; let $b\mathbf{T}_1$ be this unique pre-image of $b$. If, instead, $b$ has no pre-image in $A$ under $\mathbf{S}$, then choose some element of $A$ (it doesn't matter which; there is at least one element of $A$ available to be chosen since $A \neq \varnothing$), and let $b\mathbf{T}_1$ be this element of $A$. (To readers acquainted with the Axiom of Choice: note the use of it in the preceding sentence.) We have now selected, for each $b \in B$, an element $b\mathbf{T}_1 \in A$; hence, $\mathbf{T}_1$ maps $B \to A$. To show that $\mathbf{ST}_1 = \mathbf{I}_A$, we must show that $a(\mathbf{ST}_1) = a\mathbf{I}_A$ for all $a \in A$. But if $a \in A$, then $a\mathbf{S}$ is some element $b \in B$; thus, $a$ is a pre-image of $b$ under $\mathbf{S}$, so that, by definition $a = b\mathbf{T}_1$. Hence,

$$a(\mathbf{ST}_1) = (a\mathbf{S})\mathbf{T}_1 = b\mathbf{T}_1 = a = a\mathbf{I}_A$$

so that $\mathbf{ST}_1 = \mathbf{I}_A$ as desired. Conversely, if there exists a mapping $\mathbf{T}_1$ such that $\mathbf{ST}_1 = \mathbf{I}_A$, then the sequence of implications

$$a_1\mathbf{S} = a_2\mathbf{S} \Rightarrow (a_1\mathbf{S})\mathbf{T}_1 = (a_2\mathbf{S})\mathbf{T}_1 \Rightarrow a_1(\mathbf{ST}_1) = a_2(\mathbf{ST}_1)$$
$$\Rightarrow a_1\mathbf{I}_A = a_2\mathbf{I}_A \Rightarrow a_1 = a_2$$

shows that **S** is one-to-one. (The symbol "$\Rightarrow$" means "implies".)

*Proof of* (*b*). Assume **S** is onto. This means that each element $b \in B$ has at least one pre-image under **S**; we choose one such pre-image (it doesn't matter which one) and call it $b\mathbf{T}_2$. This defines a mapping $\mathbf{T}_2 : B \to A$. Since $b\mathbf{T}_2$ is a pre-image of $b$ under **S**, we have

$$b(\mathbf{T}_2\mathbf{S}) = (b\mathbf{T}_2)\mathbf{S} = b = b\mathbf{I}_B \qquad (\text{all } b \in B),$$

so that $\mathbf{T}_2\mathbf{S} = \mathbf{I}_B$. Conversely, if $\mathbf{T}_2$ exists such that $\mathbf{T}_2\mathbf{S} = \mathbf{I}_B$, then

$$b = b\mathbf{I}_B = b(\mathbf{T}_2\mathbf{S}) = (b\mathbf{T}_2)\mathbf{S} \qquad (\text{all } b \in B);$$

hence every $b \in B$ is the image of something under **S**, so that **S** is onto. ∎

Parts (a) and (b) of Figure 1.3.1 illustrate (respectively) parts (a) and (b) of Theorem 1.3.6. In both cases, solid arrows indicate **S**; dotted arrows indicate $\mathbf{T}_1$ and $\mathbf{T}_2$. In Figure 1.3.1(a), $b_1$ and $b_5$ have no pre-images under **S**, so we may choose $b_1\mathbf{T}_1$ and $b_5\mathbf{T}_1$ in any way that we wish; we have, in fact, chosen $b_1\mathbf{T}_1 = a_2$ and $b_5\mathbf{T}_1 = a_3$. The element $b_3$, on the other hand, has a unique pre-image $a_3$ under **S**; hence, $b_3\mathbf{T}_1 = a_3$. In Figure 1.3.1(b) we could have chosen $b_1\mathbf{T}_2$ to be either of the two pre-images of $b_1$ under **S**; we have, in fact, chosen $b_1\mathbf{T}_2 = a_1$, but the choice $b_1\mathbf{T}_2 = a_2$ would have been just as good. This illustrates the fact that the mapping $\mathbf{T}_2$ of Theorem 1.3.6(b) is *not*, in general, unique. Similarly, the mapping $\mathbf{T}_1$ of Theorem 1.3.6(a) is not normally unique.

Figure 1.3.1

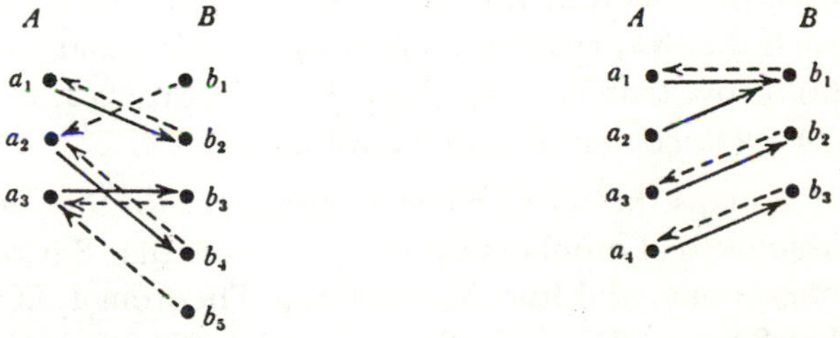


Solid arrows indicate **S**
Dotted arrows indicate $\mathbf{T}_1$
**S** is one-to-one
$\mathbf{ST}_1 = \mathbf{I}_A$
(a)

Solid arrows indicate **S**
Dotted arrows indicate $\mathbf{T}_2$
**S** is onto
$\mathbf{T}_2\mathbf{S} = \mathbf{I}_B$
(b)

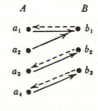

*However*, if $\mathbf{S}$ is *both* one-to-one and onto, then the mappings $\mathbf{T}_1$ and $\mathbf{T}_2$ of Theorem 1.3.6 are not merely unique but equal (see Exercise 5 below); thus in this case there is a unique mapping $\mathbf{T} : B \to A$ such that

$$\mathbf{ST} = \mathbf{I}_A; \qquad \mathbf{TS} = \mathbf{I}_B.$$

This mapping $\mathbf{T}$ is called the *inverse* of $\mathbf{S}$ and is denoted $\mathbf{S}^{-1}$. More generally, a mapping $\mathbf{T}_1$, satisfying Theorem 1.3.6(a), is called a *right inverse* of $\mathbf{S}$; a mapping $\mathbf{T}_2$, satisfying Theorem 1.3.6(b), is called a *left inverse* of $\mathbf{S}$. Theorem 1.3.6 may thus be restated as follows:

> *A mapping is one-to-one if and only if it has a right inverse; a mapping is onto if and only if it has a left inverse.*

The concepts of identity and inverse will be discussed further in Section 1.5.

### Exercises

(Asterisks denote results that will be needed later.)

1. Fill in the reason for the last step in the proof of Theorem 1.3.2, namely the step $[a(\mathbf{ST})]\mathbf{U} = a[(\mathbf{ST})\mathbf{U}]$.

*2. Prove Theorem 1.3.5.

3. Answer the "Question" in the paragraph which follows Theorem 1.3.5.

4. Let $\mathbf{R}^+$ be the set of all positive real numbers; let $\mathbf{S} : \mathbf{R}^+ \to \mathbf{R}^+$ be defined by

   $$x\mathbf{S} = 1/(3x + 4) \qquad (\text{all } x \in \mathbf{R}^+).$$

   Is $\mathbf{S}$ one-to-one? Is $\mathbf{S}$ onto? If a mapping $\mathbf{T}$ exists satisfying either half of Theorem 1.3.6, find *one* such mapping $\mathbf{T}$ explicitly.

*5. If $\mathbf{S} : A \to B$ is a mapping that is *both* one-to-one and onto, then by Theorem 1.3.6 there exist mappings $\mathbf{T}_1$, $\mathbf{T}_2$ of $B \to A$ such that $\mathbf{ST}_1 = \mathbf{I}_A$ and $\mathbf{T}_2\mathbf{S} = \mathbf{I}_B$. In this situation,

   (a) Prove that $\mathbf{T}_1 = \mathbf{T}_2$. (*Hint*: Use Theorems 1.3.5 and 1.3.2.)

   (b) Deduce that $\mathbf{T}_1$ and $\mathbf{T}_2$ are unique. (SOL)

6. If $\mathbf{S}$ maps the set of all real numbers into the set of all *non-negative* real numbers by $x\mathbf{S} = x^2$, show that $\mathbf{S}$ is onto but not one-to-one, and find $\mathbf{T}_2$ satisfying Theorem 1.3.6(b).

*7. Let $\mathbf{S} : A \to B$ and $\mathbf{T} : B \to C$, so that $\mathbf{ST} : A \to C$.

   (a) Prove: if $\mathbf{S}$, $\mathbf{T}$ are both one-to-one, then so is $\mathbf{ST}$.

   (b) Prove: if $\mathbf{S}$, $\mathbf{T}$ are both onto, then so is $\mathbf{ST}$.

*8. If **I** is the identity mapping of $A \to A$, prove that **I** is one-to-one and onto.

*9. Prove (a) that the mapping $\mathbf{T}_1$ of Theorem 1.3.6(a) is onto, and (b) that the mapping $\mathbf{T}_2$ of Theorem 1.3.6(b) is one-to-one. (SOL)

## 1.4 Relations; equivalence relations

If $a$, $b$ are any real numbers, then the statement "$a < b$" ("$a$ is less than $b$") is either true or false. If $x$, $y$ are any human beings, then the statement "$x$ is married to $y$" is either true or false. In general, suppose that $A$ is any set and that "$a \sim b$" is any statement such that, for each $a \in A$ and $b \in A$, "$a \sim b$" is either true or false. Then "$\sim$" is said to be a *relation on A*; we further say that *a is related to b*, if $a \sim b$ (that is, if the statement "$a \sim b$" is true. Falsity of the statement "$a \sim b$" is, of course, written "$a \not\sim b$".) "$<$" is a relation on the set of all real numbers, and $8 < 10$ but $9 \not< 7$. Marriage (more precisely, the phrase "is married to") is a relation on the set of all human beings.

**1.4.1 Definition.** A relation "$\sim$" on $A$ is called an *equivalence relation on A* if it has the following three properties:

(a) For all $a \in A$, $a \sim a$. ("Reflexive property")

(b) Whenever $a$, $b$ are elements of $A$ such that $a \sim b$, then $b \sim a$. ("Symmetric property")

(c) Whenever $a$, $b$, $c$ are elements of $A$ such that $a \sim b$ and $b \sim c$, then $a \sim c$. ("Transitive property")

The relation "$<$" is transitive, but it is neither reflexive nor symmetric. The relation of marriage is symmetric, but it is neither reflexive nor transitive. On the other hand, if $A$ is the set of all human beings and if "$a \sim b$" means "$a$ was born in the same year as $b$", then you should be able to see easily that "$\sim$" is reflexive, symmetric, and transitive, so that "$\sim$" is an equivalence relation.

Let us pursue this last example further. Suppose that "$a \sim b$" means "$a$ was born in the same year as $b$", and suppose that we let $S_x$ be the set of all people who were born in year $x$; for example, $S_{1945}$ is the set of all people born in 1945. It is clear that the sets $S_x$ have the following properties: (1) every human being is in one and only one set $S_x$; (2) for any two human beings $a$ and $b$, $a \sim b$ if and only if $a$, $b$ belong to the *same* set $S_x$. (Stop and convince yourself of these two facts before reading further.) Moreover, if $a$ is any element of a particular set $S_x$ (so that $a$ was born in year $x$), then the elements of the set $S_x$ are precisely the people related to $a$ via the relation "$\sim$".

The existence of sets $S_x$ with the properties described above is typical of equivalence relations, as we now show.

1.4.2 **Theorem.** Let "$\sim$" be an equivalence relation on the set $A$. For each $a \in A$, let

$$[a] = \{x \in A : \quad x \sim a\};$$

that is, $[a]$ is the set of all elements that are related to $a$. The sets $[a]$ (where $a \in A$) are called *equivalence classes*, and the following statements are true:

(a) Every element $a$ of $A$ belongs to one and only one equivalence class, namely, the class $[a]$.

(b) For any $a \in A$ and $b \in A$, $a \sim b$ if and only if $a, b$ belong to the same equivalence class.

Before proving Theorem 1.4.2, we call attention to Figure 1.4.1 which illustrates the situation pictorially. Dots represent the elements of $A$, the regions into which $A$ is divided represent the different equivalence classes, and line segments are drawn between elements which are related to each other. Any two elements in the same class are related, but elements of different classes are not related (Theorem 1.4.2(b)). Thus, any two points in the same region are joined by a line segment, but no two points in different regions are. Since each

Figure 1.4.1

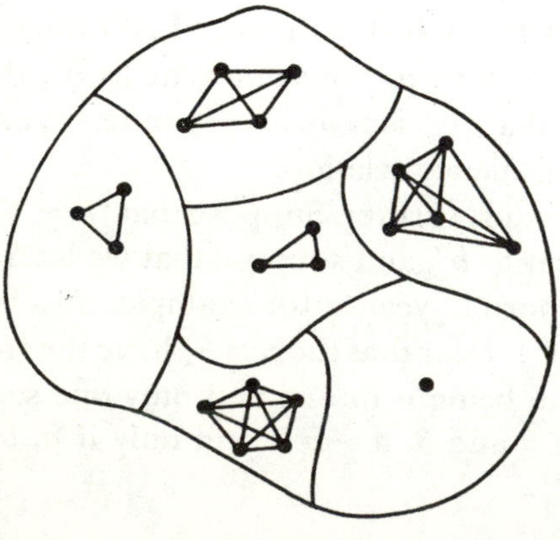

$a \in A$ belongs to only one equivalence class, no point lies in two regions (i.e., no two of the regions overlap). I have always found it useful to think of equivalence relations in terms of such a picture.

(In the example preceding Theorem 1.4.2, the equivalence classes are the sets $S_x$; for example, if Joe Doe was born in 1945, then [Joe Doe] = $S_{1945}$.)

We now prove Theorem 1.4.2.

*Proof of 1.4.2(a).* Let $a \in A$. Since "$\sim$" is reflexive, $a \sim a$, and hence $a \in [a]$. To show that $[a]$ is the *only* equivalence class that $a$ belongs to, we must show that any class which contains $a$ equals the class $[a]$. That is, assuming that

(1.4.3) $\quad a \in [b]$,

we must show that $[b] = [a]$. Since two sets are equal if and only if they have the same elements, our job is to show that every element of $[b]$ is an element of $[a]$ and vice versa.

Let $x \in [b]$; then $x \sim b$ by definition. Similarly, 1.4.3 implies that $a \sim b$. Hence $b \sim a$ (why?). Since $x \sim b$ and $b \sim a$, it follows that $x \sim a$ (why?); hence, $x \in [a]$. This shows that every element of $[b]$ is an element of $[a]$. The proof that every element of $[a]$ is an element of $[b]$ is similar; we leave it to you.

*Proof of 1.4.2(b).* By 1.4.2(a) (just proved), $[b]$ is the unique equivalence class to which $b$ belongs. Hence $a, b$ belong to the same class if and only if $a \in [b]$; that is, if and only if $a \sim b$. ▮

Part (a) of Theorem 1.4.2 can be restated as follows: *the equivalence classes are mutually disjoint, and their union is A.*

A particularly important equivalence relation, called "congruence", is introduced in Exercises 4 and 5 following this section. *You should do these exercises if you have not seen them before.* The subject of congruence will be discussed further in Section 1.13.

**Exercises**

1. Complete the proof of Theorem 1.4.2(a) by doing the part that was left to you.
2. Find a relation: (a) which is symmetric and transitive, but not reflexive; (b) which is reflexive, but is neither symmetric nor transitive.

3. If "$\sim$" is the relation defined on the set of all human beings by:

   $a \sim b$ if $a$, $b$ have a common ancestor,

   then is the relation "$\sim$" reflexive, symmetric, or transitive?

*4. Let us define a relation "$\equiv$" on the set $\mathbf{Z}$ of all integers as follows: if $a, b \in \mathbf{Z}$, then "$a \equiv b$" means that $a - b$ is divisible by 5 (that is, there exists an integer $x$ such that $a - b = 5x$). Prove that "$\equiv$" is an equivalence relation on $\mathbf{Z}$. (The usual properties of integers may be assumed.) (SOL)

*5. Let $m$ be any fixed positive integer. Show that the result of the preceding exercise remains valid if "5" is replaced by "$m$".

6. Let $A$, $B$ be sets; let $\mathbf{T} : A \to B$ be a mapping. Let "$\sim$" be the relation on $A$ defined by:

   $a_1 \sim a_2 \Leftrightarrow a_1\mathbf{T} = a_2\mathbf{T}$.

   (The symbol $\Leftrightarrow$ means "if and only if".)

   (a) Prove that "$\sim$" is an equivalence relation on $A$.

   (b) By the result of part (a) (and by Theorem 1.4.2), the set $A$ is partitioned into equivalence classes. Let $\mathcal{S}$ be the set of all equivalence classes; and for each $a \in A$, let $[a]$ denote the equivalence class containing $a$. We define a mapping $\mathbf{T}^* : \mathcal{S} \to B$ by the equation

   $[a]\mathbf{T}^* = a\mathbf{T}$ (all $a \in A$).

   Prove that the mapping $\mathbf{T}^*$ is *well defined*; that is, the image under $\mathbf{T}^*$ of any equivalence class depends only on the equivalence class and not on the particular element $a$ which represents the equivalence class.

   (c) Show that the mapping $\mathbf{T}^*$ is one-to-one.

## 1.5 Binary operations

If $m$, $n$ are integers, we can form their "sum" $m + n$, which is again an integer. If $\mathbf{S}$, $\mathbf{T}$ are mappings of $A \to A$, we can form their "product" $\mathbf{ST}$, which is again a mapping of $A \to A$. In general, if $E$ is any set, a *binary operation* "$*$" *on* $E$ is a rule by which, given any two elements $a$, $b$ in $E$ (not necessarily different), we obtain an element $a * b$ of $E$. Thus "+" (addition) is an example of a binary operation on the set of all integers; multiplication (right composition) is a binary operation on the set of all mappings of $A \to A$. Multiplication is *not* a binary operation on the set of all mappings of $A \to B$ if $B \neq A$, since if **S**, **T** are arbitrary mappings of $A \to B$ the product **ST** is not necessarily defined (why not?).

1.5.1 **Definition.** Let "$*$" be a binary operation on a set $E$. An element $e \in E$ is said to be a *left identity* of $E$ (with respect to the operation "$*$") if

$$e * x = x \qquad (\text{all } x \in E).$$

An element $e \in E$ is (similarly) called a *right identity* of $E$ if

$$x * e = x \qquad (\text{all } x \in E).$$

An element $e \in E$ is called an *identity* if it is both a left identity and a right identity.

**Examples**

1. If $A$ is any set and $E$ is the set of all subsets of $A$, then the empty set $\varnothing$ is an identity of $E$ with respect to the operation "$\cup$" (union). $E$ also has an identity with respect to the operation "$\cap$"; what is it?
2. The number 0 is an identity with respect to ordinary addition on **Z**; the number 1 is an identity with respect to ordinary multiplication on **Z**.
3. With respect to the operation "$-$" (subtraction) on **Z**, 0 is a right identity but not a left identity. It is true that $x - 0 = x$ for all $x$, but not that $0 - x = x$ for all $x$.
4. It is possible for a set to have *more than one* left identity; for example, if we define an operation "$*$" on **Z** by

   $$a * b = b \qquad (\text{all } a, b \in \mathbf{Z}),$$

   then *every* element $a \in \mathbf{Z}$ is a left identity of **Z** with respect to "$*$". However, if a set has both a left *and* a right identity, then this sort of pathological behavior cannot occur, as the following theorem shows.

1.5.2 **Theorem.** Let "$*$" be a binary operation on $E$, and suppose that $E$ has both a left identity and a right identity with respect to "$*$". Then

(a) $E$ has only one left identity with respect to "$*$".

(b) $E$ has only one right identity with respect to "$*$".

(c) The left identity equals the right identity (and hence it is an identity).

*Proof.* $E$ has a left identity $e$ and a right identity $f$, by assumption. Since $e$ is a *left* identity, $e * f = f$; since $f$ is a *right* identity, $e * f = e$. Since "things equal to the same thing are equal to each other" (remember that catechism?), $e = f$, proving part (c). If $e'$ is another left identity of $E$, the same argument shows that $e' = f$; hence, $e' = e$, proving part (a). The proof of part (b) is similar. ∎

1.5.3 **Corollary.** If $E$ has an identity with respect to "$*$", then the identity is unique.

1.5.4 **Definition.** A binary operation "$*$" on $E$ is *associative* if

$$a * (b * c) = (a * b) * c \qquad (\text{all } a, b, c \in E).$$

Addition of numbers, intersection of sets, and multiplication of mappings are all examples of associative operations. Subtraction of numbers is *not* associative; in general, $(a - b) - c$ does not equal $a - (b - c)$.

If a binary operation "$*$" on $E$ is associative, we can write $a * b * c$ without ambiguity, since both possible ways of inserting parentheses give the same result. In this situation, we can also write "products" of *more than three elements* without parentheses. For example, if $a$, $b$, $c$, $d$ are four elements of $E$, there are three possible ways of inserting parentheses in $a * b * c * d$, namely, $a * (b * c * d)$, $(a * b) * (c * d)$, and $(a * b * c) * d$. To show that these three expressions represent the same element, just let $r = c * d$ and $s = a * b$ and apply the associative law (for *three* elements) several times in succession:

$$\begin{aligned} a * (b * c * d) &= a * (b * r) = (a * b) * r = (a * b) * (c * d) \\ &= s * (c * d) = (s * c) * d = (a * b * c) * d. \end{aligned}$$

Similarly, we can then go on to prove that products of five elements, six elements, and so on, are independent of where we insert the parentheses. (A general proof would be by induction; see Sec. 1.12.) In short, *if products of three elements are associative, products of n elements are associative.*

1.5.5 **Definition.** Let $E$ have an identity $e$ with respect to a binary operation "$*$" on $E$. If the elements $a, b \in E$ satisfy the equation

$$a * b = e,$$

then $a$ is called a *left inverse of* $b$ and $b$ is called a *right inverse of* $a$ (with respect to "$*$"). If $x$ is both a left inverse and a right inverse of $y$, then $x$ is called an *inverse* of $y$.

1.5.6 **Theorem.** If $E$ has an identity $e$ with respect to "$*$" and if $a$, $b \in E$, then $a$ is an inverse of $b$ if and only if $b$ is an inverse of $a$.

This theorem follows at once from Definition 1.5.5; we omit details. The following theorem is less trivial, but not much less:

1.5.7 **Theorem.** Assume that the binary operation "$*$" on $E$ is associative and that $E$ has an identity $e$ with respect to "$*$". If an element $a \in E$ has both a left inverse and a right inverse in $E$ with respect to "$*$", then

(a) the left inverse of $a$ is unique;

(b) the right inverse of $a$ is unique;

(c) the left inverse and right inverse of $a$ are equal, and hence $a$ has an inverse.

We omit the proof of Theorem 1.5.7 since it is essentially the same as the solution to Section 1.3, Exercise 5, which you can find in Appendix G.

1.5.8 **Corollary.** Let $E$ and "$*$" satisfy the assumptions of Theorem 1.5.7 (i.e., "$*$" is associative and $E$ has an identity). If an element $a \in E$ has an inverse with respect to "$*$", then this inverse is unique.

The inverse of $a$ is normally denoted $a^{-1}$ (with *one exception*, to be discussed in Sec. 1.7).

1.5.9 **Theorem.** Assume that the binary operation "$*$" on $E$ is associative, and that $E$ has an identity $e$ with respect to "$*$". Let $a, b \in E$.

(a) If $a^{-1}$ exists in $E$, then $(a^{-1})^{-1}$ exists and equals $a$.

(b) If $a^{-1}$ and $b^{-1}$ exist in $E$, then $(a * b)^{-1}$ exists and equals $b^{-1} * a^{-1}$.

*Proof.* Theorem 1.5.6 and Corollary 1.5.8 imply that if $a^{-1} = b$, then $b^{-1} = a$; that is, $(a^{-1})^{-1} = a$, proving part (a). As for part (b), the idea of the proof is quite simple: if we wish to show that the inverse of $x$

is $y$, just show that $x * y = y * x = e$. Here we wish to show that the inverse of $a * b$ is $b^{-1} * a^{-1}$, so we must verify the equations

$$(a * b) * (b^{-1} * a^{-1}) = e$$
$$(b^{-1} * a^{-1}) * (a * b) = e.$$

To verify the first of these, we have

$$\begin{aligned}(a * b) * (b^{-1} * a^{-1}) &= a * (b * b^{-1}) * a^{-1} = a * e * a^{-1} \\ &= a * a^{-1} = e.\end{aligned}$$

The other verification is similar. ▮

Observe that in Theorem 1.5.9(b) the inverse of the product is the product of the inverses *in opposite order*. The order matters, since, in general, $a * b \neq b * a$. If it happens that $a * b = b * a$ for all $a, b \in E$ (where "$*$" is a binary operation on $E$), then the operation "$*$" is said to be *commutative*. Even if "$*$" is not commutative, it may happen that $a * b = b * a$ for certain *particular* elements $a$, $b$; in this case, we say that these elements $a$, $b$ *commute* with each other.

### Exercises

1. Answer the question "why not?" immediately preceding Definition 1.5.1.
2. Interpret binary operations as mappings.

## 1.6 Groups

1.6.1 **Definition.** Let "$*$" be a binary operation on a set $G$. $G$ is said to be a *group* under the operation "$*$" if
(a) "$*$" is associative;
(b) $G$ has an identity with respect to "$*$";
(c) Every element of $G$ has an inverse in $G$ with respect to "$*$".

Note that a group must satisfy *four* conditions, not just three: in addition to (a), (b), and (c), the fourth condition is that "$*$" be a binary operation on $G$. That is, for every $a$, $b \in G$, the element $a * b$ is defined *and belongs to* $G$. To illustrate: if "$*$" is ordinary multiplication, the set $S = \{1/3, 1, 3\}$ satisfies (a), (b), and (c); however, since $3 \cdot 3 = 9$ is not in $S$, multiplication is *not* a binary operation *on* $S$ and hence $S$ is not a group.

(We say that a set $S$ is *closed* under an operation "$*$" if for every $a$,

$b \in S$, $a * b$ belongs to $S$. This is actually the same as saying that "$*$" is a binary operation on $S$. In the example above, the set $S = \{1/3, 1, 3\}$ is not closed under multiplication.)

Some familiar examples of groups are (1) the set of all real numbers under addition, (2) the set of all integers under addition, (3) the set of all *nonzero* real numbers under multiplication. (Which part of Definition 1.6.1 would fail in example (3) if zero were not excluded?) The following example may be somewhat less familiar.

1.6.2 **Example.** If $A$ is any set, let $\mathbf{S}_A$ be the set of all mappings $\mathbf{f}$ of $A \to A$ such that $\mathbf{f}$ is one-to-one and onto. In symbols,

$$\mathbf{S}_A = \{\mathbf{f} : \mathbf{f} \text{ maps } A \to A, \mathbf{f} \text{ is 1–1 and onto}\}.$$

If $\mathbf{f}, \mathbf{g} \in \mathbf{S}_A$, then the product $\mathbf{fg}$ is also a mapping of $A \to A$ (see Sec. 1.3), and $\mathbf{fg}$ is one-to-one and onto by Section 1.3, Exercise 7. Hence $\mathbf{fg} \in \mathbf{S}_A$, showing that $\mathbf{S}_A$ is closed under multiplication. Furthermore, multiplication of mappings is associative (Theorem 1.3.2), the mapping $\mathbf{I}_A$ is an identity of $\mathbf{S}_A$ (Theorem 1.3.5 and Sec. 1.3, Exercise 8), and the existence of inverses follows from Section 1.3, Exercises 5 and 9. Thus, $\mathbf{S}_A$ is a group under multiplication. We will do more with this group in Section 5.1.

A group $G$ under "$*$" is called an *abelian group* if the operation "$*$" on $G$ is commutative; that is, if $a * b = b * a$ for all $a, b \in G$. The group $\mathbf{S}_A$ of Example 1.6.2 is *not* abelian unless $A$ has no more than two elements. Even if $A$ has only three elements (say, $a_1$, $a_2$, $a_3$), the following example shows that $\mathbf{S}_A$ is not abelian: define $\mathbf{f} : A \to A$ by

$$a_1\mathbf{f} = a_2\,; \qquad a_2\mathbf{f} = a_3\,; \qquad a_3\mathbf{f} = a_1$$

and define $\mathbf{g} : A \to A$ by

$$a_1\mathbf{g} = a_2\,; \qquad a_2\mathbf{g} = a_1\,; \qquad a_3\mathbf{g} = a_3.$$

Then $\mathbf{f}, \mathbf{g}$ belong to $\mathbf{S}_A$. The reader can verify that

$$a_1(\mathbf{fg}) = a_1\,; \qquad a_1(\mathbf{gf}) = a_3$$

and hence $\mathbf{fg} \neq \mathbf{gf}$.

1.6.3 **Theorem.** Let $G$ be a group under the operation $*$; let $a$, $b$, $c$ be elements of $G$. Then:

(a) If $a * c = b * c$, then $a = b$;
(b) If $a * b = a * c$, then $b = c$.

Statements (a) and (b) are called the right and left *cancellation laws*, for obvious reasons. The cancellation laws do not necessarily hold if $G$ is not a group. For example, in the ring $\mathbf{Z}_{(6)}$ of integers modulo 6 (see Sec. 1.13), $[4] \cdot [2] = [1] \cdot [2]$ but $[4] \neq [1]$.

To prove 1.6.3(a), let $e$ be the identity of $G$; then we have

$$a * c = b * c \qquad \text{(given)}$$
$$(a * c) * c^{-1} = (b * c) * c^{-1} \qquad \text{(why?)}$$
$$a * (c * c^{-1}) = b * (c * c^{-1}) \qquad \text{(why?)}$$
$$a * e = b * e \qquad \text{(why?)}$$
$$a = b \qquad \text{(why?)}$$

You should fill in the reasons for the steps listed above. (Refer to specific parts of Def. 1.6.1.) Of course, 1.6.3(b) is proved similarly. ∎

**Subgroups**

Let $G$ be a group and let $S$ be a subset of $G$. $S$ is called a *subgroup* of $G$ if $S$ is itself a group under the same binary operation as in $G$. For example, $\mathbf{Z}$ (the integers) is a group under addition; the set of all even integers is a subgroup of $\mathbf{Z}$. The set of all odd integers is not closed under addition ($1 + 1$ is not odd, for example) and hence cannot be a subgroup. The set of all nonnegative integers is closed under addition, but it is not a subgroup of $\mathbf{Z}$ since it fails to contain the inverses of all of its elements; for example, it contains 2 but not $-2$.

How does one determine whether a given subset of a group is a subgroup? In answering this question it is useful to observe that the four defining properties of a group are of two essentially different types, *assertions of existence* and *assertions of equality*. The closure property of a group $G$, for instance, asserts that for each $a, b \in G$ "there exists" an element $a * b$ in $G$; likewise, 1.6.1(b) and 1.6.1(c) assert the existence of certain elements of $G$. The associative property, on the other hand, is an "assertion of equality"; it tells us that two elements already known to exist in $G$ (namely, $(a * b) * c$ and $a * (b * c)$) are equal. Now suppose $S$ is a subset of the group $G$, and suppose we are able to establish that the three "assertions of existence" (closure, identity, inverses) are valid in $S$. *Then the associative law ("assertion of equality") holds in $S$ automatically, and $S$ is a subgroup.* Indeed, if $a$, $b$, $c$ are any elements of $S$, then the elements $(a * b) * c$ and $a * (b * c)$ belong to $S$ ($S$ being closed) and are equal in $G$ (by the associative law in $G$); hence they are the same element and must be equal in $S$. Let us state our result as a theorem.

**1.6.4 Theorem.** If $G$ is a group under the operation $*$ (with identity $e$) and if $S$ is a subset of $G$, then $S$ is a subgroup of $G$ if and only if the following three statements hold:

(a) For all $a, b \in S$, $a * b \in S$.

(b) $e \in S$.

(c) For all $a \in S$, $a^{-1} \in S$.

Statements (a), (b), and (c) of 1.6.4 are of course the three "assertions of existence". In each case we already know that the necessary elements exist in $G$; what is required is that these elements *belong to* $S$.

*Remarks:* 1. An argument similar to the above shows that any other "assertion of equality" valid in $G$ must be valid in $S$. For example, if the commutative law holds in $G$, then it holds in $S$ too. (Thus, if $G$ is an abelian group, any subgroup of $G$ must be abelian.)

2. Condition (b) in Theorem 1.6.4 may be replaced by the weaker statement, "$S$ is nonempty." (Indeed, if $S$ has at least one element $a$, then $a^{-1} \in S$ by 1.6.4 (c) and hence $e = a * a^{-1}$ belongs to $S$ by 1.6.4(a).) However, this fact has little practical value, since in most cases of interest the simplest way of proving $S$ nonempty is to show that $e \in S$.

3. The following objection to 1.6.4(b) may have occurred to the (alert) reader: even if $e \in S$, might not $S$ contain some other element that is an *identity for* $S$ though not an identity for $G$? The answer is no; for if $S$ had an identity $e_S$ different from the identity $e_G$ of $G$, then we would have

$$\begin{aligned} e_S * e_S &= e_S \qquad \text{(since } e_S \text{ is the identity of } S\text{)} \\ &= e_S * e_G \qquad \text{(since } e_G \text{ is the identity of } G\text{)} \end{aligned}$$

and, hence, $e_S = e_G$ by Theorem 1.6.3, contrary to assumption. A similar objection to Theorem 1.6.4(c) may be disposed of by similar reasoning.

**Exercises**

1. The set of all integers is *not* a group under multiplication. Which part of Definition 1.6.1 fails?

*2. Let $G$ be a group under $*$ and let $a, b, c \in G$. Prove that $a * b = c$ if and only if $a = c * b^{-1}$. (Note that there are *two* things to prove: each equation must be deduced from the other one.)

3. In showing that a subgroup $S$ of $G$ cannot have an identity different from the identity of $G$ (cf. Remark 3, above), why would it *not* suffice simply to refer to Corollary 1.5.3?

## 1.7 Additive notations

In mathematics one comes across all sorts of binary operations on various sets. Some of these operations are called "multiplication" (multiplication of numbers; multiplication of mappings); others are called "addition" (addition of numbers; addition of matrices; addition of vectors); others have still other names. The purpose of this section is to describe certain notational conventions that occur when the operation under consideration is called "addition," and to compare these conventions with those used in connection with binary operations other than addition.

**Plus sign**

An operation called "addition" is always represented by the plus sign "+". Conversely, an operation denoted by "+" is always called "addition". The element $x + y$ is called the *sum* of $x$ and $y$.

**Identity element**

If a set $S$ has an identity element with respect to an operation of addition on $S$, the identity element is called the *zero element* of $S$ and is usually denoted by the symbol 0 ( or $0_S$, if we want to be really precise). On the other hand, an identity element with respect to an operation other than addition (for example, multiplication) is never denoted "0"; it is normally denoted by symbols such as $e$, $I$, or 1.

**Inverses**

Suppose the set $S$ has an identity element with respect to a binary operation, and suppose the element $a \in S$ has an inverse with respect to the operation. The inverse of $a$ is denoted $a^{-1}$, *except when the operation is addition*. When the operation is addition, the inverse of $a$ is denoted $-a$ ("minus $a$"). (If the operation is not known or not specified, the notation $a^{-1}$ is used.)

The additive conventions regarding identity and inverse elements are of course motivated by the facts about ordinary numbers: the number zero *is* the identity with respect to addition of numbers, and the number $-a$ *is* the inverse of the number $a$ with respect to addition. In fact, if the two equations in Definition 1.5.1 are translated into additive notation, they become

$$0 + x = x$$
$$x + 0 = x$$

which we recognize as being true for numbers. Similarly, the defining equations for an inverse,

$$a * a^{-1} = a^{-1} * a = e$$

(using nonadditive notation), are transformed into

$$a + (-a) = (-a) + a = 0$$

when the operation is addition and additive notations are used.

**Subtraction**

In a set with an operation of addition, the expression "$x + (-y)$" is usually abbreviated "$x - y$" (to read "$x$ minus $y$"); thus, subtraction is simply a special case of addition. The expression $x - y$ is called the *difference* of $x$ and $y$.

If a binary operation of addition is defined on a set $S$, there is no necessity for $S$ to be a group under this operation, much less an abelian group. However, it turns out that in most cases of interest, sets on which addition is defined are actually abelian groups. In an abelian group, the hypotheses of Theorem 1.5.9 are satisfied, and all elements possess inverses. Similarly, the hypotheses of Theorem 1.6.3 are satisfied. By translating Theorems 1.5.9 and 1.6.3 into additive notation, we obtain the following theorem:

**1.7.1** **Theorem.** Let $E$ be an abelian group under addition. Then for all $a, b, c \in E$,

(a) $-(-a) = a$.
(b) $-(a + b) = -a - b$.
(c) If $a + c = b + c$, then $a = b$.
(d) If $a + b = a + c$, then $b = c$.

(Actually, Theorem 1.5.9(b) becomes $-(a + b) = -b - a$ in additive notation, but the "abelian" hypothesis allows us to replace $-b - a$ by $-a - b$.)

We close this section with a trivial but useful observation. As seen above, the defining equation for an inverse reads (in part)

$$a + (-a) = 0$$

when additive notation is used; by definition of subtraction, the latter equation is abbreviated $a - a = 0$. Now suppose that $x, y$ are elements of an additive group. If $x = y$, then $x - y = 0$. Conversely, if $x - y = 0$, then $x - y = 0 = y - y$, and, hence, by the cancellation law, $x = y$. In short, *two elements (of an additive group) are equal if and only if their difference is zero*. This observation (which we shall call the *Difference Principle*) will arise in proofs from time to time.

## 1.8 Homomorphisms and isomorphisms

Suppose $S$ and $S'$ are sets on which the operation of addition is defined. Let $\mathbf{T}$ be a mapping of $S \to S'$, and denote the image of any element $x$ (under $\mathbf{T}$) by $x'$. The mapping $\mathbf{T}$ is said to *preserve addition* if it has the property that whenever $x + y = z$ (in $S$), then $x' + y' = z'$ (in $S'$).

Let us say the same thing in different notation. Since $x' = x\mathbf{T}$, $y' = y\mathbf{T}$, $z' = z\mathbf{T} = (x + y)\mathbf{T}$, the equation $x' + y' = z'$ can be rewritten

**(1.8.1)** $x\mathbf{T} + y\mathbf{T} = (x + y)\mathbf{T}.$

Thus, $\mathbf{T}$ preserves addition, if 1.8.1 holds for all $x, y \in S$.

A mapping which preserves addition is called an *additive homomorphism*. Of course, there is nothing special about addition; the same discussion applies to any binary operation. For example, a multiplicative homomorphism would satisfy the equation

**(1.8.2)** $(x\mathbf{T})(y\mathbf{T}) = (xy)\mathbf{T}$ (all $x, y$).

More generally, if $*$ is a binary operation on $S$ and $\circ$ is a binary operation on $S'$, a mapping $\mathbf{T} : S \to S'$ which satisfies the equation

**(1.8.3)** $(x\mathbf{T}) \circ (y\mathbf{T}) = (x * y)\mathbf{T}$ (all $x, y \in S$)

may be called a "($*$, $\circ$)-homomorphism."

If $S$ and $S'$ are groups, a *group homomorphism* of $S \to S'$ is simply a mapping $S \to S'$ which preserves whatever the group operation happens to be. If, for example, $S$ and $S'$ are groups under addition, a "group homomorphism" of $S \to S'$ would mean an additive homomorphism; if $S$ is a group under $*$ and $S'$ is a group under $\circ$, "group homomorphism" would mean ($*$, $\circ$)-homomorphism.

**Examples**

1. Let $S$ be the set of all real numbers and take the operation $*$ on $S$ to be addition; let $S'$ be the set of all *positive* real numbers and take the operation $\circ$ on $S'$ to be multiplication. If $\mathbf{T} : S \to S'$ is defined by

   $x\mathbf{T} = 2^x$ (all $x \in S$)

   then $\mathbf{T}$ is a ($*$, $\circ$)-homomorphism; indeed, 1.8.3 reduces to the equation

   $2^x 2^y = 2^{x+y}$

which we recognize as true. In this example, **T** is actually a group homomorphism since $S$ and $S'$ are groups under addition and multiplication respectively.

2. If $\mathbf{T} : \mathbf{Z} \to \mathbf{Z}$ is defined by $n\mathbf{T} = 3n$ (all $n \in \mathbf{Z}$), then **T** fails to preserve multiplication:

$$(x\mathbf{T})\,(y\mathbf{T}) = (3x)\,(3y) = 9xy$$
$$(xy)\mathbf{T} = 3(xy) \neq 9xy$$

so that equation 1.8.2 is not satisfied. However, **T** does preserve addition:

$$x\mathbf{T} + y\mathbf{T} = 3x + 3y$$
$$(x + y)\mathbf{T} = 3(x + y) = 3x + 3y.$$

Since **Z** is a group under addition, **T** is a group homomorphism.

1.8.4 **Theorem.** Let $S$ and $S'$ be groups having the respective identities $e$, $e'$. If $\mathbf{T} : S \to S'$ is a group homomorphism, then

(a) $e\mathbf{T} = e'$;

(b) $(a^{-1})\mathbf{T} = (a\mathbf{T})^{-1}$ (all $a \in S$).

*Proof.* Let $*$, $\circ$ denote the respective binary operations on $S$, $S'$. Since **T** is a homomorphism, equation 1.8.3 holds; since $e$ is the only element of $S$ that we know anything about, it is natural to apply 1.8.3 to $e$. Choosing both $x$ and $y$ to be $e$ in 1.8.3, we obtain

$$\begin{aligned}(e\mathbf{T}) \circ (e\mathbf{T}) &= (e * e)\mathbf{T} \\ &= e\mathbf{T} \qquad (\text{since } e * e = e) \\ &= (e\mathbf{T}) \circ e' \qquad (\text{since } e' \text{ is the identity of } S').\end{aligned}$$

Canceling $e\mathbf{T}$ from both sides (Theorem 1.6.3(b)), we get $e\mathbf{T} = e'$, proving 1.8.4(a).

Statement 1.8.4(b) asserts that $(a^{-1})\mathbf{T}$ is the inverse of $a\mathbf{T}$. This statement is, of course, equivalent to the pair of equations

$$[(a^{-1})\mathbf{T}] \circ (a\mathbf{T}) = e'$$
$$(a\mathbf{T}) \circ [(a^{-1})\mathbf{T}] = e'.$$

The first of these equations is verified as follows:

$$\begin{aligned}[(a^{-1})\mathbf{T}] \circ (a\mathbf{T}) &= (a^{-1} * a)\mathbf{T} \qquad (\text{by } 1.8.3) \\ &= e\mathbf{T} \\ &= e' \qquad (\text{by } 1.8.4(\text{a})).\end{aligned}$$

The other equation is verified similarly. ▮

*Remark*: If the sets $S$ and $S'$ in Theorem 1.8.4 are groups under *addition*, then the two parts of Theorem 1.8.4 become

(1.8.5) $0\mathbf{T} = 0; \qquad (-a)\mathbf{T} = -(a\mathbf{T}) \qquad (\text{all } a \in S)$

using additive notation.

**Isomorphisms**

An *isomorphism* is a homomorphism that is one-to-one and onto. If $\mathbf{T} : S \to S'$ is a $(*, \circ)$-isomorphism, then we say that $S$ *is isomorphic to* $S'$, or that $S$ and $S'$ are *isomorphic*. In this situation, the mapping of each element $x \in S$ into its image $x' \in S'$ is a one-to-one correspondence; moreover, whenever $x * y = z$, then $x' \circ y' = z'$. Thus the structure of the set $S$ with respect to the operation "$*$" is exactly the same as the structure of $S'$ with respect to the operation "$\circ$"; whatever properties $S$ has with respect to "$*$", $S'$ must have the same properties with respect to "$\circ$". For example, if $S$ is commutative under $*$, $S'$ is commutative under $\circ$. If $S$ has exactly two elements $x$ such that $x^{-1} = x$, the same must be true of $S'$. In fact, a common way of showing that two given sets are *not* isomorphic is to exhibit a property of one set which the other set does not have. (To show that two sets $S$, $S'$ *are* isomorphic, we must usually find a specific isomorphism $\mathbf{T} : S \to S'$. This is not always trivial; there will be many mappings of $S \to S'$, and sometimes we may have to try several of them before finding one that works. See Exercise 1 at the end of this section, for example.)

1.8.6 **Theorem.** Let $*$, $\circ$ be binary operations on $S$, $S'$ respectively. If $\mathbf{T} : S \to S'$ is a $(*, \circ)$-isomorphism, then $\mathbf{T}^{-1} : S' \to S$ is a $(\circ, *)$-isomorphism.

*Proof.* Since $\mathbf{T}$ is one-to-one and onto, the results of Section 1.3 (specifically, Theorem 1.3.6 and Exercises 5 and 9) imply that there is a unique mapping $\mathbf{T}^{-1} : S' \to S$ such that $\mathbf{T}\mathbf{T}^{-1} = \mathbf{I}_S$ and $\mathbf{T}^{-1}\mathbf{T} = \mathbf{I}_{S'}$; moreover, $\mathbf{T}^{-1}$ is one-to-one and onto. Hence, $\mathbf{T}^{-1}$ will be an isomorphism provided that it is a homomorphism; that is, provided that

(1.8.7) $(w\mathbf{T}^{-1}) * (z\mathbf{T}^{-1}) = (w \circ z)\mathbf{T}^{-1} \qquad (\text{all } w, z \in S').$

To prove 1.8.7, let $w$, $z$ be elements of $S'$. If we let $x = w\mathbf{T}^{-1}$ and $y = z\mathbf{T}^{-1}$, then $w = x\mathbf{T}$ and $z = y\mathbf{T}$ (*why*?). Since $\mathbf{T}$ is a $(*, \circ)$-homomorphism, equation 1.8.3 gives

$$(x\mathbf{T}) \circ (y\mathbf{T}) = (x * y)\mathbf{T};$$

that is, $w \circ z = [(w\mathbf{T}^{-1}) * (z\mathbf{T}^{-1})]\mathbf{T}$. Applying $\mathbf{T}^{-1}$ to both sides,

$$\begin{aligned}(w \circ z)\mathbf{T}^{-1} &= [(w\mathbf{T}^{-1}) * (z\mathbf{T}^{-1})]\mathbf{T}\mathbf{T}^{-1} \\ &= [(w\mathbf{T}^{-1}) * (z\mathbf{T}^{-1})]\mathbf{I}_S \\ &= (w\mathbf{T}^{-1}) * (z\mathbf{T}^{-1})\end{aligned}$$

which is 1.8.7. ▮

**Exercises**

1. Show that the *additive* group of integers modulo 4 (cf. Sec. 1.13) is isomorphic to the *multiplicative* group $\{1, -1, i, -i\}$, where as usual $i$ denotes a complex number whose square is $-1$. (For a discussion of complex numbers, see Sec. 1.15.)
2. Let $\mathbf{R}^+$ denote the set of all positive real numbers.
   (a) The familiar equation
   $\sqrt{xy} = \sqrt{x}\sqrt{y} \qquad (\text{all } x, y \in \mathbf{R}^+)$
   implies that a certain mapping of $\mathbf{R}^+ \to \mathbf{R}^+$ is a homomorphism. What is the mapping, and what operation on $\mathbf{R}^+$ does the mapping preserve? Is $\mathbf{R}^+$ a group under this operation?
   (b) What facts about square roots result from applying Theorem 1.8.4 to the mapping of part (a)? (ANS)
3. Generalize the preceding exercise. (*Hint*: $\sqrt{x}$ is the same as $x^{1/2}$.)
4. Let $S$, $S'$, $S''$ be sets on which (respectively) binary operations "$*$", "$\circ$", "$\square$" are defined. If $\mathbf{T}_1$ is a $(*, \circ)$-homomorphism of $S \to S'$ and $\mathbf{T}_2$ is a $(\circ, \square)$-homomorphism of $S' \to S''$, prove that the mapping $\mathbf{T}_1\mathbf{T}_2$ of $S \to S''$ is a $(*, \square)$-homomorphism. (SOL)
5. Let "$*$", "$\circ$" be binary operations on $S$, $S'$, respectively; let $\mathbf{T}$ be a $(*, \circ)$-homomorphism of $S \to S'$, and assume that $\mathbf{T}$ is *onto*. Prove:
   (a) If $S$ has an identity $e$ with respect to $*$, then $e\mathbf{T}$ is an identity of $S'$ with respect to $\circ$. (*Note*: This result is stronger than Theorem 1.8.4(a) in that here we do not require either $S$ or $S'$ to be a group; but the result is weaker than 1.8.4(a) in that we require $\mathbf{T}$ to be onto.) (SOL)
   (b) If $*$ is commutative, then $\circ$ is commutative.
   (c) If $*$ is associative, then $\circ$ is associative.
   (d) If $S$ is a group under "$*$", then $S'$ is a group under "$\circ$".
6. Let $S$ be the set of integers modulo 6. Find a mapping of $S \to S$

which preserves multiplication but which does *not* map the multiplicative identity of $S$ into itself. Can you find more than one such mapping? (*Note*: Such a mapping cannot be onto, by part (a) of the preceding exercise.) (ANS)

*7. Let $G$ be a group under "$*$"; let $g$ be a *fixed* element of $G$, and let us define a mapping $\mathbf{T} : G \to G$ by

$$\mathbf{T} : x \to g^{-1} * x * g \qquad (\text{all } x \in G).$$

Prove that $\mathbf{T}$ is an isomorphism.

## 1.9 Rings and integral domains

We now consider sets on which not just one but *two* binary operations (specifically, operations of addition and multiplication) exist simultaneously.

**1.9.1 Definition.** A *ring* is a set $S$ such that

(a) Two binary operations, called *addition* (denoted "+") and *multiplication* (denoted by juxtaposition) are defined on $S$.

(b) $S$ is an abelian group under addition.

(c) Multiplication on $S$ is associative.

(d) For all $a, b, c \in S$, the following two equations hold:

$$a(b + c) = ab + ac \qquad (\text{"left distributive law"})$$
$$(a + b)c = ac + bc \qquad (\text{"right distributive law"}).$$

Observe that part (b) of Definition 1.9.1 concerns addition alone, part (c) concerns multiplication alone, and part (d) involves both operations.

The set of all integers is a ring; so is the set of all real numbers. The integers modulo $m$ form a ring (Sec. 1.13). The ring of $n$ by $n$ matrices with real coefficients (Sec. 4.4) is quite important for the study of linear transformations and other topics in linear algebra. Many further examples of rings could be given.

Since a ring $S$ is an abelian group under addition, Theorem 1.7.1 and Corollaries 1.5.3 and 1.5.8 are all applicable to addition in $S$; 1.5.3 and 1.5.8 also apply to multiplication in $S$ if an identity exists. Hence, we have the following theorem:

**1.9.2 Theorem.** Let $S$ be a ring. Then

(a) $-(-a) = a \qquad (\text{all } a \in S)$.

(b) $-(a + b) = -a - b \qquad (\text{all } a, b \in S)$.

(c) Whenever $a + c = b + c$ $(a, b, c \in S)$, then $a = b$; whenever $a + b = a + c$ $(a, b, c \in S)$, then $b = c$.

(d) The zero element of $S$ is unique.

(e) The additive inverse of each element of $S$ is unique.

(f) If $S$ has an identity for multiplication, it is unique; if an element of $S$ has an inverse with respect to multiplication, the inverse is unique.

In connection with Theorem 1.9.2(f), note that a ring $S$ need not have an identity for multiplication. If an identity for multiplication does exist, it is called the *unity* of $S$, and $S$ is then said to be a *ring with unity*.

It is useful to interpret the distributive laws (Def. 1.9.1(d)) as homomorphism equations. Let $c$ be a fixed element of a ring $S$, and let us define a mapping $\mathbf{T}: S \to S$ by

$$x\mathbf{T} = xc \qquad (\text{all } x \in S).$$

*Then* $\mathbf{T}$ *is an additive homomorphism*, since equation 1.8.1 reduces to the right distributive law $xc + yc = (x + y)c$. If we now apply equation 1.8.5 (valid since $S$ is a group under addition), we obtain

$$0c = 0; \qquad (-a)c = -(ac) \qquad (\text{all } a \in S).$$

Similarly, the *left* distributive law implies that the mapping $\mathbf{T}': S \to S$ defined by

$$x\mathbf{T}' = cx \qquad (\text{all } x \in S)$$

is an additive homomorphism, and equation 1.8.5 then gives

$$c0 = 0; \qquad c(-a) = -(ca) \qquad (\text{all } a \in S).$$

Since the element $c$ was arbitrary, we see that we have proved the following theorem:

**1.9.3** **Theorem.** Let $S$ be a ring. Then for all elements $a$, $c$ in $S$,

(a) $0c = c0 = 0$.

(b) $(-a)c = -(ac)$.

(c) $c(-a) = -(ca)$.

The properties stated in Theorem 1.9.3 are undoubtedly familiar to you from high-school algebra; the theorem asserts that these properties are valid, not just for numbers, but for elements of any ring. A few more such properties of rings are collected in the following theorem.

**1.9.4** **Theorem.** Let $S$ be a ring. Then for all elements $a$, $b$, $c$ of $S$,

(a) $(-a)(-b) = ab$.

(b) $a(b - c) = ab - ac$.
(c) $(a - b)c = ac - bc$.

This theorem is an easy corollary of the preceding one. To prove (a), for example, let $x = -b$; then

$$\begin{aligned}(-a)(-b) &= (-a)x && (x = -b, \text{ given})\\ &= a(-x) && (\text{by Theorem 1.9.3})\\ &= a[-(-b)] && (x = -b, \text{ given})\\ &= ab && (\text{Theorem 1.9.2(a)}).\end{aligned}$$

Similarly, parts (b) and (c) follow from Theorem 1.9.3 and the definition of subtraction; we leave them as exercises. ▮

1.9.5 **Theorem.** Assume that the ring $S$ has a unity $e$. Then the following statements hold:
(a) $(-e)a = a(-e) = -a$ (all $a \in S$).
(b) If $S$ has more than one element, then $e \neq 0$.

We leave part (a) as an exercise. If $e = 0$, then for all $x \in S$ we have

$$x = ex = 0x = 0$$

so that $S$ has only the one element 0; this proves (b). ▮

In a ring with unity, an element is called *inversible* if it has an inverse with respect to multiplication.

1.9.6 **Theorem.** Let $S$ be a ring with unity $e$, and let $S^*$ be the set of all inversible elements of $S$. Then $S^*$ is a group under multiplication.

*Proof.* If $a, b \in S^*$, then $a^{-1}, b^{-1}$ exist in $S$; hence $(ab)^{-1} = b^{-1}a^{-1}$ exists (Theorem 1.5.9(b)); hence $ab \in S^*$. This shows that $S^*$ is closed under multiplication. Multiplication is associative in $S^*$, since it is associative in $S$. Since $e$ is its own inverse (why?), $e \in S^*$, and $S^*$ thus has an identity. Finally, if $a \in S^*$ then $a^{-1}$ exists, hence $(a^{-1})^{-1}$ exists (Theorem 1.5.9(a)), hence $a^{-1} \in S^*$. ▮

By Definition 1.9.1, addition is automatically commutative in any ring, but multiplication may or may not be commutative. A *commutative ring* is defined to be a ring in which *multiplication* is commutative. Some of the most important rings in mathematics (such as the ring of $n$ by $n$ matrices to be introduced in Chapter 4) are noncommutative.

A *ring homomorphism* is a mapping from one ring into another which preserves both addition and multiplication. In particular, any ring homomorphism $\mathbf{T} : S \to S'$ is an (additive) group homomorphism, so that by equation 1.8.5 we have $0\mathbf{T} = 0$ and $(-a)\mathbf{T} = -(a\mathbf{T})$ (all $a \in S$).

If $A$ is a subset of a ring $S$, then $A$ is called a *subring* of $S$ if $A$ is itself a ring under the same operations of addition and multiplication as are already defined in $S$. The necessary and sufficient conditions for a subset $A$ of $S$ to be a subring are as follows:

(a) $A$ is closed under addition.
(b) $A$ is closed under multiplication.
(c) $0 \in A$.
(d) For all $a \in A$, $-a \in A$.

The proof is like that of Theorem 1.6.4; it is based on the fact that the only parts of Definition 1.9.1 which are "assertions of existence" are the closure properties and the existence of the additive identity and additive inverses. All of the other defining properties of a ring are "assertions of equality" which hold in $A$ automatically.

An *integral domain* is a commutative ring with unity, having more than one element, which has the additional property that the product of any two nonzero elements of the ring is nonzero. (Equivalently, whenever $xy = 0$, then either $x$ or $y$ is zero.) The ring of integers is an integral domain; so is the ring of real numbers. The ring of all polynomials with real coefficients (properties of which will be needed when we study eigenvalues) is an integral domain. A few general properties of integral domains are given in the exercises at the end of this section and the next section.

**Exercises**

*1. Prove Theorem 1.9.4, parts (b) and (c). (SOL)

*2. Prove Theorem 1.9.5(a).

*3. In a commutative ring, prove that $(a + b)(a - b) = aa - bb$ (the right side is usually written $a^2 - b^2$). Why is commutativity needed?

*4. We say that $x$ is a *square root* of $y$ if $x^2 = y$. Prove that in an *integral domain* an element can have at most two square roots: more precisely, if $y$ has a square root $x$, then the only square roots of $y$ are $x$ and $-x$. (*Hint*: Use Exercise 3.) (SOL)

5. Let $A$ be a subring of a ring $S$.
(a) Give an example to show that the rings $A$, $S$ may have *different* unities, even if $A \neq \{0\}$.
(b) On the other hand, the zero elements (additive identities) of $A$ and $S$ must coincide. Why? Why are the situations different for addition and multiplication?

## 1.10 Fields

In the ring of rational numbers, every nonzero number has an inverse with respect to multiplication. (Example: the inverse of 3/2 is 2/3.) In the ring **Z** of all integers, this is no longer true; for example, the integer 5 has no multiplicative inverse in **Z** since its only possible inverse, namely, 1/5, does not belong to **Z**. Systems like the rational numbers, in which the nonzero elements all have inverses, are sufficiently important to be given a name.

**1.10.1 Definition.** A *field* is a commutative ring with unity, having more than one element, in which every element other than zero has a multiplicative inverse.

We shall normally denote the unity of a field $F$ by "1", or "$1_F$" if we wish to emphasize which field we are discussing. This notation is of course motivated by the fact that the *number* 1 is the unity of the most "familiar" field, the field of real numbers. Other examples of fields are the field of all rational numbers, the field of all complex numbers, the field of integers modulo 5. The ring of integers is not a field (what part of Definition 1.10.1 fails?).

**1.10.2 Theorem.** The following statements hold in any field $F$:

(a) $1 \neq 0$.

(b) 0 does not have an inverse with respect to multiplication.

(c) The nonzero elements of $F$ form a group under multiplication.

(d) If $a \neq 0$ and $ab = ac$, then $b = c$.

(e) If $a \neq 0$ and $ba = ca$, then $b = c$.

*Proof.* (a) is immediate from Theorem 1.9.5(b). If 0 had a multiplicative inverse $x$, then $1 = 0x = 0$ contradicting (a); hence 0 has no such inverse, and (b) holds. Part (c) is then immediate from Theorem 1.9.6. The proofs of (d) and (e) are similar to that of Theorem 1.6.3 and we leave them to you (the assumption $a \neq 0$ is needed so that $a^{-1}$ will exist). ∎

*Remark*: By 1.10.2(c), the set of nonzero elements of $F$ is closed under multiplication; that is, if $a \neq 0$ and $b \neq 0$, then $ab \neq 0$. Thus, *every field is an integral domain.*

**1.10.3 Definition.** (Definition of division) If $F$ is a field and $a, b \in F$, we define $a/b$ to be $ab^{-1}$ provided that $b^{-1}$ exists; that is, provided that $b \neq 0$.

The "usual" laws concerning the arithmetic of fractions are valid in fields. We will not attempt to list *all* of these laws, but some of them are given in the next theorem.

**1.10.4 Theorem.** For all elements $a, b, c, d$ of a field $F$,

(a) $\dfrac{a}{b} + \dfrac{c}{d} = \dfrac{ad + bc}{bd}$ (if $b \neq 0$, $d \neq 0$).

(b) $\left(\dfrac{a}{b}\right) \cdot \left(\dfrac{c}{d}\right) = \dfrac{ac}{bd}$ (if $b \neq 0$, $d \neq 0$).

(c) $\dfrac{ac}{bc} = \dfrac{a}{b}$ (if $b \neq 0$, $c \neq 0$).

(d) $\dfrac{(a/b)}{(c/d)} = \dfrac{ad}{bc}$ (if $b, c, d$ are $\neq 0$).

(e) $\dfrac{1}{a} = a^{-1}$ (if $a \neq 0$).

(f) $-\left(\dfrac{a}{b}\right) = \dfrac{-a}{b} = \dfrac{a}{-b}$ (if $b \neq 0$).

We leave the proof of Theorem 1.10.4 as an exercise. The proofs of parts (a) through (e) are straightforward applications of Definition 1.10.3. The proof of part (f) is similar in spirit to the proofs of Theorems 1.5.9(b) and 1.8.4(b). (*Remark*: Regarding the requirement that denominators be nonzero, note that by Theorem 1.10.2(c), the product or quotient of nonzero elements is always nonzero. Thus, if $c$ and $d$, for instance, are nonzero, then so are $cd$ and $c/d$.)

**Exercises**

*1. Prove Theorem 1.10.4.

2. Verify, by checking each part of the definition of "field", that the ring of integers modulo 2 and the ring of integers modulo 3 are fields. What about the integers modulo 4?

3. Prove that parts (d) and (e) of Theorem 1.10.2 remain true in any integral domain. (*Hint*: Use the Difference Principle.) (SOL)

4. Show that the situation of Section 1.9, Exercise 5(a) cannot occur if the ring $S$ is an integral domain. (*Hint*: Use the result of the preceding exercise.) (SUG)

*5. Let $F, F'$ be fields, let $\mathbf{T}: F \to F'$ be a ring homomorphism, and assume that not every element of $F$ is mapped into 0 by $\mathbf{T}$. Prove:

(a) $\mathbf{T}$ maps the unity of $F$ into the unity of $F'$. (SOL)

(b) $(a^{-1})\mathbf{T} = (a\mathbf{T})^{-1}$ for all $a \in F$, $a \neq 0$. (SUG)

*Note*: This situation is covered neither by Section 1.8, Exercise 5 (we are not assuming here that $\mathbf{T}$ is onto) nor by Theorem 1.8.4 ($F$ is not a group under multiplication); hence, a new proof is required.

6. Let $F$ be a field and let $S$ be a subset of $F$. Determine, by an argument like that used for subgroups of a group (Sec. 1.6), what properties must be verified to show that $S$ is a subfield of $F$. (ANS)

## 1.11 Ordered integral domains

In the ring of integers modulo 6, the concepts of "positive" and "negative" are meaningless (why?), whereas in the field of real numbers such concepts are quite important. In this section we study systems (specifically, integral domains) in which the concepts of "positive" and "negative" are defined and the "usual" laws governing these concepts are valid.

**1.11.1 Definition.** An integral domain $D$ is called an *ordered integral domain* (abbreviated OID) if there is a subset $D^+$ of $D$ (called the set of *positive* elements of $D$) and a subset $D^-$ of $D$ (called the set of *negative* elements of $D$) having the following four properties:

(a) Every element of $D$ is positive, negative, or zero.

(b) The zero element of $D$ is not positive.

(c) If $x \in D^-$, then $-x \in D^+$.

(d) $D^+$ is closed under addition and multiplication; that is, if $x$ and $y$ are positive, then so are $x + y$ and $xy$.

The ring of integers is an OID; so is the field of real numbers. On the other hand, the field of complex numbers is not an OID.

In 1.11.1(a) it is *not* assumed that the three possibilities "positive", "negative", and "zero" are mutually exclusive. However, exclusivity can be *proved*, as follows: by 1.11.1(b) no element $x$ can be both positive and zero. If $x$ were both negative and zero, then by 1.11.1(c) the element $-x$ would be both positive and zero, R.A.A. Finally, if $x$ were both positive and negative, then by 1.11.1(c) the elements $x$ and $-x$ would both be positive; but then the element $x + (-x) = 0$ would be positive by 1.11.1(d), R.A.A.

In any OID, the converse of 1.11.1(c) holds: if $-x \in D^+$, then $x \in D^-$. (Method of proof: if $x$ is either positive or zero, we obtain a contradiction.) Thus, *an element of an OID is negative if and only if its additive inverse is positive.*

In an OID we *define* the statements

$a < b$ ($a$ is less than $b$)
$b > a$ ($b$ is greater than $a$)

to mean that $b - a$ is positive. By applying 1.11.1(a) to the element $b - a$, we see that for all elements $a$ and $b$ either $a < b$ or $a > b$ or $a = b$; these three possibilities are mutually exclusive. The notation

$a \leq b$ (or $b \geq a$)

means that either $a < b$ or $a = b$.

All of the usual laws governing inequalities are valid in OID's. We collect several such laws in the following theorem. (Laws involving fractions are not listed, since fractions need not exist in integral domains. However, see Exercise 7 at the end of this section.)

**1.11.2 Theorem.** For all elements $a, b, c, d$ of an ordered integral domain $D$, the following statements are true:

(a) $a > 0 \Leftrightarrow a$ is positive. ("$\Leftrightarrow$" means "if and only if".)
(b) $a < 0 \Leftrightarrow a$ is negative.
(c) If $a < b$ and $b < c$, then $a < c$.
(d) If $a < b$, then $-a > -b$.
(e) If $a < b$, then $a + c < b + c$.
(f) If $a < b$ and $c < d$, then $a + c < b + d$.
(g) If $a$ is positive and $b$ is negative, then $ab$ is negative.
(h) If $a$ and $b$ are negative, then $ab$ is positive.
(i) If $a < b$ and $c \in D^+$, then $ac < bc$.
(j) If $a < b$ and $c \in D^-$, then $ac > bc$.
(k) If $a \neq 0$, then $a^2$ is positive. (*Note*: $a^2$ means $a \cdot a$.)
(l) The unity of $D$ is positive.
(m) If $a$ and $ab$ are positive, then $b$ is positive.
(n) If $a$ is positive and $ab$ is negative, then $b$ is negative.
(o) If $a \in D^+$ and $ab < ac$, then $b < c$.

(p) If $a \in D^-$ and $ab < ac$, then $b > c$.
(q) If $a < b$ and $c < d$ and $a$, $b$, $c$, $d$ are all positive, then $ac < bd$.
(r) If $a, b \in D^+$ and $a^2 \leq b^2$, then $a \leq b$.

Many of these laws can be proved directly from the definitions. To prove (e), for example, we observe that $a < b$ means that $b - a \in D^+$; similarly, $a + c < b + c$ means that

$$(b + c) - (a + c) \in D^+.$$

But $(b + c) - (a + c) = b - a$ (why?); hence, (e) reduces to the statement, "If $b - a \in D^+$, then $b - a \in D^+$", which is obviously true. Similarly, (i) is equivalent to the statement, "If $b - a \in D^+$ and $c \in D^+$, then $bc - ac \in D^+$", which is true by 1.11.1(d) since $bc - ac = (b - a)c$. Parts (a)–(d) and (f) of Theorem 1.11.2 are equally direct and are left to you.

To prove (g), we use the fact (previously established) that an element is negative if and only if its additive inverse is positive. Thus, since $b$ is assumed negative, $-b$ is positive; hence, $a(-b)$ is positive by 1.11.1(d); hence, $-(ab)$ is positive (Theorem 1.9.3); hence, $ab$ is negative. We leave the proofs of (h) and (j) to you.

To prove (k), use 1.11.1(a): $a$ must be either positive or negative. If $a \in D^+$, then $a^2$ is positive by 1.11.1(d); if $a \in D^-$, then $a^2$ is positive by 1.11.2(h). Thus, in both cases, we obtain the desired result. In particular, since the unity is equal to its own square (and is nonzero by Theorem 1.9.5), (l) is a special case of (k).

The proof of (m) is indirect: we consider what happens if $b$ is *not* positive. If $b$ is negative, then by (g) $ab$ is negative, R.A.A.; if $b = 0$, then (by Theorem 1.9.3) $ab = 0$, R.A.A. Hence, $b$ is positive, since this is the only remaining possibility. The proofs of (n), (o), and (p) are similar. (q) can be proved by using (i) twice in succession, and (r) follows from (q) using an indirect proof; see if you can do these yourself. ▮

### Absolute values in an OID

Let $D$ be an OID, and suppose $y$ is an element of $D$ such that $y$ has a square root $x$ in $D$. By Section 1.9, Exercise 4, $x$ and $-x$ are the only square roots of $y$. If $y \neq 0$, then $x \neq 0$ (*why*?) and hence one of the two elements $x$, $-x$ is positive and one is negative. If $y = 0$, then $x = -x = 0$. In either case, *$y$ has exactly one nonnegative square root*, which we denote $\sqrt{y}$ or $y^{1/2}$. Note: The symbol $\sqrt{\phantom{x}}$ does *not* mean "square root of"; it means "nonnegative square root of".

For any $x \in D$, we can thus define $|x|$ (the "absolute value of $x$") by the formula

(1.11.3) $$|x| = \sqrt{x^2};$$

that is $|x|$ is the nonnegative square root of $x^2$. Equivalently, $|x|$ is either $x$ or $-x$, whichever is nonnegative.

The basic properties of absolute values in an OID are the following:

(1.11.4)
(a) $-|x| \leqslant x \leqslant |x|$
(b) $|x| = |-x|$
(c) $|x|^2 = x^2$
(d) $|xy| = |x| \cdot |y|$
(e) $|x + y| \leqslant |x| + |y|$.

We omit the detailed proofs of these properties but indicate the methods. (b) and (c) follow immediately from equation 1.11.3; (d) is a consequence of 1.11.3 and Exercise 4 below; (a) is proved by considering the two possible cases $x \geqslant 0$ and $x < 0$ separately. To prove (e), it suffices to show that $|x + y|^2 \leqslant (|x| + |y|)^2$ (see Theorem 1.11.2(r)), and the latter can be proved by expanding both sides (using 1.11.4(c) on the left side) and then using other parts of 1.11.4.

**Ordered fields**

An *ordered field* is a field $D$ satisfying 1.11.1. Since every field is automatically an integral domain (Sec. 1.10), every ordered field is automatically an ordered integral domain. The field of real numbers is an example of an ordered field.

**Exercises**

*1. Finish the proof of Theorem 1.11.2 by proving parts (a) through (d), (f), (h), (j), and (n) through (r).

2. State and prove theorems, analogous to various parts of Theorem 1.11.2, in which one or more of the symbols "<" are replaced by "≤". For which parts of 1.11.2 can this be done? (ANS)

3. In an OID, prove that if $a$ is positive and if $b \geqslant c$, then $a + b > c$.

*4. If $a$, $b$ are elements of an OID such that both $a$ and $b$ have square roots, prove that $\sqrt{a} \cdot \sqrt{b} = \sqrt{ab}$. (SOL)

*5. Prove the five parts of 1.11.4 in detail.

6. Suppose that $D$ is a *commutative ring with unity* (not necessarily an integral domain) which possesses subsets $D^+$, $D^-$ satisfying

conditions (a)–(d) of Definition 1.11.1. Prove that the product of any two nonzero elements of $D$ is nonzero. (It follows that if $D$ has more than one element, $D$ is an OID.)

*7. Prove that the following statements hold in all *ordered fields*.
(a) If $a > 0$, then $a^{-1} > 0$. (SUG)
(b) If $a, b$ are positive, then $a/b$ is positive.
(c) If $a < b$ and $c > 0$, then $a/c < b/c$.
(d) If $a < b$ and $a, b, c$ are all positive, then $c/a > c/b$. (SUG)

## 1.12 Integers and induction

In preceding sections we have often referred to the integers, tacitly assuming that the properties of integers are known to the reader. In this section we shall indicate how these properties may be developed axiomatically.

1.12.1 **Axiom.** (Axiom for the Integers) The set of all integers, denoted $\mathbf{Z}$, is an ordered integral domain satisfying the following additional condition (*Axiom of Induction*):

Whenever $S$ is a set such that
(a) the integer 1 (the unity of $\mathbf{Z}$) belongs to $S$; and
(b) whenever a positive integer $k$ belongs to $S$, then $k + 1 \in S$;
then every positive integer belongs to $S$ (that is, $\mathbf{Z}^+ \subseteq S$).

*Remark*: Assuming as known only the most basic notions of set theory, it is possible to *construct* a system $\mathbf{Z}$ satisfying Axiom 1.12.1. For a nice account of the construction (starting from scratch!), see [6].

As indicated in 1.12.1(a), the unity of $\mathbf{Z}$ is denoted 1. The integer $1 + 1$ is denoted 2; the integer $2 + 1$ is denoted 3; and so forth. (Since it is clear how to continue in this manner, we shall henceforth assume that the usual notation for integers has been established. Thus, for example, we know what is meant by "1726".)

The following argument ought to make the Axiom of Induction intuitively plausible. Let $S$ be a set satisfying conditions (a) and (b) of Axiom 1.12.1. By (a), $1 \in S$; by Theorem 1.11.2(l), 1 is positive. Hence, by (b) the integer $1 + 1$ $(=2)$ belongs to $S$. Since 2 is also positive (why?), it then follows from (b) that the integer $2 + 1$ $(=3)$ belongs to $S$. Since 3 is also positive (why?), it follows similarly that $4 \in S$. Continuing in this manner, we see that every positive integer belongs to $S$, as desired. (The phrase "continuing in this manner" is your clue that the argument is informal and nonrigorous!)

**Example**

An integer is called *even* if it is a multiple of 2 (i.e., has the form $2m$ for some $m \in \mathbf{Z}$), *odd* if it has the form $2m + 1$ for some $m \in \mathbf{Z}$. Prove that every positive integer is either even or odd.

*Solution.* Let $E$ be the set of all even integers, $D$ the set of all odd integers, $S = E \cup D$. We shall use the Axiom of Induction to show that all positive integers lie in $S$.

(a) Since $1 = 2 \cdot 0 + 1$ is odd, $1 \in D$ and, hence, $1 \in S$.

(b) Suppose $k \in S$. Then either $k \in E$ or $k \in D$. In the case $k \in E$, we have $k = 2m$ for some $m \in \mathbf{Z}$; hence $k + 1 = 2m + 1 \in D$; hence, $k + 1 \in S$. In the case $k \in D$, we have $k = 2m + 1$ (some $m \in \mathbf{Z}$); hence, $k + 1 = 2m + 1 + 1 = 2m + 2 = 2(m + 1)$ which is a multiple of 2; hence, $k + 1$ lies in $E$; hence, $k + 1$ also lies in $S$. Thus, we have shown in all cases that $k \in S \Rightarrow k + 1 \in S$.

(c) By (a) and (b), it follows from the Axiom of Induction that every positive integer is in $S$, that is, in $E \cup D$. Thus, every positive integer is either even or odd. ∎

The Axiom of Induction is often expressed in the following equivalent form, called the *Principle of Induction*:

> Suppose we have, corresponding to each positive integer $n$, a statement $\mathcal{S}(n)$. Assume further that
>
> **(1.12.2)** (a) the statement $\mathcal{S}(1)$ is true;
>
> (b) whenever $k \in \mathbf{Z}^+$ and $\mathcal{S}(k)$ is true, then $\mathcal{S}(k + 1)$ is true.
>
> Then $\mathcal{S}(n)$ is true for all positive integers $n$.

It is not hard to see that 1.12.2 is a consequence of 1.12.1. We leave it to you to supply the argument. Proofs based on 1.12.2 are usually called *proofs by induction*.

**Example**

Show that the equation

$$2(1 + 2 + \cdots + n) = n(n + 1)$$

holds for all positive integers $n$.

(Note: The precise definition of expressions like "$1 + 2 + \cdots + n$" appears later in the section. For now, regard this example only as an illustration of a method of proof.)

*Solution.* Let $\mathscr{S}(n)$ be the statement "$2(1 + \cdots + n) = n(n + 1)$". For $n = 1$ the equation reduces to $2(1) = 1(2)$ which is clearly true (multiplication in $\mathbf{Z}$ being commutative); thus, 1.12.2(a) holds. To prove 1.12.2(b), suppose $k$ is a positive integer for which $\mathscr{S}(k)$ is true; that is,

$$2(1 + \cdots + k) = k(k + 1).$$

We then have

$$\begin{aligned} 2(1 + \cdots + (k + 1)) &= 2[(1 + \cdots + k) + (k + 1)] \\ &= 2(1 + \ldots + k) + 2(k + 1) \qquad \text{(why?)} \\ &= k(k + 1) + 2(k + 1) \qquad (\text{since } \mathscr{S}(k) \text{ is true}) \\ &= (k + 2)(k + 1) \qquad \text{(why?)} \\ &= (k + 1)(k + 2) \\ &= (k + 1)[(k + 1) + 1] \end{aligned}$$

which is precisely the statement $\mathscr{S}(k + 1)$. Thus, we have shown that $\mathscr{S}(k)$ implies $\mathscr{S}(k + 1)$, that is, 1.12.2(b) holds. It follows, by the Principle of Induction, that $\mathscr{S}(n)$ is true for all positive integers $n$; that is, $2(1 + \cdots + n) = n(n + 1)$ for all such $n$. ∎

**1.12.3 Theorem.** (a) Let $n$ be an integer. Then $n$ is positive if and only if $n \geqslant 1$. (b) 1 is the least positive integer.

*Proof.* For the sake of variety we appeal to 1.12.1 rather than 1.12.2. Let

$$S = \{n \in \mathbf{Z} : n \geqslant 1\}.$$

($S$ is the set of all integers which are greater than or equal to 1). Since $1 \geqslant 1$, $1 \in S$. Also, if $k$ is any positive integer in $S$ then $k$ is *a fortiori* positive; hence, $k + 1 > 1$; hence, $k + 1 \in S$. It follows from Axiom 1.12.1 that all positive integers belong to $S$; that is, $n \geqslant 1$ whenever $n$ is a positive integer. Conversely, if $n \geqslant 1$, then the fact that 1 is positive (Theorem 1.11.2(l)) implies that $n$ must be positive. Theorem 1.12.3(a) follows; and 1.12.3(b) is an immediate consequence of 1.12.3(a). ∎

**1.12.4 Theorem.** There do not exist integers $n$, $x$ such that $n < x < n + 1$. (In other words, for each integer $n$, the next larger integer is $n + 1$.)

*Proof.* If $n < x < (n + 1)$, then $(n - n) < (x - n) < ((n + 1) - n)$; that is, $0 < (x - n) < 1$. Hence, the integer $x - n$ is positive but less than 1, contradicting 1.12.3(b). ▮

1.12.5 **Theorem.** If $S$ is any nonempty subset of $\mathbf{Z}^+$, then $S$ contains a least element (i.e., there is a unique element of $S$ which is less than all other elements of $S$).

*Proof.* Let $\mathcal{S}(n)$ be the statement: "Every subset of $\mathbf{Z}^+$ which contains $n$ contains a least element." If we can show that $\mathcal{S}(n)$ is true for all $n \in \mathbf{Z}^+$, the theorem will be proved; indeed, since we have assumed that $S$ is nonempty, $S$ must contain *some* positive integer $n$, and then the truth of the statement $\mathcal{S}(n)$ implies that $S$ has a least element.

Certainly $\mathcal{S}(1)$ is true: if any subset $S$ of $\mathbf{Z}^+$ contains 1, then 1 itself is the least element of $S$ by Theorem 1.12.3(b). Now assume that $k$ is a positive integer for which $\mathcal{S}(k)$ is true; it remains only to show that $\mathcal{S}(k + 1)$ is true. Let $S$ be a subset of $\mathbf{Z}^+$ which contains the integer $k + 1$. If $k \in S$, then the truth of $\mathcal{S}(k)$ implies that $S$ has a least element. If on the other hand $k \notin S$, then the truth of $\mathcal{S}(k)$ implies that the set

$$T = S \cup \{k\}$$

has a least element. If the least element of $T$ lies in $S$, it is automatically the least element of $S$ (why?). If, instead, the least element of $T$ is $k$ (the only remaining possibility!), then all elements of $S$ are $> k$, and hence are $\geqslant k + 1$ (Theorem 1.12.4), so that $k + 1$ itself (which by assumption lies in $S$) is the least element of $S$. We have thus shown *in all possible cases* that any subset of $\mathbf{Z}^+$ containing $k + 1$ must have a least element. Hence, $\mathcal{S}(k + 1)$ is true, completing our proof. ▮

Theorem 1.12.5 would become false if we replaced $\mathbf{Z}^+$ by $\mathbf{Z}$; for example, the set of all odd integers (positive and negative) is a nonempty subset of $\mathbf{Z}$ which contains no least element. However, we can still generalize Theorem 1.12.5 to some extent. For the generalization, we need a definition: let $S$ be a subset of an ordered integral domain $D$. An element $c \in D$ is called a *lower bound of* $S$ if $c$ is less than or equal to every element of $S$. (Similarly, $c$ is an *upper bound* of $S$ if $c \geqq n$ for all $n \in S$.) In the set $\mathbf{Z}$, for example, 4 is a lower bound of the set $\{6, 7, 8\}$; $-2$ is a lower bound of the set $\{-2, 0, 2\}$; 7 is an upper bound of the set $\{1, 2, 3\}$. Now suppose $S$ is any subset of $\mathbf{Z}$. In view of Theorem 1.12.3, it is clear that $S$ will be a subset of $\mathbf{Z}^+$ (i.e., all elements of $S$ will be positive) if and only if $S$ has 1 as a lower bound. Hence, Theorem 1.12.5 may be restated as follows: *if $S$ is a nonempty subset of* **Z** *which has 1 as a lower bound, then $S$ has a least element*. We now state our generalization of this result.

**1.12.6** **Theorem.** Let $S$ be a nonempty subset of **Z**. If $S$ has a lower bound in **Z**, then $S$ contains a least element.

*Proof.* The statement "$S$ has a lower bound in **Z**" means that every element of $S$ is greater than or equal to some fixed integer $c$. By adding $1 - c$ to every element of $S$, we obtain a new nonempty set $T$ all of whose elements are greater than or equal to $c + (1 - c) = 1$; that is, all elements of $T$ are *positive*. Hence, by Theorem 1.12.5, the set $T$ contains a least element, say $t_0$. The element $t_0 - (1 - c)$ is then the least element of $S$ (*why*?), as desired. ∎

As a numerical illustration of the proof just completed, suppose $S = \{-5, 2, 9\}$. Certainly the integer $c = -8$ is a lower bound of $S$ in **Z**. By adding $1 - c = 1 - (-8) = 9$ to each element of $S$, we obtain the set

$$T = \{4, 11, 18\}$$

all of whose elements are positive. The least element of $T$ is 4; the integer $4 - 9$ $(= -5)$ is the least element of $S$.

The technique used to prove Theorem 1.12.6 is a fairly common one: first prove a special case of the desired result, then *use the special case in proving the general case*; equivalently, reduce the general case to the special case. (In the proof above, for instance, we established the desired property of a set having the arbitrary lower bound $c$ by relating this set to a set having the specific lower bound 1.) You may remember seeing the same technique used in elementary calculus; for example, one proves the Mean Value Theorem by reducing it to its special case, Rolle's Theorem.

Just as Theorem 1.12.6 generalizes 1.12.5, so there are analogous generalizations of the Axiom of Induction and the Principle of Induction. The generalization of the Principle of Induction may be stated as follows:

(1.12.7) Let $c$ be a fixed integer. Suppose we have, corresponding to each integer $n \geqslant c$, a statement $\mathscr{S}(n)$. Assume further that

(a) The statement $\mathscr{S}(c)$ is true;

(b) Whenever $k$ is an integer $\geqslant c$ such that $\mathscr{S}(k)$ is true, then $\mathscr{S}(k + 1)$ is true.

Then $\mathscr{S}(n)$ is true for all integers $n \geqslant c$.

The derivation of 1.12.7 from 1.12.2 is like that of 1.12.6 from 1.12.5.

Even 1.12.7 is not the strongest possible form of the Principle of Induction. The following form of the Principle says even more:

> Let $c$ be a fixed integer. Suppose we have, corresponding to each integer $n \geqslant c$, a statement $\mathscr{S}(n)$. Assume further that
>
> (a) The statement $\mathscr{S}(c)$ is true;
>
> **(1.12.8)** (b) Whenever $k$ is an integer $\geqslant c$ such that the statements $\mathscr{S}(n)$ are true for all integers $n$ such that $c \leqslant n \leqslant k$, then $\mathscr{S}(k + 1)$ is true.
>
> Then $\mathscr{S}(n)$ is true for all integers $n \geqslant c$.

To illustrate how 1.12.8 strengthens 1.12.7, suppose we wish to prove that a certain statement about integers is true for all integers greater than or equal to 4. According to either 1.12.7 or 1.12.8, there are two steps in the proof. The first step is to show that the statement is true for the integer 4. The second step, if we use 1.12.7, is to prove the statement for $k + 1$ under the assumption that it holds for $k$ (also assuming $k \geqslant 4$). If we use 1.12.8, though, our task is slightly easier: when proving the statement for $k + 1$, we are allowed to assume its validity, not just for the integer $k$, *but for all integers from 4 to k, inclusive.* In some proofs by induction we can get by with 1.12.7; but in some proofs, the stronger form 1.12.8 is needed.

We have not yet shown that 1.12.8 is actually valid; let us do so now. Assume that we have statements $\mathscr{S}(n)$ $(n \geqslant c)$ satisfying (a) and (b) of 1.12.8. Let $S$ be the set of all integers $n$ such that $n \geqslant c$ and $\mathscr{S}(n)$ is *not* true. Suppose first that $S$ is nonempty. Since $c$ is a lower bound of $S$, Theorem 1.12.6 implies that $S$ has a least element $s_0$. Since $s_0 \in S$, $\mathscr{S}(s_0)$ is not true; hence, $s_0 \neq c$ by 1.12.8(a); hence, $s_0 \geqslant c + 1$ by 1.12.4; hence, $s_0 - 1 \geqslant c$. For all integers $n$ such that $c \leqslant n \leqslant s_0 - 1$, $n \notin S$ and, hence, $\mathscr{S}(n)$ is true. Hence, by 1.12.8(b), the statement $\mathscr{S}(s_0)$ is true, R.A.A. Since this contradiction arose from the assumption that $S \neq \varnothing$, we conclude that $S$ is empty. But this means that $\mathscr{S}(n)$ is true for all integers $n \geqslant c$, as desired. ▮

**Definitions by induction**

The various forms of the Principle of Induction (1.12.2, 1.12.7, 1.12.8) are used primarily as tools for proving theorems. Analogous tools are available for the purpose of making definitions. For instance, suppose we wish to define a quantity $f(n)$ for all *positive* integers $n$. One valid way of doing so is to perform the following two operations:

(1.12.9)
(a) Define $f(1)$.
(b) Define $f(k + 1)$ in terms of $f(k)$ ($k \in \mathbf{Z}^+$). (That is, define $f(k + 1)$ under the assumption that $k$ is a positive integer for which $f(k)$ already exists.)

A definition constructed in this manner is called a *definition by induction*, or *recursive definition*. As an exercise, you yourself should try to formulate the more general definition-by-induction procedures which correspond to 1.12.7 and 1.12.8 in the same way that 1.12.9 corresponds to 1.12.2.

Although "definition by induction" looks very much like proof by induction, its validity is much harder to establish; a correct proof of validity is quite tricky. If you are ambitious, you can find a correct proof in [11, Section 2.7, paragraph 7J]. (Incorrect "proofs" are also available in print; cf. Exercise 1 at the end of this section.)

**1.12.10 Example.** (Definition of positive integral exponents) Let "*" be an associative binary operation on a set $S$, and let $a$ be any element of $S$. We define

$$\begin{aligned} a^1 &= a \\ a^{k+1} &= a^k * a \qquad (\text{all } k \in \mathbf{Z}^+). \end{aligned} \tag{1.12.11}$$

According to 1.12.9, we have now defined $a^n$ for all positive integers $n$.

**1.12.12 Example.** Let $S$ be a set on which an associative operation of *addition* is defined, and suppose that to each positive integer $k$ there corresponds an element $a_k$ of $S$. The expression

$$\sum_{i=1}^{n} a_i$$

(a "sum of $n$ terms") is defined for all positive integers $n$ by the equations

$$\begin{aligned} \sum_{i=1}^{1} a_i &= a_1 \\ \sum_{i=1}^{k+1} a_i &= \left(\sum_{i=1}^{k} a_i\right) + a_{k+1} \qquad (\text{all } k \in \mathbf{Z}^+). \end{aligned} \tag{1.12.13}$$

The expression $\sum_{i=1}^{n} a_i$ is also written $a_1 + \cdots + a_n$.

*Remark*: If the set $S$ in Example 1.12.12 has an additive identity "0", we define

$$\sum_{i=1}^{0} a_i$$

(a sum consisting of *no* terms, the "empty sum") to be 0. The reason for doing so is that under this definition of the empty sum, the second equation 1.12.13 remains true when $k = 0$.

**1.12.14 Example.** If an associative operation *other than addition* (for instance, multiplication) is defined on $S$, we define the "product"

$$\prod_{i=1}^{n} a_i \qquad (a_i \in S)$$

in a manner analogous to our definition of $\Sigma_{i=1}^{n} a_i$ in Example 1.12.12. The "empty product" (product of no terms) is defined to be the identity of $S$ if an identity exists; this is, of course, analogous to the definition of "empty sum" given above.

**1.12.15 Example.** If $n$ is a nonnegative integer, the product $\Pi_{i=1}^{n} i$ (the product of all integers from 1 to $n$) is denoted "$n!$" and is called *n factorial*. Thus, $0! = 1! = 1$, and $(k + 1)! = (k!)(k + 1)$ for all positive integers $k$.

**Properties of exponents**

Let "$*$" be an associative binary operation on a set $S$. Using 1.12.11, the following properties of exponents can be proved for all $a, b \in S$ and all $m, n \in \mathbf{Z}^+$:

(1.12.16)
- (a) $a^{m+n} = a^m * a^n$.
- (b) $(a^m)^n = a^{mn}$.
- (c) If $a$ commutes with $b$ (that is, $a * b = b * a$), then $a$ commutes with $b^n$.
- (d) If $a$ commutes with $b$, then $(a * b)^n = a^n * b^n$.
- (e) If $S$ has an identity $e$, then $e^n = e$.

All five of the statements 1.12.16 can be proved by induction; see if you can do them. When proving (a) and (b), regard $m$ as fixed and use induction on $n$ (that is, in proving (a), let $\mathscr{S}(n)$ be the statement "$a^{m+n} = a^m * a^n$"; in proving (b), let $\mathscr{S}(n)$ be the statement "$(a^m)^n = a^{mn}$"). In case you are lazy, the proofs of (a) through (d) are given in [11].

Again let "$*$" be an associative operation on $S$, and assume that $S$ has an identity $e$ with respect to this operation. We then define

(1.12.17) $a^0 = e \qquad (\text{all } a \in S)$.

In addition, if $a$ is an element of $S$ which possesses an inverse $a' \in S$ with respect to the operation "$*$", it can be proved that

**(1.12.18)** $(a')^n = (a^n)'$ (all $n \in \mathbf{Z}^+$)

where the prime symbol $'$ denotes inverse. (See Exercise 2 below.) The quantity 1.12.18 is denoted $a^{-n}$, giving us a definition of *negative exponents*. Thus,

**(1.12.19)** $a^{-n} = (a')^n = (a^n)'$ (all $n \in \mathbf{Z}^+$)

provided that $a$ has an inverse. Note that when $n = 1$, equations 1.12.19 reduce to $a^{-1} = a'$, which is consistent with the standard notation for inverses which we have used throughout.

The equations 1.12.19 hold not only when $n$ is a positive integer but also when $n$ is zero or a negative integer. The proof for $n = 0$ is easy. In order to prove 1.12.19 for negative integers $n$, it is useful to set $r = -n$; since $r$ is then positive, 1.12.19 holds for $r$ in place of $n$, and we have

$$a^{-n} = a^r;$$

$$(a')^n = (a')^{-r} = [\,(a')']^r \qquad \text{(by 1.12.19 with } r \text{ in place of } n \text{ and } a' \text{ in place of } a\text{)}$$

$$= a^r \qquad \text{(by Theorem 1.5.9(a));}$$

$$(a^n)' = (a^{-r})' = [\,(a^r)']' \qquad \text{(by 1.12.19 with } r \text{ in place of } n\text{)}$$

$$= a^r \qquad \text{(Theorem 1.5.9(a)).}$$

Since all sides of 1.12.19 are equal to the same thing, 1.12.19 is established for negative $n$.

It can now be shown that the five equations 1.12.16 hold for *all* integers $m$ and $n$, whether positive, negative, or zero (with the proviso that only elements having inverses may be raised to negative powers). In general, the method of proof is to introduce the additive inverses of any negative exponents that occur (just as we introduced $r = -n$ in the proof above) and then use the fact that 1.12.16 is already known for *positive* exponents and 1.12.19 is known for *all* exponents. In the proofs of 1.12.16(a) and 1.12.16(b), several cases occur, depending on the signs of $m$, $n$, and (in 1.12.16(a)) $m + n$. You should try at least some of these proofs as exercises.

**Integral multiples**

If $S$ is an abelian group under addition, we define $na$ ($n \in \mathbf{Z}$, $a \in S$) in a manner exactly analogous to the definition of $a^n$ in a set with an operation other than addition. Thus, the additive analog of 1.12.11 is

$$\text{(1.12.20)}\quad \begin{aligned} 1a &= a \\ (k+1)a &= ka + a \qquad (k \in \mathbf{Z}^+), \end{aligned}$$

the analogs of 1.12.17 and 1.12.19 are

$$\text{(1.12.21)}\quad \begin{aligned} 0a &= 0_S \\ (-n)a &= n(-a) = -(na) \end{aligned}$$

and the analogs of parts (a), (b), (d), (e) of 1.12.16 are

$$\text{(1.12.22)}\quad \begin{aligned} (m+n)a &= ma + na \\ n(ma) &= (mn)a \\ n(a+b) &= na + nb \\ n0_S &= 0_S. \end{aligned}$$

In 1.12.21 and 1.12.22 the subscript $S$ indicates the zero element of $S$, as opposed to the zero integer which we write without a subscript.

Although equations 1.12.20–1.12.22 look very much like properties of rings, they are not quite the same as ring properties since two different sets are now involved: the elements $a$, $b$ belong to $S$ while the integers $m$, $n$ belong to $\mathbf{Z}$. It *may* happen, of course, that $S = \mathbf{Z}$; in this case, the fact that $\mathbf{Z}$ is a ring implies that 1.12.20 and 1.12.21 hold for ordinary multiplication in $\mathbf{Z}$, so that the definition of "$na$" as given above is consistent with ordinary multiplication in $\mathbf{Z}$.

**Exercises**

1. In McCoy, [10, Section 3.3], a brief argument is given to justify the method of recursive definition (definition by induction) as applied specifically to the definition of positive exponents. Look up this argument and find a logical fallacy in it. (ANS)

*2. Prove equation 1.12.18. (*Suggestion*: This equation asserts that a certain element is the inverse of another element. Imitate previous proofs of such assertions. Do *not* use induction.) (SUG)

3. Prove by induction that the following statements hold for all $n \in \mathbf{Z}^+$:
   (a) $2^n > n$. (SOL)
   (b) $6(1^2 + 2^2 + \cdots + n^2) = n(n+1)(2n+1)$.

*4. Prove by induction that the following statement holds for all $n \in \mathbf{Z}^+$: if $a_1, \ldots, a_n$ are elements of an OID and we let $S_n = a_1^2 + \cdots + a_n^2$, then
   (a) $S_n \geq 0$;

(b) If at least one of the elements $a_1, \ldots, a_n$ is nonzero, then $S_n > 0$.

*5. State the generalized definition-by-induction procedure which corresponds to 1.12.7 in the same way that 1.12.9 corresponds to 1.12.2.

*6. Let $S$ be a set on which there is an associative operation of addition. Let $m$ be a fixed integer, and assume that to each integer $k \geqslant m$ there corresponds an element $a_k$ of $S$. Using the procedure from the preceding exercise, define

$$\sum_{i=m}^{n} a_i$$

for all integers $n \geqslant m$. (ANS)

*7. Let $S$, $m$, $a_i$ be as in the preceding exercise. If $r$ and $n$ are integers such that $m \leqslant r < n$, prove that

$$\sum_{i=m}^{r} a_i + \sum_{i=r+1}^{n} a_i = \sum_{i=m}^{n} a_i.$$

(*Suggestion*: Regard $r$ as fixed and use induction on $n$. The first step is to prove that the equation holds when $n = r + 1$; then go on from there.)

*8. Prove the *generalized distributive law*

$$\sum_{i=1}^{n} (ca_i) = c\left(\sum_{i=1}^{n} a_i\right) \qquad (\text{all } n \in \mathbf{Z}^+)$$

where $c$ and the elements $a_i$ are assumed to belong to a ring $S$. (The process of replacing the left side by the right side is usually called "factoring out $c$".) (SOL)

*9. If $a_i$ $(1 \leqslant i \leqslant n)$ and $b_i$ $(1 \leqslant i \leqslant n)$ are elements of an additive abelian group, prove that

$$\sum_{i=1}^{n} (a_i + b_i) = \sum_{i=1}^{n} a_i + \sum_{i=1}^{n} b_i.$$

*10. If $x_{ij}$ $(1 \leqslant i \leqslant m,\ 1 \leqslant j \leqslant n)$ are $mn$ elements of an additive abelian group, prove that

$$\sum_{i=1}^{m}\left(\sum_{j=1}^{n} x_{ij}\right) = \sum_{j=1}^{n}\left(\sum_{i=1}^{m} x_{ij}\right).$$

In other words, it is permissible to "reverse the order of summation". (*Suggestion*: Regard $m$ as fixed and use induction on $n$.)

11. Using the definitions given in the section, show that for all positive integers $n$, $a^n$ means the same thing as $\Pi_{i=1}^{n} a$ and $na$ means the same thing as $\Sigma_{i=1}^{n} a$.

## 1.13 The integers modulo *m*

Let $m$ be a fixed positive integer. If $a, b \in \mathbf{Z}$, we say that *$a$ is congruent to $b$ modulo $m$*, written

$$a \equiv b \pmod{m},$$

if there exists an integer $x$ such that $a - b = mx$ (that is, if $a - b$ is divisible by $m$). By Section 1.4, Exercise 5, congruence modulo $m$ is an equivalence relation on $\mathbf{Z}$. We shall denote by $[a]_{(m)}$ the equivalence class which contains the element $a$; that is, the set $[a]_{(m)}$ consists of all integers which are congruent to $a$, modulo $m$. We denote by $\mathbf{Z}_{(m)}$ the set of all equivalence classes; that is,

$$\mathbf{Z}_{(m)} = \{[a]_{(m)} : a \in \mathbf{Z}\}.$$

**1.13.1 Theorem.** If $a \equiv a' \pmod{m}$ and $b \equiv b' \pmod{m}$, then
(a) $a + b \equiv a' + b' \pmod{m}$;
(b) $ab \equiv a'b' \pmod{m}$.

*Proof.* By hypothesis, $a - a' = mx$ and $b - b' = my$ for some $x, y \in \mathbf{Z}$. Hence,

$$\begin{aligned}(a + b) - (a' + b') &= (a - a') + (b - b')\\ &= mx + my = m(x + y)\end{aligned}$$

proving 1.13.1(a). To prove 1.13.1(b), it will suffice (since congruence is transitive) to prove that $ab \equiv a'b$ and $a'b \equiv a'b' \pmod{m}$. But

$$\begin{aligned}ab - a'b &= (a - a')b = (mx)b = m(xb)\\ a'b - a'b' &= a'(b - b') = a'(my) = m(a'y)\end{aligned}$$

and the result follows. ∎

It follows from Theorem 1.13.1(a) that if $S = [a]_{(m)}$ and $T = [b]_{(m)}$ are any two elements of $\mathbf{Z}_{(m)}$, and if we add any element $a'$ of $S$ to any element $b'$ of $T$, the sum $a' + b'$ will belong to the class $U = [a + b]_{(m)}$. This class $U$ is thus the same no matter which elements $a' \in S$, $b' \in T$ are chosen; that is, *$U$ depends only on $S$ and $T$*, not on $a'$ and $b'$. It follows that if we define $U$ to be the *sum* of $S$ and $T$, we have a well-defined binary operation of addition on $\mathbf{Z}_{(m)}$:

**(1.13.2)** $[a]_{(m)} + [b]_{(m)} = [a + b]_{(m)}$

for all $a, b$. By similar reasoning (using Theorem 1.13.1(b)), the binary operation of multiplication on $\mathbf{Z}_{(m)}$ defined by

**(1.13.3)** $[a]_{(m)}[b]_{(m)} = [ab]_{(m)}$

is well defined.

**1.13.4 Theorem.** The equivalence classes $[0]_{(m)}, [1]_{(m)}, \ldots, [m-1]_{(m)}$ are distinct and are the only elements of $\mathbf{Z}_{(m)}$. In other words, given any integer $n$, $n$ is congruent (modulo $m$) to exactly one integer $r$ such that $0 \leqslant r \leqslant m - 1$.

*Sketch of proof.* We assume $m > 1$ (if $m = 1$, then *all* integers are congruent to 0 and the result is trivial). Suppose $a, b$ are two distinct integers ($0 \leqslant a < m, 0 \leqslant b < m$) such that $[a]_{(m)} = [b]_{(m)}$. Without loss of generality we may assume $a < b$. Then $b \equiv a \pmod{m}$ so that $b - a = mx$; since $b - a$ and $m$ are positive, so is $x$ (Theorem 1.11.2(m)); hence, $x \geqslant 1$; hence, $mx \geqslant m > b \geqslant b - a = mx$, R.A.A. This shows that the classes $[0]_{(m)}, \ldots, [m-1]_{(m)}$ are distinct. If $n \in \mathbf{Z}^+$, the fact that $[n]_{(m)}$ equals one of the classes $[0]_{(m)}, \ldots, [m-1]_{(m)}$ may be proved by induction on $n$ (using the obvious fact that $[m]_{(m)} = [0]_{(m)}$). The same result for negative $n$ follows from the result for positive $n$, since if $n < 0$, then $[n]_{(m)} = [n - nm]_{(m)}$ and $n - nm > 0$. ∎

**Example**

The distinct elements of $\mathbf{Z}_{(6)}$ are [0], [1], [2], [3], [4], [5] (omitting the subscript ($m$) to save space). Every element of $\mathbf{Z}_{(6)}$ is equal to one of these six elements; for example, since $20 - 2 = 18 = 6 \cdot 3$, we have $20 \equiv 2 \pmod{6}$ and, hence,

$$[20] = [2].$$

Similarly, $[2] \cdot [5] = [10] = [4]$; $[3] + [4] = [1]$; etc.

**1.13.5 Theorem.** $\mathbf{Z}_{(m)}$ is a commutative ring under the operations of addition and multiplication defined by 1.13.2 and 1.13.3. The zero element of $\mathbf{Z}_{(m)}$ is $[0]_{(m)}$; the unity of $\mathbf{Z}_{(m)}$ is $[1]_{(m)}$; the additive inverse of any element $[a]_{(m)} \in \mathbf{Z}_{(m)}$ is $[-a]_{(m)}$.

*Proof.* Each of the needed properties of $\mathbf{Z}_{(m)}$ follows easily from the analogous property of $\mathbf{Z}$; we shall give the proof of only one property (to show how it's done) and will leave the others as exercises.

Let us prove the left distributive law. For any elements $[a]_{(m)}$, $[b]_{(m)}$, $[c]_{(m)}$ in $\mathbf{Z}_{(m)}$ we have

$$\begin{aligned}
[a]_{(m)}([b]_{(m)} + [c]_{(m)}) &= [a]_{(m)}[b + c]_{(m)} && \text{(by 1.13.2)}\\
&= [a(b + c)]_{(m)} && \text{(by 1.13.3)}\\
&= [ab + ac]_{(m)} && \text{(by the distributive law in } \mathbf{Z})\\
&= [ab]_{(m)} + [ac]_{(m)} && \text{(by 1.13.2)}\\
&= [a]_{(m)}[b]_{(m)} + [a]_{(m)}[c]_{(m)} && \text{(by 1.13.3)}
\end{aligned}$$

which is the desired result. ∎

*Remark*: Theorem 1.13.4 implies that $\mathbf{Z}_{(m)}$ has exactly $m$ elements (the concept of the "number of elements" in a set will be defined formally in Section 1.14); thus there exists a ring having any specified number of elements. In particular, the two distinct elements of $\mathbf{Z}_{(2)}$ are $[0]_{(2)}$ and $[1]_{(2)}$. It is easy to see that $[0]_{(2)}$ is precisely the set of all even integers and that $[1]_{(2)}$ is the set of odd integers (proof?); thus every integer is either even or odd, but no integer is both even and odd. (This strengthens the result of the first example in Section 1.12.)

It can be shown that if $p > 1$ is a prime (that is, an integer whose only integral divisors are $\pm 1$ and $\pm p$) then the ring $\mathbf{Z}_{(p)}$ is actually a field. A proof may be found in McCoy [10], Chapter 5. In particular, $\mathbf{Z}_{(2)}$ is a field; thus a field exists which has only two elements. (By definition, every field must have *at least* two elements, the zero and the unity.)

**Exercises**

1. Write down the addition table and multiplication table for the ring $\mathbf{Z}_{(6)}$, using only the elements [0], [1], ..., [5].
2. Show that Theorem 1.13.4 is equivalent to the following statement (usually called the *division algorithm*): if $m \in \mathbf{Z}^+$, then every integer $x$ can be expressed uniquely in the form

   $x = qm + r$

   where $q \in \mathbf{Z}$, $r \in \mathbf{Z}$, $0 \leq r < m$.
3. Prove the remaining parts of Theorem 1.13.5.

## 1.14 Finite and infinite sets

Consider a set $S$ consisting of the following elements: a triangle, a circle, a square, an asterisk, and a plus sign. When we say that this set

$S$ has *five elements,* we mean that the integers from 1 to 5 can be "attached" to the elements of $S$ in such a way that each of these integers corresponds to exactly one element of $S$ and vice versa. In other words, *we have a one-to-one correspondence between the set S and the set of all integers from 1 to 5.* Figure 1.14.1 exhibits two of the many possible ways of setting up such a correspondence.

Since there is nothing special about the number 5, it is clear that the preceding paragraph can be generalized. Before doing so, we introduce two items of terminology. First: if $A$ and $B$ are sets, we say that $A$ is *equipotent* to $B$ (*notation:* $A \sim B$) if there exists a one-to-one mapping of $A$ onto $B$. (Many authors use the word "equivalent" instead of "equipotent".) Second: if $n$ is a nonnegative integer, we define

$$E_n = \{x \in \mathbf{Z}^+ : 1 \leqslant x \leqslant n\};$$

that is, $E_n$ is the set of all positive integers from 1 to $n$. (Note that $E_0$ is the empty set.) The italicized statement in the preceding paragraph can now be rephrased as: "$S$ is equipotent to $E_5$." Similarly, if $n$ is any nonnegative integer, we define the phrase

*S has n elements*

to mean that $S$ is equipotent to $E_n$. (Equivalently, $E_n$ is equipotent to $S$; cf. Lemma 1.14.2 below.) It will be shown later (Theorem 1.14.11) that the number of elements in $S$ is *unique*; that is, $S$ cannot be equipotent to both $E_n$ and $E_m$ if $n$ and $m$ are distinct nonnegative integers.

In particular, the set $E_n$ itself has $n$ elements; the empty set $E_0$ has 0 elements. The ring of integers modulo $m$ has exactly $m$ elements, since by Theorem 1.13.4 the mapping of $E_m \to \mathbf{Z}_{(m)}$ defined by

$$x \to [x - 1]_{(m)}$$

is one-to-one and onto.

If $S$ has $n$ elements, these elements may be labeled $a_1, ..., a_n$. Indeed, since there is a one-to-one mapping $\mathbf{T}$ of $E_n$ onto $S$, we can simply let $a_i$ denote the image of $i$ under $\mathbf{T}$, for all $i \in E_n$. The mapping $\mathbf{T}$ is not unique, of course; in fact, there are $n!$ such mappings (Exercise 6 below). Choosing a different mapping of $E_n \to S$ in place of $\mathbf{T}$

Figure 1.14.1

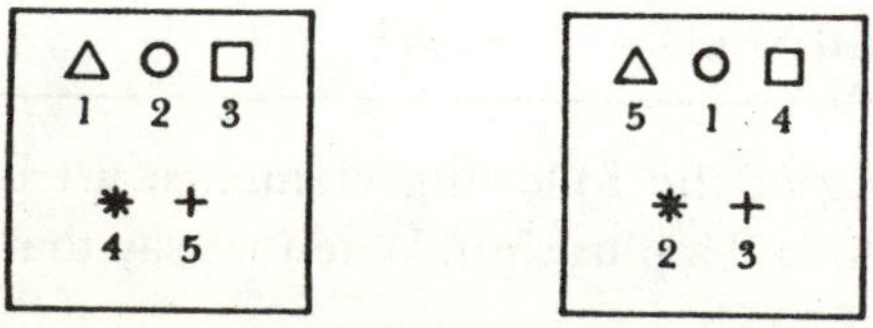

amounts to the same thing as writing the elements of $S$ in a different order.

**1.14.1 Theorem.** If $S$ is any set, then the following two statements are equivalent (that is, either one is true if and only if the other is true):

(a) There exists an integer $n \geq 0$ such that $S$ has $n$ elements.

(b) Every one-to-one mapping of $S \to S$ is onto.

The set $S$ is said to be *finite* if statements 1.14.1(a) and 1.14.1(b) are true (equivalently, if either one of them is true). Statement (a) is, of course, the "usual" definition of finiteness. Statement (b), which may be unfamiliar to you (unless you have had a course in set theory), shows that the concept of finiteness may be defined without mentioning integers at all! A set is called *infinite* if it is not finite.

To prove Theorem 1.14.1 we need a sequence of lemmas. For convenience in stating them, we shall call a set $S$ "$\alpha$-finite" if it satisfies 1.14.1(a), "$\beta$-finite" if it satisfies 1.14.1(b). Proofs of most of the lemmas are omitted; you should do them as exercises.

**1.14.2 Lemma.** "Equipotence" is an equivalence relation. That is,

(a) Every set $S$ is equipotent to itself;

(b) If $S \sim T$, then $T \sim S$;

(c) If $R \sim S$ and $S \sim T$, then $R \sim T$.

**1.14.3 Lemma.** The empty set is both $\alpha$-finite and $\beta$-finite.

**1.14.4 Lemma.** Suppose that $n$ is a nonnegative integer, $S \sim E_n$, and $x$ is an element not in $S$. Then $S \cup \{x\} \sim E_{n+1}$.

**1.14.5 Lemma.** If $S$ is $\beta$-finite and if $x$ is an element not in $S$, then $S \cup \{x\}$ is $\beta$-finite.

*Proof.* Let $R = S \cup \{x\}$ and let $\mathbf{T}: R \to R$ be one-to-one. We must prove that $\mathbf{T}$ is onto.

*Case 1. $s\mathbf{T} \in S$ for all $s \in S$.* In this case, $\mathbf{T} \upharpoonright S$ (the restriction of $\mathbf{T}$ to $S$) is a one-to-one mapping of $S$ into $S$; hence, it maps $S$ *onto* $S$, since $S$ is $\beta$-finite. It follows that $x\mathbf{T} \notin S$ (why?); hence, $x\mathbf{T} = x$; hence, $\mathbf{T}$ is onto (why?).

*Case 2. $s_1\mathbf{T} = x$ for some $s_1 \in S$.* Then $x\mathbf{T} \neq x$; hence $x\mathbf{T} = s_2$ for some $s_2 \in S$. Define $\mathbf{T}' : R \to R$ by

$$s_1\mathbf{T}' = s_2$$
$$x\mathbf{T}' = x$$
$$s\mathbf{T}' = s\mathbf{T} \qquad \text{(all elements } s \neq s_1 \text{ in } S).$$

Since $\mathbf{T}'$ is obtained from $\mathbf{T}$ simply by interchanging the images of $s_1$ and $x$, it is clear that $\mathbf{T}'$ is still one-to-one. Moreover, since $\mathbf{T}'$ maps $x \to x$ it must map elements of $S$ into elements of $S$. Hence, $\mathbf{T}'$ is onto by the result of Case 1. Since $\mathbf{T}$ has the same range as $\mathbf{T}'$, it follows that $\mathbf{T}$ is onto. ▮

**1.14.6 Lemma.** For every integer $n \geqslant 0$, $E_n$ is $\beta$-finite.
(*Method*: Use induction on $n$ together with previous lemmas.)

**1.14.7 Lemma.** $\mathbf{Z}^+$ is not $\beta$-finite.

*Proof.* The mapping $x \to x + 1$ of $\mathbf{Z}^+ \to \mathbf{Z}^+$ is one-to-one but not onto (why not?). ▮

**1.14.8 Lemma.** If $S$ is equipotent to $S'$ and $S$ is $\beta$-finite, then $S'$ is $\beta$-finite.

*Sketch of proof.* By hypothesis there exists a one-to-one mapping $\mathbf{M}$ of $S$ onto $S'$. Let $\mathbf{T}$ be any one-to-one mapping of $S' \to S'$. Then $\mathbf{MTM}^{-1}$ is a one-to-one mapping of $S \to S$ (show this!). Since $S$ is $\beta$-finite, $\mathbf{MTM}^{-1}$ is onto. Hence, $\mathbf{T}$ is onto (why?) as desired. ▮

**1.14.9 Lemma.** Any subset of a $\beta$-finite set is $\beta$-finite.

**1.14.10 Lemma.** For every set $S$, either $S$ is $\alpha$-finite or $\mathbf{Z}^+$ is equipotent to some subset of $S$.

*Proof.* Assume $S$ is not $\alpha$-finite. Then any $\alpha$-finite subset $R$ of $S$ is a *proper* subset of $S$; hence, for any such $R$ we may choose (by the Axiom of Choice!) an element $x_R \in S$ such that $x_R \notin R$. In particular, the empty set is $\alpha$-finite and hence we have an element $x_\varnothing \in S$.

If $R$ is $\alpha$-finite, then $R \cup \{x_R\}$ is $\alpha$-finite by Lemma 1.14.4. Hence, for each $n \in \mathbf{Z}^+$ we can define an $\alpha$-finite subset $R_n$ of $S$ by means of the following *definition by induction*:

$$R_1 = \{x_\varnothing\}$$
$$R_{k+1} = R_k \cup \{x_{R_k}\} \qquad \text{(all } k \in \mathbf{Z}^+).$$

It can now be proved that

$$x_{R_m} \in R_n \qquad \text{whenever } n > m$$

(prove this by induction on $n$, keeping $m$ fixed). In particular, if $n > m$, then $x_{R_m} \neq x_{R_n}$ (since $x_{R_m} \in R_n$ but $x_{R_n} \notin R_n$). Hence, the mapping

$$n \to x_{R_n}$$

is a *one-to-one* mapping of $\mathbf{Z}^+$ onto the set

$$S' = \{x_{R_n} : n \in \mathbf{Z}^+\}.$$

Thus, $\mathbf{Z}^+ \sim S'$, as desired. ∎

It is now easy to complete the proof of Theorem 1.14.1. If $S$ is $\alpha$-finite, then $E_n \sim S$ for some $n$; hence, $S$ is $\beta$-finite by Lemmas 1.14.6 and 1.14.8. Conversely, suppose $S$ is $\beta$-finite. If $\mathbf{Z}^+$ were equipotent to a subset $S'$ of $S$, then $S'$ would be $\beta$-finite (Lemma 1.14.9) and, hence, $\mathbf{Z}^+$ would be $\beta$-finite, contradicting 1.14.7. Hence, $S$ is $\alpha$-finite by 1.14.10. ∎

**1.14.11 Theorem.** If $S$ is finite, then the integer $n$ which satisfies Theorem 1.14.1(a) is unique. In other words, if $n$ and $m$ are nonnegative integers such that $S \sim E_n$ and $S \sim E_m$ then $n = m$.

*Proof.* Suppose instead that $n \neq m$; without loss of generality we may assume that $m < n$. Since $S \sim E_n$ and $S \sim E_m$, Lemma 1.14.2 gives $E_n \sim E_m$; thus, there is a one-to-one mapping $\mathbf{T}$ of $E_n$ onto $E_m$. But $E_m$ is a proper subset of $E_n$; thus, $\mathbf{T}$ is a one-to-one mapping of $E_n \to E_n$ which is not onto $E_n$. Hence, $E_n$ is not $\beta$-finite, contradicting Lemma 1.14.6. ∎

**Exercises**

1. (a) Prove Lemmas 1.14.2, 1.14.3, 1.14.4, and 1.14.6.
   (b) Prove Lemma 1.14.9. (SOL)

*2. If $A$ and $B$ are disjoint finite sets having $m$ elements and $n$ elements (respectively), prove that $A \cup B$ has $m + n$ elements.

3. Prove: if $A$ and $B$ are finite sets (not necessarily disjoint) then $A \cup B$ is finite.

*4. Prove: if $n$ is a positive integer and $A_1, \ldots, A_n$ are $n$ mutually disjoint finite sets each having $m$ elements, then $A_1 \cup \cdots \cup A_n$ has $mn$ elements. (*Suggestion*: Use induction on $n$.)

*5. Let $S$ be a finite set of $n$ elements, let $A$ be a subset of $S$ having $m$ elements, and let $A'$ be the set of all elements of $S$ which are not in $A$. Prove: $A'$ has $n - m$ elements. (SOL)

*6. Prove: if $n$ is a positive integer and the sets $A$, $B$ each have $n$ elements, then there are exactly $n!$ one-to-one mappings of $A$ onto $B$. (*Suggestion*: Use induction on $n$.) (SOL)

*7. Prove: a finite set of $n$ elements has exactly $2^n$ subsets. (*Suggestion*: Use induction on $n$.)

8. Prove: every OID has infinitely many elements. (SUG)

## 1.15 Rational, real, and complex numbers

It is well known that once we have the system of integers, we can then *construct* the systems of rational, real, and complex numbers without assuming their existence in advance. We shall not perform the constructions here; what we *shall* do is summarize, mostly without proof, those properties of rational, real, and complex numbers which we will need later. References are given below to books in which the missing constructions and proofs may be found.

We use the following notations, which are fairly standard:

**Q** = set of all rational numbers
**R** = set of all real numbers
**C** = set of all complex numbers.

These notations will remain in force throughout the book. Of course, we continue to denote the set of all integers by **Z**. The sets **Q**, **R**, and **C** are fields, and

$$\mathbf{Z} \subset \mathbf{Q} \subset \mathbf{R} \subset \mathbf{C}.$$

**Q** is the smallest field which contains **Z**; in fact, every rational number is a quotient of two integers. For a good account of the construction of **Q**, see McCoy [10], Section 5.4. (The construction depends only on the fact that **Z** is an integral domain; no other properties of **Z** are used. Hence, by the same procedure we may construct a field containing any given integral domain; that is, *every integral domain is a subring of some field*. This fact will be needed later on.)

The field **R** of real numbers is an ordered field that is characterized by the following additional property, called the *completeness property*: if $S$ is any nonempty subset of **R** such that $S$ has an upper bound in **R**, then $S$ has a least upper bound in **R**. (The phrase "upper bound" was defined in Sec. 1.12.) There are many other equivalent forms of the completeness property, some of which are more intuitive than the form in which we have stated it; however, our form is perhaps the most common. For a construction of **R**, see either Hamilton-Landin

[6] (Chapters 4–5), Stoll [14] (Secs. 3.5 and 3.6), or Mostow [11] (Secs. 15.6, 4.3, and 4.4). A nice account of the construction in the first edition of [10] has been deleted from subsequent editions.

The most important algebraic property of **R** which will be needed later is the fact that *every positive real number has a real square root.* More generally, if $n$ is a positive integer, then any positive real number has a real $n$th root; for a proof (based on the completeness property as stated above), see Mostow [11], Section 4.4, Theorem 4.3.

In Appendix D we shall need to refer to the following property of **R**: *if* $\mathbf{T} : \mathbf{R} \to \mathbf{R}$ *is a ring homomorphism, then either* $r\mathbf{T} = 0$ *(all* $r \in \mathbf{R}$*) or else* $\mathbf{T} = \mathbf{I}$. (In words: the only ring homomorphisms of $\mathbf{R} \to \mathbf{R}$ are the zero mapping and the identity mapping.) We sketch the proof as follows: let $\mathbf{T} : \mathbf{R} \to \mathbf{R}$ be a homomorphism which is not the zero mapping. By Section 1.10, Exercise 5, we have $1\mathbf{T} = 1$; also, if $k\mathbf{T} = k$, then $(k + 1)\mathbf{T} = k\mathbf{T} + 1\mathbf{T} = k + 1$. Hence, by induction,

**(1.15.1)** $$n\mathbf{T} = n$$

whenever $n$ is a *positive integer*. By applying equation 1.8.5, we then establish 1.15.1 whenever $n$ is zero or a negative integer; thus, 1.15.1 holds for all $n \in \mathbf{Z}$. Since $(n^{-1})\mathbf{T} = (n\mathbf{T})^{-1}$ (Sec. 1.10, Exercise 5), it easily follows that 1.15.1 holds for all *rational* numbers $n$. Suppose now that there exists a real number $x$ such that $x\mathbf{T} = y \neq x$. We may assume without loss of generality that $y < x$ (replace $x$ by $-x$ if necessary). By McCoy [10], Theorem 6.5, there exists a rational number $q$ between $y$ and $x$:

$$y < q < x.$$

Since $x - q$ is positive, it has a square root $c = (x - q)^{1/2}$. Then

$$\begin{aligned}(c\mathbf{T})^2 &= (c\mathbf{T})\,(c\mathbf{T}) = (cc)\mathbf{T} = (x - q)\mathbf{T} \\ &= x\mathbf{T} - q\mathbf{T} = y - q < 0\end{aligned}$$

contradicting Theorem 1.11.2(k). The contradiction arose from assuming $x\mathbf{T} \neq x$; hence, $x\mathbf{T} = x$ for all real $x$, and $\mathbf{T} = \mathbf{I}$. ▮

The field **C** of *complex numbers* contains an element $i$ such that $i^2 = -1$; every complex number $z$ can be expressed in the form

**(1.15.2)** $$z = a + bi \qquad (a \in \mathbf{R},\ b \in \mathbf{R}).$$

(For the construction of **C**, see [10] or [11].) The numbers $a$, $b$ in 1.15.2 are called the *real part* of $z$ and the *imaginary part* of $z$, respectively; in symbols,

$$a = \operatorname{Re} z$$
$$b = \operatorname{Im} z.$$

Note that the "imaginary part" of $z$ is a *real* number.

*Remark*: The word "imaginary" is not intended to cast doubt on the existence of the elements under discussion, any more than the word "real" is intended to suggest that nonreal complex numbers somehow do not "really" exist. A century or so ago, the words "imaginary" and "real" may have had such unfortunate connotations, but today we should regard them as purely technical terms. They are certainly not the only words which mean different things in mathematics and English.

If $z = a + bi$ as in 1.15.2, then the number $\bar{z} = a - bi$ is called the *complex conjugate* of $z$. The mapping $z \to \bar{z}$ is an isomorphism of the field **C** onto itself; thus, the conjugate of a sum is the sum of the conjugates, and similarly for products. In symbols,

$$\overline{z + w} = \bar{z} + \bar{w}$$
$$\overline{zw} = \bar{z}\,\bar{w}.$$

You should be able to verify these equations yourself by direct computation. You should also be able to verify that

$$\text{(1.15.3)} \quad \begin{aligned} z + \bar{z} &= 2 \operatorname{Re} z \\ z - \bar{z} &= 2i \operatorname{Im} z. \end{aligned}$$

Unlike the fields **Q** and **R**, the field **C** is not an *ordered* field (for example, the equation $1^2 + i^2 = 0$ contradicts Section 1.12, Exercise 4); however, we can still define the concept of *absolute value* in **C**. The definition is

$$\text{(1.15.4)} \quad \begin{aligned} |z| = |a + bi| &= (a^2 + b^2)^{1/2} \\ &= ((\operatorname{Re} z)^2 + (\operatorname{Im} z)^2)^{1/2} \end{aligned}$$

where $z = a + bi$ as in 1.15.2. That is, $|z|$ is the *real, nonnegative square root* of the real nonnegative number $a^2 + b^2$. Moreover, $a^2 + b^2 = 0$ only if $a = b = 0$ (Sec. 1.12, Exercise 4); hence,

$$|z| = 0 \Leftrightarrow z = 0.$$

Also, since $z\bar{z} = (a + bi)(a - bi) = a^2 + b^2$, 1.15.4 may be restated in the form

$$\text{(1.15.5)} \quad |z| = (z\bar{z})^{1/2}.$$

This is *not* the same as the definition of absolute value in OID's, which was $|x| = (x^2)^{1/2}$ (equation 1.11.3). However, 1.11.3 and 1.15.5 agree when $z = x$ is *real*, since a real number is its own complex conjugate.

From 1.15.5 we have

$$\text{(1.15.6)} \quad \begin{aligned} |zw| &= ((zw)(\overline{zw}))^{1/2} = (zw\bar{z}\bar{w})^{1/2} = ((z\bar{z})(w\bar{w}))^{1/2} \\ &= (z\bar{z})^{1/2}(w\bar{w})^{1/2} \qquad \text{(Sec. 1.11, Exercise 4)} \\ &= |z| \cdot |w|. \end{aligned}$$

Thus, the absolute value of a product is the product of the absolute values.

**Exercises**

1. One often identifies the field of complex numbers with the $xy$ plane by associating the complex number $a + bi$ with the point $(a, b)$ in the plane. Using this identification, interpret the absolute value of a complex number geometrically.

*2. Prove that for all $z$, $w$ in **C**,

$$\text{Re}(z + w) = \text{Re } z + \text{Re } w$$
$$\text{Im}(z + w) = \text{Im } z + \text{Im } w.$$

*3. Prove that for all $z \in \mathbf{C}$, $z = \bar{\bar{z}}$ (the conjugate of the conjugate is the original number).

## 1.16 Polynomials

In this section, we summarize those properties of polynomials that will be needed in later chapters. We shall omit many of the proofs; they may be found in [10], Chapter 9, or in [11], Chapter 6 (except for the Fundamental Theorem of Algebra, for which we give other references below).

Let $T$ be a ring, let $S$ be a subring of $T$, and let $x$ be an element of $T$ such that *$x$ commutes with all elements of $S$*. By a *polynomial in $x$ over $S$*, we mean an element $f \in T$ that can be expressed in the form

$$f = a_0 + a_1x + a_2x^2 + \cdots + a_nx^n \quad (a_0, \ldots, a_n \in S) \tag{1.16.1}$$

for some integer $n \geqslant 0$. (If $n = 0$, then $f = a_0$ is called a *constant polynomial*; thus the constant polynomials are simply the elements of $S$.) It is not hard to show that the sum or product of any two elements of the form 1.16.1 is itself expressible in the form 1.16.1; in fact, the set of all polynomials in $x$ over $S$ is a subring of $T$. This subring will be denoted $S[x]$. Clearly, $S \subseteq S[x]$.

The element $x \in T$ is called (an) *indeterminate over $S$* if (1) $x$ commutes with all elements of $S$, and (2) the expression 1.16.1 is never zero unless all $a_i = 0$ ($i = 0, \ldots, n$). For example, the real number $\pi$ is indeterminate over the field **Q** of rational numbers; the proof is quite difficult. On the other hand, $\sqrt{3}$ is not indeterminate over **Q**, since

$$-3 + (\sqrt{3})^2 = 0.$$

If $x$ is indeterminate over $S$, the ring $S[x]$ is called the *ring of polynomials in one indeterminate over S*, or more simply the *ring of polynomials over S*. This ring is unique in the sense that if $x$ and $y$ are both indeterminate over $S$, then the rings $S[x]$ and $S[y]$ are isomorphic (see 1.16.4 below, and the associated discussion).

Let $x$ be indeterminate over $S$. If $f$ is any element of $S[x]$, the elements $a_i$ in the expression 1.16.1 are called the *coefficients* of $f$; specifically, $a_i$ is the *coefficient of* $x^i$ *in* $f$. The coefficients of $f$ are essentially unique, in the sense that if

$$f = a_0 + \cdots + a_nx^n = b_0 + \cdots + b_rx^r \qquad (a_i \in S,\ b_j \in S) \tag{1.16.2}$$

with $n \leqslant r$, then

$$\begin{aligned} a_i &= b_i \qquad (\text{all } i,\ 0 \leqslant i \leqslant n) \\ b_j &= 0 \qquad (\text{all } j > n). \end{aligned} \tag{1.16.3}$$

The proof of 1.16.3 is based on the Difference Principle (cf. Sec. 1.7). If 1.16.2 holds, then

$$(a_0 - b_0) + (a_1 - b_1)x + \cdots + (a_n - b_n)x^n + \sum_{j>n} b_jx^j = f - f = 0.$$

Since $x$ is indeterminate, it follows that $a_0 - b_0 = 0, \ldots, a_n - b_n = 0$, $b_j = 0$ $(j > n)$, and hence 1.16.3 holds.

It is not hard to show that given any ring $S$ with unity, there exists a ring $T$ having the same unity and containing an indeterminate over $S$; equivalently, *there exists a polynomial ring in one indeterminate over S*. To do this somewhat informally, simply let $T$ $(= S[x])$ be the set of all "formal expressions" of the form 1.16.1 and define addition and multiplication of such "expressions" in the natural way. For a more rigorous construction, see Mostow [11], Section 6.2, Theorem 2.5.

Let $x$ be indeterminate over $S$. If $S'$ is any ring containing $S$, and if $c$ is an element of $S'$ which commutes with all elements of $S$, then the mapping $\mathbf{M}_c$ defined by

$$\mathbf{M}_c : a_0 + a_1x + \cdots + a_nx^n \to a_0 + a_1c + \cdots + a_nc^n \qquad (a_i \in S) \tag{1.16.4}$$

maps $S[x]$ onto $S[c]$. ($\mathbf{M}_c$ is well defined due to the uniqueness of the coefficients of a given polynomial.) Moreover, it is not hard to show that $\mathbf{M}_c$ preserves addition and multiplication (try this as an exercise) and, hence, is a ring homomorphism. If $c$ is also indeterminate over $S$, then $\mathbf{M}_c$ is one-to-one (*why?*) and, hence, the rings $S[x]$ and $S[c]$ are isomorphic. (This justifies our previous assertion that the ring of polynomials in one indeterminate over $S$ is unique.)

The homomorphism $\mathbf{M}_c$ is called the *substitution mapping* because the right side of 1.16.4 is obtained from the left side by substituting $c$ for $x$. Indeed, the fact that 1.16.4 is a homomorphism enables us to justify substitution in polynomial equations in general. (*Example*: Since $x^2 - 4 = (x - 2)(x + 2)$, the fact that $\mathbf{M}_c$ preserves multiplication implies that $c^2 - 4 = (c - 2)(c + 2)$.)

Many (perhaps most) authors use functional notation for polynomials; thus an expression like 1.16.1 would be denoted $f(x)$ instead of $f$. Similarly, the left and right sides of 1.16.4 would be denoted $f(x)$ and $f(c)$, respectively. However, you should *not* think of polynomials as being functions. There are indeed such things as "polynomial functions" (mappings) which *correspond* to polynomials; specifically, if $f(x)$ is a polynomial in the indeterminate $x$ over $S$, then the corresponding "polynomial function" $\mathscr{F}$ is a mapping of $S \to S$ which is defined by the rule

$$\mathscr{F} : c \to f(c) \qquad (\text{all } c \in S).$$

In calculus, we differentiate polynomial functions, not polynomials.

If $x$ is indeterminate over $S$ and $f$ is a nonzero element of $S[x]$, then at least one of the coefficients of $f$ must be nonzero. By the *degree* of $f$ (*notation*: $\deg f$) we mean the largest $n$ such that the coefficient of $x^n$ in $f$ is nonzero. (If this coefficient is equal to 1, $f$ is called *monic*.) Thus, for example, the polynomial $x^3 - 2x$ has degree 3 and is monic; the polynomial $2x^3 - 2x$ has degree 3 and is not monic. Similarly,

$$\deg(3 - x - x^2) = 2$$
$$\deg(3) = 0$$
$$\deg(2 - x^2 + 3x^4) = \deg(2 - x^2 + 3x^4 + 0x^5) = 4.$$

The polynomials of degree zero are precisely the constant polynomials (except for the zero polynomial whose degree is undefined).

1.16.5 **Theorem.** Let $D$ be an integral domain and let $x$ be an indeterminate over $D$. Then

(a) $\deg(fg) = \deg f + \deg g$ for all nonzero elements $f$, $g$ in $D[x]$;
(b) $D[x]$ is an integral domain.

*Proof.* $D[x]$ is clearly a commutative ring with unity. If $\deg f = n$ and $\deg g = m$, then we may write $f = a_0 + \cdots + a_n x^n$, $g = b_0 + \cdots + b_m x^m$ with $a_n \neq 0$, $b_m \neq 0$. Since $D$ is an integral domain, $a_n b_m \neq 0$. But it is easy to see that $a_n b_m$ is the coefficient of $x^{n+m}$ in $fg$ and that no higher power of $x$ appears in $fg$. Hence, $\deg(fg) = n + m$, proving

1.16.5(a). Moreover, since $a_n b_m \neq 0$, $fg \neq 0$ whenever $f$ and $g$ are nonzero; hence, $D[x]$ is an integral domain. ▮

In particular, if $F$ is a field, then $F[x]$ is an integral domain. For the remainder of this section, we will consider only the case of a polynomial ring over a field; thus, $F$ will denote a field and $F[x]$ will denote the ring of all polynomials in one indeterminate over $F$.

If $f \in F[x]$, an element $c \in F$ is called a *root* of $f$ if $f(c) = 0$. (For example, the polynomial $x^2 + 1$ has no roots in $\mathbf{R}$, but it does have the complex root $i$ since $i^2 + 1 = 0$.) If $F = \mathbf{Q}$ (the rational numbers), there is a standard method for finding all rational roots of $f$; see [10], Section 9.6. Of course, the irrational roots of $f$ (if any) cannot be found by this method.

If $f, g \in F[x]$, $g$ is said to be a *factor* of $f$ (or $f$ is said to be *divisible by* $g$) if $f = gh$ for some $h \in F[x]$. It is proved in both [10] and [11] that if $f \in F[x]$ and $c \in F$, then *$c$ is a root of $f$ if and only if the polynomial $x - c$ is a factor of $f$.* In this situation it is possible that some higher power of $x - c$ may also be a factor of $f$. If $m$ is the largest integer such that $(x - c)^m$ is a factor of $f$, then $c$ is said to be a root of $f$ of *multiplicity* $m$. For example, since

$$x^3 - x^2 - 8x + 12 = (x - 2)^2 (x + 3),$$

the polynomial $f = x^3 - x^2 - 8x + 12$ has 2 as a root of multiplicity 2, whereas $-3$ is a root of $f$ of multiplicity 1.

A polynomial $f \in F[x]$ is said to be *prime*, or *irreducible*, if $f$ is nonconstant (i.e., has degree $> 0$) and cannot be expressed as the product of two nonconstant polynomials. *Examples*: $x^2 + 1$ is prime in $\mathbf{R}[x]$, but is not prime in $\mathbf{C}[x]$, since over the field of complex numbers $x^2 + 1 = (x + i)(x - i)$. The polynomial $x^4 + 1$ is prime over the field of rational numbers, but not over the field of real numbers since

$$x^4 + 1 = (x^2 - \sqrt{2}\,x + 1)(x^2 + \sqrt{2}\,x + 1).$$

As shown in [10] and [11], every nonconstant polynomial $f \in F[x]$ can be expressed as a product of (one or more) *prime* polynomials:

$$\text{(1.16.6)} \quad f = \prod_{i=1}^{r} p_i \qquad (p_i \in F[x],\ p_i \text{ prime}).$$

(The $p_i$ are not necessarily distinct.) Moreover, the expression 1.16.6 for $f$ is unique except for (1) the order in which the factors $p_i$ are written, and (2) constant factors (for example, $p_1 p_2 = (3p_1)(\frac{1}{3}p_2)$). It easily follows that any prime polynomial in $F[x]$ which is a factor of $f$ must be a constant multiple of one of the $p_i$. In particular, if $c$ is a root of $f$, then one of the $p_i$ must equal $x - c$ (or $x - c$ times a constant). Since the number of such factors $p_i$ clearly cannot exceed the degree of $f$ (cf. Theorem 1.16.5(a)), it follows that *a polynomial of degree n in* $F[x]$ *has no more than n distinct roots.*

If the $p_i$ in 1.16.6 are all of degree 1, we say that $f$ *splits* (over $F$). For example, $x^2 + 1$ does not split over **R** but does split over **C**. More generally, given any field $F$ and any nonconstant polynomial $f \in F[x]$, it is always possible to find a larger field $F'$ containing $F$ such that $f$ splits over $F'$. For one method of constructing such a field $F'$, see [10], Section 10.3.

In the case where $F$ is the field of real or complex numbers, some special results are known. One of the most important of these results is the so-called *Fundamental Theorem of Algebra* (F.T.O.A.), which asserts that *every nonconstant polynomial in* **C**$[x]$ *has a complex root*. In particular, every prime polynomial $p \in$ **C**$[x]$ has a root $c \in$ **C**; hence, every such $p$ has a factor $x - c$ of degree one; hence, every such $p$ must itself be of degree 1, since otherwise it would have a nontrivial factorization and would not be prime. We conclude that *every nonconstant polynomial in* **C**$[x]$ *splits over* **C**.

The traditional proof of the F.T.O.A. is based on the theory of complex integration. However, proofs of a more elementary nature exist. For instance, the proof in [9], pp. 35–38, assumes no prior knowledge except for (1) facts previously stated in this section and the preceding section, and (2) some standard facts about continuous functions, only one of which is not normally covered in a first course in calculus: namely, the fact (proved in [8] and in many advanced calculus texts) that a continuous real-valued function $f(x, y)$ defined on a closed disk $\{(x, y) : x^2 + y^2 \leq r^2\}$ in the $xy$ plane must take on a minimum value at some point in the disk.

(The reader familiar with DeMoivre's Theorem–more precisely, with its corollary that all complex numbers have $n$th roots–should be able to follow a much shorter proof which is similar in spirit to the proof in [9] but occupies only fourteen lines; it appears in the *American Mathematical Monthly*, October, 1976, p. 647.)

An important consequence of the F.T.O.A. is that every prime polynomial over the field of *real* numbers must have degree 1 or 2 (see [10] or [11] for the derivation). In particular, if $f \in$ **R**$[x]$ has odd degree, then its prime factors cannot all be of even degree and, hence, $f$ must have a factor of degree 1, say, $ax - b$; but then $b/a$ is a root of $f$ (*why?*). In short, *every polynomial of odd degree in* **R**$[x]$ *has a real root.*

For an alternate proof of this fact, consider the graph of the equation $y = \mathcal{F}(x)$, where $\mathcal{F}$ is a *polynomial function* of odd degree $n$, say

$$\mathcal{F}(x) = a_0 + \cdots + a_n x^n$$

(we change our point of view here and regard $x$ as a real variable rather than an indeterminate). Whenever $|x|$ is sufficiently large, $\mathcal{F}(x)$ has the same sign as $a_n x^n$; since $n$ is odd, this sign is not the same when $x < 0$ as when $x > 0$. Hence, $\mathcal{F}(x)$ is positive for some values of $x$ and negative for others. Therefore, the graph must cross the $x$ axis somewhere, say, at $x = c$; and then $c$ is a root of $f$.

Any polynomial in $\mathbf{R}[x]$ is also an element of $\mathbf{C}[x]$, since $\mathbf{R} \subseteq \mathbf{C}$. If $f \in \mathbf{R}[x]$ and if $c$ is a root of $f$ over $\mathbf{C}$, then $\bar{c}$ (the complex conjugate of $c$) is also a root of $f$ over $\mathbf{C}$. In fact, $\bar{c}$ and $c$ *have the same multiplicity* as roots of $f$ over $\mathbf{C}$. For a proof, see Exercise 1 below.

**Exercise**

1. Let $f$ be a polynomial with *real* coefficients ($f \in \mathbf{R}[x]$), and suppose $b_0, b_1, \ldots, b_k, c$ are *complex* numbers such that

   $$f = (x - c)^m(b_0 + b_1 x + \cdots + b_k x^k).$$

   Prove that also

   $$f = (x - \bar{c})^m(\bar{b}_0 + \bar{b}_1 x + \cdots + \bar{b}_k x^k).$$

   Deduce that $c$ and $\bar{c}$ have the same multiplicity as roots of $f$ over $\mathbf{C}$.

# 2 Vector spaces

## 2.1 Definition of vector space

If you have taken a course in elementary physics (or elementary calculus) using vector methods, you probably remember that the objects **u** and **v** in Figure 2.1.1 were called *vectors*. That is, a vector was a directed line segment, or "arrow", with two arrows having equal lengths and the same direction considered to be equal as vectors. Thus, the arrows **Ar** *AB* and **Ar** *CD* in Figure 2.1.1 represent the same vector **u**; the arrows **Ar** *DE* and **Ar** *FG* represent the same vector **v**. (**Ar** *AB* denotes the *arrow from A to B*; cf. Sec. 3.1.) If you do not like to think of distinct segments as being "equal", you can consider equivalence classes of segments rather than the segments themselves.

Two vectors **u**, **v** (of the type under discussion) can be added to obtain a vector **u** + **v**, in the manner shown in Figure 2.1.1. Thus, *addition is a binary operation on the set of all vectors*. (You have already encountered the idea of a binary operation in Chapter 1 or in a course in abstract algebra.) Similarly, we may multiply a number by a vector to obtain a new vector; the segments **Ar** *FG* and **Ar** *KL* in Figure 2.1.1 illustrate how to do so. Multiplication of a number by a vector is called *scalar multiplication*; in fact, numbers themselves are usually called

Figure 2.1.1

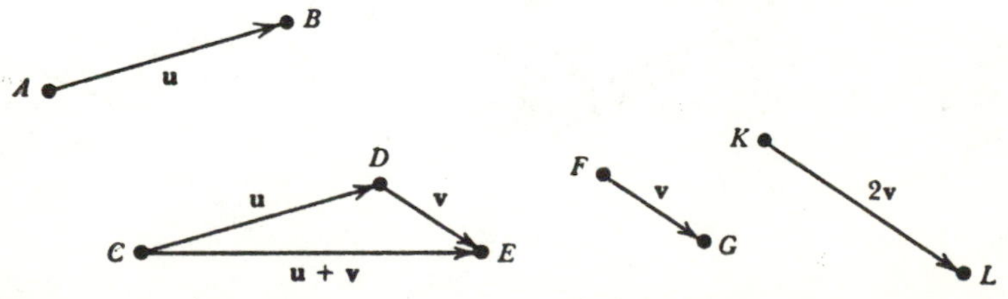

*scalars* when scalar multiplication is being discussed. (Scalar multiplication is not quite a "binary operation" in the strict sense of Section 1.5, since the two elements being multiplied do not belong to the same set.) The geometric effect of scalar multiplication, at least when the scalar is positive, is to multiply the length of the original vector by the scalar while leaving the direction unchanged.

In addition to their geometric significance, vector addition and scalar multiplication have useful algebraic properties. For example, addition is associative (Fig. 2.1.2(a)) and commutative (Fig. 2.1.2(b)); in fact, the set of all vectors is an abelian group under addition (what is the zero vector? What segment(s) in Figure 2.1.1 could be used to represent the vector $-\mathbf{u}$?). Scalar multiplication satisfies the distributive law $a(\mathbf{u} + \mathbf{v}) = a\mathbf{u} + a\mathbf{v}$ (where $a$ is a scalar and $\mathbf{u}$ and $\mathbf{v}$ are vectors) and various other laws.

Our object in this section is to generalize the foregoing ideas by defining the concept of "vector space". A vector space is a set consisting of certain elements called "vectors"; these vectors can be added, and certain elements called "scalars" can be multiplied by them. However, the "scalars" are no longer required to be numbers; they may be elements of any field. (Stop here and review the definition of "field" from Section 1.10.) Likewise, the "vectors" need not be directed line segments; in fact, they need have no geometric significance whatever. What the vectors in a "vector space" do have in common with the vectors of Figure 2.1.1 is the list of *algebraic* properties which they satisfy: addition makes the vector space an abelian group, scalar multiplication satisfies the distributive law, and so forth. In fact, instead of saying "and so forth", it might be a good idea to exhibit the whole list of properties by giving the actual definition.

Figure 2.1.2

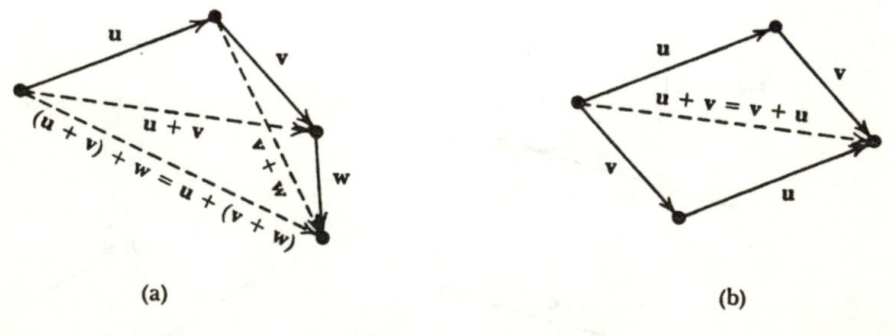

2.1.1 **Definition.** Let $F$ be a field. A *vector space over* $F$ is a set $V$ (whose elements are called *vectors*) such that:

(a) A binary operation of addition is defined on $V$; that is, for every $\mathbf{u} \in V$ and every $\mathbf{v} \in V$ there exists a unique element $\mathbf{u} + \mathbf{v}$ belonging to $V$.

(b) For every $a \in F$ and every $\mathbf{v} \in V$, there exists a unique element $a\mathbf{v}$ belonging to $V$. (This operation, which multiplies an element of $F$ by an element of $V$ to obtain an element of $V$, is called *scalar multiplication*, and the elements of $F$ are called *scalars*.)

(c) $V$ is an abelian group under addition.

(d) Scalar multiplication has the following properties: for all $a$, $b$ in $F$ and all $\mathbf{u}, \mathbf{v}$ in $V$,

(i) $a(\mathbf{u} + \mathbf{v}) = a\mathbf{u} + a\mathbf{v}$

(ii) $(a + b)\mathbf{u} = a\mathbf{u} + b\mathbf{u}$

(iii) $(ab)\mathbf{u} = a(b\mathbf{u})$

(iv) $1_F\mathbf{u} = \mathbf{u}$.

As a matter of terminology, we say that $V$ is "closed under addition" if 2.1.1(a) holds, "closed under scalar multiplication" if 2.1.1(b) holds. (Compare with the general definition of "closed" at the beginning of Sec. 1.6.)

We recall from Section 1.6 the meaning of property 2.1.1(c): addition is associative and commutative, there exists a "zero vector" $\mathbf{0}$ (identity for addition), and every vector $\mathbf{u}$ has an additive inverse $-\mathbf{u}$. That is, for all vectors $\mathbf{u}$, $\mathbf{v}$, $\mathbf{w}$, we have

$$\begin{aligned} (\mathbf{u} + \mathbf{v}) + \mathbf{w} &= \mathbf{u} + (\mathbf{v} + \mathbf{w}) \\ \mathbf{u} + \mathbf{v} &= \mathbf{v} + \mathbf{u} \\ \mathbf{u} + \mathbf{0} &= \mathbf{0} + \mathbf{u} = \mathbf{u} \\ \mathbf{u} + (-\mathbf{u}) &= (-\mathbf{u}) + \mathbf{u} = \mathbf{0}. \end{aligned} \tag{2.1.2}$$

As in Chapter 1, the expression $\mathbf{u} + (-\mathbf{v})$ is abbreviated $\mathbf{u} - \mathbf{v}$, and $1_F$ denotes the unity of the field $F$.

As suggested earlier, the "vectors" of Figure 2.1.1 do satisfy the conditions of Definition 2.1.1 and thus constitute a vector space. There are many other important vector spaces; in fact, the importance of vector spaces to mathematics can hardly be overestimated. The elements of a vector space may be directed line segments (more precisely, equivalence classes of such segments); they may be matrices; they may be points. The set $\mathscr{C}$ of all continuous real-valued functions on the interval $(0, \infty)$ is a vector space under the "usual" definitions of addition and scalar multiplication; that is, if $f$ and $g$ are functions, then $f + g$ and $af$ ($a \in \mathbf{R}$) are defined by the equations

$$(2.1.3)\qquad \begin{aligned}(f + g)(x) &= f(x) + g(x) \qquad (\text{all } x \in \mathbf{R});\\ (af)(x) &= a(f(x)) \qquad (\text{all } x \in \mathbf{R}).\end{aligned}$$

If $S$ is the subset of $\mathscr{C}$ consisting of only those functions $y = y(x)$ which satisfy the differential equation

$$x^2y'' + xy' = y \qquad (x > 0),$$

then $S$ is also a vector space. Not all of these examples will be pursued further; our purpose in stating them here is simply to suggest the great variety of mathematical objects to which vector-space theory can be applied. Certain specific vector spaces *will* be studied in more detail; one of the most important ones will be introduced in this section (Theorem 2.1.5 below).

In addition to the properties listed in Definition 2.1.1, certain other properties of vector spaces which are easy consequences of Definition 2.1.1 will be used repeatedly in what follows. We collect these properties in a theorem.

**2.1.4** **Theorem.** Let $V$ be a vector space over $F$. The following statements hold for all $a, b \in F$ and for all vectors $\mathbf{u}, \mathbf{u}_i, \mathbf{v}, \mathbf{w} \in V$.

(a) $-(-\mathbf{u}) = \mathbf{u}$.
(b) $-(\mathbf{u} + \mathbf{v}) = -\mathbf{u} - \mathbf{v}$.
(c) $0\mathbf{u} = \mathbf{0}$.
(d) $a\mathbf{0} = \mathbf{0}$.
(e) $-(a\mathbf{u}) = a(-\mathbf{u}) = (-a)\mathbf{u}$.
(f) $(-a)(-\mathbf{u}) = a\mathbf{u}$.
(g) $a(\mathbf{u} - \mathbf{v}) = a\mathbf{u} - a\mathbf{v}$.
(h) $(a - b)\mathbf{u} = a\mathbf{u} - b\mathbf{u}$.
(i) $a(\mathbf{u}_1 + \cdots + \mathbf{u}_n) = a\mathbf{u}_1 + \cdots + a\mathbf{u}_n$.
(j) $(-1_F)\mathbf{u} = -\mathbf{u}$.
(k) If $\mathbf{u} + \mathbf{v} = \mathbf{u} + \mathbf{w}$, then $\mathbf{v} = \mathbf{w}$.
(l) If $a\mathbf{u} = \mathbf{0}$, then $a = 0$ or $\mathbf{u} = \mathbf{0}$.
(m) If $a \neq 0$ and $a\mathbf{u} = a\mathbf{v}$, then $\mathbf{u} = \mathbf{v}$.
(n) If $\mathbf{u} \neq \mathbf{0}$ and $a\mathbf{u} = b\mathbf{u}$, then $a = b$.

*Proof and remarks.* $\mathbf{0}$ of course denotes the zero vector, whereas $0$ denotes the zero scalar. Properties (a), (b), and (k) have already been proved for additive abelian groups (Theorem 1.7.1), so that they automatically hold in vector spaces.

Parts (c), (d), and (e) of Theorem 2.1.4 may be proved by imitating the proof of Theorem 1.9.3. (We cannot directly *apply* 1.9.3 since $V$ is not a ring; properties 2.1.1(d), though analogous to ring properties, are *not* ring properties since the scalars and vectors belong to different sets.) The only major change in the proof is that the homomorphisms $\mathbf{T} : S \to S$, $\mathbf{T}' : S \to S$ must be replaced by homomorphisms of $F \to V$ and $V \to V$ respectively. Parts (f), (g), (h), and (j) of Theorem 2.1.4 may be deduced from parts (a) and (e) exactly as Theorems 1.9.4 and 1.9.5(a) followed from Theorem 1.9.3.

If $a\mathbf{u} = \mathbf{0}$ but $a \neq 0$, then $a$ has a multiplicative inverse $a^{-1} \in F$ since $F$ is a field. We then have

$$\mathbf{u} = 1_F\mathbf{u} = (aa^{-1})\mathbf{u} = a^{-1}(a\mathbf{u}) = a^{-1}(\mathbf{0}) = \mathbf{0}$$

using various parts of Definition 2.1.1 as well as 2.1.4(d). This proves part (l) of Theorem 2.1.4.

The proof of part (i) is by induction (the case $n = 2$ is Definition 2.1.1(d) (i)) and we leave it to you as an exercise.

To prove part (n), we use the Difference Principle (cf. end of Section 1.7). Since $a\mathbf{u} = b\mathbf{u}$, we have $a\mathbf{u} - b\mathbf{u} = \mathbf{0}$; hence, $(a - b)\mathbf{u} = \mathbf{0}$ by part (h); hence, $a - b = 0$ or $\mathbf{u} = \mathbf{0}$ by part (l). Since $\mathbf{u} \neq \mathbf{0}$ is given, we must have $a - b = 0$; hence, $a = b$. The proof of part (m) is similar. ▮

(*Question*: Can you see a shorter proof of part (m)? Why would this shorter method *not* work for part (n)?)

The *vector space of n-tuples*, to be introduced in the next theorem, is one of the most important of all vector spaces; we shall pay a great deal of attention to it. The space of $n$-tuples will be fundamental to much of the geometry in this book, and to much of the linear algebra as well. Points in the plane or in space, directed line segments, rows of matrices, solutions of linear equations – all of these different types of mathematical objects can be represented by $n$-tuples, as we will eventually show.

**2.1.5** **Theorem.** Let $F$ be any field and let $n$ be any positive integer. Let $F_n$ denote the set of all $n$-tuples $(a_1, ..., a_n)$ such that $a_1, ..., a_n$ belong to $F$. Then $F_n$ is a vector space over $F$, if addition of $n$-tuples and multiplication of a scalar by an $n$-tuple are defined (respectively) by

$$\text{(2.1.6)}\quad \begin{aligned} &\text{(a) } (a_1, a_2, \ldots, a_n) + (b_1, b_2, \ldots, b_n) \\ &\quad = (a_1 + b_1, a_2 + b_2, \ldots, a_n + b_n); \\ &\text{(b) } c(a_1, a_2, \ldots, a_n) = (ca_1, ca_2, \ldots, ca_n). \end{aligned}$$

*Proof.* It is clear from 2.1.6 that the sum of two elements of $F_n$ is an element of $F_n$, and that a scalar times an element of $F_n$ equals an element of $F_n$. Thus, parts (a) and (b) of Definition 2.1.1 hold. Verification of parts (c) and (d) consists of a total of eight properties (remembering that 2.1.1(c) consists of the four properties listed in 2.1.2). We give just one of the eight verifications here, to show how it's done; you should do the others yourself. Let us verify the left distributive law (Definition 2.1.1(d), part (i)). Remembering that the "vectors" in this situation are $n$-tuples, we must show that

$$\text{(2.1.7)}\quad c[(a_1, \ldots, a_n) + (b_1, \ldots, b_n)] = c(a_1, \ldots, a_n) + c(b_1, \ldots, b_n)$$

where $c$, $a_i$, $b_i$ are elements of $F$. We shall prove 2.1.7 by showing that both sides are equal to the same $n$-tuple. Considering the left side of 2.1.7 first, we have

$$\begin{aligned} c[(a_1, \ldots, a_n) + (b_1, \ldots, b_n)] &= c(a_1 + b_1, \ldots, a_n + b_n) && \text{(by 2.1.6(a))} \\ &= (c(a_1 + b_1), \ldots, c(a_n + b_n)) && \text{(by 2.1.6(b))} \\ &= (ca_1 + cb_1, \ldots, ca_n + cb_n), \end{aligned}$$

where in the last step we have used the distributive law *in the field F*. Similarly, the right side of 2.1.7 is

$$\begin{aligned} c(a_1, \ldots, a_n) + c(b_1, \ldots, b_n) &= (ca_1, \ldots, ca_n) + (cb_1, \ldots, cb_n) && \text{(by 2.1.6(b))} \\ &= (ca_1 + cb_1, \ldots, ca_n + cb_n) && \text{(by 2.1.6(a)).} \end{aligned}$$

Since both sides of 2.1.7 have been reduced to the same $n$-tuple, 2.1.7 is true. ("Things equal to the same thing are equal to each other"!) ∎

Notice that in the preceding argument we used the left distributive law for fields to prove the analogous law in $F_n$. Similarly, each part of Definition 2.1.1(c) and 2.1.1(d) for $F_n$ is a consequence of the analogous property of fields; in particular, (0, 0, ..., 0) is the zero vector in $F_n$ and the additive inverse of $(a_1, a_2, ..., a_n)$ is $(-a_1, -a_2, ..., -a_n)$. We leave the detailed verifications to you.

The vector space of $n$-tuples (especially when $F = \mathbf{R}$, the field of real numbers) has two important geometric interpretations, which we now describe briefly.

### The "point" interpretation

If you have taken a course in analytic geometry, you are acquainted with the idea of assigning coordinates $(x, y)$ to points in the plane. Once such coordinates are assigned, each point in the plane corresponds to an ordered pair of real numbers; that is, to an element of the vector space $\mathbf{R}_2$. Similarly, if we assign coordinates $(x, y, z)$ to points in ordinary 3-dimensional space, we obtain a correspondence between points in 3-space and elements of $\mathbf{R}_3$. More generally, we could define "real $n$-space" to be the vector space $\mathbf{R}_n$, with a "point" in $\mathbf{R}_n$ defined to be an $n$-tuple $(a_1, a_2, ..., a_n)$. It is only in the cases $n = 2$ and $n = 3$, however, that realistic pictures or graphs can be drawn to illustrate the vector-space concepts geometrically.

Figure 2.1.3 illustrates what addition and scalar multiplication look like in the vector space $\mathbf{R}_2$, when elements $(x, y)$ of $\mathbf{R}_2$ are interpreted as points. If $P$, $Q$ are the points $(1, 3)$ and $(5, 2)$ respectively, then from the definition of addition in $\mathbf{R}_2$, we have

$$P + Q = (1, 3) + (5, 2) = (1 + 5, 3 + 2) = (6, 5).$$

Figure 2.1.3

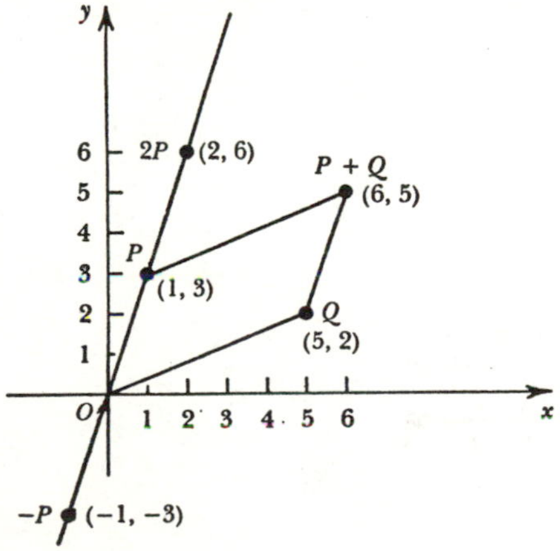

Similarly, from the definition of scalar multiplication in $\mathbf{R}_2$, we have

$$2P = 2(1, 3) = (2 \cdot 1, 2 \cdot 3) = (2, 6).$$

Geometrically, $P + Q$ is the fourth vertex of the parallelogram whose other vertices are $P$, the origin, and $Q$ (in that order). The point $2P$ lies on the line through $P$ and the origin and is twice as far as $P$ from the origin; the point $-P$ lies on the same line but on the opposite side of the origin; and so forth.

**The "arrow" interpretation**

Here we return to the idea of directed line segments ("arrows") with which we began this section. For simplicity, we take $F = \mathbf{R}$ and $n = 2$, so that our vector space is the space $\mathbf{R}_2$ of all ordered pairs of real numbers. If $A(a_1, a_2)$ and $B(b_1, b_2)$ are two points in the plane, we associate with the directed segment **Ar** $AB$ the ordered pair (vector) $\mathbf{u} = (b_1 - a_1, b_2 - a_2) \in \mathbf{R}_2$. The numbers $b_1 - a_1$ and $b_2 - a_2$ which are the coordinates of $\mathbf{u}$ can be thought of as distances traveled in the $x$ direction and $y$ direction (respectively) by a particle moving from $A$ to $B$. Observe that since in the vector space $\mathbf{R}_2$ we have

$$(b_1 - a_1, b_2 - a_2) = (b_1, b_2) - (a_1, a_2),$$

identification of points with ordered pairs gives us the equation $\mathbf{u} = B - A$, where $\mathbf{u}$ is the vector associated with **Ar** $AB$.

The "arrow" interpretation is illustrated in Figure 2.1.4. There the directed segment **Ar** $AB$ corresponds to the vector $\mathbf{u} = (6 - 2, 4 - 3) = (4, 1)$, while the vector associated with **Ar** $BC$ is $\mathbf{v} = (11 - 6, 2 - 4)$

Figure 2.1.4

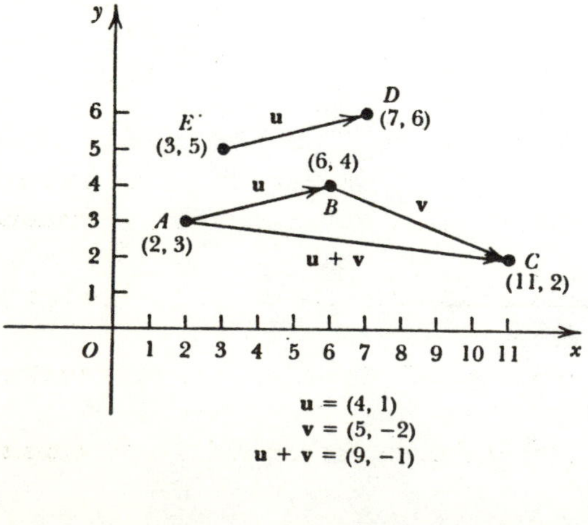

= (5, −2). The segment **Ar** *ED*, which has the same length and direction as **Ar** *AB*, is associated with the same vector **u**, since

$$(7 - 3, 6 - 5) = (4, 1) = (6 - 2, 4 - 3).$$

Observe that the sum of the vectors corresponding to **Ar** *AB* and **Ar** *BC* equals the vector corresponding to **Ar** *AC*: (4, 1) + (5, −2) = (9, −1). A similar statement holds for any three points *A, B, C*.

Both the "point" and "arrow" interpretations of *n*-tuples (especially when $n = 2$ or 3) are quite useful in illustrating and motivating various aspects of vector-space theory, and we shall refer to them often in what follows. However, it must always be kept in mind that *not every vector space consists of n-tuples*; there are many other vector spaces besides $F_n$. In proving theorems about vector spaces, or in doing exercises, be careful not to represent your vectors as *n*-tuples unless that's actually what they are.

**Exercises**

(Reminder: A star (asterisk) next to an exercise means that the result of the exercise will be needed later.)

1. Using the "arrow" interpretation in $\mathbf{R}_2$, draw a picture illustrating the left distributive law of scalar multiplication.
2. Answer both parts of the "Question" following the proof of Theorem 2.1.4.

*3. Finish the proof of Theorem 2.1.5 by verifying the seven remaining parts of Definition 2.1.1(c)–2.1.1(d). In particular, show that the zero vector is (0, 0, ..., 0) and that the additive inverse of $(a_1, a_2, \ldots, a_n)$ is $(-a_1, -a_2, \ldots, -a_n)$.

*4. In the vector space $F_n$, show that

$$(a_1, \ldots, a_n) - (b_1, \ldots, b_n) = (a_1 - b_1, \ldots, a_n - b_n).$$

(For the general definition of subtraction, see Sec. 1.7.)

*5. Prove Theorem 2.1.4(i) by induction.

6. Let $F$ and $F'$ be fields such that $F$ is a subfield of $F'$.

(a) Show that $F'$ is a vector space over $F$, if addition and scalar multiplication are taken to be the addition and multiplication already defined in $F'$.

(b) Show that $F$ need *not* be a vector space over $F'$. What part of Definition 2.1.1 fails?

(c) If $V$ is a vector space over $F'$, show that $V$ is also a vector space over $F$.

7. Let $F$ be a field; let $F[x]$ be the set consisting of all polynomials

in the indeterminate $x$ with coefficients in $F$ (cf. Sec. 1.16). Using the obvious definitions of addition and scalar multiplication, show that $F[x]$ is a vector space over $F$.

## 2.2 Subspaces

Throughout this section, $F$ will denote a field.

If $V$ is a vector space over $F$, and if $S$ is a subset of $V$, a question which sometimes arises is whether or not $S$ is itself a vector space. In other words, if we take the given operations of addition and scalar multiplication on $V$ but look only at what these operations do to elements of $S$, does $S$ (in place of $V$) satisfy all parts of Definition 2.1.1? If the answer is "yes", $S$ is called a *subspace* of $V$.

**2.2.1** **Definition.** If $V$ is a vector space over $F$ and $S$ is a subset of $V$, then $S$ is called a *subspace* of $V$ if $S$ itself is a vector space over $F$ (under the same operations as are given in $V$).

Definition 2.2.1 is, of course, completely analogous to the definitions of "subgroup" and "subring" given in Sections 1.6 and 1.9.

As an example, let $S$ be the set of all *rational* "solutions" $(x, y, z)$ of the equation

**(2.2.2)** $2x - 3y + 4z = 6.$

Since the elements of $S$ are ordered triples of rational numbers, $S$ is a subset of the vector space $\mathbf{Q}_3$ over the field $\mathbf{Q}$ of rational numbers. However, $S$ is not a *subspace* of $\mathbf{Q}_3$. To show that $S$ is not a subspace we need merely observe that $(3, 0, 0)$ and $(1, 0, 1)$ are solutions of equation 2.2.2 and thus belong to $S$, while their sum $(4, 0, 1)$ is not a solution and thus is not in $S$. In other words, $S$ is not closed under addition; it does not satisfy part (a) of Definition 2.1.1.

On the other hand, the set of rational solutions of

**(2.2.3)** $2x - 3y + 4z = 0$

does turn out to be a subspace of $\mathbf{Q}_3$, as we will show below. (*Question*: What do you think is the significant difference between 2.2.2 and 2.2.3?)

Let us now return to the general situation involving a subset $S$ of an arbitrary vector space $V$. By definition, $S$ is a subspace of $V$ if and only if all parts of Definition 2.1.1 hold for $S$ in place of $V$. Now, Definition 2.1.1 consists of a total of *ten* statements (counting parts

(c) and (d) as four statements each). However, it is not necessary to actually verify all ten statements for $S$, since those statements which are "assertions of equality" (as opposed to "assertions of existence") hold in $S$ automatically. (If you have not already read the discussion in Section 1.6 regarding the two types of assertions, go back and read it now before continuing further.) If we examine our ten vector-space properties to see which ones are "assertions of existence" and which are "assertions of equality", we obtain the following classification:

**Assertions of existence**

1. Closure under addition.
2. Closure under scalar multiplication.
3. Existence of identity for addition.
4. Existence of inverses for addition.

**Assertions of equality**

1. Addition is commutative ($\mathbf{u} + \mathbf{v} = \mathbf{v} + \mathbf{u}$ for all $\mathbf{u}, \mathbf{v}$).
2. Addition is associative ($(\mathbf{u} + \mathbf{v}) + \mathbf{w} = \mathbf{u} + (\mathbf{v} + \mathbf{w})$).
3. $a(\mathbf{u} + \mathbf{v}) = a\mathbf{u} + a\mathbf{v}$.
4. $(a + b)\mathbf{u} = a\mathbf{u} + b\mathbf{u}$.
5. $(ab)\mathbf{u} = a(b\mathbf{u})$.
6. $1_F\mathbf{u} = \mathbf{u}$.

Thus, in showing that a subset $S$ of $V$ is a subspace of $V$, there are only four things to prove rather than ten. Even one of these four is superfluous; indeed, the fourth assertion (inverses for addition) is a special case of the second (closure under scalar multiplication) since for any $\mathbf{s} \in S$ we have $-\mathbf{s} = (-1_F)\mathbf{s}$. We hence have the following theorem:

2.2.4 **Theorem.** Let $V$ be a vector space over $F$, and let $S$ be a subset of $V$. Then $S$ is a subspace of $V$ if and only if it has the following three properties:

(a) Whenever $\mathbf{s} \in S$ and $\mathbf{t} \in S$, then $\mathbf{s} + \mathbf{t} \in S$. ($S$ is closed under addition.)

(b) Whenever $a \in F$ and $\mathbf{s} \in S$, then $a\mathbf{s} \in S$. ($S$ is closed under scalar multiplication.)

(c) The zero vector of $V$ belongs to $S$.

2.2.5 **Example.** Let $S$ be the set of all rational solutions $(x, y, z)$ of equation 2.2.3. Since the elements of $S$ are ordered triples of rational numbers, $S$ is a subset of $\mathbf{Q}_3$. We shall show that $S$ is a *subspace* of $\mathbf{Q}_3$ by verifying the three parts of Theorem 2.2.4.

(a) *Closure under addition.* Let $\mathbf{s}$, $\mathbf{t}$ be any elements of $S$. Then $\mathbf{s} = (x_1, y_1, z_1)$, $\mathbf{t} = (x_2, y_2, z_2)$ where both of these ordered triples are solutions of 2.2.3; that is,

$$\begin{aligned} 2x_1 - 3y_1 + 4z_1 &= 0. \\ 2x_2 - 3y_2 + 4z_2 &= 0. \end{aligned} \tag{2.2.6}$$

To show that the triple $\mathbf{s} + \mathbf{t} = (x_1 + x_2, y_1 + y_2, z_1 + z_2)$ belongs to $S$, we must show that it too is a solution of 2.2.3; that is, we must show that

$$2(x_1 + x_2) - 3(y_1 + y_2) + 4(z_1 + z_2) = 0. \tag{2.2.7}$$

But if we simply add the two equations 2.2.6 and then rearrange terms (using properties of the field $\mathbf{Q}$), we obtain 2.2.7 as desired.

(b) *Closure under scalar multiplication.* Let $\mathbf{s} = (x_1, y_1, z_1)$ be any element of $S$ and let $a$ be a scalar (rational number). Since $\mathbf{s} \in S$, we have $2x_1 - 3y_1 + 4z_1 = 0$. Multiplying by $a$ and rearranging terms, we obtain

$$2(ax_1) - 3(ay_1) + 4(az_1) = 0$$

which shows that the ordered triple $a\mathbf{s} = (ax_1, ay_1, az_1)$ belongs to $S$.

(c) Finally, the zero vector $(0, 0, 0)$ is clearly a solution of 2.2.3 and hence belongs to $S$. ∎

There is a useful way of "adding" two subspaces of a vector space $V$. More generally, any two *subsets* of $V$ can be added: if $A$ and $B$ are arbitrary subsets of $V$ (not necessarily subspaces), we define

$$A + B = \{\mathbf{a} + \mathbf{b} : \mathbf{a} \in A, \mathbf{b} \in B\}. \tag{2.2.8}$$

In other words, a vector lies in the subset $A + B$ if and only if it can be expressed as the sum of a vector in $A$ and a vector in $B$. For example, if $V = \mathbf{R}_2$ and $A$ consists of the two vectors $(1, 0)$ and $(2, 2)$ while $B$ consists of the vectors $(3, 1)$ and $(2, -1)$, then

$$A + B = \{(4, 1), (3, -1), (5, 3)\}.$$

Sums of arbitrary subsets of $V$ are usually of no special interest, but sums of *subspaces* arise often in connection with various topics in linear algebra and geometry; in Chapters 8 and 9, the concept of the sum of subspaces will be indispensable.

**2.2.9 Theorem.** If $S$ and $T$ are subspaces of the vector space $V$, then the set $S + T$ is a subspace of $V$.

*Proof.* (a) We show first that $S + T$ is closed under addition. If $\mathbf{u}$, $\mathbf{v}$ are any elements of $S + T$, then by definition of $S + T$ we may write

$$\mathbf{u} = \mathbf{s}_1 + \mathbf{t}_1 \qquad (\mathbf{s}_1 \in S, \mathbf{t}_1 \in T)$$
$$\mathbf{v} = \mathbf{s}_2 + \mathbf{t}_2 \qquad (\mathbf{s}_2 \in S, \mathbf{t}_2 \in T).$$

Adding (and rearranging terms), we get $\mathbf{u} + \mathbf{v} = (\mathbf{s}_1 + \mathbf{s}_2) + (\mathbf{t}_1 + \mathbf{t}_2)$. But $S$, being a subspace, is closed under addition and, hence, $\mathbf{s}_1 + \mathbf{s}_2 \in S$; similarly, $\mathbf{t}_1 + \mathbf{t}_2 \in T$. Thus, we have expressed $\mathbf{u} + \mathbf{v}$ as the sum of an element of $S$ and an element of $T$, so that $\mathbf{u} + \mathbf{v} \in S + T$.

(b) The proof that $S + T$ is closed under scalar multiplication is similar; we leave it to you.

(c) Since $\mathbf{0} \in S$ and $\mathbf{0} \in T$, we have $\mathbf{0} + \mathbf{0} \in S + T$; that is, $\mathbf{0} \in S + T$. This completes the proof. ∎

*Remarks:* 1. If a set $A$ consists of only one element (say, $A = \{\alpha\}$), the sum $A + B = \{\alpha\} + B$ is usually abbreviated $\alpha + B$; similarly, the sum $B + A = B + \{\alpha\}$ is abbreviated $B + \alpha$. Thus,

$$\begin{aligned} \alpha + B &= \{\alpha + \mathbf{b}: \mathbf{b} \in B\} \\ B + \alpha &= \{\mathbf{b} + \alpha: \mathbf{b} \in B\}. \end{aligned} \tag{2.2.10}$$

Of course, the two sets $\alpha + B$ and $B + \alpha$ are the same (why?). As an example, in $\mathbf{R}_2$ we have

$$(2, 3) + \{(-1, 0), (5, 1)\} = \{(1, 3), (7, 4)\}.$$

2. It is not hard to show that addition of subsets of $V$, as defined by 2.2.8, satisfies the following laws (see Exercise 11 below): for all subsets $A$, $B$, $C$ of $V$,

$$\begin{aligned} (A + B) + C &= A + (B + C) \\ A + B &= B + A \\ A + \{\mathbf{0}\} &= A. \end{aligned} \tag{2.2.11}$$

These are three of the four conditions which define an abelian group. Unfortunately, the fourth condition (existence of additive inverses) does not hold, and hence any properties of abelian groups which depend on additive inverses cannot be applied to addition of subsets. For example, subtraction is not defined: if $A + B = A + C$, we cannot subtract $A$ from both sides and conclude that $B = C$. (To put it another way, the cancellation law is false.) However, since *vectors* have additive inverses, it is perfectly legitimate to subtract a *single vector* from both sides of a set equation: if $A + \mathbf{v} = B + \mathbf{v}$ (using the notation of 2.2.10), then $A = B$.

3. Just as 2.2.8 defines the sum of *two* subsets of $V$, so the equation

$$A_1 + \cdots + A_n = \{\mathbf{a}_1 + \cdots + \mathbf{a}_n : \mathbf{a}_1 \in A_1, \ldots, \mathbf{a}_n \in A_n\}$$

may be taken as the definition of the sum of $n$ subsets. Alternatively, the sum of $n$ subsets may be defined by induction, as in Example 1.12.12, Section 1.12. It is not hard to show that these two ways of defining $A_1 + \cdots + A_n$ are equivalent.

**Exercises**

1. In Example 2.2.5 it was shown that the set of rational solutions $(x, y, z)$ of equation 2.2.3 is a subspace of $\mathbf{Q}_3$. Why would the argument fail if equation 2.2.2 were used instead of 2.2.3?

*2. (a) If $S$ and $T$ are subspaces of a vector space $V$, prove that their intersection $S \cap T$ is a subspace of $V$. (SOL)
(b) More generally, prove that the intersection of any number of subspaces of $V$ (even infinitely many) is always a subspace of $V$.

3. Let $F$ be a field, let $a_1, \ldots, a_n$ be fixed elements of $F$, and let $S$ be the set of all solutions $(x_1, \ldots, x_n) \in F_n$ of the equation
$$a_1x_1 + \cdots + a_nx_n = 0.$$
Prove that $S$ is a subspace of $F_n$. (This generalizes Example 2.2.5.)

4. More generally, if $S$ is the set of all solutions $(x_1, \ldots, x_n) \in F_n$ of the system
$$\begin{aligned} a_{11}x_1 + \cdots + a_{1n}x_n &= 0 \\ a_{21}x_1 + \cdots + a_{2n}x_n &= 0 \\ &\cdots \\ a_{m1}x_1 + \cdots + a_{mn}x_n &= 0 \end{aligned}$$
(where the $a$'s are fixed elements of $F$), prove that $S$ is a subspace of $F_n$. (*Suggestion*: Use the results of preceding exercises.) (SOL)

5. Using the "point" interpretation of the vector space $\mathbf{R}_2$ (cf. Sec. 2.1), which of the following subsets of $\mathbf{R}_2$ are subspaces?
(a) The line segment from $(1, 1)$ to $(-1, -1)$. (Interpret this as "the set of all points $(x, y)$ which lie on the line segment from $(1, 1)$ to $(-1, -1)$".)
(b) The line $2x + 3y = 0$.
(c) The line $2x + 3y = 1$.
(d) The line $x = 4$.
(e) The circle $x^2 - 2x + y^2 = 0$.
(f) The union of the $x$ axis and the $y$ axis (that is, the set of all points at least one of whose coordinates is zero).

(g) $\mathbf{R}_2$ itself. (ANS)

6. Which of the following subsets of $\mathbf{R}_3$ are subspaces? Describe each subset geometrically, using the "point" interpretation of $\mathbf{R}_3$.
   (a) The set of all elements of the form $(a, a, a)$ where $a \in \mathbf{R}$ (that is, the set of points having three equal coordinates).
   (b) The set of all elements of the form $(a, b, a + b)$, where $a, b \in \mathbf{R}$.
   (c) The set of all elements of the form $(c, 2c, 1)$, where $c \in \mathbf{R}$.
   (d) The set of all elements of the form $(c, 2c, 0)$, where $c \in \mathbf{R}$.
   (e) The set of all elements of the form $(c, c^2, 0)$, where $c \in \mathbf{R}$.
7. Let $\mathbf{R}[x]$ be the vector space of all polynomials in one indeterminate over $\mathbf{R}$ (cf. Sec. 2.1, Exercise 7). Which of the following subsets of $\mathbf{R}[x]$ are subspaces of $\mathbf{R}[x]$?
   (a) The set of all constant polynomials.
   (b) The set of all polynomials of degree 2, together with the zero polynomial.
   (c) The set of all polynomials of degree 2 or less, together with the zero polynomial.
   (d) The set of all polynomials of degree 2 or more, together with the zero polynomial.
   (e) The set of all polynomials having the number 4 as a root.
   (f) The set of all polynomials having the polynomial $x^2 + 2$ as a factor.
   (g) The set of all polynomials which contain only even powers of $x$ (that is, polynomials of the form $a_0 + a_2x^2 + a_4x^4 + \cdots + a_{2n}x^{2n}$). (ANS)

*8. If $V$ is any vector space, prove that $\{\mathbf{0}\}$ is a subspace of $V$.

9. Complete the proof of Theorem 2.2.9 by showing that $S + T$ is closed under scalar multiplication.
10. Suppose that $S$ is a subspace of $\mathbf{R}_2$, and that the vector $(1, 2)$ lies in $S$. What other vectors (at least) must $S$ contain? Why? Draw a figure using the "point" interpretation of $\mathbf{R}_2$.

The next four exercises (11–14) refer to addition of subsets of a vector space, as defined by 2.2.8.

*11. Prove that the three equations 2.2.11 hold for all subsets $A$, $B$, $C$ of a vector space $V$.

*12. If $S$ is any subset of $V$ and $T$ is a *subspace* of $V$, prove that $S \subseteq S + T$. (SOL)

*13. If $A$, $B$ are arbitrary subsets of the vector space $V$ and if $\mathbf{u}$ is any element of $V$, prove that
$(\mathbf{u} + A) \cap (\mathbf{u} + B) = \mathbf{u} + (A \cap B)$.

14. Let $B$, $C$, $D$, $E$ be the subsets of $\mathbf{R}_2$ which are described (respectively) in parts (b), (c), (d), (e) of Exercise 5 above.
(a) Show that $B + C = C$. (SOL)
(b) Show that $B + D = \mathbf{R}_2$.
(c) Describe the set $D + E$ both algebraically and geometrically. (ANS)

## 2.3 The subspace spanned by a set of vectors

Throughout this section, $V$ will denote a vector space over a field $F$.

Suppose that $S$ is a subspace of $V$, and that $\mathbf{u}_1, \mathbf{u}_2, \ldots, \mathbf{u}_n$ are certain vectors belonging to $S$. In view of Theorem 2.2.4, certain other vectors must then belong to $S$ automatically. For instance, since $S$ is closed under scalar multiplication, all vectors of the form $a\mathbf{u}_1$ ($a \in F$) must belong to $S$; likewise, all vectors of the form $b\mathbf{u}_2$ ($b \in F$). Since $S$ is closed under addition, it then follows that all vectors of the form $a\mathbf{u}_1 + b\mathbf{u}_2$ ($a, b \in F$) belong to $S$. By extending this argument a little further, we see that all vectors of the form

$$a_1\mathbf{u}_1 + a_2\mathbf{u}_2 + \cdots + a_n\mathbf{u}_n \qquad (a_1, \ldots, a_n \in F) \tag{2.3.1}$$

must belong to $S$. Vectors of the form 2.3.1 are called *linear combinations of* $\mathbf{u}_1, \mathbf{u}_2, \ldots, \mathbf{u}_n$ *over* $F$. (The phrase "over $F$" is usually omitted unless more than one field of scalars is under discussion.) What we have shown is that any subspace which contains given vectors $\mathbf{u}_1, \ldots, \mathbf{u}_n$ must also contain all linear combinations of those vectors.

Although the preceding discussion involved only finitely many $\mathbf{u}$'s, the infinite case is not essentially different. If $A$ is any subset of $V$ (finite or infinite), by a "linear combination of elements of $A$" we mean any expression of the form 2.3.1 such that all of the $\mathbf{u}$'s in the expression belong to $A$; that is, any expression of the form

$$a_1\mathbf{u}_1 + \cdots + a_n\mathbf{u}_n \qquad (a_1, \ldots, a_n \in F;\ \mathbf{u}_1, \ldots, \mathbf{u}_n \in A). \tag{2.3.2}$$

We even allow $n$ to be zero in 2.3.2; in this case, 2.3.2 becomes a sum having no terms, the "empty sum". By convention (see Sec. 1.12) an empty sum always equals the additive identity, which in a vector space is the vector $\mathbf{0}$. Thus, no matter what the set $A$ consists of (even if $A$

itself is empty!), $\mathbf{0}$ is a linear combination of the elements of $A$. It has become fairly common practice to let $[A]$ denote the set of all linear combinations of elements of $A$. (If $A$ consists of only finitely many vectors $\mathbf{u}_1, ..., \mathbf{u}_n$, it is usual to write $[A]$ as $[\mathbf{u}_1, ..., \mathbf{u}_n]$ instead of the more rigorous notation $[\{\mathbf{u}_1, ..., \mathbf{u}_n\}]$.) In particular, we see from the discussion above that if $\varnothing$ denotes the empty set, then

(2.3.3) $$[\varnothing] = \{\mathbf{0}\};$$

in words, the "set of all linear combinations of vectors in the empty set" equals the (nonempty) set consisting of the single vector $\mathbf{0}$. Note the different kinds of brackets in 2.3.3: the brackets [ ] mean "the set of all linear combinations of", while the brackets { } mean "the set consisting of".

2.3.4 **Example.** A linear combination of a single vector is simply a scalar multiple of that vector. In $\mathbf{R}_2$, for instance, the set $[(1, 3)]$ consists of all vectors of the form $a(1, 3)$ $(a \in \mathbf{R})$, that is, all vectors of the form $(a, 3a)$. Equivalently, $[(1, 3)]$ is the set of all vectors $(x, y)$ which satisfy the equation $y = 3x$. Geometrically (using the "point" interpretation of $\mathbf{R}_2$) this set is the line which passes through the origin and the point (1, 3).

2.3.5 **Theorem.** Let $A$ be any subset of the vector space $V$; let $[A]$ denote the set of all linear combinations of elements of $A$. Then

(a) $[A]$ is a subspace of $V$;

(b) $A \subseteq [A]$;

(c) If $S$ is any subspace of $V$ such that $A \subseteq S$, then $[A] \subseteq S$.

*Proof.* Part (c) of 2.3.5 asserts that any subspace containing a given set of vectors must contain all linear combinations of these vectors. This has already been established when $A$ is a finite set; the proof when $A$ is infinite is the same. (The empty linear combination causes no trouble since every subspace must contain $\mathbf{0}$.) The proof of 2.3.5(b) is trivial: if $\mathbf{u} \in A$, then $\mathbf{u} = 1\mathbf{u}$ has the form 2.3.2 (with $n = 1$) and hence belongs to $[A]$. To prove 2.3.5(a), we will show that $[A]$ satisfies the three conditions of Theorem 2.2.4. *Closure under addition*: if $\mathbf{u}, \mathbf{v}$ are any elements of $[A]$, then we may write

$$\mathbf{u} = a_1\mathbf{u}_1 + \cdots + a_n\mathbf{u}_n \qquad (a_i \in F, \mathbf{u}_i \in A);$$
$$\mathbf{v} = b_1\mathbf{v}_1 + \cdots + b_m\mathbf{v}_m \qquad (b_j \in F, \mathbf{v}_j \in A).$$

Adding, we get

$$\mathbf{u} + \mathbf{v} = a_1\mathbf{u}_1 + \cdots + a_n\mathbf{u}_n + b_1\mathbf{v}_1 + \cdots + b_m\mathbf{v}_m.$$

This is still of the form 2.3.2 since all of the **u**'s and **v**'s are in $A$; hence, $\mathbf{u} + \mathbf{v} \in [A]$. *Closure under scalar multiplication*: proof is similar and we leave it to you. Since we have observed earlier that $[A]$ contains $\mathbf{0}$ (the "empty sum"!), the proof is complete. ∎

*Remark*: It follows from Theorem 2.3.5 that $[A]$ *is the smallest subspace of $V$ which contains $A$*. Indeed, by 2.3.5(a) and 2.3.5(b), $[A]$ is a subspace of $V$ which contains $A$; and by 2.3.5(c), $[A]$ is the smallest such subspace since any other such subspace contains $[A]$.

**2.3.6** **Definition.** For any subset $A$ of $V$, the subspace $[A]$ is called the subspace *spanned by A* (or *generated by A*).

**2.3.7** **Example.** We use the "point" interpretation of $n$-tuples (Sec. 2.1) to identify points in 3-space with elements of $\mathbf{R}_3$. If $P$, $Q$ are two points in $\mathbf{R}_3$ which do not lie on the same line through the origin, then the scalar multiples of $P$ constitute the line $[P]$ through $P$ and $O$, while the scalar multiples of $Q$ constitute the line $[Q]$ through $Q$ and $O$ (Fig. 2.3.1). The elements of the subspace $[P, Q]$ are the linear combinations $aP + bQ$, all of which lie in the plane determined by the two lines $[P]$ and $[Q]$. Conversely, given any point $X$ in this plane, it is easy to see from Figure 2.3.1 how to locate a point $bQ$ on $[Q]$ and a point $aP$ on $[P]$ such that $X = aP + bQ$. We thus see that the subspace $[P, Q]$ spanned by $P$ and $Q$ is the entire plane determined by the lines $[P]$, $[Q]$ (equivalently, the plane determined by the three points $O$, $P$, $Q$).

**2.3.8** **Example.** Show by purely algebraic reasoning that the subspace of $\mathbf{R}_2$ which is spanned by the two vectors (2, 5) and (1, 3) is the whole space $\mathbf{R}_2$.

Figure 2.3.1

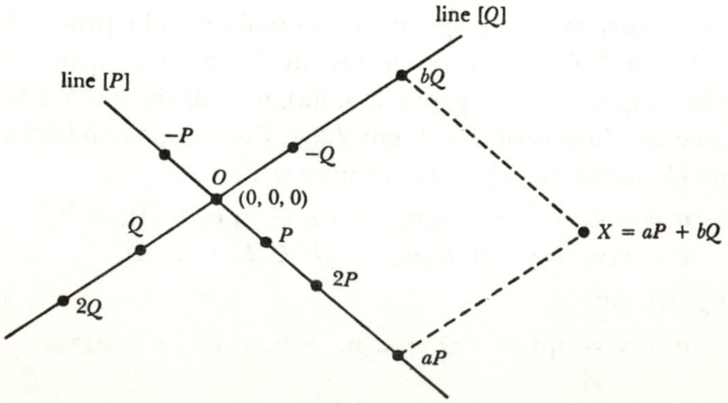

*Solution.* We must show that any vector $(a, b)$ in $\mathbf{R}_2$ is a linear combination of $(2, 5)$ and $(1, 3)$; that is, we must find scalars $x$, $y$ such that

**(2.3.9)** $(a, b) = x(2, 5) + y(1, 3).$

Using the definitions of addition and scalar multiplication in $\mathbf{R}_2$, we see that equation 2.3.9 is the same as $(a, b) = (2x + y, 5x + 3y)$, which in turn is equivalent to the pair of equations

**(2.3.10)**
$$\begin{aligned} 2x + y &= a \\ 5x + 3y &= b. \end{aligned}$$

Solving for $x$ and $y$ in terms of $a$ and $b$, we obtain

**(2.3.11)** $x = 3a - b; \qquad y = 2b - 5a.$

(Once the solution 2.3.11 is obtained, it should be checked by substituting 2.3.11 back into 2.3.10.) Hence, for any $(a, b)$, there exist scalars $x$, $y$ (depending on $a$, $b$) satisfying 2.3.9, as we wished to show. (For a second solution to this problem, see the Exercises at the end of this section; for a third solution, see Sec. 2.6. It is worth noting that in mathematics, as elsewhere, there is often more than one way of solving a problem.) ▮

*Remark*: The minimum number of vectors required to span a given subspace is called the "dimension" of the subspace (see Sec. 2.6, Definition 2.6.7 and Exercise 9). The reason why is suggested by Example 2.3.4 (in which one vector spanned a line) and Examples 2.3.7 and 2.3.8 (in which two vectors spanned a plane). It should be noted in connection with these examples that not every line or plane is a subspace; indeed, a line or plane which does not pass through the origin cannot be a subspace since it does not contain $\mathbf{0}$. Of course, if the vector space $V$ is not $\mathbf{R}_2$ or $\mathbf{R}_3$, the subspaces of $V$ need not have geometric interpretations at all; they may be purely algebraic objects, as is the case in Section 2.2, Exercise 7.

Some simple facts about subspaces spanned by subsets are collected in the following two theorems.

**2.3.12** **Theorem.** Let $V$ be a vector space over $F$. If $U$ is a subspace of $V$, then $[U] = U$.

**2.3.13** **Theorem.** Let $V$ be a vector space over $F$. If $A$, $B$, $C$ are any subsets of $V$, the following statements are true:

(a) $[[A]] = [A]$.
(b) If $A \subseteq [B]$, then $[A] \subseteq [B]$.
(c) If $A \subseteq B$, then $[A] \subseteq [B]$.

(d) If $A \subseteq [B]$ and $B \subseteq [A]$, then $[A] = [B]$.
(e) If $A \subseteq [B]$ and $B \subseteq [C]$, then $A \subseteq [C]$.

*Proofs.* Theorem 2.3.12 should be obvious; indeed, since $U$ is a *subspace*, the smallest subspace containing $U$ is $U$ itself; that is, $[U] = U$. (If this proof was too informal for you, here's a sketch of a more formal one: $U \subseteq [U]$ by Theorem 2.3.5(b) and $[U] \subseteq U$ by Theorem 2.3.5(c).)

Theorem 2.3.13(a) is obtained by setting $U = [A]$ in Theorem 2.3.12; Theorem 2.3.13(b) is obtained by setting $S = [B]$ in Theorem 2.3.5(c). (Remember that $[A]$ and $[B]$ are subspaces.) Theorem 2.3.13(c) follows from 2.3.13(b), using the fact that $B \subseteq [B]$. Theorem 2.3.13(d) follows from 2.3.13(b): if $A \subseteq [B]$ and $B \subseteq [A]$, then by 2.3.13(b) we have $[A] \subseteq [B]$ and $[B] \subseteq [A]$, so that $[A] = [B]$. Part (e) is proved similarly; we omit the details. ∎

If the vectors in the set $A$ are denoted $\mathbf{u}_1, \mathbf{u}_2, \ldots$, and the vectors in $B$ are denoted $\mathbf{v}_1, \mathbf{v}_2, \ldots$, we may rephrase Theorem 2.3.13(d) as follows: *If the* **u**'s *are linear combinations of the* **v**'s *and the* **v**'s *are linear combinations of the* **u**'s, *then the subspace spanned by the* **u**'s *equals the subspace spanned by the* **v**'s. This principle is sometimes useful in proving that two subspaces are equal.

**2.3.14 Example.** In $\mathbf{R}_3$, show that $[(1, 2, 3), (0, 4, 1)] = [(1, 6, 4), (1, -2, 2)]$.

*Solution.* Using the principle stated in the preceding paragraph, we need only show that (1, 2, 3) and (0, 4, 1) can be expressed as linear combinations of (1, 6, 4) and (1, −2, 2), and vice versa. That is, we must find scalars $a, b, c, d, e, f, g, h$ satisfying the equations

$$\begin{aligned} (1, 2, 3) &= a(1, 6, 4) + b(1, -2, 2) \\ (0, 4, 1) &= c(1, 6, 4) + d(1, -2, 2) \\ (1, 6, 4) &= e(1, 2, 3) + f(0, 4, 1) \\ (1, -2, 2) &= g(1, 2, 3) + h(0, 4, 1). \end{aligned} \tag{2.3.15}$$

Each of these four equations can be solved for the unknown scalars just as we solved equation 2.3.9. For example, the first of the four equations 2.3.15 is equivalent to the system

$$\begin{aligned} a + b &= 1 \\ 6a - 2b &= 2 \\ 4a + 2b &= 3 \end{aligned}$$

which has the solution $a = 1/2$, $b = 1/2$. The other equations may be solved similarly. (Be sure to check each solution by substitution.) ▮

**2.3.16 Theorem.** Let $\mathbf{u}_1, \ldots, \mathbf{u}_n$ be elements of a vector space $V$. Then:

(a) The subspace $[\mathbf{u}_1, \ldots, \mathbf{u}_n]$ is unchanged if one vector $\mathbf{u}_i$ is replaced by $\mathbf{u}_i + a\mathbf{u}_j$, where $a$ is a scalar and $j$ is an index different from $i$. That is,

$$[\mathbf{u}_1, \ldots, \mathbf{u}_{i-1}, \mathbf{u}_i, \mathbf{u}_{i+1}, \ldots, \mathbf{u}_n] = [\mathbf{u}_1, \ldots, \mathbf{u}_{i-1}, \mathbf{u}_i + a\mathbf{u}_j, \mathbf{u}_{i+1}, \ldots, \mathbf{u}_n].$$

(b) The subspace $[\mathbf{u}_1, \ldots, \mathbf{u}_n]$ is unchanged if one vector $\mathbf{u}_i$ is replaced by $c\mathbf{u}_i$, where $c$ is a scalar and $c \neq 0$. That is,

$$[\mathbf{u}_1, \ldots, \mathbf{u}_i, \ldots, \mathbf{u}_n] = [\mathbf{u}_1, \ldots, c\mathbf{u}_i, \ldots, \mathbf{u}_n].$$

*Proof.* Using the same principle as before, to prove (a) we must show that the vectors

$$\mathbf{u}_1, \ldots, \mathbf{u}_{i-1}, \mathbf{u}_i, \mathbf{u}_{i+1}, \ldots, \mathbf{u}_n \tag{2.3.17}$$

are linear combinations of the vectors

$$\mathbf{u}_1, \ldots, \mathbf{u}_{i-1}, \mathbf{u}_i + a\mathbf{u}_j, \mathbf{u}_{i+1}, \ldots, \mathbf{u}_n \tag{2.3.18}$$

and vice versa. The equations which express the vectors 2.3.17 as linear combinations of the vectors 2.3.18 are

$$\begin{aligned}
\mathbf{u}_1 &= \mathbf{u}_1 \\
&\ldots \\
\mathbf{u}_{i-1} &= \mathbf{u}_{i-1} \\
\mathbf{u}_i &= (\mathbf{u}_i + a\mathbf{u}_j) - a(\mathbf{u}_j) \\
\mathbf{u}_{i+1} &= \mathbf{u}_{i+1} \\
&\ldots \\
\mathbf{u}_n &= \mathbf{u}_n.
\end{aligned}$$

Conversely, it is trivial that the vectors 2.3.18 are linear combinations of the vectors 2.3.17. This proves (a). Note where the assumption $j \neq i$ is needed: since all of the original **u**'s *except* $\mathbf{u}_i$ are among the vectors 2.3.18, it follows that $\mathbf{u}_j$ is one of the vectors 2.3.18, so that the right side of the expression

$$\mathbf{u}_i = (\mathbf{u}_i + a\mathbf{u}_j) - a(\mathbf{u}_j)$$

is indeed a linear combination of the vectors 2.3.18, as we want it to be.

The proof of (b) is similar, and we leave it as an exercise. ▮

The principal use of Theorem 2.3.16 will occur later on, in connection with matrices (cf. Theorem 5.5.6). In addition, examples such as 2.3.14 can be simplified by using Theorem 2.3.16:

(2.3.19)

$$[(1, 2, 3), (0, 4, 1)] = [(1, 2, 3), (1, 6, 4)] \quad \text{(replacing } \mathbf{u}_2 \text{ by } \mathbf{u}_2 + \mathbf{u}_1\text{)}$$

$$[(1, 2, 3), (1, 6, 4)] = [(1/2, -1, 1), (1, 6, 4)] \quad \text{(replacing } \mathbf{u}_1 \text{ by } \mathbf{u}_1 - \mathbf{u}_2/2\text{)}$$

$$[(1/2, -1, 1), (1, 6, 4)] = [(1, -2, 2), (1, 6, 4)] \quad \text{(replacing } \mathbf{u}_1 \text{ by } 2\mathbf{u}_1\text{)}$$

and hence $[(1, 2, 3), (0, 4, 1)] = [(1, 6, 4), (1, -2, 2)]$. (The sequence of steps in 2.3.19 is admittedly not very systematic. For a more systematic form of the same argument, see Sec. 5.6, Example 5.6.5.)

At the end of Section 2.2, we defined the sum of two subsets of $V$. The connection between that concept and the "span" concept is given by the following theorem:

2.3.20 **Theorem.** If $A$ and $B$ are any subsets of the vector space $V$, then

$$[A] + [B] = [A \cup B].$$

(Recall that $A \cup B$ is the "union" of $A$ and $B$, consisting of all elements of $A$ together with all elements of $B$.).

*Proof.* Since $A \subseteq A \cup B$, $[A] \subseteq [A \cup B]$; similarly, $[B] \subseteq [A \cup B]$. Since $[A \cup B]$ is a subspace, it is closed under addition; it follows that $[A] + [B] \subseteq [A \cup B]$. (Does the last step need further justification? Think about it.)

Conversely, we must show that $[A \cup B] \subseteq [A] + [B]$. But we know that $A \subseteq [A]$, and $[A] \subseteq [A] + [B]$ (Sec. 2.2, Exercise 12); hence, $A \subseteq [A] + [B]$. Similarly, $B \subseteq [A] + [B]$. Hence, $A \cup B \subseteq [A] + [B]$, by definition of union. Moreover, $[A] + [B]$ is a subspace by Theorem 2.2.9. Hence, $[A \cup B] \subseteq [A] + [B]$ by Theorem 2.3.5(c). Since we have now shown that each of the two sets $[A] + [B]$ and $[A \cup B]$ is a subset of the other, Theorem 2.3.20 follows. ▮

2.3.21 **Corollary.** If $S$ is a subspace of $V$, then $S + S = S$.

*Proof.* Set $A = B = S$ in Theorem 2.3.20 and use Theorem 2.3.12. ▮

As an illustration of Theorem 2.3.20, consider the situation of Example 2.3.7 (Fig. 2.3.1). If we let $A = \{P\}$ (the set consisting of the single point $P$) and $B = \{Q\}$, Theorem 2.3.20 reduces to $[P] + [Q] = [P, Q]$. As shown in Example 2.3.7, $[P, Q]$ is the plane determined by the lines $[P]$ and $[Q]$. In other words, the sum of two distinct lines through the origin is the plane which they determine.

**Exercises**

1. Complete the proof of Theorem 2.3.5(a) by showing that $[A]$ is closed under scalar multiplication.
2. In $\mathbf{R}_3$, let $\mathbf{u} = (1, 2, 3)$, $\mathbf{v} = (0, 1, 2)$, $\mathbf{w} = (3, 1, -1)$.
   (a) Express vector $(8, 5, 2)$ as a linear combination of $\mathbf{u}$, $\mathbf{v}$, and $\mathbf{w}$. Is there more than one such expression for $(8, 5, 2)$? (ANS)
   (b) Can an arbitrary vector $(a, b, c) \in \mathbf{R}_3$ be expressed as a linear combination of $\mathbf{u}$, $\mathbf{v}$, and $\mathbf{w}$? (*Suggestion*: Try to do so as in Example 2.3.8, and see what happens.) Which vectors in $\mathbf{R}_3$ can be so expressed? What does this tell you about the subspace $[\mathbf{u}, \mathbf{v}, \mathbf{w}]$? (ANS)
*3. Prove Theorem 2.3.16(b). Why must we assume that $c \neq 0$?
4. (a) In the vector space $F_2$, verify that $(a, b) = a(1, 0) + b(0, 1)$ for any scalars $a$, $b$. Deduce that $[(1, 0), (0, 1)] = F_2$.
   (b) State and prove an analogous result for $F_3$; for $F_n$.
5. In the vector space $\mathbf{R}_2$, use Theorem 2.3.16 to show that
   $[(2, 5), (1, 3)] = [(1, 0), (0, 1)]$.
   (Note that this, together with Exercise 4(a), provides a second solution to Example 2.3.8.)

## 2.4 Linear dependence and independence

The concepts of linear dependence and independence, which we are about to define, are very important in both algebra and geometry. (They also have applications in analysis; see, for example, Appendix E, Theorem B.) Throughout this section, $V$ will denote a vector space over the field $F$.

**2.4.1 Definition.** (a) Let $n$ be a positive integer and let $\mathbf{u}_1, \ldots, \mathbf{u}_n$ be (not necessarily distinct) vectors in $V$. The ordered $n$-tuple $(\mathbf{u}_1, \ldots, \mathbf{u}_n)$ is said to be *linearly dependent* (more strictly, *linearly dependent over F*) if there exist scalars $a_1, \ldots, a_n$ in $F$, *not all of which are zero*, such that

(2.4.2) $$a_1\mathbf{u}_1 + \cdots + a_n\mathbf{u}_n = \mathbf{0}.$$

(b) Let $S$ be a subset of $V$. We say that the set $S$ is *linearly dependent* if there exist *distinct* vectors $\mathbf{u}_1, \ldots, \mathbf{u}_n \in S$ and scalars $a_1, \ldots, a_n$ not all of which are zero, such that equation 2.4.2 holds. In other words, a set of vectors is linearly dependent if and only if some $n$-tuple of *distinct* vectors in the set is linearly dependent.

(c) "Linearly independent" means "not linearly dependent". Thus, the $n$-tuple $(\mathbf{u}_1, \ldots, \mathbf{u}_n)$ is linearly independent if and only if the following statement is true: whenever equation 2.4.2 holds, then $a_1 = \cdots = a_n = 0$. Similarly, a subset $S$ of $V$ is linearly independent if and only if the following is true: whenever $\mathbf{u}_1, \ldots, \mathbf{u}_n$ are distinct vectors in $S$ such that 2.4.2 holds, then all of the scalars $a_1, \ldots, a_n$ are zero.

*Remarks*: 1. The adverb "linearly" in Definition 2.4.1 is often omitted to save space, although sometimes we need it in order to distinguish linear dependence from other kinds of dependence, for example, barycentric dependence (to be defined in Sec. 3.5) or algebraic dependence (which will not be defined in this book).

2. Note the distinction made between *ordered n-tuples* of vectors and *sets* of vectors. Definition 2.4.1 easily implies that the $n$-tuple $(\mathbf{u}_1, \ldots, \mathbf{u}_n)$ is linearly independent if and only if all of the vectors $\mathbf{u}_1, \ldots, \mathbf{u}_n$ are distinct *and* the set $\{\mathbf{u}_1, \ldots, \mathbf{u}_n\}$ is independent. (We leave the formal proof of this assertion as an exercise.) If any two of $\mathbf{u}_1, \ldots, \mathbf{u}_n$ are equal, then the $n$-tuple $(\mathbf{u}_1, \ldots, \mathbf{u}_n)$ is dependent regardless of the status of the set $\{\mathbf{u}_1, \ldots, \mathbf{u}_n\}$. As a matter of terminology, statements like "the vectors $\mathbf{u}_1, \ldots, \mathbf{u}_n$ are independent" will be understood to refer to the $n$-tuple $(\mathbf{u}_1, \ldots, \mathbf{u}_n)$, *not* the set $\{\mathbf{u}_1, \ldots, \mathbf{u}_n\}$. For example, if

$$\mathbf{u}_1 = (1, 0, 2)$$
$$\mathbf{u}_2 = (1, 0, 2)$$

in $\mathbf{R}_3$, then the set $\{\mathbf{u}_1\}$ is independent and, hence, so is the set $\{\mathbf{u}_1, \mathbf{u}_2\}$ (since it is the same set!); on the other hand, the pair $(\mathbf{u}_1, \mathbf{u}_2)$ is dependent. In this case, it would be correct to say "the vectors $\mathbf{u}_1, \mathbf{u}_2$ are dependent".

**2.4.3** **Example.** Determine whether the vectors $\mathbf{u}_1, \mathbf{u}_2, \mathbf{u}_3$ in $\mathbf{R}_3$ are dependent or independent, where

$$\mathbf{u}_1 = (1, 2, 3); \qquad \mathbf{u}_2 = (0, 1, 2); \qquad \mathbf{u}_3 = (3, 1, -1).$$

*Solution.* We must determine whether there exist scalars $a_1, a_2, a_3$, not all zero, which satisfy the equation

$$a_1(1, 2, 3) + a_2(0, 1, 2) + a_3(3, 1, -1) = (0, 0, 0) \tag{2.4.4}$$

(in which case $\mathbf{u}_1$, $\mathbf{u}_2$, $\mathbf{u}_3$, are dependent), or whether the only solution of 2.4.4 is $a_1 = a_2 = a_3 = 0$ (in which case $\mathbf{u}_1$, $\mathbf{u}_2$, $\mathbf{u}_3$ are independent). By using the definitions of addition and scalar multiplication in $\mathbf{R}_3$, we can reduce the vector equation 2.4.4 to a system of three linear equations:

$$\begin{aligned} a_1 \qquad\quad + 3a_3 &= 0 \\ 2a_1 + a_2 + a_3 &= 0 \\ 3a_1 + 2a_2 - a_3 &= 0. \end{aligned} \tag{2.4.5}$$

By solving these equations for the "unknowns" $a_1$, $a_2$, $a_3$, we find that any scalars $a_i$ which satisfy the conditions

$$a_1 = -3a_3; \qquad a_2 = 5a_3$$

will satisfy 2.4.5; in particular, we may choose $a_1 = -3$, $a_2 = 5$, $a_3 = 1$. Since not all of these scalars are zero, the vectors $\mathbf{u}_1$, $\mathbf{u}_2$, $\mathbf{u}_3$ are dependent. ∎

(If we change the vector $\mathbf{u}_3$ in this example from (3, 1, −1) to (4, 1, −1), we obtain in place of 2.4.5 the equations

$$\begin{aligned} a_1 \qquad\quad + 4a_3 &= 0 \\ 2a_1 + a_2 + a_3 &= 0 \\ 3a_1 + 2a_2 - a_3 &= 0. \end{aligned}$$

Here the only solution is $a_1 = a_2 = a_3 = 0$; hence, the vectors

$$(1, 2, 3), (0, 1, 2), (4, 1, -1)$$

are independent.)

*Comment (author's apology)*: In the example above, as well as in previous examples in Section 2.3, it has been necessary to solve systems of linear equations (such as 2.4.5 and 2.3.10) or to do other kinds of arithmetical calculations (as in 2.3.19). The mechanics of actually performing such calculations will be discussed systematically in Sections 5.6 and 5.10; my failure to do so now is based on (1) my desire to wait until matrices are available, (2) a wish not to distract your attention from other matters, and (3) my belief that you can find ways of doing these calculations yourself. (*Exercise*: Justify part (3) of the preceding sentence.)

The next theorem gives an alternate condition for linear dependence which is sometimes useful.

**2.4.6 Theorem.** A subset $S$ of $V$ is linearly dependent if and only if some vector $\mathbf{u} \in S$ is equal to a linear combination of the other vectors in $S$. Similarly, an $n$-tuple $(\mathbf{u}_1, \ldots, \mathbf{u}_n)$ is dependent if and only if, for some $i$ $(1 \leq i \leq n)$, the vector $\mathbf{u}_i$ is a linear combination of the vectors

$$\mathbf{u}_j : 1 \leq j \leq n, \qquad j \neq i.$$

We give the proof for $n$-tuples (the proof for subsets is essentially the same). If $(\mathbf{u}_1, ..., \mathbf{u}_n)$ is dependent, then $\mathbf{u}_1, ..., \mathbf{u}_n$ satisfy an equation of the form 2.4.2 in which at least one of the $a$'s, say $a_i$, is not zero. Subtracting all of the terms except $a_i\mathbf{u}_i$ from both sides of 2.4.2, we get

$$a_i\mathbf{u}_i = -\sum_{j \neq i} a_j\mathbf{u}_j$$

(the sum on the right is the sum of the $n - 1$ terms corresponding to the $n - 1$ possible values of $j$ from 1 to $n$, excluding $i$). Multiplying both sides by $a_i^{-1}$ (which exists since $a_i \neq 0$), we get

$$\mathbf{u}_i = -\sum_{j \neq i} (a_i^{-1}a_j)\mathbf{u}_j$$

which expresses $\mathbf{u}_i$ as a linear combination of the vectors $\mathbf{u}_j$, $j \neq i$. Conversely, suppose some $\mathbf{u}_i$ is a linear combination of the vectors $\mathbf{u}_j$ $(j \neq i)$, say,

$$\mathbf{u}_i = \sum_{j \neq i} b_j\mathbf{u}_j.$$

Then

$$1\mathbf{u}_i - \sum_{j \neq i} b_j\mathbf{u}_j = \mathbf{0}.$$

The latter equation has the form 2.4.2, and *not all of the scalars are zero* since $1 \neq 0$ by Theorem 1.10.2(a). Hence, the $n$-tuple $(\mathbf{u}_1, ..., \mathbf{u}_n)$ is dependent. ∎

**Some special cases**

1. If $\mathbf{u}$, $\mathbf{v}$ are any two distinct vectors in $V$, Theorem 2.4.6 asserts that the set $\{\mathbf{u}, \mathbf{v}\}$ is dependent if and only if one of these two vectors is a scalar multiple of the other one. (Interpret this geometrically!) Thus, for instance, the set $\{(1,3), (2, 6)\}$ is dependent but the set $\{(1, 3), (2, 5)\}$ is independent.

2. *Any set containing the zero vector is dependent*, since $\mathbf{0}$ can always be expressed as a linear combination of the remaining vectors by choosing all of the scalars in the linear combination to be zero. For example, $\{\mathbf{0}, \mathbf{v}, \mathbf{w}\}$ is dependent since $\mathbf{0} = 0\mathbf{v} + 0\mathbf{w}$. Even if $\mathbf{0}$ is the only vector in the set, we can still regard $\mathbf{0}$ as the *empty* linear combination of the "other vectors" since there are no other vectors. (Recall from Section

2.3 that the empty sum equals zero!) Thus, in particular, the set $\{\mathbf{0}\}$, consisting of *just* the zero vector, is dependent.

3. Conversely, a set $\{\mathbf{u}\}$ consisting of a single *nonzero* vector is always *independent*; for since $\mathbf{u} \neq \mathbf{0}$, $\mathbf{u}$ does not equal the empty sum and, hence, is not a "linear combination of the other vectors"!

4. *The empty set is independent*; indeed, both Definition 2.4.1(b) and Theorem 2.4.6 require a dependent set to contain at least one vector.

**2.4.7 Theorem.** If $\mathbf{u}_1, ..., \mathbf{u}_n$ are any vectors in $V$, then the following statements are equivalent:

(a) The $n$-tuple $(\mathbf{u}_1, ..., \mathbf{u}_n)$ is independent.

(b) Whenever

$$\sum_{i=1}^{n} a_i\mathbf{u}_i = \sum_{i=1}^{n} b_i\mathbf{u}_i,$$

where the $a$'s and $b$'s are scalars, then $a_i = b_i$ for all $i$ $(1 \leqslant i \leqslant n)$.

(Roughly speaking, Theorem 2.4.7 says that we may "equate coefficients" in any equation involving independent vectors.)

*Proof that (a) ⇒ (b).* Assume $(\mathbf{u}_1, ..., \mathbf{u}_n)$ is independent. If

$$\sum_{i=1}^{n} a_i\mathbf{u}_i = \sum_{i=1}^{n} b_i\mathbf{u}_i,$$

then

$$\sum_{i=1}^{n} (a_i - b_i)\mathbf{u}_i = \mathbf{0};$$

hence, $a_i - b_i = 0$ for each $i$ (Definition 2.4.1(c) ); hence, $a_i = b_i$ for each $i$. (Note how this argument generalizes the proof of Theorem 2.1.4(n).)

*Proof that (b) ⇒ (a).* Assume statement (b) is true. If $a_1, ..., a_n$ are any scalars such that equation 2.4.2 holds, then we have

$$\sum_{i=1}^{n} a_i\mathbf{u}_i = \mathbf{0} = \sum_{i=1}^{n} 0\mathbf{u}_i.$$

Since 2.4.7(b) is assumed true, it follows that $a_i = 0$ for all $i$; hence, $(\mathbf{u}_1, ..., \mathbf{u}_n)$ is independent by Definition 2.4.1(c). ∎

**2.4.8 Theorem.** Let $A$, $B$ be subsets of the vector space $V$.

(a) If $A \subseteq B$ and $A$ is dependent, then $B$ is dependent.

(b) If $A \subseteq B$ and $B$ is independent, then $A$ is independent.

*Proof.* Assume $A \subseteq B$. If $A$ is dependent, then by Theorem 2.4.6 some vector $\mathbf{u} \in A$ is a linear combination of other vectors $\mathbf{v}_1, \ldots, \mathbf{v}_k$ in $A$:

$$\mathbf{u} = a_1\mathbf{v}_1 + \cdots + a_k\mathbf{v}_k. \tag{2.4.9}$$

But since $A \subseteq B$ the vectors $\mathbf{u}, \mathbf{v}_1, \ldots, \mathbf{v}_k$ also lie in $B$; hence, 2.4.9 expresses an element of $B$ as a linear combination of other elements of $B$, so that $B$ is dependent. This proves 2.4.8(a). Since 2.4.8(b) is the contrapositive of 2.4.8(a), 2.4.8(b) follows without further proof. (In logic, the contrapositive of the statement "If $x$, then $y$" is the statement "If not $y$, then not $x$"; a statement is true if and only if its contrapositive is true.) ∎

To rephrase Theorem 2.4.8, a subset of an independent set is independent; an overset of a dependent set is dependent.

We close this section with the following theorem, which will be used repeatedly.

**2.4.10 Theorem.** Let $V$ be a vector space. If $A$ is an independent subset of $V$ and $\mathbf{v}$ is an element of $V$ such that the set $A \cup \{\mathbf{v}\}$ is dependent, then $\mathbf{v} \in [A]$. (In other words: if an independent set becomes dependent when one additional vector is added to it, then the additional vector is a linear combination of the vectors in the original set.)

Similarly, if $\mathbf{u}_1, \ldots, \mathbf{u}_n, \mathbf{v}$ are elements of $V$ such that the $n$-tuple $(\mathbf{u}_1, \ldots, \mathbf{u}_n)$ is independent but $(\mathbf{u}_1, \ldots, \mathbf{u}_n, \mathbf{v})$ is dependent, then $\mathbf{v} \in [\mathbf{u}_1, \ldots, \mathbf{u}_n]$.

We give the proof for $n$-tuples (the proof for subsets is similar). The assumption that $(\mathbf{u}_1, \ldots, \mathbf{u}_n, \mathbf{v})$ is dependent implies that some equation of the form

$$a_1\mathbf{u}_1 + \cdots + a_n\mathbf{u}_n + b\mathbf{v} = \mathbf{0} \tag{2.4.11}$$

holds, where not all of the scalars in 2.4.11 are zero. If $b \neq 0$, then we can solve 2.4.11 for $\mathbf{v}$ (just as we solved 2.4.2 for $\mathbf{u}_i$ in the first half of the proof of Theorem 2.4.6), obtaining

$$\mathbf{v} = -b^{-1}(a_1\mathbf{u}_1 + \cdots + a_n\mathbf{u}_n)$$

so that $\mathbf{v} \in [\mathbf{u}_1, \ldots, \mathbf{u}_n]$ as desired. On the other hand, if $b = 0$, then 2.4.11 reduces to

$$a_1\mathbf{u}_1 + \cdots + a_n\mathbf{u}_n = \mathbf{0} \tag{2.4.12}$$

and since not all of the scalars in 2.4.11 are zero it follows that not all of the scalars in 2.4.12 are zero, contradicting the assumption that $(\mathbf{u}_1, \ldots, \mathbf{u}_n)$ is independent. ∎

*Remark*: Since any statement implies its contrapositive, Theorem 2.4.10 implies that if $A$ is independent and $\mathbf{u} \notin [A]$, then $A \cup \{\mathbf{u}\}$ is independent. This tells us how to extend a given independent set to a larger independent set: simply adjoin any vector which is not already a linear combination of the vectors in the original set. For example, suppose $\mathbf{v}_1 = (1, 0, 1)$ and $\mathbf{v}_2 = (2, 3, 0)$ (in $\mathbf{R}_3$) and we wish to find $\mathbf{v}_3$ such that $\{\mathbf{v}_1, \mathbf{v}_2, \mathbf{v}_3\}$ is an independent set. (It is clear that $\{\mathbf{v}_1, \mathbf{v}_2\}$ is independent.) Since the most general vector in $[\mathbf{v}_1, \mathbf{v}_2]$ has the form

$$a(1, 0, 1) + b(2, 3, 0) = (a + 2b, 3b, a)$$

(in which the first coordinate equals the third coordinate plus two-thirds of the second coordinate), we need only select any vector *not* of this form as our vector $\mathbf{v}_3$. For instance, since $0 \neq 1 + (2/3) \cdot 4$, the vector $\mathbf{v}_3 = (0, 4, 1)$ is not in $[\mathbf{v}_1, \mathbf{v}_2]$ and, hence, $\{\mathbf{v}_1, \mathbf{v}_2, \mathbf{v}_3\}$ is independent.

**Exercises**

1. Let $V$ be a vector space. As shown in this section, the smallest subset of $V$ (namely, $\varnothing$) is linearly independent. Show that the largest subset of $V$ (namely, $V$ itself) is linearly dependent.
2. In each of the following cases, determine whether the given subset $S$ of $\mathbf{R}_n$ is dependent or independent.
   (a) $n = 3$; $S = \{(1, 1, 1), (1, -1, 3), (2, 5, -1)\}$.
   (b) $n = 3$; $S = \{(1, 0, -1), (2, 3, 0), (0, 1, 4)\}$.
   (c) $n = 4$; $S = \{(2, 1, 0, 1), (3, 1, 1, 4), (1, -1, 3, 1)\}$.
   (d) $n = 4$; $S = \{(1, 3, -2, 4), (2, 4, -3, 6), (-3, 5, -1, 2)\}$. (ANS)
3. (a) If $\mathbf{v}_1 = (1, 1, 3)$ and $\mathbf{v}_2 = (2, 2, 1)$ ( in $\mathbf{R}_3$), find a vector $\mathbf{v}_3$ in $\mathbf{R}_3$ such that the set $\{\mathbf{v}_1, \mathbf{v}_2, \mathbf{v}_3\}$ is independent.
   (b) With $\mathbf{v}_1, \mathbf{v}_2, \mathbf{v}_3$ as above, can be a vector $\mathbf{v}_4$ be found in $\mathbf{R}_3$ such that $\{\mathbf{v}_1, \mathbf{v}_2, \mathbf{v}_3, \mathbf{v}_4\}$ is independent? Why or why not? (ANS)
4. Using the "point" interpretation of $n$-tuples, interpret geometrically: (a) the dependence or independence of two vectors (points) in $\mathbf{R}_3$; (b) the dependence or independence of three vectors in $\mathbf{R}_3$. Do the same thing using the "arrow" interpretation of $n$-tuples.

*5. It was stated in this section that an $n$-tuple $(\mathbf{u}_1, ..., \mathbf{u}_n)$ is independent if and only if both of the following statements hold:

(a) The set $\{\mathbf{u}_1, ..., \mathbf{u}_n\}$ is independent;
(b) The vectors $\mathbf{u}_1, ..., \mathbf{u}_n$ are all distinct.
Give a formal proof.

6. If $\mathbf{C}$ is the field of complex numbers, $\mathbf{C}_2$ is a vector space over $\mathbf{C}$. We may also consider $\mathbf{C}_2$ as a vector space over $\mathbf{R}$ if we take only real numbers as our scalars. Find a subset of $\mathbf{C}_2$ which is linearly independent over $\mathbf{R}$ but linearly dependent over $\mathbf{C}$. (ANS)
7. Can a subset of $\mathbf{C}_2$ be linearly dependent over $\mathbf{R}$ while being linearly independent over $\mathbf{C}$? Why or why not? (ANS)
8. Let $F$ be any field, and let $\mathbf{u} = (a, b)$ and $\mathbf{v} = (c, d)$ be any two elements of the vector space $F_2$. Prove that the vectors $\mathbf{u}$, $\mathbf{v}$ are dependent if and only if $ad - bc = 0$. (This result will be generalized in Sec. 5.6.)

## 2.5 A basic theorem

Let $V$ be a vector space over a field $F$. As we have seen, any subset $A$ of $V$ spans a subspace $[A]$ of $V$; in general, the larger the set $A$, the larger the subspace $[A]$. In particular, the more vectors $A$ contains, the more likely it is that $A$ will span the entire vector space $V$.

Another probability which appears to increase as $A$ gets larger is the probability that $A$ will be linearly dependent. The smallest possible subset of $V$ (namely, $\varnothing$) is independent; the largest possible subset (namely, $V$) is dependent. Enlarging an independent set sometimes makes it dependent, but if a dependent set is enlarged it remains dependent (Theorem 2.4.8).

Roughly speaking, then, "large" subsets are more likely to span $V$; "large" subsets are more likely to be dependent. The connection between these two ideas is made precise by the following theorem, which asserts that any "sufficiently large" subset of $V$ *must* be dependent.

**2.5.1 Theorem.** Let $V$ be a vector space over $F$, and suppose there exists a finite subset $A$ of $V$ such that $[A] = V$; say $A$ has $n$ elements.

Then any set of more than $n$ vectors in $V$ is linearly dependent.

To illustrate Theorem 2.5.1: since $F_2 = [(1, 0), (0, 1)]$ (Sec. 2.3, Exercise 4), any three (or more) vectors in $F_2$ are dependent. Similarly,

$F_3 = [(1, 0, 0), (0, 1, 0), (0, 0, 1)]$ and, hence, any set of four or more vectors in $F_3$ is dependent. More generally, $F_n$ is spanned by $n$ vectors (Theorem 2.6.2 in the next section) and, hence, any $n + 1$ vectors in $F_n$ are dependent.

The proof of Theorem 2.5.1, to which we devote the rest of this section, is the most difficult proof in Chapter 2. If you find it heavy going, do not worry too much; the proofs in the remaining sections of this chapter will be easier!

*Proof of Theorem 2.5.1.* We may write

$$A = \{\mathbf{v}_1, \ldots, \mathbf{v}_n\}$$

where $n$ is the number of vectors in $A$. Let us (for the duration of this proof only) call a set "large" if it has more than $n$ elements; Theorem 2.5.1 asserts that every large subset of $V$ is dependent. The proof will be indirect: we shall assume there exists a large *independent* subset of $V$ and shall derive a contradiction from this assumption.

For each large independent set $S$, let $e(S)$ be the number of elements in the set $S \cap A$; that is, the number of elements which $S$ has in common with $A$. Since the number $e(S)$ cannot exceed $n$ (why not?), there are only finitely many possibilities for $e(S)$. Hence, assuming there exists at least one large independent subset of $V$, we may choose $S$ such that

(a) $S$ is a large independent set;
(b) The number $e(S)$ is as large as possible, subject to (a).

For this particular set $S$, let $e(S) = r$. In other words, $S \cap A$ has $r$ elements, and *no other large independent set has more than $r$ elements in common with $A$.*

By renumbering the vectors $\mathbf{v}_1, \ldots, \mathbf{v}_n$ in $A$ if necessary, we may assume that the $r$ vectors in $S \cap A$ are $\mathbf{v}_1, \ldots, \mathbf{v}_r$. Thus

$$S = \{\mathbf{v}_1, \ldots, \mathbf{v}_r, \mathbf{u}_{r+1}, \ldots, \mathbf{u}_n, \mathbf{u}_{n+1}, \ldots\}$$

for certain vectors $\mathbf{u}_j$ (remember that $S$ has at least $n + 1$ elements since $S$ is large). We now let

$$S_0 = \{\mathbf{v}_1, \ldots, \mathbf{v}_r, \mathbf{u}_{r+1}, \ldots, \mathbf{u}_n\}.$$

Since the vectors $\mathbf{v}_1, \ldots, \mathbf{v}_r$ belong to $S_0$, they clearly belong to $[S_0]$ as well. We will now show that $\mathbf{v}_{r+1}, \ldots, \mathbf{v}_n$ (the remaining vectors in $A$) also belong to $[S_0]$. Indeed, if $i$ is any index such that $r + 1 \leqslant i \leqslant n$, consider the set $T_i = S_0 \cup \{\mathbf{v}_i\}$ which has $n + 1$ elements. Since $T_i$ is "large" and has $r + 1$ elements in common with $A$ (namely, $\mathbf{v}_1, \ldots, \mathbf{v}_r, \mathbf{v}_i$), the last sentence of the preceding paragraph implies that

$T_i$ cannot be independent; that is, $T_i$ is dependent. On the other hand, $S$ has been assumed independent and, therefore, the subset $S_0$ of $S$ is also independent (Theorem 2.4.8). We can now apply Theorem 2.4.10: since $S_0$ is independent and $T_i = S_0 \cup \{\mathbf{v}_i\}$ is dependent, it follows that $\mathbf{v}_i \in [S_0]$ as claimed. We have, thus, established the fact that the vectors $\mathbf{v}_1, ..., \mathbf{v}_n$ all lie in $[S_0]$; that is,

$$A \subseteq [S_0].$$

Hence, $[A] \subseteq [S_0]$ (Theorem 2.3.5). But $[A]$ is the entire space $V$ by hypothesis; thus every vector in $V$ lies in $[S_0]$. In particular,

**(2.5.2)** $\mathbf{u}_{n+1} \in [S_0]$.

But $\mathbf{u}_{n+1}$ and the elements of $S_0$ are distinct vectors in $S$; thus, 2.5.2 asserts that one of the vectors in $S$ is a linear combination of other vectors in $S$. But then $S$ is dependent by Theorem 2.4.6. Since $S$ was originally chosen to be independent, we have our desired contradiction. ▮

Theorem 2.5.1 may be rephrased as follows: *an independent subset of V cannot have more elements than a set which spans V.* A set which is independent *and* spans $V$ is called a "basis" of $V$; in the next section, in which we begin the study of bases, Theorem 2.5.1 and its consequences will play a major role.

## 2.6 Basis and dimension

**2.6.1** **Definition.** Let $V$ be a vector space over $F$.

(a) A subset $B$ of $V$ is called a *basis* of $V$ if it has both of the following properties:

(i) $B$ is linearly independent;

(ii) $[B] = V$; that is, $B$ spans $V$.

(b) If $\mathbf{u}_1, ..., \mathbf{u}_n$ are elements of $V$, the ordered $n$-tuple $(\mathbf{u}_1, ..., \mathbf{u}_n)$ is called an *ordered basis* of $V$ if

(i) $(\mathbf{u}_1, ..., \mathbf{u}_n)$ is linearly independent, and

(ii) $[\mathbf{u}_1, ..., \mathbf{u}_n] = V$.

It follows at once from Definition 2.6.1 (and Sec. 2.4, Exercise 5) that the $n$-tuple $(\mathbf{u}_1, ..., \mathbf{u}_n)$ is an ordered basis of $V$ if and only if $\mathbf{u}_1, ..., \mathbf{u}_n$ are all distinct and the set $\{\mathbf{u}_1, ..., \mathbf{u}_n\}$ is a basis of $V$.

A vector space $V$ will, in general, have many different bases, and there is usually no reason to single out any one basis of $V$ as "better" than any other. In the space $F_n$, however, one particular basis *is* often singled out, namely, the so-called *canonical basis* which we describe in the following theorem.

**2.6.2 Theorem.** In the vector space $F_n$, let

$$\begin{aligned} \mathbf{e}_1 &= (1, 0, 0, ..., 0) \\ \mathbf{e}_2 &= (0, 1, 0, ..., 0) \\ \mathbf{e}_3 &= (0, 0, 1, ..., 0) \\ &\cdots \\ \mathbf{e}_n &= (0, 0, 0, ..., 1). \end{aligned}$$

That is, for each $i$ ($1 \leqslant i \leqslant n$) we define $\mathbf{e}_i$ to be the element of $F_n$ whose $i$th coordinate is 1 and whose other coordinates are zero.

Then the $n$-tuple $(\mathbf{e}_1, ..., \mathbf{e}_n)$ is an ordered basis of $F_n$, called the canonical basis of $F_n$.

*Proof.* Using the definitions of addition and scalar multiplication in $F_n$ it is easy to verify that

**(2.6.3)** $$(c_1, c_2, ..., c_n) = c_1\mathbf{e}_1 + c_2\mathbf{e}_2 + \cdots + c_n\mathbf{e}_n$$

(do so!); hence, any $n$-tuple $(c_1, ..., c_n)$ is a linear combination of $\mathbf{e}_1, ..., \mathbf{e}_n$, so that $[\mathbf{e}_1, ..., \mathbf{e}_n] = F_n$. Also, since the zero vector in $F_n$ is $(0, 0, ..., 0)$, the expression 2.6.3 is the zero vector only when all of the $c_i$ are zero; hence $(\mathbf{e}_1, ..., \mathbf{e}_n)$ is independent. ∎

The following characterization of bases is sometimes used in place of Definition 2.6.1.

**2.6.4 Theorem.** Let $V$ be a vector space over $F$. If $B$ is a subset of $V$, the following two statements are equivalent:

(a) $B$ is a basis of $V$;

(b) Each vector in $V$ can be expressed uniquely (i.e., in one and only one way) as a linear combination of the elements of $B$.

(*Note*: Uniqueness is not contradicted by the fact that, say, $2\mathbf{u}_1 + 5\mathbf{u}_2 = 2\mathbf{u}_1 + 5\mathbf{u}_2 + 0\mathbf{u}_3$, since two expressions which coincide except for the inclusion or omission of terms with zero coefficients are considered to be the same.)

*Proof.* Since $[B]$ is the set of all linear combinations of elements of $B$, every element of $V$ can be expressed as such a linear combination if and only if $[B] = V$. The *uniqueness* of the expression for each vector is equivalent to the statement that $B$ is independent (cf. Theorem 2.4.7). Hence, 2.6.4(b) holds if and only if $B$ spans $V$ and is independent, that is, if and only if $B$ is a basis of $V$. ∎

By Theorem 2.6.4, if $(\mathbf{u}_1, \mathbf{u}_2, ..., \mathbf{u}_n)$ is an ordered basis of $V$ then every vector $\mathbf{v} \in V$ can be expressed in exactly one way in the form

**(2.6.5)** $$\mathbf{v} = c_1\mathbf{u}_1 + \cdots + c_n\mathbf{u}_n \qquad (c_i \in F).$$

The uniquely determined scalars $c_i$ in 2.6.5 are called the *components* of the vector $\mathbf{v}$ with respect to the ordered basis $(\mathbf{u}_1, \mathbf{u}_2. ..., \mathbf{u}_n)$. For example, equation 2.6.3 shows that $c_1, ..., c_n$ are the components of $(c_1, ..., c_n)$ with respect to the canonical basis of $F_n$. Similarly, the scalars $x$ and $y$ in equation 2.3.9 are the components of the vector $(a, b)$ with respect to the ordered basis $((2, 5), (1, 3))$ of $\mathbf{R}_2$.

The idea of components with respect to a basis can be interpreted geometrically. Consider the situation of Example 2.3.7 (Fig. 2.3.1); let $V$ be the plane $[P, Q]$ spanned by the points (vectors) $P$ and $Q$ in that example. Since $P$ and $Q$ lie on different lines through $O$, neither point is a scalar multiple of the other and, hence, the set $\{P, Q\}$ is independent. Since $V = [P, Q]$, it follows that $\{P, Q\}$ is a basis of $V$. If we take the lines $[P]$, $[Q]$ to be new "coordinate axes" in the plane $V$, and if we further determine the scales along these new axes by taking $P$ and $Q$ to be our unit points (so that $P$, for instance, has coordinates $(1, 0)$ with respect to the "new" axes and scale), it is not hard to see that the coordinates of the point $X = aP + bQ$ with respect to the new axes and scale are precisely the scalars $a$, $b$; these scalars are, of course, the components of $X$ with respect to the ordered basis $(P, Q)$. In other words, the components of a vector with respect to a basis may be interpreted as the coordinates of that vector (point) with respect to the axes (and scales) determined by the basis.

In the example just discussed, each of the (two) basis vectors determined one axis; hence, there were as many axes altogether as there were elements of the basis. On the other hand, you are probably accustomed to thinking of the number of coordinate axes needed in a space as being equal to the dimension of that space. It will probably not surprise you, therefore, when we define the *dimension* of a vector space $V$ to be the number of elements in a basis of $V$. First, however, we must show that this number does not depend on which basis we choose; it is uniquely determined by $V$.

**2.6.6** **Theorem.** Let $V$ be a vector space over $F$, and suppose that $V$ has a finite basis consisting of exactly $n$ elements. Then:

(a) Every basis of $V$ has exactly $n$ elements.

(b) If $A$ is any independent subset of $V$, then $A$ has at most $n$ elements.

(c) If $B$ is any subset of $V$ such that $[B] = V$, then $B$ has at least $n$ elements.

*Proof.* By Theorem 2.5.1, *an independent subset of $V$ cannot have more elements than a set which spans $V$*. By assumption, $V$ has a basis $S$ consisting of $n$ elements. Since $S$ is independent, 2.6.6(c) follows; since $S$ spans $V$, 2.6.6(b) follows. Finally, since a basis is independent *and* spans $V$, 2.6.6(a) follows from (b) and (c). ▮

**2.6.7 Definition.** Under the hypotheses of Theorem 2.6.6, the number $n$ (which by Theorem 2.6.6(a) depends only on $V$) is called the *dimension* of $V$ (more precisely, the dimension of $V$ over $F$) and is denoted dim $V$, or $\dim_F(V)$. A vector space $V$ which does not have a finite basis is said to have *infinite dimension* (in symbols, dim $V = \infty$; see Exercise 6 at the end of the section).

Two special cases of importance are contained in the following theorem.

**2.6.8 Theorem.** (a) The dimension of the vector space $F_n$ (over $F$) is $n$.

(b) If $V$ is any vector space, the one-dimensional subspaces of $V$ are precisely the subspaces of the form $[\mathbf{u}]$, where $\mathbf{u}$ is a nonzero vector in $V$.

2.6.8(a) follows from the fact that the canonical basis of $F_n$ has $n$ vectors (Theorem 2.6.2); 2.6.8(b) follows from the fact that the set $\{\mathbf{u}\}$ is independent if and only if $\mathbf{u} \neq \mathbf{0}$ (Sec. 2.4). ▮

*Remark*: Theorem 2.6.6 applies only to a vector space which is known to have a finite basis. It is possible to show that every vector space has a basis (finite or infinite) and that Theorem 2.6.6 still holds in the infinite case if $n$ is an "infinite cardinal number"; however, a full discussion of this would require a greater background in set theory and foundations of mathematics than we have assumed you to have. (The proof that every vector space has a basis, for example, requires Zorn's Lemma.) Fortunately, the cases covered by Theorem 2.6.6 suffice for most applications, and we will not need the stronger result in this book.

There is a third way of characterizing bases, in addition to Definition

2.6.1 and Theorem 2.6.4. In order to state this third characterization (which appears in Corollary 2.6.10 below), we must first explain the meaning of the word "maximal" as applied to sets having a given property. Suppose that by a "$P$-set" we mean a set that has a certain property $P$. By a *maximal P-set*, we mean a set $S$ such that

(a) $S$ is a $P$-set (i.e., $S$ has property $P$);
(b) $S$ is not a proper subset of any other $P$-set.

For example, if $A$ is a set of vectors in a vector space, a "maximal independent subset of $A$" is a subset $S$ of $A$ such that (a) $S$ is independent, and (b) any subset of $A$ which properly contains $S$ is dependent. Figure 2.6.1 illustrates the concept of "maximal $P$-set" pictorially: dots represent $P$-sets, and lines indicate set inclusion (for instance, the line connecting $C$ to $F$ indicates that $F$ is a subset of $C$; similarly, $E$ is a subset of $D$). The maximal $P$-sets in Figure 2.6.1 are $A$, $B$, $C$, and $D$.

**2.6.9 Theorem.** Let $V$ be a vector space and let $A$ be a subset of $V$ which spans $V$; that is, $[A] = V$. Then any maximal independent subset of $A$ is a basis of $V$.

*Proof.* Let $B$ be a maximal independent subset of $A$. If $\mathbf{v}$ is any element of $A$ which is not in $B$, then $B \cup \{\mathbf{v}\}$ is dependent (by definition of "maximal"); hence, $\mathbf{v} \in [B]$ by Theorem 2.4.10. It follows that all elements of $A$ belong to $[B]$; that is, $A \subseteq [B]$. Hence, $[A] \subseteq [B]$ (Theorem 2.3.5); that is, $V \subseteq [B]$, so that $B$ spans $V$. Thus, $B$ is a basis of $V$ by Definition 2.6.1. ▮

**2.6.10 Corollary.** Every maximal independent subset of $V$ is a basis of $V$.

Figure 2.6.1

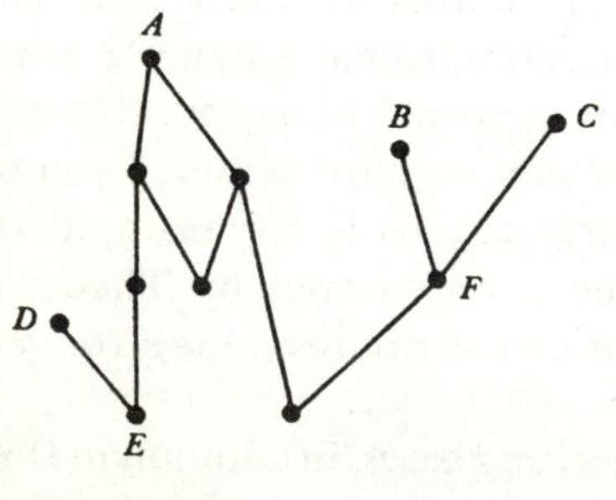

*Proof.* Let $A = V$ in Theorem 2.6.9. ▮

**2.6.11 Example.** In $\mathbf{R}_5$, find a maximal independent subset of the set

$$A = \{(1, 2, 0, -2, 1), (3, 0, -4, 4, 3), (0, 3, 2, -5, 0), (0, 0, 0, 1, 2), (2, 1, -2, 3, 6)\}.$$

*Solution.* Let the five vectors in $A$ be labeled $\mathbf{u}_1, \ldots, \mathbf{u}_5$ (in the order in which they are listed). Since $\mathbf{u}_1 \neq \mathbf{0}$, the subset $\{\mathbf{u}_1\}$ is independent. Enlarging this subset by adding $\mathbf{u}_2$ to it, we find that the subset $\{\mathbf{u}_1, \mathbf{u}_2\}$ is still independent (*why*?). Enlarging again by adding $\mathbf{u}_3$, we find that $\{\mathbf{u}_1, \mathbf{u}_2, \mathbf{u}_3\}$ is *dependent* (this requires computation, as in Example 2.4.3); hence, we discard $\mathbf{u}_3$ and try $\mathbf{u}_4$ instead. The set $\{\mathbf{u}_1, \mathbf{u}_2, \mathbf{u}_4\}$ turns out to be independent. Enlarging it by adding $\mathbf{u}_5$, we find that $\{\mathbf{u}_1, \mathbf{u}_2, \mathbf{u}_4, \mathbf{u}_5\}$ is dependent; hence, we discard $\mathbf{u}_5$. The set

$$B = \{\mathbf{u}_1, \mathbf{u}_2, \mathbf{u}_4\}$$

is a maximal independent subset of $A$; for not only is it independent, but our argument implies that if any one of the other vectors in $A$ ($\mathbf{u}_3$ or $\mathbf{u}_5$) is added to $B$, the resulting set is dependent.

Of course, $\{\mathbf{u}_1, \mathbf{u}_2, \mathbf{u}_4\}$ is not the only maximal independent subset of $A$. If instead of considering the vectors $\mathbf{u}_1, \ldots, \mathbf{u}_5$ *in that order*, we had taken them in the order $\mathbf{u}_5, \mathbf{u}_4, \mathbf{u}_1, \mathbf{u}_3, \mathbf{u}_2$, we would have come out with the maximal independent subset $C = \{\mathbf{u}_5, \mathbf{u}_4, \mathbf{u}_1\}$. Note that the sets $B$ and $C$ have the same number of elements. This is no accident; indeed, by Theorem 2.6.9 any maximal independent subset of $A$ is a basis of $[A]$ and, hence, the number of elements in it is the (uniquely determined) dimension of $[A]$. ▮

**2.6.12 Theorem.** If $V$ has finite dimension $n$, then

(a) Every subset of $V$ which spans $V$ includes a basis of $V$. (We say that "$A$ includes $B$" if $B \subseteq A$.)

(b) Every $n$-element subset of $V$ which spans $V$ *is* a basis of $V$. (By an "$n$-element set" we, of course, mean a set having exactly $n$ elements.)

*Proof.* Let $A$ be a subset of $V$ such that $[A] = V$. Since an independent subset of $A$ can have at most $n$ elements (Theorem 2.6.6(b)), and since $A$ has at least one independent subset (namely, $\varnothing$), it follows that among all independent subsets of $A$ there exists one (say, $B$) having the greatest number of elements. Clearly, $B$ is then a *maximal* independent subset of $A$; by Theorem 2.6.9, $B$ is a basis of $V$. We have thus shown that $A$ includes a basis $B$ of $V$, proving 2.6.12(a). Moreover, $B$ has $n$ elements by Theorem 2.6.6; if $A$ has $n$ elements also, it follows from $A \supseteq B$ that $A = B$, so that $A$ is a basis of $V$, proving 2.6.12(b). ∎

**2.6.13 Theorem.** Suppose that $A$ is a finite subset of $V$ such that $[A] = V$, and suppose $S$ is an independent subset of $V$. Then there exists a subset $A_0$ of $A$, such that $S \cup A_0$ is a basis of $V$.

*Proof.* An argument like that used above (in the proof of 2.6.12(a)) shows that among all independent sets of the form $S \cup A_0$ ($A_0 \subseteq A$) there exists one for which $A_0$ has the greatest number of elements. For this choice of $A_0$, $S \cup A_0$ is a maximal independent subset of $S \cup A$. Since $[A] = V$, certainly $[S \cup A] = V$; hence, $S \cup A_0$ is a basis of $V$ by Theorem 2.6.9. (*Remark*: $A_0$ could be the empty set, in which case $S$ itself is a basis of $V$.) ∎

**2.6.14 Corollary.** If $V$ has finite dimension $n$, then

(a) Every independent subset of $V$ is included in a basis of $V$.

(b) Every $n$-element independent subset of $V$ *is* a basis of $V$.

*Proof.* By hypothesis, $V$ has a finite basis $A$; in particular, $A$ is a finite set which spans $V$. Hence, Theorem 2.6.13 is applicable, and we see that any independent set $S$ is included in some basis $S \cup A_0$, proving 2.6.14(a). Moreover, $S \cup A_0$ has $n$ elements (Theorem 2.6.6); if $S$ has $n$ elements also, it follows that $S = S \cup A_0$ and, hence, $S$ is a basis, proving 2.6.14(b). ∎

Observe that the two parts of 2.6.14 parallel the two parts of 2.6.12. Every independent set can be enlarged to a basis; every set which spans $V$ can be reduced (diminished) to a basis. Every set of either type (independent or spanning $V$) is a basis if it has the right number of elements. Thus, for example, if we wish to know whether $\{(0, 1, 1), (1, 0, 2), (1, 1, 1)\}$ is a basis of $\mathbf{R}_3$, we need only determine whether the set is independent. The set $\{(0, 1, 1), (2, 0, 3)\}$, on the other hand, cannot be a basis of $\mathbf{R}_3$, since it does not have exactly 3 elements. The set $\{(2, 5), (1, 3)\}$ in $\mathbf{R}_2$ has exactly 2 elements and is independent, and hence is a basis of $\mathbf{R}_2$. (Note that this provides still another solution of Example 2.3.8.)

A useful special case of Theorem 2.6.13 is obtained by taking $V$ to be $F_n$ and $A$ to be the canonical basis of $F_n$. In this case, Theorem 2.6.13 reduces to the following:

**2.6.15 Theorem.** If $S$ is an independent subset of $F_n$ having exactly $k$ elements, then $S$ can be enlarged to a basis of $F_n$ by adjoining exactly $n - k$ of the vectors $\mathbf{e}_1, \ldots, \mathbf{e}_n$ (where the $\mathbf{e}_i$ are as defined in Theorem 2.6.2).

For example, if we wish to extend the set $S = \{(1, 4, 2), (0, 2, 1)\}$ (which is clearly independent) to a basis of $\mathbf{R}_3$, we need only adjoin to $S$ one of the three vectors $(1, 0, 0)$, $(0, 1, 0)$, $(0, 0, 1)$. If we try adjoining $(1, 0, 0)$, we find that the enlarged set is dependent; however, if we try adjoining $(0, 1, 0)$, we find that the enlarged set

$$\{(1, 4, 2), (0, 2, 1), (0, 1, 0)\}$$

is independent and, hence, is a basis of $\mathbf{R}_3$.

In subsequent chapters it will often be necessary to deal with subspaces of $F_n$, and we will need to know that every such subspace has a finite basis. The theorem which assures us of this is the following:

**2.6.16 Theorem.** Let $V$ be a vector space having finite dimension $n$. If $U$ is any subspace of $V$, then:

(a) $U$ has a finite basis, and $\dim U \leq n$.
(b) $\dim U = n$ only if $U = V$.
(c) Every basis of $U$ can be extended to a basis of $V$.

*Proof.* By Theorem 2.6.6, any independent set in $V$ has at most $n$ elements. Since any independent subset of $U$ is automatically an independent subset of $V$, it follows that any independent subset of $U$ has at most $n$ elements. Since $U$ has at least one independent subset (namely $\varnothing$), it follows that among all independent subsets of $U$, there exists one (say, $B$) having the greatest number of elements. $B$ is then a maximal independent subset of $U$ and, hence, is a basis of $U$. (The preceding argument should look familiar by now!) *We have thus shown that a basis of $U$ exists.* To prove the rest of the theorem, let $S$ be *any* basis of $U$; then $S$ is an independent subset of $U$ and, hence, also of $V$, so that by 2.6.14(a) we can extend $S$ to a basis $S^*$ of $V$, proving 2.6.16(c). Since $S^*$ has $n$ elements (Theorem 2.6.6(a)) and since $S \subseteq S^*$, it follows that $S$ has at most $n$ elements so that $\dim U \leq n$, proving 2.6.16(a). Finally, if $S$ itself has $n$ elements, then we must have $S = S^*$, from which $[S] = [S^*]$, that is, $U = V$, proving 2.6.16(b). ∎

If $S$ and $T$ are subspaces of $V$, then $S \cap T$ and $S + T$ are subspaces also (see Sec. 2.2, Theorem 2.2.9 and Exercise 2). There is a nice result connecting the dimensions of $S \cap T$ and $S + T$, which we now prove.

**2.6.17 Theorem.** Let $V$ be a vector space and let $S$ and $T$ be finite-dimensional subspaces of $V$. Then $S \cap T$ and $S + T$ are finite-dimensional, and

$$\dim(S \cap T) + \dim(S + T) = \dim S + \dim T. \tag{2.6.18}$$

*Proof.* By Theorem 2.6.16 (applied to $S \cap T$ as a subspace of $S$), $S \cap T$ has a finite basis $\{\mathbf{u}_1, ..., \mathbf{u}_r\}$ which can be extended to a basis $A_1 = \{\mathbf{u}_1, ..., \mathbf{u}_r, \mathbf{v}_1, ..., \mathbf{v}_s\}$ of $S$; we may obviously assume that the vectors $\mathbf{u}_1, ..., \mathbf{u}_r, \mathbf{v}_1, ..., \mathbf{v}_s$ are all distinct. Similarly, $\{\mathbf{u}_1, ..., \mathbf{u}_r\}$ can be extended to a basis $A_2 = \{\mathbf{u}_1, ..., \mathbf{u}_r, \mathbf{w}_1, ..., \mathbf{w}_t\}$ of $T$, where the $\mathbf{u}$'s and $\mathbf{w}$'s are all distinct. If we can show that the $(r + s + t)$-tuple

$$B = (\mathbf{u}_1, ..., \mathbf{u}_r, \mathbf{v}_1, ..., \mathbf{v}_s, \mathbf{w}_1, ..., \mathbf{w}_t)$$

is an *ordered* basis of $S + T$, the theorem will be proved since, then, $\mathbf{u}_1, ..., \mathbf{u}_r, \mathbf{v}_1, ..., \mathbf{v}_s, \mathbf{w}_1, ..., \mathbf{w}_t$ are necessarily distinct and 2.6.18 reduces to the true equation $r + (r + s + t) = (r + s) + (r + t)$.

Since $S = [A_1]$ and $T = [A_2]$, it follows from Theorem 2.3.20 that

$$S + T = [A_1 \cup A_2] = [B];$$

hence, to prove that $B$ is an ordered basis of $S + T$ we need only show that $B$ is independent. Suppose that

$$\sum_{i=1}^{r} a_i\mathbf{u}_i + \sum_{j=1}^{s} b_j\mathbf{v}_j + \sum_{k=1}^{t} c_k\mathbf{w}_k = \mathbf{0}, \tag{2.6.19}$$

where the $a$'s, $b$'s, and $c$'s are scalars. Letting $\mathbf{x}$ be the sum of the terms $a_i\mathbf{u}_i$ and $b_j\mathbf{v}_j$, we have

$$\mathbf{x} = \sum_{i=1}^{r} a_i\mathbf{u}_i + \sum_{j=1}^{s} b_j\mathbf{v}_j = -\sum_{k=1}^{t} c_k\mathbf{w}_k. \tag{2.6.20}$$

Since the right side of 2.6.20 belongs to $T$ and the left side belongs to $S$, it follows that $\mathbf{x}$ belongs to both $S$ and $T$; that is, $\mathbf{x} \in S \cap T$. Hence, $\mathbf{x}$ is a linear combination of the $\mathbf{u}$'s:

$$\mathbf{x} = \sum_{i=1}^{r} d_i\mathbf{u}_i. \tag{2.6.21}$$

Subtracting 2.6.20 from 2.6.21, we get

$$\text{(2.6.22)} \quad \sum_{i=1}^{r} (d_i - a_i)\mathbf{u}_i - \sum_{j=1}^{s} b_j\mathbf{v}_j = \sum_{i=1}^{r} d_i\mathbf{u}_i + \sum_{k=1}^{t} c_k\mathbf{w}_k = \mathbf{x} - \mathbf{x} = \mathbf{0}.$$

Since the $(r + t)$-tuple $(\mathbf{u}_1, ..., \mathbf{u}_r, \mathbf{w}_1, ..., \mathbf{w}_t)$ is independent, 2.6.22 implies that all of the $d$'s and $c$'s are zero; similarly, independence of $(\mathbf{u}_1, ..., \mathbf{u}_r, \mathbf{v}_1, ..., \mathbf{v}_s)$ implies that all of the scalars $d_i - a_i$ and $b_j$ are zero. It follows that all of the $a$'s, $b$'s and $c$'s (i.e., all of the scalars in 2.6.19) are zero and, hence, the $(r + s + t)$-tuple $B$ is independent as desired. ∎

Theorem 2.6.17 has geometric applications; for example, see Exercise 11 below. In Section 3.4, a generalization of Theorem 2.6.17 will play a key role in proving such statements as, "Two nonintersecting lines in the same plane are parallel", "Two nonintersecting planes in the same 3-space are parallel", and the like.

**Exercises**

1. Verify equation 2.6.3.
2. Which of the following subsets of $\mathbf{R}_3$ are bases of $\mathbf{R}_3$? (ANS)
   (a) $\{(1, 2, 3), (4, 5, 6), (7, 8, 9)\}$.
   (b) $\{(2, 0, 3), (0, 1, 1), (0, 1, 2)\}$.
   (c) $\{(1, 0, 1), (0, 1, 1), (1, 1, 0), (1, 1, 1)\}$.
3. Extend the set $\{(2, 1, -1, 0), (3, -2, 2, 0)\}$ to a basis of $\mathbf{R}_4$.
4. In each of the following, find a maximal independent subset of the given set of vectors. (In all cases, the vectors belong to $\mathbf{R}_5$.) (ANS)
   (a) $\{(1, 0, 1, 0, 2), (3, 0, 3, 0, 6), (0, 0, 1, 1, 1), (2, 0, 0, -2, 2)\}$.
   (b) $\{(1, 0, 1, 0, 2), (5, 1, 0, 1, 7), (3, 1, -2, 1, 3), (3, 2, -7, 2, 0), (1, 1, 1, 1, 1)\}$.
   (c) $\{(0, 0, 0, 0, 0), (1, 1, 1, 1, 1), (2, 2, 2, 2, 2)\}$.

*5. (a) If $V$ is any vector space, show that $\varnothing$ (the empty set) is a basis of the subspace $\{\mathbf{0}\}$ of $V$.
   (b) Deduce that a vector space has dimension 0 if and only if it consists of just the zero vector.

6. Prove: a vector space $V$ which does not have a finite basis must contain an independent subset consisting of infinitely many

vectors. (This is why we say that such a space $V$ has "infinite dimension", cf. Definition 2.6.7.)

7. Exhibit an (infinite) basis of the vector space $\mathbf{R}[x]$ of all polynomials with real coefficients. (Use Theorem 2.6.4.)
8. Referring to Section 2.2, Exercises 6 and 7, find a basis for each of the subsets (in those exercises) which are subspaces. (ANS)
9. Prove the following result, which parallels Corollary 2.6.10: every minimal set which spans $V$ is a basis of $V$.
10. Prove the converse of Corollary 2.6.10: every basis of $V$ is a maximal independent subset of $V$.
11. (a) Using Theorem 2.6.17 and other results, prove that if $V$ is a 3-dimensional vector space and $S$ and $T$ are 2-dimensional subspaces of $V$, then either $S = T$ or $\dim(S \cap T) = 1$. (What "other results" are needed?) (SOL)
(b) Interpret the result of (a) geometrically. (SOL)
(c) What can be said about the intersection of a 1-dimensional and a 2-dimensional subspace (still assuming $\dim V = 3$)? Give a proof.
12. Prove that any vector space with more than two elements has more than one basis.
13. The pair ((2, 3), (4, 7)) is an ordered basis of $\mathbf{R}_2$. Compute the components of each of the following vectors with respect to this basis: (1, 1), (0, 2), $(a, b)$. (ANS)
14. Derive a formula for the components of an arbitrary vector $(x, y)$ with respect to an arbitrary ordered basis $((a, b), (c, d))$ of $\mathbf{R}_2$, where $ad - bc \neq 0$ (cf. Sec. 2.4, Exercise 8). (ANS)
15. Find a basis for $\mathbf{C}$ (the complex numbers) as a vector space over $\mathbf{R}$. What is the dimension of $\mathbf{C}$ over $\mathbf{R}$?
16. Let $F$, $F'$ be fields such that $F$ is a subfield of $F'$, and let $V$ be a vector space over $F'$. As observed previously (Section 2.1, Exercise 6), both $F'$ and $V$ are vector spaces over $F$. Suppose that $(\alpha_1, \ldots, \alpha_m)$ is an ordered basis of $F'$ over $F$ and that $(\mathbf{v}_1, \ldots, \mathbf{v}_n)$ is an ordered basis of $V$ over $F'$. Prove that the $mn$ elements

$$\alpha_i \mathbf{v}_j \qquad (1 \leqslant i \leqslant m;\ 1 \leqslant j \leqslant n)$$

are distinct and constitute a basis of $V$ over $F$. (The products $\alpha_i \mathbf{v}_j$, of course, exist since $\alpha_i \in F'$ and $V$ is a vector space over $F'$.) What does this tell you about the dimension of $V$ over $F$?

17. Let $W_1, \ldots, W_r$ be finite-dimensional subspaces of a vector space $V$. For each $i$, let

$$W_i^* = W_1 + \cdots + W_{i-1} + W_{i+1} + \cdots + W_r;$$

that is, $W_i^*$ is the sum of the $r - 1$ subspaces $W_j$ such that $1 \leqslant j \leqslant r, j \neq i$. Prove: if $W_i \cap W_i^* = \{0\}$ for all $i$ $(1 \leqslant i \leqslant r)$, then

$$\dim(W_1 + \cdots + W_r) = \dim W_1 + \cdots + \dim W_r.$$

(*Remark*: The case $r = 2$ follows readily from Theorem 2.6.17, and the general case can be done by induction on $r$. However, be careful – experience has shown that even good students often do this proof erroneously.)

## 2.7 Cosets and quotient spaces

Throughout this section, $V$ will denote a vector space over $F$ and $W$ will denote a subspace of $V$.

If $\mathbf{u}, \mathbf{v}$ are any elements of $V$, we say that $\mathbf{u}$ *is congruent to* $\mathbf{v}$ *modulo* $W$ (*notation*: $\mathbf{u} \equiv \mathbf{v} \pmod{W}$) if $\mathbf{u} - \mathbf{v} \in W$. The concept of congruence is a simple one, but it has important uses in both algebra and geometry. In this section we shall develop some properties of congruence modulo a subspace; these properties are quite analogous to the properties of "congruence modulo an integer" which were discussed in Section 1.13. Before reading further, you should review the material in Section 1.4 on equivalence relations and equivalence classes.

**2.7.1 Theorem.** If $W$ is a subspace of $V$, then the relation of congruence (modulo $W$) is an equivalence relation on $V$.

*Proof.* We must show that for all $\mathbf{u}, \mathbf{v}, \mathbf{w}$ in $V$,

(a) $\mathbf{u} \equiv \mathbf{u} \pmod{W}$;

(2.7.2) (b) If $\mathbf{u} \equiv \mathbf{v} \pmod{W}$, then $\mathbf{v} \equiv \mathbf{u} \pmod{W}$;

(c) If $\mathbf{u} \equiv \mathbf{v}$ and $\mathbf{v} \equiv \mathbf{w} \pmod{W}$, then $\mathbf{u} \equiv \mathbf{w} \pmod{W}$.

We shall prove (c) and shall leave (a) and (b) as exercises. Assume $\mathbf{u} \equiv \mathbf{v}$ and $\mathbf{v} \equiv \mathbf{w} \pmod{W}$. By definition, this means that

(2.7.3) $\mathbf{u} - \mathbf{v} \in W; \quad \mathbf{v} - \mathbf{w} \in W.$

To prove $\mathbf{u} \equiv \mathbf{w} \pmod{W}$, we must show that

(2.7.4) $\mathbf{u} - \mathbf{w} \in W.$

But $\mathbf{u} - \mathbf{w} = (\mathbf{u} - \mathbf{v}) + (\mathbf{v} - \mathbf{w})$; and the subspace $W$ is closed under addition. Hence, 2.7.3 implies 2.7.4, q.e.d. ∎

Since we now have an equivalence relation on $V$, Theorem 1.4.2 tells us that each vector $\mathbf{v} \in V$ belongs to a unique "equivalence class", namely, the set $\mathbf{v}_*$ consisting of all elements $\mathbf{x}$ which are related to $\mathbf{v}$. (We use the notation $\mathbf{v}_*$ here in place of the notation $[\mathbf{v}]$ used in Theorem 1.4.2, since in this chapter we have defined $[\mathbf{v}]$ to mean something else.) Now the condition for $\mathbf{x}$ to be related (congruent) to $\mathbf{v}$ is that $\mathbf{x} - \mathbf{v} \in W$; that is, $\mathbf{x} - \mathbf{v} = \mathbf{w}$ for some $\mathbf{w} \in W$; that is, $\mathbf{x} = \mathbf{v} + \mathbf{w}$ for some $\mathbf{w} \in W$. Hence, $\mathbf{v}_*$ is the set of all elements of the form $\mathbf{v} + \mathbf{w}$ ($\mathbf{w} \in W$); in the notation of equation 2.2.10, $\mathbf{v}_* = \mathbf{v} + W$. We state this result as a theorem.

2.7.5 **Theorem.** Let $W$ be a subspace of $V$. If $\mathbf{v}$ is any element of $V$, then (with respect to the equivalence relation of congruence modulo $W$) the unique equivalence class to which $\mathbf{v}$ belongs is the set

$$\mathbf{v}_* = \{\mathbf{v} + \mathbf{w} : \mathbf{w} \in W\} = \mathbf{v} + W.$$

2.7.6 **Definition.** The equivalence classes $\mathbf{v}_* = \mathbf{v} + W$ are called *cosets of W* (more precisely, *cosets of W in V*).

Note that since every element of $V$ belongs to one and only one coset of $W$, *two distinct cosets of W have no elements in common.*

Some elementary properties of cosets are summarized in the following theorem.

2.7.7 **Theorem.** The following statements hold for all elements $\mathbf{u}$, $\mathbf{v}$ of $V$ and all subspaces $W$, $W_1$, $W_2$ of $V$.

(a) $\mathbf{v} \in \mathbf{u} + W$ if and only if $\mathbf{u} + W = \mathbf{v} + W$.

(b) $\mathbf{v} \in W$ if and only if $\mathbf{v} + W = W$.

(c) $\mathbf{0} \in \mathbf{u} + W$ if and only if $\mathbf{u} + W = W$.

(d) If $\mathbf{u} + W_1 \subseteq \mathbf{v} + W_2$, then $W_1 \subseteq W_2$.

(e) If $\mathbf{u} + W_1 \subset \mathbf{v} + W_2$, then $W_1 \subset W_2$.

(f) If $\mathbf{u} + W_1 = \mathbf{v} + W_2$, then $W_1 = W_2$. (In other words, the same set cannot be a coset of two different subspaces.)

(g) $\mathbf{u}$, $\mathbf{v}$ belong to the same coset of $W$ if and only if $\mathbf{u} - \mathbf{v} \in W$.

*Proof.* Since $\mathbf{v} + W$ is the unique coset of $W$ which contains $\mathbf{v}$, part (a) of Theorem 2.7.7 is obvious. By letting $\mathbf{u} = \mathbf{0}$ in (a), we obtain (b); by letting $\mathbf{v} = \mathbf{0}$ in (a), we obtain (c).

If $\mathbf{u} + W_1 \subseteq \mathbf{v} + W_2$, then by subtracting $\mathbf{u}$ we get

$$W_1 \subseteq (\mathbf{v} - \mathbf{u}) + W_2.$$

(*Question*: Does this step need justification?) Since $\mathbf{0} \in W_1$ (why?), it follows that $\mathbf{0} \in (\mathbf{v} - \mathbf{u}) + W_2$; hence, $(\mathbf{v} - \mathbf{u}) + W_2 = W_2$ by (c), so that $W_1 \subseteq W_2$, proving (d). The proofs of (e) and (f) are similar. Finally, by Theorem 1.4.2 the vectors $\mathbf{u}$, $\mathbf{v}$ belong to the same equivalence class (coset of $W$) if and only if they are equivalent (congruent modulo $W$), and by definition $\mathbf{u} \equiv \mathbf{v} \pmod{W}$ means that $\mathbf{u} - \mathbf{v} \in W$. ▮

A geometric interpretation of cosets is suggested by Figure 2.7.1, in which we take $V$ to be $\mathbf{R}_2$ and $W$ to be the one-dimensional subspace $[\mathbf{v}]$, where $\mathbf{v} = (1, 2)$. If $\mathbf{u} = (2, 1)$, the effect of adding $\mathbf{u}$ to each element of $W$ is to move each element (point) two units to the right and one unit up, and thus to "translate" the line $W$ into the line $L$. That is, $\mathbf{u} + W = L$; a coset of $W$ is simply a line parallel to $W$. The fact that each element of $V$ belongs to a unique coset of $W$ can now be interpreted as the familiar geometric statement: "Through each point there is a unique line parallel to the line $W$." Similarly, if $V$ were $\mathbf{R}_3$ and the subspace $W$ were two-dimensional (a plane through the origin), the cosets of $W$ would be the planes parallel to $W$.

Figure 2.7.1

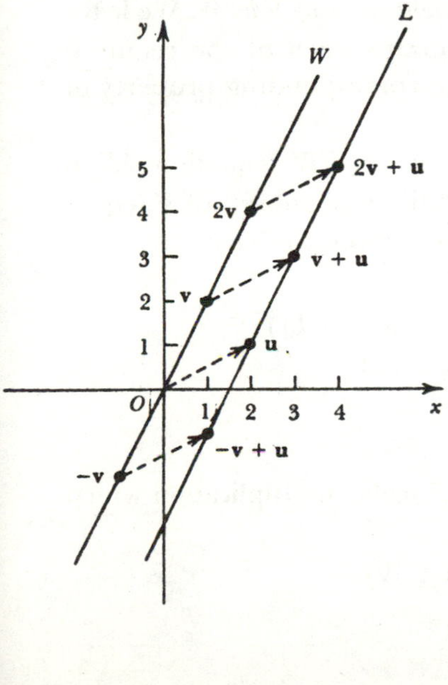

2.7.8 **Theorem.** Let $W$ be a subspace of $V$. If $\mathbf{u} \equiv \mathbf{u}'$ and $\mathbf{v} \equiv \mathbf{v}'$ (mod $W$), then

(a) $\mathbf{u} + \mathbf{v} \equiv \mathbf{u}' + \mathbf{v}'$ (mod $W$);

(b) $c\mathbf{u} \equiv c\mathbf{u}'$ (mod $W$) for all scalars $c$.

*Proof.* Exercise. (This theorem is analogous to Theorem 1.13.1.)

We now let $V/W$ denote the set of all cosets of $W$ in $V$:

$$V/W = \{\mathbf{v}_* : \mathbf{v} \in V\} = \{\mathbf{v} + W : \mathbf{v} \in V\}.$$

An argument just like that in Section 1.13, using Theorem 2.7.8 in place of Theorem 1.13.1, now shows that the operations of addition and scalar multiplication on $V/W$ defined by

(2.7.9) $\mathbf{u}_* + \mathbf{v}_* = (\mathbf{u} + \mathbf{v})_* \qquad (\text{all } \mathbf{u}, \mathbf{v} \in V)$

(2.7.10) $c\mathbf{u}_* = (c\mathbf{u})_* \qquad (\text{all } \mathbf{u} \in V, c \in F)$

are well defined.

2.7.11 **Theorem.** If $W$ is a subspace of $V$, then $V/W$ is a vector space over $F$ under the operations of addition and scalar multiplication defined by 2.7.9 and 2.7.10. The zero element (additive identity) of $V/W$ is the coset $\mathbf{0}_*$; the additive inverse of any coset $\mathbf{v}_*$ is the coset $(-\mathbf{v})_*$.

The vector space $V/W$ is called the *quotient space of V by W*. We leave the proof of Theorem 2.7.11 as an exercise; each of the required properties of $V/W$ follows easily from the corresponding property of $V$.

*Remarks*: 1. The definition of addition in $V/W$ (equation 2.7.9) agrees with our previous definition of addition of subsets of $V$ (equation 2.2.8). Indeed, using 2.2.8 as our definition, we have

$$\begin{aligned} \mathbf{u}_* + \mathbf{v}_* &= (\mathbf{u} + W) + (\mathbf{v} + W) \\ &= (\mathbf{u} + \mathbf{v}) + (W + W) \qquad (\text{by } 2.2.11) \\ &= (\mathbf{u} + \mathbf{v}) + W \qquad (\text{by } 2.3.21) \\ &= (\mathbf{u} + \mathbf{v})_* \end{aligned}$$

which coincides with 2.7.9.

2. On the other hand, the definition of scalar multiplication which would be analogous to 2.2.8, namely,

(2.7.12) $cA = \{c\mathbf{v} : \mathbf{v} \in A\} \qquad (c \in F, A \subseteq V),$

is *not* consistent with 2.7.10 (see Exercise 4 below). For this reason, *we shall never use 2.7.12 in this book.*

3. Since $\mathbf{0}_* = \mathbf{0} + W = W$, $W$ itself is a coset of $W$, and we see from Theorem 2.7.11 that $W$ is the "zero element" of vector space $V/W$. (The latter fact could also be deduced from 2.3.21 in view of Remark 1 above.)

**2.7.13 Example.** Let $V = \mathbf{R}_2$, $\mathbf{v} = (1, 2)$, $W = [\mathbf{v}]$ (as in Fig. 2.7.1). Show that $V/W$ has dimension 1.

*Solution.* We will show that the element $(1, 0)_*$ constitutes a basis of $V/W$. Since $(1, 0) \notin W$, $(1, 0)_* \neq W$; that is, $(1, 0)_*$ is not the zero element of $V/W$ and, hence, the set $\{(1, 0)_*\}$ is independent. To complete the proof, we must show that $(1, 0)_*$ spans $V/W$; that is, given any vector $(a, b)_* \in V/W$, we must show that

**(2.7.14)** $(a, b)_* = x(1, 0)_*$

for some scalar $x$. Now equation 2.7.14 is equivalent to $(a, b)_* = (x, 0)_*$, that is, to $(a, b) - (x, 0) \in W$ (Theorem 2.7.7(g)), that is, to $(a - x, b) \in W$, that is, to

**(2.7.15)** $(a - x, b) = y(1, 2) \qquad (\text{some } y \in \mathbf{R}).$

But equation 2.7.15 can be solved for $x$ and $y$ (do so!); the value of $x$ thus obtained is a solution of 2.7.14, and the desired conclusion follows. ∎

The preceding example is illustrated in Figure 2.7.2. Any coset of

Figure 2.7.2

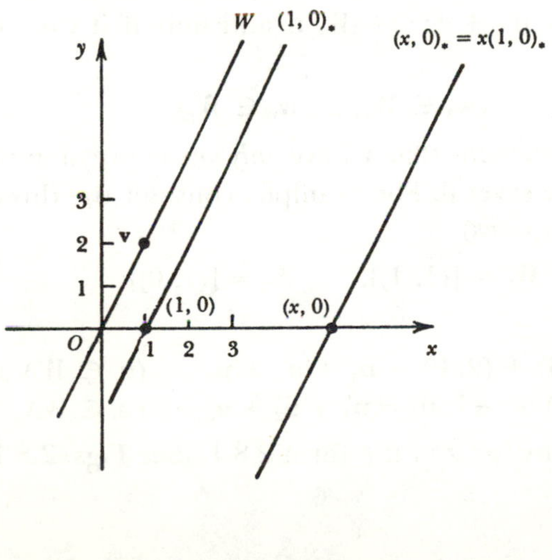

$W$ is a line parallel to $W$; this line must cross the $x$ axis at some point $(x, 0)$ and, hence, is a scalar multiple of the coset $(1, 0)_*$.

**Exercises**

*1. Prove parts (a) and (b) of 2.7.2.

*2. Prove Theorem 2.7.8.

*3. Prove Theorem 2.7.11.

4. Give an example to show that 2.7.12 does not always coincide with 2.7.10.

5. As shown in this section, $W$ is the additive identity of the vector space $V/W$. On the other hand, by 2.2.11 we have $\{\mathbf{0}\} + A = A$ for all subsets $A$ of $V$. In view of the fact that $W \neq \{\mathbf{0}\}$ in general, why does this not contradict Corollary 1.5.3 on the uniqueness of an identity?

6. Let $V = \mathbf{R}_3$, $\mathbf{v} = (2, 1, 3) \in V$, $W = [\mathbf{v}]$. Show that the set $\{(1, 0, 0)_*, (0, 1, 0)_*\}$ is a basis for $V/W$. Interpret geometrically.

7. Generalizing the preceding exercise, let $V$ be a vector space of dimension $n$ and let $W$ be a subspace of $V$ of dimension $m$. Prove that $\dim(V/W) = n - m$. (*Suggestion*: Extend an ordered basis $(\mathbf{u}_1, \ldots, \mathbf{u}_m)$ of $W$ to an ordered basis $(\mathbf{u}_1, \ldots, \mathbf{u}_m, \mathbf{u}_{m+1}, \ldots, \mathbf{u}_n)$ of $V$, then show that $((\mathbf{u}_{m+1})_*, \ldots, (\mathbf{u}_n)_*)$ is an ordered basis of $V/W$.

## 2.8 Direct sums of subspaces

Let $W_1, \ldots, W_r$ be subspaces of a vector space $V$. By definition, a vector $\mathbf{v} \in V$ belongs to the sum $W_1 + \cdots + W_r$ if and only if it can be expressed in the form

**(2.8.1)** $$\mathbf{v} = \mathbf{w}_1 + \cdots + \mathbf{w}_r \qquad (\mathbf{w}_1 \in W_1, \ldots, \mathbf{w}_r \in W_r).$$

However, there is no requirement that $\mathbf{v}$ have *only one* expression of the form 2.8.1; it may have several. For example, consider the three lines (one-dimensional subspaces)

$$W_1 = [(-1, 2)]; \qquad W_2 = [(1, 1)]; \qquad W_3 = [(1, 0)]$$

in $\mathbf{R}_2$. If $\mathbf{v} = (2, 9)$, then

$$\mathbf{v} = (-3, 6) + (3, 3) + (2, 0) = \mathbf{u}_1 + \mathbf{u}_2 + \mathbf{u}_3 \qquad (\mathbf{u}_i \in W_i)$$
$$\mathbf{v} = (-2, 4) + (5, 5) + (-1, 0) = \mathbf{u}_1' + \mathbf{u}_2' + \mathbf{u}_3' \qquad (\mathbf{u}_i' \in W_i)$$

are two different expressions for $\mathbf{v}$ in the form 2.8.1. (See Figs. 2.8.1 and 2.8.2, respectively.)

On the other hand, if $W_1$ and $W_2$ are, respectively, a line and a plane in $\mathbf{R}_3$ which meet only at the origin, then the vectors $\mathbf{w}_1$ and $\mathbf{w}_2$ in any expression of the form

$$\mathbf{v} = \mathbf{w}_1 + \mathbf{w}_2 \qquad (\mathbf{w}_i \in W_i)$$

are uniquely determined by $\mathbf{v}$; see Figure 2.8.3.

**2.8.2** **Definition.** If $W_1, \ldots, W_r$ are subspaces of $V$, the sum $W_1 + \cdots + W_r$ is called a *direct sum*, and is denoted

$$W_1 \oplus \cdots \oplus W_r, \qquad \text{or} \qquad \oplus \sum_{i=1}^{r} W_i\,,$$

if every vector $\mathbf{v}$ in $W_1 + \cdots + W_r$ has only one expression of the form 2.8.1; that is, if the vectors $\mathbf{w}_1, \ldots, \mathbf{w}_r$ in 2.8.1 are uniquely determined by $\mathbf{v}$. In this situation, the subspaces $W_1, \ldots, W_r$ are said to be *independent*.

According to Definition 2.8.2, then, the sum of the two subspaces $W_i$ in Figure 2.8.3 is direct, but the sum of the three subspaces $W_1$, $W_2$, $W_3$ shown in Figures 2.8.1 and 2.8.2 is not direct.

The direct sum concept will be needed later in connection with

Figure 2.8.1

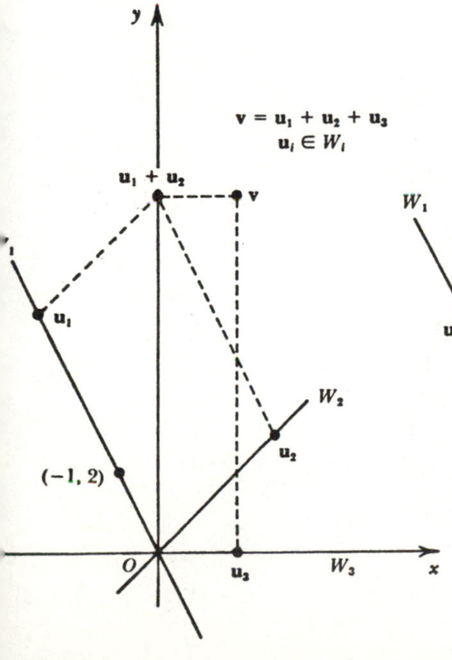


Figure 2.8.2

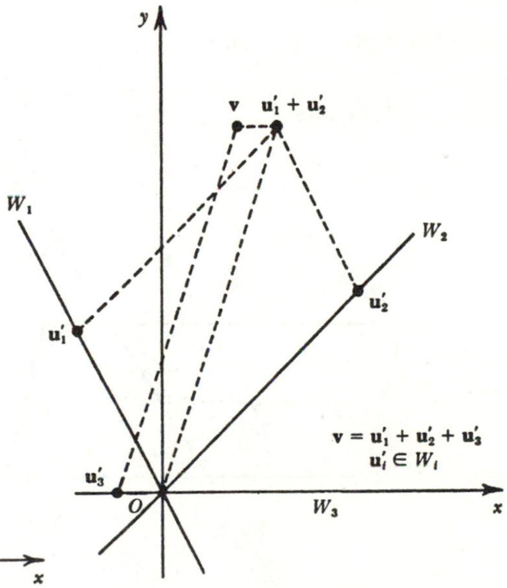

various topics in both algebra and geometry, such as isometries (Secs. 6.5 and 9.4), orthogonality (Sec. 3.8), and the Jordan form (Chap. 8).

**2.8.3** **Theorem.** Let $W_1, \ldots, W_r$ be subspaces of $V$, and assume each $W_i$ is finite dimensional. Then the following six statements are equivalent; that is, if any one of them is true, they all are.

(a) The sum $W_1 + \cdots + W_r$ is direct; that is, every vector $\mathbf{v}$ in $W_1 + \cdots + W_r$ can be expressed in the form 2.8.1 in only one way.

(b) Whenever $\mathbf{w}_i \in W_i$ $(i = 1, \ldots, r)$ and $\mathbf{w}_1 + \cdots + \mathbf{w}_r = \mathbf{0}$, then $\mathbf{w}_1 = \cdots = \mathbf{w}_r = \mathbf{0}$.

(c) For each $i$ $(1 \leq i \leq r)$, $W_i \cap W_i^* = \{\mathbf{0}\}$, where

$$W_i^* = W_1 + \cdots + W_{i-1} + W_{i+1} + \cdots + W_r = \sum_{j \neq i} W_j.$$

(d) $\dim(W_1 + \cdots + W_r) = \dim W_1 + \cdots + \dim W_r$.

(e) For every choice of bases $B_i$ for $W_i$ $(i = 1, \ldots, r)$, the sets $B_1, \ldots, B_r$ are mutually disjoint and their union is a basis of $W_1 + \cdots + W_r$.

(f) For some choice of bases $B_i$ for $W_i$ $(i = 1, \ldots, r)$, the sets $B_1, \ldots, B_r$ are mutually disjoint and their union is a linearly independent set.

The proof will consist of showing that (a) ⇒ (b) ⇒ (c) ⇒ (d) ⇒ (e) ⇒ (f) ⇒ (a).

*Proof that (a) ⇒ (b).* Exercise.

Figure 2.8.3

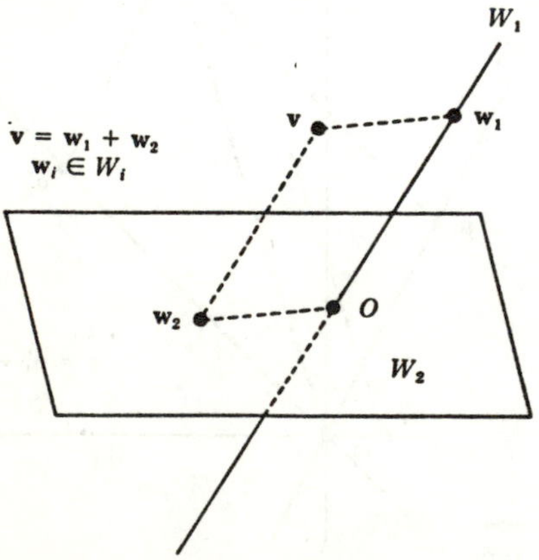

*Proof that* (*b*) ⇒ (*c*). Exercise.

*Proof that* (*c*) ⇒ (*d*). For $r = 1$ this is trivial. For $r = 2$ the statement "(c) ⇒ (d)" asserts that if $W_1 \cap W_2 = \{\mathbf{0}\}$, then $\dim(W_1 + W_2) = \dim W_1 + \dim W_2$; this assertion is true by Theorem 2.6.17. For $r > 2$ we use induction: assume that (c) implies (d), whenever $r = k$. To prove the same implication when $r = k + 1$, suppose that (c) holds for the $k + 1$ subspaces $W_1, \ldots, W_{k+1}$. Then each of these $k + 1$ subspaces has zero intersection with the sum of the other $k$ subspaces; in particular,

$$W_{k+1} \cap (W_1 + \cdots + W_k) = \{\mathbf{0}\}.$$

Hence by the result already proved when $r = 2$,

$$\dim[(W_1 + \cdots + W_k) + W_{k+1}] = \dim(W_1 + \cdots + W_k) + \dim W_{k+1}. \tag{2.8.4}$$

Also, since each $W_i$ has zero intersection with the sum of the other $k$ subspaces, each $W_i$ must also have zero intersection with the sum of any $k - 1$ of the others; this implies that (c) holds for the $k$ subspaces $W_1, \ldots, W_k$. Since we have assumed (c) ⇒ (d) when $r = k$, it follows that (d) holds for the subspaces $W_1, \ldots, W_k$. That is,

$$\dim(W_1 + \cdots + W_k) = \dim W_1 + \cdots + \dim W_k. \tag{2.8.5}$$

Adding $\dim W_{k+1}$ to both sides of 2.8.5 and then using 2.8.4, we get

$$\dim (W_1 + \cdots + W_k + W_{k+1}) = \dim W_1 + \cdots + \dim W_k + \dim W_{k+1}$$

which shows that (d) holds when $r = k + 1$. Thus, we have shown that (c) implies (d) when $r = k + 1$, as desired.

(*Note*: The preceding proof is the solution to Exercise 17, Section 2.6. Any "proof" which looks significantly simpler than the above is probably fallacious!)

*Proof that* (*d*) ⇒ (*e*). Let $\dim W_i = m_i$, and let $B_i$ be any basis of $W_i$ ($i = 1, \ldots, r$). Since any vector in $W_1 + \cdots + W_r$ is a sum of vectors from the individual subspaces $W_i$, which in turn are linear combinations of the vectors in $B_i$, it follows that the union of the $B_i$ *spans* $W_1 + \cdots + W_r$. Moreover, assuming (d), the dimension of $W_1 + \cdots + W_r$ equals $m_1 + \cdots + m_r$ which equals the total number of vectors in $B_1, \ldots, B_r$. Hence, the $B_i$ are disjoint by 2.6.6(c), and their union is a basis of $W_1 + \cdots + W_r$ by 2.6.12(b).

*Proof that (e) ⇒ (f).* Obvious.

*Proof that (f) ⇒ (a).* Assume that (f) is true; that is, there exist bases

$$B_i = \{\mathbf{v}_1^{(i)}, \mathbf{v}_2^{(i)}, \dots\}$$

for $W_i$ ($i = 1, \dots, r$), such that the vectors $\mathbf{v}_j^{(i)}$ (all $i, j$) are independent. Now suppose an element $\mathbf{v} \in W_1 + \cdots + W_r$ has *two* expressions in the form 2.8.1:

$$\mathbf{v} = \mathbf{w}_1 + \cdots + \mathbf{w}_r = \mathbf{w}_1' + \cdots + \mathbf{w}_r' \qquad (\mathbf{w}_i \in W_i, \mathbf{w}_i' \in W_i;\ 1 \leqslant i \leqslant r). \tag{2.8.6}$$

Since each $B_i$ is a basis of the corresponding $W_i$, we have

$$\mathbf{w}_i = \sum_j a_j^{(i)} \mathbf{v}_j^{(i)}; \qquad \mathbf{w}_i' = \sum_j b_j^{(i)} \mathbf{v}_j^{(i)} \tag{2.8.7}$$

for certain scalars $a_j^{(i)}$, $b_j^{(i)}$. Substituting the expressions 2.8.7 into equation 2.8.6, we get

$$\mathbf{v} = \sum_i \sum_j a_j^{(i)} \mathbf{v}_j^{(i)} = \sum_i \sum_j b_j^{(i)} \mathbf{v}_j^{(i)}.$$

Since the $\mathbf{v}_j^{(i)}$ (all $i, j$) are independent, Theorem 2.4.7 implies that $a_j^{(i)} = b_j^{(i)}$ (all $i$, $j$). Hence, $\mathbf{w}_i = \mathbf{w}_i'$ by 2.8.7. Hence, the two expressions 2.8.6 for $\mathbf{v}$ are the same, which proves statement (a). ∎

**Examples**

In Figures 2.8.1–2.8.2, each of the lines $W_1$, $W_2$, $W_3$ has dimension 1 but the plane $\mathbf{R}_2$ in which they lie has dimension 2; thus condition (d) of Theorem 2.8.3 fails to hold and the sum $W_1 + W_2 + W_3$ is not direct. However, in Figure 2.8.3 the sum $W_1 + W_2$ is direct since condition (c) holds: $W_1 \cap W_2 = \{\mathbf{0}\}$.

(Note that in one of these examples we determined directness via condition (c); in the other we used condition (d). In the proofs of certain theorems later on, we shall use conditions (e) and (f). It is convenient to have several conditions to choose from!)

The implication (f) ⇒ (a) can be stated in the following slightly modified form:

2.8.8 **Theorem.** Let $W_1, \dots, W_r, Y$ be subspaces of $V$; let $W_i = [B_i]$ ($i = 1, \dots, r$). If the sets $B_1, \dots, B_r$ are disjoint and their union is a basis of $Y$, then $Y = W_1 \oplus \cdots \oplus W_r$.

*Proof.* By Theorem 2.3.20, we have

$$\begin{aligned} Y &= [B_1 \cup \cdots \cup B_r] = [B_1] + \cdots + [B_r] \\ &= W_1 + \cdots + W_r. \end{aligned} \tag{2.8.9}$$

Since the union of the $B_i$ is independent, each set $B_i$ is independent (Theorem 2.4.8(b)) and hence is a basis of $W_i$, so that 2.8.3(f) holds. Hence, 2.8.3(a) holds (note that the proof of the implication (f) ⇒ (a) did not really depend on finite dimensionality); that is, the sum 2.8.9 is direct.

**Exercises**

*1. Complete the proof of Theorem 2.8.3 by showing that (a) ⇒ (b) and (b) ⇒ (c).

2. In $\mathbf{R}_4$, let

$$W_1 = \{(x, y, z, w) : x + y = z + w = 0\}$$
$$W_2 = \{(x, y, z, w) : y + z = x + w = 0\}.$$

$W_1$ and $W_2$ are both subspaces of $\mathbf{R}_4$ (*why*?). Determine whether or not the sum $W_1 + W_2$ is direct. (ANS)

3. Same question for the sum $W_1 + W_2 + W_3$, where

$$W_1 = [(1, 1, 0, 0)]; \qquad W_2 = [(1, 0, 1, 0)];$$
$$W_3 = [(1, 0, 0, 1), (0, 2, 3, 0)].$$ (ANS)

4. If $\mathbf{u}_1, \ldots, \mathbf{u}_r$ are elements of a vector space $V$, show that the $r$-tuple $(\mathbf{u}_1, \ldots, \mathbf{u}_r)$ is linearly independent if and only if the subspaces $[\mathbf{u}_1], \ldots, [\mathbf{u}_r]$ are independent in the sense of Definition 2.8.2.

5. Let $V$ be finite dimensional and $W$, $X$ be subspaces of $V$ such that $V = W + X$. Prove: there exists a subspace $X^*$ of $X$ such that $V = W \oplus X^*$.

# 3 Some geometry in vector spaces

## 3.0 Introduction

In the preceding chapter, geometric illustrations were given for several algebraic concepts. The illustrations were informal and nonrigorous; our theoretical development would have been logically complete (though poorly motivated) without them. It was assumed that the concepts of point, line, plane, and others, which were mentioned in the illustrations, were known to the student, and we assumed their properties without proof.

In the present chapter, we change our point of view towards geometry: we will *define* words like "point", "line", "distance", "angle", and so forth, without assuming any of the postulates of Euclidean geometry; some of these "postulates" will in fact be proven as theorems. Of course, we will try to formulate our definitions in ways that seem geometrically reasonable; we will not hesitate to use pictures and informal discussions to relate our formal development to the ideas about geometry which you already possess. Our object is not to develop *all* of standard Euclidean geometry (we will fall far short of that!), but to give you some idea of the power of vector methods and to lay a foundation for what follows.

Much of what we do in this chapter will take place in an arbitrary vector space. However, at times we shall restrict our considerations to vector spaces over the field **R** (and sometimes specifically to the space $\mathbf{R}_n$), especially when discussing such things as length, angle, and distance, which are not always meaningful over more general fields.

## 3.1 Points and arrows

In discussing the "point" interpretation of $n$-tuples in Section 2.1, we saw that it makes sense to identify points in the plane with elements of the vector space $\mathbf{R}_2$, points in ordinary 3-space with elements of $\mathbf{R}_3$, and so on. Let us now do this formally.

**3.1.1 Definition.** By *real n-space*, we mean the vector space $\mathbf{R}_n$; by a *point* in real $n$-space, we mean an element of $\mathbf{R}_n$. That is, a point in real $n$-space is an $n$-tuple $(a_1, ..., a_n)$, where the $a_i$ are real numbers. The numbers $a_i$ are called the *coordinates* of the point $(a_1, ..., a_n)$.

More generally, if $V$ is an arbitrary vector space, a *point in V* will mean an element of $V$.

When we are thinking of elements of a vector space $V$ as points, we shall denote them by capital letters ($P$, $Q$, $A$, $B$, etc.); when we think of them as algebraic objects (elements of a vector space) we shall continue to denote them $\mathbf{u}$, $\mathbf{v}$, etc. Of course, expressions such as

| | |
|---|---|
| $P + \mathbf{u}$ | (a point plus a vector) |
| $P + Q$ | (a point plus a point) |
| $aP$ | (a scalar times a point) |
| $\mathbf{u} = P$ | (a vector equals a point) |

make sense since $\mathbf{u}$, $P$, $Q$ all belong to $V$. For example, if $P$, $Q$ are the points (1, 3) and (5, 2) in $\mathbf{R}_2$, then $P + Q$ is the point (6, 5) (see Fig. 2.1.3, Sec. 2.1); if $A$ is the point (2, 3) and $\mathbf{u}$ is the vector (4, 1), then $A + \mathbf{u}$ is the point (6, 4) (see Fig. 2.1.4).

The vector $\mathbf{0}$, when regarded as a point, will be called the *origin* and will be denoted by the letter $O$.

Our next definition formalizes the "arrow interpretation" which we introduced informally in Section 2.1.

**3.1.2 Definition.** Let $A$, $B$ be points in a vector space $V$. By the *arrow from A to B*, also called the *arrow with initial point A and terminal point B* (and denoted **Ar** $AB$), we mean the vector $B - A$ in $V$. In the vector space $\mathbf{R}_n$ (or $F_n$) this means that if $A = (a_1, ..., a_n)$ and $B = (b_1, ..., b_n)$, then

(3.1.3) $$\textbf{Ar } AB = B - A = (b_1 - a_1, b_2 - a_2, ..., b_n - a_n)$$

(cf. Sec. 2.1, Exercise 4). In Figure 3.1.1, for example, **Ar** $AB = B - A = (6, 4) - (2, 3) = (4, 1)$.

It is clear that **Ar** $AB$ can equal **Ar** $CD$ even if $A$, $B$, $C$, $D$ are all different points. In fact, if we wish to represent a given vector $\mathbf{u}$ as an arrow **Ar** $AB$, the initial point $A$ may be chosen arbitrarily; however, once the initial point is chosen, the terminal point $B$ is uniquely determined. Let us state this as a formal theorem.

**3.1.4 Theorem.** Let $V$ be a vector space. Given any vector $\mathbf{u} \in V$ and any point $A \in V$, there is a unique point $B \in V$ such that $\mathbf{u} = \mathbf{Ar}\, AB$.

*Proof.* The equation $\mathbf{u} = \mathbf{Ar}\, AB$ is the same as $\mathbf{u} = B - A$, which in turn is equivalent to the equation $\mathbf{u} + A = B$ (Sec. 1.6, Exercise 2); thus the unique point $B$ satisfying the theorem is $\mathbf{u} + A$. ∎

**3.1.5 Theorem.** For all points $A, B, C$ in $V$, $\mathbf{Ar}\, AB + \mathbf{Ar}\, BC = \mathbf{Ar}\, AC$. (See Fig. 3.1.1.)

*Proof.* $\mathbf{Ar}\, AB + \mathbf{Ar}\, BC = (B - A) + (C - B) = C - A = \mathbf{Ar}\, AC$. (Note at which step we used the properties 2.1.1(c) of addition in vector spaces.) ∎

### Distance and length in real *n*-space

You have probably seen (at least in 2 and 3 dimensions) the formula expressing the distance between two points in terms of the coordinates of those points. Since our approach in this chapter assumes no formal knowledge of geometry, we shall take the aforementioned formula as our *definition* of distance, and shall generalize the definition to real *n*-space. (Taking the formula as a definition is sort of a cop-out, but we have to start *somewhere*.)

Figure 3.1.1

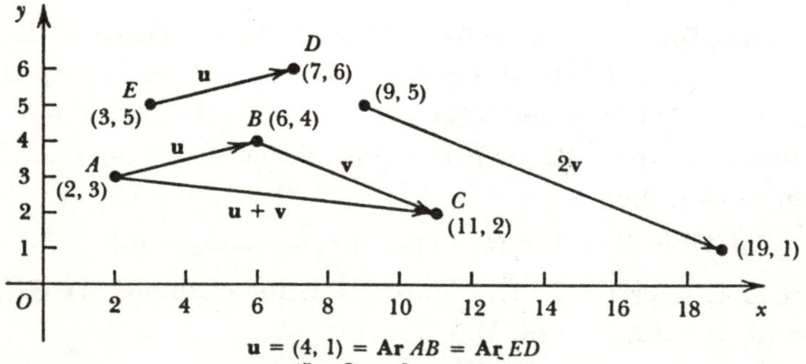


$\mathbf{u} = (4, 1) = \mathbf{Ar}\, AB = \mathbf{Ar}\, ED$
$\mathbf{v} = (5, -2)$; $2\mathbf{v} = (10, -4)$
$\mathbf{u} + \mathbf{v} = (9, -1)$

3.1.6 **Definition.** If $A = (a_1, ..., a_n)$ and $B = (b_1, ..., b_n)$ are points in $\mathbf{R}_n$, the *distance from A to B* is the number

$$d(A, B) = ((b_1 - a_1)^2 + \cdots + (b_n - a_n)^2)^{1/2}.$$

(Note that the right side exists, since in the field **R** every nonnegative number has a square root (Sec. 1.15). Since this property of **R** does not hold in arbitrary fields $F$, Definition 3.1.6 *cannot* be extended to the vector space $F_n$.)

The simplest properties of distance in $\mathbf{R}_n$ are summarized in the next theorem.

3.1.7 **Theorem.** For all points $A$, $B$ in $\mathbf{R}_n$,

(a) $d(A, B) = d(B, A)$.

(b) $d(A, B) = 0$ if and only if $A = B$.

(c) $d(A, B) = d(O, B - A)$.

*Proof.* Part (a) follows from the fact that $(b_i - a_i)^2 = (a_i - b_i)^2$, part (c) follows from the equation

$$b_i - a_i = (b_i - a_i) - 0$$

and part (b) follows from Section 1.12, Exercise 4. ∎

If **Ar** $AB$ and **Ar** $ED$ are two arrows which represent the same vector $\mathbf{u} \in \mathbf{R}_n$ (that is, **Ar** $AB$ = **Ar** $ED = \mathbf{u}$), Figure 3.1.1 suggests that the two arrows have the same "length"; that is, the distance from $A$ to $B$ equals the distance from $E$ to $D$. The latter statement is true; its proof is as follows: since the equation $\mathbf{u}$ = **Ar** $AB$ = **Ar** $ED$ is the same as $\mathbf{u} = B - A = D - E$, 3.1.7(c) gives

$$d(A, B) = d(O, B - A) = d(O, D - E) = d(E, D).$$

We have thus established the following theorem (3.1.8) and have motivated the following definition (3.1.9):

3.1.8 **Theorem.** If $\mathbf{u}$ = **Ar** $AB$ is a vector in $\mathbf{R}_n$, then the number $d(A, B)$ depends only on the vector $\mathbf{u}$ and not on the particular choice of points $A$, $B$.

3.1.9 **Definition.** The number $d(A, B)$ in Theorem 3.1.8 is called the *length* of $\mathbf{u}$, and is denoted $|\mathbf{u}|$.

As a corollary, we have

**3.1.10 Theorem.** If $\mathbf{u} = (a_1, \ldots, a_n)$ is any vector in $\mathbf{R}_n$, then

(a) $|\mathbf{u}| = (a_1^2 + \cdots + a_n^2)^{1/2}$.

(b) $|\mathbf{u}| = 0$ if and only if $\mathbf{u} = \mathbf{0}$.

*Proof.* Since $\mathbf{u} = \mathbf{u} - \mathbf{0}$, 3.1.9 gives $|\mathbf{u}| = d(\mathbf{0}, \mathbf{u})$. Hence, 3.1.10(a) follows easily from Definition 3.1.6, and 3.1.10(b) is the same as 3.1.7(b). ∎

We can now justify the assertion, made at the beginning of Chapter 2 in conjunction with Figure 2.1.1, regarding the effect of scalar multiplication on the length of a vector.

**3.1.11 Theorem.** If $\mathbf{u} \in \mathbf{R}_n$ and $c \in \mathbf{R}$, then

$$|c\mathbf{u}| = |c| \cdot |\mathbf{u}|.$$

That is, if a vector is multiplied by $c$, its length is multiplied by $|c|$.

*Proof.* Exercise. (Use Theorem 3.1.10(a) and equation 2.1.6(b).)

**Parallel vectors**

A glance at Figure 3.1.1 suggests that two arrows which represent the same vector $\mathbf{u}$ are parallel; more generally, that multiplication by a scalar results in an arrow parallel to the original arrow ($2\mathbf{v}$ is parallel to $\mathbf{v}$). This motivates the following definition:

**3.1.12 Definition.** Let $V$ be a vector space. If $\mathbf{u}, \mathbf{v} \in V$, we say that $\mathbf{u}$ *is parallel to* $\mathbf{v}$ (*notation*: $\mathbf{u} \parallel \mathbf{v}$) if $\mathbf{u}$ is a scalar multiple of $\mathbf{v}$; that is, if $\mathbf{u} \in [\mathbf{v}]$.

Note that unlike the concepts of distance and length, the concept of parallelism is meaningful in any vector space, not just in $\mathbf{R}_n$.

Some properties of parallelism are collected in the following theorem.

**3.1.13 Theorem.** For all vectors $\mathbf{u}, \mathbf{v}, \mathbf{w}$ in $V$,

(a) $\mathbf{u} \parallel \mathbf{u}$.

(b) If $\mathbf{u} \neq \mathbf{0}$ and $\mathbf{u} \parallel \mathbf{v}$, then $\mathbf{v} \parallel \mathbf{u}$.

(c) If $\mathbf{u} \parallel \mathbf{v}$ and $\mathbf{v} \parallel \mathbf{w}$, then $\mathbf{u} \parallel \mathbf{w}$.

(d) The set of all vectors in $V$ which are parallel to the fixed vector $\mathbf{v}$ is precisely the subspace $[\mathbf{v}]$; in particular, if $\mathbf{u}_1$ and $\mathbf{u}_2$ are parallel to $\mathbf{v}$, then any linear combination of $\mathbf{u}_1$ and $\mathbf{u}_2$ is parallel to $\mathbf{v}$.

(e) If $\mathbf{u} \neq \mathbf{0}$, then $\mathbf{u} \parallel \mathbf{v}$ if and only if $[\mathbf{u}] = [\mathbf{v}]$.

Statement (d) is just a restatement of Definition 3.1.12. To prove (e), we observe first that

$$\mathbf{u} \parallel \mathbf{v} \Leftrightarrow \mathbf{u} \in [\mathbf{v}] \Leftrightarrow [\mathbf{u}] \subseteq [\mathbf{v}]$$

by Theorem 2.3.13(b). If $\mathbf{u} \neq \mathbf{0}$, then the subspace $[\mathbf{u}]$ is one dimensional (Theorem 2.6.8); since the subspace $[\mathbf{v}]$ is *at most* one dimensional (why?), Theorem 2.6.16 implies that we can have $[\mathbf{u}] \subseteq [\mathbf{v}]$ if and only if $[\mathbf{u}] = [\mathbf{v}]$, proving (e). We leave as an exercise the proof of statements (a), (b), (c); see Exercise 5 below. (These three statements tell us that parallelism is an equivalence relation on the set of all *nonzero* elements of $V$. If the zero vector is included, parallelism is not quite symmetric, since $\mathbf{0}$ is parallel to every vector but not conversely.) ∎

To illustrate the concepts of this section (and to show the power of the vector-space approach in solving geometric problems), we shall prove that in the vector space $\mathbf{R}_n$, *the lengths of opposite sides of a parallelogram are equal.* To say the same thing more formally: if $A$, $B$, $C$, $D$ are distinct points such that $\mathbf{Ar}\ AB \parallel \mathbf{Ar}\ DC$ and $\mathbf{Ar}\ AD \parallel \mathbf{Ar}\ BC$ (see Fig. 3.1.2), then $|\mathbf{Ar}\ AB| = |\mathbf{Ar}\ DC|$ and $|\mathbf{Ar}\ AD| = |\mathbf{Ar}\ BC|$. Actually, this statement is not quite true (for a counterexample, see Fig. 3.1.3); to make it true, we must assume in addition that $\mathbf{Ar}\ AB$ and $\mathbf{Ar}\ AD$ are not parallel. Under this assumption, $\mathbf{Ar}\ AB$ and $\mathbf{Ar}\ AD$ are independent (Exercise 4 below). Since $\mathbf{Ar}\ DC \parallel \mathbf{Ar}\ AB$, we may write $\mathbf{Ar}\ DC = a(\mathbf{Ar}\ AB)$ for some scalar $a \in \mathbf{R}$; similarly, $\mathbf{Ar}\ BC = b(\mathbf{Ar}\ AD)$ for some $b \in \mathbf{R}$. Then

$$\begin{aligned}
\mathbf{Ar}\ AC &= \mathbf{Ar}\ AB + \mathbf{Ar}\ BC = \mathbf{Ar}\ AB + b(\mathbf{Ar}\ AD)\\
\mathbf{Ar}\ AC &= \mathbf{Ar}\ AD + \mathbf{Ar}\ DC = \mathbf{Ar}\ AD + a(\mathbf{Ar}\ AB)\\
&= a(\mathbf{Ar}\ AB) + \mathbf{Ar}\ AD
\end{aligned}$$

and, hence,

$$\mathbf{Ar}\ AB + b(\mathbf{Ar}\ AD) = a(\mathbf{Ar}\ AB) + \mathbf{Ar}\ AD.$$

Since $\mathbf{Ar}\ AB$, $\mathbf{Ar}\ AD$ are independent, we may equate coefficients in the last equation (Theorem 2.4.7); thus, $a = 1$, $b = 1$. Hence, $\mathbf{Ar}\ DC$

Figure 3.1.2

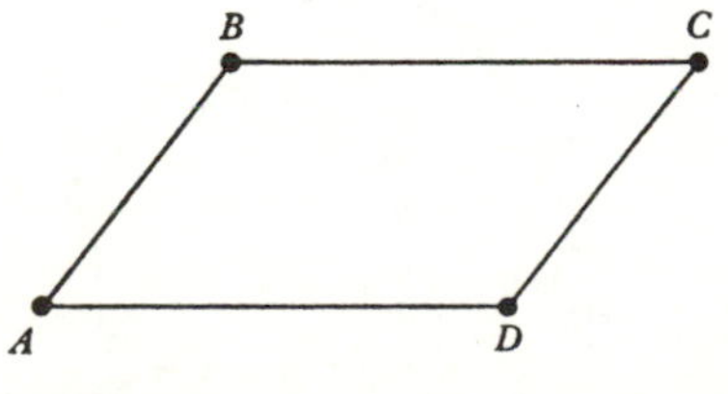


Figure 3.1.3

= **Ar** $AB$ and **Ar** $BC$ = **Ar** $AD$, from which certainly $|\mathbf{Ar}\ DC| = |\mathbf{Ar}\ AB|$ and $|\mathbf{Ar}\ BC| = |\mathbf{Ar}\ AD|$, q.e.d. (Note that this argument used no fact that has not been formally established; for example, the concept of "angle" was not used.)

According to our definition of parallelism, the arrow **Ar** $AB$ in Figure 3.1.1 is parallel to both **Ar** $ED$ and **Ar** $DE$. We can make a finer distinction by saying that $\mathbf{u}$ has *the same direction* as $\mathbf{v}$ if $\mathbf{u} = c\mathbf{v}$ for some *positive* scalar $c$; similarly, $\mathbf{u}$ has the *opposite direction* from $\mathbf{v}$ if $\mathbf{u} = c\mathbf{v}$, $c < 0$. (This makes sense, of course, only if the field of scalars is an *ordered* field; in particular, it makes sense in $\mathbf{R}_n$.) Thus, in Figure 3.1.1, **Ar** $AB$ has the same direction as **Ar** $ED$, the opposite direction from **Ar** $DE$. It is now easy to show that two nonzero elements of $\mathbf{R}_n$ are equal if and only if they have the same length and direction (justifying an informal statement at the beginning of Sec. 2.1). Indeed, if $\mathbf{u}$, $\mathbf{v}$ have the same direction, then $\mathbf{u} = c\mathbf{v}$ with $c > 0$; if also $\mathbf{u}$, $\mathbf{v}$ have the same length, then by Theorem 3.1.11 we have

$$|c| \cdot |\mathbf{v}| = |c\mathbf{v}| = |\mathbf{u}| = |\mathbf{v}|;$$

hence, $|c| = 1$; hence, $c = 1$ (since $c > 0$); hence, $\mathbf{u} = \mathbf{v}$. (For a generalization of this result, see Exercise 9 below.)

### Exercises

(In the exercises below, points and vectors are assumed to belong to an arbitrary vector space $V$ unless the exercise states otherwise.)

1. Prove that for any points $P$ and $Q$, $-$**Ar** $PQ$ = **Ar** $QP$ and **Ar** $PP$ = $\mathbf{0}$.

*2. If **Ar** $AB = \mathbf{u}$ and **Ar** $AC = \mathbf{v}$, prove that **Ar** $CB = \mathbf{u} - \mathbf{v}$. (See Figure 3.1.4.)

3. Prove that if **Ar** $AB$ = **Ar** $CD$, then **Ar** $AC$ = **Ar** $BD$. What familiar theorem of high school geometry is this result related to? Draw a figure to illustrate.

Figure 3.1.4

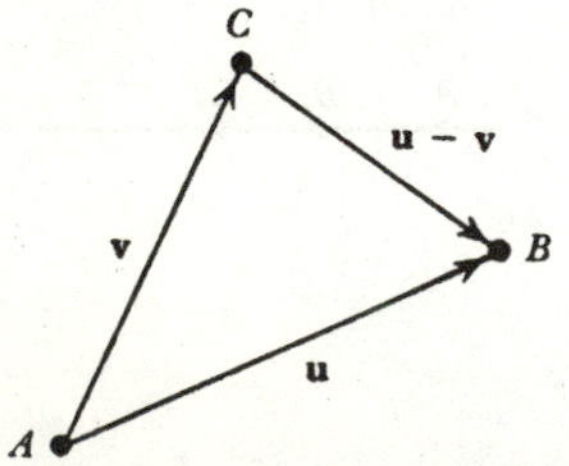

*4. Prove: if $\mathbf{u} \neq \mathbf{0}$, then $\mathbf{v} \parallel \mathbf{u}$ if and only if the ordered pair $(\mathbf{u}, \mathbf{v})$ is linearly dependent. (SOL)

*5. Show that parts (a), (b), and (c) of Theorem 3.1.13 are special cases of certain theorems from Sections 2.3–2.4. Which theorems?

(These three parts of 3.1.13 can, of course, be proved directly, without reference to any theorems; however, the connection with the theorems in question is worth noticing.) (ANS)

6. If $A = (1, 3)$, $B = (5, 2)$ and $C = (3, 0)$ are three consecutive vertices of a parallelogram in $\mathbf{R}_2$, find the fourth vertex. (Formally, this means: find $D$ such that $\mathbf{Ar}\ AB \parallel \mathbf{Ar}\ DC$ and $\mathbf{Ar}\ AD \parallel \mathbf{Ar}\ BC$.)

*7. Prove Theorem 3.1.11.

8. If "$\mathbf{u} \sim \mathbf{v}$" means "$\mathbf{u}$ has the same direction as $\mathbf{v}$" (see the last paragraph of the section), show that "$\sim$" is an equivalence relation on the set of all nonzero elements of $\mathbf{R}_n$. (This result enables us to define a "direction" to be an equivalence class of nonzero vectors.)

9. Prove: if the nonzero vectors $\mathbf{u}, \mathbf{v}$ in $\mathbf{R}_n$ have the same direction and if the length of $\mathbf{v}$ is $c$ times the length of $\mathbf{u}$, then $\mathbf{v} = c\mathbf{u}$.

10. The vector space $\mathbf{R}_1$ is essentially the same as $\mathbf{R}$ itself; in this case, vectors may be identified with real numbers. Thus, if $x \in \mathbf{R}\ (= \mathbf{R}_1)$, the notation $|x|$ may stand for either the length of (the vector) $x$ or the absolute value of (the number) $x$. Discuss the consistency of this.

## 3.2 Lines and line segments

According to Euclid, two points determine a unique line. How do they do so? That is, if $A$ and $B$ are distinct points and $L$ is the line which they "determine", of what, precisely, does $L$ consist? To put it another way, what condition must a point $P$ satisfy in order to lie on $L$? One of the possible answers to this question is suggested by Figure 3.2.1: the arrow $\mathbf{Ar}\ AP$ must be parallel to $\mathbf{Ar}\ AB$. We therefore make the following definition.

3.2.1 **Definition.** Let $V$ be a vector space. If $A, B$ are distinct points in $V$, the *line determined by $A$ and $B$* (denoted $L(A, B)$) is the set of all points $P \in V$ such that $\mathbf{Ar}\ AP \parallel \mathbf{Ar}\ AB$.

As a matter of terminology, if a point $P$ belongs to a line $L$, we shall say that $P$ *lies on* $L$, and that $L$ goes *through* $P$.

As suggested above, Definition 3.2.1 is not the only possible definition of "line". In Section 2.7 we interpreted lines as *cosets of one-dimensional subspaces*; as shown below (Theorem 3.2.5(a)), this interpretation and Definition 3.2.1 agree. In the vector space $\mathbf{R}_n$ (in which "distance" is defined), still a third definition is possible: $P$ lies on $L(A, B)$ if and only if the largest of the three distances $d(A, B)$, $d(A, P)$, $d(B, P)$ is the sum of the other two (see Sec. 3.7, Exercise 16).

It is obvious that the points $A$ and $B$ themselves lie on the line $L(A, B)$ (we leave a formal proof of this as an exercise). The fact that $L(A, B)$ is the *unique* line through $A$ and $B$ will be a corollary of the next theorem.

**3.2.2 Theorem.** Let $A$, $B$ be distinct points in the vector space $V$, and let $A'$, $B'$ be any points on the line $L(A, B)$. Then

(a) $\mathbf{Ar}\ A'B' \parallel \mathbf{Ar}\ AB$;

(b) If $A' \neq B'$, then $L(A', B') = L(A, B)$.

You should draw a figure to convince yourself that Theorem 3.2.2 is true. To prove the theorem formally, let $A', B' \in L(A, B)$. By Section 3.1, Exercise 2 (or by Definition 3.1.2),

$$\mathbf{Ar}\ A'B' = \mathbf{Ar}\ AB' - \mathbf{Ar}\ AA'.$$

Since both $\mathbf{Ar}\ AB'$ and $\mathbf{Ar}\ AA'$ are parallel to $\mathbf{Ar}\ AB$ by Definition 3.2.1, it follows from 3.1.13(d) that $\mathbf{Ar}\ A'B' \parallel \mathbf{Ar}\ AB$, proving 3.2.2(a). Now assume also that $A' \neq B'$. If $P$ is any point on $L(A', B')$, then by definition $\mathbf{Ar}\ A'P \parallel \mathbf{Ar}\ A'B'$. Since also $\mathbf{Ar}\ A'B' \parallel \mathbf{Ar}\ AB$ (just shown), Theorem 3.1.13(c) implies that $\mathbf{Ar}\ A'P \parallel \mathbf{Ar}\ AB$. Since also $\mathbf{Ar}\ AA' \parallel \mathbf{Ar}\ AB$, Theorem 3.1.13(d) implies that the vector

$$\mathbf{Ar}\ AA' + \mathbf{Ar}\ A'P = \mathbf{Ar}\ AP$$

is parallel to $\mathbf{Ar}\ AB$; hence, by definition $P \in L(A, B)$. A similar

Figure 3.2.1

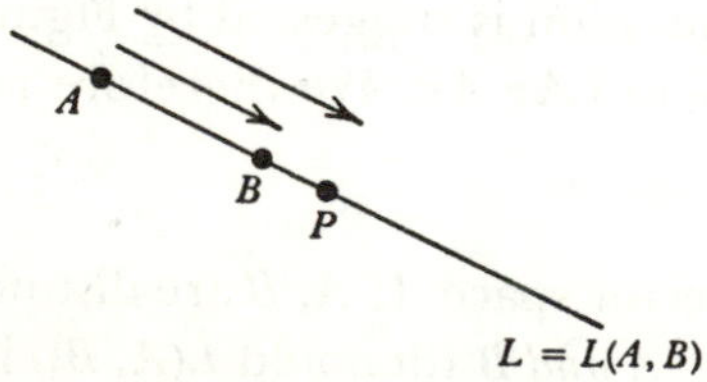

argument (which we leave as an exercise) shows that every point on $L(A, B)$ lies on $L(A', B')$. Hence, $L(A', B') = L(A, B)$, proving 3.2.2(b). ∎

**3.2.3 Corollary.** (a) Given any two distinct points $C, D$ in $V$, $L(C, D)$ is the *unique* line through $C$ and $D$. (b) Two distinct lines can intersect in at most one point.

*Proof.* If $L(A, B)$ is any other line through $C$ and $D$, then 3.2.2(b) (with $A' = C$, $B' = D$) implies that $L(C, D) = L(A, B)$, proving 3.2.3(a). As for 3.2.3(b), if $L_1$ and $L_2$ are lines which intersect in more than one point (say, $L_1 \cap L_2$ contains the two distinct points $C$ and $D$), then by 3.2.3(a) both $L_1$ and $L_2$ are equal to $L(C, D)$ and thus are not distinct lines. ∎

In Section 2.7 we interpreted lines as being cosets of one-dimensional subspaces; more precisely, the cosets of a given one-dimensional subspace $W$ were interpreted as being the lines parallel to $W$. This interpretation will be justified formally in the next theorem; but first we need a definition of "parallel lines".

**3.2.4 Definition.** The line $L(A, B)$ is *parallel* to the line $L(C, D)$, if **Ar** $AB \parallel$ **Ar** $CD$.

Figure 3.2.2 should convince you that Definition 3.2.4 is reasonable; however, the fact that Definition 3.2.4 is *legal* requires proof. The problem is that parallelism of lines ought to depend only on the lines, not on which points are chosen to determine the lines. Thus, if $L(A, B) = L(A', B')$ and $L(C, D) = L(C', D')$ (see Fig. 3.2.2), we must show that **Ar** $AB \parallel$ **Ar** $CD$ if and only if **Ar** $A'B' \parallel$ **Ar** $C'D'$. The proof is as follows: since 3.2.2(a) implies that

$$\mathbf{Ar}\ AB \parallel \mathbf{Ar}\ A'B',$$

it follows from 3.1.13 (parts (b) and (c)) that

$$\mathbf{Ar}\ AB \parallel \mathbf{Ar}\ CD \Leftrightarrow \mathbf{Ar}\ A'B' \parallel \mathbf{Ar}\ CD.$$

Figure 3.2.2

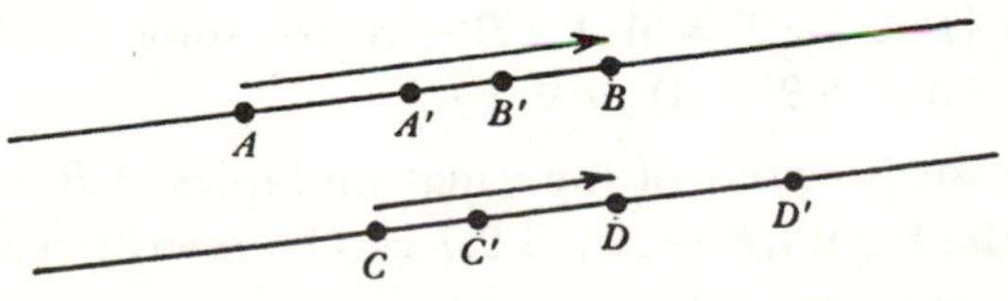

Similarly, we have **Ar** $CD \parallel$ **Ar** $C'D'$ and, hence,

$$\mathbf{Ar}\, A'B' \parallel \mathbf{Ar}\, CD \Leftrightarrow \mathbf{Ar}\, A'B' \parallel \mathbf{Ar}\, C'D'.$$

Hence, **Ar** $AB \parallel$ **Ar** $CD \Leftrightarrow$ **Ar** $A'B' \parallel$ **Ar** $C'D'$, as desired.

*Remarks*: (1) According to our definition, *every line is parallel to itself*, since by 3.1.13 every vector is parallel to itself. (2) It follows from the first three parts of 3.1.13 that parallelism is an equivalence relation on the set of all lines.

3.2.5 **Theorem.** The following assertions hold in any vector space $V$:

(a) A subset of $V$ is a line if and only if it is a coset of a one-dimensional subspace of $V$.

(b) Each line $L(A, B)$ is a coset of *only one* subspace of $V$, namely, the subspace spanned by the vector **Ar** $AB$.

(c) If $L$ is a coset of the subspace $W$ and if $\mathbf{v} \in L$, then $L = W + \mathbf{v}$.

(d) If $L$ is a coset of the subspace $W$ and if $\mathbf{v} \in L$, then $\mathbf{u} \in L \Leftrightarrow \mathbf{u} - \mathbf{v} \in W$.

(e) The lines through the origin are precisely the one-dimensional subspaces of $V$.

(f) Two lines are parallel if and only if they are cosets of the same subspace.

(g) If $W$ is a one-dimensional subspace (i.e., a line through the origin), the lines parallel to $W$ are precisely the cosets of $W$.

*Proof.* By 2.6.8(b) the one-dimensional subspaces of $V$ are precisely the sets $[\mathbf{u}]$, where $\mathbf{u} \neq \mathbf{0}$; hence, the cosets of one-dimensional subspaces are the sets of the form

$$\mathbf{v} + [\mathbf{u}] \qquad (\mathbf{v}, \mathbf{u} \in V;\ \mathbf{u} \neq \mathbf{0}). \tag{3.2.6}$$

On the other hand, if $A$, $B$ are distinct points in $V$ (and if $F$ denotes the field of scalars), then

$$\begin{aligned} L(A, B) &= \{P \in V : \mathbf{Ar}\, AP \parallel \mathbf{Ar}\, AB\} && \text{(by Def. 3.2.1)} \\ &= \{P \in V :\ (P - A) \parallel (B - A)\} && \text{(by Def. 3.1.2)} \\ &= \{P \in V : P - A = t(B - A) \text{ for some } t \in F\} && \\ & && \text{(by Def. 3.1.12)} \\ &= \{P \in V : P = A + t(B - A) \text{ for some } t \in F\} && \\ &= \{A + t(B - A) : t \in F\}. && \end{aligned} \tag{3.2.7}$$

Since $[B - A]$ consists precisely of the scalar multiples of $B - A$ (that is, the vectors of the form $t(B - A)$), 3.2.7 can be rewritten as

$$L(A, B) = A + [B - A] = A + [\mathbf{Ar}\, AB]. \tag{3.2.7'}$$

Since the condition $A \neq B$ in Definition 3.2.1 is equivalent to $B - A \neq \mathbf{0}$, it is clear that the sets 3.2.6 are the same as the sets 3.2.7′ (let $\mathbf{v} = A$, $\mathbf{u} = B - A$, $B = \mathbf{u} + \mathbf{v}$). This proves parts (a) and (b) of Theorem 3.2.5, noting that the uniqueness assertion in part (b) follows from Theorem 2.7.7(f).

Parts (c) and (d) of 3.2.5 follow at once from 2.7.7(a) and 2.7.7(g). In particular, take $\mathbf{v} = \mathbf{0}$ in 3.2.5(c); this implies that if $\mathbf{0} \in L$ then $L = W$, from which 3.2.5(e) easily follows.

By Definition 3.2.4 and Theorem 3.1.13(e), $L(A, B)$ is parallel to $L(C, D)$ if and only if $[\mathbf{Ar}\, AB] = [\mathbf{Ar}\, CD]$; hence, 3.2.5(f) follows from 3.2.5(b). Finally, since any subspace $W$ is a coset of itself, 3.2.5(g) is a special case of 3.2.5(f). ∎

**3.2.8** **Corollary.** (a) If $L = L(A, B)$ is any line in $V$ and $P$ is any point in $V$, then there is a unique line $L^*$ through $P$ which is parallel to $L$. In fact, $L^* = [B - A] + P$.

(b) Two distinct parallel lines do not meet; that is, their intersection is empty.

*Proof.* By 3.2.5(b), the line $L = L(A, B)$ is a coset of $[B - A]$; hence, by 3.2.5(f), the lines parallel to $L$ are the cosets of $[B - A]$. By Theorem 2.7.5 the unique coset of $[B - A]$ containing $P$ is $P + [B - A]$, proving 3.2.8(a). Since distinct cosets of the same subspace have no elements in common, 3.2.8(b) follows at once from 3.2.5(f). ∎

*Remark*: In Section 3.4 (after "plane" has been defined) we shall show that two distinct lines are parallel *if and only if* they lie in the same plane and do not meet. Our definition of parallel lines is thus consistent with the standard high-school definition. (See Theorem 3.4.9.)

**Equations of a line**

By 3.2.7, the general point $P$ on the fixed line $L(A, B)$ is

(3.2.9) $$P = A + t(B - A) \qquad (t \in F);$$

as $t$ varies, the point $P$ varies along the line. If we let $a = 1 - t$ and $b = t$, equation 3.2.9 becomes

(3.2.10) $$P = aA + bB \qquad (a, b \in F;\ a + b = 1).$$

Conversely, if $a$ and $b$ are any scalars *whose sum is 1*, we can obtain 3.2.9 from 3.2.10 by letting $t = b$ (so that $1 - t = a$). Thus, either 3.2.9 or 3.2.10 may be used as the *equation of the line* $L(A, B)$; that is, the equation which a point $P$ satisfies if and only if $P$ lies on the line. Note that 3.2.10 is symmetric with respect to $A$ and $B$.

In the vector space $F_n$, points can be represented as $n$-tuples and equation 3.2.9 can be put in a more familiar form. If the variable point $P$ has coordinates $(x_1, \dots, x_n)$ and if we let

$$A = (a_1, \dots, a_n); \qquad \mathbf{Ar}\, AB = B - A = (c_1, \dots, c_n),$$

then 3.2.9 becomes

$$(x_1, \dots, x_n) = (a_1, \dots, a_n) + t(c_1, \dots, c_n)$$

which is equivalent to the system of equations

$$\begin{aligned} x_1 &= a_1 + c_1 t \\ x_2 &= a_2 + c_2 t \\ &\dots \\ x_n &= a_n + c_n t \end{aligned} \qquad (c_1, \dots, c_n \text{ not all zero}). \tag{3.2.11}$$

(The $c_i$ are not all zero since $B - A \neq \mathbf{0}$.) You probably recognize 3.2.11 as the *parametric equations of a line*.

For fixed $A$ and $B$, the scalar $t$ in 3.2.9 is uniquely determined by $P$. Indeed, if

$$P = A + t_1(B - A) = A + t_2(B - A),$$

then subtraction of $A$ from both sides gives $t_1(B - A) = t_2(B - A)$; since $A \neq B$, Theorem 2.1.4(n) implies that $t_1 = t_2$.

In view of the connection between 3.2.9 and 3.2.10, the uniqueness of $t$ implies that the scalars $a$ and $b$ in 3.2.10 are also uniquely determined by $P$. We call $a$ and $b$ the *barycentric coordinates of P with respect to A and B*. The concept of barycentric coordinates will be generalized to higher dimensions in Section 3.6.

**Line segments**

Let us examine the geometric significance of the scalar $t$ in equation 3.2.9. To begin with, 3.2.9 can be rewritten as $P - A = t(B - A)$, or

$$\mathbf{Ar}\, AP = t(\mathbf{Ar}\, AB). \tag{3.2.12}$$

In the case $V = \mathbf{R}_n$, Theorem 3.1.11 then gives

$$|\mathbf{Ar}\, AP| = |t| \cdot |\mathbf{Ar}\, AB|; \tag{3.2.13}$$

that is, $|t|$ is the ratio of the lengths of **Ar** $AP$ and **Ar** $AB$. The *sign* of $t$ is also significant. When $t > 0$, **Ar** $AP$ has the same direction as **Ar** $AB$ (by 3.2.12) and, thus, $P$ and $B$ lie on the same side of $A$. If $0 < t < 1$, then **Ar** $AP$ is shorter than **Ar** $AB$ (by 3.2.13) so that $P$ lies *between* $A$ and $B$, whereas if $t > 1$, then **Ar** $AP$ is longer than **Ar**

$AB$ and, hence, $P$ is farther from $A$ than $B$ is. As $t$ increases from 0 to 1, $P$ moves continuously from $A$ to $B$. Figure 3.2.3 shows the positions of points $P$ corresponding to various values of $t$.

The above considerations, some of which are informal, motivate the following formal definition:

**3.2.14 Definition.** Let $V$ be a vector space over the field $\mathbf{R}$, let $A$, $B$ be distinct points in $V$, and let $P$ be a point on the line $L(A, B)$. $P$ is said to lie *between* $A$ and $B$ if the scalar $t$ in equation 3.2.12 (3.2.9) satisfies $0 < t < 1$. We say that $P$ is *weakly between $A$ and $B$* if $0 \leqslant t \leqslant 1$ (i.e., if $P$ either lies between $A$ and $B$ or equals $A$ or $B$). By the *line segment from $A$ to $B$* (which we denote $\overline{AB}$) we mean the set of all points which are weakly between $A$ and $B$; that is,

$$\overline{AB} = \{P : \mathbf{Ar}\ AP = t(\mathbf{Ar}\ AB),\ 0 \leqslant t \leqslant 1\}. \tag{3.2.15}$$

The points $A$ and $B$ are called the *endpoints* of the segment $\overline{AB}$ (cf. Example 3.2.20 below). In the case $V = \mathbf{R}_n$ we further define the *length* of the segment $\overline{AB}$ (denoted $|\overline{AB}|$) to be the length of the vector $\mathbf{Ar}$ $AB$.

Figure 3.2.4 shows the positions of some points $P \in L(A, B)$ corresponding to various values of the *barycentric coordinates* $a$ and $b$ of equation 3.2.10. (As before, $a$ and $b$ are related to $t$ by the equations $a = 1 - t$, $b = t$. Compare Figs. 3.2.3 and 3.2.4.) In the case $V = \mathbf{R}_n$, the ratio $|b/a|$ is the ratio between the lengths of $\mathbf{Ar}\ AP$ and $\mathbf{Ar}\ BP$ (see Exercise 8, this section). Thus, for example, when $b = 1/4$ and $a = 3/4$, then the arrow $\mathbf{Ar}\ AP$ is one-third as long as $\mathbf{Ar}\ BP$; when $b = a = 1/2$, then $|\mathbf{Ar}\ AP| = |\mathbf{Ar}\ BP|$ so that $P$ is equidistant from $A$ and $B$ (in other words, $P$ is the *midpoint* of the segment $\overline{AB}$). It is clear that $0 \leqslant t \leqslant 1$ if and only if both $a$ and $b$ are nonnegative; thus, the definition of the line segment $\overline{AB}$ may be rewritten in the form

$$\overline{AB} = \{P : P = aA + bB,\ a \geqslant 0,\ b \geqslant 0,\ a + b = 1\}. \tag{3.2.16}$$

Definition 3.2.14 is not the only way of formally defining "line segment" and "between". For two alternatives, both of which should appear geometrically reasonable, see Exercises 9 and 10 at the end of this section.

By convention, the *line segment from $A$ to $A$* is defined to be the set consisting of the single point $A$. It is easy to see that this convention is consistent with 3.2.15 and 3.2.16.

The following examples exhibit some of the things we can do with lines and line segments using our algebraic approach. Further examples will occur in the exercises.

**3.2.17 Example.** Let $L_1$ and $L_2$ be distinct lines in $\mathbf{R}_n$ which meet at the point $P$. Let $A$, $B$ be points on $L_1$ different from $P$, and let $C$, $D$ be points on $L_2$ different from $P$; assume also that $A$ lies between $P$ and $B$, and that $C$ lies between $P$ and $D$ (see Fig. 3.2.5). Prove that $\mathbf{Ar}\, AC \parallel \mathbf{Ar}\, BD$ if and only if the two ratios

$$|\overline{PA}|/|\overline{PB}|; \qquad |\overline{PC}|/|\overline{PD}|$$

are equal. (You probably recognize this as a fact about "similar triangles".)

Figure 3.2.3

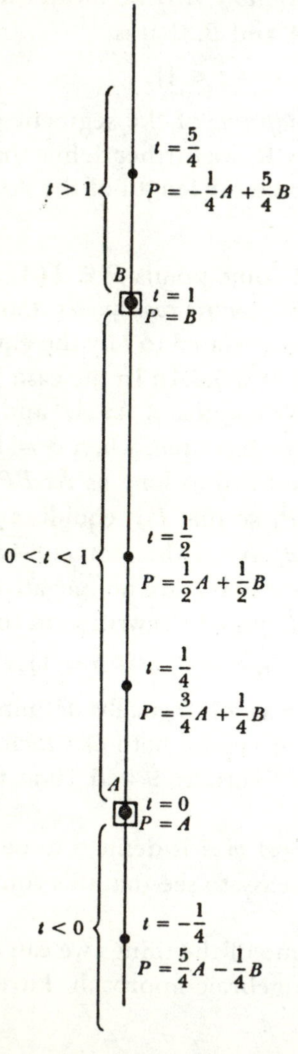


Figure 3.2.4

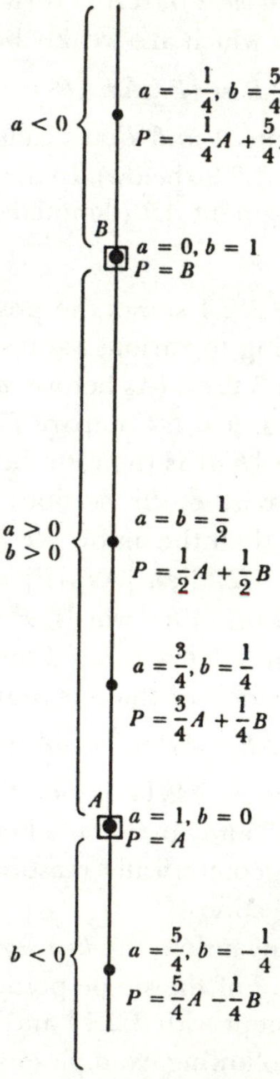

*Solution.* Since $A$ is between $P$ and $B$, we have

$$\mathbf{Ar}\ PA = a(\mathbf{Ar}\ PB); \qquad 0 < a < 1.$$

Similarly,

$$\mathbf{Ar}\ PC = b(\mathbf{Ar}\ PD); \qquad 0 < b < 1.$$

It follows from Theorem 3.1.11 that

$$|\overline{PA}|/|\overline{PB}| = a; \qquad |\overline{PC}|/|\overline{PD}| = b. \tag{3.2.18}$$

If the two ratios are equal than $a = b$ and we have

$$\begin{aligned}\mathbf{Ar}\ AC = \mathbf{Ar}\ PC - \mathbf{Ar}\ PA &= b(\mathbf{Ar}\ PD) - a(\mathbf{Ar}\ PB)\\ &= a(\mathbf{Ar}\ PD) - a(\mathbf{Ar}\ PB)\\ &= a(\mathbf{Ar}\ PD - \mathbf{Ar}\ PB) = a(\mathbf{Ar}\ BD)\end{aligned}$$

so that $\mathbf{Ar}\ AC \parallel \mathbf{Ar}\ BD$ by definition 3.2.12. Conversely, if $\mathbf{Ar}\ AC \parallel \mathbf{Ar}\ BD$, then $\mathbf{Ar}\ AC = c(\mathbf{Ar}\ BD)$ for some scalar $c$, and we have

$$\begin{aligned}b(\mathbf{Ar}\ PD) - a(\mathbf{Ar}\ PB) &= \mathbf{Ar}\ PC - \mathbf{Ar}\ PA = \mathbf{Ar}\ AC\\ &= c(\mathbf{Ar}\ BD)\\ &= c(\mathbf{Ar}\ PD - \mathbf{Ar}\ PB)\\ &= c(\mathbf{Ar}\ PD) - c(\mathbf{Ar}\ PB).\end{aligned} \tag{3.2.19}$$

Since $L_1 \neq L_2$, the points $P$, $B$, $D$ are not all on the same line; thus, by Exercise 3 (below) the vectors $\mathbf{Ar}\ PD$, $\mathbf{Ar}\ PB$ are linearly independent. Hence, by Theorem 2.4.7 we may equate coefficients on both sides of equation 3.2.19. Equating coefficients of $\mathbf{Ar}\ PD$, $b = c$; equating coefficients of $\mathbf{Ar}\ PB$, $a = c$. Hence $a = b$ and the two ratios 3.2.18 are equal. ∎

**3.2.20 Example.** Prove that if the line segments $\overline{AB}$ and $\overline{CD}$ are equal (i.e., consist of the same points), then either $A = C$ and $B = D$, or $A = D$ and $B = C$. (This shows that the endpoints of a line segment are uniquely determined by the segment, except for their order.)

Figure 3.2.5

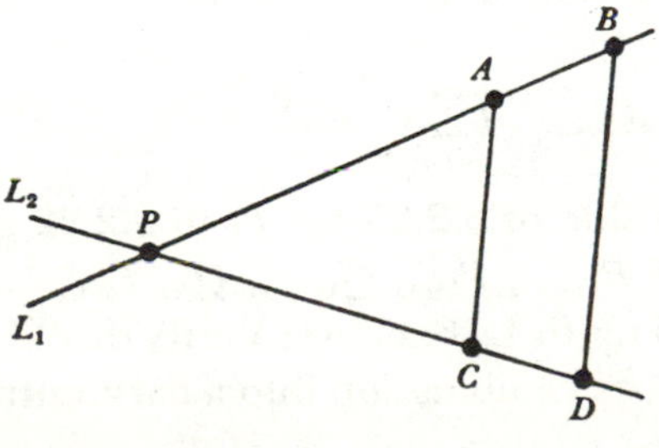

*Solution.* If $A = B$ this is trivial, so assume $A \neq B$. Since $C \in \overline{CD} = \overline{AB}$ and $D \in \overline{CD} = \overline{AB}$, we have

$$C = cA + dB \qquad (c \geqslant 0, d \geqslant 0, c + d = 1)$$
$$D = eA + fB \qquad (e \geqslant 0, f \geqslant 0, e + f = 1).$$

Similarly, since $A \in \overline{AB} = \overline{CD}$, we have

$$\begin{aligned} A &= aC + bD \qquad (a \geqslant 0, b \geqslant 0, a + b = 1) \\ &= a(cA + dB) + b(eA + fB) \qquad (3.2.21) \\ &= (ac + be)A + (ad + bf)B. \end{aligned}$$

Moreover, the coefficients of $A$ and $B$ on the right side of 3.2.21 add up to 1, since

$$a(c + d) + b(e + f) = a + b = 1.$$

Hence, by uniqueness of the barycentric coordinates,

$$ac + be = 1; \qquad ad + bf = 0.$$

Since all scalars are nonnegative, $ad + bf = 0$ implies that $a = 0$ or $d = 0$; that is, $A = D$ or $C = A$; that is, $A$ is one of the two points $C$, $D$. By a similar argument, $B$ is one of the two points $C$, $D$. The desired result follows. ∎

**Further remarks**

1. Two line segments are said to be *parallel* if the corresponding vectors are parallel; that is, we say that $\overline{AB} \parallel \overline{CD}$ if **Ar** $AB \parallel$ **Ar** $CD$. It is clear from the result of Example 3.2.20 that parallelism of segments is well defined (i.e., it depends only on the segments as sets of points, not on how we express the segments in terms of endpoints).

2. It was no accident that the coefficients of $A$ and $B$ on the right side of 3.2.21 added up to 1. More generally, if we have an equation of the form

$$A = a_1P_1 + \cdots + a_nP_n \qquad (3.2.22)$$

where $a_1 + \cdots + a_n = 1$, we say that 3.2.22 expresses $A$ as a *barycentric combination* of the points $P_1, \ldots, P_n$. If in turn one of the $P$'s is expressed as a barycentric combination of points $Q_1, \ldots, Q_m$, say

$$P_k = c_1Q_1 + \cdots + c_mQ_m \qquad \left(\sum_i c_i = 1\right), \qquad (3.2.23)$$

then substitution of the right side of 3.2.23 for $P_k$ in 3.2.22 gives an expression for $A$ in terms of $P_1, \ldots, P_{k-1}, Q_1, \ldots, Q_m, P_{k+1}, \ldots, P_n$ in which the coefficients still add up to 1. (Exercise: Verify this!) In other words, substituting a barycentric combination into a barycentric combination yields a barycentric combination. The concept of barycentric combination will arise again in Section 3.6.

**Exercises**

*1. Prove formally that the points $A$ and $B$ lie on the line $L(A, B)$. (We assume $A \neq B$.)

2. Complete the proof of Theorem 3.2.2(b) by showing that every point on $L(A, B)$ lies on $L(A', B')$.

*3. Let $V$ be a nonzero vector space (i.e., $V$ has at least one element other than $\mathbf{0}$). Prove that three points $A, B, C$ in $V$ are "collinear" (lie on the same line) if and only if the vectors $\mathbf{Ar}\,AB$, $\mathbf{Ar}\,AC$ are linearly dependent. To which exercise in Section 3.1 is this related? (SOL)

4. In Section 3.1 it was shown that if $A, B, C, D$ are distinct points such that $\mathbf{Ar}\,AB \parallel \mathbf{Ar}\,DC$, $\mathbf{Ar}\,AD \parallel \mathbf{Ar}\,BC$, and $\mathbf{Ar}\,AB$ is not parallel to $\mathbf{Ar}\,AD$, then $\mathbf{Ar}\,AB = \mathbf{Ar}\,DC$ and $\mathbf{Ar}\,AD = \mathbf{Ar}\,BC$. Show that the hypothesis "$\mathbf{Ar}\,AB$ is not parallel to $\mathbf{Ar}\,AD$" may be replaced by "$A, B, C, D$ are not all on the same line". (SOL)

In the remaining exercises of this section, all points are assumed to belong to $\mathbf{R}_n$.

5. Prove that the line segments $\overline{AB}$ and $\overline{BA}$ are equal (that is, they consist of the same points). (SUG)

6. Is the line (in $\mathbf{R}_2$) through (3, 4) and (5, 1) parallel to the line through (−1, 3) and (3, −3)? Why?

7. If $A = (2, 4)$ and $B = (10, 0)$ (in $\mathbf{R}_2$), compute the coordinates of the point $P = \frac{3}{4}A + \frac{1}{4}B$ by direct substitution. Then verify by means of a graph (on graph paper!) that the position of $P$ relative to $A$ and $B$ agrees with Figure 3.2.4.

*8. Prove formally that for all points $P$ satisfying equation 3.2.10, $|b/a| = |\mathbf{Ar}\,AP|/|\mathbf{Ar}\,BP|$ whenever $a \neq 0$. (*Suggestion*: Express both $\mathbf{Ar}\,AP$ and $\mathbf{Ar}\,BP$ as scalar multiples of $\mathbf{Ar}\,AB$.)

9. Prove formally that a point $P$ on $L(A, B)$ is weakly between $A$ and $B$ (according to Definition 3.2.14) if and only if $|\mathbf{Ar}\,AP| + |\mathbf{Ar}\,PB| = |\mathbf{Ar}\,AB|$. Does this result seem geometrically reasonable? Draw a figure to illustrate. (*Note*: As will be shown later in Sec. 3.7, Exercise 14, the hypothesis $P \in L(A, B)$ is superfluous.)

10. Prove formally that a point $P$ on $L(A, B)$ is between $A$ and $B$ if and only if the vectors $\mathbf{Ar}\,AP$ and $\mathbf{Ar}\,BP$ have *opposite*

*directions* (as defined in the last paragraph of Sec. 3.1). Does this result seem geometrically reasonable? Draw a figure.

11. Let $P_1$, $P_2$, $P_3$ be distinct points on the line $L(A, B)$; then by 3.2.10 we can write $P_i = a_iA + b_iB$ where $a_i + b_i = 1$ ($i = 1, 2, 3$). Prove that the following statements are equivalent:
(a) $P_2$ is between $P_1$ and $P_3$.
(b) $a_2$ is between $a_1$ and $a_3$ (that is, $a_1 < a_2 < a_3$ or $a_1 > a_2 > a_3$).
(c) $b_2$ is between $b_1$ and $b_3$. (SOL)
12. Let $L$ be a line in $\mathbf{R}_n$. Use the result of the preceding exercise to prove the following statements:
(a) Given three distinct points on $L$, one of them (and only one) is between the other two.
(b) If $A, B, C, D$ are distinct points on $L$ such that $B \in \overline{AC}$ and $C \in \overline{AD}$, then $C \in \overline{BD}$ and $B \in \overline{AD}$. (Draw a figure to illustrate.)
(c) If $A, B, C, D$ are distinct points on $L$ such that $B \in \overline{AC}$ and $C \in \overline{BD}$, then $B \in \overline{AD}$ and $C \in \overline{AD}$. (Draw a figure to illustrate.)
13. Let $A, B, P$ be distinct points. (*Note*: Here we have not assumed that $P \in L(A, B)$.) Prove: $P$ is between $A$ and $B$ if and only if the line segment $\overline{AB}$ is the union of the segments $\overline{AP}$ and $\overline{PB}$. Does this result seem geometrically reasonable? Draw a figure.
14. Prove formally: if $A, B$ are distinct points, then
(a) the point $P = \frac{1}{2}A + \frac{1}{2}B$ (the "midpoint of $\overline{AB}$") is equidistant from $A$ and $B$;
(b) no other point on $L(A, B)$ is equidistant from $A$ and $B$. (SUG)
15. Let $A, B, C$ be noncollinear points. Prove that the three medians of the triangle $ABC$ meet at the point $P = \frac{1}{3}A + \frac{1}{3}B + \frac{1}{3}C$, and that $P$ is twice as far from each vertex as from the midpoint of the opposite side. Draw a figure to illustrate. (We leave it to you to supply the obvious formal definitions of terms like "median", "vertex", "opposite side", etc.)
16. Draw a figure showing that the conclusion of Example 3.2.17 is no longer true if the assumption, "$A$ is between $P$ and $B$", is omitted. Where does the proof break down?
17. Let $L_1$, $L_2$, $L_3$ be distinct lines which intersect at a point $P$. Let $A_i$, $B_i$ be points different from $P$ on $L_i$ ($i = 1, 2, 3$). Prove: if $\mathbf{Ar}\, A_1A_2 \parallel \mathbf{Ar}\, B_1B_2$ and $\mathbf{Ar}\, A_2A_3 \parallel \mathbf{Ar}\, B_2B_3$, then $\mathbf{Ar}\, A_1A_3 \parallel \mathbf{Ar}\, B_1B_3$. Draw a figure to illustrate. (SOL)

18. Show that the conclusion of the preceding exercise remains valid if "intersect at a point $P$" is replaced by "are parallel". Draw a figure to illustrate. (SUG)
19. The general point in $\mathbf{R}_2$ is often written $(x, y)$ instead of $(x_1, x_2)$; in this notation, equations 3.2.11 in the case $n = 2$ reduce to

$$(*) \qquad \begin{aligned} x &= a_1 + c_1 t \\ y &= a_2 + c_2 t. \end{aligned}$$

Show that every pair of equations (*) is equivalent to a single equation having one of the two forms

$y = mx + b; \qquad x = c$

and that conversely every equation of either of the latter two forms is equivalent to a pair of equations (*). (The purpose of this exercise is to show that the equations of lines in real 2-space are what you always thought they were.)
20. Referring to the preceding exercise, the line $y = mx + b$ is said to have *slope* $m$ while the line $x = c$ is said to have *slope* $\infty$. Prove: two lines in $\mathbf{R}_2$ are parallel if and only if they have the same slope. (*Suggestion*: For fixed $m$, show that the line $y = mx$ is a subspace and that the line $y = mx + b$ is a coset of this subspace; etc.)
21. If $P$ is a point on a line $L$, then $P$ divides $L$ into two "half lines." How might one define "half line" formally? (This will be done in Sec. 3.9.)
22. Let $L_1$, $L_2$, $L_3$ be three distinct lines in $\mathbf{R}_2$ which intersect at a point $P$, and let $Q_2$ be a point different from $P$ on $L_2$ (see Figure 3.2.6). Prove: if $k$ is any given positive real number,

Figure 3.2.6

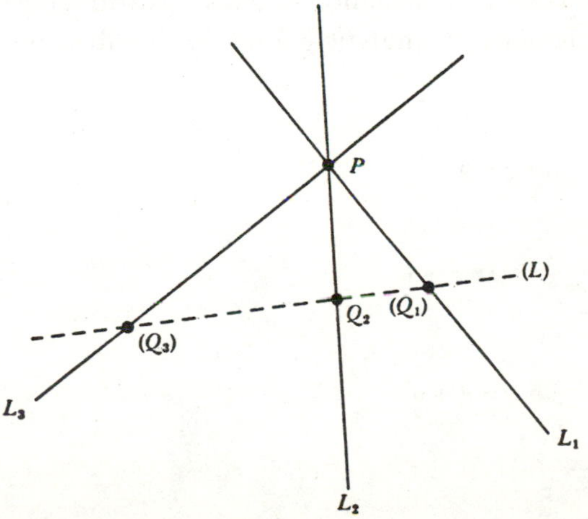

then there exists a line $L \neq L_2$ which passes through $Q_2$ and intersects $L_1$, $L_3$ at points $Q_1$, $Q_3$ (respectively) such that $Q_2$ is between $Q_1$ and $Q_3$ and

$$\frac{|\overline{Q_1Q_2}|}{|\overline{Q_2Q_3}|} = k.$$

## 3.3 Translations

Let $V$ be a vector space and let $\mathbf{u}$ be a fixed vector in $V$. If $A$, $B$ are points such that $\mathbf{Ar}\,AB = \mathbf{u}$, then by definition $\mathbf{u} = B - A$ and, hence, $B = A + \mathbf{u}$. In other words, the terminal point of any arrow which represents $\mathbf{u}$ is obtained by adding $\mathbf{u}$ to the initial point. (See Fig. 3.3.1.)

We can say the same thing in terms of mappings. Let $\mathscr{T}_{\mathsf{u}}$ be the mapping of $V \to V$ defined by

$$\mathscr{T}_{\mathsf{u}} : \mathbf{x} \to \mathbf{x} + \mathbf{u} \qquad (\text{all } \mathbf{x} \in V).$$

(*Note*: To facilitate typesetting, the subscript letter appears in sans serif type "u" instead of in boldface type "**u**"; similar changes in notation apply also to other subscript letters that would normally appear boldface.) Then the initial point of any arrow representing $\mathbf{u}$ is mapped by $\mathscr{T}_{\mathsf{u}}$ into the terminal point of that arrow. In real $n$-space, all such arrows have the same length and direction, so that $\mathscr{T}_{\mathsf{u}}$ has the effect of moving all points in $\mathbf{R}_n$ the same distance in the same direction.

**3.3.1 Definition.** A *translation of V* is a mapping of the form $\mathscr{T}_{\mathsf{u}}$ for some $\mathbf{u} \in V$.

*Remark*: The concept of "translation of axes", as you have probably come across it in elementary analytic geometry, involves moving the

Figure 3.3.1

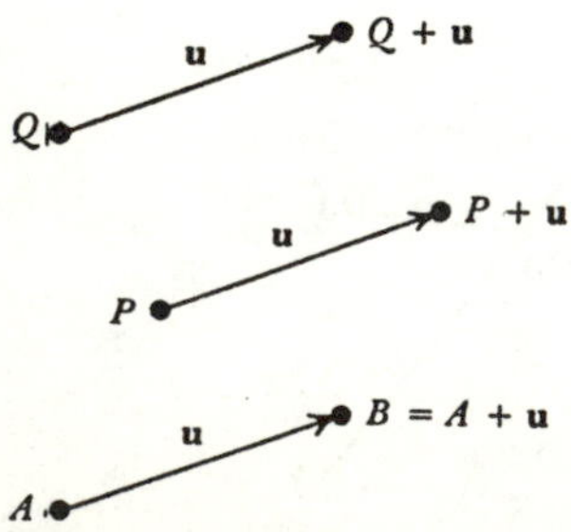

coordinate axes (equivalently, choosing a new origin) while the points remain fixed. A "translation" as defined by 3.3.1 does the opposite: it moves the points while the axes remain fixed. The relation between the two concepts is illustrated in Figure 3.3.2: a translation of *axes* in a given direction has the same effect on the coordinates of a point as a translation of *points* the same distance in the *opposite* direction. In this section we discuss only translations of points, not change of axes. (Change of axes will be discussed in Sec. 3.10.)

Under the mapping $\mathcal{T}_{\mathbf{u}}$, the image of any element $\mathbf{x} \in V$ is $\mathbf{x} + \mathbf{u}$; hence, the image of any subspace $W$ of $V$ is the coset $W + \mathbf{u}$. In other words, *the cosets of a subspace $W$ of $V$ are precisely the images of $W$ under translations of $V$.* We have already seen this idea illustrated in Figure 2.7.1, in which $W$ and its coset $L = W + \mathbf{u}$ are lines; $L$ is the image of $W$ under the translation $\mathcal{T}_{\mathbf{u}}$. (Compare Figs. 2.7.1 and 3.3.1.) This interpretation of cosets enables us to restate Theorem 3.2.5(a) as follows:

**3.3.2 Theorem.** A subset of $V$ is a line if and only if it is the image of a one-dimensional subspace under a translation.

The notion of parallelism may also be restated in terms of translations:

**3.3.3 Theorem.** Two lines $L_1$, $L_2$ are parallel if and only if $L_2$ is the image of $L_1$ under a translation.

*Proof.* Exercise. (Use Theorem 3.2.5(f) and the definition of translation.)

Figure 3.3.2

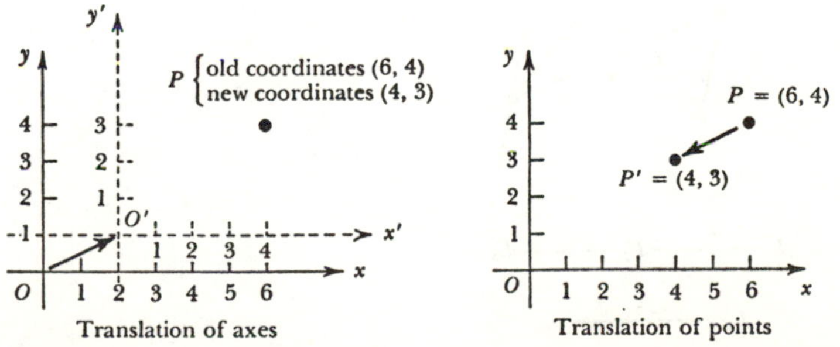

In Section 1.3 we defined *multiplication of mappings*. The definition may be applied in particular to translations:

$$\mathbf{x}(\mathcal{T}_\mathbf{u}\mathcal{T}_\mathbf{v}) = (\mathbf{x}\mathcal{T}_\mathbf{u})\mathcal{T}_\mathbf{v} \qquad (\text{all } \mathbf{x} \in V; \text{ all } \mathbf{u}, \mathbf{v} \in V).$$

Since the effect of applying $\mathcal{T}_\mathbf{u}$ is to add $\mathbf{u}$, the above equation is the same as

$$\begin{aligned}\mathbf{x}(\mathcal{T}_\mathbf{u}\mathcal{T}_\mathbf{v}) &= (\mathbf{x} + \mathbf{u}) + \mathbf{v}\\ &= \mathbf{x} + (\mathbf{u} + \mathbf{v})\\ &= \mathbf{x}\mathcal{T}_{\mathbf{u}+\mathbf{v}}.\end{aligned}$$

Since this holds for *all* $\mathbf{x} \in V$, it follows that

(3.3.4) $\mathcal{T}_\mathbf{u}\mathcal{T}_\mathbf{v} = \mathcal{T}_{\mathbf{u}+\mathbf{v}}.$

That is, *the product of translations is a translation*. This is illustrated in Figure 3.3.3: the vectors $\mathbf{u}$, $\mathbf{v}$, $\mathbf{u} + \mathbf{v}$ are represented by arrows, and the line $L_1$ is mapped into the line $L_3$ either by the translation $\mathcal{T}_{\mathbf{u}+\mathbf{v}}$ (dotted arrows) or by the product $\mathcal{T}_\mathbf{u}\mathcal{T}_\mathbf{v}$ ($\mathcal{T}_\mathbf{u}$ followed by $\mathcal{T}_\mathbf{v}$). The lines $L_1$, $L_2$, $L_3$ are all parallel by Theorem 3.3.3.

Does the form of equation 3.3.4 look familiar? It should; it's a homomorphism equation. Indeed, let $\tau(V)$ be the set of all translations of $V$, and let $\phi$ be the mapping of $V \to \tau(V)$ defined by

Figure 3.3.3

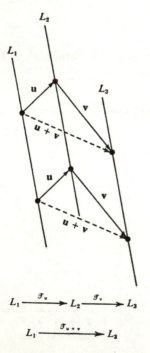

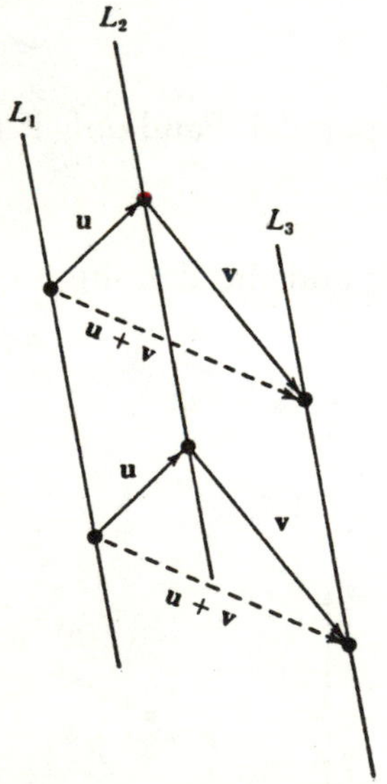


$$L_1 \xrightarrow{\mathcal{T}_\mathbf{u}} L_2 \xrightarrow{\mathcal{T}_\mathbf{v}} L_3$$

$$L_1 \xrightarrow{\mathcal{T}_{\mathbf{u}+\mathbf{v}}} L_3$$

$$\phi : \mathbf{u} \to \mathcal{T}_{\mathbf{u}} \qquad (\text{all } \mathbf{u} \in V).$$

Then 3.3.4 can be rewritten $(\mathbf{u}\phi)(\mathbf{v}\phi) = (\mathbf{u} + \mathbf{v})\phi$, which is precisely the condition for $\phi$ to be a homomorphism (see Sec. 1.8). Moreover, $\phi$ is clearly onto; and since

$$\mathbf{u} \neq \mathbf{v} \Rightarrow \mathbf{0}\mathcal{T}_{\mathbf{u}} \neq \mathbf{0}\mathcal{T}_{\mathbf{v}} \Rightarrow \mathcal{T}_{\mathbf{u}} \neq \mathcal{T}_{\mathbf{v}}$$

$\phi$ is one-to-one. Hence, $\phi$ is an isomorphism of $V$ (under addition) onto $\tau(V)$ (under multiplication). In particular, since $V$ is an abelian group, so is $\tau(V)$ (Sec. 1.8, Exercise 5); since $\mathbf{0}$ is the identity of $V$, $\mathcal{T}_{\mathbf{0}}$ is the identity of $\tau(V)$ (Theorem 1.8.4(a)). The latter is, of course, easy to prove directly: since

$$\mathbf{x}\mathcal{T}_{\mathbf{0}} = \mathbf{x} + \mathbf{0} = \mathbf{x}$$

for all $\mathbf{x}$, it follows that $\mathcal{T}_{\mathbf{0}}$ is the identity mapping $\mathbf{I}$. Similarly, since $\mathcal{T}_{-\mathbf{u}}\,\mathcal{T}_{\mathbf{u}} = \mathcal{T}_{-\mathbf{u}+\mathbf{u}} = \mathcal{T}_{\mathbf{0}} = \mathbf{I}$, $\mathcal{T}_{-\mathbf{u}}$ is the inverse of $\mathcal{T}_{\mathbf{u}}$; this could also be deduced from Theorem 1.8.4(b). Summarizing our results, we have

**3.3.5 Theorem.** The set $\tau(V)$, consisting of all translations of $V$, is an abelian group under multiplication of mappings; this group is isomorphic to the additive group $V$. The identity of $\tau(V)$ is the identity mapping $\mathbf{I} = \mathcal{T}_{\mathbf{0}}$; the inverse of $\mathcal{T}_{\mathbf{u}}$ is $\mathcal{T}_{-\mathbf{u}}$.

**Exercises**

*1. Prove Theorem 3.3.3.

2. By Theorem 3.3.5 (or equation 3.3.4), any two translations of $V$ commute with each other. Show by an example that this is not true of arbitrary mappings of $V \to V$. (You may take $V = \mathbf{R}_2$.)

*3. Prove that any translation of $\mathbf{R}_n$ *preserves distance*; that is, given any two points in $\mathbf{R}_n$, the distance between them equals the distance between their images under the translation. Draw a figure to illustrate.

## 3.4 Planes and flats

Given any two distinct points in a plane, the entire line through them lies in the plane. Also, if a point $P$ and a line $L$ lie in a given plane, so does the line through $P$ parallel to $L$. (Don't worry about the definition of "plane"; we're being informal.) We have just stated two properties of planes. However, planes are not the only objects having these two properties. It is clear, for example, that the first two sentences of this paragraph would remain true if the word "plane" were replaced by "line". They would also remain true if "plane" were replaced by "vector space" (or even if "plane" were replaced by "point"!). It is useful to give a name to sets having these properties.

3.4.1 **Definition.** Let $V$ be a vector space over a field $F$. By a *flat* in $V$ we mean a nonempty subset $S$ of $V$ having the following two properties:

(a) Whenever $A$, $B$ are distinct points in $S$, then $L(A, B) \subseteq S$.

(b) Whenever a point $P$ and a line $L$ are contained in $S$, then $S$ contains the line through $P$ which is parallel to $L$. (Note that there is exactly one such line through $P$, by Corollary 3.2.8(a).)

The next theorem shows that, under rather general conditions, property 3.4.1(b) is actually superfluous; that is, if 3.4.1(a) holds, then so does 3.4.1(b) automatically. In addition, the theorem shows that flats can be characterized *algebraically* in a manner similar to the way Theorem 3.2.5(a) (or Theorem 3.3.2) characterizes lines.

3.4.2 **Theorem.** Let $V$ be a vector space over $F$.

(a) A subset $S$ of $V$ is a flat if and only if it is a coset of some subspace of $V$; equivalently, if and only if $S$ is the image of some subspace under a translation.

(b) If $1_F + 1_F \neq 0_F$, then any nonempty subset $S$ of $V$ which satisfies 3.4.1(a) will also satisfy 3.4.1(b) and is, hence, a flat.

We shall postpone the proof of Theorem 3.4.2 until the end of this section, so that we may first study the consequences of the theorem. Even before giving a proof, however, the following comment may be worth making. If we compare Definition 3.4.1 (the characterization of "flat") with Theorem 2.2.4 (the characterization of "subspace"), we see that the two are essentially similar in nature: both assert that if certain elements belong to the set in question then so do certain other elements. Hence, the fact that flats are closely related to subspaces (Theorem 3.4.2(a)) is not totally surprising. (Because of this close relationship, flats are sometimes called *affine subspaces*. In this book, however, the word "subspace" will be used only in the sense of Definition 2.2.1. Other common synonyms for the word "flat" are "linear variety" and "linear manifold".)

To begin our discussion of flats, we observe that if $S$ is a flat then the subspace $W$ of which $S$ is a coset is *uniquely determined by* $S$ (Theorem

2.7.7(f)); this unique subspace $W$ is called the *direction space* of $S$. (The phrase "direction space" is probably due to M. Hausner; cf. [7].) It is natural to define the *dimension* of a flat to be the dimension of its direction space (and we do so); thus Theorem 3.2.5(a) may be restated as, "line = one-dimensional flat". We define a *plane* to be a two-dimensional flat. By a *k-flat* we will mean a $k$-dimensional flat. It is common to define a *hyperplane in V* to be a flat whose dimension is one less than that of $V$; thus, in $\mathbf{R}_3$ the hyperplanes are the planes, while in $\mathbf{R}_4$ the hyperplanes are the 3-flats. The 0-flats are precisely the points (more rigorously, they are the sets consisting of just one point); indeed, the only 0-dimensional subspace of $V$ is $\{\mathbf{0}\}$ (Sec. 2.6, Exercise 5) and the cosets of this subspace are the sets of the form $\mathbf{u} + \{\mathbf{0}\} = \{\mathbf{u}\}$ which consist of just one element $\mathbf{u}$.

A *subflat* of a flat $S$ is, of course, a subset of $S$ which is itself a flat. Later we shall need the following result: if $S_1$ is a subflat of a finite-dimensional flat $S_2$, then $\dim S_1 \leqslant \dim S_2$, with the dimensions being equal only if $S_1 = S_2$. This result follows from the similar theorem for subspaces (Theorem 2.6.16) together with parts (d) and (e) of Theorem 2.7.7.

We have seen (Theorem 3.2.5(f)) that two lines are parallel if and only if they are cosets of the same subspace $W$. Figure 3.4.1 suggests, similarly, that two planes $S_1$ and $S_2$ will be parallel if they are cosets of the same subspace. It is natural to define parallelism of $k$-flats (for arbitrary $k$) in an analogous manner:

Figure 3.4.1

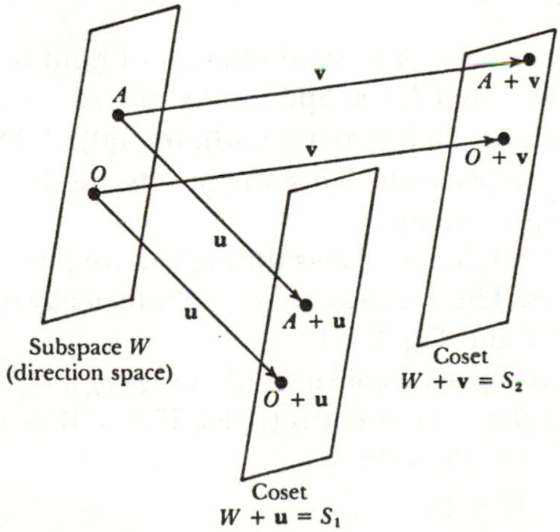

**3.4.3 Definition.** Two flats of the same dimension in $V$ are said to be *parallel* if they are cosets of the same subspace; that is, if they have the same direction space.

Note that 3.4.3 defines parallelism only for flats of the same dimension. How would you define parallelism of two flats of different dimensions, such as a line and a plane? (The general definition is given in Exercise 1 at the end of this section; try to think of a suitable definition yourself before looking at the exercise!)

Several elementary properties of flats are summarized in the following theorem.

**3.4.4 Theorem.** The following statements hold in any vector space $V$.

(a) If a flat $S$ has direction space $W$ and if $\mathbf{v} \in S$, then

$$S = W + \mathbf{v}.$$

(b) If a flat $S$ has direction space $W$ and if $\mathbf{v} \in S$, then

$$\mathbf{u} \in S \Leftrightarrow \mathbf{u} - \mathbf{v} \in W.$$

(c) The flats through the origin are precisely the subspaces of $V$.

(d) The $k$-flats parallel to a given $k$-dimensional subspace $W$ are the cosets of $W$.

(e) If $S$ is a $k$-flat with direction space $W$ and if $P$ is any point, then there is a unique $k$-flat $S^*$ through $P$ which is parallel to $S$. In fact, $S^* = W + P$.

(f) Two distinct parallel $k$-flats have empty intersection.

(g) Two $k$-flats $S_1$, $S_2$ are parallel if and only if $S_2$ is the image of $S_1$ under a translation.

Each part of Theorem 3.4.4 is a generalization of (and is proved like) an earlier theorem about lines. Specifically, the seven parts of 3.4.4 are generalizations of 3.2.5 (parts (c), (d), (e), (g)), 3.2.8 (parts (a) and (b)), and 3.3.3, respectively. Since the proofs are the same as before, we need not repeat them.

As a special case of 3.4.4(c), the planes through $O$ are precisely the 2-dimensional subspaces. This fact was suggested informally in Chapter 2 (see Example 2.3.7 and Fig. 2.3.1).

Theorem 3.4.4(g) has a partial converse: *the image of a k-flat under a translation is always a k-flat*. The proof is trivial: if $S = W + \mathbf{u}$ is a $k$-flat with direction space $W$, then the set

$$S\mathcal{T}_{\mathbf{v}} = S + \mathbf{v} = W + (\mathbf{u} + \mathbf{v})$$

is a flat having the same direction space and, hence, the same dimenson.

**Intersections of flats**

If two lines in a plane do not meet, they are parallel; but two lines in 3-space may be nonparallel and still fail to meet. On the other hand, the intersection of two *planes* in 3-space is empty only if the planes are parallel. It is possible for the intersection of *three* planes $S_1$, $S_2$, $S_3$ in 3-space to be empty even if no two of the planes are parallel; but this can occur only if the three lines $S_1 \cap S_2$, $S_1 \cap S_3$, $S_2 \cap S_3$ are parallel (see Fig. 3.4.2).

The statements in the preceding paragraph are not quite as miscellaneous as they seem; in fact, we shall prove a general theorem on intersections of flats (Theorem 3.4.8 below) which implies all of these statements as special cases. The proof is based upon the following preliminary result (a generalization of Theorem 2.6.17) concerning intersections of subspaces.

3.4.5 **Theorem.** Let $V$ be a vector space of finite dimension $n$. Let $W_1, \ldots, W_k$ be subspaces of $V$, and let $e_i = n - \dim W_i$ ($i = 1, \ldots, k$). Then

Figure 3.4.2

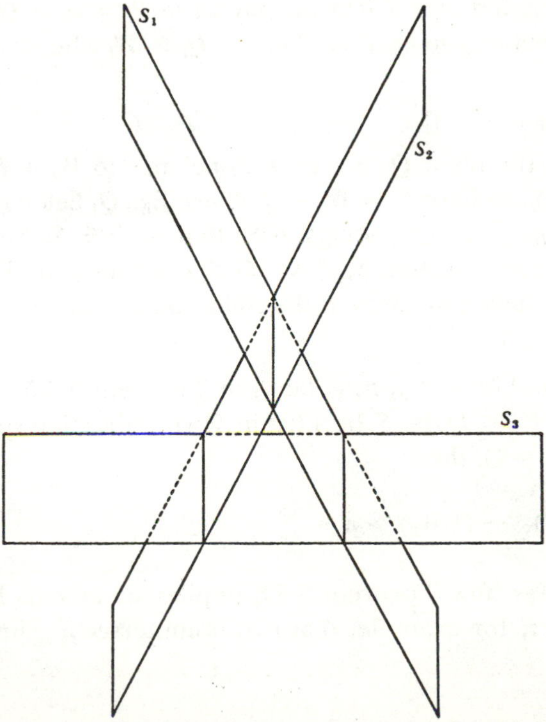

**(3.4.6)** $\dim(W_1 \cap \cdots \cap W_k) \geq n - \sum_{i=1}^{k} e_i.$

Moreover, the following two statements are equivalent:

**(3.4.7)**
(a) $\dim(W_1 \cap \cdots \cap W_k) = n - \sum_{i=1}^{k} e_i;$
(b) For all $i$ $(i = 1, \ldots, k)$, $W_i + (\bigcap_{j \neq i} W_j) = V$

where in 3.4.7(b), $j$ runs over all indices from 1 to $k$ except $i$.

For example, if $W_1$, $W_2$, $W_3$ are subspaces of dimensions 7, 8, 9 (respectively) in a 10-dimensional vector space $V$, then the $e_i$ in Theorem 3.4.5 are 3, 2, 1, and the theorem asserts that $\dim(W_1 \cap W_2 \cap W_3) \geq 4$, with equality only if

$$\begin{aligned} W_1 + (W_2 \cap W_3) &= W_2 + (W_1 \cap W_3) \\ &= W_3 + (W_1 \cap W_2) = V. \end{aligned}$$

The proof of Theorem 3.4.5 is given at the end of this section. Assuming that Theorem 3.4.5 has been established, let $S_i$ be a flat having direction space $W_i$ $(i = 1, \ldots, k)$ and suppose that 3.4.7(b) holds (equivalently, 3.4.7(a) holds). For each $i$ we may write $S_i = W_i + P_i$ for some point $P_i$; and by 3.4.7(b) we may write $P_i = \mathbf{w}_i + Q_i$ where $\mathbf{w}_i \in W_i$ and $Q_i$ belongs to $\bigcap_{j \neq i} W_j$. That is, $Q_i \in W_j$ whenever $j \neq i$. Let

$$Q = Q_1 + \cdots + Q_k.$$

For each fixed $i$, the point $Q_i = -\mathbf{w}_i + P_i$ belongs to $W_i + P_i = S_i$; hence, by 3.4.4(a), we have $S_i = W_i + Q_i$. Since $\sum_{j \neq i} Q_j$ belongs to $W_i$, it follows that $(\sum_{j \neq i} Q_j) + Q_i$ belongs to $S_i$; that is, $Q \in S_i$. Since this holds for all $i$, the intersection $S_1 \cap \cdots \cap S_k$ contains $Q$ and, hence, is nonempty. We have thus proved the following theorem:

**3.4.8 Theorem.** Let $V$, $W_i$, $n$, $e_i$ be as in Theorem 3.4.5, and for each $i$ $(1 \leq i \leq k)$, let $S_i$ be a flat in $V$ with direction space $W_i$. If $S_1 \cap \cdots \cap S_k = \varnothing$, then

$$\dim(W_1 \cap \cdots \cap W_k) > n - \sum_{i=1}^{k} e_i.$$

Let us now show how Theorem 3.4.8 implies statements like the ones made earlier; for example, that two nonintersecting lines in a plane (or two nonintersecting planes in 3-space) are parallel. More generally, we can prove the following:

**3.4.9 Theorem.** Let $V$ be a vector space. If $S_1$, $S_2$ are distinct $k$-flats in $V$, then $S_1$ and $S_2$ are parallel if and only if they lie in the same $(k + 1)$-flat and have empty intersection.

*Proof.* Assume $S_1$, $S_2$ lie in the same $(k + 1)$-flat $S$ and that $S_1 \cap S_2 = \varnothing$. Let $W_1$, $W_2$, $W$ be the respective direction spaces of $S_1$, $S_2$, $S$; in particular, $S = W + \mathbf{v}$ for some $\mathbf{v} \in V$. Since $S_i \subseteq S$, Theorem 2.7.7(d) implies that $W_i \subseteq W$; also, since $S_i \subseteq W + \mathbf{v}$ it follows that $S_i - \mathbf{v} \subseteq W$. Moreover, $S_1 \cap S_2 = \varnothing$ implies $(S_1 - \mathbf{v}) \cap (S_2 - \mathbf{v}) = \varnothing$ (*why*?). Hence, we can apply Theorem 3.4.8 with $V$ replaced by $W$, $S_i$ by $S_i - \mathbf{v}$, $n$ by $k + 1$. We obtain

$$e_i = (k + 1) - \dim W_i = (k + 1) - k = 1 \qquad (i = 1, 2)$$

and, hence, by Theorem 3.4.8

$$\dim(W_1 \cap W_2) > (k + 1) - (1 + 1),$$

that is,

$$\dim(W_1 \cap W_2) \geqslant k. \tag{3.4.10}$$

But $\dim W_1 = \dim W_2 = k$, and $W_1 \cap W_2$ is a subspace of both $W_1$ and $W_2$. Hence, 3.4.10 is possible only if $W_1 = W_1 \cap W_2 = W_2$, so that $S_1$ and $S_2$ have the same direction space and are hence parallel.

Conversely, suppose $S_1 \| S_2$. Then certainly $S_1 \cap S_2 = \varnothing$, by 3.4.4(f). Moreover, by definition the parallel flats $S_1$ and $S_2$ have the same direction space $W_0$; say $S_1 = W_0 + \mathbf{v}_1$, $S_2 = W_0 + \mathbf{v}_2$. Thus

$$S_1 = W_0 + \mathbf{v}_1 = W_0 + (\mathbf{v}_1 - \mathbf{v}_2) + \mathbf{v}_2 \subseteq [W_0, \mathbf{v}_1 - \mathbf{v}_2] + \mathbf{v}_2$$
$$S_2 = W_0 + \mathbf{v}_2 \subseteq [W_0, \mathbf{v}_1 - \mathbf{v}_2] + \mathbf{v}_2$$

so that $S_1$ and $S_2$ both lie in the flat $S = [W_0, \mathbf{v}_1 - \mathbf{v}_2] + \mathbf{v}_2$. Since $\dim W_0 = k$, $\dim S \leqslant k + 1$ (why?); hence, $\dim S = k + 1$ (not $k$, since $S$ is strictly bigger than $S_1$); that is, $S_1$ and $S_2$ lie in the same $(k + 1)$-flat, as desired. ∎

The basic theorem regarding flats having *non*empty intersection is the following:

**3.4.11 Theorem.** Let $W_1, \ldots, W_k$ be subspaces of $V$; for each $i$ $(1 \leqslant i \leqslant k)$, let $S_i$ be a flat in $V$ with direction space $W_i$. If $S_1 \cap \cdots \cap S_k$ is nonempty, then $S_1 \cap \cdots \cap S_k$ is a flat whose direction space is $W_1 \cap \cdots \cap W_k$.

*Proof.* By hypothesis, $S_1 \cap \cdots \cap S_k$ contains at least one vector $\mathbf{v}$. Then $\mathbf{v} \in S_i$ (all $i$) and, hence, $S_i = W_i + \mathbf{v}$. If $\mathbf{u} \in S_1 \cap \cdots \cap S_k$, then for each $k$ we have $\mathbf{u} \in S_i = W_i + \mathbf{v}$; hence, $\mathbf{u} - \mathbf{v} \in W_i$; hence, $\mathbf{u} - \mathbf{v} \in W_1 \cap \cdots \cap W_k$; hence

$$\mathbf{u} \in (W_1 \cap \cdots \cap W_k) + \mathbf{v}.$$

Conversely, any vector in $(W_1 \cap \cdots \cap W_k) + \mathbf{v}$ lies in $W_i + \mathbf{v}$, hence, in $S_i$ (all $i$), hence, in $S_1 \cap \cdots \cap S_k$. Thus,

$$S_1 \cap \cdots \cap S_k = (W_1 \cap \cdots \cap W_k) + \mathbf{v}$$

which is a coset of $W_1 \cap \cdots \cap W_k$. This is the desired result. ∎

*Remark*: Theorem 3.4.11 shows that the intersection of any number of flats $S_i$ is always a flat (if nonempty). This result remains true even when the number of flats $S_i$ is infinite; the proof is the same.

3.4.12 **Example.** In a 3-dimensional vector space $V$, prove that the intersection of two nonparallel planes is a line.

*Solution.* If the planes $S_1$, $S_2$ are not parallel, then their direction spaces $W_1$, $W_2$ are not equal; hence, $\dim(W_1 \cap W_2) = 1$ by Section 2.6, Exercise 11(a). Also, $S_1 \cap S_2 \neq \varnothing$ by 3.4.9; hence, by 3.4.11, $S_1 \cap S_2$ is a coset of $W_1 \cap W_2$ and thus is a line. ∎

3.4.13 **Example.** Let $S_1$, $S_2$, $S_3$ be planes in a 3-dimensional vector space $V$. If $S_1 \cap S_2 \cap S_3$ is empty but no two of the $S_i$ are parallel, prove that the three lines $S_1 \cap S_2$, $S_1 \cap S_3$, $S_2 \cap S_3$ are parallel. (This is the statement asserted earlier, which Fig. 3.4.2 illustrates.)

*First solution.* The intersections $S_i \cap S_j$ are indeed lines by Example 3.4.12. Suppose two of them (say, $S_1 \cap S_2$ and $S_1 \cap S_3$) are not parallel. Since these two lines lie in the same plane $S_1$, they have a point in common (Theorem 3.4.9), contradicting the assumption that $S_1 \cap S_2 \cap S_3 = \varnothing$.

*Second solution.* Theorem 3.4.8 is applicable with $n = 3$, $e_i = 3 - 2 = 1$ $(i = 1, 2, 3)$, $n - \Sigma e_i = 0$; thus

(3.4.14) $\dim(W_1 \cap W_2 \cap W_3) \geq 1.$

But each of the subspaces $W_1 \cap W_2$, $W_1 \cap W_3$, $W_2 \cap W_3$ contains $W_1 \cap W_2 \cap W_3$, and each is one dimensional, by Example 3.4.12. Hence, 3.4.14 is possible only if

(3.4.15) $W_1 \cap W_2 \cap W_3 = W_1 \cap W_2 = W_1 \cap W_3 = W_2 \cap W_3.$

Now no two of the $S_i$ are parallel, by assumption; hence, the intersections $S_i \cap S_j$ are nonempty by 3.4.9. Hence, 3.4.11 and 3.4.15 imply that $S_1 \cap S_2$, $S_1 \cap S_3$, $S_2 \cap S_3$ are cosets of the same subspace and thus are parallel. (This solution, though longer than the first solution, has some interest as a further application of Theorems 3.4.8 and 3.4.11.) ▮

**The proofs of Theorems 3.4.5 and 3.4.2**

When we stated Theorems 3.4.2 and 3.4.5, we postponed their proofs. Let us give these proofs now.

*Proof of 3.4.2(a).* Let $W$ be any subspace and let $S = W + \mathbf{u}$ be any coset of $W$. We must show that $S$ satisfies 3.4.1(a) and 3.4.1(b). To prove 3.4.1(a), let $A, B \in S$ with $A \neq B$. Then $A = \mathbf{w}_1 + \mathbf{u}$, $B = \mathbf{w}_2 + \mathbf{u}$ (where $\mathbf{w}_1, \mathbf{w}_2 \in W$). For any scalar $t$, we then have

$$A + t(B - A) = (1 - t)\mathbf{w}_1 + t\mathbf{w}_2 + \mathbf{u} \in W + \mathbf{u}.$$

that is, $L(A, B) \subseteq S$. To prove 3.4.1(b), let $P \in S$ and $L(A, B) \subseteq S$. By Corollary 3.2.8(a), the line through $P$ parallel to $L(A, B)$ is the line

$$L^* = [B - A] + P.$$

Also, since $A, B \in S$, we have $B - A \in W$ (Theorem 3.2.5(d)), and since $P \in S$, we have $S = W + P$. Hence,

$$L^* = [B - A] + P \subseteq W + P = S$$

proving 3.4.1(b). We have thus shown that any coset of a subspace is a flat. (Note that a coset $W + \mathbf{u}$ is automatically nonempty, since it contains $\mathbf{u}$.)

Conversely, let $S$ be a flat. Since $S \neq \varnothing$, we may choose an element $\mathbf{u} \in S$. Let $W = S - \mathbf{u}$; then $S = W + \mathbf{u}$, and we must show that $W$ is a subspace. By Theorem 2.2.4 there are three things to verify:

(a) *$W$ contains* $\mathbf{0}$. This is easy, since $\mathbf{0} = \mathbf{u} - \mathbf{u} \in S - \mathbf{u} = W$.

(b) *$W$ is closed under scalar multiplication.* Let $\mathbf{w} \in W$ and let $t$ be a scalar. If $\mathbf{w} = \mathbf{0}$, then trivially $t\mathbf{w} = \mathbf{0} \in W$. If instead $\mathbf{w} \neq \mathbf{0}$, then the vectors $\mathbf{u}$ and $\mathbf{w} + \mathbf{u}$ are different; since $\mathbf{u} \in S$ and $\mathbf{w} + \mathbf{u} \in W + \mathbf{u} = S$, it follows that $S$ contains the line $L(\mathbf{u}, \mathbf{w} + \mathbf{u})$. In particular, $S$ contains the element

$$(1 - t)\mathbf{u} + t(\mathbf{w} + \mathbf{u})$$

and, hence, $W = S - \mathbf{u}$ contains the vector

$$(1 - t)\mathbf{u} + t(\mathbf{w} + \mathbf{u}) - \mathbf{u} = t\mathbf{w}$$

as desired.

(c) *W is closed under addition.* Let $\mathbf{w}_1, \mathbf{w}_2 \in W$; we must show that $\mathbf{w}_1 + \mathbf{w}_2 \in W$. As before, the vectors $\mathbf{u}$, $\mathbf{w}_i + \mathbf{u}$ belong to $S$. Since $S$ is a flat, it follows, first that $S$ contains the line $L = L(\mathbf{u}, \mathbf{w}_1 + \mathbf{u})$, and then that $S$ contains the line $L^*$ through $\mathbf{w}_2 + \mathbf{u}$ which is parallel to $L$. By Corollary 3.2.8(a), $L^* = [\mathbf{w}_1] + \mathbf{w}_2 + \mathbf{u}$. Hence, $S$ contains the vector $\mathbf{w}_1 + \mathbf{w}_2 + \mathbf{u}$, so that $W = S - \mathbf{u}$ contains $\mathbf{w}_1 + \mathbf{w}_2$ as desired. This completes the proof of Theorem 3.4.2(a). ▮

*Proof of 3.4.2(b).* Assume that $S$ is nonempty and satisfies 3.4.1(a). As above, choose $\mathbf{u} \in S$ and write $W = S - \mathbf{u}$, $S = W + \mathbf{u}$. We wish to show that $S$ is a flat; by 3.4.2(a) (which we have just proved!), it will suffice to show that $W$ is a subspace. The arguments given above, showing that $\mathbf{0} \in W$ and that $W$ is closed under scalar multiplication, are still valid since they used only 3.4.1(a), not 3.4.1(b). The proof of closure under addition used 3.4.1(b) and hence is no longer valid, but a different proof of the same result can be given using the assumption that $1_F + 1_F \neq 0_F$. The argument is as follows: let $\mathbf{w}_1, \mathbf{w}_2 \in W$. Then $\mathbf{w}_1 + \mathbf{u}$, $\mathbf{w}_2 + \mathbf{u}$ lie in $S$, so that by 3.4.1(a), $S$ contains the line $L(\mathbf{w}_1 + \mathbf{u}, \mathbf{w}_2 + \mathbf{u})$. It follows from 3.2.10, with $a = b = \frac{1}{2}$ (division by 2 being allowed since $2_F \neq 0_F$!), that $S$ contains the element

$$\tfrac{1}{2}(\mathbf{w}_1 + \mathbf{u}) + \tfrac{1}{2}(\mathbf{w}_2 + \mathbf{u}).$$

Hence, $W = S - \mathbf{u}$ contains the vector

$$\mathbf{v} = \tfrac{1}{2}(\mathbf{w}_1 + \mathbf{u}) + \tfrac{1}{2}(\mathbf{w}_2 + \mathbf{u}) - \mathbf{u} = \tfrac{1}{2}(\mathbf{w}_1 + \mathbf{w}_2).$$

Since $W$ is closed under scalar multiplication, it follows that $W$ contains the vector $2\mathbf{v} = \mathbf{w}_1 + \mathbf{w}_2$. Hence, $W$ is closed under addition and is a subspace, as desired. ▮

*Proof of 3.4.5.* The formula

$$\begin{aligned} &\dim(W_1 \cap \cdots \cap W_k) \\ &= n - \sum_{i=1}^{k} e_i + \sum_{j=1}^{k-1} [n - \dim((W_1 \cap \cdots \cap W_j) + W_{j+1})] \end{aligned} \tag{3.4.16}$$

may be established by induction on $k$, using Theorem 2.6.17 to show that if 3.4.16 holds for $k$ subspaces then it holds for $k + 1$ subspaces. (We leave the details as an exercise; the argument is straightforward.) It follows at once that 3.4.6 holds, and that 3.4.7(a) holds if and only if

$$(W_1 \cap \cdots \cap W_j) + W_{j+1} = V \qquad (\text{all } j,\ 1 \leq j \leq k-1). \tag{3.4.17}$$

In particular, 3.4.7(a) implies that

$$(W_1 \cap \cdots \cap W_{k-1}) + W_k = V. \tag{3.4.18}$$

If the subspaces $W_i$ are permuted, the truth or falsity of 3.4.7(a) is clearly unaffected; since we have shown that 3.4.7(a) implies 3.4.18, it follows that 3.4.7(a) must also imply 3.4.7(b). Conversely, if 3.4.7(b) holds, then for any $j$ ($1 \leq j < k$), we have

$$\begin{aligned} V &= W_{j+1} + (W_1 \cap \cdots \cap W_j \cap W_{j+2} \cap \cdots \cap W_k) \\ &\subseteq W_{j+1} + (W_1 \cap \cdots \cap W_j) \end{aligned}$$

which implies 3.4.17 and hence 3.4.7(a). ▮

**Fields of characteristic 2**

A field $F$ is said to have *characteristic 2* (*notation*: char $F = 2$) if $1_F + 1_F = 0_F$. (Examples of fields of characteristic 2: (1) the integers modulo 2; (2) the field of four elements given in McCoy [10], Sec. 2.3, Example 7.) Theorem 3.4.2(b) is only one of many geometric theorems in which it is necessary to specify that the field of scalars does not have characteristic 2. To show that 3.4.2(b) is actually false without this specification, let $F$ be the field of integers modulo 2 and consider the vector space $V = F_2$. Since $F$ has just two elements (0 and 1, with $1 + 1 = 0$), $V$ has just four points. Figure 3.4.3 is a diagram of the points and lines in $V$. It is easily seen from the figure that if $S$ is a subset of $V$ consisting of exactly three points, then $S$ satisfies 3.4.1(a) but does not satisfy 3.4.1(b). Hence, 3.4.2(b) does not hold in $V$.

**Exercises**

*1. If $h \leq k$ and if $S_1$, $S_2$ are (respectively) an $h$-flat and a $k$-flat in $V$, prove that the following two statements are equivalent:
(a) $S_1$ is parallel to some $h$-flat contained in $S_2$.
(b) The direction space of $S_1$ is contained in the direction space of $S_2$.

We say that $S_1$ is *parallel* to $S_2$ if (a) holds; equivalently, if (b) holds. Does this definition agree with your geometric intuition? Note that this definition reduces to Definition 3.4.3

Figure 3.4.3

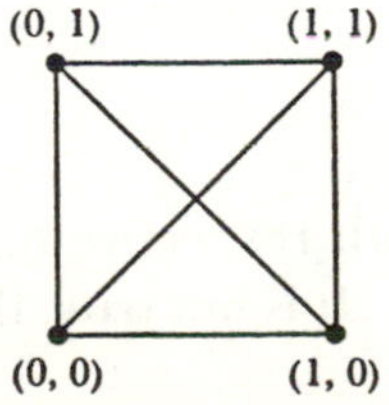

if $h = k$. Note also that statement (a) (or (b)) clearly implies that $h \leqslant k$; thus, according to our definition, a flat can be parallel only to a flat of the same or higher dimension. (If $h < k$ and $S_1$ is parallel to $S_2$, we do *not* say that $S_2$ is parallel to $S_1$.) (SOL)

*2. Suppose that the $h$-flat $S_1$ is parallel to the $k$-flat $S_2$, where $h \leqslant k$. Prove: either $S_1 \subseteq S_2$ or $S_1 \cap S_2 = \varnothing$. Does this agree with your geometric intuition? (Think of the case $h = 1$, $k = 2$.) (SOL)

*3. Let $V$ be a finite-dimensional vector space; let $S$ be any flat in $V$ and let $H$ be a hyperplane in $V$.
   (a) Prove: if $S \cap H$ is empty, then $S$ is parallel to $H$. Draw a figure illustrating this result in the case where $V = \mathbf{R}_3$ and $S$ is a line. (SOL)
   (b) Prove: if $S \cap H$ is nonempty, then either $S \subseteq H$ or dim $(S \cap H) = \dim S - 1$. (SOL)
   (c) Deduce from (a) and (b): if $S$ is not parallel to $H$, then $S \cap H$ is a flat whose dimension equals dim $S - 1$.

*4. Let $S_1$, $S_2$ be flats in $V$, and suppose that every line in $S_1$ is parallel to at least one line in $S_2$. Prove: $S_1$ is parallel to $S_2$. (In particular, this implies that dim $S_1 \leqslant$ dim $S_2$, cf. Exercise 1.) (SOL)

5. Prove that if two distinct lines intersect, then they lie in the same plane. Generalize, if possible.

6. Let $V$ be a finite-dimensional vector space, and let $H_1$, $H_2$, $H_3$ be three hyperplanes in $V$. Prove: if $H_1 \cap H_2 \cap H_3$ is empty but no two of $H_1$, $H_2$, $H_3$ are parallel, then the intersections $H_1 \cap H_2$, $H_1 \cap H_3$, $H_2 \cap H_3$ are parallel $(n - 2)$-flats. (This generalizes Example 3.4.13.)

*7. Using suitable results from Chapter 2, prove: if $h < k \leqslant$ dim $V$, then every $h$-flat in $V$ is contained in some $k$-flat in $V$.

*8. Prove: if the $k$-flat $S$ is not a subspace, then the subspace spanned by $S$ has dimension $k + 1$.

9. Prove the following partial converse of Theorem 3.4.8: if $V$, $W_i$, $n$, $e_i$ are as in Theorem 3.4.5 and if

$$\dim (W_i \cap \cdots \cap W_k) > n - \sum_{i=1}^{k} e_i,$$

then there exist flats $S_i$ $(1 \leqslant i \leqslant k)$ with respective direction spaces $W_i$ such that $S_1 \cap \cdots \cap S_k = \varnothing$. (It is not true, though, that $S_1 \cap \cdots \cap S_k = \varnothing$ for *all* choices of flats having direction spaces $W_i$.)

10. Interpret Exercise 9 in the case where $V = \mathbf{R}_3$, $k = 2$, and $\dim W_1 = \dim W_2 = 1$. Draw a figure to illustrate.
11. Fill in the details of the proof of equation 3.4.16.
12. Let $f(x_1, ..., x_n)$ be a function of $n$ real variables defined by

$$f(x_1, ..., x_n) = \sum_{i=1}^{n} \sum_{j=1}^{n} a_{ij}x_ix_j + \sum_{i=1}^{n} b_ix_i + c$$

where the $a_{ij}$, $b_i$, $c$ are real constants. (Such functions will arise in Chapter 7.) Prove: If $f(x_1, ..., x_n) \geq 0$ for all real numbers $x_1, ..., x_n$, then the set

$S = \{(x_1, ..., x_n) : f(x_1, ..., x_n) = 0\}$

is a flat in $\mathbf{R}_n$ provided that $S \neq \varnothing$.

(*Suggestion*: Using results from Section 1.16, show that a real quadratic polynomial in one variable which has two distinct roots must take on both positive and negative values; then use the latter to show that $S$ satisfies 3.4.1(a).)

13. (a) Let $a_1, ..., a_n$, $c$ be fixed elements of a field $F$ and let $S$ be the set of all solutions $(x_1, ..., x_n) \in F_n$ of the equation

$a_1x_1 + \cdots + a_nx_n = c.$

Prove that if $S \neq \varnothing$, then $S$ is a flat in $F_n$.

(b) State and prove a similar result concerning the set of all solutions of a system of $m$ such equations.

(Compare with Sec. 2.2, Exercises 3 and 4.)

## 3.5 The flat determined by a set of points

Given two distinct points $P$, $Q$ in a vector space $V$, the line $L(P, Q)$ is the smallest flat which contains both $P$ and $Q$. More generally, given any subset $\mathscr{P}$ of $V$, there exists a smallest flat containing $\mathscr{P}$, as the following theorem shows.

**3.5.1 Theorem.** Let $\mathscr{P}$ be any nonempty subset of the vector space $V$. Then there exists a unique flat $\langle \mathscr{P} \rangle$ in $V$, called the *flat determined by* $\mathscr{P}$, such that

(a) $\mathscr{P} \subseteq \langle \mathscr{P} \rangle$;

(b) Whenever $S$ is a flat in $V$ such that $\mathscr{P} \subseteq S$, then $\langle \mathscr{P} \rangle \subseteq S$. (Thus, $\langle \mathscr{P} \rangle$ is the smallest flat containing $\mathscr{P}$.)

*Proof.* Let the flats which contain $\mathscr{P}$ be denoted $S_i$, where $i$ runs over a (possibly infinite) set of indices. (Note that there is at least one $S_i$, namely, $V$ itself.) Now define $\langle \mathscr{P} \rangle$ to be the intersection of *all* of the flats $S_i$; thus, a point belongs to $\langle \mathscr{P} \rangle$ if and only if it belongs to every $S_i$. Since $\mathscr{P} \subseteq S_i$ for every $i$, $\mathscr{P} \subseteq \langle \mathscr{P} \rangle$; and whenever $S$ is a flat in $V$ such that $\mathscr{P} \subseteq S$, then $S$ is one of the $S_i$ and, hence, $\langle \mathscr{P} \rangle \subseteq S$. Thus, 3.5.1(a) and 3.5.1(b) hold. In particular, 3.5.1(a) implies that $\langle \mathscr{P} \rangle \neq \varnothing$; since $\langle \mathscr{P} \rangle$ is an intersection of flats, it follows that $\langle \mathscr{P} \rangle$ is a flat (cf. the "Remark" following the proof of Theorem 3.4.11). Finally, if $\langle \mathscr{P} \rangle^*$ were a flat having the same properties as $\langle \mathscr{P} \rangle$, then by 3.5.1(b) we have $\langle \mathscr{P} \rangle \subseteq \langle \mathscr{P} \rangle^*$ and $\langle \mathscr{P} \rangle^* \subseteq \langle \mathscr{P} \rangle$, hence, $\langle \mathscr{P} \rangle^* = \langle \mathscr{P} \rangle$; thus, the flat determined by $\mathscr{P}$ is unique. ∎

If $\mathscr{P}$ is a set consisting of only finitely many points $P_0, P_1, \ldots, P_k$, then $\langle \mathscr{P} \rangle$ may be written $\langle P_0, P_1, \ldots, P_k \rangle$ and may be called the *flat determined by* $P_0, P_1, \ldots, P_k$. The flat determined by a single point is just the point itself (more rigorously, it is the set consisting of the point). If $P$ and $Q$ are distinct points, then the flat $\langle P, Q \rangle$ determined by $P$ and $Q$ is the line $L(P, Q)$ (to prove this formally, use Definition 3.4.1 and Theorem 3.2.2(b)).

The (geometric) concept of the flat determined by a set of points is somewhat analogous to the (algebraic) concept of the subspace spanned by a set of vectors. In each case we are concerned with the smallest set of a certain type (flat, subspace) which contains given elements. In fact, the points which determine a flat are closely related to the vectors which span the direction space of that flat, as the following theorem shows.

**3.5.2** **Theorem.** If $S = \langle P_0, P_1, \ldots, P_k \rangle$ is the flat determined by the points $P_0, \ldots, P_k$, then the direction space of $S$ is the subspace $[P_1 - P_0, \ldots, P_k - P_0]$ spanned by the $k$ vectors $\mathbf{v}_i = \mathbf{Ar}\, P_0P_i = P_i - P_0$ $(1 \leqslant i \leqslant k)$. In particular, $\dim S \leqslant k$.

*Proof.* Let $W$ be the direction space of $S$, and let $W^* = [P_1 - P_0, \ldots, P_k - P_0]$. Since $P_0 \in S$, $S = W + P_0$. Since the $P_i$ all lie in $S$, the vectors $P_i - P_0$ all lie in $S - P_0 = W$; hence $W^* \subseteq W$ by Theorem 2.3.5(c). Conversely, since the vectors $\mathbf{0}, P_1 - P_0, \ldots, P_k - P_0$ all lie in $W^*$, the points $P_0, P_1, \ldots, P_k$ all lie in $W^* + P_0$. Hence, $S \subseteq W^* + P_0$ (Theorem 3.5.1(b)), so that $W = S - P_0 \subseteq W^*$. It follows that $W = W^* = [P_1 - P_0, \ldots, P_k - P_0]$ as desired. (The statement about the dimension follows at once, using either 2.6.12(a) or 2.5.1.) ∎

*Remark*: By comparing Theorem 3.5.2 with equation 3.2.7′, we obtain a second proof that $\langle P, Q \rangle = L(P, Q)$ whenever $P, Q$ are distinct points.

By Theorem 3.5.2, $k + 1$ points determine a flat of dimension at most $k$ (e.g., two points determine a line; three points determine at most a plane). The following theorem tells us under what circumstances the dimension will be *exactly* $k$.

**3.5.3 Theorem.** If $S = \langle P_0, ..., P_k \rangle$, then the following statements are equivalent:

(a) $\dim S = k$.

(b) The $k$ vectors $P_1 - P_0, ..., P_k - P_0$ are linearly independent.

(c) The points $P_0, ..., P_k$ are not all contained in the same flat of dimension less than $k$.

(d) No one of the points $P_i$ $(0 \leqslant i \leqslant k)$ lies in the flat determined by the other $k$ points.

Moreover, if $\dim V \geqslant k - 1$, then statements (a) through (d) are also equivalent to the following additional statement:

(e) The points $P_0, ..., P_k$ are not all contained in the same $(k - 1)$-flat.

(*Note*: In order for 3.5.3(d) to be meaningful in the case $k = 0$, we must argue as follows: since the flat determined by the empty set is undefined, it is true that "$P_0$ does not lie in the flat determined by the other 0 points" since there is no such flat! In other words, 3.5.3(d) may be regarded as vacuously true when $k = 0$. It is easily seen that the other four statements are also true when $k = 0$; hence, in the proof of the theorem we may assume $k \geqslant 1$.)

*Proof when $k \geqslant 1$.* Exercise 7 of the preceding section implies that (c) and (e) are equivalent when $\dim V \geqslant k - 1$. Hence, it will suffice to prove that statements (a) through (d) are equivalent. We shall do this by showing that (b) $\Rightarrow$ (a) $\Rightarrow$ (c) $\Rightarrow$ (d) $\Rightarrow$ (b).

Since $S$ has direction space $[P_1 - P_0, ..., P_k - P_0]$, it is clear that (b) $\Rightarrow$ (a). Since $S$ is the *smallest* flat containing the $P_i$, (a) $\Rightarrow$ (c). Since the flat determined by only $k$ points has dimension less then $k$, (c) $\Rightarrow$ (d). Finally, assume (d) and suppose (b) is false; that is, suppose the vectors $P_1 - P_0, ..., P_k - P_0$ are dependent. Then by 2.4.6, one of these vectors lies in the subspace spanned by the others. To simplify the notation, we may assume that

$$P_k - P_0 \in [P_1 - P_0, ..., P_{k-1} - P_0].$$

Adding $P_0$ to both sides,

$$P_k \in [P_1 - P_0, ..., P_{k-1} - P_0] + P_0.$$

But Theorems 3.5.2 and 3.4.4(a) imply that

$$\langle P_0, ..., P_{k-1} \rangle = [P_1 - P_0, ..., P_{k-1} - P_0] + P_0.$$

Hence, $P_k \in \langle P_0, ..., P_{k-1} \rangle$, contradicting (d). It follows that (d) $\Rightarrow$ (b). ∎

The cases $k = 1$, $k = 2$ of Theorem 3.5.3 reduce to familiar statements of Euclidean geometry. Thus, for $k = 2$, the theorem says that three points $P_0$, $P_1$, $P_2$ determine a plane (3.5.3(a)) if and only if they are not collinear (3.5.3(e)); equivalently, if and only if the vectors $\mathbf{Ar}\ P_0P_1$, $\mathbf{Ar}\ P_0P_2$ are linearly independent (3.5.3(b)). (Note that this gives us a solution of Sec. 3.2, Exercise 3.) Similarly, when $k = 1$, the theorem says that two points determine a line (3.5.3(a)) if and only if they are distinct (3.5.3(d) or 3.5.3(e)).

**3.5.4 Definition.** The ordered $(k + 1)$-tuple $(P_0, ..., P_k)$ is called *barycentrically independent* (abbreviated "*b*-independent") if any one of the statements 3.5.3(a) through 3.5.3(d) holds (if one of them holds, they all do). Of course, "barycentrically dependent" (*b*-dependent) means "not *b*-independent". Thus, the points $P_0, ..., P_k$ are barycentrically dependent if some one of them belongs to the flat determined by the others.

(Note how this parallels the characterization of "linearly dependent" (Theorem 2.4.6): the vectors $\mathbf{u}_1, ..., \mathbf{u}_n$ are *linearly* dependent if one of them belongs to the *subspace spanned* by the others.)

Since statements (a), (c), (e) of 3.5.3 obviously depend only on the *set* $\{P_0, ..., P_k\}$ and not on the order in which the $P_i$ are written, the same must be true of statement (b). That is, the linear independence or dependence of the $k$ vectors $\mathbf{Ar}\ P_0P_1, ..., \mathbf{Ar}\ P_0P_k$ is unaffected if the points $P_0, P_1, ..., P_k$ are permuted. Taking $k = 3$, for example, we see that if any one of the four ordered triples

$$(\mathbf{Ar}\ P_0P_1, \mathbf{Ar}\ P_0P_2, \mathbf{Ar}\ P_0P_3); (\mathbf{Ar}\ P_1P_0, \mathbf{Ar}\ P_1P_2, \mathbf{Ar}\ P_1P_3);$$
$$(\mathbf{Ar}\ P_3P_0, \mathbf{Ar}\ P_3P_1, \mathbf{Ar}\ P_3P_2); (\mathbf{Ar}\ P_2P_0, \mathbf{Ar}\ P_2P_1, \mathbf{Ar}\ P_2P_3)$$

is linearly independent, then each of the other three triples is linearly independent also.

By Definition 3.5.4, any $k + 1$ *b*-independent points determine a $k$-flat. Conversely, we have

3.5.5 **Theorem.** If $S$ is any $k$-flat, then there exist $k + 1$ $b$-independent points $P_0, ..., P_k$ such that $S = \langle P_0, ..., P_k \rangle$.

*Proof.* Choose $P_0 \in S$; then $S = W + P_0$ where $W$ is the direction space of $S$. Since dim $W = k$, $W$ has a basis $\{\mathbf{u}_1, ..., \mathbf{u}_k\}$. Let $P_i = \mathbf{u}_i + P_0$ $(i = 1, ..., k)$. By 3.5.2 the flat $\langle P_0, ..., P_k \rangle$ has direction space

$$[P_1 - P_0, ..., P_k - P_0] = [\mathbf{u}_1, ..., \mathbf{u}_k] = W$$

and, hence, $\langle P_0, ..., P_k \rangle = W + P_0 = S$. The $P_i$ are $b$-independent since 3.5.3(a) holds. ∎

3.5.6 **Theorem.** The points $P_0, ..., P_k$ are $b$-dependent if and only if there exist scalars $a_0, ..., a_k$, *not all zero*, such that

(a) $a_0P_0 + \cdots + a_kP_k = \mathbf{0}$, and

(b) $a_0 + \cdots + a_k = 0$.

We leave the proof of Theorem 3.5.6 as an exercise (use 3.5.3(b) as the condition for $b$-independence). Note that if 3.5.6(b) were deleted, we would have the condition for *linear* dependence; thus, $b$-dependence implies linear dependence but not conversely.

We include the following example as an application of Theorem 3.5.6.

3.5.7 **Example.** (*Fano's Axiom*) Given four distinct points $A$, $B$, $C$, $D$ in a plane (and assuming the field of scalars does not have characteristic 2), prove that the line through some two of these points intersects the line through the other two points. (*Note*: Figure 3.4.3 shows that the assertion is false without the exclusion of fields of characteristic 2.)

*Solution.* For $\langle A, B\rangle$ to meet $\langle C, D\rangle$ (say), it suffices that some point on $\langle A, B\rangle$ coincide with some point on $\langle C, D\rangle$. That is, it suffices that

$$\alpha A + \beta B = \gamma C + \delta D; \qquad \alpha + \beta = \gamma + \delta = 1 \tag{3.5.8}$$

for some scalars $\alpha, \beta, \gamma, \delta$. Transposing terms, we see that 3.5.8 is the same as

$$\begin{aligned} &\alpha A + \beta B - \gamma C - \delta D = \mathbf{0}; \qquad \alpha + \beta - \gamma - \delta = 0; \\ &\alpha + \beta = 1. \end{aligned} \tag{3.5.9}$$

The form of the latter equations suggests use of Theorem 3.5.6.

Certainly $A, B, C, D$ are $b$-dependent (four points in a flat of dimension less than 3); hence, by 3.5.6, there exist scalars $a, b, c, d$ such that

$$aA + bB + cC + dD = \mathbf{0}; \qquad a + b + c + d = 0;$$
$a, b, c, d$ not all zero.

If $a + b$ is not zero (say, $a + b = k$), then 3.5.9 is obtained by letting $\alpha = a/k$, $\beta = b/k$, $\gamma = -c/k$, $\delta = -d/k$, and we conclude that $\langle A, B\rangle$ meets $\langle C, D\rangle$. Likewise, if $a + c \neq 0$, then a similar argument shows that $\langle A, C\rangle$ meets $\langle B, D\rangle$; and so on. In other words, it suffices to show that *the sum of some two of the scalars a, b, c, d is nonzero*. We leave this last detail to you (Exercise 7 below); note that this is where we must use our assumption that the field does not have characteristic 2. ∎

The construction of proofs using Theorem 3.5.6 sometimes requires quite a bit of ingenuity and/or persistence. If you are ambitious, you might try Exercise 10 below.

**Exercises**

Many of the theorems about linear dependence carry over to barycentric dependence with little change (and often with the same proofs). Exercises 1–5 below are examples of this.

1. Prove that any subset of a barycentrically independent set is barycentrically independent.
*2. Prove: if $P_0, \ldots, P_k$ are barycentrically independent but $P_0, \ldots, P_k, Q$ are barycentrically dependent, then $Q \in \langle P_0, \ldots, P_k\rangle$. (Here the proof is *not* the same as before.) (SUG)
3. Prove: if $S$ is the flat determined by a set of $m$ points, then any set of more than $m$ points in $S$ is barycentrically dependent.
4. Prove: if $S = \langle P_0, \ldots, P_k\rangle$, then any maximal $b$-independent subset of $\{P_0, \ldots, P_k\}$ also determines $S$. (This result is the analog of 2.6.9.)
*5. Prove: if $A$ is any set of points such that the flat $\langle A\rangle$ is finite-dimensional, then any $b$-independent subset of $A$ can be extended to a $b$-independent set that determines $\langle A\rangle$. (Compare with 2.6.14(a).) (SUG)
6. Prove Theorem 3.5.6.
7. Let $F$ be a field which does not have characteristic 2, and let $a_1, \ldots, a_k$ ($k \geqslant 3$) be elements of $F$. Prove: if the $a_i$ are not all zero, then the sum of some two of them is nonzero. (This

result was used in the solution to Example 3.5.7.) Why is it necessary to assume that $F$ does not have characteristic 2?

8. Find the dimension of each of the following flats in $\mathbf{R}_5$:
   (a) $\langle$ (0, −1, 2, 4, 1), (1, 0, 3, 5, 2), (2, −1, 1, 6, 2), (1, 2, 6, 5, 3) $\rangle$
   (b) $\langle$ (1, 0, −1, 2, 0), (3, −1, 2, 0, 1), (0, 0, 0, 3, −2), (2, −1, 3, 1, −1), (1, 0, 5, 2, 0) $\rangle$
   (c) $\langle$ (1, 0, 0, 1, 0), (2, 1, 1, 2, 1), (3, 2, 2, 3, 2), (4, 3, 3, 4, 3) $\rangle$. (ANS)
9. In each of the following, $S$ and $T$ are flats in $\mathbf{R}_4$. Determine whether $S$ is parallel to $T$.
   (a) $S = \langle$ (1, 0, 1, 0), (3, −1, 0, 1), (1, 3, −1, 2) $\rangle$
   $T = \langle$ (0, 5, 2, 1), (−2, 6, 3, 0), (−2, 3, 5, −2) $\rangle$.
   (b) $S = \langle$ (1, 0, 1, 2), (4, 3, 0, 0) $\rangle$
   $T = \langle$ (0, 0, 0, 0), (1, 4, 0, 2), (−2, 1, 1, 4) $\rangle$. (ANS)
10. Let $A, B, C, D$ be four points in a plane; assume that no three of these points are collinear and that no two of the lines which they determine are parallel. Let $E, F, G$ be (respectively) the points of intersection of $\langle A, B \rangle$ and $\langle C, D \rangle$, $\langle B, C \rangle$ and $\langle A, D \rangle$, $\langle A, C \rangle$ and $\langle B, D \rangle$ (see Figure 3.5.1). Assuming the field of scalars does not have characteristic 2, prove that the points $E, F, G$ are not collinear. (SOL)

(*Remark*: If we pretend that two parallel lines in a given plane meet at a "point at infinity", and that the set of all such "points at infinity" is a "line at infinity", then Example 3.5.7 (Fano's Axiom) actually becomes a special case of Exercise 10; indeed, the statement that *E, F, G do not all lie on the line at infinity* is equivalent to 3.5.7. For a rigorous treatment of this sort of thing, see Secs. 7.6 and 7.7; in particular, see Exercise 6, Sec. 7.7.)

Figure 3.5.1

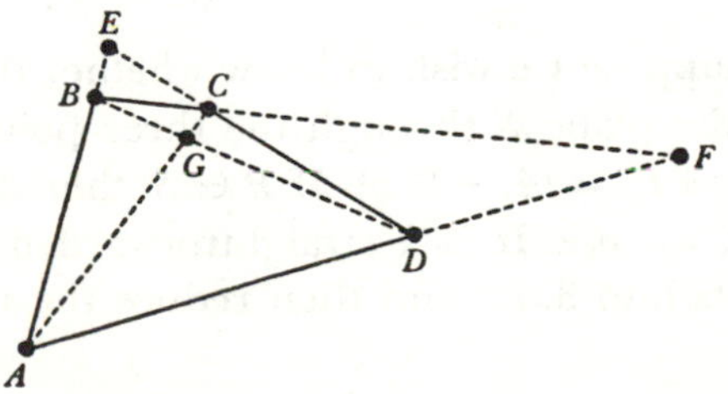

## 3.6 Barycentric combinations and coordinates

We showed in Section 3.2 that the general point on the line $L(A, B)$ can be uniquely expressed in the form $aA + bB$, where $a + b = 1$. We now generalize this result to arbitrary flats.

**3.6.1 Theorem.** Let $P_0, \ldots, P_k$ be points in a vector space $V$, and let $S = \langle P_0, \ldots, P_k \rangle$. Then a point $P \in V$ lies in $S$ if and only if there exist scalars $a_0, \ldots, a_k$ such that

(3.6.2) $$P = a_0P_0 + \cdots + a_kP_k; \qquad a_0 + \cdots + a_k = 1.$$

Moreover, if $P_0, \ldots, P_k$ are $b$-independent, then the scalars $a_i$ in 3.6.2 are uniquely determined by $P$.

*Proof.* As we saw in the preceding section, $S = [P_1 - P_0, \ldots, P_k - P_0] + P_0$; hence, the general point $P$ in $S$ has the form

(3.6.3) $$P = P_0 + b_1 (P_1 - P_0) + \cdots + b_k (P_k - P_0).$$

If we let

$$a_1 = b_1, \ldots, a_k = b_k, \qquad a_0 = 1 - \sum_{i=1}^{k} b_i$$

then $a_0 + \cdots + a_k = 1$ and 3.6.3 coincides with 3.6.2. (Conversely, starting from 3.6.2 we obtain 3.6.3 by letting $b_1 = a_1, \ldots, b_k = a_k$.) Moreover, if the $P_i$ are $b$-independent, then the vectors $P_1 - P_0$, $\ldots, P_k - P_0$ are linearly independent, so that the $b_i$ are uniquely determined by $P$ (Theorem 2.4.7) and hence so are the $a_i$. ∎

**3.6.4 Definition.** A point $P$ which satisfies 3.6.2 is called a *barycentric combination* (abbreviated *b-combination*) of $P_0, \ldots, P_k$. If $P_0, \ldots, P_k$ are $b$-independent, the (unique) scalars $a_i$ in 3.6.2 are called the *barycentric coordinates of $P$ with respect to the points $P_0, \ldots, P_k$*. Note that in a $b$-combination the scalars must add up to 1 (unlike a linear combination, in which the scalars are unrestricted).

As an example in $\mathbf{R}_3$, suppose we wish to know whether the point $P = (-5, 5, -6)$ lies in the plane $S$ through the three points $P_0 = (0, 1, 2)$, $P_1 = (3, 1, 0)$, and $P_2 = (4, -1, 5)$. If $P \in S$, then 3.6.2 has at least one solution $(a_0, a_1, a_2)$. It is a straightforward matter to substitute the given points into 3.6.2 and then reduce the result to the system

$$\begin{aligned} 3a_1 + 4a_2 &= -5 \\ a_0 + a_1 - a_2 &= 5 \\ 2a_0 + 5a_2 &= -6 \\ a_0 + a_1 + a_2 &= 1. \end{aligned}$$

These four equations in three unknowns do have the solution $a_0 = 2$, $a_1 = 1$, $a_2 = -2$; hence, $P \in S$. (In the case of a flat in $\mathbf{R}_n$ determined by $k + 1$ points $P_i$, there would be $n + 1$ equations in $k + 1$ unknowns.) Note that in this problem there is no need to determine whether the solution $(a_0, a_1, a_2)$ is unique.

Observe how Theorem 3.6.1 parallels Theorem 2.3.5: the smallest (flat/subspace) containing given elements is the set of all (linear/barycentric) combinations of these elements. In general, statements involving the concepts of

flat; barycentric combination; barycentric independence

often parallel those involving the concepts of

subspace; linear combination; linear independence.

To illustrate this remark further, compare Theorem 2.4.10 with Section 3.5, Exercise 2. (Also see Definition 3.5.4 and the accompanying comments.)

We shall later have occasion to refer to the following special case of Theorem 3.6.1 (which partially parallels Theorem 2.6.4!):

**3.6.5 Theorem.** If $P_0, \ldots, P_n$ are $n + 1$ $b$-independent points in a vector space $V$ of dimension $n$, then every point $P \in V$ is uniquely expressible as a barycentric combination of $P_0, \ldots, P_n$.

*Proof.* By 3.6.1 this is certainly true of all points $P$ which lie in the flat $S = \langle P_0, \ldots, P_n \rangle$. But $S$ has dimension $n$ (Theorem 3.5.3) and, hence, is all of $V$. ∎

### The signs of the barycentric coordinates

In Section 3.2 we saw that if $A$ and $B$ are distinct points in $\mathbf{R}_n$ and if $P = aA + bB$ is a point on the line $L(A, B)$ $(a + b = 1)$, then the signs of $a$ and $b$ determine in which part of the line $P$ lies (see Fig. 3.2.4). In the case of higher-dimensional flats the signs of the barycentric coordinates have similar significance. We shall now discuss this significance in detail in the case of a plane determined by three points. Throughout this discussion we assume that we are working in the vector space $\mathbf{R}_n$ (real $n$-space), since in more general fields the concepts of "positive" and "negative" are not always definable.

Let $S$ be a plane in $\mathbf{R}_n$; then by Theorem 3.5.5 there are three $b$-independent points $P_0$, $P_1$, $P_2$ which determine $S$. The points $P_i$ determine the three lines

$$L_0 = \langle P_1, P_2\rangle; \qquad L_1 = \langle P_0, P_2\rangle; \qquad L_2 = \langle P_0, P_1\rangle$$

as shown in Figure 3.6.1. If

$$\textbf{(3.6.6)} \quad P = a_0P_0 + a_1P_1 + a_2P_2 \qquad (a_0 + a_1 + a_2 = 1)$$

is any point in the plane $S$, then a necessary and sufficient condition for $P$ to lie on one of the three lines $L_i$ is that the corresponding barycentric coordinate $a_i$ is zero. For example, if $a_2 = 0$, then $P = a_0P_0 + a_1P_1$ with $a_0 + a_1 = 1$; hence, $P$ belongs to $\langle P_0,P_1\rangle = L_2$. Conversely, if $P \in L_2$, then $P = a_0'P_0 + a_1'P_1 (a_0' + a_1' = 1)$, so that by the uniqueness of the $a_i$ we must have $a_0 = a_0'$, $a_1 = a_1'$, $a_2 = 0$.

If $a_2 \neq 0$ (i.e., if $P \notin L_2$), then $a_2$ is either positive or negative. We claim that *all points $P$ such that $a_2 > 0$ lie on one side of $L_2$, and all points $P$ such that $a_2 < 0$ lie on the other side of $L_2$.* To say the same thing a bit more formally: let $P = a_0P_0 + a_1P_1 + a_2P_2$ and $Q = b_0P_0 + b_1P_1 + b_2P_2$, where $a_0 + a_1 + a_2 = b_0 + b_1 + b_2 = 1$. What we actually assert is that (1) if $a_2$, $b_2$ have the same sign, then the segment $\overline{PQ}$ does not intersect $L_2$, but that (2) if $a_2$, $b_2$ have opposite signs, then $\overline{PQ}$ intersects

Figure 3.6.1

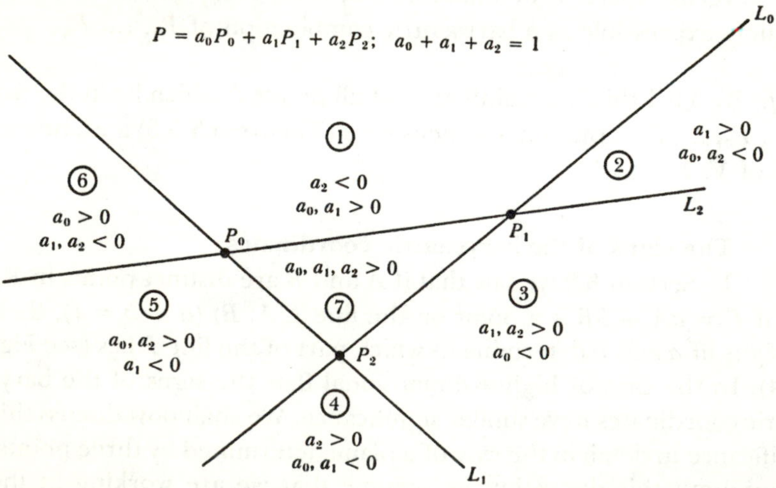

$L_2$. These two assertions can be proved formally (see Exercises 1–4, this section); informally, it is clear that they imply our assertion (above, in italics) about "sides" of $L_2$. Moreover, it is easy to tell which side of $L_2$ is which: since

$$P_2 = 0P_0 + 0P_1 + 1P_2; \qquad 0 + 0 + 1 = 1$$

it follows that the side of $L_2$ on which $a_2 > 0$ must be the side containing the point $P_2$. Thus, in Figure 3.6.1, we have $a_2 > 0$ for all points in regions 3, 4, 5, 7, while $a_2 < 0$ for all points in regions 1, 2, 6. By applying a similar argument to the lines $L_0$ and $L_1$, we easily determine the precise combination of signs corresponding to each of the seven regions in Figure 3.6.1. For example, the point $2P_0 - 3P_1 + 2P_2$ lies in region 5. In general, if $P$ is any point satisfying 3.6.6, the coordinate $a_i$ will be positive if $P$ lies on the same side of $L_i$ as $P_i$, negative if $P$ and $P_i$ lie on opposite sides of $L_i$. (Note that since each of $a_0$, $a_1$, $a_2$ can be positive or negative, there are *eight* possible combinations of signs; however, only seven of these combinations can actually occur on the plane $S$, since the condition $a_0 + a_1 + a_2 = 1$ implies that not all $a_i$ are negative. This accounts for there being only seven regions in Fig. 3.6.1.)

**Comments on the preceding discussion**

1. The fact that points corresponding to $a_2 > 0$ and $a_2 < 0$ lie on opposite sides of $L_2$ can also be deduced from the Intermediate Value Theorem for continuous functions. (Look up this theorem in any calculus book.) The argument, roughly, is that $a_2$ cannot move continuously from a negative value at $P$ to a positive value at $Q$ without passing through the "intermediate" value 0; and as soon as $a_2 = 0$ we have a point on $L_2$, showing that the segment $\overline{PQ}$ meets $L_2$. It is not difficult to make this argument rigorous using elementary facts about continuity.

2. The region in which the $a_i$ are all positive (region 7) is precisely the "inside" of the triangle $P_0P_1P_2$. In this connection, see Exercise 6 below.

3. This whole discussion can be generalized to higher dimensions. In the discussion of the general $k$-flat $\langle P_0, \ldots, P_k \rangle$, the $(k - 1)$-flats determined by $k$ of the points $P_i$ play the same role as was played by the lines $L_i$ in the case $k = 2$. You should try to work out the details yourself (see Exercise 7 below).

### Miscellaneous applications of barycentric coordinates

**3.6.7 Example.** Let $A$, $B$, $C$ be noncollinear points in $\mathbf{R}_n$. If the numbers in Figure 3.6.2 represent distances (for example, $d(B, C) = 8$), express $P$ as a barycentric combination of $A$, $B$, $C$.

*Solution.* The results of Section 3.2 imply that

$$D = \frac{1}{3}A + \frac{2}{3}B$$
$$E = \frac{7}{8}B + \frac{1}{8}C$$
$$F = \frac{1}{2}B + \frac{1}{2}C$$
$$G = \frac{1}{3}A + \frac{2}{3}C.$$

Since $P$ lies on both $\langle D, E \rangle$ and $\langle F, G \rangle$, we have

$$\begin{aligned} P &= tD + (1-t)E \\ &= \frac{1}{3}tA + \left(\frac{2}{3}t + \frac{7}{8}(1-t)\right)B + \frac{1}{8}(1-t)\,C \\ P &= rF + (1-r)G \\ &= \frac{1}{3}(1-r)A + \frac{1}{2}rB + \left(\frac{1}{2}r + \frac{2}{3}(1-r)\right)C. \end{aligned} \tag{3.6.8}$$

Each of the two equations 3.6.8 expresses $P$ as a barycentric combination of $A$, $B$, $C$. (Recall from the end of Section 3.2 that substituting a $b$-combination into a $b$-combination yields a $b$-combination!) Since $A$, $B$, $C$ are $b$-independent, it follows that we may equate coefficients.

Figure 3.6.2

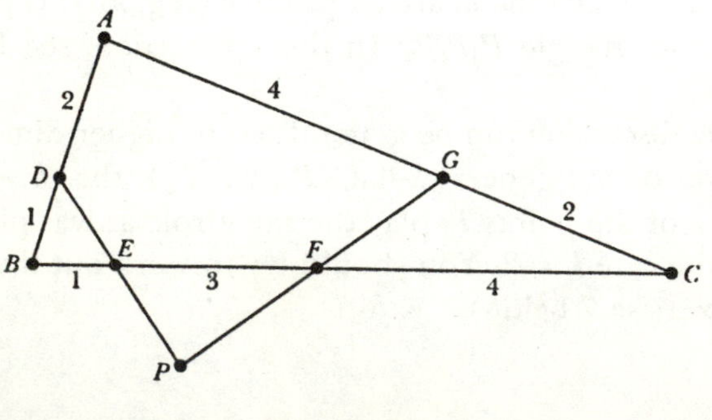

Equating coefficients of $A$, we get $t = 1 - r$; equating coefficients of $C$, we get

$$\frac{1}{8}r = \frac{1}{8}(1-t) = \frac{1}{2}r + \frac{2}{3}(1-r)$$

from which $r = 16/7$, $t = 1 - r = -9/7$. Hence, 3.6.8 reduces to

$$P = -\frac{3}{7}A + \frac{8}{7}B + \frac{2}{7}C$$

which is the desired expression.

**3.6.9** **Example.** Let $A_1, B_1, C_1, D_1$ be distinct points on a line $L_1$ in $\mathbf{R}_n$; let $P \in \mathbf{R}_n$ be a point not on $L_1$. Assume that the line $L_2$ (not passing through $P$) intersects the lines $\langle P, A_1\rangle, ..., \langle P, D_1\rangle$ at points $A_2, ..., D_2$, respectively (see Fig. 3.6.3). Prove that

$$\frac{|\overline{A_1B_1}|/|\overline{B_1D_1}|}{|\overline{A_1C_1}|/|\overline{C_1D_1}|} = \frac{|\overline{A_2B_2}|/|\overline{B_2D_2}|}{|\overline{A_2C_2}|/|\overline{C_2D_2}|}. \tag{3.6.10}$$

(The property to be proved is called the *invariance of the cross-ratio*; stated more simply, it asserts that if the numbers $\alpha$, $\beta$, $\gamma$ in Fig. 3.6.4

Figure 3.6.3

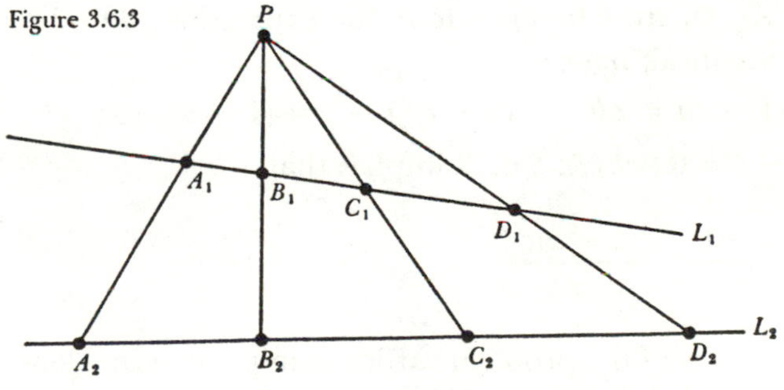


Figure 3.6.4

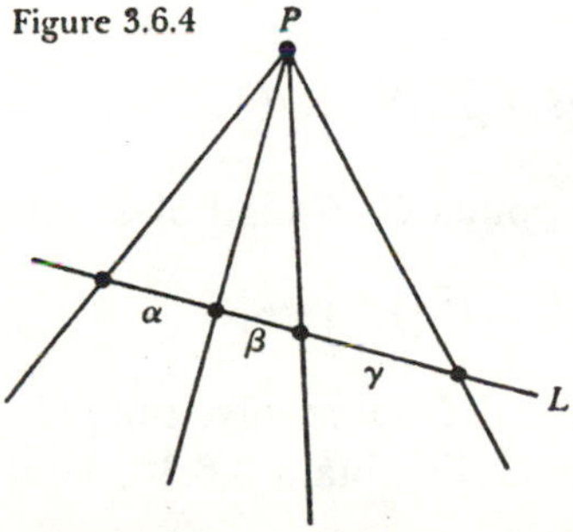

represent distances, then the ratio

$$\left(\frac{\alpha}{\beta+\gamma}\right)\Big/\left(\frac{\alpha+\beta}{\gamma}\right)$$

depends only on the four lines through $P$, not on the line $L$ which crosses them.)

*Solution.* The method will be to express $B_1$ and $C_1$ as barycentric combinations of $P$, $A_2$, and $D_2$ in two different ways, and equate coefficients.

Since $B_1 \in \langle P, B_2\rangle$, we can write $B_1 = rP + (1 - r)B_2$; since $B_2 \in \langle A_2, D_2\rangle$, we have $B_2 = tA_2 + (1 - t)D_2$. By substitution, this gives

**(3.6.11)** $B_1 = rP + (1 - r)tA_2 + (1 - r)(1 - t)D_2,$

which expresses $B_1$ as a barycentric combination of $P, A_2, D_2$. Similarly, we have

$$B_1 = aA_1 + (1 - a)D_1; \qquad A_1 = bA_2 + (1 - b)P;$$
$$D_1 = cD_2 + (1 - c)P;$$

substitution then gives

**(3.6.12)** $B_1 = [a(1 - b) + (1 - a)(1 - c)]P + abA_2 + (1 - a)cD_2.$

Since $P, A_2, D_2$ are $b$-independent, the expressions 3.6.11 and 3.6.12 must be identical; hence,

**(3.6.13)** $(1 - r)t = ab; \qquad (1 - r)(1 - t) = (1 - a)c.$

Since $1 - r \neq 0$ (*why*?), 3.6.13 implies that

**(3.6.14)** $$\frac{1-t}{t} = \frac{(1-a)c}{ab}$$

(the hypotheses of the problem suffice to imply that the denominators are nonzero). Using Section 3.2, Exercise 8, we see that 3.6.14 is the same as

**(3.6.15)** $|\overline{A_2B_2}|/|\overline{B_2D_2}| = (|\overline{A_1B_1}|/|\overline{B_1D_1}|) \cdot |c/b|.$

Similar reasoning applied to the points $C_i$ instead of $B_i$ gives

**(3.6.16)** $|\overline{A_2C_2}|/|\overline{C_2D_2}| = (|\overline{A_1C_1}|/|\overline{C_1D_1}|) \cdot |c/b|$

(note that the definitions of $b$ and $c$ did not involve the points $B_i$, $C_i$). Comparing 3.6.15 with 3.6.16, we easily obtain 3.6.10. ∎

**3.6.17** **Example.** Let hypotheses and notations be as in Example 3.6.9. Assume further that $B_1$ is between $A_1$ and $C_1$, $C_1$ is between $B_1$ and $D_1$, and $C_2$ is between $B_2$ and $D_2$ (see Fig. 3.6.3). Prove that $A_2$ cannot lie between $B_2$ and $D_2$. (*Note*: $A_2$ can lie on *either* side of the segment $\overline{B_2D_2}$; Figs. 3.6.3 and 3.6.5 show the two possibilities.)

*Solution.* Suppose $A_2$ is between $B_2$ and $C_2$, and let $\alpha_i, \beta_i, \gamma_i$ $(i = 1, 2)$ be the distances indicated in Figure 3.6.6. Equation 3.6.10 then becomes

$$\frac{\alpha_1\gamma_1}{(\alpha_1 + \beta_1)(\beta_1 + \gamma_1)} = \frac{\alpha_2\gamma_2}{(\alpha_2 + \beta_2 + \gamma_2)\beta_2}.$$

Equation 3.6.10 remains true if the roles of $A_i$, $B_i$ are reversed (why?); hence

$$\frac{\alpha_1\gamma_1}{(\alpha_1 + \beta_1 + \gamma_1)\beta_1} = \frac{\alpha_2\gamma_2}{(\alpha_2 + \beta_2)(\beta_2 + \gamma_2)}.$$

It follows that

$$\text{(3.6.18)} \quad \frac{(\beta_1 + \alpha_1)(\beta_1 + \gamma_1)}{(\alpha_1 + \beta_1 + \gamma_1)\beta_1} = \frac{(\alpha_2 + \beta_2 + \gamma_2)\beta_2}{(\alpha_2 + \beta_2)(\beta_2 + \gamma_2)}.$$

But the left side of 3.6.18 is $> 1$ while the right side is $< 1$, R.A.A. Similarly, if $A_2$ were between $C_2$ and $D_2$, then the left and right sides

Figure 3.6.5

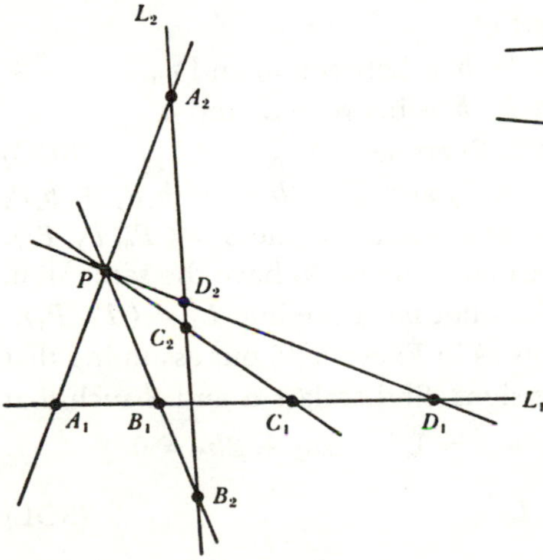


Figure 3.6.6

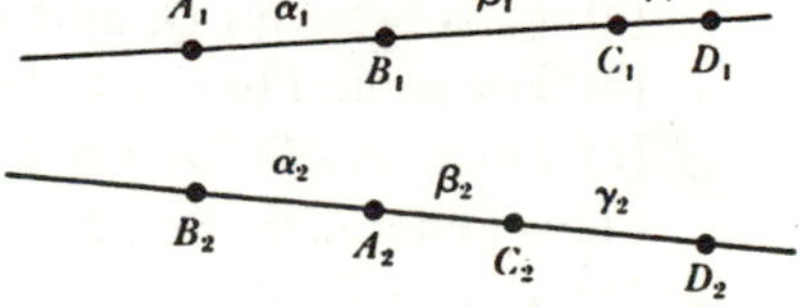

of 3.6.10 would be $< 1$ and $> 1$, respectively, R.A.A. Hence, $A_2$ lies neither on $\overline{B_2C_2}$ nor on $\overline{C_2D_2}$, so that $A_2 \notin \overline{B_2D_2}$. ∎

(*Remark*: In the above solution we have tacitly used certain properties of "betweenness" which are established in some of the exercises at the end of Section 3.2. Which exercises?)

As the material of this section suggests, barycentric coordinates can be usefully applied to many types of geometric problems. Barycentric coordinates also have applications in physics; for a nice (partially informal) treatment of the *center of mass* along such lines, the interested reader is referred to Hausner [7, Chapter 1]. For further geometric examples involving barycentric coordinates, see [7, Sections 3.6 through 3.8]; also see the Exercises below.

### Exercises

In all of the exercises below, all points, lines, etc., are assumed to lie in $\mathbf{R}_n$ unless stated otherwise.

1. Let $P_0, \ldots, P_k$ be *b*-independent points and let $Q_1, Q_2, Q_3$ be *collinear* points which lie in the flat $\langle P_0, \ldots, P_k \rangle$. We can therefore write

$$Q_1 = \sum_{i=0}^{k} a_i P_i; \qquad Q_2 = \sum_{i=0}^{k} b_i P_i; \qquad Q_3 = \sum_{i=0}^{k} c_i P_i$$

where $\Sigma a_i = \Sigma b_i = \Sigma c_i = 1$. Prove that the following statements are equivalent:

(a) $Q_2$ is between $Q_1$ and $Q_3$.
(b) For some $i$ $(0 \leq i \leq k)$, $b_i$ is between $a_i$ and $c_i$.
(c) For every $i$ $(0 \leq i \leq k)$, $b_i$ is between $a_i$ and $c_i$.
(This generalizes Sec. 3.2, Exercise 11.) (SUG)

2. Let $P = a_0P_0 + a_1P_1 + a_2P_2$ and $Q = b_0P_0 + b_1P_1 + b_2P_2$ (where $\Sigma a_i = \Sigma b_i = 1$) be points in the plane $S = \langle P_0, P_1, P_2 \rangle$. Deduce from Exercise 1 that if $a_2$ and $b_2$ have the same sign, then the segment $\overline{PQ}$ does not meet the line $L_2 = \langle P_0, P_1 \rangle$.

3. Using the same notations as in Exercise 2, but assuming that $a_2$ and $b_2$ have opposite signs, find scalars $\alpha$ and $\beta$ such that

$$\alpha > 0, \qquad \beta > 0, \qquad \alpha + \beta = 1, \qquad \alpha a_2 + \beta b_2 = 0.$$

Deduce that $\overline{PQ}$ meets $L_2$. (SOL)

4. Referring to the seven numbered "regions" of Figure 3.6.1 (corresponding to the seven possible combinations of signs of the $a_i$), deduce from the preceding exercises that the line segment joining any two points in the same region lies entirely within that region, and that the line segment joining any two points in different regions crosses one of the lines $L_i$.
5. (For readers who know some topology.) Again referring to the seven regions in Figure 3.6.1, prove that
   (a) each region is an open subset of the plane $S$;
   (b) each region is connected;
   (c) the union of any two of the regions is disconnected.

   (*Note*: In proving (a) you may need to use the fact that if $L$ is a line and $P$ is a point not on $L$ then there exists a positive shortest distance from $P$ to $L$. This fact will be derived in a later section of this chapter; specifically, see Theorem 3.8.23.)
6. A set of points in $\mathbf{R}_n$ is said to be *bounded* if the distance between any two points in the set is less than some fixed number $M$ depending only on the set. Referring to the seven regions of Figure 3.6.1, prove formally that region 7 (in which all $a_i > 0$) is bounded but that each of the other six regions is unbounded. (This is why region 7 is called the "inside" of the triangle $P_0P_1P_2$.)
7. Can you generalize Exercises 2–6 to higher dimensions?
8. In $\mathbf{R}_4$, let $S$ be the plane determined by the three points (2, 2, 0, 1), (1, 1, 2, 1), (5, 3, 0, 2). Determine which, if any, of the following points lie on this plane: (2, 0, 6, 2), (0, 1, 1, 1), (3, 3, 2, 2), (3, 5, 4, −2). (ANS)
9. In $\mathbf{R}_3$, let $P_0 = (-1, 0, 1)$, $P_1 = (4, 3, 2)$, $P_2 = (0, -2, 6)$, $P_3 = (5, 0, -2)$.

   (a) Prove that $P_0, P_1, P_2, P_3$ are $b$-independent.
   (b) Determine whether the point $(1, 5/2, -1)$ lies inside, outside, or on the tetrahedron whose vertices are $P_0$, $P_1$, $P_2$, $P_3$. (SOL)
10. Let $S$ be a nonempty subset of an arbitrary vector space $V$ over any field $F$ (in this exercise only, we do not assume $V = \mathbf{R}_n$). Prove that $S$ is a flat if and only if it satisfies the following condition: whenever $A, B, C \in S$, then all $b$-combinations of $A, B, C$ lie in $S$. (In other words, $S$ is closed under formation

of $b$-combinations of *three* points. The similar statement for *two* points will not suffice to make $S$ a flat if $F$ has characteristic 2.)

11. Let $P_0$, $P_1$, $P_2$ be $b$-independent points in real 3-space ($\mathbf{R}_3$), and let $S$ be the plane $\langle P_0, P_1, P_2 \rangle$.
(a) Prove that $P_0$, $P_1$, $P_2$ are *linearly* independent (as elements of the vector space $\mathbf{R}_3$) if and only if $S$ does not pass through the origin. (SUG)
(b) Assume $S$ does not pass through the origin. Deduce (using a result from Chapter 2) that any point $P \in \mathbf{R}_3$ can be expressed uniquely as a *linear* (not necessarily barycentric) combination $P = a_0P_0 + a_1P_1 + a_2P_2$, where the $a_i$ are scalars.
(c) Assume $S$ does not pass through the origin. Using the notation of part (b), show (informally, if you wish) that the points $P$ for which $a_0 + a_1 + a_2 > 1$ lie on one side of $S$ while the points $P$ for which $a_0 + a_1 + a_2 < 1$ lie on the other side of $S$.
(d) Still using the same notation, let $k$ be a fixed real number. Prove that the set of all points $P$ for which $a_0 + a_1 + a_2 = k$ is a plane parallel to $S$. (We still assume $S$ does not pass through the origin.) (SUG)
12. Let $A$, $B$, $C$ be noncollinear points, and let $P$ lie in the plane $\langle A, B, C\rangle$; then $P = aA + bB + cC$ where $a + b + c = 1$. Prove: if $k$ is a fixed real number, then the set of all points $P$ for which $a = k$ is a line $L_k$ parallel to the line $\langle B, C\rangle$. Draw a figure exhibiting the lines $L_k$ for $k = 0, 1/2, 1$. In general, how does the position of $L_k$ vary as $k$ increases? (SUG)
13. Let $P = aA + bB + cC$ with $a + b + c = 1$, as in the preceding exercise. Obtain a condition on the numbers $a$, $b$, $c$ which is equivalent to the statement: "$P$ lies on the line which passes through $C$ and the midpoint of $\overline{AB}$."
14. In Figure 3.6.7, $B$ is the midpoint of $\overline{AD}$ and $M$ is the midpoint of $\overline{BC}$. (We assume $A$, $B$, $C$ are noncollinear.)
(a) Express $P$ as a barycentric combination of $A$, $B$, $C$.
(b) If the segment $\overline{PC}$ has length 3, how long is $\overline{PD}$?
(c) If the segment $\overline{PM}$ has length 2, how long is $\overline{AM}$? (ANS)
15. Suppose $A$, $B$, $C$, $D$ are points such that $aA + bB + cC + dD = 0$, $a + b + c + d = 0$, and suppose the scalars $a$, $b$, $c$, $d$ are

*all* nonzero. Prove: if not all four of the points are collinear, then no three are collinear.

16. If a point $P$ lies inside a triangle $ABC$ (cf. Exercise 6), show that $\overline{PA}$ is not parallel to $\overline{BC}$.
17. Let $A$, $B$, $C$ be noncollinear points; let $D$ be between $A$ and $B$, and let $E$ be between $A$ and $C$ (see Fig. 3.6.8). Prove: the segments $\overline{DC}$ and $\overline{BE}$ intersect, and the point of intersection lies inside the triangle $ABC$.
18. In Figure 3.6.9, $A$, $B$, $C$ are noncollinear and $D$, $E$ are points on the segments $\overline{AB}$, $\overline{AC}$, respectively (with $D$ and $E$ different from $A$, $B$, $C$); $P$ is the intersection of $\overline{DC}$ and $\overline{BE}$, and $M$ is the intersection of $\langle B, C\rangle$ and $\langle A, P\rangle$. Prove: if **Ar** $DE \parallel$ **Ar** $BC$, then $M$ is the midpoint of $\overline{BC}$.

(Note that the existence of the points $P$ and $M$ is assured by the two preceding exercises.)

Figure 3.6.7

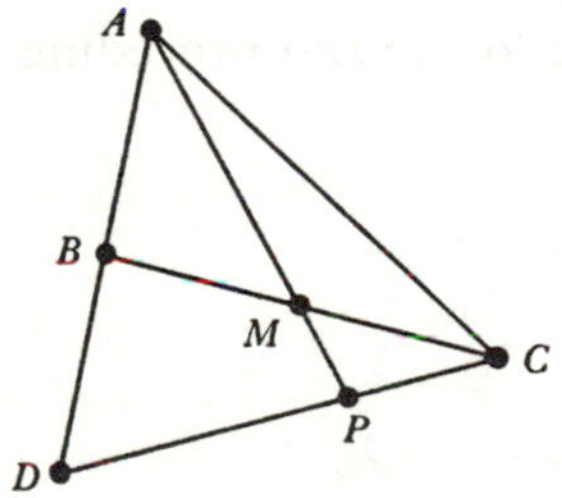


Figure 3.6.8

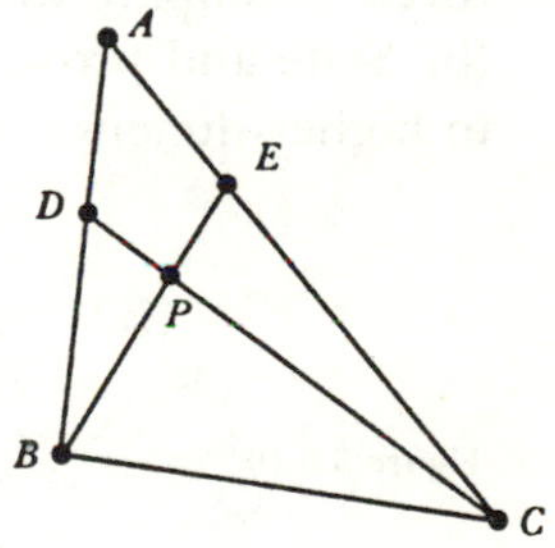


Figure 3.6.9

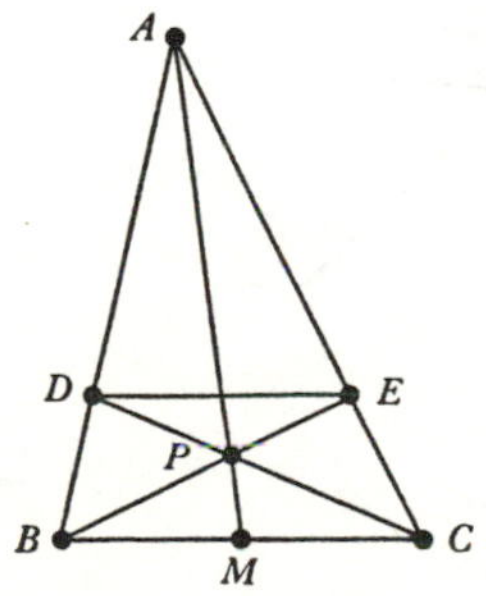

19. Assume that $A_1, A_2, A_3, A_4$ are distinct equally spaced points on a line ("equally spaced" means that $|\overline{A_1A_2}| = |\overline{A_2A_3}| = |\overline{A_3A_4}|$), $P$ is a point not on that line, and $L$ is a line in the plane $\langle P, A_1, A_2, A_3, A_4\rangle$. Assume further that $L$ is parallel to (but distinct from) the line $\langle P, A_1\rangle$. The line $L$ intersects $\langle P, A_2\rangle$, $\langle P, A_3\rangle$, and $\langle P, A_4\rangle$ in points $B_2, B_3, B_4$, respectively (see Fig. 3.6.10). Prove that

    $$|\overline{B_2B_3}|/|\overline{B_3B_4}| = 3.$$

    (*Suggestion*: It might be useful to regard this as a limiting case of Example 3.6.9, even though we need not use limits to obtain the solution.) (SOL)
20. Generalize Example 3.6.17.
21. (a) Let $V$ be any vector space, not necessarily $\mathbf{R}_n$, and let $A$, $B$, $C$, $D$, $E$ be five points in the same 3-flat in $V$ which do not all lie in the same plane. Assuming the field of scalars does not have characteristic 2, prove that the line through some two of these points intersects the plane through the other three. (Compare with Example 3.5.7.)
    (b) State and prove a generalization of the preceding result to higher-dimensional flats.

Figure 3.6.10

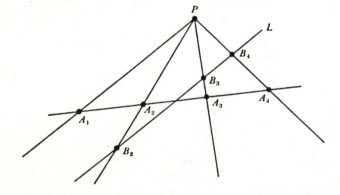

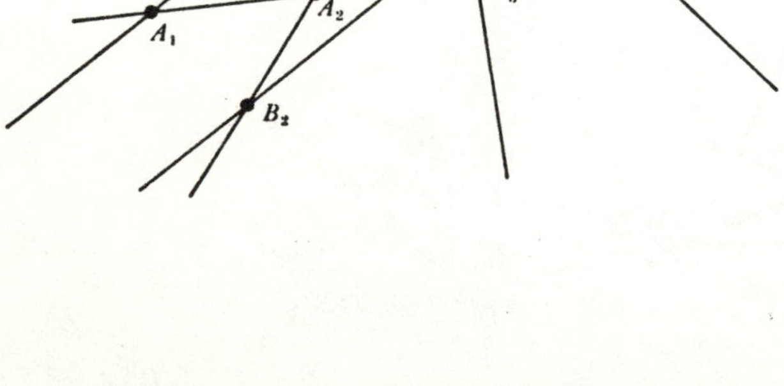

## 3.7 The dot product; real Euclidean spaces

Let $F$ be a field. If $\mathbf{u} = (a_1, a_2, ..., a_n)$ and $\mathbf{v} = (b_1, b_2, ..., b_n)$ are any elements of the vector space $F_n$, we define

$$\text{(3.7.1)} \qquad \mathbf{u} \cdot \mathbf{v} = a_1b_1 + a_2b_2 + \cdots a_nb_n = \sum_{i=1}^{n} a_ib_i;$$

that is, $\mathbf{u} \cdot \mathbf{v}$ is the sum of the products of corresponding coordinates of $\mathbf{u}$ and $\mathbf{v}$. The quantity $\mathbf{u} \cdot \mathbf{v}$ is called the *dot product* of $\mathbf{u}$ and $\mathbf{v}$. It is essential to remember that *the dot product of two vectors is a scalar, not a vector*. Some simple properties of the dot product are collected in the following theorem.

**3.7.2 Theorem.** If $F$ is any field, the following three statements hold for all $c \in F$ and all $\mathbf{u}, \mathbf{v}, \mathbf{w} \in F_n$:

(a) $\mathbf{u} \cdot \mathbf{v} = \mathbf{v} \cdot \mathbf{u}$

(b) $\mathbf{u} \cdot (\mathbf{v} + \mathbf{w}) = (\mathbf{u} \cdot \mathbf{v}) + (\mathbf{u} \cdot \mathbf{w})$

(c) $c(\mathbf{u} \cdot \mathbf{v}) = (c\mathbf{u}) \cdot \mathbf{v} = \mathbf{u} \cdot (c\mathbf{v})$.

All parts of Theorem 3.7.2 are easy consequences of 3.7.1; we leave the proofs as exercises.

Although the dot product is defined in terms of coordinates and has some algebraic uses (in Chapter 4 we will use it to simplify the description of matrix multiplication), its principal applications in this book will be to geometry in real $n$-space. The dot product in $\mathbf{R}_n$ is, of course, called the *real dot product*. In addition to the properties listed in Theorem 3.7.2, we will need two further properties of the real dot product, both of which follow at once from Theorem 3.1.10:

**3.7.3 Theorem.** If $\mathbf{u}$ is any vector in $\mathbf{R}_n$, then

(a) $\mathbf{u} \cdot \mathbf{u} = |\mathbf{u}|^2$

(b) $\mathbf{u} \cdot \mathbf{u} = 0$ if and only if $\mathbf{u} = \mathbf{0}$.

We emphasize that Theorem 3.7.3 is valid only over the field $\mathbf{R}$, in contrast to 3.7.2 which holds over arbitrary fields. (The vector $\mathbf{u} = (1, i)$ in $\mathbf{C}_2$ provides a counterexample to 3.7.3(b) over the field $\mathbf{C}$.) For the rest of this section we shall assume that our field of scalars is $\mathbf{R}$.

It is useful to generalize the definition of the real dot product as follows: let $E$ be any vector space over $\mathbf{R}$. If there exists a function $p : E \times E \to \mathbf{R}$ such that the laws

(3.7.4)
(a) $p(\mathbf{u}, \mathbf{v}) = p(\mathbf{v}, \mathbf{u})$
(b) $p(\mathbf{u}, \mathbf{v} + \mathbf{w}) = p(\mathbf{u}, \mathbf{v}) + p(\mathbf{u}, \mathbf{w})$
(c) $cp(\mathbf{u}, \mathbf{v}) = p(c\mathbf{u}, \mathbf{v}) = p(\mathbf{u}, c\mathbf{v})$
(d) $p(\mathbf{u}, \mathbf{u}) \geqslant 0$
(e) $p(\mathbf{u}, \mathbf{u}) = 0 \Leftrightarrow \mathbf{u} = \mathbf{0}$

hold for all $\mathbf{u}, \mathbf{v}, \mathbf{w} \in E$ and $c \in \mathbf{R}$, then $E$ is called a *real Euclidean space* (or *real inner product space*) and the function $p$ is called a *real inner product* on $E$. For any vector $\mathbf{u} \in E$, we then define the *norm* (*length*) *of* $\mathbf{u}$ *with respect to* $p$ (denoted $\|\mathbf{u}\|_p$, or simply $\|\mathbf{u}\|$) to be $(p(\mathbf{u}, \mathbf{u}))^{1/2}$. The *distance* $d(\mathbf{u}, \mathbf{v})$ between any two elements $\mathbf{u}, \mathbf{v}$ of $E$ is defined to be $\|\mathbf{u} - \mathbf{v}\|$; similarly, the length $|\overline{PQ}|$ of a line segment $\overline{PQ}$ ($P, Q \in E$) is defined to be $\|\mathbf{Ar}\ PQ\| = \|Q - P\| = d(P, Q)$. Theorems 3.7.2 and 3.7.3 of course imply that the space $\mathbf{R}_n$ (or, more generally, any subspace of $\mathbf{R}_n$) is a real Euclidean space under the inner product $p(\mathbf{u}, \mathbf{v}) = \mathbf{u} \cdot \mathbf{v}$, and that the norm of any vector $\mathbf{u}$ with respect to this inner product is the length of $\mathbf{u}$. (The formula $d(\mathbf{u}, \mathbf{v}) = \|\mathbf{u} - \mathbf{v}\|$ is essentially the same as Definition 3.1.9.) In what follows, we shall, whenever possible, state our results in terms of arbitrary real Euclidean spaces; all such results will apply in particular to $\mathbf{R}_n$ and subspaces of $\mathbf{R}_n$. (For an important example of a real Euclidean space of an entirely different type, see Exercise 17 at the end of this section.)

The formula $\|\mathbf{u}\| = (p(\mathbf{u}, \mathbf{u}))^{1/2}$ (or 3.7.3(a)) provides the connection between inner products and geometry; it shows that lengths are expressible in terms of inner products. Conversely, *inner products are expressible in terms of lengths*, as shown by part (c) of the next theorem.

**3.7.5 Theorem.** For all vectors $\mathbf{u}, \mathbf{v}$ in a real Euclidean space, the following equations hold:

(a) $\|\mathbf{u} + \mathbf{v}\|^2 = \|\mathbf{u}\|^2 + \|\mathbf{v}\|^2 + 2p(\mathbf{u}, \mathbf{v})$.
(b) $\|\mathbf{u} - \mathbf{v}\|^2 = \|\mathbf{u}\|^2 + \|\mathbf{v}\|^2 - 2p(\mathbf{u}, \mathbf{v})$.
(c) $p(\mathbf{u}, \mathbf{v} = \frac{1}{2} ( \|\mathbf{u}\|^2 + \|\mathbf{v}\|^2 - \|\mathbf{u} - \mathbf{v}\|^2)$.
(d) $\|t\mathbf{u}\| = |t| \cdot \|\mathbf{u}\|$ for all $t \in \mathbf{R}$.

*Proof.* $\|\mathbf{u} + \mathbf{v}\|^2 = p(\mathbf{u} + \mathbf{v}, \mathbf{u} + \mathbf{v}) = p(\mathbf{u},\mathbf{u}) + p(\mathbf{v}, \mathbf{v}) + 2p(\mathbf{u}, \mathbf{v})$ (by 3.7.4(a) and 3.7.4(b)), which is the same as 3.7.5(a). The proof of (b) is similar, and (c) is immediate from (b). We leave the proof of (d) as an exercise. ∎

In connection with Theorem 3.7.5, we recall from Section 3.1 (Exercise 2 and Theorem 3.1.5) that the arrow representations of $\mathbf{u} + \mathbf{v}$ and $\mathbf{u} - \mathbf{v}$ are related to those of $\mathbf{u}$ and $\mathbf{v}$ in the manner shown in Figure 3.7.1. Thus, each of the first three parts of 3.7.5 relates $p(\mathbf{u}, \mathbf{v})$ to the lengths of the three sides of a certain triangle. As a consequence, we can sometimes establish properties of triangles (or other figures, for that matter) by use of the inner product. The following example illustrates the sort of thing that can be done.

**3.7.6 Example.** In a real Euclidean space, prove that if two medians of a triangle have equal length then the triangle is isosceles.

*Solution.* Let the two equal-length medians of triangle $ABC$ be $\overline{BP}$ and $\overline{CQ}$, as shown in Figure 3.7.2, and let $\mathbf{Ar}\ AQ = \mathbf{u}$, $\mathbf{Ar}\ AP = \mathbf{v}$. Since $Q$, $P$ are the midpoints of $\overline{AB}$ and $\overline{AC}$, it follows that

$$\mathbf{Ar}\ AB = 2\mathbf{u}; \qquad \mathbf{Ar}\ AC = 2\mathbf{v}$$

(formal proof: $Q = \frac{1}{2}(A + B)$, so that $B - A = 2(Q - A)$, i.e., $\mathbf{Ar}\ AB = 2 \cdot \mathbf{Ar}\ AQ = 2\mathbf{u}$; and similarly for $\mathbf{Ar}\ AC$). Hence

$$\mathbf{Ar}\ CQ = \mathbf{Ar}\ AQ - \mathbf{Ar}\ AC = \mathbf{u} - 2\mathbf{v}; \qquad \mathbf{Ar}\ BP = \mathbf{v} - 2\mathbf{u}.$$

By assumption, the lengths of these medians are equal: $\|\mathbf{u} - 2\mathbf{v}\| = \|\mathbf{v} - 2\mathbf{u}\|$. Squaring both sides and then expanding via Theorem 3.7.5(b), we get

$$\|\mathbf{u}\|^2 + 4\|\mathbf{v}\|^2 - 4p(\mathbf{u}, \mathbf{v}) = \|\mathbf{v}\|^2 + 4\|\mathbf{u}\|^2 - 4p(\mathbf{u}, \mathbf{v}) \tag{3.7.7}$$

(note where we have used 3.7.4(c) and 3.7.5(d)). Cancelling equal terms from both sides of 3.7.7, we get $3\|\mathbf{v}\|^2 = 3\|\mathbf{u}\|^2$. Hence, $\|\mathbf{v}\| = \|\mathbf{u}\|$, $\|2\mathbf{v}\| = \|2\mathbf{u}\|$, $|\overline{AC}| = |\overline{AB}|$ which is the desired result.

Figure 3.7.1

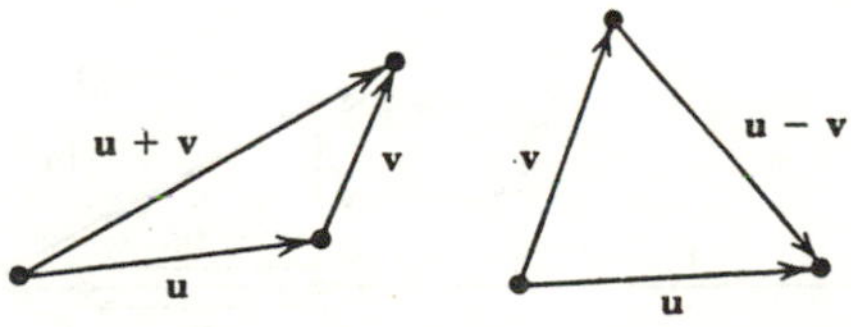


(a) (b)

Figure 3.7.2

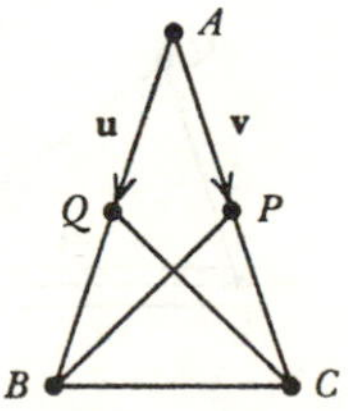

Inner products are closely related to the concept of *perpendicularity*. Let $E$ be a real Euclidean space, let $L = L(A, B)$ be a line in $E$, let $Q$ be a point in $E$, and let us write $\mathbf{v} = \mathbf{Ar}\, AQ$, $\mathbf{u} = \mathbf{Ar}\, AB$ as in Figure 3.7.3(a). For any point $P \neq A$ on $L$, we have $\mathbf{Ar}\, AP = t\mathbf{u}$ for some scalar $t \neq 0$ (cf. equation 3.2.12). By Theorem 3.7.5(b), we have

(3.7.8) $$|\overline{PQ}|^2 = \|\mathbf{v} - t\mathbf{u}\|^2 = \|\mathbf{v}\|^2 + \|t\mathbf{u}\|^2 - 2t \cdot p(\mathbf{u}, \mathbf{v}).$$

If it happens that $p(\mathbf{u}, \mathbf{v}) = 0$, then the last term in 3.7.8 drops out, and since $\|t\mathbf{u}\|^2 > 0$ it follows that $|\overline{PQ}|^2 > \|\mathbf{v}\|^2$, $|\overline{PQ}| > \|\mathbf{v}\| = |\overline{QA}|$. In other words, if $p(\mathbf{u}, \mathbf{v}) = 0$, then $\overline{QA}$ is the shortest segment from $Q$ to any point on $L$; equivalently, the distance from $Q$ to $A$ is the *shortest distance from $Q$ to $L$*. This result, which is illustrated in Figure 3.7.3(b), clearly suggests that in this situation we should call the vectors $\mathbf{u}$ and $\mathbf{v}$ *perpendicular* and should call the segment $\overline{QA}$ the *perpendicular segment from $Q$ to $L$*. We thus make the following definitions:

**3.7.9 Definition.** In a real Euclidean space $E$,

(a) Two vectors $\mathbf{u}$, $\mathbf{v}$ are *perpendicular*, or *orthogonal* (*notation*: $\mathbf{u} \perp \mathbf{v}$) if $p(\mathbf{u}, \mathbf{v}) = 0$.

(b) Two flats $S_1$, $S_2$ which have a point $A$ in common are said to be *perpendicular* (*at $A$*) if

$$\mathbf{Ar}\, AP_1 \perp \mathbf{Ar}\, AP_2 \text{ for all points } P_1 \in S_1,\ P_2 \in S_2.$$

(See Fig. 3.7.4(a) to make this definition seem reasonable.)

(c) If $A$ is a point on a flat $S$, a line segment $\overline{QA}$ is called a *perpendicular segment from $Q$ to $S$* if $\mathbf{Ar}\, AQ \perp \mathbf{Ar}\, AP$ for all points $P \in S$. (See Fig. 3.7.4(b) to make this definition seem reasonable.)

Note that the word "orthogonal" is *not* used in parts (b) and (c) of the definition. Roughly speaking, geometric objects (flats or line seg-

Figure 3.7.3

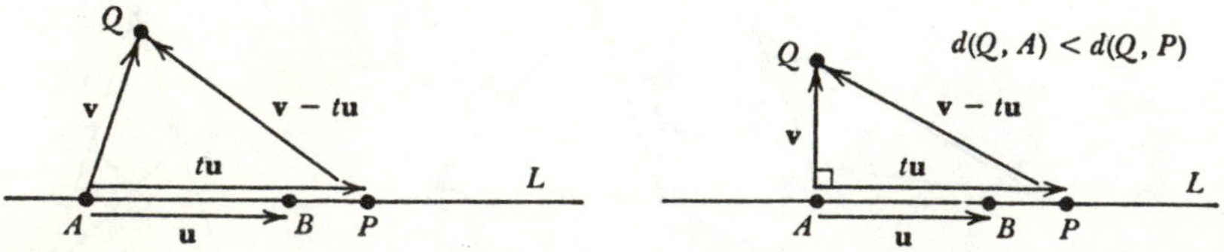


(a) (b)

ments) may be "perpendicular", but only algebraic objects (e.g., vectors) can be "orthogonal". The distinction is not hard and fast, but after a while you should become accustomed to normal usage.

The "shortest-distance" result, which we obtained above, can now be rephrased as follows: if $Q$ is a point and $L$ is a line in $E$, then any perpendicular segment from $Q$ to $L$ is shorter than all other segments from $Q$ to $L$. A similar result holds for flats (see Fig. 3.7.4(b)): *if $S$ is a flat, $A$ is a point on $S$, and $\overline{QA}$ is a perpendicular segment from $Q$ to $S$, then $d(Q, A) < d(Q, P)$ for all points $P \neq A$ on $S$.* (Proof: apply the result already obtained for lines to the line $L = \langle A, P\rangle$.) In particular, this shows that a perpendicular segment from $Q$ to $S$, if it exists, must be *unique*, since if there were two such segments, then each would be shorter than the other.

In connection with Definition 3.7.9(c), note that if $A \in S$ and $W$ is the direction space of $S$, then

$$P \in S \Leftrightarrow \mathbf{Ar}\, AP \in W$$

(Theorem 3.4.4(b)). Hence, the conditions for $\overline{QA}$ to be the perpendicular segment from $Q$ to $S$ are

**(3.7.10)**
(a) $A \in S$
(b) $\mathbf{Ar}\, AQ$ is orthogonal to all vectors in the direction space $W$ of $S$.

Let us now show that the perpendicular segment from a point $Q$ to a line $L = L(A, B)$ always exists. (The analogous result for an arbitrary flat instead of a line will be proved in the next section.) As

Figure 3.7.4

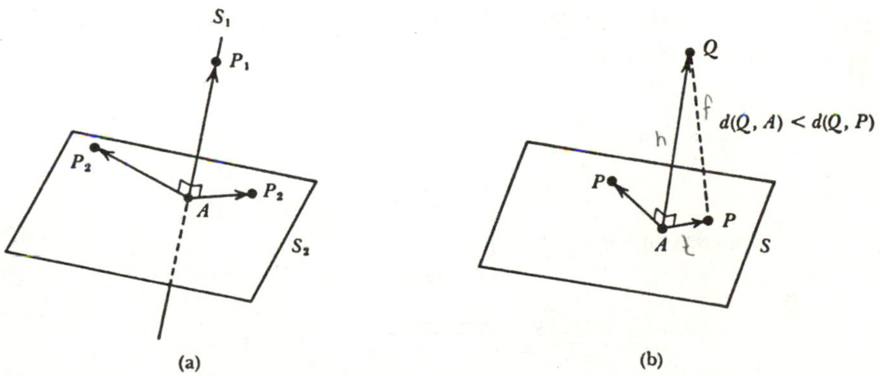


(a) (b)

before, let $\mathbf{u} = \mathbf{Ar}\ AB$ and $\mathbf{v} = \mathbf{Ar}\ AQ$ (see Fig. 3.7.5.). If we can find a scalar $t$ such that

**(3.7.11)** $p(\mathbf{u}, \mathbf{v} - t\mathbf{u}) = 0,$

then the desired perpendicular segment $\overline{QP}$ is found by setting $P = A + t\mathbf{u}$ (see Fig. 3.7.5); indeed, 3.7.11 implies that $\mathbf{Ar}\ PQ = \mathbf{v} - t\mathbf{u}$ is perpendicular to $k\mathbf{u}$ for every real $k$, and, hence, to $\mathbf{Ar}\ PP'$ for every point $P'$ on $L$. Thus, we need only solve 3.7.11 for $t$. But 3.7.11 is the same as

$$0 = p(\mathbf{u}, \mathbf{v}) - t \cdot p(\mathbf{u}, \mathbf{u}) = p(\mathbf{u}, \mathbf{v}) - t\|\mathbf{u}\|^2$$

which clearly has the solution

**(3.7.12)** $$t = \frac{p(\mathbf{u}, \mathbf{v})}{\|\mathbf{u}\|^2} = \frac{p(\mathbf{u}, \mathbf{v})}{p(\mathbf{u}, \mathbf{u})}.$$

Note that the denominator is nonzero by 3.7.4(e).

**3.7.13** **Example.** If $L$ is the line in $\mathbf{R}_3$ whose parametric equations (see 3.2.11) are

**(3.7.14)** $$\begin{aligned} x_1 &= 3 + t \\ x_2 &= -2 + 3t \\ x_3 &= 1 - 2t \end{aligned}$$

and if $Q$ is the point $(-1, 14, -5)$, find the point $P$ on $L$ such that $\overline{QP}$ is the perpendicular segment from $Q$ to $L$.

*Solution.* We may take the vector $\mathbf{u} = \mathbf{Ar}\ AB$ to be $(1, 3, -2)$ (these are the coefficients of $t$ in the equations 3.7.14; compare with the derivation of 3.2.11). The point $A$ may be chosen to be any point on $L$; for example, setting $t = 0$ we find that the point

$$A = (3, -2, 1)$$

lies on $L$. We then have, using the same notations as in Figure 3.7.5 and equation 3.7.12,

Figure 3.7.5

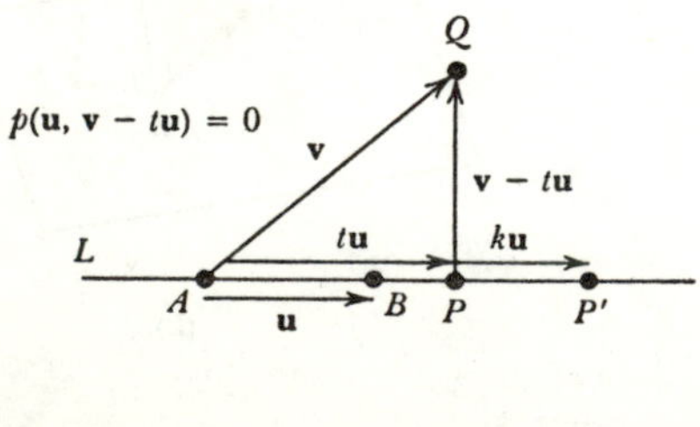

$$\mathbf{v} = \mathbf{Ar}\, AQ = Q - A = (-4, 16, -6)$$

$$t = \frac{\mathbf{u} \cdot \mathbf{v}}{\mathbf{u} \cdot \mathbf{u}} = \frac{(1, 3, -2) \cdot (-4, 16, -6)}{(1, 3, -2) \cdot (1, 3, -2)} = \frac{56}{14} = 4$$

$$P = A + t\mathbf{u} = (3, -2, 1) + 4(1, 3, -2) = (7, 10, -7).$$

The desired point is (7, 10, −7). ∎

**Further results and remarks**

If $\mathbf{u}, \mathbf{v}$ are vectors such that $\mathbf{u} \perp \mathbf{v}$ (i.e., $p(\mathbf{u}, \mathbf{v}) = 0$), Theorem 3.7.5(a) reduces to $\|\mathbf{u} + \mathbf{v}\|^2 = \|\mathbf{u}\|^2 + \|\mathbf{v}\|^2$. This is precisely the *Pythagorean Theorem* (as a glance at Fig. 3.7.6 should convince you): if two sides of a triangle are perpendicular, then the sum of the squares of their lengths equals the square of the length of the remaining side. The following is a generalization:

**3.7.15** **Theorem.** (Pythagorean Theorem for $k$ vectors) If $\mathbf{u}_1, \ldots, \mathbf{u}_k$ are mutually perpendicular vectors in a real Euclidean space $E$, then

$$\|\mathbf{u}_1 + \cdots + \mathbf{u}_k\|^2 = \|\mathbf{u}_1\|^2 + \cdots + \|\mathbf{u}_k\|^2.$$

This has already been established in the case $k = 2$. The proof in the general case is similar; we leave it to you. Incidentally, the truth of the Pythagorean Theorem is hardly surprising, since our very definition of distance (Definition 3.1.6) was essentially a special case of this theorem.

If we apply the Pythagorean Theorem to the situation of Figure 3.7.5, we get

$$\|\mathbf{v}\|^2 = \|t\mathbf{u}\|^2 + \|\mathbf{v} - t\mathbf{u}\|^2$$

and in particular $\|t\mathbf{u}\|^2 \leq \|\mathbf{v}\|^2$; taking square roots, we have

**(3.7.16)** $|t| \cdot \|\mathbf{u}\| \leq \|\mathbf{v}\|$.

Figure 3.7.6

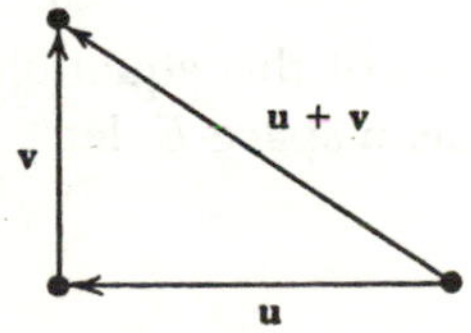

Multiplying both sides of 3.7.16 by $\|\mathbf{u}\|$ and then substituting the expression 3.7.12 for $t$, we get the *Cauchy-Schwarz inequality*

$$(3.7.17)\qquad |p(\mathbf{u}, \mathbf{v})| \leqslant \|\mathbf{u}\| \cdot \|\mathbf{v}\|.$$

Although the argument leading to 3.7.17 assumed $\mathbf{u} \neq \mathbf{0}$, 3.7.17 also holds when $\mathbf{u} = \mathbf{0}$ since $p(\mathbf{0}, \mathbf{v}) = 0$ for all $\mathbf{v} \in E$ (this follows from 3.7.4(b) just as 1.9.3(a) followed from 1.9.1(d)). Thus, 3.7.17 is valid for *all* $\mathbf{u}, \mathbf{v} \in E$.

The interpretation of 3.7.17 in the case $E = \mathbf{R}_n$ is of some interest; if we substitute 3.7.1 for $p(\mathbf{u}, \mathbf{v})$ and then square both sides of 3.7.17, we obtain

$$(3.7.18)\qquad \left(\sum_{i=1}^{n} a_i b_i\right)^2 \leqslant \left(\sum_{i=1}^{n} a_i^2\right)\left(\sum_{i=1}^{n} b_i^2\right)$$

for all real numbers $a_1, \ldots, a_n, b_1, \ldots, b_n$. This is a good illustration of the power of vector methods; it would be rather difficult to prove an inequality like 3.7.18 without the use of vectors. (Try doing so!)

An inequality closely related to 3.7.17 is the *triangle inequality*, which asserts that in any Euclidean space $E$,

$$(3.7.19)\qquad \|\mathbf{u} + \mathbf{v}\| \leqslant \|\mathbf{u}\| + \|\mathbf{v}\| \qquad (\text{all } \mathbf{u}, \mathbf{v} \in E).$$

Geometrically, this says that the length of any side of a triangle cannot exceed the sum of the lengths of the other two sides (see Fig. 3.7.1(a)). To prove 3.7.19, we use 3.7.5(a):

$$\begin{aligned} \|\mathbf{u} + \mathbf{v}\|^2 &= \|\mathbf{u}\|^2 + \|\mathbf{v}\|^2 + 2p(\mathbf{u}, \mathbf{v}) \\ &\leqslant \|\mathbf{u}\|^2 + \|\mathbf{v}\|^2 + 2\|\mathbf{u}\| \cdot \|\mathbf{v}\| \qquad (\text{by } 3.7.17) \\ &= (\|\mathbf{u}\| + \|\mathbf{v}\|)^2 \end{aligned}$$

and by taking square roots we obtain the desired result.

## Exercises

*1. Prove Theorems 3.7.2 and 3.7.5(d).

2. By adding equations 3.7.5(a) and 3.7.5(b), we obtain the equation

$$\|\mathbf{u} + \mathbf{v}\|^2 + \|\mathbf{u} - \mathbf{v}\|^2 = 2\|\mathbf{u}\|^2 + 2\|\mathbf{v}\|^2.$$

Find a simple geometric interpretation of this equation.

*3. (a) For any subset $S$ of a real Euclidean space $E$, let

$$S^{\perp} = \{\mathbf{u} \in E : \mathbf{u} \perp \mathbf{s} \text{ for all } \mathbf{s} \in S\}.$$

Prove that $S^{\perp}$ is a subspace of $E$.

(b) Let $F$ be any field. If we replace $E$ by $F_n$ and use $\mathbf{u} \cdot \mathbf{v} = 0$ in place of $p(\mathbf{u}, \mathbf{v}) = 0$ as the definition of perpendicularity, show that the result of part (a) remains valid.

4. Interpret Exercises 3 and 4 of Section 2.2 in the light of the preceding exercise.
5. Let $A$, $B$ be points in $\mathbf{R}_n$. If $M$ is the midpoint of the line segment $\overline{AB}$, the set

   $S = \{P \in \mathbf{R}_n : \text{Ar } MP \perp \text{Ar } AB\}$

   is called the *perpendicular bisector* of the segment $\overline{AB}$ (see Fig. 3.7.7). Prove:

   (a) $S$ is a flat; in fact, $S = [\text{Ar } AB]^{\perp} + M$.

   (b) $S$ is the largest flat which is perpendicular to the line $\langle A, B\rangle$ at $M$.
6. Let $\mathbf{u}, \mathbf{v} \in \mathbf{R}_n$.

   (a) Prove: $|\mathbf{u}| = |\mathbf{v}|$ if and only if $(\mathbf{u} + \mathbf{v}) \perp (\mathbf{u} - \mathbf{v})$.

   (b) Deduce from part (a): the diagonals of a rhombus are mutually perpendicular. Draw a figure to illustrate.

   (c) Deduce from part (a): the line joining the midpoint of a chord of a circle to the center of the circle is perpendicular to the chord. (Equivalently: the perpendicular bisector of a chord of a circle passes through the center of the circle.) Draw a figure to illustrate.

   (d) Deduce from part (a): the perpendicular bisector of a segment $\overline{AB}$ is equal to the set of all points which are equidistant from $A$ and $B$. Draw a figure to illustrate.

Figure 3.7.7

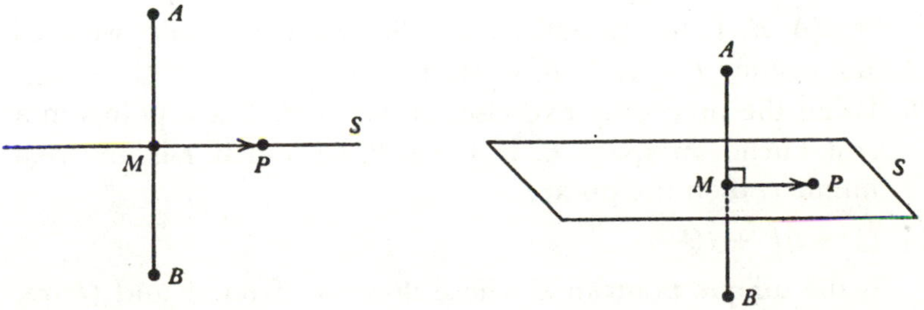


(a) The perpendicular bisector of $\overline{AB}$ in $\mathbf{R}_2$ (b) The perpendicular bisector of $\overline{AB}$ in $\mathbf{R}_3$

7. (a) In $\mathbf{R}_3$, find the perpendicular segment from the point $(-1, 2, -8)$ to the line through the points $(1, 0, 2)$ and $(3, -2, 3)$. (ANS)
(b) In $\mathbf{R}_4$, find the perpendicular segment from the point $(6, 3, 1, 4)$ to the line through the points $(1, 3, 2, 1)$ and $(7, 0, -1, 4)$. (ANS)
8. If $y = m_1x + b_1$ and $y = m_2x + b_2$ are the equations of two lines in $\mathbf{R}_2$ (see Section 3.2, Exercise 19), prove that the two lines are perpendicular if and only if $m_1m_2 = -1$.
9. (a) If two flats in a real Euclidean space $E$ are perpendicular (as in Definition 3.7.9(b)), prove that they have only one point in common.
(b) Show that the result of part (a) is no longer valid if $E$ is replaced by $F_n$ (where $F$ is an arbitrary field) and $\mathbf{u} \perp \mathbf{v}$ is defined to mean $\mathbf{u} \cdot \mathbf{v} = 0$. Why does the proof fail?
10. How would you define perpendicularity of two planes in real 3-space? (This case is not covered by Definition 3.7.9(b), since two such planes have more than one point in common.)
*11. Prove Theorem 3.7.15. (The case $k = 2$ has already been proved.)
12. According to an old Greek proverb, "a straight line is the shortest distance between two points". Although the proverb is technically incorrect (a distance is a number while a line is a set of points), there is a correct idea behind it. What result of this section could be used to justify the proverb?
13. By examining the proof of the inequality 3.7.19 in detail, show that equality holds ($\|\mathbf{u} + \mathbf{v}\| = \|\mathbf{u}\| + \|\mathbf{v}\|$) if and only if either $\mathbf{u} = \mathbf{0}$ or $\mathbf{v} = t\mathbf{u}$ for some scalar $t \geqslant 0$. Interpret geometrically.
14. Using the result of the preceding exercise, show that if $A$, $B$, $P$ are points in a real Euclidean space with $A \neq B$, then $P$ lies on the segment $\overline{AB}$ if and only if $d(A, P) + d(P, B) = d(A, B)$. (This strengthens Sec. 3.2, Exercise 9 since we need not assume $P \in L(A, B)$ to start with.)
15. Using the preceding exercise, prove: if $P$, $Q$ are points in a real Euclidean space $E$, if $k = d(P, Q)$ and if $t$ is any real number, then the point

$$(1 - t)P + tQ$$

is the unique point in $E$ whose distances from $P$ and $Q$ are, respectively, $|t| \cdot k$ and $|1 - t| \cdot k$. (Consider separately the cases $0 \leqslant t \leqslant 1$, $t < 0$, $t > 1$.)

16. Deduce from Exercise 14 that if $A$, $B$ are distinct points in $E$, then a point $P \in E$ lies on the line $L(A, B)$ if and only if one of the three numbers

$$d(A, B),\ d(A, P),\ d(B, P)$$

is the sum of the other two. (Cf. Sec. 3.2, Exercise 12(a).)

17. Let $E$ be the set of all continuous real-valued functions whose domain is the closed interval $[0, 1]$. ($E$ is a vector space over $\mathbf{R}$ under the "usual" definitions of addition and scalar multiplication; cf. equations 2.1.3.) Show that the function $p: E \times E \to \mathbf{R}$ defined by

$$p(f, g) = \int_0^1 f(x)g(x)\,dx$$

is a real inner product on $E$. (This exercise, of course, requires some knowledge of the basic properties of continuous functions and definite integrals.)

## 3.8 Orthogonal sets

Throughout this section, $E$ will denote a real Euclidean space with inner product $p$.

A set $S$ of vectors in $E$ is called an *orthogonal set* if any two distinct vectors in $S$ are orthogonal. An *orthonormal set* is an orthogonal set each of whose elements has norm 1. "Orthogonal $k$-tuple" and "orthonormal $k$-tuple" are defined similarly (cf. the distinction made between sets and $n$-tuples in Sec. 2.4). The phrase "$\mathbf{u}_1, \ldots, \mathbf{u}_k$ are orthogonal (orthonormal)" will be taken to mean that the *k-tuple* $(\mathbf{u}_1, \ldots, \mathbf{u}_k)$ is orthogonal (orthonormal). Thus, $\mathbf{u}_1, \ldots, \mathbf{u}_k$ are orthogonal if and only if $p(\mathbf{u}_i, \mathbf{u}_j) = 0$ whenever $i \neq j$; and $\mathbf{u}_1, \ldots, \mathbf{u}_k$ are orthonormal if and only if

$$\textbf{(3.8.1)}\qquad p(\mathbf{u}_i, \mathbf{u}_j) = \begin{Bmatrix} 1 & \text{if} & i = j \\ 0 & \text{if} & i \neq j \end{Bmatrix}$$

for all $i, j$ $(1 \leq i \leq k;\ 1 \leq j \leq k)$. In this section we shall prove the *existence* of certain orthogonal and orthonormal sets (e.g., if $\dim E < \infty$, then $E$ has an orthonormal basis) and shall obtain various consequences therefrom.

We begin with the following fundamental result:

3.8.2 **Theorem.** Any orthogonal $k$-tuple of *nonzero* vectors in $E$ is linearly independent.

*Proof.* Assume that the $k$-tuple $(\mathbf{u}_1, ..., \mathbf{u}_k)$ is orthogonal and that $\mathbf{u}_i \neq \mathbf{0}$ (all $i$). If $a_1, ..., a_k$ are scalars such that

$$a_1\mathbf{u}_1 + a_2\mathbf{u}_2 + \cdots + a_k\mathbf{u}_k = \mathbf{0},$$

then by taking the inner product of both sides with $\mathbf{u}_1$ and applying 3.7.4, we obtain

$$a_1p(\mathbf{u}_1, \mathbf{u}_1) + a_2p(\mathbf{u}_2, \mathbf{u}_1) + ... + a_kp(\mathbf{u}_k, \mathbf{u}_1) = p(\mathbf{0}, \mathbf{u}_1) = 0. \tag{3.8.3}$$

Since $p(\mathbf{u}_2, \mathbf{u}_1) = \cdots = p(\mathbf{u}_k, \mathbf{u}_1) = 0$, equation 3.8.3 reduces to

$$a_1p(\mathbf{u}_1, \mathbf{u}_1) = 0. \tag{3.8.4}$$

Since $\mathbf{u}_1$ is nonzero, $p(\mathbf{u}_1, \mathbf{u}_1) \neq 0$ (*why*?); hence, 3.8.4 implies that $a_1 = 0$. A similar argument shows that all of the $a_i$ are zero; hence, the $\mathbf{u}_i$ are independent. ∎

*Remark*: Theorem 3.8.2 has a familiar interpretation in the case $k = 2$: two perpendicular nonzero vectors cannot be parallel. (Cf. Sec. 3.1, Exercise 4.) What is the geometric interpretation when $k = 3$?

We now consider questions involving existence. Given a nonzero vector in the plane $\mathbf{R}_2$, we can always find (i.e., there always exists) a second nonzero vector in $\mathbf{R}_2$ which is perpendicular to the first one. Similarly, given two mutually perpendicular vectors in $\mathbf{R}_3$, there exists a third vector perpendicular to the first two. The following theorem generalizes these remarks to arbitrary real Euclidean spaces.

3.8.5 **Theorem.** Let $\mathbf{w}_1, ..., \mathbf{w}_k$ be orthogonal nonzero vectors in $E$, and let $\mathbf{w}^*$ be any vector in $E$. Then the vector

$$\mathbf{w}_{k+1} = \mathbf{w}^* - \sum_{i=1}^{k} \left( \frac{p(\mathbf{w}^*, \mathbf{w}_i)}{p(\mathbf{w}_i, \mathbf{w}_i)} \right) \mathbf{w}_i \tag{3.8.6}$$

is orthogonal to each of the vectors $\mathbf{w}_1, ..., \mathbf{w}_k$; moreover,

$$[\mathbf{w}_1, ..., \mathbf{w}_k, \mathbf{w}_{k+1}] = [\mathbf{w}_1, ..., \mathbf{w}_k, \mathbf{w}^*]. \tag{3.8.7}$$

(Note that since the $\mathbf{w}_i$ are nonzero, the denominators in 3.8.6 are nonzero.)

*Proof.* 3.8.7 is obtained by applying Theorem 2.3.16(a) $k$ times in succession. As for the rest, if we take the inner product of both sides of 3.8.6 with the vector $\mathbf{w}_1$ and use 3.7.4, we obtain

(3.8.8) $$p(\mathbf{w}_{k+1}, \mathbf{w}_1) = p(\mathbf{w}^*, \mathbf{w}_1) - \sum_{i=1}^{k} \left(\frac{p(\mathbf{w}^*, \mathbf{w}_i)}{p(\mathbf{w}_i, \mathbf{w}_i)}\right) p(\mathbf{w}_i, \mathbf{w}_1).$$

Since $\mathbf{w}_1, \ldots, \mathbf{w}_k$ are assumed orthogonal, the factors $p(\mathbf{w}_i, \mathbf{w}_1)$ on the right side of 3.8.8 are zero for $i = 2, \ldots, k$. Hence, all terms in the sum vanish except for $i = 1$, and 3.8.8 reduces to

$$p(\mathbf{w}_{k+1}, \mathbf{w}_1) = p(\mathbf{w}^*, \mathbf{w}_1) - \left(\frac{(\mathbf{w}^*, \mathbf{w}p)}{p(\mathbf{w}_1, \mathbf{w}_1)}\right) p(\mathbf{w}_1, \mathbf{w}_1)$$
$$= p(\mathbf{w}^*, \mathbf{w}_1) - p(\mathbf{w}^*, \mathbf{w}_1) = 0$$

so that $\mathbf{w}_{k+1} \perp \mathbf{w}_1$. By exactly the same argument, $\mathbf{w}_{k+1}$ is perpendicular to $\mathbf{w}_2, \ldots, \mathbf{w}_k$. ∎

The scalars $p(\mathbf{w}^*, \mathbf{w}_i)/p(\mathbf{w}_i, \mathbf{w}_i)$ which appear in equation 3.8.6 are not new; we came across them earlier in equation 3.7.12. In fact, if we think of the arrow $t\mathbf{u} = \mathbf{Ar}\, AP$ in Figure 3.7.5 as being the "projection of $\mathbf{v}$ on $\mathbf{u}$", then by changing notation and using 3.7.12 we see that the vector

$$\left(\frac{p(\mathbf{w}^*, \mathbf{w}_i)}{p(\mathbf{w}_i, \mathbf{w}_i)}\right) \mathbf{w}_i$$

is the projection of $\mathbf{w}^*$ on $\mathbf{w}_i$. Theorem 3.8.5 thus has a simple interpretation: if we subtract from a vector $\mathbf{w}^*$ the projections of $\mathbf{w}^*$ on orthogonal vectors $\mathbf{w}_1, \ldots, \mathbf{w}_k$, we get a vector perpendicular to $\mathbf{w}_1, \ldots, \mathbf{w}_k$. (Try to visualize this in the case $k = 2$.) The result is *not* true if the original vectors $\mathbf{w}_1, \ldots, \mathbf{w}_k$ are not orthogonal.

Suppose that $\mathbf{w}_1, \ldots, \mathbf{w}_k$ are orthogonal nonzero vectors in $E$. If $E$ has finite dimension, then the set $\{\mathbf{w}_1, \ldots, \mathbf{w}_k\}$ can be extended to an *orthogonal basis* of $E$. Indeed, letting $n = \dim E$, $E$ has a basis $\{\mathbf{u}_1^*, \ldots, \mathbf{u}_n^*\}$. By subtracting from $\mathbf{u}_1^*$ its projections on $\mathbf{w}_1, \ldots, \mathbf{w}_k$, we get a vector

$$\mathbf{u}_1 = \mathbf{u}_1^* - \sum_{i=1}^{k} \left(\frac{p(\mathbf{u}_1^*, \mathbf{w}_i)}{p(\mathbf{w}_i, \mathbf{w}_i)}\right) \mathbf{w}_i$$

which by Theorem 3.8.5 is orthogonal to $\mathbf{w}_1, \ldots, \mathbf{w}_k$. Similarly, by subtracting from $\mathbf{u}_2^*$ its projections on the *nonzero* vectors among $\mathbf{w}_1, \ldots, \mathbf{w}_k, \mathbf{u}_1$ (note that $\mathbf{u}_1$ may be zero), we get a vector $\mathbf{u}_2$ orthogonal to all of these nonzero vectors. But $\mathbf{u}_2$ is also orthogonal to the zero vector; thus $\mathbf{u}_2$ is orthogonal to $\mathbf{w}_1, \ldots, \mathbf{w}_k, \mathbf{u}_1$. Continuing in the same manner, we obtain $\mathbf{u}_3$ by subtracting from $\mathbf{u}_3^*$ its projections on the nonzero vectors among $\mathbf{w}_1, \ldots, \mathbf{w}_k, \mathbf{u}_1, \mathbf{u}_2$; and so on. After $n$ steps, we end up with an orthogonal set $\{\mathbf{w}_1, \ldots, \mathbf{w}_k, \mathbf{u}_1, \ldots, \mathbf{u}_n\}$; let $\mathbf{w}_1, \ldots, \mathbf{w}_k, \mathbf{u}_{i_1}, \ldots, \mathbf{u}_{i_m}$ be the *nonzero* vectors in this set. Then $\mathbf{w}_1, \ldots, \mathbf{w}_k, \mathbf{u}_{i_1}, \ldots, \mathbf{u}_{i_m}$ are independent (Theorem 3.8.2), and

$$\begin{aligned}[\mathbf{w}_1, \ldots, \mathbf{w}_k, \mathbf{u}_{i_1}, \ldots, \mathbf{u}_{i_m}] &= [\mathbf{w}_1, \ldots, \mathbf{w}_k, \mathbf{u}_1, \ldots, \mathbf{u}_n] \\ &= [\mathbf{w}_1, \ldots, \mathbf{w}_k, \mathbf{u}_1^*, \ldots, \mathbf{u}_n^*] \qquad \text{(by 3.8.7 applied } n \text{ times)} \\ &\supseteq [\mathbf{u}_1^*, \ldots, \mathbf{u}_n^*] = E.\end{aligned}$$

Hence, $\{\mathbf{w}_1, \ldots, \mathbf{w}_k, \mathbf{u}_{i_1}, \ldots, \mathbf{u}_{i_m}\}$ is the desired orthogonal basis of $E$. (Note that this implies that $m = n - k$.) We state our result as a theorem.

**3.8.9 Theorem.** If $E$ has finite dimension, then every orthogonal set of nonzero vectors in $E$ can be extended to an orthogonal basis of $E$.

**3.8.10 Corollary.** If $E$ has finite dimension, then $E$ has an orthogonal basis.

*Proof.* The empty set is (trivially) an orthogonal set of nonzero vectors in $E$ and hence can be extended to an orthogonal basis of $E$ by 3.8.9. ∎

The process described above, by which an orthogonal set is extended to an orthogonal basis, is known as the *Gram-Schmidt process*. In this process, the set $\{\mathbf{u}_1^*, \ldots, \mathbf{u}_n^*\}$ (notations as above) need not actually be a basis of $E$; it need only *span* $E$ (independence of the $\mathbf{u}_i^*$ was never used).

**3.8.11 Example.** Find an orthogonal basis of the subspace $W = [(1, 2, 1, 2), (0, 1, 1, 1), (2, 1, 0, -1)]$ of $\mathbf{R}_4$.

*Solution.* We use the notations of the proof of Theorem 3.8.9, with $E = W$, $k = 0$, and with

$$\mathbf{u}_1^* = (1, 2, 1, 2); \qquad \mathbf{u}_2^* = (0, 1, 1, 1); \qquad \mathbf{u}_3^* = (2, 1, 0, -1).$$

We then compute

$$\begin{aligned}\mathbf{u}_1 &= \mathbf{u}_1^* = (1, 2, 1, 2); \\ \mathbf{u}_2 &= \mathbf{u}_2^* - \left(\frac{\mathbf{u}_2^* \cdot \mathbf{u}_1}{\mathbf{u}_1 \cdot \mathbf{u}_1}\right)\mathbf{u}_1 = \left(\frac{-1}{2}, 0, \frac{1}{2}, 0\right).\end{aligned}$$

It is convenient, before going further, to get rid of the fractions by replacing $\mathbf{u}_2$ by $2\mathbf{u}_2$ (such a substitution is permissible since it affects neither the orthogonality of the vectors nor the space which they span); thus, our new $\mathbf{u}_2$ is

$\mathbf{u}_2 = (-1, 0, 1, 0)$.

Similarly, using $\mathbf{u}_1$ and the new $\mathbf{u}_2$, we obtain

$$\mathbf{u}_3 = \mathbf{u}_3^* - \left( \frac{\mathbf{u}_3^* \cdot \mathbf{u}_1}{\mathbf{u}_1 \cdot \mathbf{u}_1} \right) \mathbf{u}_1 - \left( \frac{\mathbf{u}_3^* \cdot \mathbf{u}_2}{\mathbf{u}_2 \cdot \mathbf{u}_2} \right) \mathbf{u}_2 = \left( \frac{4}{5}, \frac{3}{5}, \frac{4}{5}, -\frac{7}{5} \right).$$

We may then multiply by 5 to eliminate fractions. Our final orthogonal basis of $W$ is

(3.8.12) $\{(1, 2, 1, 2), (-1, 0, 1, 0), (4, 3, 4, -7)\}$. ▮

**3.8.13 Example.** Extend the set 3.8.12 to an orthogonal basis of $\mathbf{R}_4$.

*Solution.* Again we use the notations of the proof of Theorem 3.8.9, this time with $E = \mathbf{R}_4$, $\mathbf{u}_i^* = \mathbf{e}_i$, $\mathbf{w}_1 = (1, 2, 1, 2)$, $\mathbf{w}_2 = (-1, 0, 1, 0)$, $\mathbf{w}_3 = (4, 3, 4, -7)$, $k = 3$. Then

$$\mathbf{u}_1 = \mathbf{u}_1^* - \sum_{i=1}^{3} \left( \frac{\mathbf{u}_1^* \cdot \mathbf{w}_i}{\mathbf{w}_i \cdot \mathbf{w}_i} \right) \mathbf{w}_i = \frac{1}{9}\left(2, -3, 2, 1\right).$$

Multiplying by 9 to eliminate fractions, we now have an orthogonal set

(3.8.14) $\{\mathbf{w}_1, \mathbf{w}_2, \mathbf{w}_3, \mathbf{u}_1\} = \{ (1, 2, 1, 2), (-1, 0, 1, 0), (4, 3, 4, -7), (2, -3, 2, 1) \}$.

Since this set contains four vectors, it is already the desired basis of $\mathbf{R}_4$; there is no need to compute $\mathbf{u}_2$, $\mathbf{u}_3$, $\mathbf{u}_4$. In fact, these additional vectors are necessarily zero (*why*?); for example,

$$\begin{aligned}
\mathbf{u}_2 &= \mathbf{u}_2^* - \sum_{i=1}^{3} \left( \frac{\mathbf{u}_2^* \cdot \mathbf{w}_i}{\mathbf{w}_i \cdot \mathbf{w}_i} \right) \mathbf{w}_i - \left( \frac{\mathbf{u}_2^* \cdot \mathbf{u}_1}{\mathbf{u}_1 \cdot \mathbf{u}_1} \right) \mathbf{u}_1 \\
&= (0, 1, 0, 0) - \frac{2}{10}\left(1, 2, 1, 2\right) \\
&\quad - \frac{3}{90}\left(4, 3, 4, -7\right) + \frac{3}{18}\left(2, -3, 2, 1\right) \\
&= (0, 0, 0, 0).
\end{aligned}$$ ▮

Given any vector $\mathbf{u} \neq \mathbf{0}$ in $E$, $\|\mathbf{u}\| \neq 0$ and, hence, the vector $\mathbf{u}' = \mathbf{u}/\|\mathbf{u}\|$ is defined; moreover, $\|\mathbf{u}'\| = 1$ (why?). Thus, any set $S$ of nonzero orthogonal vectors can be transformed into an *orthonormal* set $S'$ by dividing each vector in $S$ by its norm. If $S$ was a basis of $E$, $S'$ will also be a basis (*why*?). Hence, 3.8.9 and 3.8.10 remain true with the word "orthogonal" replaced by "orthonormal". We state this result as a theorem.

3.8.15 **Theorem.** If $E$ has finite dimension, then (a) $E$ has an orthonormal basis; (b) every orthonormal subset of $E$ can be extended to an orthonormal basis of $E$.

In practice, when computing an orthonormal basis it may be best, in order to simplify the arithmetic (especially when fractions are involved), to save the "normalization" (replacement of $\mathbf{u}$ by $\mathbf{u}/\|\mathbf{u}\|$) until the very end, rather than normalizing each vector as you go along. If you wish, for instance, to obtain an orthonormal basis for the space $W$ of Example 3.8.11, you would proceed as before to obtain the *orthogonal* basis 3.8.12 and *then* normalize each vector to obtain the orthonormal basis

$$\left\{(1/\sqrt{10}, 2/\sqrt{10}, 1/\sqrt{10}, 2/\sqrt{10}), (-1/\sqrt{2}, 0, 1/\sqrt{2}, 0),\right.$$
$$\left.(4/\sqrt{90}, 3/\sqrt{90}, 4/\sqrt{90}, -7/\sqrt{90})\right\}.$$

(On the other hand, by normalizing each vector in the orthogonal basis as you go along, you eliminate the denominators in formula 3.8.6; if the numbers in question are complicated decimals and you're using a computer, this may be the better way.)

The following two theorems on orthonormal sets will be needed later; we state them now for future reference.

3.8.16 **Theorem.** The canonical basis of $\mathbf{R}_n$ is an orthonormal set.

3.8.17 **Theorem.** If $\mathbf{u}_1, \ldots, \mathbf{u}_k$ are orthonormal vectors in $E$, and if $a_1, \ldots, a_k, b_1, \ldots, b_k$ are any scalars, then

$$p\left(\sum_{i=1}^{k} a_i\mathbf{u}_i, \sum_{j=1}^{k} b_j\mathbf{u}_j\right) = a_1b_1 + \cdots + a_kb_k.$$

We leave the proofs of 3.8.16 and 3.8.17 as exercises (use equation 3.8.1). As a special case of 3.8.17, if the vectors $\mathbf{u}_1, \ldots, \mathbf{u}_k$ are orthonormal, then

$$\|a_1\mathbf{u}_1 + \cdots + a_k\mathbf{u}_k\| = (a_1^2 + \cdots + a_k^2)^{1/2}.$$

(Interpret the latter geometrically!)

If $W$ is any subspace of the real Euclidean space $E$, we define the *orthogonal complement of* $W$ *(in* $E$*)* to be the set

$$W^\perp = \{\mathbf{v} \in E : \mathbf{v} \perp \mathbf{w} \text{ for all } \mathbf{w} \in W\}.$$

The next theorem should suggest to you a reason for using the word "complement" to describe the set $W^\perp$.

**3.8.18 Theorem.** If $E$ has finite dimension $n$ and if $W$ is any subspace of $E$, then

(a) $W^\perp$ is a subspace of $E$, and $E = W \oplus W^\perp$.

(b) $\dim W + \dim W^\perp = n$. In fact, any orthogonal basis $\{\mathbf{w}_1, ..., \mathbf{w}_k\}$ of $W$ can be extended to an orthogonal basis $\{\mathbf{w}_1, ..., \mathbf{w}_k, \mathbf{w}_{k+1}, ..., \mathbf{w}_n\}$ of $E$, and then $\{\mathbf{w}_{k+1}, ..., \mathbf{w}_n\}$ is necessarily an orthogonal basis of $W^\perp$.

(c) Every vector $\mathbf{u} \in E$ can be expressed in one and only one way in the form $\mathbf{u} = \mathbf{w} + \mathbf{v}$ where $\mathbf{w} \in W$ and $\mathbf{v} \in W^\perp$.

(d) $(W^\perp)^\perp = W$.

*Proof.* We note that $W$ must have an orthogonal basis by 3.8.10, and that by 3.8.9 we can extend any such basis $\{\mathbf{w}_1, ..., \mathbf{w}_k\}$ to an orthogonal basis $\{\mathbf{w}_1, ..., \mathbf{w}_k, \mathbf{w}_{k+1}, ..., \mathbf{w}_n\}$ of $E$. Any element $\mathbf{w}$ of $E$ can then be written in the form

$$\mathbf{w} = c_1\mathbf{w}_1 + \cdots + c_k\mathbf{w}_k + c_{k+1}\mathbf{w}_{k+1} + \cdots + c_n\mathbf{w}_n$$

and we have

$$p(\mathbf{w}, \mathbf{w}_i) = p\left(\sum_{j=1}^{n} c_j\mathbf{w}_j, \mathbf{w}_i\right) = \sum_{j=1}^{n} c_j p(\mathbf{w}_j, \mathbf{w}_i) = c_i p(\mathbf{w}_i, \mathbf{w}_i).$$

Hence, $\mathbf{w}$ is orthogonal to $\mathbf{w}_1, ..., \mathbf{w}_k$ if and only if $c_1, ..., c_k$ are zero, that is, if and only if $\mathbf{w} \in [\mathbf{w}_{k+1}, ..., \mathbf{w}_n]$. It follows that

$$W^\perp = [\mathbf{w}_{k+1}, ..., \mathbf{w}_n].$$

Part (b) of Theorem 3.8.18 follows at once, and parts (a) and (c) follow from the results of Section 2.8 (see Theorems 2.8.8 and 2.8.3). We leave the proof of (d) as an exercise (Exercise 4 at the end of this section).

*Example*: In $\mathbf{R}_3$, if $W$ is a plane through the origin, then $W^\perp$ is the line through the origin which is perpendicular to this plane. Note that $\dim W^\perp = 3 - 2 = 1$, by Theorem 3.8.18(b).

*Remark*: The vectors $\mathbf{w}$, $\mathbf{v}$ of 3.8.18(c) can be found explicitly if $\mathbf{u}$ is given. Indeed, if we let

$$\text{(3.8.19)} \quad \mathbf{w} = \sum_{i=1}^{k} \left(\frac{p(\mathbf{u}, \mathbf{w}_i)}{p(\mathbf{w}_i, \mathbf{w}_i)}\right) \mathbf{w}_i; \qquad \mathbf{v} = \mathbf{u} - \mathbf{w}$$

then $\mathbf{w} \in W$, $\mathbf{v} \in W^\perp$ by Theorem 3.8.5, and $\mathbf{u} = \mathbf{w} + \mathbf{v}$.

3.8.20 **Corollary.** If $\mathbf{w}_1, ..., \mathbf{w}_k$ are nonzero orthogonal vectors and if

$$\mathbf{u} = \sum_{i=1}^{k} c_i \mathbf{w}_i$$

is any linear combination of $\mathbf{w}_1, ..., \mathbf{w}_k$, then $c_i = p(\mathbf{u}, \mathbf{w}_i)/p(\mathbf{w}_i, \mathbf{w}_i)$.

*Proof.* Let $W = [\mathbf{w}_1, ..., \mathbf{w}_k]$; then $\mathbf{u} \in W$ and, hence, in 3.8.18(c), we must have $\mathbf{u} = \mathbf{w}$, $\mathbf{v} = \mathbf{0}$ by uniqueness. Since $\mathbf{w}$ is given by 3.8.19, the result follows. ▮

The next example exhibits an interesting application of the concept of orthogonal complement.

3.8.21 **Example.** (Solving Homogeneous Linear Equations) Find all real solutions $(x_1, x_2, x_3, x_4)$ of the system

$$\begin{aligned} x_1 + 2x_2 + x_3 + 2x_4 &= 0 \\ x_2 + x_3 + x_4 &= 0 \\ 2x_1 + x_2 \qquad - x_4 &= 0. \end{aligned} \qquad (3.8.22)$$

*Solution.* The set of all solutions of this system is precisely the orthogonal complement in $\mathbf{R}_4$ of the subspace

$$W = [\ (1, 2, 1, 2), (0, 1, 1, 1), (2, 1, 0, -1)\ ]$$

(*why*?). We have already seen (Examples 3.8.11 and 3.8.13) that the set 3.8.12 is an orthogonal basis of $W$ which can be extended to one of $\mathbf{R}_4$ by adjoining the vector $(2, -3, 2, 1)$; hence, the latter vector constitutes a basis of $W^\perp$ by 3.8.18(b). We conclude that the general solution of 3.8.22 has the form

$$(2c, -3c, 2c, c) \qquad (c \in \mathbf{R}).$$

(*Note*: A more efficient way of solving linear equations will be described in Sec. 5.10.) ▮

In Section 3.7 we defined the *perpendicular segment from a point to a flat*. The *existence* of such a perpendicular segment is a consequence of Theorem 3.8.18(c), as we now show.

3.8.23 **Theorem.** If $E$ has finite dimension, $Q$ is any point in $E$, and $S$ is any flat in $E$, then there exists a perpendicular segment from $Q$ to $S$.

*Proof.* Let $W$ be the direction space of $S$ and let us choose a fixed point $P_0$ on $S$; then $S = W + P_0$. Now apply 3.8.18(c) to the vector $\mathbf{u} = \mathbf{Ar}\, P_0Q$; we obtain $\mathbf{u} = \mathbf{w} + \mathbf{v}$ where $\mathbf{w} \in W$ and $\mathbf{v}$ is perpendicular to all vectors in $W$ (see Fig. 3.8.1). Let $A = P_0 + \mathbf{w}$; then

$$A \in P_0 + W = S$$

so that 3.7.10(a) holds; and

$$\mathbf{Ar}\, AQ = \mathbf{Ar}\, P_0Q - \mathbf{Ar}\, P_0A = \mathbf{u} - \mathbf{w} = \mathbf{v}$$

so that 3.7.10(b) holds. Hence, $\overline{QA}$ is the perpendicular segment from $Q$ to $S$. ∎

We have already shown that the perpendicular segment $\overline{QA}$ from $Q$ to $S$ is unique; in fact, we showed that $\overline{QA}$ is the shortest segment from $Q$ to any point on $S$. The point $A$ is called the *perpendicular projection of $Q$ on $S$*; also, if $\mathbf{v} = \mathbf{Ar}\, AQ$, then the point $Q' = A - \mathbf{v}$ is called the *reflection point of $Q$ through $S$*. Physically, we may think of $Q'$ as the mirror image of $Q$ where $S$ is regarded as the mirror (see Fig. 3.8.2). Note that $A$ is the midpoint of $\overline{QQ'}$, since

Figure 3.8.1

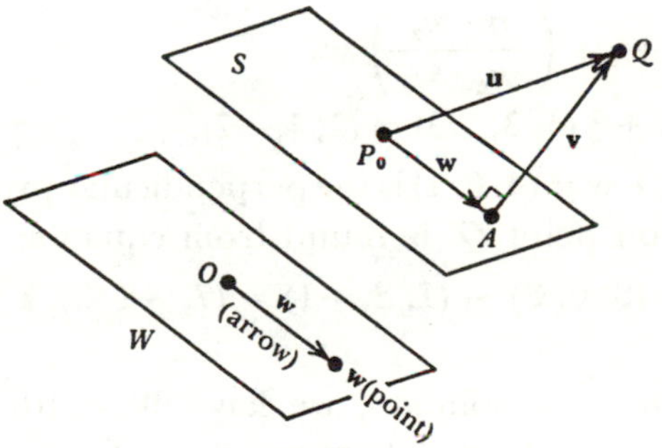


Figure 3.8.2

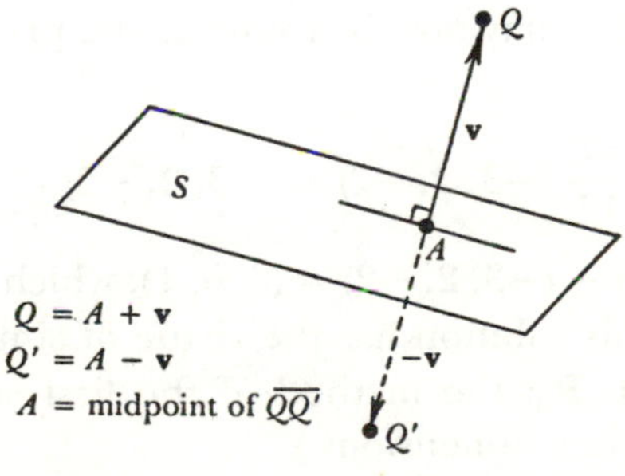

(3.8.24) $\frac{1}{2}(Q + Q') = \frac{1}{2}((A + \mathbf{v}) + (A - \mathbf{v})) = A.$

Also, solving 3.8.24 for $Q'$ yields

(3.8.25) $Q' = 2A - Q.$

If $Q$ lies on $S$, then $Q = A = Q'$.

**3.8.26 Example.** In $\mathbf{R}_3$, find the perpendicular projection of the point $Q = (1, 2, -1)$ on the plane $S = \langle(2, -1, 3), (2, 0, 4), (4, 2, 3)\rangle$. Also, find the reflection point of $Q$ through $S$.

*First solution.* The given plane $S$ has direction space $W = [(0, 1, 1), (2, 3, 0)]$. The basis of $W$ must first be orthogonalized; you should be able to apply the method of Example 3.8.11 to obtain the orthogonal basis $\{\mathbf{w}_1, \mathbf{w}_2\}$ of $W$, where

$$\mathbf{w}_1 = (0, 1, 1); \qquad \mathbf{w}_2 = (4, 3, -3).$$

Using the notations of Figure 3.8.1, we may choose $P_0 = (2, -1, 3)$, so that

$$\mathbf{u} = \mathbf{Ar}\, P_0Q = (1, 2, -1) - (2, -1, 3) = (-1, 3, -4).$$

The vector $\mathbf{w}$ of Figure 3.8.1 is then obtained from 3.8.19:

$$\mathbf{w} = \left(\frac{\mathbf{u}\cdot\mathbf{w}_1}{\mathbf{w}_1\cdot\mathbf{w}_1}\right)\mathbf{w}_1 + \left(\frac{\mathbf{u}\cdot\mathbf{w}_2}{\mathbf{w}_2\cdot\mathbf{w}_2}\right)\mathbf{w}_2$$
$$= -\tfrac{1}{2}(0, 1, 1) + \tfrac{1}{2}(4, 3, -3) = (2, 1, -2).$$

Finally, the point $A = P_0 + \mathbf{w} = (4, 0, 1)$ is the perpendicular projection of $Q$ on $S$. The reflection point $Q'$ is found from equation 3.8.25:

$$Q' = 2A - Q = (8, 0, 2) - (1, 2, -1) = (7, -2, 3). \blacksquare$$

*Second solution.* As in the first solution, we have $W = [(0, 1, 1), (2, 3, 0)]$ and $\mathbf{u} = (-1, 3, -4)$. We now borrow a result from Chapter 5 (see Section 5.4, Exercise 7, and Section 5.9, Exercise 8): the cross product $\mathbf{w}' = (0, 1, 1) \times (2, 3, 0) = (-3, 2, -2)$ constitutes a basis of $W^{\perp}$. The vector $\mathbf{v}$ of Figure 3.8.1 may now be found as the projection of $\mathbf{u}$ on $\mathbf{w}'$ (*why*?); that is,

$$\mathbf{v} = \left(\frac{\mathbf{u}\cdot\mathbf{w}'}{\mathbf{w}'\cdot\mathbf{w}'}\right)\mathbf{w}' = \frac{17}{17}(-3, 2, -2) = (-3, 2, -2).$$

Hence, $A = Q - \mathbf{v} = (1, 2, -1) - (-3, 2, -2) = (4, 0, 1)$, which agrees with our previous answer. (This solution has the virtue of simplicity, but the method only works in $\mathbf{R}_3$; the method of the first solution carries over unchanged to higher dimensions.)

*Remark*: If $Q$ is a point in $E$, $S$ is a flat, and $\mathbf{u}$ is any vector in $E$, then the statements

$$A \in S; \qquad Q - A \perp P - A \text{ for all points } P \in S$$

are equivalent (respectively) to the statements

$$A + \mathbf{u} \in S + \mathbf{u}; \qquad (Q + \mathbf{u}) - (A + \mathbf{u}) \perp (P + \mathbf{u}) - (A + \mathbf{u}) \text{ for all points } P + \mathbf{u} \in S + \mathbf{u}.$$

It thus follows from Definition 3.7.9(c) that

(3.8.27) $A$ is the perpendicular projection of $Q$ on $S$ $\Leftrightarrow A + \mathbf{u}$ is the perpendicular projection of $Q + \mathbf{u}$ on $S + \mathbf{u}$.

Statement 3.8.27 is illustrated by Figure 3.8.3.

**Reflections**

Let $E$ have finite dimension and let $S$ be a flat in $E$. The mapping of $E \to E$ which assigns to each point $Q$ in $E$ its reflection point through $S$ is called the *reflection* (*of* $E$) *through* $S$, and we shall denote this mapping $\mathcal{M}_S$. (Think of the letter $\mathcal{M}$ as standing for "mirror".) In the notations of Figure 3.8.2, $\mathcal{M}_S$ maps $Q \to Q'$, or

(3.8.28) $\mathcal{M}_S : A + \mathbf{v} \to A - \mathbf{v}$.

Equation 3.8.28 holds whenever $A$ belongs to $S$ and $\mathbf{v}$ is perpendicular to all vectors in the direction space of $S$. Note that if $\mathbf{v}$ has this property, then so does $-\mathbf{v}$, so that $\mathcal{M}_S$ maps $A - \mathbf{v} \to A + \mathbf{v}$. It follows at once that

(3.8.29) $\mathcal{M}_S\mathcal{M}_S = \mathbf{I}$;

that is, the square of any reflection is the identity mapping. (Equivalently, any reflection is its own inverse.)

One of the important properties of a reflection is that it preserves distance; that is, if $P$ and $Q$ are any points and $P'$, $Q'$ are their respective

Figure 3.8.3

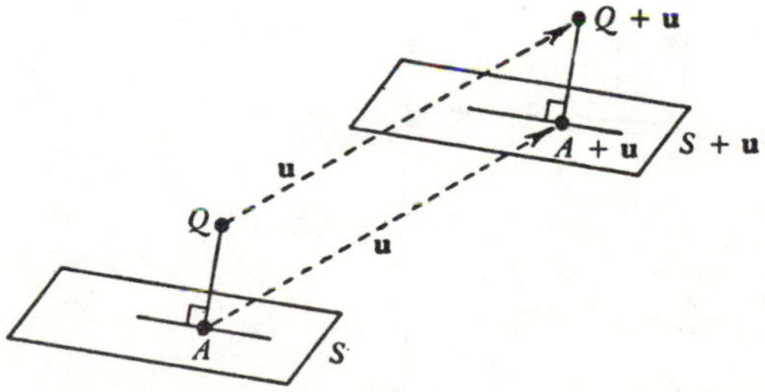

images under the reflection, then $d(P, Q) = d(P', Q')$ (see Fig. 3.8.4). We leave the formal proof to you (Exercise 11 below). A mapping (of one real Euclidean space into another) which preserves distance is called an *isometry*; in later chapters we shall study the isometries of $\mathbf{R}_n$ in considerable detail. Rotations and translations are other examples of isometries. It is a rather surprising fact that *every* isometry of $\mathbf{R}_n$ can be expressed as a product of reflections (see the exercises following Sec. 9.4).

The following theorem shows that any reflection through a flat is equal to a reflection through a subspace, followed by a translation; thus, when studying reflections, it often suffices to consider only reflections through subspaces.

Figure 3.8.4

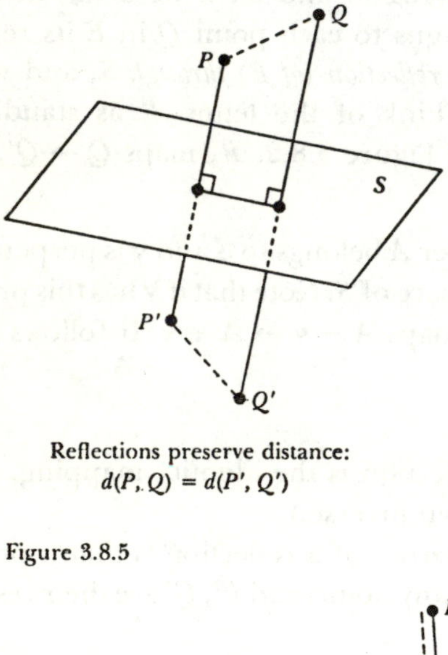


Reflections preserve distance:
$d(P, Q) = d(P', Q')$

Figure 3.8.5

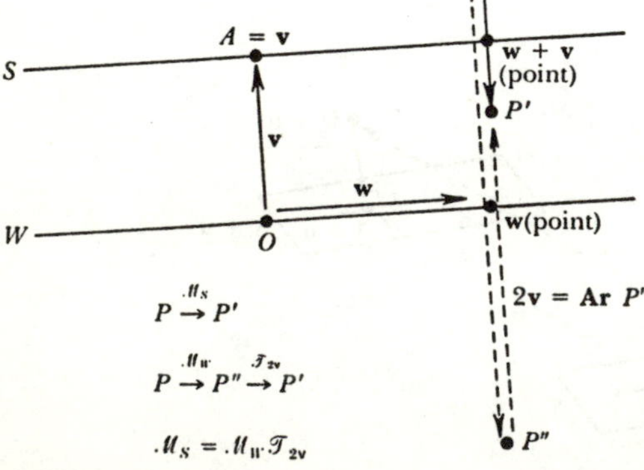

3.8.30 **Theorem.** Assume $\dim E < \infty$. If $S$ is a flat in $E$ with direction space $W$, then

(a) $S \cap W^{\perp}$ consists of a single element $\mathbf{v}$;

(b) For this vector $\mathbf{v}$, we have $\mathcal{M}_S = \mathcal{M}_W \mathcal{T}_{2\mathbf{v}}$, $\mathcal{T}_{2\mathbf{v}} = \mathcal{M}_W \mathcal{M}_S$.

*Proof* (cf. Fig. 3.8.5). By 3.7.10, a point belongs to $S \cap W^{\perp}$ if and only if it is the perpendicular projection $A$ of $O$ (the origin) on $S$; hence, 3.8.30(a) holds with $\mathbf{v} = A$ (see Fig. 3.8.5). To prove 3.8.30(b), let $P$ be any point in $E$ and let $\mathbf{w}$ be the perpendicular projection of $P$ on $W$. Then $P - \mathbf{w} \in W^{\perp}$; hence, $P - (\mathbf{w} + \mathbf{v}) \in W^{\perp}$; hence, $\mathbf{w} + \mathbf{v}$ is the perpendicular projection of $P$ on $W + \mathbf{v} = S$. Thus, we have

$$P\mathcal{M}_W = 2\mathbf{w} - P$$

$$P\mathcal{M}_S = 2(\mathbf{w} + \mathbf{v}) - P = (2\mathbf{w} - P) + 2\mathbf{v} = (P\mathcal{M}_W)\,\mathcal{T}_{2\mathbf{v}}.$$

Since $P$ was arbitrary, it follows that $\mathcal{M}_S = \mathcal{M}_W \mathcal{T}_{2\mathbf{v}}$. Solving for $\mathcal{T}_{2\mathbf{v}}$, we obtain

$$\mathcal{T}_{2\mathbf{v}} = \mathcal{M}_W^{-1}\mathcal{M}_S = \mathcal{M}_W \mathcal{M}_S. \ \blacksquare$$

3.8.31 **Remark.** An alternate proof of 3.8.30(a) goes as follows: by Theorem 3.8.18, we have $E = W \oplus W^{\perp}$ and $\dim W + \dim W^{\perp} = n$. Hence, $W \cap W^{\perp} = \{\mathbf{0}\}$ and

$$\dim (W \cap W^{\perp}) = 0 = n - (n - \dim W) - (n - \dim W^{\perp}).$$

Hence, Theorem 3.4.8 implies that $S \cap W^{\perp}$ is nonempty. Theorem 3.4.11 then implies that $S \cap W^{\perp}$ is a coset of $W \cap W^{\perp} = \{\mathbf{0}\}$ and thus consists of a single point. (By a similar argument, if $S'$ is any coset of $W^{\perp}$, then $S \cap S'$ is a single point.) $\blacksquare$

Another result relating reflections to translations is

3.8.32 **Theorem.** Assume $\dim E < \infty$. For any flat $S$ in $E$ and any vector $\mathbf{u} \in E$, $\mathcal{M}_{S+\mathbf{u}} = \mathcal{T}_{-\mathbf{u}}\,\mathcal{M}_S \mathcal{T}_{\mathbf{u}}$.

This theorem is illustrated by Figure 3.8.6 when $S$ is a line; as seen

Figure 3.8.6

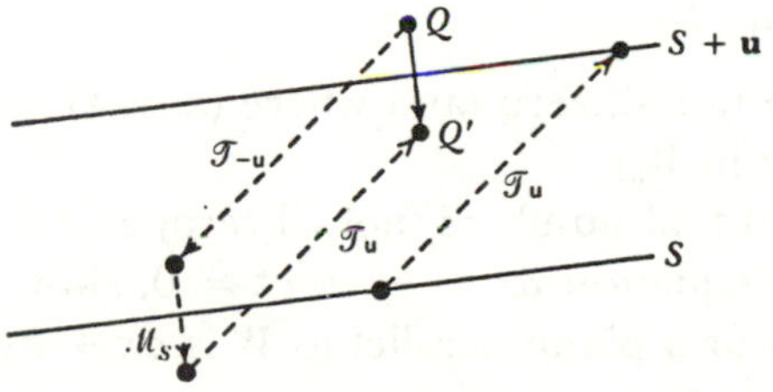


Solid arrow represents $\mathcal{M}_{S+\mathbf{u}}$

from the figure, both $\mathcal{M}_{S+\mathbf{u}}$ and $\mathcal{T}_{-\mathbf{u}}\mathcal{M}_S\mathcal{T}_{\mathbf{u}}$ map $Q \to Q'$. To prove the theorem formally, let $Q$ be any point in $E$. If $A$ is the perpendicular projection of $Q$ on $S + \mathbf{u}$, then, by 3.8.27, the point $A - \mathbf{u}$ is the perpendicular projection of $Q - \mathbf{u}$ on $S$. By 3.8.25, we thus have

$$Q\mathcal{M}_{S+\mathbf{u}} = 2A - Q$$
$$(Q - \mathbf{u})\mathcal{M}_S = 2(A - \mathbf{u}) - (Q - \mathbf{u}) = (2A - Q) - \mathbf{u}.$$

Hence,

$$Q\mathcal{T}_{-\mathbf{u}}\mathcal{M}_S\mathcal{T}_{\mathbf{u}} = (Q - \mathbf{u})\mathcal{M}_S + \mathbf{u} = 2A - Q = Q\mathcal{M}_{S+\mathbf{u}}$$

and 3.8.32 follows.

**Exercises**

1. Find an orthogonal basis of each of the following subspaces of $\mathbf{R}_4$:
   (a) $[(3, 0, 1, 1), (4, 2, 0, -1), (-5, 0, 3, 1)]$. (ANS)
   (b) $W^\perp$, if $W = [(2, 1, -1, 1)]$. (SOL)
   (c) The set $S$ of all solutions $(x, y, z, w)$ of the system
   $$3x \quad + 6z + 5w = 0$$
   $$x + y - 2z - \quad w = 0.$$ (SOL)
2. In $\mathbf{R}_5$, find an orthogonal basis of $W$, where
   $W = [(1, 1, 1, 1, 1), (4, 5, 2, -2, 1)]$.
*3. Prove Theorems 3.8.16 and 3.8.17.
*4. Prove Theorem 3.8.18(d). (SUG)
5. Let $E$ be the (infinite-dimensional) space described in Exercise 17, Section 3.7. Find a subspace $W$ of $E$ such that statements (c) and (d) of 3.8.18 are false. (This shows that the hypothesis of finite dimensionality in 3.8.18 is necessary.) (SUG)
6. Deduce from Theorem 3.8.18 that the general equation of a plane through the origin in $\mathbf{R}_3$ is
   $$ax + by + cz = 0$$
   where $a, b, c$ are not all zero (and where $(x, y, z)$ represents the general point in $\mathbf{R}_3$).
7. If $a, b, c$ are fixed real numbers (not all zero) and if $W$ is the plane in $\mathbf{R}_3$ with equation $ax + by + cz = 0$, show that the general equation of a plane parallel to $W$ is $ax + by + cz = d$, where $d$ is a constant.
8. Generalizing the preceding exercise, show that the general

equation of a hyperplane in $\mathbf{R}_n$ is $a_1x_1 + \cdots + a_nx_n = c$, where $a_1, \ldots, a_n, c$ are constants and not all of the $a_i$ are zero.

9. In each case, find the reflection point of $Q$ through the given plane. (ANS)
   (a) $Q = (6, 7, 1)$; plane $2x + 3y - z = 4$ in $\mathbf{R}_3$.
   (b) $Q = (-2, 4, -5)$; plane $\langle(1, 0, 1), (3, 1, 0), (2, 2, 3)\rangle$ in $\mathbf{R}_3$.
   (c) $Q = (1, 7, -1, 5)$; plane $\langle(1, 0, 0, 1), (0, 1, 2, 0), (1, 1, 1, 2)\rangle$ in $\mathbf{R}_4$.
10. In $\mathbf{R}_4$, show that the intersection of the two hyperplanes
$$x_1 + x_2 - x_3 - 2x_4 = -6$$
$$3x_1 - 2x_2 + 2x_3 - 6x_4 = 2$$
is a plane; then find the reflection point of $Q = (3, 0, 0, 0)$ through this plane. (ANS)

In Exercises 11–20, $E$ denotes a finite-dimensional real Euclidean space.

*11. If $S$ is any flat in $E$, prove that the reflection $\mathcal{M}_S$ preserves distance; that is, $d(P, Q) = d(P\mathcal{M}_S, Q\mathcal{M}_S)$ for all points $P, Q \in E$.
*12. Show that a point $P \in E$ is left fixed (i.e., mapped into itself) by the reflection $\mathcal{M}_S$ if and only if $P \in S$.
*13. If $W$ is a *subspace* of $E$, prove that
$$\mathcal{M}_W : \mathbf{w} \to \mathbf{w} \quad (\text{all } \mathbf{w} \in W).$$
$$\mathcal{M}_W : \mathbf{v} \to -\mathbf{v} \quad (\text{all } \mathbf{v} \in W^\perp).$$
(SUG)
*14. Generalize Theorem 3.8.30(b) as follows: if $S$ is a flat in $E$ having direction space $W$, and if $\mathbf{x} \in W^\perp$, then $\mathcal{M}_{S+\mathbf{x}} = \mathcal{M}_S\mathcal{T}_{2\mathbf{x}}$ and $\mathcal{T}_{2\mathbf{x}} = \mathcal{M}_S\mathcal{M}_{S+\mathbf{x}}$. (SOL)
15. Deduce from the preceding exercise: if $0 \leq k < \dim E$, then every translation of $E$ is the product of two reflections through $k$-flats.
16. (a) Prove: for any flat $S$ in $E$ and any vector $\mathbf{u} \in E$, the inverse of $\mathcal{M}_S\mathcal{T}_\mathbf{u}$ is $\mathcal{M}_{S+\mathbf{u}}\,\mathcal{T}_{-\mathbf{u}}$. (*Suggestion*: Use 3.8.32.) Draw a figure to illustrate.
    (b) Deduce: if $\mathbf{u}$ belongs to the direction space of $S$, then the inverse of $\mathcal{M}_S\mathcal{T}_\mathbf{u}$ is $\mathcal{M}_S\mathcal{T}_{-\mathbf{u}}$.
*17. Assume that $S$ is a flat, $W$ is the direction space of $S$, and $\mathbf{w} \in W$. Prove that the reflection $\mathcal{M}_S$ and the translation $\mathcal{T}_\mathbf{w}$ commute with each other; that is, $\mathcal{M}_S\mathcal{T}_\mathbf{w} = \mathcal{T}_\mathbf{w}\mathcal{M}_S$. Draw a figure to illustrate.
18. If $W_1, \ldots, W_k$ are subspaces of $E$, prove that
$$(W_1 + \cdots + W_k)^\perp = W_1^\perp \cap \cdots \cap W_k^\perp.$$

19. Let $W_1, \ldots, W_k$ be *mutually orthogonal* subspaces of $E$; that is, if $i \neq j$, then

$$\mathbf{w}_i \perp \mathbf{w}_j \qquad (\text{all } \mathbf{w}_i \in W_i,\ \mathbf{w}_j \in W_j).$$

Prove that if $S_i$ is any flat with direction space $W_i^\perp$ ($i = 1, \ldots, k$), then $S_i \cap \cdots \cap S_k$ is nonempty (and, hence, is a coset of $W_1^\perp \cap \cdots \cap W_k^\perp$). (This generalizes Remark 3.8.31.) (SUG)

20. By the *distance from a point P to a line L* (denoted $d(P, L)$), we mean the shortest distance from $P$ to any point on $L$; equivalently, the length of the perpendicular segment from $P$ to $L$. Let $L_1, L_2$ be lines in $E$, and suppose that the distances from points on $L_2$ to the line $L_1$ are bounded; that is, there exists a constant $c$ such that

$$d(P, L_1) \leqslant c \qquad (\text{all } P \in L_2).$$

Prove that $L_1$ and $L_2$ are parallel. (You may need to use the fact that a nonconstant polynomial in one variable with real coefficients is unbounded.)

21. Let $A, B, C, D$ be distinct points in the vector space $\mathbf{R}_2$, and assume that the opposite sides of quadrilateral $ABCD$ have equal lengths; that is, $|\mathbf{Ar}\, AB| = |\mathbf{Ar}\, DC|$ and $|\mathbf{Ar}\, AD| = |\mathbf{Ar}\, BC|$.

(a) Prove that the vector $\mathbf{Ar}\, AD - \mathbf{Ar}\, BC$ is orthogonal to both $\mathbf{Ar}\, AC$ and $\mathbf{Ar}\, BD$. (SOL)

(b) Deduce that either $\mathbf{Ar}\, AD = \mathbf{Ar}\, BC$ or $\mathbf{Ar}\, AC \mid \mathbf{Ar}\, BD$. Draw figures to illustrate both possibilities. (SOL)

## 3.9 Cosines, sines, and angles

Throughout this section, $E$ will denote a real Euclidean space with inner product $p$.

Let $\mathbf{u}$, $\mathbf{v}$ be any nonzero vectors in $E$. By Theorem 3.1.4, $\mathbf{u}$ and $\mathbf{v}$ can be represented by arrows having the same "initial point" $A$; say $\mathbf{u} = \mathbf{Ar}\, AB$, $\mathbf{v} = \mathbf{Ar}\, AQ$. As shown in Section 3.7, if

**(3.9.1)** $$t = p(\mathbf{u}, \mathbf{v})/\|\mathbf{u}\|^2,$$

then the vector $\mathbf{Ar}\, AP = t\mathbf{u}$ is perpendicular to the vector $\mathbf{Ar}\, PQ = \mathbf{v} - t\mathbf{u}$ (see Fig. 3.9.1(a), which coincides with Fig. 3.7.5 from Sec. 3.7). If $t > 0$, the arrows $\mathbf{u}$ and $t\mathbf{u}$ have the same direction (Fig. 3.9.1(a));

in this case, it is natural to define the *cosine* of the pair (**u**, **v**) (*notation*: cos(**u**, **v**)) to be the ratio $|\overline{AP}|/|\overline{AQ}|$, in agreement with the familiar right-triangle definition of cosines. Since $|\overline{AP}| = \|t\mathbf{u}\| = t\|\mathbf{u}\|$ and $|\overline{AQ}| = \|\mathbf{v}\|$, our definition becomes

$$\text{(3.9.2)} \quad \cos(\mathbf{u}, \mathbf{v}) = \frac{t\|\mathbf{u}\|}{\|\mathbf{v}\|}.$$

If instead $t < 0$, then **u** and $t\mathbf{u}$ have opposite directions (Fig. 3.9.1(b)) and the angle between **u** and **v** is obtuse; in this case, the usual convention that the cosine is negative remains consistent with 3.9.2. Hence, we take 3.9.2 to be our definition of the cosine *in all cases*. Observe that by 3.9.2, cos(**u**, **v**) depends only on **u** and **v**, not on the choice of "initial point" $A$.

By substituting 3.9.1 into 3.9.2, we obtain

$$\text{(3.9.3)} \quad \cos(\mathbf{u}, \mathbf{v}) = \left(\frac{p(\mathbf{u}, \mathbf{v})}{\|\mathbf{u}\|^2}\right)\frac{\|\mathbf{u}\|}{\|\mathbf{v}\|} = \frac{p(\mathbf{u}, \mathbf{v})}{\|\mathbf{u}\| \cdot \|\mathbf{v}\|},$$

which can be rewritten in the familiar form

$$\text{(3.9.4)} \quad p(\mathbf{u}, \mathbf{v}) = \|\mathbf{u}\| \cdot \|\mathbf{v}\| \cos(\mathbf{u}, \mathbf{v}).$$

If we then substitute 3.9.4 into 3.7.5(b), we get

$$\text{(3.9.5)} \quad \|\mathbf{u} - \mathbf{v}\|^2 = \|\mathbf{u}\|^2 + \|\mathbf{v}\|^2 - 2\|\mathbf{u}\| \cdot \|\mathbf{v}\| \cos(\mathbf{u}, \mathbf{v}).$$

You will probably recognize 3.9.5 as the *Law of Cosines* by examining Figure 3.7.1(b).

Similarly, it is natural to define the *sine* of the pair (**u**, **v**) to be the ratio $|\overline{PQ}|/|\overline{AQ}|$, where $P$, $A$, $Q$ are as above; that is,

$$\text{(3.9.6)} \quad \sin(\mathbf{u}, \mathbf{v}) = \|\mathbf{v} - t\mathbf{u}\|/\|\mathbf{v}\| = \left\|\mathbf{v} - \left(\frac{p(\mathbf{u}, \mathbf{v})}{\|\mathbf{u}\|^2}\right)\mathbf{u}\right\|/\|\mathbf{v}\|.$$

Figure 3.9.1

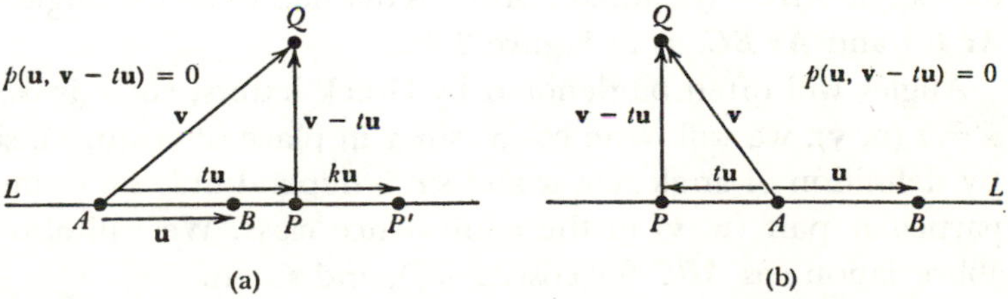

(Again, this depends only on **u** and **v**, not on $A$.) Since $\mathbf{Ar}\,AP \perp \mathbf{Ar}\,PQ$, we have $\|\mathbf{Ar}\,AP\|^2 + \|\mathbf{Ar}\,PQ\|^2 = \|\mathbf{Ar}\,AQ\|^2$ by Theorem 3.7.15; dividing both sides by $\|\mathbf{Ar}\,AQ\|^2$, we get

$$\cos^2(\mathbf{u}, \mathbf{v}) + \sin^2(\mathbf{u}, \mathbf{v}) = 1 \tag{3.9.7}$$

as expected. Conversely, we have the following theorem:

**3.9.8 Theorem.** If $\dim E \geqslant 2$, and if $a, b$ are any real numbers such that

$$a^2 + b^2 = 1; \qquad b \geqslant 0,$$

then there exist nonzero vectors $\mathbf{u}, \mathbf{v} \in E$ such that

$$a = \cos(\mathbf{u}, \mathbf{v}); \qquad b = \sin(\mathbf{u}, \mathbf{v}).$$

*Proof.* By the results of the preceding section, there exist two orthonormal vectors $\mathbf{w}_1, \mathbf{w}_2$ in $E$ since $\dim E \geqslant 2$. Let $\mathbf{u} = \mathbf{w}_1$, $\mathbf{v} = a\mathbf{w}_1 + b\mathbf{w}_2$; then $\cos(\mathbf{u}, \mathbf{v}) = a$ by 3.9.3 and 3.8.17, and then $\sin(\mathbf{u}, \mathbf{v}) = b$ by 3.9.7. ∎

Note that although the cosine may be either positive or negative, the sine is necessarily nonnegative. (Equivalently, we restrict our angles to the range from 0° to 180°. When *oriented* angles are defined in Sec. 6.2, this restriction will be removed.)

By this point it may have occurred to you that the word "angle" has not yet been formally defined. One possible approach is the following: start by calling two pairs of nonzero vectors $(\mathbf{u}, \mathbf{v})$ and $(\mathbf{u}', \mathbf{v}')$ "angularly equivalent" if $\cos(\mathbf{u}, \mathbf{v}) = \cos(\mathbf{u}', \mathbf{v}')$. (Informally, this is motivated by the fact that two angles between 0° and 180° are equal if and only if they have the same cosine.) It is obvious that "angular equivalence" is an equivalence relation. We then call the equivalence classes "angles"; specifically, the *angle between* **u** *and* **v** (denoted $\angle(\mathbf{u}, \mathbf{v})$) is defined to be the equivalence class to which the pair $(\mathbf{u}, \mathbf{v})$ belongs. Similarly, if $A, B, C$ are points in $E$ with $A \neq B$ and $B \neq C$, the "angle $ABC$" (denoted $\angle ABC$) is defined to be the angle between $\mathbf{Ar}\,BA$ and $\mathbf{Ar}\,BC$, as in Figure 3.9.2.

Angles will often be denoted by Greek letters, such as $\alpha, \beta, \gamma$. If $\alpha = \angle(\mathbf{u}, \mathbf{v})$, we will write $\cos\alpha$, $\sin\alpha$ in place of $\cos(\mathbf{u}, \mathbf{v})$, $\sin(\mathbf{u}, \mathbf{v})$. By definition of angle, $\cos\alpha$ and $\sin\alpha$ depend only on $\alpha$, not on the particular pair $(\mathbf{u}, \mathbf{v})$ in the equivalence class. We will also use the abbreviation $\cos ABC$ for $\cos(\angle ABC)$, and so on.

It is well known that we can associate with each angle a real number, called the number of *radians* in the given angle. (There is also a second such number, the number of *degrees*.) The best way to do this rigorously involves calculus and thus lies outside the scope of this book. (Briefly, the method is as follows: we use the limit concept (equivalently, the definite integral) to define "arc length" and then define radian measure in terms of arc length along a unit circle.) Some familiarity with facts about degrees and radians will be assumed informally, but we will not depend on such knowledge to prove numbered theorems.

### Half lines

If $A'$ lies between $A$ and $B$, and $C$ lies between $C'$ and $B$ (Fig. 3.9.3), it is intuitively evident that $\angle A'BC' = \angle ABC$. Facts such as these may be stated conveniently in terms of the concept of "half line," which we now define.

**3.9.9 Definition.** Let $P$ and $Q$ be distinct points in $E$. The *half line* (or *ray*) *from P through Q*, denoted $H(P, Q)$, is the set of all points $R$ such that $\mathbf{Ar}\ PR = t \cdot \mathbf{Ar}\ PQ$ for some scalar $t \geq 0$. (See Fig. 3.9.4.) The point $P$ is called the *initial point* of $H(P, Q)$.

Figure 3.9.2

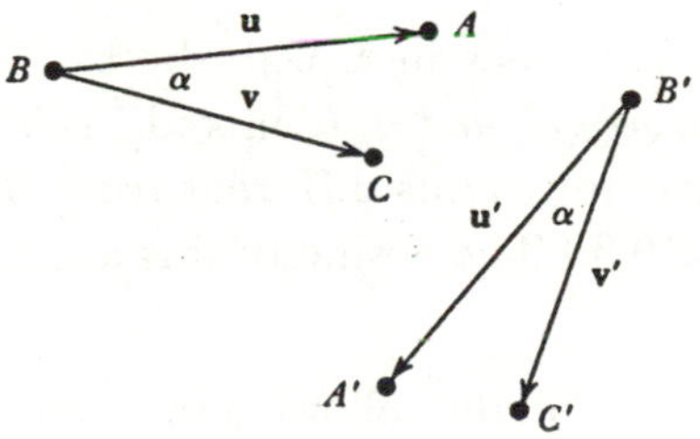


Equal angles:
$\alpha = \angle(\mathbf{u}, \mathbf{v}) = \angle ABC$
$= \angle(\mathbf{u}', \mathbf{v}')$
$= \angle A'B'C'$

Figure 3.9.3

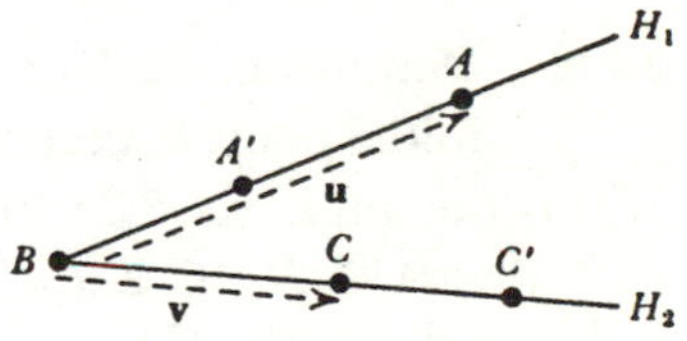


$\angle ABC = \angle A'BC'$
$= \angle(H_1, H_2)$
$= \angle(\mathbf{u}, \mathbf{v})$

Figure 3.9.4

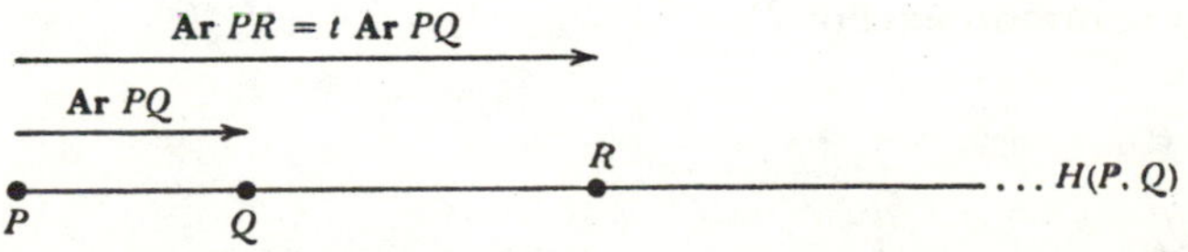

The condition $t \geqslant 0$ in Definition 3.9.9 is equivalent to specifying that **Ar** $PR$ have the same direction as **Ar** $PQ$ (unless $R = P$, which is the case when $t = 0$.) For those who prefer barycentric coordinates, an argument like those used in Section 3.2 easily implies that

**(3.9.10)** $H(P, Q) = \{aP + bQ : a + b = 1, b \geqslant 0\}.$

By comparing 3.9.10 with 3.2.16 and 3.2.10, we see that

$$\overline{PQ} \subseteq H(P, Q) \subseteq L(P, Q).$$

The initial point $P$ is uniquely determined by the half line $H(P, Q)$, in the sense that if $H(P, Q) = H(P', Q')$ then $P = P'$ (Theorem 3.9.13(a) below); this justifies the use of the word "the" before the phrase "initial point" in the last sentence of Definition 3.9.9. On the other hand, the point $Q$ is not uniquely determined by $H(P, Q)$; in fact, by Theorem 3.9.13(b) (see below), we get the same ray if $Q$ is replaced by any other point (except $P$) on $H(P, Q)$.

**3.9.11** **Theorem.** If $A' \in H(B, A)$ and $C' \in H(B, C)$, with $A' \neq B$ and $C' \neq B$, then $\angle A'BC' = \angle ABC$.

Theorem 3.9.11 is exhibited pictorially in Figure 3.9.3; we leave its proof as an exercise. In view of 3.9.11, the following definition is unambiguous:

**3.9.12** **Definition.** If $H_1$ and $H_2$ are half lines in $E$ with the same initial point $B$, then the *angle between $H_1$ and $H_2$* (denoted $\angle(H_1, H_2)$) is the angle $\angle ABC$ where $A$, $C$ are any points different from $B$ such that $A \in H_1$, $C \in H_2$. (See Fig. 3.9.3.) The cosine of this angle is denoted $\cos(H_1, H_2)$.

Some properties of half lines are collected in the following theorem:

**3.9.13** **Theorem.** (a) If $H(P, Q) = H(P', Q')$, then $P = P'$. (That is, the initial point of a given ray is unique.)

(b) If $Q'$ lies on $H(P, Q)$ and $Q' \neq P$, then $H(P, Q) = H(P, Q')$. (See Fig. 3.9.5.)

(c) Two distinct half lines with the same initial point $P$ have no point in common except $P$.

Figure 3.9.5

(d) If $P$ is the initial point of a half line $H$ and if $k$ is any positive real number, then there is a unique point $Q$ on $H$ such that $d(P, Q) = k$.

(e) Two rays $H_1, H_2$ with the same initial point are equal if and only if $\cos(H_1, H_2) = 1$.

To prove 3.9.13(a), assume $H(P, Q) = H(P', Q')$. Since $P \in H(P, Q)$, we have $P \in H(P', Q')$ and, hence,

$$P = (1-b)P' + bQ' \qquad (b \geqslant 0).$$

By similar reasoning,

$$P' = (1-c)P + cQ \qquad (c \geqslant 0).$$

Then by substitution,

$$2P - P' = (1-2b)P' + 2bQ' \in H(P', Q') = H(P, Q);$$
$$2P - P' = 2P - (1-c)P - cQ = (1+c)P - cQ.$$

Hence, by uniqueness of the barycentric coordinates on the line $\langle P, Q\rangle$ we must have $-c \geqslant 0$. But $c \geqslant 0$; hence, $c = 0$ and $P' = (1-c)P = P$.

We leave the proofs of parts (b), (c), (d) of Theorem 3.9.13 as exercises. To prove part (e), let $H_1 = H(B, A)$, $H_2 = H(B, C)$, $\mathbf{u} = \mathbf{Ar}\ BA$, $\mathbf{v} = \mathbf{Ar}\ BC$ as in Figure 3.9.3. Then

$$H_1 = H_2 \Leftrightarrow C \in H_1 \qquad \text{(by 3.9.13(b))}$$
$$\Leftrightarrow \mathbf{v} = t\mathbf{u} \qquad \text{for some } t > 0 \qquad \text{(by Def. 3.9.9).}$$

*If* $\mathbf{v} = t\mathbf{u}$ with $t > 0$, then

$$\cos(H_1, H_2) = \cos(\mathbf{u}, \mathbf{v}) = \cos(\mathbf{u}, t\mathbf{u}) = 1$$

(cf. Exercise 1 at the end of this section). Conversely, if $\cos(\mathbf{u}, \mathbf{v}) = 1$ then

$$p(\mathbf{u}, \mathbf{v}) = \|\mathbf{u}\| \cdot \|\mathbf{v}\| \qquad \text{(by 3.9.4)}$$
$$\sin(\mathbf{u}, \mathbf{v}) = 0 \qquad \text{(by 3.9.7)}$$

and, hence,

$$0 = \|\mathbf{v}\| \sin(\mathbf{u}, \mathbf{v}) = \left\|\mathbf{v} - \left(\frac{p(\mathbf{u}, \mathbf{v})}{\|\mathbf{u}\|^2}\right)\mathbf{u}\right\| \qquad \text{(by 3.9.6)}$$

from which

$$\mathbf{v} = \left(\frac{p(\mathbf{u}, \mathbf{v})}{\|\mathbf{u}\|^2}\right)\mathbf{u} = \left(\frac{\|\mathbf{u}\| \cdot \|\mathbf{v}\|}{\|\mathbf{u}\|^2}\right)\mathbf{u} = \left(\frac{\|\mathbf{v}\|}{\|\mathbf{u}\|}\right)\mathbf{u},$$

that is, $\mathbf{v} = t\mathbf{u}$, where $t > 0$; hence, $H_1 = H_2$. ∎

Figure 3.9.6

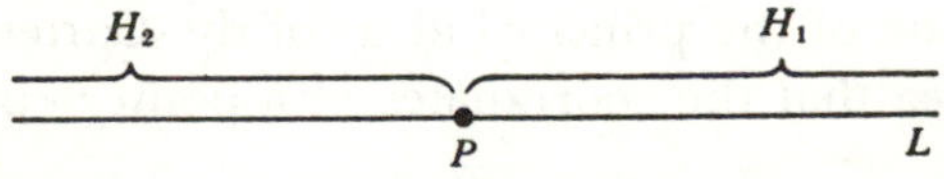

It is intuitively clear that any point on a line "divides" the line into two half lines (see Fig. 3.9.6). A rigorous statement of this fact is as follows:

**3.9.14 Theorem.** Let $L$ be a line in $E$ and let $P$ be a point on $L$. Then there are exactly two distinct rays having initial point $P$ which are contained in $L$; and $L$ is the union of these two half lines.

We leave the proof as an exercise.

**Exercises**

*1. For any scalar $t > 0$ and any nonzero vectors $\mathbf{u}$, $\mathbf{v}$, prove:
(a) $\cos(\mathbf{u}, \mathbf{u}) = 1$.
(b) $\cos(\mathbf{u}, \mathbf{v}) = \cos(t\mathbf{u}, \mathbf{v}) = \cos(\mathbf{v}, \mathbf{u})$.
(c) $\cos(-\mathbf{u}, \mathbf{v}) = -\cos(\mathbf{u}, \mathbf{v})$.

2. (a) How would you formally define a "right angle"? A "straight angle"?
(b) In high-school geometry we learn that "a straight angle consists of two right angles". Justify this statement formally. Draw a figure to illustrate.

*3. Prove 3.9.10.

*4. Prove Theorem 3.9.11.

*5. Prove parts (b), (c), (d) of Theorem 3.9.13. Draw figures to illustrate parts (c) and (d).

*6. Prove Theorem 3.9.14.

## 3.10 Orthogonal coordinate systems in $\mathbf{R}_n$

In a first course in analytic geometry, the usual approach is the so-called Cartesian approach: geometric concepts (point, line, distance, angle) are regarded as already known, so that we have a pre-existing geometric system upon which we then superimpose coordinates ($n$-tuples). In this book (specifically, in the present chapter), we do things the other way around, starting with $n$-tuples and using them to formally *define* point, line, length, and so on, after which the usual geometric "postulates" can be proved as theorems (see, e.g., Corollaries 3.2.3(a) and 3.2.8(a)). The object of this section is to incorporate into our formal approach one of the principal ideas of the Cartesian approach, namely, the idea that the coordinates of a point depend on a choice of coordinate axes. For simplicity, we shall restrict our considerations to the space $\mathbf{R}_n$.

To begin, we observe that to choose a set of mutually perpendicular axes in $\mathbf{R}_n$ really amounts to choosing an "origin" (the point where the axes meet) and a set of $n$ mutually orthogonal vectors of length 1. Figure 3.10.1 illustrates this in $\mathbf{R}_2$: if $P_0$ (the "origin") and $\mathbf{u}_1$, $\mathbf{u}_2$ are given, then the points $Q_i = P_0 + \mathbf{u}_i$ are determined, giving us not only the lines $L_1$ and $L_2$ but the positive direction along each line (the direction from $P_0$ to $Q_i$). Hence, we make the following definition:

**3.10.1 Definition.** By an *orthogonal coordinate system* in $\mathbf{R}_n$, we mean an $(n + 1)$-tuple of the form $K = (P_0; \mathbf{u}_1, ..., \mathbf{u}_n)$ where $P_0$ is a fixed point in $\mathbf{R}_n$ and $\{\mathbf{u}_1, ..., \mathbf{u}_n\}$ is an orthonormal basis of $\mathbf{R}_n$. The $n$ lines $L_i = P_0 + [\mathbf{u}_i]$ are called the *axes* of the coordinate system. By the *positive direction along* $L_i$, we mean the direction of the vector $\mathbf{u}_i$. (See Section 3.1, Exercise 8 for the definition of "direction".)

Let $K = (P_0; \mathbf{u}_1, ..., \mathbf{u}_n)$ be an orthogonal coordinate system in $\mathbf{R}_n$. Let $Q_i = P_0 + \mathbf{u}_i$; then the axes of $K$ are the lines

$$L_i = P_0 + [\mathbf{u}_i] = \langle P_0, Q_i \rangle.$$

Let $S_i$ be the flat determined by $P_0$ and all of the $Q$'s except $Q_i$; for example,

$$S_2 = \langle P_0, Q_1, Q_3, Q_4, ..., Q_n \rangle.$$

Equivalently, $S_i$ is the smallest flat containing all of the coordinate

Figure 3.10.1

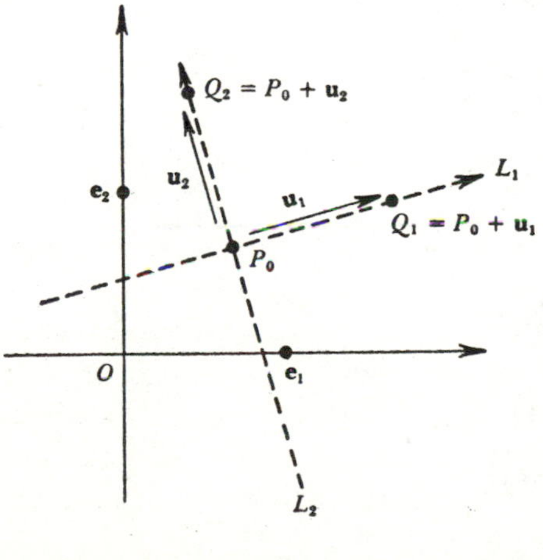

axes of $K$ except $L_i$. By Theorem 3.5.2, the direction space of $S_i$ is spanned by the set consisting of all of the **u**'s except $\mathbf{u}_i$; hence,

$$\dim S_i = n - 1.$$

The axis $L_i$ is, of course, perpendicular to $S_i$ at $P_0$. (In the special case $n = 2$, we have $L_1 = S_2$ and $L_2 = S_1$; see Fig. 3.10.2.) If we now let $P = (x_i, \ldots, x_n)$ be any point in $\mathbf{R}_n$, the *coordinates of P in the system K* (also called the coordinates of $P$ with respect to the axes $L_i$) are the numbers $x'_1, \ldots, x'_n$ defined as follows: for each $i$, let $A_i$ be the perpendicular projection of $P$ on $S_i$ (cf. Section 3.8). Then

**(3.10.2)** $x'_i = \pm d(A_i, P) = \pm|\overline{A_iP}|$

where the sign is plus if the vectors $\mathbf{u}_i$, **Ar** $A_iP$ have the same direction, minus if they have opposite directions. (If **Ar** $A_iP = \mathbf{0}$ the sign is immaterial.) The number $x'_i$ is called the *directed distance* from $A_i$ (or $S_i$) to $P$; it equals the actual distance except for sign. For example, the point $P$ in Figure 3.10.2 has coordinates $(5, -2\frac{1}{2})$ with respect to the

Figure 3.10.2

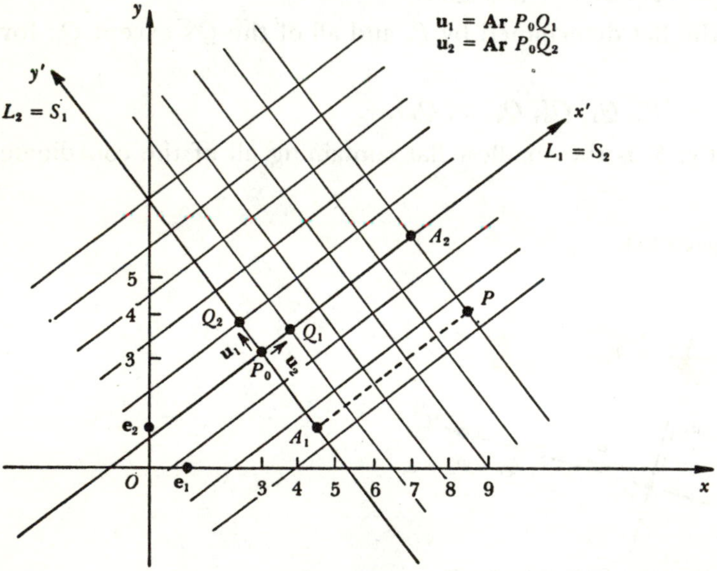


Coordinates of $P$ are:

$$(x, y) = \left(8\frac{1}{2}, 4\right)$$

$$(x', y') = \left(5, -2\frac{1}{2}\right)$$

axes $L_1$, $L_2$; the same point $P$ has coordinates $(8\frac{1}{2}, 4)$ with respect to the "usual" axes $[\mathbf{e}_1]$, $[\mathbf{e}_2]$.

**3.10.3 Theorem.** If $K = (P_0; \mathbf{u}_1, ..., \mathbf{u}_n)$ is an orthogonal coordinate system in $\mathbf{R}_n$ and if $P \in \mathbf{R}_n$, then the following statements are equivalent:

(a) The coordinates of $P$ in the system $K$ are $x_1', ..., x_n'$.

(b) $\mathbf{Ar}\ P_0P = x_1'\mathbf{u}_1 + \cdots + x_n'\mathbf{u}_n$.

We leave the proof as an exercise. In particular, if $K_0$ is the "usual" coordinate system $(O; \mathbf{e}_1, ..., \mathbf{e}_n)$, then the coordinates of any $n$-tuple $(x_1, ..., x_n)$ in the system $K_0$ are precisely $x_1, ..., x_n$, since

$$(x_1, ..., x_n) = x_1\mathbf{e}_1 + \cdots + x_n\mathbf{e}_n.$$

Thus, coordinates in the system $K_0$ coincide with coordinates as defined in Definition 3.1.1. More generally, the coordinates of $P$ in the system $K = (P_0; \mathbf{u}_1, ..., \mathbf{u}_n)$ are the components of $\mathbf{Ar}\ P_0P$ with respect to the ordered basis $(\mathbf{u}_1, ..., \mathbf{u}_n)$.

The next theorem tells us how to actually compute the coordinates of a given point in a given orthogonal coordinate system.

**3.10.4 Theorem.** Let $K = (P_0; \mathbf{u}_1, ..., \mathbf{u}_n)$ be an orthogonal coordinate system in $\mathbf{R}_n$; let $P \in \mathbf{R}_n$ and let $\mathbf{v} = \mathbf{Ar}\ P_0P$. If $x_1', ..., x_n'$ are the coordinates of $P$ in the system $K$, then

$$x_i' = \mathbf{v} \cdot \mathbf{u}_i \qquad (i = 1, ..., n).$$

*Proof.* By 3.10.3, $\mathbf{v} = \Sigma_i x_i'\mathbf{u}_i$. Since the $\mathbf{u}_i$ are orthogonal, we have

$$x_i' = \frac{\mathbf{v} \cdot \mathbf{u}_i}{\mathbf{u}_i \cdot \mathbf{u}_i}$$

by Corollary 3.8.20; and since the $\mathbf{u}_i$ are orthonormal, $\mathbf{u}_i \cdot \mathbf{u}_i = 1$. The result follows.

**3.10.5 Example.** Let $K$ be the coordinate system in $\mathbf{R}_2$ whose axes are the lines $L_1$, $L_2$ of Figure 3.10.2. Here, $\mathbf{u}_1 = (\frac{4}{5}, \frac{3}{5})$, $\mathbf{u}_2 = (-\frac{3}{5}, \frac{4}{5})$, $P_0 = (3, 3)$. If we take $P = (8\frac{1}{2}, 4)$ as in the Figure, then

$$\mathbf{v} = \mathbf{Ar}\ P_0P = \left(5\frac{1}{2}, 1\right)$$

$$x' = \mathbf{v} \cdot \mathbf{u}_1 = \left(5\frac{1}{2}, 1\right) \cdot \left(\frac{4}{5}, \frac{3}{5}\right) = 5$$

$$y' = \mathbf{v} \cdot \mathbf{u}_2 = \left(5\frac{1}{2}, 1\right) \cdot \left(-\frac{3}{5}, \frac{4}{5}\right) = -2\frac{1}{2}$$

as expected.

If $x'_1, \ldots, x'_n$ are the coordinates of a point $P$ in a given coordinate system $K$, it will be convenient to write

$$P = (x'_1, \ldots, x'_n)_K.$$

In particular, if $K_0 = (O; \mathbf{e}_1, \ldots, \mathbf{e}_n)$ (as above), then

$$(x_1, \ldots, x_n) = (x_1, \ldots, x_n)_{K_0}$$

for all $(x_1, \ldots, x_n) \in \mathbf{R}_n$. If $K$ and $P$ are as in Example 3.10.5 (Fig. 3.10.2), then

$$P = \left(8\frac{1}{2}, 4\right) = \left(8\frac{1}{2}, 4\right)_{K_0} = \left(5, -2\frac{1}{2}\right)_K.$$

If our "new" coordinate system $K$ has the form $(P_0; \mathbf{e}_1, \ldots, \mathbf{e}_n)$, where the $\mathbf{e}_i$ are the elements of the canonical basis, the corresponding transformation of axes and coordinates is called a *translation of axes*. In this case, the new axes are parallel to the "usual" axes; and if $P_0 = (a_1, \ldots, a_n)$, the old and new coordinates of any point $P = (x_1, \ldots, x_n) = (x'_1, \ldots, x'_n)_K$ are related by the equations

$$x'_i = x_i - a_i$$

(cf. 3.10.3(b) or 3.10.4). Figure 3.10.3 shows a translation of axes in $\mathbf{R}_2$ for which $P_0 = (2, 1)$; if we choose $P = (6, 4)$, then the $K$ coordinates of $P$ are $(x', y') = (6 - 2, 4 - 1) = (4, 3)$.

Figure 3.10.3

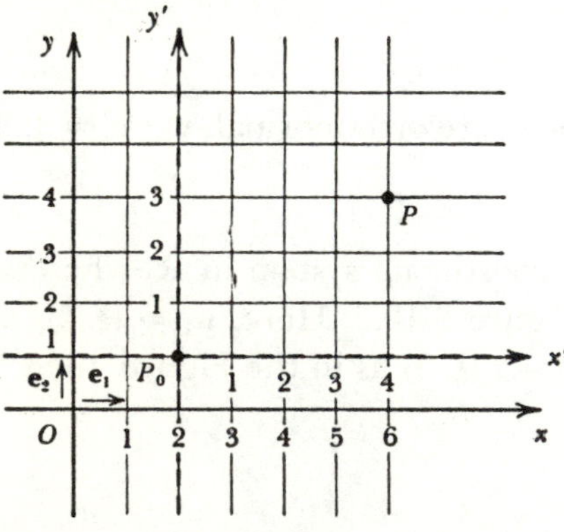


$P_0 = (2, 1)$
$K = (P_0; \mathbf{e}_1, \mathbf{e}_2)$
$K_0$ coordinates of $P$ are (6, 4)
$K$ coordinates of $P$ are (4, 3)

**Exercises**

1. Let $x$, $y$ represent coordinates with respect to the "usual" coordinate system $(O; \mathbf{e}_1, \mathbf{e}_2)$ in $\mathbf{R}_2$. Let $x'$, $y'$ represent coordinates in the system $K = (P_0; \mathbf{u}_1, \mathbf{u}_2)$ determined by taking the lines $2y = x + 3$ and $y = 14 - 2x$ as the $x'$ axis and $y'$ axis, respectively, where the positive directions of the new axes are as shown in Figure 3.10.4.
   (a) Find the point $P_0$ and the orthonormal vectors $\mathbf{u}_1$, $\mathbf{u}_2$. (ANS)
   (b) Find the coordinates of the point $P = (2, 0)$ in the system $K$. (ANS)
   (c) Draw an accurate graph of your results on graph paper.

*2. Let $K$ be an orthogonal coordinate system in $\mathbf{R}_n$, and let

$$P = (x_1', \ldots, x_n')_K; \qquad Q = (y_1', \ldots, y_n')_K.$$

Show that $d(P, Q) = ((x_1' - y_1')^2 + \cdots + (x_n' - y_n')^2)^{1/2}$. In other words, the distance formula remains valid if $K$ coordinates are used in place of "usual" coordinates.

*3. Prove Theorem 3.10.3.

Figure 3.10.4

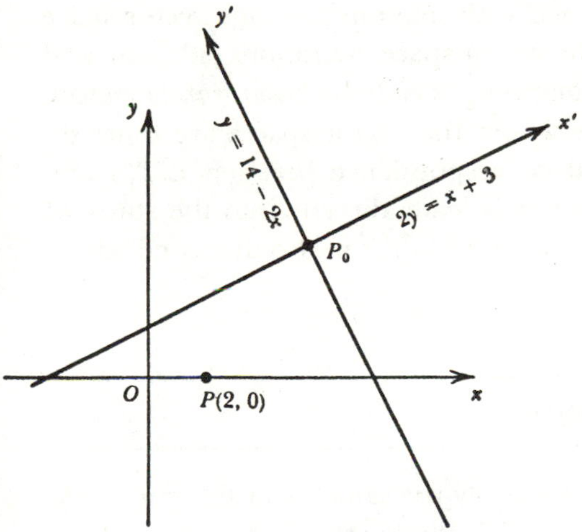

# 4 Linear transformations and matrices

## 4.0 Introduction

In studying mathematical systems, we are often concerned with those mappings of a system into itself (more generally, of one system into another) which preserve certain properties of the system. For example, if a binary operation "*" is defined on a set $S$, it is natural to study the mappings of $S \to S$ which preserve the operation "*". (In Section 1.8 we called such mappings *homomorphisms*.) Similarly, we can study mappings of geometric systems which preserve one or more geometric properties (e.g., angles, length, collinearity).

This chapter will be concerned with mappings of one vector space into another which preserve the vector space operations (addition and scalar multiplication); such a mapping is called a *linear transformation*, or L.T. for short. In the case where the vector spaces are finite dimensional, there is a natural correspondence between L.T.'s and matrices, so that the study of L.T.'s leads directly into the study of matrices. The geometric significance of L.T.'s is also discussed where appropriate.

## 4.1 Linear transformations

Let $V$ and $V'$ be vector spaces (possibly the same) over the same field $F$, and let $\mathbf{T}$ be a mapping of $V \to V'$. We say that $\mathbf{T}$ *preserves addition* if, whenever $\mathbf{u}, \mathbf{v}$ are elements of $V$ and $\mathbf{u}', \mathbf{v}'$ are their images under $\mathbf{T}$ (respectively), then the image of $\mathbf{u} + \mathbf{v}$ is $\mathbf{u}' + \mathbf{v}'$. That is, $\mathbf{T}$ preserves addition if

**(4.1.1)** $(\mathbf{u} + \mathbf{v})\mathbf{T} = \mathbf{u}\mathbf{T} + \mathbf{v}\mathbf{T} \qquad (\text{all } \mathbf{u}, \mathbf{v} \in V).$

(We introduced this terminology in Sec. 1.8, which you should review at the present time. Also review Sec. 1.2 on "Mappings".) Similarly, **T** *preserves scalar multiplication* if the image of $c\mathbf{v}$ is $c\mathbf{v}'$ for all $c \in F$ and all $\mathbf{v} \in V$; that is, if

**(4.1.2)** $(c\mathbf{v})\mathbf{T} = c(\mathbf{v}\mathbf{T}) \qquad (\text{all } \mathbf{v} \in V, c \in F).$

**4.1.3** **Definition.** Let $V$, $V'$ be vector spaces over the field $F$. A mapping $\mathbf{T} : V \to V'$ is called a *linear transformation* (abbreviated L.T.) if it preserves addition and scalar multiplication; that is, if it satisfies equations 4.1.1 and 4.1.2.

Note that in the terminology of Section 1.8, a linear transformation is simply a vector-space homomorphism.

Some basic properties of linear transformations are collected in the following theorem:

**4.1.4** **Theorem.** Let $V$, $V'$ be vector spaces over $F$; let $\mathbf{T} : V \to V'$ be a linear transformation. Then

(a) $\mathbf{0T} = \mathbf{0}$.

(b) $(-\mathbf{v})\mathbf{T} = -(\mathbf{vT}) \qquad (\text{all } \mathbf{v} \in V)$.

(c) $\left(\sum_{i=1}^{k} c_i\mathbf{v}_i\right)\mathbf{T} = \sum_{i=1}^{k} c_i(\mathbf{v}_i\mathbf{T}) \qquad (\text{all } k \in \mathbf{Z}^+, c_i \in F, \mathbf{v}_i \in V)$.

(d) If $(\mathbf{v}_1, ..., \mathbf{v}_k)$ is a linearly dependent $k$-tuple of vectors in $V$, then the $k$-tuple $(\mathbf{v}_1\mathbf{T}, ..., \mathbf{v}_k\mathbf{T})$ is linearly dependent.

Parts (a) and (b) of 4.1.4 have already been proved in Chapter 1 (translate Theorem 1.8.4 into additive notation). We leave parts (c) and (d) as exercises (use induction on $k$ to prove (c), then use (a) and (c) in proving (d)). ∎

Theorem 4.1.4(c) can be restated as follows: if an L.T. maps $\mathbf{v}_i \to \mathbf{v}'_i$, then it maps

$$\sum_{i=1}^{k} c_i\mathbf{v}_i \to \sum_{i=1}^{k} c_i\mathbf{v}'_i.$$

In other words, L.T.'s *preserve linear combinations*. This is one of the reasons why we call them "linear" transformations. (Another reason is that they preserve collinearity; see Exercise 10, this section.) Similarly, 4.1.4(d) may be restated as, "L.T.'s preserve linear dependence". (They do *not*, in general, preserve linear *independence*; cf. Exercise 6(b).)

If $S$ is a subset of $V$ and $\mathbf{T}$ is a mapping of $V \to V'$, the *image of $S$ under* $\mathbf{T}$ (denoted $S\mathbf{T}$) is the set of all vectors which are images (under $\mathbf{T}$) of elements of $S$. That is,

$$S\mathbf{T} = \{\mathbf{s}\mathbf{T} : \mathbf{s} \in S\}.$$

Similarly, if $S'$ is a subset of $V'$, the *pre-image of $S'$ under* $\mathbf{T}$ is the set of all pre-images of elements of $S'$; that is, it is the set

$$\{\mathbf{s} \in V : \mathbf{s}\mathbf{T} \in S'\}.$$

For example, let $V = V' = \mathbf{R}_2$ and let $\mathbf{T}$ map $(x, y) \to (x, 0)$ for all $(x, y) \in \mathbf{R}_2$ (geometrically, $\mathbf{T}$ is the perpendicular projection onto the $x$ axis; see Fig. 4.1.1). If $S$ is the shaded region in Figure 4.1.1, the image of $S$ under $\mathbf{T}$ is the line segment from $(a, 0)$ to $(b, 0)$; the pre-image of this line segment is the entire vertical strip between the lines $x = a$ and $x = b$. The pre-image of the origin is the $y$ axis; the pre-image of the line $x = y$ is also the $y$ axis (*why*?). The image of the $y$ axis is the origin; the image of any nonempty subset of the $y$ axis is also the origin. The pre-image of the shaded region $S$ is the empty set, since no vector has its image in $S$.

**4.1.5 Theorem.** Let $\mathbf{T} : V \to V'$ be a linear transformation.

(a) If $S$ is any subspace of $V$, then the image of $S$ under $\mathbf{T}$ is a subspace of $V'$.

(b) If $S'$ is any subspace of $V'$, then the pre-image of $S'$ under $\mathbf{T}$ is a subspace of $V$.

Figure 4.1.1

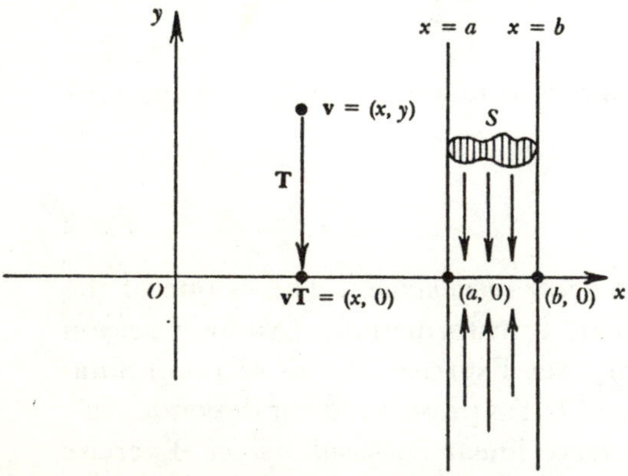


arrows indicate the effects of **T**

We leave the proof of Theorem 4.1.5 as an exercise. Two special cases of Theorem 4.1.5 (namely, the image of $V$ and the pre-image of $\{\mathbf{0}\}$) turn out to be particularly important, and we will study them in more detail in Section 4.6.

Although L.T.'s are defined algebraically, the preservation of algebraic properties is closely related to the preservation of geometric properties; as a result, certain types of geometrically defined transformations turn out to be linear provided that they leave the origin fixed (by 4.1.4(a), any L.T. must map $\mathbf{0} \to \mathbf{0}$). The following theorem is a case in point. (For another, see Theorem 4.9.9.)

**4.1.6 Theorem.** Let $E_1$, $E_2$ be real Euclidean spaces, and let $\mathbf{T}$ be a mapping of $E_1 \to E_2$ which preserves distance; that is, for all points $P$, $Q$ in $E_1$,

**(4.1.7)** $$d(P, Q) = d(P\mathbf{T}, Q\mathbf{T}).$$

Then $\mathbf{T}$ is a linear transformation provided that $\mathbf{0T} = \mathbf{0}$.

Recall that a mapping satisfying 4.1.7 is called an *isometry* (Sec. 3.8); thus Theorem 4.1.6 states that any isometry which leaves the origin fixed is linear. To prove the theorem, we note that 4.1.7 is equivalent to the equation

**(4.1.8)** $$\|\mathbf{u} - \mathbf{v}\| = \|\mathbf{uT} - \mathbf{vT}\| \qquad (\text{all } \mathbf{u}, \mathbf{v} \in E_1).$$

Setting $\mathbf{v} = \mathbf{0}$ in 4.1.8 and using the hypothesis $\mathbf{0T} = \mathbf{0}$, we obtain

**(4.1.9)** $$\|\mathbf{u}\| = \|\mathbf{uT}\| \qquad (\text{all } \mathbf{u} \in E_1).$$

Now for any vectors $\mathbf{u}, \mathbf{v}, \mathbf{w}$ in $E_1$, the identity

**(4.1.10)** $$\|\mathbf{u} + \mathbf{v} - \mathbf{w}\|^2 = \|\mathbf{u} - \mathbf{w}\|^2 + \|\mathbf{v} - \mathbf{w}\|^2 - \|\mathbf{u} - \mathbf{v}\|^2 + \|\mathbf{u}\|^2 + \|\mathbf{v}\|^2 - \|\mathbf{w}\|^2$$

is valid (proof: expand both sides using Theorem 3.7.5). By 4.1.8 and 4.1.9, every term on the right side of 4.1.10 remains unchanged if we replace $\mathbf{u}, \mathbf{v}, \mathbf{w}$ by $\mathbf{uT}, \mathbf{vT}, \mathbf{wT}$, respectively. Hence, the left side of 4.1.10 must remain unchanged also. That is,

$$\|\mathbf{u} + \mathbf{v} - \mathbf{w}\|^2 = \|\mathbf{uT} + \mathbf{vT} - \mathbf{wT}\|^2.$$

Setting $\mathbf{w} = \mathbf{u} + \mathbf{v}$, this becomes

**(4.1.11)** $$0 = \|\mathbf{uT} + \mathbf{vT} - (\mathbf{u} + \mathbf{v})\mathbf{T}\|^2.$$

Since only the zero vector has length zero, it follows from 4.1.11 that

$$\mathbf{uT} + \mathbf{vT} - (\mathbf{u} + \mathbf{v})\mathbf{T} = \mathbf{0}$$

and thus $(\mathbf{u} + \mathbf{v})\mathbf{T} = \mathbf{uT} + \mathbf{vT}$, proving that $\mathbf{T}$ preserves addition. The proof that $\mathbf{T}$ preserves scalar multiplication is similar, using the identity

$$\|a\mathbf{u} - \mathbf{w}\|^2 = (1 - a)\,(\|\mathbf{w}\|^2 - a\|\mathbf{u}\|^2) + a\|\mathbf{u} - \mathbf{w}\|^2 \tag{4.1.12}$$

in place of 4.1.10. ∎

It follows from Theorem 4.1.6, for example, that a rotation about the origin in $\mathbf{R}_2$, or a rotation about an axis through the origin in $\mathbf{R}_3$, is a linear transformation. Similarly, if $W$ is a subspace of a real Euclidean space, the reflection through $W$ is linear. (Recall that reflections were defined in Sec. 3.8. Rotations will be treated formally in Chap. 6.)

**Exercises**

*1. Prove parts (c) and (d) of Theorem 4.1.4.

*2. Let $V$, $V'$ be vector spaces over $F$. Prove: a mapping $\mathbf{T} : V \to V'$ is linear *if and only if*

$$(a\mathbf{u} + b\mathbf{v})\mathbf{T} = a(\mathbf{uT}) + b(\mathbf{vT}) \qquad (\text{all } a, b \in F;\ \text{all } \mathbf{u}, \mathbf{v} \in V).$$

3. Let $F$ be any field.
(a) If $a$, $b$, $c$, $d$ are any scalars (elements of $F$) and if $\mathbf{T} : F_2 \to F_2$ is defined by the equation

$$(x, y)\mathbf{T} = (ax + cy, bx + dy) \qquad (\text{all } x, y \in F),$$

prove that $\mathbf{T}$ is a linear transformation.
(b) If $a, b, c, d, e, f$ are scalars and $\mathbf{T} : F_2 \to F_2$ is defined by

$$(x, y)\mathbf{T} = (ax + cy + e, bx + dy + f),$$

prove that $\mathbf{T}$ is *not* linear unless $e = f = 0$.
(c) See if you can generalize part (a) to mappings of $F_n \to F_n$ (or even $F_n \to F_m$).

*4. Let $m$, $n$ be positive integers with $m \leqslant n$. If $\mathbf{T} : F_n \to F_m$ is defined by

$$(a_1, \ldots, a_m, a_{m+1}, \ldots, a_n)\mathbf{T} = (a_1, \ldots, a_m),$$

prove that $\mathbf{T}$ is linear.

5. Let $\mathbf{u}_1, \ldots, \mathbf{u}_n$ be elements of a vector space $V$ over $F$. If $\mathbf{T} : F_n \to V$ is defined by

$$(a_1, \ldots, a_n)\mathbf{T} = a_1\mathbf{u}_1 + \cdots + a_n\mathbf{u}_n,$$

prove that $\mathbf{T}$ is linear.

6. Theorem 4.1.4(d) asserts that L.T.'s preserve linear dependence.
   (a) Prove the following partial converse: if a linear transformation is *one-to-one* (cf. Sec. 1.2), then it preserves linear independence; that is, the images of independent vectors are independent.
   (b) Show by an example that the result of part (a) is false if the "one-to-one" hypothesis is deleted.

*7. Prove Theorem 4.1.5. (*Caution*: *Do not use bases* in your proof.)

8. If $\mathbf{T} : \mathbf{R}_3 \to \mathbf{R}_3$ is defined by $(x, y, z)\mathbf{T} = (x, y, x + y)$, find:
   (a) the image of the plane $2x = y$ (cf. Sec. 3.8, Exercise 6);
   (b) the pre-image of the line $\langle(1, 2, 6), (3, 0, 0)\rangle$. (ANS)

*9. If $\mathbf{T} : V \to V'$ is a linear transformation and $A$ and $B$ are arbitrary subsets of $V$, prove that $(A + B)\mathbf{T} = A\mathbf{T} + B\mathbf{T}$. (Thus, 4.1.1. holds for sets in place of vectors.)

10. Let $\mathbf{T} : V \to V'$ be a linear transformation.
   (a) If $W$ is a subspace of $V$ of dimension $k$, prove that $W\mathbf{T}$ (which is a subspace of $V'$ by 4.1.5) has dimension $\leqslant k$.
   (b) If $S$ is a flat in $V$, prove that $S\mathbf{T}$ is a flat in $V'$, and that if $\dim S = k$, then $\dim S\mathbf{T} \leqslant k$.
   (c) Deduce from (b) that $\mathbf{T}$ maps collinear points into collinear points; that is, if $P$, $Q$, $R$ lie on the same line in $V$, then $P\mathbf{T}$, $Q\mathbf{T}$, $R\mathbf{T}$ lie on the same line in $V'$. (Here we assume that $V'$ has dimension greater than zero.)

11. For any complex number $z$, let $z^*$ denote the complex conjugate of $z$. Let $\mathbf{T} : \mathbf{C}_2 \to \mathbf{C}_2$ be defined by
   $(z_1, z_2)\mathbf{T} = (z_1^*, z_2^*)$.
   (a) Show that $\mathbf{T}$ is *not* a linear transformation. (SUG)
   (b) Show that $\mathbf{T}$ *does* map collinear points into collinear points. (*Suggestion*: Use 3.2.10 as the equation of a line.) (SOL)

*12. Verify equations 4.1.10 and 4.1.12.

13. Prove that an isometry (a mapping $\mathbf{T}$ satisfying equation 4.1.7) is necessarily one-to-one.

14. Let $E_1$, $E_2$, $E_3$ be real Euclidean spaces. If $\mathbf{T} : E_1 \to E_2$ and $\mathbf{S} : E_2 \to E_3$ are isometries, prove that the product $\mathbf{TS}$ (cf. Sec. 1.3) is an isometry.

15. Let $E_1$, $E_2$ be a real Euclidean spaces. Using Theorem 4.1.6, prove that every isometry of $E_1 \to E_2$ can be expressed as a product $\mathbf{ST}$ where $\mathbf{S}$ is a *linear* isometry of $E_1 \to E_2$ and $\mathbf{T}$ is a translation of $E_2$.

16. Let $E_1$, $E_2$ be real Euclidean spaces; let $\mathbf{T}$ be a mapping of $E_1 \to E_2$. Prove that the following two statements are equivalent:
(a) $\mathbf{T}$ is an isometry and $\mathbf{0T} = \mathbf{0}$.
(b) $\mathbf{T}$ preserves inner products; that is, $p(\mathbf{u}, \mathbf{v}) = p(\mathbf{uT}, \mathbf{vT})$ for all $\mathbf{u}, \mathbf{v}$ in $E_1$.
17. Construct an alternate proof of Theorem 4.1.6 based on the result of Section 3.7, Exercise 15.
18. Let $V$ be any vector space and let $\mathbf{S} : V \to V$ be a linear transformation. Prove that for all $\mathbf{v} \in V$,
$\mathcal{T}_{\mathbf{v}}\mathbf{S} = \mathbf{S}\mathcal{T}_{\mathbf{vS}}$,
where $\mathcal{T}_{\mathbf{v}}$ denotes the translation $\mathbf{x} \to \mathbf{x} + \mathbf{v}$ ( as in Sec. 3.3.).

## 4.2 L.T.'s, bases, and matrices

Let $F$ be a field. Exercise 3(a) of the preceding section (rephrased slightly) asserts that if $a, b, c, d \in F$ and if

$$\begin{aligned} x' &= ax + cy \\ y' &= bx + dy, \end{aligned} \tag{4.2.1}$$

then the mapping $(x, y) \to (x', y')$ is a linear transformation of $F_2 \to F_2$. Interestingly, the converse is true: *every* L.T. of $F_2 \to F_2$ can be described by equations of the form 4.2.1, where $(x', y')$ denotes the image of $(x, y)$. Moreover, the result generalizes to arbitrary finite-dimensional vector spaces (equations 4.2.14 below). The key to the proof lies in the following theorem, which asserts that a linear transformation of $V \to V'$ is uniquely determined by what it does to a basis of $V$.

**4.2.2 Theorem.** Let $V$, $V'$ be vector spaces over $F$. Suppose that $V$ has a finite (ordered) basis $(\mathbf{v}_1, \ldots, \mathbf{v}_n)$, and let $\mathbf{v}'_1, \ldots, \mathbf{v}'_n$ be any elements of $V'$ (not necessarily distinct).

Then there exists one and only one linear transformation $\mathbf{T} : V \to V'$ such that

$$\mathbf{v}_i\mathbf{T} = \mathbf{v}'_i \qquad (i = 1, \ldots, n). \tag{4.2.3}$$

*Proof.* If $\mathbf{T} : V \to V'$ is an L.T. satisfying 4.2.3, then 4.1.4(c) implies that

$$\text{(4.2.4)}\quad \mathbf{T} : \sum_{i=1}^{n} c_i\mathbf{v}_i \to \sum_{i=1}^{n} c_i\mathbf{v}'_i$$

for all scalars $c_1, ..., c_n$. Since $\{\mathbf{v}_1, ..., \mathbf{v}_n\}$ is a basis, every element of $V$ can be expressed in the form

$$\text{(4.2.5)}\quad \sum_{i=1}^{n} c_i\mathbf{v}_i \qquad (c_i \in F)$$

and, hence, 4.2.4 determines **T** completely. We have thus shown that if there exists a linear transformation **T** satisfying 4.2.3, then **T** is unique. Conversely, to show that **T** actually exists, simply define **T** by 4.2.4; the definition is unambiguous since the expression of any element of $V$ in the form 4.2.5 is unique (Theorem 2.6.4). We leave it as an exercise to show that the mapping **T** defined by 4.2.4 is actually linear and satisfies 4.2.3. This completes the proof. ∎

We can now determine the equations of the most general linear transformation of $V \to W$, where $V$ and $W$ are any two finite-dimensional vector spaces over $F$. Let $\dim V = m$ and $\dim W = n$, so that $V$ has an ordered basis $(\mathbf{v}_1, ..., \mathbf{v}_m)$ and $W$ has an ordered basis $(\mathbf{w}_1, ..., \mathbf{w}_n)$. Suppose **T** is any linear transformation of $V \to W$. For each vector $\mathbf{v}_i$ in the basis, $\mathbf{v}_i\mathbf{T}$ belongs to $W$ and, hence, can be expressed (uniquely) as a linear combination of $\mathbf{w}_1, ..., \mathbf{w}_n$; that is,

$$\text{(4.2.6)}\quad \begin{aligned} \mathbf{v}_i\mathbf{T} &= a_{i1}\mathbf{w}_1 + a_{i2}\mathbf{w}_2 + \cdots + a_{in}\mathbf{w}_n \\ &= \sum_{j=1}^{n} a_{ij}\mathbf{w}_j \qquad (i = 1, ..., m). \end{aligned}$$

(Notice that each scalar has *two* subscripts: $a_{ij}$ is the coefficient of $\mathbf{w}_j$ in the expression for $\mathbf{v}_i\mathbf{T}$.) There are $m$ equations 4.2.6 (one for each $i$); if we write out each of the $m$ equations separately, we have

$$\text{(4.2.7)}\quad \begin{aligned} \mathbf{v}_1\mathbf{T} &= a_{11}\mathbf{w}_1 + a_{12}\mathbf{w}_2 + \cdots + a_{1n}\mathbf{w}_n \\ \mathbf{v}_2\mathbf{T} &= a_{21}\mathbf{w}_1 + a_{22}\mathbf{w}_2 + \cdots + a_{2n}\mathbf{w}_n \\ &\cdots \\ \mathbf{v}_m\mathbf{T} &= a_{m1}\mathbf{w}_1 + a_{m2}\mathbf{w}_2 + \cdots + a_{mn}\mathbf{w}_n. \end{aligned}$$

The scalars $a_{ij}$ are, of course, uniquely determined by **T**. Conversely, given any set of scalars $a_{ij}$ $(1 \leq i \leq m,\ 1 \leq j \leq n)$, Theorem 4.2.2 tells us that there is a unique linear transformation $\mathbf{T} : V \to W$ satisfying 4.2.7.

It is common practice to write the scalars $a_{ij}$ as a rectangular array:

$$(4.2.8) \quad A = \begin{bmatrix} a_{11} & a_{12} & \cdots & a_{1n} \\ a_{21} & a_{22} & \cdots & a_{2n} \\ & & \cdots & \\ a_{m1} & a_{m2} & \cdots & a_{mn} \end{bmatrix}.$$

Such an array is called an $m \times n$ ("*m by n*") *rectangular matrix*, or simply an $m \times n$ *matrix* (more precisely, an $m \times n$ *matrix over F*, indicating that the entries $a_{ij}$ belong to $F$). The matrix 4.2.8 has $m$ rows and $n$ columns, and may be denoted by a single capital letter $A$. If $m = n$, the matrix $A$ is said to be a *square* matrix. The results of the preceding paragraph can now be stated in the following form:

**4.2.9 Theorem.** Let $V$ and $W$ be vector spaces over $F$, having finite dimensions $m$, $n$, respectively. Given any fixed ordered bases $(\mathbf{v}_1, ..., \mathbf{v}_m)$ of $V$ and $(\mathbf{w}_1, ..., \mathbf{w}_n)$ of $W$, there is a one-to-one correspondence between the set of all linear transformations of $V \to W$ and the set of all $m \times n$ matrices over $F$; each linear transformation corresponds to a unique matrix and vice versa. Specifically, the linear transformation $\mathbf{T}$ corresponds to the matrix 4.2.8 where the $a_{ij}$ are the unique scalars satisfying 4.2.7 (or 4.2.6).

The above correspondence between L.T.'s and matrices is extremely important and will arise repeatedly; the equations which describe the correspondence (4.2.6 or 4.2.7) *should be memorized.* The matrix $A$ which corresponds to $\mathbf{T}$ is called the *matrix of* $\mathbf{T}$ with respect to the given ordered bases $(\mathbf{v}_1, ..., \mathbf{v}_m)$ and $(\mathbf{w}_1, ..., \mathbf{w}_n)$. In the case where the two vector spaces $V$ and $W$ are the same (i.e., there is only one vector space under discussion), it is usual to take the two bases to be the same also ($\mathbf{v}_i = \mathbf{w}_i$ for all $i$); we then speak of the matrix of the transformation $\mathbf{T} : V \to V$ with respect to a given basis of $V$.

(If this whole discussion seems abstract and you feel that a numerical example is called for, be patient; examples appear later in the section.)

**Notation and terminology**

The element which appears in the $i$th row, $j$th column of a matrix $A$ will be called the $(i, j)$ entry in $A$ and will be denoted $A_{ij}$. (It is usual for the first subscript to denote the row while the second denotes the column.) In 4.2.8, for example, $A_{ij} = a_{ij}$ for all $i$ and $j$. Similarly, if $A$ is the $2 \times 3$ matrix

$$\begin{bmatrix} 1 & 2 & 6 \\ 3 & 7 & 5 \end{bmatrix},$$

then $A_{13} = 6$, $A_{21} = 3$, and so forth. (If the subscript $i$ or $j$ has more than one digit, it is necessary to separate $i$ from $j$ by a comma; thus, the entry in the second row, twelfth column of $A$ would be denoted $A_{2,12}$ rather than $A_{212}$.) Still another common convention is to abbreviate 4.2.8 by

**(4.2.10)** $A = (a_{ij})$

indicating that the entries in $A$ are the scalars $a_{ij}$. The notation 4.2.10 is sometimes inadequate, however. For example, we may wish to discuss a matrix $M$ whose entry in the $i$th row, $j$th column is not $a_{ij}$ but $a_{ji}$; in this situation, to write $M = (a_{ji})$ would give the (erroneous) impression that $a_{ji}$ appears in the $j$th row, $i$th column. It is better to write

$$M_{ij} = a_{ji}$$

to indicate that the entry in the $i$th row, $j$th column of $M$ is $a_{ji}$.

If $A$ is an $m \times n$ matrix over $F$, the $n$-tuple $(a_{i1}, ..., a_{in})$, which consists of the entries in the $i$th row of $A$, may of course be regarded as an element of the vector space $F_n$; when so doing, we refer to this $n$-tuple as the $i$th *row vector* of $A$. Similarly, the $j$th *column vector* of $A$ is the $m$-tuple $(a_{1j}, ..., a_{mj})$ regarded as an element of $F_m$.

The equations of $\mathbf{T}$, which we have written in the form 4.2.7, can also be written in a form which generalizes 4.2.1. Still using the notations of Theorem 4.2.9, we can write the most general element of $V$ in the form

$$x_1\mathbf{v}_1 + \cdots + x_m\mathbf{v}_m$$

and we may denote the image of this element by $x'_1\mathbf{w}_1 + \cdots + x'_n\mathbf{w}_n$:

**(4.2.11)** $$(x_1\mathbf{v}_1 + \cdots + x_m\mathbf{v}_m)\mathbf{T} = x'_1\mathbf{w}_1 + \cdots + x'_n\mathbf{w}_n = \sum_{j=1}^{n} x'_j\mathbf{w}_j.$$

A second expression for $(x_1\mathbf{v}_1 + \cdots + x_m\mathbf{v}_m)\mathbf{T}$ as a linear combination of the $\mathbf{w}$'s is obtained using the fact that $\mathbf{T}$ is linear:

$$(x_1\mathbf{v}_1 + \cdots + x_m\mathbf{v}_m)\mathbf{T} = \left(\sum_{i=1}^{m} x_i\mathbf{v}_i\right)\mathbf{T}$$

$$= \sum_{i=1}^{m} x_i\,(\mathbf{v}_i\mathbf{T}) = \sum_{i=1}^{m} x_i \left( \sum_{j=1}^{n} a_{ij}\mathbf{w}_j \right)$$

**(4.2.12)** (by 4.2.6)

$$= \sum_{j=1}^{n} \left( \sum_{i=1}^{m} a_{ij}x_i \right) \mathbf{w}_j$$

(cf. Sec. 1.12, Exercise 10),

the purpose of the last step being to exhibit explicitly the coefficient of each $\mathbf{w}_j$. Since the $\mathbf{w}$'s are independent and the expressions 4.2.11 and 4.2.12 are equal, Theorem 2.4.7 allows us to *equate coefficients*. Doing so, we obtain

$$x'_j = \sum_{i=1}^{m} a_{ij}x_i = a_{1j}x_1 + a_{2j}x_2 + \cdots + a_{mj}x_m \qquad (j = 1, \ldots, n). \tag{4.2.13}$$

Writing out each equation 4.2.13 separately, we have

$$\begin{aligned} x'_1 &= a_{11}x_1 + a_{21}x_2 + \cdots + a_{m1}x_m \\ x'_2 &= a_{12}x_1 + a_{22}x_2 + \cdots + a_{m2}x_m \\ &\cdots \\ x'_n &= a_{1n}x_1 + a_{2n}x_2 + \cdots + a_{mn}x_m. \end{aligned} \tag{4.2.14}$$

Whereas 4.2.7 expresses the image of each basis vector of $V$ in terms of the basis vectors of $W$, 4.2.14 expresses the *components* of the image of *any* vector in terms of the components of the given vector. (By "components" we of course mean components with respect to the given bases.) In the special case where $V = W = F_2$ and our basis is the canonical basis $\{\mathbf{e}_1, \mathbf{e}_2\}$, 4.2.14 reduces to 4.2.1.

Observe that the arrangement of the scalars $a_{ij}$ is not the same in 4.2.14 as in 4.2.7. The scalar which appeared in 4.2.7 as the *j*th entry in the *i*th row now appears in 4.2.14 as the *i*th entry in the *j*th row (equivalently, as the *j*th entry in the *i*th *column*). The matrix which is obtained from a given matrix $A$ by changing rows to columns (equivalently, changing columns to rows) is called the *transpose* of $A$ and shall be denoted $A^t$; thus, if $A$ is the matrix 4.2.8, then

$$A^t = \begin{bmatrix} a_{11} & a_{21} \cdots & a_{m1} \\ a_{12} & a_{22} \cdots & a_{m2} \\ & \cdots & \\ a_{1n} & a_{2n} \cdots & a_{mn} \end{bmatrix}.$$

If $A$ is an $m \times n$ matrix, $A^t$ is an $n \times m$ matrix. The entry in the $i$th row, $j$th column of $A$ is the same as the entry in the $j$th row, $i$th column of $A^t$; that is,

$$A_{ij} = (A^t)_{ji}.$$

If $A$ represents the arrangement of the scalars in 4.2.7, then $A^t$ represents their arrangement in 4.2.14. Remember that it is $A$, not $A^t$, which is called the matrix of $\mathbf{T}$. (At least, this is the convention in all books which use right-hand notation for mappings, as this book does. In a book using left-hand notation, $A^t$ would be the matrix of $\mathbf{T}$. See the discussion in Sec. 1.2, second paragraph.)

We have shown that any L.T. of $V \to W$ (where $V$, $W$ are as above) has equations of the form 4.2.14. (This is yet another reason why L.T.'s are called "linear": they can be described by means of linear equations, at least in the finite-dimensional case.) Conversely, *any mapping of $V \to W$ whose equations are of the form 4.2.14 must be linear.* The proof is simple: if the mapping $\mathbf{S}$ has the equations 4.2.14, let $\mathbf{T}$ be the *linear* mapping satisfying 4.2.7. Our argument then shows that $\mathbf{T}$ satisfies 4.2.14; hence, $\mathbf{T} = \mathbf{S}$; hence, $\mathbf{S}$ is linear.

**4.2.15 Example.** In $\mathbf{R}_3$, let $L$ be a line through the origin and let $H$ be a plane through the origin which does not contain $L$. For each point $\mathbf{v} \in \mathbf{R}_3$, let $\mathbf{vT}$ be the point at which $H$ intersects the line through $\mathbf{v}$ parallel to $L$ (see Fig. 4.2.1). Prove that $\mathbf{T}$, as thus defined, is a linear transformation of $\mathbf{R}_3 \to \mathbf{R}_3$.

*Solution.* If $L = [\mathbf{u}_1]$ and $H = [\mathbf{u}_2, \mathbf{u}_3]$, then $(\mathbf{u}_1, \mathbf{u}_2, \mathbf{u}_3)$ is an ordered basis of $\mathbf{R}_3$ (why?). If $\mathbf{v} = x_1\mathbf{u}_1 + x_2\mathbf{u}_2 + x_3\mathbf{u}_3$, the line through $\mathbf{v}$ parallel to $L$ is

Figure 4.2.1

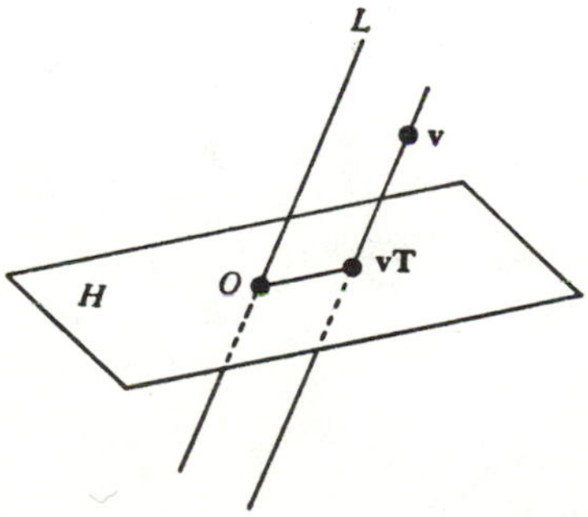

$$L + \mathbf{v} = [\mathbf{u}_1] + \mathbf{v} = [\mathbf{u}_1] + x_2\mathbf{u}_2 + x_3\mathbf{u}_3;$$

this line intersects $H$ precisely in the point $\mathbf{vT} = x_2\mathbf{u}_2 + x_3\mathbf{u}_3$. (*Note*: The fact that the intersection is a single point could also be deduced from Sec. 3.4, Exercise 3(c).) In other words, the components $(x_1', x_2', x_3')$ of $\mathbf{vT}$ with respect to the basis $(\mathbf{u}_1, \mathbf{u}_2, \mathbf{u}_3)$ are given by

$$\begin{aligned} x_1' &= 0 \\ x_2' &= \quad x_2 \\ x_3' &= \qquad x_3. \end{aligned}$$

Since these are linear equations of the form 4.2.14, $\mathbf{T}$ is linear. ▮

**4.2.16 Example.** Let $\mathbf{T} : \mathbf{R}_2 \to \mathbf{R}_2$ be the reflection through the line $y = x/2$. Find the matrix of $\mathbf{T}$ with respect to the canonical basis $(\mathbf{e}_1, \mathbf{e}_2)$ of $\mathbf{R}_2$. Also find the equations of $\mathbf{T}$ in the form 4.2.14.

*Solution.* Note that the equation $y = x/2$ does represent a line (Sec. 3.2, Exercise 19). We have already observed that $\mathbf{T}$ is linear (see comment following the proof of Theorem 4.1.6); hence, $\mathbf{T}$ will be completely determined if we can find the images of the basis vectors $\mathbf{e}_1$ and $\mathbf{e}_2$. One way to find $\mathbf{e}_1\mathbf{T}$ is to first find the perpendicular projection $P$ of $\mathbf{e}_1$ on the line $y = x/2$ (see Fig. 4.2.2) and then use formula 3.8.25:

$$\mathbf{e}_1\mathbf{T} = 2P - \mathbf{e}_1 \tag{4.2.17}$$

to obtain $\mathbf{e}_1\mathbf{T}$. (For two other ways of finding $\mathbf{e}_1\mathbf{T}$, see Exercises 3 and 4 at the end of this section.) The point $P$ can be found by the method

Figure 4.2.2

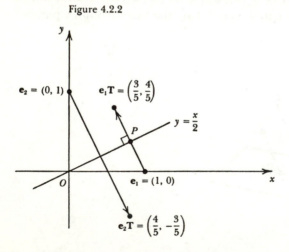

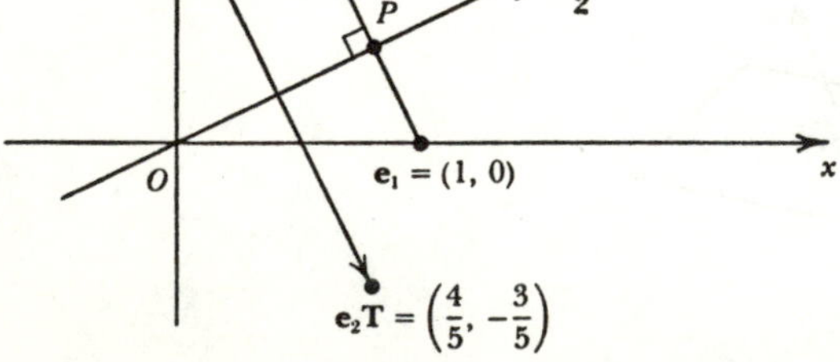

of Example 3.7.13 with $Q = (1,0)$, $A = (0,0)$, $\mathbf{u} = (2, 1)$. (The more general method of Example 3.8.26 would be used if **T** were a reflection through a higher-dimensional flat such as a plane.) We leave it to you to do the computations, which give the result

$$P = \left(\frac{4}{5}, \frac{2}{5}\right)$$

from which 4.2.17 gives

$$\mathbf{e}_1\mathbf{T} = \left(\frac{3}{5}, \frac{4}{5}\right) = \frac{3}{5}\mathbf{e}_1 + \frac{4}{5}\mathbf{e}_2.$$

A similar argument gives

$$\mathbf{e}_2\mathbf{T} = \left(\frac{4}{5}, -\frac{3}{5}\right) = \frac{4}{5}\mathbf{e}_1 - \frac{3}{5}\mathbf{e}_2.$$

These are the equations of the form 4.2.7, and from them we immediately conclude that the matrix of **T** is

$$A = \begin{bmatrix} \frac{3}{5} & \frac{4}{5} \\ \frac{4}{5} & -\frac{3}{5} \end{bmatrix}. \tag{4.2.18}$$

In this case $A^t = A$, and the equations 4.2.14 are

$$x' = \frac{3}{5}x + \frac{4}{5}y \qquad y' = \frac{4}{5}x - \frac{3}{5}y \tag{4.2.19}$$

where $(x', y') = x'\mathbf{e}_1 + y'\mathbf{e}_2$ is the image of $(x, y) = x\mathbf{e}_1 + y\mathbf{e}_2$ under **T**. ▮

It should be emphasized that the matrix of **T** depends on the original choice of basis or bases, not merely on **T**. For example, if in Example 4.2.16 we use the ordered basis ((1, 1), (0, 2)) in place of the canonical basis, we obtain

$$(1, 1)\mathbf{T} = \left(\frac{7}{5}, \frac{1}{5}\right) = \frac{7}{5}(1, 1) - \frac{3}{5}(0, 2)$$

$$(0, 2)\mathbf{T} = \left(\frac{8}{5}, -\frac{6}{5}\right) = \frac{8}{5}(1, 1) - \frac{7}{5}(0, 2).$$

(Note that after finding the images of the basis vectors, we must then express those images as linear combinations of the basis vectors.) Hence, the matrix of **T** with respect to the basis ((1, 1), (0, 2)) is

$$A' = \begin{bmatrix} \frac{7}{5} & -\frac{3}{5} \\ \frac{8}{5} & -\frac{7}{5} \end{bmatrix}$$

which is *not the same* as the matrix $A$ in 4.2.18. (However, the two matrices $A$ and $A'$ do have some properties in common; for example, they have the same determinant. See Chap. 5, Theorem 5.6.11.)

4.2.20 **Example.** Let $\mathbf{T} : \mathbf{R}_2 \to \mathbf{R}_2$ be the mapping which rotates each point counterclockwise about the origin through an angle $\theta$. The effect of $\mathbf{T}$ on the vectors $\mathbf{e}_1$ and $\mathbf{e}_2$ (as well as on an arbitrary vector $(x, y)$) is shown in Figure 4.2.3. Since $\mathbf{T}$ is distance-preserving (why?) and maps $O \to O$, $\mathbf{T}$ is linear. Looking at the two right triangles shown in Figure 4.2.3, both triangles have hypotenuse of length 1 and, hence, their other sides have lengths $\cos\theta$ and $\sin\theta$; it follows easily that $\mathbf{e}_1\mathbf{T}$ is the point $(\cos\theta, \sin\theta)$ and $\mathbf{e}_2\mathbf{T}$ is the point $(-\sin\theta, \cos\theta)$. That is,

$$\mathbf{e}_1\mathbf{T} = (\cos\theta)\,\mathbf{e}_1 + (\sin\theta)\,\mathbf{e}_2$$
$$\mathbf{e}_2\mathbf{T} = (-\sin\theta)\,\mathbf{e}_1 + (\cos\theta)\,\mathbf{e}_2.$$

Hence, the matrix of $\mathbf{T}$ with respect to the canonical basis is

$$A = \begin{bmatrix} \cos\theta & \sin\theta \\ -\sin\theta & \cos\theta \end{bmatrix}. \qquad (4.2.21)$$

By taking the transpose of $A$, we obtain the "equations of $\mathbf{T}$":

Figure 4.2.3

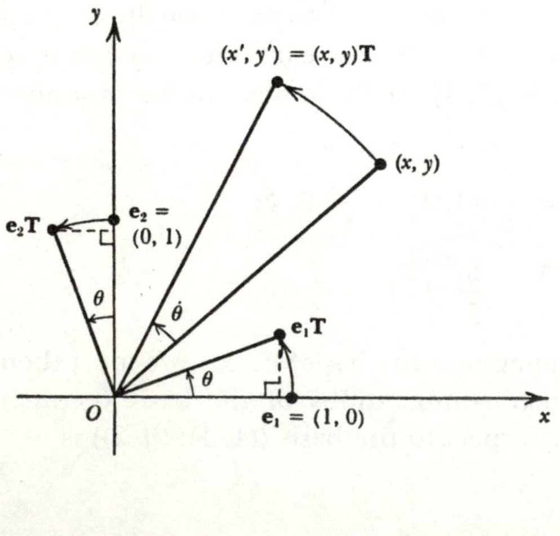

$$x' = x\cos\theta - y\sin\theta$$
(4.2.22)
$$y' = x\sin\theta + y\cos\theta.$$

(These differ from the equations of "rotation of axes" in standard calculus and analytic geometry texts in that the roles of $(x, y)$ and $(x', y')$ are reversed; see, for example, [13], Sec. 8.9. The reason for the difference is that "rotation of axes" involves moving the axes while keeping the points stationary, whereas here we are rotating the *points* while keeping the *axes* stationary.)

In connection with Example 4.2.20, we recall that two real numbers can be represented as the cosine and sine of an oriented angle if and only if the sum of their squares is 1. In other words, a $2 \times 2$ matrix represents a rotation about $O$ (in $\mathbf{R}_2$) if and only if the matrix has the form

$$\begin{bmatrix} a & b \\ -b & a \end{bmatrix}$$

where $a^2 + b^2 = 1$. (This result has been derived informally; where was the argument informal? The same result will be obtained rigorously in Sec. 6.1.)

We conclude this section with an important special case of Theorem 4.2.2. Let us take the space $V'$ of that theorem to be $F_n$, and let the vectors $\mathbf{v}_1', \dots, \mathbf{v}_n'$ be the elements of the canonical basis: $\mathbf{v}_i' = \mathbf{e}_i$. Equation 4.2.4 then becomes

(4.2.23) $$\mathbf{T} : \sum_{i=1}^{n} c_i\mathbf{v}_i \rightarrow (c_1, \dots, c_n).$$

It is easy to see that the mapping $\mathbf{T}$ defined by 4.2.23 is one-to-one and onto (*why*?). In Section 1.8, we defined an *isomorphism* to be a homomorphism which is one-to-one and onto. Our result may thus be stated as follows:

4.2.24 **Theorem.** If $V$ is any vector space of finite dimension $n$ over $F$, then $V$ is isomorphic to the vector space $F_n$. In fact, if $(\mathbf{v}_1, \dots, \mathbf{v}_n)$ is any ordered basis of $V$, the mapping 4.2.23 is an isomorphism of $V \rightarrow F_n$.

For example, if $V$ is the set of all real-valued functions $y = y(x)$ which satisfy the differential equation

(4.2.25) $$x^2y'' + xy' = y \qquad (x > 0),$$

a general result in the theory of differential equations (see Appendix E, Theorem C) implies that $V$ is a vector space of dimension 2 over $\mathbf{R}$. It is easily verified that the functions $y_1 = x$ and $y_2 = 1/x$ are solutions of 4.2.25, and these two functions are clearly linearly independent (why?); hence, they constitute a basis of $V$, so that the general solution of 4.2.25 is given by

$$y = ax + \frac{b}{x} \qquad (a, b \in \mathbf{R}).$$

Under the isomorphism of $V \to \mathbf{R}_2$ described by Theorem 4.2.24, the function $y = ax + b/x$ (an element of $V$) corresponds to the element $(a, b)$ of $\mathbf{R}_2$.

**Exercises**

*1. Complete the proof of Theorem 4.2.2 by showing that if $\mathbf{T}$ is the mapping defined by equation 4.2.4, then $\mathbf{T}$ is linear and satisfies 4.2.3.

2. Verify the values of $\mathbf{e}_1\mathbf{T}$ and $\mathbf{e}_2\mathbf{T}$ which are stated in Example 4.2.16. (Use the method suggested in the Example.)

3. In Example 4.2.16, the vector $\mathbf{e}_1\mathbf{T}$ must satisfy the following two conditions: (a) $|\mathbf{e}_1\mathbf{T}| = 1$ (why?); (b) the line through the points $\mathbf{e}_1$ and $\mathbf{e}_1\mathbf{T}$ is perpendicular to the line $y = x/2$. Express these two conditions as equations and then solve for the two coordinates of $\mathbf{e}_1\mathbf{T}$. Find $\mathbf{e}_2\mathbf{T}$ by a similar method.

4. In Example 4.2.16, let $\theta$ be the acute angle between the $x$ axis and the line $y = x/2$. Show that the coordinates of $\mathbf{e}_1\mathbf{T}$ are $(\cos 2\theta, \sin 2\theta)$; hence, find $\mathbf{e}_1\mathbf{T}$ by finding $\cos\theta$ and $\sin\theta$ and then using the double-angle formulas of trigonometry. Find $\mathbf{e}_2\mathbf{T}$ by a similar method. (This exercise is "informal"; why?)

5. With respect to the mapping $\mathbf{T}$ of Example 4.2.16, use equations 4.2.19 to plot (on graph paper) various points $P = (x, y)$ and their images $P\mathbf{T} = (x', y')$. Does each image point $P\mathbf{T}$ appear to be the reflection point of $P$ through the line $y = x/2$?

6. In the matrix 4.2.18, the entries $a_{12}$ and $a_{21}$ are equal. Determine whether or not the matrix of the reflection through an arbitrary line $y = mx$ in $\mathbf{R}_2$ has the same property. (Use the canonical basis.)

7. *Informally*, derive the "addition formulas" of trigonometry from equations 4.2.22.

8. Let $m$ be a fixed number different from zero. For each point $P = (a, b)$ in $\mathbf{R}_2$, let $P' = P\mathbf{T}$ be the unique point on the line $y = m(x - a)$ such that the vector $\mathbf{Ar}\ PP'$ is parallel to the $x$ axis (see Fig. 4.2.4). Also, let us write $(x, y)\mathbf{T} = (x', y')$.
   (a) *Without assuming that* $\mathbf{T}$ *is linear*, find $x'$ and $y'$ in terms of $x$ and $y$.
   (b) Deduce from (a) that $\mathbf{T}$ is linear, and determine the matrix of $\mathbf{T}$ with respect to the canonical basis. (ANS)
9. Let $m$ be a fixed number; for each point $P \in \mathbf{R}_2$, let $P\mathbf{T}$ be the perpendicular projection of $P$ on the line $y = mx$ (see Fig. 4.2.5). Show that $\mathbf{T}$ is linear and find the matrix of $\mathbf{T}$ with respect to the canonical basis. (ANS)
10. (*Generalization of the preceding exercise*) Let $E$ be a real Euclidean space of finite dimension $n$ and let $W$ be a subspace of $E$. For each point $P \in E$, let $P\mathbf{T}$ be the perpendicular projection of $P$ on $W$. Show that $\mathbf{T}$ is linear. ($\mathbf{T}$ is called the *perpendicular projection mapping of E on W*.) (SOL)
11. Find the matrix (with respect to the canonical basis of $\mathbf{R}_3$) of the reflection through the plane $x - y + 2z = 0$ in $\mathbf{R}_3$. (Cf. Example 3.8.26.) (ANS)
12. If the linear transformation $\mathbf{T} : \mathbf{R}_2 \to \mathbf{R}_2$ maps $(1, 2) \to (7, 1)$ and $(3, -2) \to (5, -5)$, find the matrix of $\mathbf{T}$ with respect to the canonical basis. (SOL)
13. If the linear transformation $\mathbf{T} : \mathbf{R}_2$ maps $(1, 1) \to (3, 0)$ and $(2, -1) \to (0, 3)$, find the matrix of $\mathbf{T}$ : (a) with respect to the canonical basis; (b) with respect to the ordered basis $((1, 1), (2, -1))$. (ANS)

Figure 4.2.4

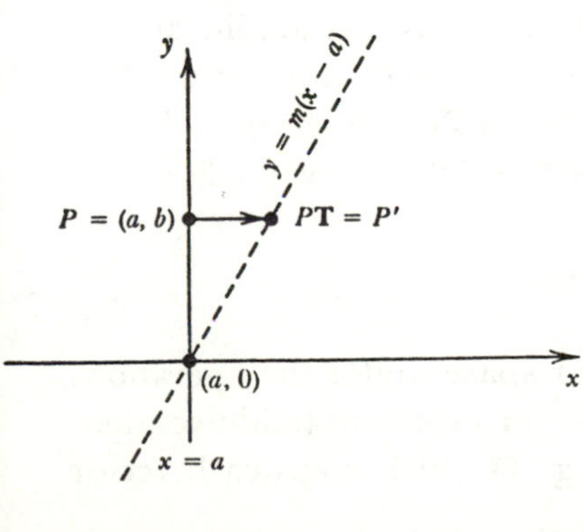


Figure 4.2.5

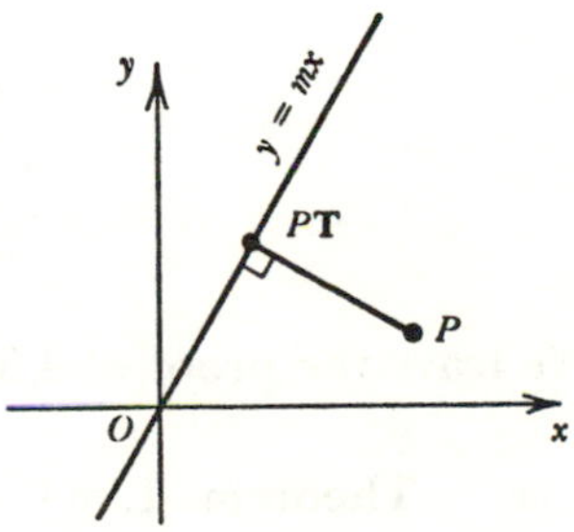

## 4.3 Algebraic operations on L.T.'s

If $V$, $V'$ are vector spaces over $F$ and if

$$\text{Lin}(V, V')$$

denotes the set of all linear transformations of $V \to V'$, it is possible to make Lin($V$, $V'$) itself a vector space. That is, we can define *addition of linear transformations* in such a way that the sum of two L.T.'s is itself an L.T.; similarly, we can multiply a scalar by an L.T. with the product again being an L.T. The necessary definitions are as follows:

**4.3.1 Definition.** Let $V$, $V'$ be vector spaces over $F$. If $\mathbf{S}$, $\mathbf{T}$ are any elements of Lin($V$, $V'$) and $c$ is any scalar, then $\mathbf{S} + \mathbf{T}$ is the mapping of $V \to V'$ defined by

$$\mathbf{v}(\mathbf{S} + \mathbf{T}) = \mathbf{v}\mathbf{S} + \mathbf{v}\mathbf{T} \qquad (\text{all } \mathbf{v} \in V) \tag{4.3.2}$$

and $c\mathbf{T}$ is the mapping of $V \to V'$ defined by

$$\mathbf{v}(c\mathbf{T}) = c(\mathbf{v}\mathbf{T}) \qquad (\text{all } \mathbf{v} \in V). \tag{4.3.3}$$

It seems hardly necessary to note that the right sides of 4.3.2 and 4.3.3 do indeed belong to $V'$, so that $\mathbf{S} + \mathbf{T}$ and $c\mathbf{T}$ are well-defined mappings of $V \to V'$. To show that these mappings are actually *linear* (i.e., belong to Lin($V$, $V'$)), we must verify the equations

$$(a\mathbf{u} + b\mathbf{v})(\mathbf{S} + \mathbf{T}) = a[\mathbf{u}(\mathbf{S} + \mathbf{T})] + b[\mathbf{v}(\mathbf{S} + \mathbf{T})] \tag{4.3.4}$$

$$(a\mathbf{u} + b\mathbf{v})(c\mathbf{T}) = a[\mathbf{u}(c\mathbf{T})] + b[\mathbf{v}(c\mathbf{T})] \tag{4.3.5}$$

for all $\mathbf{u}, \mathbf{v} \in V$ and all $a, b \in F$ (cf. Sec. 4.1, Exercise 2). The proof of 4.3.4 goes as follows:

$$\begin{aligned}
(a\mathbf{u} + b\mathbf{v})(\mathbf{S} + \mathbf{T}) &= (a\mathbf{u} + b\mathbf{v})\mathbf{S} + (a\mathbf{u} + b\mathbf{v})\mathbf{T} \quad (\text{by 4.3.2})\\
&= [a(\mathbf{u}\mathbf{S}) + b(\mathbf{v}\mathbf{S})] + [a(\mathbf{u}\mathbf{T}) + b(\mathbf{v}\mathbf{T})]\\
&\qquad (\text{since } \mathbf{S} \text{ and } \mathbf{T} \text{ themselves are linear})\\
&= [a(\mathbf{u}\mathbf{S}) + a(\mathbf{u}\mathbf{T})] + [b(\mathbf{v}\mathbf{S}) + b(\mathbf{v}\mathbf{T})] \quad (\text{why?})\\
&= a(\mathbf{u}\mathbf{S} + \mathbf{u}\mathbf{T}) + b(\mathbf{v}\mathbf{S} + \mathbf{v}\mathbf{T}) \quad (\text{why?})\\
&= a(\mathbf{u}(\mathbf{S} + \mathbf{T})) + b(\mathbf{v}(\mathbf{S} + \mathbf{T})) \quad (\text{by 4.3.2}).
\end{aligned}$$

We leave the proof of 4.3.5 as an exercise.

**4.3.6 Theorem.** Lin($V$, $V'$) is a vector space under the operations defined by 4.3.2 and 4.3.3. The zero element (additive identity) of Lin($V$, $V'$) is the "zero mapping" $\mathbf{O}$ which maps each vector

$\mathbf{u} \in V$ into the vector $\mathbf{0} \in V'$. The additive inverse of any element $\mathbf{T} \in \text{Lin}(V, V')$ is the mapping $-\mathbf{T}$ which is defined by

$$\mathbf{v}(-\mathbf{T}) = -(\mathbf{vT}) \qquad (\text{all } \mathbf{v} \in V).$$

*Proof.* By the preceding discussion, Lin($V$, $V'$) is closed under addition and scalar multiplication. The eight properties which comprise parts (c) and (d) of Definition 2.1.1 must also be verified; we give just one of the verifications here, leaving the others as exercises (see Exercise 1 at the end of this section). Specifically, we shall give the proof that

**(4.3.7)** $$c(\mathbf{S} + \mathbf{T}) = c\mathbf{S} + c\mathbf{T}$$

whenever $c \in F$ and $\mathbf{S}$, $\mathbf{T}$ belong to Lin($V$, $V'$) (this is part (d)(i) of Definition 2.1.1). Since both sides of 4.3.7 are mappings with domain $V$, they will be equal if they do the same thing to each vector $\mathbf{v} \in V$. But

$$\begin{aligned} \mathbf{v}[c(\mathbf{S} + \mathbf{T})] &= c[\mathbf{v}(\mathbf{S} + \mathbf{T})] && \text{(by 4.3.3)} \\ &= c[\mathbf{vS} + \mathbf{vT}] && \text{(by 4.3.2)} \\ &= c(\mathbf{vS}) + c(\mathbf{vT}) && \text{(why?);} \\ \mathbf{v}(c\mathbf{S} + c\mathbf{T}) &= \mathbf{v}(c\mathbf{S}) + \mathbf{v}(c\mathbf{T}) && \text{(by 4.3.2)} \\ &= c(\mathbf{vS}) + c(\mathbf{vT}) && \text{(why?),} \end{aligned}$$

and 4.3.7 follows. ∎

The *product of two mappings* has already been defined in Section 1.3: if $\mathbf{S} : V \to V'$ and $\mathbf{T} : V' \to V''$, then the mapping $\mathbf{ST}$ of $V \to V''$ is defined by

$$\mathbf{v}(\mathbf{ST}) = (\mathbf{vS})\mathbf{T} \qquad (\text{all } \mathbf{v} \in V).$$

If $\mathbf{S}$ and $\mathbf{T}$ are both linear, then

$$\begin{aligned} (a\mathbf{u} + b\mathbf{v})(\mathbf{ST}) &= [(a\mathbf{u} + b\mathbf{v})\mathbf{S}]\mathbf{T} && \text{(by definition of } \mathbf{ST}) \\ &= [a(\mathbf{uS}) + b(\mathbf{vS})]\mathbf{T} && \text{(since } \mathbf{S} \text{ is linear)} \\ &= a[(\mathbf{uS})\mathbf{T}] + b[(\mathbf{vS})\mathbf{T}] && \text{(since } \mathbf{T} \text{ is linear)} \\ &= a[\mathbf{u}(\mathbf{ST})] + b[\mathbf{v}(\mathbf{ST})] && \text{(why?)} \end{aligned}$$

and, hence, $\mathbf{ST}$ is linear also. We have thus proved part (a) of the following theorem:

**4.3.8** **Theorem.** If $\mathbf{S} \in \text{Lin}(V, V')$ and $\mathbf{T} \in \text{Lin}(V', V'')$, then
(a) $\mathbf{ST} \in \text{Lin}(V, V'')$;
(b) For all scalars $c$, $c(\mathbf{ST}) = (c\mathbf{S})\mathbf{T} = \mathbf{S}(c\mathbf{T})$.

The proof of 4.3.8(b) is left as an exercise. Note that two different kinds of products are involved in 4.3.8(b): the product of two mappings and the product of a scalar and a mapping.

The case where all of the vector spaces are the same ($V = V' = V''$) is especially important: in this case, all of the L.T.'s under discussion belong to Lin($V$, $V$). By 4.3.6, Lin($V$, $V$) is a vector space (in particular, an abelian group under addition); by 4.3.8(a), Lin($V$, $V$) is closed under multiplication; by Theorem 1.3.2, multiplication in Lin($V$, $V$) is associative. It is easily verified that multiplication is also distributive:

$$\text{(4.3.9)} \quad \begin{aligned} \mathbf{S}(\mathbf{T} + \mathbf{U}) &= \mathbf{ST} + \mathbf{SU} \\ (\mathbf{S} + \mathbf{T})\mathbf{U} &= \mathbf{SU} + \mathbf{TU} \end{aligned}$$

for all **S, T, U** in Lin($V$, $V$). (Once again, we leave the proof as an exercise.) Hence, Lin($V$, $V$) is a ring. The "identity mapping" $\mathbf{I}_V$, which maps $\mathbf{v} \to \mathbf{v}$ for all $\mathbf{v} \in V$, is a unity of the ring Lin($V$, $V$); indeed, it acts as an identity for multiplication by Theorem 1.3.5, and it is linear since

$$(a\mathbf{u} + b\mathbf{v})\mathbf{I}_V = a\mathbf{u} + b\mathbf{v} = a(\mathbf{u}\mathbf{I}_V) + b(\mathbf{v}\mathbf{I}_V).$$

Let us summarize our results in the form of a theorem.

**4.3.10 Theorem.** If $V$ is any vector space, then Lin($V$, $V$) is a ring which has as its unity the identity mapping $\mathbf{I}_V$.

As in Section 1.3, we will write **I** instead of $\mathbf{I}_V$ when it is clear which vector space we are discussing.

If $V$ is a vector space, the theorems of this section tell us that Lin($V$, $V$) is both a ring and a vector space, and that ring multiplication and scalar multiplication in Lin($V$, $V$) are related by 4.3.8(b). A system having all of these properties is called an *algebra*. In this book we shall not discuss properties of algebras in general; however, there do exist algebras other than Lin($V$, $V$), and the theory of algebras plays an important role in advanced mathematics.

### Exercises

*1. Complete the proof of Theorem 4.3.6, as follows:
   (a) Fill in the missing reasons in the proof of 4.3.4.
   (b) Prove 4.3.5. Give a reason for each step.
   (c) Fill in the missing reasons in the proof of 4.3.7.
   (d) Verify the seven remaining vector-space properties from Definition 2.1.1, parts (c) and (d). In the process, show that

the mappings **O** and $-\mathbf{T}$ have the properties attributed to them by Theorem 4.3.6, and that these two mappings are in fact *linear.*

*2. Prove Theorem 4.3.8(b). Give a reason for each step.

*3. Prove 4.3.9.

4. Let $\mathbf{S} : \mathbf{R}_2 \rightarrow \mathbf{R}_2$ be the reflection through the line $y = x$, and let $\mathbf{T} : \mathbf{R}_2 \rightarrow \mathbf{R}_2$ be the counterclockwise rotation through 90° about the origin. Describe both **ST** and **TS** geometrically. Are they the same? (ANS)

5. Let $\mathbf{S} : \mathbf{R}_2 \rightarrow \mathbf{R}_2$ be the reflection through the line $x = 3$; let $\mathbf{T} : \mathbf{R}_2 \rightarrow \mathbf{R}_2$ be the reflection through the line $x = 7$. Describe **ST** and **TS**. (*Note*: These two mappings **S** and **T** are *not* linear.) (ANS)

## 4.4 Algebraic operations on matrices

In Section 4.2 we established a correspondence between L.T.'s and matrices; in Section 4.3 we introduced three algebraic operations on L.T.'s (addition, multiplication, scalar multiplication). The obvious next step is to use these operations on L.T.'s to obtain corresponding operations on matrices.

### Addition

Let $V$ be an $m$-dimensional vector space with an ordered basis $(\mathbf{v}_1, ..., \mathbf{v}_m)$ and let $W$ be an $n$-dimensional vector space with an ordered basis $(\mathbf{w}_1, ..., \mathbf{w}_n)$. With respect to these bases, let the mapping $\mathbf{S} \in \text{Lin}(V, W)$ have matrix $A$ and let $\mathbf{T} \in \text{Lin}(V, W)$ have matrix $B$. This means (cf. 4.2.6) that

$$\mathbf{v}_i\mathbf{S} = \sum_{j=1}^{n} a_{ij}\mathbf{w}_j \qquad (i = 1, ..., m)$$

$$\mathbf{v}_i\mathbf{T} = \sum_{j=1}^{n} b_{ij}\mathbf{w}_j \qquad (i = 1, ..., m)$$

where $A = (a_{ij}), B = (b_{ij})$. Let us define $A + B = (c_{ij})$ to be the matrix of the mapping $\mathbf{S} + \mathbf{T}$ (with respect to the same bases, of course); thus,

$$\text{(4.4.1)} \qquad \mathbf{v}_i(\mathbf{S} + \mathbf{T}) = \sum_{j=1}^{n} c_{ij}\mathbf{w}_j.$$

By definition of **S** + **T**, we also have

$$\mathbf{v}_i(\mathbf{S}+\mathbf{T}) = \mathbf{v}_i\mathbf{S} + \mathbf{v}_i\mathbf{T}$$

$$(4.4.2) \qquad = \sum_{j=1}^{n} a_{ij}\mathbf{w}_j + \sum_{j=1}^{n} b_{ij}\mathbf{w}_j$$

$$= \sum_{j=1}^{n} (a_{ij} + b_{ij})\mathbf{w}_j.$$

Comparing 4.4.1 with 4.4.2, and remembering that we can equate coefficients (*why*?), we obtain

$$(4.4.3) \qquad (A + B)_{ij} = c_{ij} = a_{ij} + b_{ij} = A_{ij} + B_{ij}.$$

In other words, *an entry in $A + B$ is obtained simply by adding the entries in the corresponding position in $A$ and $B$*. For example,

$$\begin{bmatrix} 2 & 3 & 0 \\ 4 & -5 & -1 \end{bmatrix} + \begin{bmatrix} 4 & 7 & 2 \\ 3 & 1 & 5 \end{bmatrix} = \begin{bmatrix} 6 & 10 & 2 \\ 7 & -4 & 4 \end{bmatrix}.$$

Observe that since dim $V = m$ and dim $W = n$, the matrices which correspond to elements of Lin($V$, $W$) are the $m \times n$ matrices, so that *the sum of two $m \times n$ matrices is an $m \times n$ matrix*. Moreover, it is clear from 4.4.3 that the sum of two matrices depends only on the matrices, *not* on the choice of bases for $V$ and $W$; in other words, addition of $m \times n$ matrices is well defined.

### Scalar multiplication

If $\mathbf{S} \in$ Lin($V$, $W$) has matrix $A$ and if $c$ is a scalar, we define the matrix $cA$ to be the matrix of the mapping $c\mathbf{S}$. An argument like that for addition leads to the equation

$$(4.4.4) \qquad (cA)_{ij} = ca_{ij} = c(A_{ij}).$$

In other words, to multiply a matrix by a scalar, *multiply each entry by that scalar*. For example,

$$5\begin{bmatrix} 2 & 3 & 0 \\ 4 & -5 & -1 \end{bmatrix} = \begin{bmatrix} 10 & 15 & 0 \\ 20 & -25 & -5 \end{bmatrix}.$$

A scalar times an $m \times n$ matrix equals an $m \times n$ matrix. Again, we see that our operation (in this case, scalar multiplication) is well defined, since the formula for $cA$ (equation 4.4.4) depends only on $c$ and $A$.

### Multiplication

The formula for matrix multiplication turns out to be more complicated. The product **ST** is defined whenever $\mathbf{S} \in$ Lin($V$, $V'$) and

$\mathbf{T} \in \mathrm{Lin}(V', V'')$. Here we assume that

$V$ has an ordered basis $(\mathbf{v}_1, \ldots, \mathbf{v}_m)$;
$V'$ has an ordered basis $(\mathbf{v}'_1, \ldots, \mathbf{v}'_n)$;
$V''$ has an ordered basis $(\mathbf{v}''_1, \ldots, \mathbf{v}''_p)$.

With respect to these bases, let $\mathbf{S}$ have matrix $A$ and let $\mathbf{T}$ have matrix $B$. (Since $\mathbf{S}$ maps $V \to V'$, $A$ is an $m \times n$ matrix; similarly, $B$ is an $n \times p$ matrix.) Then

$$\text{(4.4.5)} \qquad \mathbf{v}_i\mathbf{S} = \sum_{j=1}^{n} a_{ij}\mathbf{v}'_j \qquad (i = 1, \ldots, m)$$

$$\text{(4.4.6)} \qquad \mathbf{v}'_j\mathbf{T} = \sum_{k=1}^{p} b_{jk}\mathbf{v}''_k \qquad (j = 1, \ldots, n).$$

We define $AB$ to be the matrix of $\mathbf{ST}$; since $\mathbf{ST}$ maps $V \to V''$, $AB$ is an $m \times p$ matrix and we can denote the entries in $AB$ by $c_{ik}$ $(1 \leqslant i \leqslant m, 1 \leqslant k \leqslant p)$. By 4.2.6, we have

$$\text{(4.4.7)} \qquad \mathbf{v}_i(\mathbf{ST}) = \sum_{k=1}^{p} c_{ik}\mathbf{v}''_k;$$

by definition of $\mathbf{ST}$, we have

$$\begin{aligned} \mathbf{v}_i(\mathbf{ST}) &= (\mathbf{v}_i\mathbf{S})\mathbf{T} \\ &= \left(\sum_{j=1}^{n} a_{ij}\mathbf{v}'_j\right)\mathbf{T} && \text{(by 4.4.5)} \\ \text{(4.4.8)} \qquad &= \sum_{j=1}^{n} a_{ij}\,(\mathbf{v}'_j\mathbf{T}) && \text{(since } \mathbf{T} \text{ is linear)} \\ &= \sum_{j=1}^{n} a_{ij}\left(\sum_{k=1}^{p} b_{jk}\mathbf{v}''_k\right) && \text{(by 4.4.6)} \\ &= \sum_{k=1}^{p}\left(\sum_{j=1}^{n} a_{ij}b_{jk}\right)\mathbf{v}''_k. \end{aligned}$$

Equating coefficients in 4.4.7 and 4.4.8 (*why*?), we obtain

$$\text{(4.4.9)} \qquad (AB)_{ik} = c_{ik} = \sum_{j=1}^{n} a_{ij}b_{jk} = a_{i1}b_{1k} + a_{i2}b_{2k} + \cdots + a_{in}b_{nk}$$

which is the desired formula: it expresses the entries in $AB$ in terms of the entries in $A$ and $B$. Once again, we see that our operation on matrices is well defined. Let us also emphasize again the sizes of the matrices: $A$ is $m \times n$, $B$ is $n \times p$, $AB$ is $m \times p$. In other words, *the product $AB$ exists if and only if the number of columns of $A$ equals the number* *of rows of B. In this case, AB has the same number of rows as A and the same number of columns as B.*

*Remarks*: The method by which we derived equation 4.4.9 is essentially the same as that used to derive 4.2.13. You should compare the two derivations and note the similarities. In Section 4.8 the same method will occur again.

The right side of 4.4.9 is an expression of a type that we have seen before: a sum of $n$ products of two terms each. Indeed, in Section 3.7, we defined the *dot product of two elements of* $F_n$ to be just such an expression (equation 3.7.1). Thus, we can rewrite 4.4.9 as a dot product:

**(4.4.10)** $(AB)_{ik} = (a_{i1}, a_{i2}, \ldots, a_{in}) \cdot (b_{1k}, b_{2k}, \ldots, b_{nk}).$

Observe that the first $n$-tuple on the right side of 4.4.10 is the $i$th row vector of $A$, while the second is the $k$th column vector of $B$. Thus, we may rewrite 4.4.10 as

**(4.4.11)** $(AB)_{ik} = (i\text{th row of } A) \cdot (k\text{th column of } B).$

**4.4.12 Example.** Compute the product

$$\begin{bmatrix} 1 & 2 & 3 \\ 4 & 5 & 6 \end{bmatrix} \begin{bmatrix} 2 & -1 \\ -1 & 2 \\ 0 & 3 \end{bmatrix}.$$

*Solution.* The first matrix $A$ is a $2 \times 3$ matrix; the second matrix $B$ is $3 \times 2$; hence, the product $AB = C$ exists and is $2 \times 2$. Denoting the entries in the product by $c_{ij}$, we have

$$\begin{aligned} c_{11} &= (\text{first row of } A) \cdot (\text{first col. of } B) \\ &= (1, 2, 3) \cdot (2, -1, 0) = 0 \\ c_{12} &= (\text{first row of } A) \cdot (\text{second col. of } B) \\ &= (1, 2, 3) \cdot (-1, 2, 3) = 12 \\ c_{21} &= (\text{second row of } A) \cdot (\text{first col. of } B) \\ &= (4, 5, 6) \cdot (2, -1, 0) = 3 \\ c_{22} &= (\text{second row of } A) \cdot (\text{second col. of } B) \\ &= (4, 5, 6) \cdot (-1, 2, 3) = 24 \end{aligned}$$

and, hence,

$$\begin{bmatrix} 1 & 2 & 3 \\ 4 & 5 & 6 \end{bmatrix} \begin{bmatrix} 2 & -1 \\ -1 & 2 \\ 0 & 3 \end{bmatrix} = \begin{bmatrix} 0 & 12 \\ 3 & 24 \end{bmatrix}.$$

(After a little practice, you should be able to write down the answers to problems like this one without actually writing out the intermediate steps.) ▮

*Caution: Matrix multiplication is not commutative*; in general, $AB \neq BA$. Though this is inconvenient, it should not surprise you; indeed, we have already seen that multiplication of mappings is noncommutative (Sec. 1.3; cf. also Sec. 4.3, Exercise 4). There are other ways, too, in which matrix multiplication fails to behave as we might like; for instance, a product of two nonzero matrices can equal the zero matrix (see Sec. 4.5, Exercise 6).

Since the algebraic operations on matrices have been defined in such a way as to correspond exactly to the operations on linear transformations, it follows at once that the one-to-one correspondence between L.T.'s and matrices (Theorem 4.2.9) is an isomorphism. Specifically, we have

**4.4.13 Theorem.** For any positive integers $m$, $n$ and any field $F$, let $\mathrm{Mat}(m, n; F)$ denote the set of all $m \times n$ matrices over $F$. Then

(a) $\mathrm{Mat}(m, n; F)$ is a vector space over $F$. If $V$, $W$ are any vector spaces over $F$ of dimensions $m$, $n$, respectively, then $\mathrm{Mat}(m, n; F)$ is isomorphic to $\mathrm{Lin}(V, W)$.

(b) $\mathrm{Mat}(n, n; F)$ is a ring. If $V$ is any $n$-dimensional vector space over $F$, then $\mathrm{Mat}(n, n; F)$ is isomorphic to the ring $\mathrm{Lin}(V, V)$.

(c) If $A$, $B$ are $m \times n$ and $n \times p$ matrices (respectively), then $AB$ is defined and

$$c(AB) = (cA)B = A(cB) \qquad (\text{all } c \in F).$$

(d) If $A$, $B$, $C$ are matrices over $F$ such that the products $AB$ and $BC$ are defined, then

$$(AB)C = A(BC);$$

that is, matrix multiplication is always associative.

The four parts of 4.4.13 follow respectively from Theorems 4.3.6, 4.3.10, 4.3.8(b), and 1.3.2.

### The matrix equation of a linear transformation

The equations of a linear transformation can be expressed very nicely in terms of matrix multiplication. Let $V$, $W$ be vector spaces having ordered bases $(\mathbf{v}_1, \ldots, \mathbf{v}_m)$ and $(\mathbf{w}_1, \ldots, \mathbf{w}_n)$, respectively, and let

$A$ be the matrix (with respect to these bases) of a linear transformation $\mathbf{T}: V \to W$. If

$$\mathbf{T}: \sum_{i=1}^{m} x_i \mathbf{v}_i \to \sum_{j=1}^{n} x'_j \mathbf{w}_j,$$

the $x'_j$ are related to the $x_i$ via equations 4.2.13. Let $X$ denote the 1-by-$m$ "row matrix" whose entries are $x_1, \ldots, x_m$ (a "row matrix" is, of course, a matrix with only one row); similarly, let $X'$ be the 1-by-$n$ row matrix $[x'_1\ x'_2 \cdots x'_n]$. Since $X_{1i} = x_i$ and $X'_{1j} = x'_j$, we may rewrite 4.2.13 in the form

$$\begin{aligned} X'_{1j} &= \sum_{i=1}^{m} a_{ij} X_{1i} \\ &= \sum_{i=1}^{m} X_{1i} a_{ij} = (XA)_{1j} \qquad \text{(by 4.4.9).} \end{aligned}$$

Since this holds for all $j$, the matrices $X'$ and $XA$ have the same entries and thus are equal. We state this result as a theorem.

**4.4.14 Theorem.** Let $A$ be the matrix of the linear transformation $\mathbf{T}: V \to W$ with respect to given ordered bases of $V$ and $W$. If $X$ is the row matrix consisting of the components of an arbitrary vector $\mathbf{v} \in V$ (with respect to the given basis of $V$), and if $X'$ is the row matrix consisting of the components of the image vector $\mathbf{vT}$, then $X' = XA$ (matrix product).

In the special case where $V = F_m$, $W = F_n$ and the specified bases are the canonical bases, the components of a vector $(c_1, c_2, \ldots, c_m) \in V$ are simply the coordinates $c_1, \ldots, c_m$, and similarly for vectors in $W$. Hence, in this case, we can find the image of a given vector just by treating the vector as a $1 \times m$ matrix and multiplying it by the matrix $A$. For example, if $\mathbf{T}: \mathbf{R}_2 \to \mathbf{R}_2$ is the reflection through the line $y = x/2$ (see Example 4.2.16), $A$ is given by 4.2.18. To find the image of (1, 1), we compute the matrix product

$$[1\ \ 1] \begin{bmatrix} \frac{3}{5} & \frac{4}{5} \\ \frac{4}{5} & -\frac{3}{5} \end{bmatrix} = \begin{bmatrix} \frac{7}{5} & \frac{1}{5} \end{bmatrix};$$

thus, $(1, 1)\mathbf{T} = (7/5, 1/5)$.

**Exercises**

1. Fill in the details of the derivation of 4.4.4.
2. Given the matrices

$$A = \begin{bmatrix} 1 & 2 \\ 3 & 0 \end{bmatrix}; \qquad B = \begin{bmatrix} 4 & 1 \\ -1 & 1 \end{bmatrix},$$

compute $A + B$, $AB$, and $BA$. Does $AB$ equal $BA$? (ANS)

3. Compute $AB$ and $BA$ if

$$A = \begin{bmatrix} 3 & 0 & 1 \\ 4 & 3 & 1 \end{bmatrix}; \qquad B = \begin{bmatrix} 1 & -1 \\ 2 & 3 \\ -5 & 2 \end{bmatrix}.$$

(ANS)

4. Can it ever happen that $AB$ is defined but $BA$ is not defined? Explain.
5. If $\mathbf{T} : \mathbf{R}_2 \to \mathbf{R}_2$ is the linear transformation which has the matrix

$$\begin{bmatrix} 1 & 3 \\ 2 & -1 \end{bmatrix}$$

with respect to the canonical basis, compute the image of the vector (3, 4) under $\mathbf{T}$ in two ways: (a) using equations 4.2.14; (b) using Theorem 4.4.14. (Your answers should agree!) (ANS)

6. Let $A$, $B$ be the matrices of Exercise 3; let $\mathbf{S} : F_2 \to F_3$ be the L.T. which has matrix $A$ with respect to the canonical bases of $F_2$ and $F_3$, and let $\mathbf{T} : F_3 \to F_2$ be the L.T. which has matrix $B$.

   (a) Write the equations for $\mathbf{S}$ and $\mathbf{T}$ in the form 4.2.7.

   (b) Using $\mathbf{e}_i(\mathbf{ST}) = (\mathbf{e}_i\mathbf{S})\mathbf{T}$, obtain the similar equations for $\mathbf{ST}$ by direct substitution using part (a).

   (c) Check that your answer to (b) agrees with your answer to Exercise 3.

7. Same exercise as the preceding but with $\mathbf{S}$ and $\mathbf{T}$ replaced by the mappings of Section 4.3, Exercise 4.
8. If $A = \begin{bmatrix} a & 1 \\ 0 & a \end{bmatrix}$, compute $A^n$ ($n \in \mathbf{Z}^+$). (ANS)

*9. Let $C$ be a 1-by-1 matrix whose sole entry is $c$; let $B$ be a $1 \times n$ matrix. Prove that $cB = CB$. (Thus, we may "identify" the matrix $C$ with the scalar $c$ without being inconsistent.)

*10. If $\mathbf{u}_1, \ldots, \mathbf{u}_m$ are vectors in $F_n$, let us denote by $R(\mathbf{u}_1, \ldots, \mathbf{u}_m)$ the $m \times n$ matrix whose row vectors are $\mathbf{u}_1, \ldots, \mathbf{u}_m$. Prove: if

$\mathbf{T} : F_n \to F_p$ is a linear transformation having matrix $A$ with respect to the canonical bases of $F_n$ and $F_p$, then

$(R(\mathbf{u}_1, ..., \mathbf{u}_m))A = R(\mathbf{u}_1\mathbf{T}, ..., \mathbf{u}_m\mathbf{T})$

for all $\mathbf{u}_1, ..., \mathbf{u}_m$ in $F_n$. (SOL)

11. Verify the result of the preceding exercise in the case where

$$m = n = 2; \qquad p = 3; \qquad A = \begin{bmatrix} 3 & 0 & 1 \\ 4 & 3 & 1 \end{bmatrix}; \qquad \mathbf{u}_1 = (3, -1);$$

$\mathbf{u}_2 = (4, 3)$

by computing the matrices $(R(\mathbf{u}_1, ..., \mathbf{u}_m))A$ and $R(\mathbf{u}_1\mathbf{T}, ..., \mathbf{u}_m\mathbf{T})$ independently. (ANS)

*12. Let $A$, $B$ be, respectively, an $m \times n$ and an $n \times p$ matrix over $F$; let $A_{(i)}$ denote the $i$th row vector of $A$, and similarly for the row vectors of other matrices. Prove: if $c_1, ..., c_m$ are any scalars such that

$$\sum_{i=1}^{m} c_i A_{(i)} = \mathbf{0},$$

then $\sum_{i=1}^{m} c_i(AB)_{(i)} = \mathbf{0}$. Deduce: if the row vectors of $A$ are linearly dependent, then the row vectors of $AB$ are linearly dependent. (SUG)

13. The preceding exercise states a result involving rows. State and prove an analogous result involving columns. (Be careful!)

14. Theorem 4.4.13(d) asserts that matrix multiplication is always associative. Try to prove this *without* reference to linear transformations; that is, using only 4.4.9 as the definition of matrix multiplication.

15. For any matrix $A = (a_{ij})$ over $\mathbf{C}$ (the field of complex numbers), let $\overline{A}$ be the matrix whose $(i, j)$ entry is $\overline{a_{ij}}$ (bar denotes complex conjugate). Prove that if $A$, $B$ are matrices over $\mathbf{C}$, then
(a) $\overline{A + B} = \overline{A} + \overline{B}$ whenever $A + B$ is defined;
(b) $\overline{AB} = \overline{A}\,\overline{B}$ whenever $AB$ is defined;
(c) $\overline{cA} = \overline{c}\,\overline{A}$ for all $c \in \mathbf{C}$.

16. Generalize the preceding exercise as follows: let $F$ be a field, and suppose $a \to a^*$ is a ring homomorphism of $F \to F$. For any matrix $A = (a_{ij})$ over $F$, let $A^*$ be the matrix $(a_{ij}^*)$ whose entries are the images (under the homomorphism) of the entries in $A$. Prove that if $A$, $B$ are matrices over $F$, then
(a) $(A + B)^* = A^* + B^*$ whenever $A + B$ is defined;
(b) $(AB)^* = A^*B^*$ whenever $AB$ is defined;
(c) $(cA)^* = c^*A^*$ for all $c \in F$.

## 4.5 Miscellaneous results

In the preceding section we defined algebraic operations on matrices. In this section we collect various results related to those operations.

**The matrix of the zero mapping**

Let $V$, $W$ be vector spaces over $F$. By Theorem 4.3.6, the "zero mapping" $\mathbf{O}$ defined by

$$\mathbf{vO} = \mathbf{0} \qquad (\text{all } \mathbf{v} \in V)$$

is the zero element of the vector space Lin($V$, $W$). If $V$ and $W$ have ordered bases $(\mathbf{v}_1, \ldots, \mathbf{v}_m)$ and $(\mathbf{w}_1, \ldots, \mathbf{w}_n)$, respectively, then $\mathbf{v}_i\mathbf{O} = \mathbf{0}$ (all $i$) and, hence, clearly the scalars $a_{ij}$ in 4.2.7 are all zero. In other words, the matrix of $\mathbf{O}$ is the $m \times n$ matrix *all of whose entries are zero.* We call this matrix the *zero matrix* and denote it by the symbol 0. (This symbol is also used for the *scalar* zero, but in any given situation the meaning of the symbol will normally be clear from the context.) Since the zero mapping is the additive identity of Lin($V$, $W$), Theorem 4.4.13(a) implies that the zero matrix is the additive identity of Mat ($m$, $n$; $F$); that is,

$$0 + A = A + 0 = A$$

for all matrices $A$, where

$$0 = \begin{bmatrix} 0 & 0 & \cdots & 0 \\ 0 & 0 & \cdots & 0 \\ & & \cdots & \\ 0 & 0 & \cdots & 0 \end{bmatrix}.$$

**The matrix of the identity mapping**

By Theorem 4.3.10, the identity mapping $\mathbf{I} : V \to V$ defined by

$$\mathbf{vI} = \mathbf{v} \qquad (\text{all } \mathbf{v} \in V)$$

is the unity of the ring Lin($V$, $V$). If $V$ has finite dimension $n$ and $(\mathbf{v}_1, \ldots, \mathbf{v}_n)$ is any ordered basis of $V$, we have

$$\begin{aligned} \mathbf{v}_1\mathbf{I} &= \mathbf{v}_1 = 1\mathbf{v}_1 + 0\mathbf{v}_2 + \cdots + 0\mathbf{v}_n \\ \mathbf{v}_2\mathbf{I} &= \mathbf{v}_2 = 0\mathbf{v}_1 + 1\mathbf{v}_2 + \cdots + 0\mathbf{v}_n \\ &\cdots \\ \mathbf{v}_n\mathbf{I} &= \mathbf{v}_n = 0\mathbf{v}_1 + 0\mathbf{v}_2 + \cdots + 1\mathbf{v}_n. \end{aligned}$$

Hence, the matrix of **I** is the $n \times n$ *identity matrix*

$$I_n = \begin{bmatrix} 1 & 0 & 0 & \cdots & 0 \\ 0 & 1 & 0 & \cdots & 0 \\ 0 & 0 & 1 & \cdots & 0 \\ & & & \cdots & \\ 0 & 0 & 0 & \cdots & 1 \end{bmatrix}$$

in which the entries along the main diagonal are 1 and all other entries are zero. (The "main diagonal" of any square matrix $A = (a_{ij})$ consists of the entries $a_{11}, a_{22}, \ldots$ for which the row and column subscripts are the same.) Since the identity mapping is the unity of Lin($V$, $V$), the identity matrix $I_n$ is the unity of the ring Mat($n$, $n$; $F$). In fact, we have the following slightly stronger result, which follows from Theorem 1.3.5:

**4.5.1 Theorem.** If $F$ is a field and $m$ and $n$ are positive integers, then $I_mA = A = AI_n$ for every $m \times n$ matrix $A$ over $F$.

The entries in the matrix $I_n$ are usually denoted $\delta_{ij}$; that is,

$$(I_n)_{ij} = \delta_{ij} = \begin{Bmatrix} 1 & \text{if} & i = j \\ 0 & \text{if} & i \neq j \end{Bmatrix}. \tag{4.5.2}$$

The symbol $\delta_{ij}$ is known as the *Kronecker delta.* (L. Kronecker was a 19th-century mathematician who made some important contributions to algebra.) The Kronecker delta occurs in many mathematical situations; for example, equation 3.8.1 may be rewritten as

$$p(\mathbf{u}_i, \mathbf{u}_j) = \delta_{ij}.$$

As you might expect, we will write $I$ in place of $I_n$ when the integer $n$ is not essential to the discussion.

### Properties of the transpose

The transpose of a matrix $A$ (denoted $A^t$) was defined in Section 4.2. The following theorem relates the transpose to our algebraic operations on matrices.

**4.5.3 Theorem.** If $A$, $B$ are matrices over the field $F$, then

(a) $(A + B)^t = A^t + B^t$ whenever $A + B$ is defined.

(b) $(cA)^t = c(A^t)$ for all $c \in F$.

(c) $(AB)^t = B^tA^t$ whenever $AB$ is defined.

(d) $(A^t)^t = A$.

In other words, the transpose of a sum is the sum of the transposes; the transpose of a product is the product of the transposes *in reverse order.*

We shall leave parts (a), (b), and (d) of Theorem 4.5.3 as exercises. The proof of part (c) is as follows: for all $i$ and $k$,

$$\begin{aligned}[(AB)^t]_{ki} &= (AB)_{ik} \\ &= (i\text{th row of } A) \cdot (k\text{th column of } B) \\ &= (i\text{th column of } A^t) \cdot (k\text{th row of } B^t) \\ &= (k\text{th row of } B^t) \cdot (i\text{th column of } A^t) \\ &= (B^tA^t)_{ki}.\end{aligned}$$

Thus, the matrices $(AB)^t$, $B^tA^t$ have the same entry in each position and, hence, are equal. ∎

**Diagonal and triangular matrices**

Let $A$ be an $n \times n$ (square) matrix. By the *main diagonal* of $A$, we mean the $n$-tuple $(a_{11}, a_{22}, \ldots, a_{nn})$ which consists of those entries $a_{ij}$ for which $i = j$. Pictorially, these are the entries which lie on the line from the upper left corner of $A$ to the lower right corner of $A$. The entries *below* the main diagonal are those entries $a_{ij}$ for which $i > j$ (why?); if all such entries are zero, $A$ is called *upper triangular*. Similarly, $A$ is *lower triangular* if all entries *above* the main diagonal are zero. For example, the matrices

$$\begin{bmatrix} 1 & 2 & 3 & 4 \\ 0 & 5 & 6 & 7 \\ 0 & 0 & 8 & 9 \\ 0 & 0 & 0 & 10 \end{bmatrix}, \quad \begin{bmatrix} 3 & 0 & 0 & 0 \\ 1 & 2 & 0 & 0 \\ 3 & -1 & 0 & 0 \\ 6 & 7 & 3 & 4 \end{bmatrix}$$

are upper triangular and lower triangular, respectively. A matrix is *triangular* if it is either upper triangular or lower triangular. A square matrix is called a *diagonal matrix* if it is *both* upper triangular and lower triangular, that is, if all of its entries not on the main diagonal are zero. We shall not discuss diagonal and triangular matrices in any detail at present (though they will figure in the discussion of determinants in Chap. 5); however, some of the properties of these matrices are developed in the exercises at the end of this section.

**Submatrices**

Let $A$ be a matrix. By a *submatrix* of $A$, we mean any matrix obtained from $A$ by deleting any number of complete rows and any number of complete columns. The number of rows or columns deleted may be zero; thus, $A$ is a submatrix of itself. The empty matrix is a submatrix of every matrix.

**Block multiplication of matrices**

Let $A$ be an $m \times n$ matrix and let $B$ be an $n \times p$ matrix. Suppose also that

$$m = m_1 + m_2; \qquad n = n_1 + n_2; \qquad p = p_1 + p_2$$

where the numbers $m_i$, $n_i$, $p_i$ $(i = 1, 2)$ are nonnegative integers. If we separate the first $m_1$ rows of $A$ from the last $m_2$ rows, and the first $n_1$ columns from the last $n_2$ columns, the effect is to partition $A$ into four submatrices:

$$\text{(4.5.4)} \qquad A = \begin{array}{c} \begin{array}{cc} n_1 & n_2 \\ \text{columns} & \text{columns} \end{array} \\ \begin{bmatrix} A^{11} & A^{12} \\ A^{21} & A^{22} \end{bmatrix} \end{array} \begin{array}{l} m_1 \text{ rows} \\ m_2 \text{ rows} \end{array}$$

where each $A^{ij}$ is an $m_i \times n_j$ matrix $(i = 1, 2; j = 1, 2)$. (We use superscripts rather than subscripts since the notation $A_{ij}$ has been used to mean the $(i,j)$ entry in $A$.) Similarly, $B$ may be partitioned as follows:

$$\text{(4.5.5)} \qquad B = \begin{array}{c} \begin{array}{cc} p_1 & p_2 \\ \text{columns} & \text{columns} \end{array} \\ \begin{bmatrix} B^{11} & B^{12} \\ B^{21} & B^{22} \end{bmatrix} \end{array} \begin{array}{l} n_1 \text{ rows} \\ n_2 \text{ rows} \end{array}$$

where each $B_{ij}$ is an $n_i \times p_j$ matrix. Since the product $C = AB$ is an $m \times p$ matrix, we may similarly partition $C$ in the form

$$\text{(4.5.6)} \qquad C = \begin{array}{c} \begin{array}{cc} p_1 & p_2 \\ \text{columns} & \text{columns} \end{array} \\ \begin{bmatrix} C^{11} & C^{12} \\ C^{21} & C^{22} \end{bmatrix} \end{array} \begin{array}{l} m_1 \text{ rows} \\ m_2 \text{ rows.} \end{array}$$

The significant fact about all this is that the expression 4.5.6 for $C$ may be obtained by multiplying the expressions 4.5.4 and 4.5.5 *as if the blocks behaved like entries*. That is, we have

$$\text{(4.5.7)} \qquad C^{ik} = \sum_{j=1}^{2} A^{ij}B^{jk}$$

for each $i$ and $k$. The proof is a tedious but straightforward application of the definitions. For example, to show that

$$\text{(4.5.8)} \qquad C^{12} = A^{11}B^{12} + A^{12}B^{22}$$

we consider any fixed entry in the matrix $C^{12}$, say the entry in the

$(r, t)$ position. Letting subscripts denote position within a given matrix, we have

$$
\begin{aligned}
(C^{12})_{rt} &= C_{r,p_1+t} \qquad \text{(by definition of } C^{12}\text{)} \\
&= \sum_{s=1}^{n} A_{rs}B_{s,p_1+t} \qquad \text{(by 4.4.9)} \\
&= \sum_{s=1}^{n_1} A_{rs}B_{s,p_1+t} + \sum_{s=n_1+1}^{n} A_{rs}B_{s,p_1+t} \\
&= \sum_{s=1}^{n_1} A_{rs}B_{s,p_1+t} + \sum_{j=1}^{n_2} A_{r,j+n_1}B_{j+n_1,p_1+t} \\
&\qquad\qquad \text{(letting } s = j + n_1, j = s - n_1\text{)} \\
&= \sum_{s=1}^{n_1} (A^{11})_{rs}\,(B^{12})_{st} + \sum_{j=1}^{n_2} (A^{12})_{rj}\,(B^{22})_{jt} \qquad \text{(by 4.5.4 and 4.5.5)} \\
&= (A^{11}B^{12})_{rt} + (A^{12}B^{22})_{rt} \qquad \text{(by 4.4.9)}
\end{aligned}
$$

and 4.5.8 follows. The other cases of 4.5.7 are proved similarly.

More generally: if $A$, $B$ are $m \times n$ and $n \times p$ matrices, respectively, and if

$$
\begin{aligned}
m &= m_1 + m_2 + \cdots + m_\alpha = \sum_{i=1}^{\alpha} m_i \\
n &= n_1 + n_2 + \cdots + n_\beta = \sum_{j=1}^{\beta} n_j \\
p &= p_1 + p_2 + \cdots + p_\lambda = \sum_{k=1}^{\lambda} p_k,
\end{aligned}
$$

we obtain a partition of $A$ into $\alpha\beta$ submatrices $A^{ij}$ of sizes $m_i \times n_j$, a partition of $B$ into $\beta\lambda$ submatrices $B^{jk}$ of sizes $n_j \times p_k$, and a partition of $C = AB$ into $\alpha\lambda$ submatrices $C^{ik}$ of sizes $m_i \times p_k$; and it remains true that

$$C^{ik} = \sum_{j=1}^{\beta} A^{ij}B^{jk}$$

for all $i$ and $k$. The proof, although more complicated than the proof of 4.5.7, is similar to it in spirit; we omit details.

**Exercises**

*1. Prove Theorem 4.5.3, parts (a), (b), and (d).

2. Using the same notation as in Section 4.4, Exercise 16, prove that $(A^*)^t = (A^t)^*$ for every matrix $A$ over $F$.

3. Let $\mathbf{T} \in \text{Lin}(F_n, F_n)$; let $A$ be the matrix of $\mathbf{T}$ with respect to the canonical basis of $F_n$. Since the vectors in $F_n$ are $n$-tuples, they can be regarded as $1 \times n$ matrices (row matrices). In particular, if $\mathbf{u}, \mathbf{v} \in F_n$ and we regard $\mathbf{u}$ and $\mathbf{v}$ as row matrices, then $\mathbf{u}A\mathbf{v}^t$ is a $1 \times 1$ matrix. Show that the sole entry in the matrix $\mathbf{u}A\mathbf{v}^t$ is the scalar $(\mathbf{u}\mathbf{T}) \cdot \mathbf{v}$. Hence, show that if $A = A^t$, then $(\mathbf{u}\mathbf{T}) \cdot \mathbf{v} = \mathbf{u} \cdot (\mathbf{v}\mathbf{T})$ for all $\mathbf{u}, \mathbf{v} \in F_n$.
4. (a) How many nonempty submatrices does an $n \times n$ matrix have? (ANS)

   (b) How many square submatrices (including the empty submatrix) does an $n \times n$ matrix have? (ANS)
*5. If $B$ is a submatrix of $A$, and if certain row vectors of $B$ are linearly independent, prove that the corresponding row vectors of $A$ are linearly independent. (SOL)
6. (a) Find two $2 \times 2$ matrices $A$ and $B$ (over $\mathbf{R}$) such that $AB = 0$ but neither $A$ nor $B$ equals 0.

   (b) Find a $2 \times 2$ matrix $A$ (over $\mathbf{R}$) such that $A^2 = I$ but $A$ equals neither $I$ nor $-I$.
7. (a) Prove: if the $n \times n$ matrices $A = (a_{ij})$ and $B = (b_{ij})$ are both upper triangular, and if $C = AB = (c_{ij})$, then $C$ is upper triangular and $c_{ii} = a_{ii}b_{ii}$ $(1 \leq i \leq n)$.

   (b) State and prove a similar result for lower triangular matrices.

   (c) State and prove a similar result for diagonal matrices.
8. (a) Prove that any two $n \times n$ diagonal matrices $A$ and $B$ (over a field $F$) commute with each other: $AB = BA$.

   (b) Conversely, if $A = (a_{ij})$ is a diagonal matrix whose diagonal entries $a_{11}, a_{22}, \ldots, a_{nn}$ are *all different*, prove that every matrix which commutes with $A$ is a diagonal matrix. (SOL)
9. If $c$ is a scalar, a matrix of the form $cI$ (a scalar multiple of the identity matrix) is called a *scalar matrix*.

   (a) Prove that an $n \times n$ scalar matrix commutes with all other $n \times n$ matrices.

   (b) Prove the converse: if an $n \times n$ matrix $A$ commutes with all other $n \times n$ matrices, then $A$ is a scalar matrix. (SOL)
10. Let

$$A = \begin{bmatrix} 1 & 2 & 3 & -2 \\ 0 & 3 & -1 & -4 \\ -3 & -1 & 2 & 2 \\ 4 & 2 & 0 & 1 \end{bmatrix}, \qquad B = \begin{bmatrix} 3 & 1 & -2 & 5 \\ -2 & 0 & 2 & -3 \\ 0 & 0 & 1 & 1 \\ 1 & 2 & 1 & 2 \end{bmatrix}$$

and take $m_1 = m_2 = n_1 = n_2 = p_1 = p_2 = 2$ in equations 4.5.4 through 4.5.6. Compute $C = AB$ by block multiplication (i.e., using equation 4.5.7). Then check your answer by computing $AB$ directly from formula 4.4.11, without blocks.

11. Using the notations of equations 4.5.4 through 4.5.6 (with $C = AB$), suppose that $m_1 = n_1 = p_1$, $m_2 = n_2 = p_2$, and that $A^{12} = B^{12} = 0$. Show that $C^{12} = 0$. Can this result be generalized? (Compare with Exercise 7.)

## 4.6 The range and nullspace of a linear transformation

Let $V$ and $W$ be vector spaces over $F$ and let $\mathbf{T} \in \text{Lin}(V, W)$. By Theorem 4.1.5, the image of any subspace of $V$ under $\mathbf{T}$, or the pre-image of any subspace of $W$ under $\mathbf{T}$, is itself a subspace. Two of these subspaces, the *range* and *nullspace* of $\mathbf{T}$, are of special importance. In this section, we define these two subspaces and establish a few of their basic properties.

**4.6.1 Definition.** If $\mathbf{T} \in \text{Lin}(V, W)$, the image of $V$ under $\mathbf{T}$ is called the *range* of $\mathbf{T}$ and is denoted Ran $\mathbf{T}$; the pre-image of $\{\mathbf{0}_W\}$ under $\mathbf{T}$ is called the *nullspace* (or *kernel*) of $\mathbf{T}$ and is denoted Nul $\mathbf{T}$.

Our definition of "range" agrees with that in Section 1.2. By Theorem 4.1.5, Ran $\mathbf{T}$ is a subspace of $W$ and Nul $\mathbf{T}$ is a subspace of $V$. Ran $\mathbf{T}$ consists of all vectors of the form $\mathbf{vT}$ ($\mathbf{v} \in V$), while Nul $\mathbf{T}$ consists of all $\mathbf{v} \in V$ such that $\mathbf{vT} = \mathbf{0}$. In symbols,

$$\text{Ran } \mathbf{T} = \{\mathbf{vT} : \mathbf{v} \in V\} = V\mathbf{T} \subseteq W$$
$$\text{Nul } \mathbf{T} = \{\mathbf{v} \in V : \mathbf{vT} = \mathbf{0}\} \subseteq V.$$

**4.6.2 Example.** Let $\mathbf{T} : \mathbf{R}_3 \to \mathbf{R}_3$ be the mapping defined in Example 4.2.15 (see Fig. 4.6.1). For all $\mathbf{v} \in \mathbf{R}_3$, $\mathbf{vT}$ lies in the

Figure 4.6.1

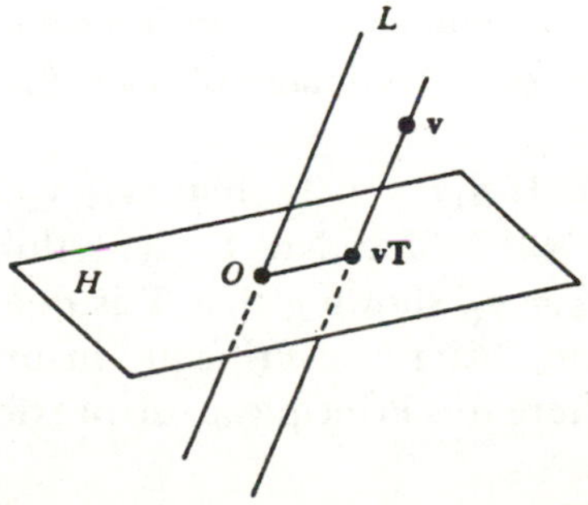

plane $H$; conversely, every point in $H$ is the image of at least one vector (namely, itself). Hence, Ran $\mathbf{T} = H$. The points $\mathbf{v}$ which are mapped into $O$ by $\mathbf{T}$ are precisely the points on the line $L$ (why?); thus, Nul $\mathbf{T} = L$. In this example, the dimensions of the range and nullspace are 2 and 1, respectively. Note that the *sum* of these dimensions (3) equals the dimension of the entire space on which $\mathbf{T}$ is defined. This is no accident; we shall prove below that the same dimension relation holds for any linear transformation.

**4.6.3** **Example.** If $\mathbf{T} \in \text{Lin}(\mathbf{R}_3, \mathbf{R}_3)$ is defined by

$$(x, y, z)\mathbf{T} = (2x - y, y + z, 4x + y + 3z),$$

find the nullspace of $\mathbf{T}$.

*Solution.* A vector $(x, y, z)$ lies in Nul $\mathbf{T}$ if and only if $(x, y, z)\mathbf{T} = (0, 0, 0)$; that is, if and only if

$$\begin{aligned} 2x - y \phantom{{}+3z} &= 0 \\ y + \phantom{3}z &= 0 \\ 4x + y + 3z &= 0. \end{aligned} \tag{4.6.4}$$

The solutions of the system 4.6.4 are the vectors of the form $(a, 2a, -2a)$ $(a \in \mathbf{R})$; hence, Nul $\mathbf{T}$ is the one-dimensional subspace $[(1, 2, -2)]$. ▮

One of the principal uses of the nullspace of a linear transformation $\mathbf{T}$ is in determining whether $\mathbf{T}$ is one-to-one. (Recall from Sec. 1.2 that a mapping is *one-to-one* if no two elements of its domain have the same image.) Specifically, we have the following theorem:

**4.6.5** **Theorem.** If $\mathbf{T} \in \text{Lin}(V, W)$, then $\mathbf{T}$ is one-to-one if and only if Nul $\mathbf{T} = \{\mathbf{0}\}$.

*Proof.* If $\mathbf{T}$ is one-to-one, then no element of $W$ can have more than one pre-image; in particular, the only pre-image of $\mathbf{0}_W$ is $\mathbf{0}_V$, so that Nul $\mathbf{T}$ consists only of $\mathbf{0}_V$.

Conversely, let Nul $\mathbf{T} = \{\mathbf{0}_V\}$. If $\mathbf{v}_1\mathbf{T} = \mathbf{v}_2\mathbf{T}$, then $(\mathbf{v}_1 - \mathbf{v}_2)\mathbf{T} = \mathbf{v}_1\mathbf{T} - \mathbf{v}_2\mathbf{T} = \mathbf{0}$ and, hence, $\mathbf{v}_1 - \mathbf{v}_2 \in$ Nul $\mathbf{T}$. Since Nul $\mathbf{T} = \{\mathbf{0}\}$, this implies that $\mathbf{v}_1 - \mathbf{v}_2 = \mathbf{0}$ and, hence, $\mathbf{v}_1 = \mathbf{v}_2$, showing that $\mathbf{T}$ is one-to-one. (Note once again the use of the Difference Principle in proving a theorem. Can you remember where this Principle arose previously?) ▮

As we see from Example 4.6.3, the process of finding the nullspace of an L.T. involves solving simultaneous linear equations. To find the *range* of an L.T., we use the following theorem:

**4.6.6** **Theorem.** If $V = [\mathbf{u}_1, ..., \mathbf{u}_n]$ and $\mathbf{T} \in \text{Lin}(V, W)$, then Ran $\mathbf{T} = [\mathbf{u}_1\mathbf{T}, ..., \mathbf{u}_n\mathbf{T}]$.

We leave the proof as an exercise. The theorem asserts that if given vectors span the domain of **T**, then their images under **T** span the range of **T**. In Example 4.6.3, for instance,

$$\begin{aligned}\text{Ran } \mathbf{T} &= [\mathbf{e}_1\mathbf{T}, \mathbf{e}_2\mathbf{T}, \mathbf{e}_3\mathbf{T}] \\ &= [(2, 0, 4), (-1, 1, 1), (0, 1, 3)].\end{aligned}$$

The set $\{(2, 0, 4), (-1, 1, 1)\}$ is a maximal independent subset of $\{(2, 0, 4), (-1, 1, 1), (0, 1, 3)\}$ and hence is a basis of Ran **T**. We conclude that Ran **T** is the two-dimensional subspace $[(2, 0, 4), (-1, 1, 1)]$. Again, as in Example 4.6.2, the dimensions of Ran **T** and Nul **T** add up to the dimension of the domain of **T**. We now prove that this dimension relation holds in general.

**4.6.7** **Theorem.** If $\mathbf{T} \in \text{Lin}(V, W)$ and if $V$ has finite dimension, then

$$\dim(\text{Nul } \mathbf{T}) + \dim(\text{Ran } \mathbf{T}) = \dim V. \tag{4.6.8}$$

In fact, if $(\mathbf{u}_1, ..., \mathbf{u}_k)$ is any ordered basis of Nul **T** and we extend this basis to an ordered basis $(\mathbf{u}_1, ..., \mathbf{u}_k, \mathbf{u}_{k+1}, ..., \mathbf{u}_n)$ of $V$, then $(\mathbf{u}_{k+1}\mathbf{T}, ..., \mathbf{u}_n\mathbf{T})$ is an ordered basis of Ran **T**.

(Note that by Theorem 2.6.16 the subspace Nul **T** does have a basis, and any such basis can indeed be extended to a basis of $V$.)

*Proof of Theorem 4.6.7.* Since the vectors $\mathbf{u}_1, ..., \mathbf{u}_k$ are in Nul **T**, their images $\mathbf{u}_1\mathbf{T}, ..., \mathbf{u}_k\mathbf{T}$ are zero. Hence, Theorem 4.6.6 implies that

$$\begin{aligned}\text{Ran } \mathbf{T} &= [\mathbf{0}, ..., \mathbf{0}, \mathbf{u}_{k+1}\mathbf{T}, ..., \mathbf{u}_n\mathbf{T}] \\ &= [\mathbf{u}_{k+1}\mathbf{T}, ..., \mathbf{u}_n\mathbf{T}].\end{aligned} \tag{4.6.9}$$

Moreover, the vectors $\mathbf{u}_{k+1}\mathbf{T}, ..., \mathbf{u}_n\mathbf{T}$ are independent; for suppose some linear combination of them were zero:

$$\sum_{i=k+1}^{n} a_i (\mathbf{u}_i\mathbf{T}) = \mathbf{0}.$$

Then by linearity of $\mathbf{T}$, we have

$$\left(\sum_{i=k+1}^{n} a_i\mathbf{u}_i\right)\mathbf{T} = \mathbf{0}$$

so that $\sum_{i=k+1}^{n} a_i\mathbf{u}_i \in \text{Nul }\mathbf{T}$. We can, thus, express $\sum_{i=k+1}^{n} a_i\mathbf{u}_i$ in terms of the basis $(\mathbf{u}_1, ..., \mathbf{u}_k)$ of Nul $\mathbf{T}$:

$$\sum_{i=k+1}^{n} a_i\mathbf{u}_i = \sum_{j=1}^{k} b_j\mathbf{u}_j = \sum_{j=1}^{k} b_j\mathbf{u}_j + 0\mathbf{u}_{k+1} + \cdots + 0\mathbf{u}_n.$$

Since $\mathbf{u}_1, ..., \mathbf{u}_n$ are independent, we may equate coefficients of $\mathbf{u}_{k+1}$, ..., $\mathbf{u}_n$ on both sides, which gives $a_{k+1} = \cdots = a_n = 0$. It follows that $\mathbf{u}_{k+1}\mathbf{T}, ..., \mathbf{u}_n\mathbf{T}$ are independent, as claimed. Since these vectors also span Ran $\mathbf{T}$ by 4.6.9, they constitute a basis of Ran $\mathbf{T}$, and the theorem follows. ∎

An important special case of Theorem 4.6.7 will be stated here for future reference. Let $W$ be a subspace of $V$, and consider the quotient space $V/W$ which consists of all cosets of $W$ in $V$ (see Theorem 2.7.11). A typical element of $V/W$ has the form

$$\mathbf{v}_* = \mathbf{v} + W$$

where $\mathbf{v}$ is an element of $V$. Let us define a mapping $\mathbf{\Psi} : V \to V/W$ by

**(4.6.10)** $\mathbf{\Psi} : \mathbf{v} \to \mathbf{v}_* \qquad (\text{all } \mathbf{v} \in V).$

Using the definitions of addition and scalar multiplication in $V/W$ (equations 2.7.9 and 2.7.10), it is an easy exercise to show that $\mathbf{\Psi}$ is linear. Since an arbitrary element $\mathbf{v}_* \in V/W$ is the image of $\mathbf{v} \in V$, $\mathbf{\Psi}$ is clearly *onto* and, hence,

$$\text{Ran }\mathbf{\Psi} = V/W.$$

As for the nullspace of $\mathbf{\Psi}$, we have

$$\begin{aligned} \mathbf{v} \in \text{Nul }\mathbf{\Psi} &\Leftrightarrow \mathbf{v\Psi} = \mathbf{0}_{V/W} \Leftrightarrow \mathbf{v}_* = \mathbf{0}_{V/W} \\ &\Leftrightarrow \mathbf{v}_* = W \qquad \text{(see the third "Remark" following Theorem 2.7.11)} \\ &\Leftrightarrow \mathbf{v} \in W \qquad \text{(by Theorem 2.7.7(b))} \end{aligned}$$

so that Nul $\mathbf{\Psi} = W$. Since $\mathbf{\Psi}$ is linear, Theorem 4.6.7 is applicable, and we obtain the following result:

**4.6.11 Theorem.** Let $V$ be a finite-dimensional vector space and let $W$ be a subspace of $V$. If an ordered basis $(\mathbf{u}_1, ..., \mathbf{u}_k)$ of $W$

is extended to an ordered basis $(\mathbf{u}_1, ..., \mathbf{u}_k, \mathbf{u}_{k+1}, ..., \mathbf{u}_n)$ of $V$, then $(\mathbf{u}_{k+1} + W, ..., \mathbf{u}_n + W)$ is an ordered basis of $V/W$.

*Concluding remark.* It is worth pointing out that at least two important theorems from previous chapters are special cases of the theorems of this section: specifically, Theorem 2.4.7 (on equating coefficients of independent vectors) is a special case of 4.6.5, and Theorem 3.8.18(b) (on the dimension of an orthogonal complement) is a special case of 4.6.8. To convince yourself, do Exercises 8 and 9 below.

**Exercises**

*1. Prove Theorem 4.6.6.

*2. Prove that the mapping $\boldsymbol{\Psi}$ defined by 4.6.10 is linear.

3. Let $\mathbf{T} : \mathbf{R}_2 \to \mathbf{R}_2$ be defined by $\mathbf{T} : (x, y) \to (x - y, x - y)$. Then $\mathbf{T}$ is linear (*why*?). Find the range and nullspace of $\mathbf{T}$ and verify equation 4.6.8. (ANS)

4. Let $\mathbf{T} \in \text{Lin}\,(\mathbf{R}_4, \mathbf{R}_3)$ be the L.T. which has the matrix

$$\begin{bmatrix} 0 & 2 & -2 \\ 3 & 1 & 5 \\ 1 & 1 & 1 \\ 0 & 2 & -2 \end{bmatrix}$$

with respect to the canonical bases. Find bases for the spaces Ran $\mathbf{T}$ and Nul $\mathbf{T}$ and verify equation 4.6.8. (ANS)

5. In each of the following, exhibit a linear transformation $\mathbf{T} : \mathbf{R}_4 \to \mathbf{R}_4$ satisfying the given condition. Describe $\mathbf{T}$ geometrically whenever possible.
   (a) Ran $\mathbf{T} = [\mathbf{e}_1, \mathbf{e}_2]$; Nul $\mathbf{T} = [\mathbf{e}_3, \mathbf{e}_4]$.
   (b) Ran $\mathbf{T} = [\mathbf{e}_1, \mathbf{e}_2]$; Nul $\mathbf{T} = [\mathbf{e}_2, \mathbf{e}_4]$. (ANS)
   (c) Ran $\mathbf{T} = [\mathbf{e}_1, \mathbf{e}_2, \mathbf{e}_4]$; Nul $\mathbf{T} = [\mathbf{e}_2]$.
   (d) Ran $\mathbf{T} = [\mathbf{e}_2]$; Nul $\mathbf{T} = [\mathbf{e}_1, \mathbf{e}_2, \mathbf{e}_4]$.

6. More generally, let $V$ and $W$ be vector spaces over a field $F$ and let $N$, $S$ be any subspaces of $V$, $W$ (respectively) such that $\dim N + \dim S = \dim V$. Show that there exists a linear transformation $\mathbf{T} : V \to W$ such that

   Nul $\mathbf{T} = N$;     Ran $\mathbf{T} = S$.

7. (a) If $\mathbf{T} \in \text{Lin}(V, V')$ and $\mathbf{S} \in \text{Lin}(V', V'')$, prove that Ran $\mathbf{TS} \subseteq$ Ran $\mathbf{S}$ and Nul $\mathbf{TS} \supseteq$ Nul $\mathbf{T}$.

(b) Deduce from part (a); if $\mathbf{T} \in \text{Lin}(V, V)$ and $k$ is a positive integer, then Ran $\mathbf{T}^{k+1} \subseteq$ Ran $\mathbf{T}^k$ and Nul $\mathbf{T}^{k+1} \supseteq$ Nul $\mathbf{T}^k$.

8. Show that Theorem 2.4.7 is really a special case of Theorem 4.6.5. (*Suggestion*: Apply 4.6.5 to the mapping $\mathbf{T}$ defined in Sec. 4.1, Exercise 5.)
9. Show that Theorem 3.8.18(b) is really a special case of equation 4.6.8. (*Suggestion*: If $W$ is any subspace of the $n$-dimensional real Euclidean space $E$, let $\mathbf{T} : E \to W$ be the perpendicular projection mapping of $E$ on $W$.) (ANS)
10. (A generalization of Theorem 4.6.5) Let $G$ and $G'$ be groups with identities $e$ and $e'$, respectively. If $\mathbf{T} : G \to G'$ is a homomorphism (see Sec. 1.8), the *kernel* of $\mathbf{T}$ (denoted Ker $\mathbf{T}$) is defined to be the set of all elements $x \in G$ such that $x\mathbf{T} = e'$. Prove that $\mathbf{T}$ is one-to-one if and only if Ker $\mathbf{T} = \{e\}$.

## 4.7 L.T.'s which are one-to-one or onto; left and right inverses

We recall that a linear transformation is an *isomorphism* if it is one-to-one and onto. Of these two properties, "one-to-oneness" is the more important; the reason is that, in a certain sense, *every* mapping may be regarded as "onto". (Specifically, if $\mathbf{T}$ maps $V$ into $W$ and $V' = V\mathbf{T}$ is the range of $\mathbf{T}$, we may think of $\mathbf{T}$ as a mapping of $V$ *onto* $V'$.) Thus, it is useful to have criteria by which we can determine whether a given L.T. is one-to-one. One such criterion appeared in the preceding section (Theorem 4.6.5). Another, which applies to *all* mappings (not just L.T.'s), was established in Section 1.3 (Theorem 1.3.6(a)): a mapping $\mathbf{T} : V \to W$ is one-to-one if and only if there is a mapping $\mathbf{T}' : W \to V$ (not necessarily unique!) such that $\mathbf{T}\mathbf{T}' = \mathbf{I}_V$. Such a mapping $\mathbf{T}'$ is called a *right inverse* of $\mathbf{T}$. Since at present the mappings which interest us are primarily the linear ones, it is natural to ask whether right inverses of L.T.'s must themselves be linear. The answer is: not *all* of the right inverses of a one-to-one L.T. need be linear, *but at least one of them is linear*. In the next theorem we establish this fact as well as several other conditions equivalent to "one-to-oneness".

**4.7.1 Theorem.** Let $V$, $W$ be vector spaces of finite dimensions $m$, $n$, respectively; let $\mathbf{T} \in \text{Lin}(V, W)$, and let $A$ be the matrix of $\mathbf{T}$ with respect to ordered bases $(\mathbf{v}_1, \ldots, \mathbf{v}_m)$ of $V$, $(\mathbf{w}_1, \ldots, \mathbf{w}_n)$ of $W$. Then the following statements are equivalent:

(a) $\mathbf{T}$ is one-to-one.

(b) Nul $\mathbf{T} = \{\mathbf{0}\}$.

(c) $\mathbf{T}$ preserves linear independence; that is, whenever $\mathbf{u}_1, \ldots, \mathbf{u}_k$ are independent vectors in $V$, then the vectors $\mathbf{u}_1\mathbf{T}, \ldots, \mathbf{u}_k\mathbf{T}$ are independent.

(d) For some ordered basis $(\mathbf{u}_1, \ldots, \mathbf{u}_m)$ of $V$, the vectors $\mathbf{u}_1\mathbf{T}, \ldots, \mathbf{u}_m\mathbf{T}$ are independent.

(e) There exists a linear transformation $\mathbf{T}' : W \to V$ such that $\mathbf{T}\mathbf{T}' = \mathbf{I}_V$.

(f) There exists a mapping $\mathbf{T}' : W \to V$ such that $\mathbf{T}\mathbf{T}' = \mathbf{I}_V$.

(g) There exists an $n \times m$ matrix $A'$ such that $AA' = I_m$.

(h) The row vectors of $A$ are linearly independent.

Note that statements (g) and (h) involve the matrix of $\mathbf{T}$ rather than $\mathbf{T}$ itself. Of course, the matrix of $\mathbf{T}$ is not uniquely determined since it depends on a choice of bases; but it is clear that if (g) and (h) hold for one matrix of $\mathbf{T}$, then they also hold for any other matrix of $\mathbf{T}$ (*why*?). As you might expect, the matrix $A'$ in (g) is called a *right inverse* of $A$.

*Proof of Theorem 4.7.1.* Since statement (g) is obviously equivalent to (e), it suffices to prove the equivalence of statements (a) through (f) and (h). To do so, we shall establish the circle of implications

$$(a) \Rightarrow (b) \Rightarrow (c) \Rightarrow (h) \Rightarrow (d) \Rightarrow (e) \Rightarrow (f) \Rightarrow (a).$$

The first of these is already known: (a) $\Rightarrow$ (b) by Theorem 4.6.5. If (b) holds and if $\mathbf{u}_1, \ldots, \mathbf{u}_k$ are independent vectors, then

$$\sum_{i=1}^{k} c_i\,(\mathbf{u}_i\mathbf{T}) = \mathbf{0} \Rightarrow \left(\sum_{i=1}^{k} c_i\mathbf{u}_i\right)\mathbf{T} = \mathbf{0}$$
$$\Rightarrow \sum_{i=1}^{k} c_i\mathbf{u}_i = \mathbf{0} \qquad \text{(by (b))}$$
$$\Rightarrow \text{all } c_i = 0$$

so that $\mathbf{u}_1\mathbf{T}, \ldots, \mathbf{u}_k\mathbf{T}$ are independent; thus (b) $\Rightarrow$ (c).

To prove the next two implications in the circle, we recall that by Theorem 4.2.24, the mapping

$$\mathbf{S} : \sum_{j=1}^{n} c_j\mathbf{w}_j \to (c_1, \ldots, c_n)$$

is an isomorphism of $W$ onto $F_n$. Applying $\mathbf{S}$ to both sides of equation 4.2.6 (Sec. 4.2), we have

$$(\mathbf{v}_i\mathbf{T})\mathbf{S} = \left(\sum_{j=1}^{n} a_{ij}\mathbf{w}_j\right)\mathbf{S} = (a_{i1}, \ldots, a_{in}) = i\text{th row of } A$$

for $i = 1, \ldots, m$. If statement (c) holds, then the vectors $\mathbf{v}_i\mathbf{T}$ are independent; since $\mathbf{S}$ is one-to-one, it follows (by the implication (a) $\Rightarrow$ (c) which we have already proved!) that the vectors $(\mathbf{v}_i\mathbf{T})\mathbf{S}$ are independent, that is, the rows of $A$ are independent, showing that (c) $\Rightarrow$ (h). Conversely, if the rows of $A$ are independent, then so are their pre-images under $\mathbf{S}$ (Theorem 4.1.4(d)); that is, the vectors $\mathbf{v}_i\mathbf{T}$ are independent, showing that (h) $\Rightarrow$ (d).

If (d) holds, then $(\mathbf{u}_1\mathbf{T}, \ldots, \mathbf{u}_m\mathbf{T})$ can be extended to an ordered basis $(\mathbf{u}_1\mathbf{T}, \ldots, \mathbf{u}_m\mathbf{T}, \mathbf{w}^*_{m+1}, \ldots, \mathbf{w}^*_n)$ of $W$. By Theorem 4.2.2 there then exists a *linear* mapping $\mathbf{T}' : W \to V$ such that

$$\mathbf{T}' : \mathbf{u}_i\mathbf{T} \to \mathbf{u}_i \qquad (1 \leq i \leq m)$$
$$\mathbf{T}' : \mathbf{w}^*_j \to \mathbf{0} \qquad (m + 1 \leq j \leq n).$$

For each $i$ $(1 \leq i \leq m)$, we then have

$$\mathbf{u}_i(\mathbf{T}\mathbf{T}') = (\mathbf{u}_i\mathbf{T})\mathbf{T}' = \mathbf{u}_i = \mathbf{u}_i\mathbf{I}_V;$$

since $\mathbf{T}\mathbf{T}'$ is linear, it follows by the "uniqueness" part of Theorem 4.2.2 that $\mathbf{T}\mathbf{T}' = \mathbf{I}_V$. Thus, (d) $\Rightarrow$ (e). The implication (e) $\Rightarrow$ (f) is trivial. Since (f) $\Rightarrow$ (a) by Theorem 1.3.6(a), our circle of implications is complete. ∎

The next theorem lists some necessary and sufficient conditions for an L.T. to be *onto*. Note how these conditions parallel parts (e) through (h) of Theorem 4.7.1.

**4.7.2** **Theorem.** Using the same notations ($V$, $W$, $\mathbf{T}$, $A$, $m$, $n$, $\mathbf{v}_i$, $\mathbf{w}_j$) as in Theorem 4.7.1, the following statements are equivalent:

(a) $\mathbf{T}$ is onto.

(b) There exists a linear transformation $\mathbf{T}' : W \to V$ such that $\mathbf{T}'\mathbf{T} = \mathbf{I}_W$.

(c) There exists a mapping $\mathbf{T}' : W \to V$ such that $\mathbf{T}'\mathbf{T} = \mathbf{I}_W$.

(d) There exists an $n \times m$ matrix $A'$ such that $A'A = I_n$.

(e) The column vectors of $A$ are linearly independent.

A mapping $\mathbf{T}'$ satisfying statement (c) is called a *left inverse* of $\mathbf{T}$; similarly, the matrix $A'$ in (d) is a *left inverse* of $A$.

*Proof of 4.7.2.* If $\mathbf{T}$ is onto, there exist vectors $\mathbf{v}^*_1, \ldots, \mathbf{v}^*_n$ in $V$ such that $\mathbf{v}^*_j\mathbf{T} = \mathbf{w}_j$ $(1 \leq j \leq n)$. By Theorem 4.2.2 there exists $\mathbf{T}' \in \text{Lin}(W, V)$ such that

$$\mathbf{w}_j\mathbf{T}' = \mathbf{v}_j^* \qquad (1 \leq j \leq n).$$

Then $\mathbf{w}_j(\mathbf{T}'\mathbf{T}) = \mathbf{v}_j^*\mathbf{T} = \mathbf{w}_j$ (all $j$) and, hence, $\mathbf{T}'\mathbf{T} = \mathbf{I}_W$, showing that (a) $\Rightarrow$ (b). Clearly (b) $\Rightarrow$ (c), and (c) $\Rightarrow$ (a) by Theorem 1.3.6(b). Thus, (a), (b), and (c) are equivalent, and evidently (b) is equivalent to (d). Finally, the equivalence of (d) and (e) is proved by applying Theorem 4.7.1 to the matrix $A^t$ instead of $A$ (remember that the transpose of a product is the product of the transposes *in reverse order*); we leave the details as an exercise. ∎

The situation is especially nice when $V$ and $W$ have the same dimension (in particular, when $V = W$). The next theorem suggests why.

**4.7.3 Theorem.** If $V$ and $W$ have the same finite dimension and $\mathbf{T} \in \mathrm{Lin}(V, W)$, then $\mathbf{T}$ is one-to-one if and only if $\mathbf{T}$ is onto.

*Proof.* $\mathbf{T}$ is one-to-one $\Leftrightarrow$ $\mathrm{Nul}\,\mathbf{T} = \{\mathbf{0}\}$

$\Leftrightarrow \dim(\mathrm{Nul}\,\mathbf{T}) = 0$ (why?)

$\Leftrightarrow \dim(\mathrm{Ran}\,\mathbf{T}) = \dim V$ (by 4.6.8)

$\Leftrightarrow \dim(\mathrm{Ran}\,\mathbf{T}) = \dim W$ (by hypothesis)

$\Leftrightarrow \mathrm{Ran}\,\mathbf{T} = W$ (by 2.6.16(b))

$\Leftrightarrow$ $\mathbf{T}$ is onto. ∎

It is important to remember the *hypothesis* of Theorem 4.7.3: $\dim V = \dim W < \infty$. The conclusion is false if $\dim V \neq \dim W$, or if the spaces are infinite dimensional (see Exercise 6 below).

Theorem 4.7.3 implies that *in the case $V = W$, all thirteen of the lettered statements in Theorems 4.7.1 and 4.7.2 are equivalent.* Moreover, in this case the mappings $\mathbf{T}'$ of 4.7.1(e) and 4.7.1(f) and the mappings $\mathbf{T}'$ of 4.7.2(b) and 4.7.2(c) are necessarily unique and equal, by Theorem 1.5.7. (The set $E$ in 1.5.7 may be taken to be the set of all mappings of $V \to V$.) Hence, we have

**4.7.4 Theorem.** If $\dim V$ is finite and $\mathbf{T} \in \mathrm{Lin}(V, V)$, then any one-sided (not necessarily linear) inverse of $\mathbf{T}$ is necessarily a two-sided *linear* inverse of $\mathbf{T}$; and if such an inverse exists, it is unique.

Translating 4.7.4 into a statement about matrices, we have at once

**4.7.5 Theorem.** If $A$, $B$ are $n \times n$ matrices (over a field $F$) such that $AB = I$, then $BA = I$.

Although 4.7.4 is false in the infinite-dimensional case, we do have the following result:

**4.7.6 Theorem.** Let $V$ be any vector space (not necessarily finite dimensional) and let $\mathbf{T} \in \text{Lin}(V, V)$. If a mapping $\mathbf{T}^{-1}: V \to V$ exists such that $\mathbf{T}^{-1}\mathbf{T} = \mathbf{T}\mathbf{T}^{-1} = \mathbf{I}$, then $\mathbf{T}^{-1}$ is linear. Such a mapping $\mathbf{T}^{-1}$ is unique if it exists, and it exists if and only if $\mathbf{T}$ is both one-to-one and onto.

*Proof.* Uniqueness of $\mathbf{T}^{-1}$ follows from 1.5.8, the assertion about existence from 1.3.6, and the linearity of $\mathbf{T}^{-1}$ from 1.8.6. (Actually, 1.8.6 only shows that $\mathbf{T}^{-1}$ preserves addition, since scalar multiplication is not strictly a binary operation *on* $V$. However, the proof for scalar multiplication is similar.) ▮

**Exercises**

1. If $A$ is an $m \times n$ matrix where $m > n$, show that $A$ cannot have a right inverse; that is, there is no $n \times m$ matrix $B$ such that $AB = I_m$. Similarly, show that $A$ cannot have a left inverse if $m < n$.
2. Find *more than one* right inverse of the matrix $A = \begin{bmatrix} 1 & 2 & 3 \\ 0 & 1 & -1 \end{bmatrix}$. (ANS)
3. In each case, determine whether the given matrix has a right inverse. (ANS)

   (a) $\begin{bmatrix} 1 & 2 & 5 & 3 \\ -2 & -4 & -10 & -6 \end{bmatrix}$ (b) $\begin{bmatrix} 1 & 1 & 0 & 1 \\ -1 & 0 & 1 & 2 \\ 2 & 1 & 0 & 5 \end{bmatrix}$

   (c) $\begin{bmatrix} 1 & 1 & 1 & 1 \\ 2 & -3 & 4 & 0 \\ 0 & -5 & 2 & -2 \end{bmatrix}$
4. In each case, determine whether the given matrix has a (two-sided) inverse. (ANS)

   (a) $\begin{bmatrix} 1 & -1 & 2 \\ 1 & 0 & 1 \\ 0 & 1 & 0 \end{bmatrix}$ (b) $\begin{bmatrix} 3 & 1 & 0 \\ -1 & 0 & 2 \\ 3 & 2 & 6 \end{bmatrix}$

*5. Carry out the suggested proof that statements (d) and (e) of Theorem 4.7.2 are equivalent.

6. Let $V$ be the set of all *infinite sequences* $(c_1, c_2, c_3, \ldots)$ where the $c_i$ are real numbers. $V$ is then a vector space over $\mathbf{R}$ under the natural definitions of addition and scalar multiplication. Find a linear transformation of $V \to V$ which is one-to-one but not onto. (This shows that Theorem 4.7.3 is false without the assumption of finite dimensionality.)

7. Interpret Section 4.4, Exercise 12, in the light of Theorem 4.7.1.
8. Let $A$, $B$ be matrices over $F$ such that $AB$ is defined. Prove: *if the row vectors of B are independent*, then any relation of dependence among the rows of $AB$ must also hold for the rows of $A$; that is, if

$$\sum_{i=1}^{m} c_i(AB)_{(i)} = \mathbf{0}$$

(where $X_{(i)}$ denotes the $i$th row vector of the matrix $X$ and where $m$ is the number of rows of $A$), then

$$\sum_{i=1}^{m} c_i A_{(i)} = \mathbf{0}.$$

(This is a partial converse of Sec. 4.4, Exercise 12.)

## 4.8 Change of basis

Let $V$ be a vector space of finite dimension $n$ over a field $F$. As we have already seen, the correspondence between Lin($V$, $V$) and Mat ($n$, $n$; $F$) depends on a choice of basis; a given L.T. may have matrix $A$ with respect to one basis of $V$ and matrix $A'$ with respect to another basis, where possibly $A \neq A'$. However, the matrices $A$ and $A'$ are not wholly unrelated; in this section, we shall derive a relation between them (equation 4.8.6 below). To keep things simple we shall stick to Lin($V$, $V$) rather than consider the more general system Lin($V$, $W$).

To fix our notation, assume that the mapping $\mathbf{T} \in \text{Lin}(V, V)$ has matrix $A = (a_{ij})$ with respect to the ordered basis $(\mathbf{v}_1, ..., \mathbf{v}_n)$ of $V$ and matrix $A' = (a'_{ij})$ with respect to the basis $(\mathbf{v}'_1, ..., \mathbf{v}'_n)$ of $V$. This means that

$$\textbf{(4.8.1)} \quad \mathbf{v}_i\mathbf{T} = \sum_{j=1}^{n} a_{ij}\mathbf{v}_j \qquad (1 \leq i \leq n);$$

$$\textbf{(4.8.2)} \quad \mathbf{v}'_j\mathbf{T} = \sum_{k=1}^{n} a'_{jk}\mathbf{v}'_k \qquad (1 \leq j \leq n).$$

Now, since $(\mathbf{v}'_1, ..., \mathbf{v}'_n)$ is a basis, every element of $V$ is a linear combination of $\mathbf{v}'_1, ..., \mathbf{v}'_n$; in particular, each $\mathbf{v}_i$ is a linear combination of the $\mathbf{v}'_j$, so that we may write

$$\textbf{(4.8.3)} \quad \mathbf{v}_i = \sum_{j=1}^{n} p_{ij}\mathbf{v}'_j \qquad (1 \leq i \leq n)$$

for suitable scalars $p_{ij}$. Similarly, each $\mathbf{v}'_j$ can be written as a linear combination of $\mathbf{v}_1, \ldots, \mathbf{v}_n$:

$$\mathbf{v}'_j = \sum_{k=1}^{n} q_{jk}\mathbf{v}_k \qquad (1 \leq j \leq n). \tag{4.8.4}$$

Let $P$ be the matrix consisting of the scalars $p_{ij}$ (that is, $P_{ij} = p_{ij}$) and similarly let $Q$ be the matrix consisting of the $q$'s ($Q_{jk} = q_{jk}$).

**4.8.5 Theorem.** Under the assumptions and notations above, $Q = P^{-1}$ and $PA' = AP$.

*Proof.* Substituting 4.8.4 into 4.8.3, we get

$$\begin{aligned}\mathbf{v}_i = \sum_{j=1}^{n} p_{ij}\mathbf{v}'_j &= \sum_{j=1}^{n} p_{ij}\left(\sum_{k=1}^{n} q_{jk}\mathbf{v}_k\right)\\ &= \sum_{k=1}^{n}\left(\sum_{j=1}^{n} p_{ij}q_{jk}\right)\mathbf{v}_k\\ &= \sum_{k=1}^{n} (PQ)_{ik}\mathbf{v}_k.\end{aligned}$$

On the other hand, we also have (trivially)

$$\begin{aligned}\mathbf{v}_i &= 0\mathbf{v}_1 + \ldots + 1\mathbf{v}_i + \ldots + 0\mathbf{v}_n \qquad \text{(all scalars zero except the coefficient of } \mathbf{v}_i\text{)}\\ &= \delta_{i1}\mathbf{v}_i + \ldots + \delta_{ii}\mathbf{v}_i + \ldots + \delta_{in}\mathbf{v}_n\\ &= \sum_{k=1}^{n} \delta_{ik}\mathbf{v}_k\\ &= \sum_{k=1}^{n} I_{ik}\mathbf{v}_k \qquad \text{(by 4.5.2).}\end{aligned}$$

Equating coefficients, we obtain $(PQ)_{ik} = I_{ik}$ for all $i$, $k$, and, hence, $PQ = I$. (This type of argument should be familiar by now!) It follows from Theorem 4.7.5 that also $QP = I$, so that $Q = P^{-1}$.

For the second half of the theorem, we apply $\mathbf{T}$ to both sides of 4.8.3:

$$\begin{aligned}\mathbf{v}_i\mathbf{T} = \left(\sum_{j=1}^{n} p_{ij}\mathbf{v}'_j\right)\mathbf{T} &= \sum_{j=1}^{n} p_{ij}\,(\mathbf{v}'_j\mathbf{T})\\ &= \sum_{j=1}^{n} p_{ij}\left(\sum_{k=1}^{n} a'_{jk}\mathbf{v}'_k\right) \qquad \text{(by 4.8.2)}\end{aligned}$$

$$= \sum_{k=1}^{n} \left( \sum_{j=1}^{n} p_{ij} a'_{jk} \right) \mathbf{v}'_k = \sum_{k=1}^{n} (PA')_{ik} \mathbf{v}'_k.$$

But we also have, by 4.8.1,

$$\mathbf{v}_i\mathbf{T} = \sum_{j=1}^{n} a_{ij}\mathbf{v}_j$$

$$= \sum_{j=1}^{n} a_{ij} \left( \sum_{k=1}^{n} p_{jk}\mathbf{v}'_k \right) \qquad \text{(using 4.8.3 with } (i, j) \text{ replaced by } (j, k))$$

$$= \sum_{k=1}^{n} \left( \sum_{j=1}^{n} a_{ij}p_{jk} \right) \mathbf{v}'_k$$

$$= \sum_{k=1}^{n} (AP)_{ik}\mathbf{v}'_k.$$

Equating coefficients, $(PA')_{ik} = (AP)_{ik}$ for all $i$, $k$ and thus $PA' = AP$. ∎

Since $P^{-1}$ exists, the equation $PA' = AP$ can be multiplied by $P^{-1}$ on the left. This gives

(4.8.6) $\quad A' = P^{-1}AP.$

(It is too bad that matrix multiplication isn't commutative; if it were, 4.8.6 would imply that $A' = AP^{-1}P = AI = A$ and the matrix of **T** would be independent of the basis. Unfortunately, this is not the case.) Still another form of the equation is obtained by multiplying 4.8.6 by $Q$ on the right and remembering that $PQ = I$; this gives $A'Q = P^{-1}A$, or

(4.8.7) $\quad A'Q = QA.$

**Example**

Let $\mathbf{T} \in \text{Lin}(\mathbf{R}_2, \mathbf{R}_2)$ be defined by $(x, y)\,\mathbf{T} = (2x - y, x + y)$. Let $A$ be the matrix of **T** with respect to the canonical basis, and let $A'$ be the matrix of **T** with respect to the basis $\{(1, 1), (2, 3)\}$. Find $P$, $Q$, $A$, $A'$ *without* using Theorem 4.8.5; then verify Theorem 4.8.5 for these matrices.

*Solution.* Since

$$\mathbf{e}_1\mathbf{T} = (1, 0)\mathbf{T} = (2, 1) = 2\mathbf{e}_1 + \mathbf{e}_2$$
$$\mathbf{e}_2\mathbf{T} = (0, 1)\mathbf{T} = (-1, 1) = -\mathbf{e}_1 + \mathbf{e}_2$$

it follows that

$$A = \begin{bmatrix} 2 & 1 \\ -1 & 1 \end{bmatrix}.$$

Similarly, from the equations

$$\begin{aligned}(1, 1)\mathbf{T} &= (1, 2) = -(1, 1) + (2, 3)\\ (2, 3)\mathbf{T} &= (1, 5) = -7(1, 1) + 4(2, 3)\end{aligned}$$

we get

$$A' = \begin{bmatrix} -1 & 1 \\ -7 & 4 \end{bmatrix}.$$

From the equations

$$\begin{aligned}(1, 1) &= 1\mathbf{e}_1 + 1\mathbf{e}_2\\ (2, 3) &= 2\mathbf{e}_1 + 3\mathbf{e}_2\end{aligned}$$

we have

$$Q = \begin{bmatrix} 1 & 1 \\ 2 & 3 \end{bmatrix}$$

(cf. 4.8.4); similarly, from the equations

$$\begin{aligned}\mathbf{e}_1 &= (1, 0) = 3(1, 1) - (2, 3)\\ \mathbf{e}_2 &= (0, 1) = -2(1, 1) + (2, 3)\end{aligned}$$

we have

$$P = \begin{bmatrix} 3 & -1 \\ -2 & 1 \end{bmatrix}.$$

By matrix multiplication.

$$PA' = \begin{bmatrix} 3 & -1 \\ -2 & 1 \end{bmatrix} \begin{bmatrix} -1 & 1 \\ -7 & 4 \end{bmatrix} = \begin{bmatrix} 4 & -1 \\ -5 & 2 \end{bmatrix}$$

$$AP = \begin{bmatrix} 2 & 1 \\ -1 & 1 \end{bmatrix} \begin{bmatrix} 3 & -1 \\ -2 & 1 \end{bmatrix} = \begin{bmatrix} 4 & -1 \\ -5 & 2 \end{bmatrix}$$

and thus $PA' = AP$ as desired. We leave it to you to verify similarly (by multiplication) that $PQ = QP = I$. ∎

Theorem 4.8.5 shows that if $P$ is a matrix which relates two bases of $V$ as in 4.8.3, then $P$ has an inverse. Conversely, we have

**4.8.8 Theorem.** Suppose that $P$ is an $n \times n$ matrix over $F$ such that $P^{-1}$ exists, and let $(\mathbf{v}_1', ..., \mathbf{v}_n')$ be an ordered basis for an $n$-dimensional vector space $V$ over $F$. If $\mathbf{v}_1, ..., \mathbf{v}_n$ are the vectors defined by 4.8.3, then $(\mathbf{v}_1, ..., \mathbf{v}_n)$ is an ordered basis of $V$.

*Proof.* By Theorem 4.2.2, there is a unique linear transformation $\mathbf{S}: V \to V$ which maps $\mathbf{v}_j' \to \mathbf{v}_j$ for all $j$. By 4.8.3, $\mathbf{S}$ has matrix $P$ with respect to the basis $(\mathbf{v}'_1, ..., \mathbf{v}'_n)$. Since $P$ has an inverse, $\mathbf{S}$ is onto; thus,

$$V = \text{Ran } \mathbf{S} = [\mathbf{v}'_1\mathbf{S}, ..., \mathbf{v}'_n\mathbf{S}] \qquad \text{(cf. Theorem 4.6.6)}$$
$$= [\mathbf{v}_1, ..., \mathbf{v}_n].$$

It follows from Theorem 2.6.12(b) that $(\mathbf{v}_1, ..., \mathbf{v}_n)$ is a basis. ▮

(*Author's remark*: The above proof was originally suggested to me by one of my students; I like its reliance on reasoning in place of computation. If you prefer computation, see Exercise 3 below for an alternate proof.)

By combining Theorems 4.8.5 and 4.8.8, we see that two $n \times n$ matrices $A$, $A'$ (over $F$) represent the same linear transformation of $V \to V$ (with respect to possibly different bases) *if and only if* there exists a matrix $P$ over $F$ such that $A' = P^{-1}AP$. Two such matrices $A$, $A'$ are said to be *similar over F*. Actually, the phrase "over $F$" is redundant, since it can be proved that similarity is *independent of the field of scalars* in the following sense: if $F$ is a subfield of some larger field $K$, if $A$ and $A'$ belong to Mat($n$, $n$; $F$), and if there exists a matrix $P$ *over K* which satisfies 4.8.6, then there will exist a matrix $P$ *over F* which satisfies 4.8.6. (The proof is far beyond the scope of this book; fortunately, we will not need the result.)

In Section 2.6 we defined the *components* of a vector with respect to a given basis. The following theorem tells how the components of a fixed vector with respect to two different bases are related.

**4.8.9 Theorem.** Let $V$ be a vector space of dimension $n$ over $F$; let $(\mathbf{v}_1, ..., \mathbf{v}_n)$ and $(\mathbf{v}'_1, ..., \mathbf{v}'_n)$ be ordered bases of $V$, and let $\mathbf{u} \in V$ be a fixed vector. Let $x_1, ..., x_n$ be the components of $\mathbf{u}$ with respect to the basis $(\mathbf{v}_1, ..., \mathbf{v}_n)$; let $x'_1, ..., x'_n$ be the components of $\mathbf{u}$ with respect to the basis $(\mathbf{v}'_1, ..., \mathbf{v}'_n)$. That is,

$$\mathbf{u} = \sum_{i=1}^{n} x_i\mathbf{v}_i = \sum_{i=1}^{n} x'_i\mathbf{v}'_i.$$

If $P$ and $Q$ are the matrices defined by equations 4.8.3 and 4.8.4 and if $X$, $X'$ are the row matrices defined by

$$X = [x_1\, x_2 \cdots x_n]; \qquad X' = [x'_1\, x'_2 \cdots x'_n],$$

then $X' = XP$ and $X = X'Q$.

We leave the proof of 4.8.9 as an exercise.

**Exercises**

1. Let $\mathbf{T} : \mathbf{R}_2 \to \mathbf{R}_2$ be defined by $(x, y)\mathbf{T} = (x - y, 2x)$. Using the notations of the section, let $\mathbf{v}_1 = \mathbf{e}_1$, $\mathbf{v}_2 = \mathbf{e}_2$, $\mathbf{v}_1' = (1, 2)$, $\mathbf{v}_2' = (3, 5)$.
(a) Compute the matrices $A$, $A'$, $P$, $Q$ independently (i.e, without using Theorem 4.8.5). (ANS)
(b) Verify by matrix multiplication that $PQ = I$, $PA' = AP$, $QA = A'Q$.
(c) Verify Theorem 4.8.9 for the vector $\mathbf{u} = (1, -1)$. (SOL)
(d) Verify Theorem 4.8.9 for the vector $\mathbf{u} = (3, 1)$.

*2. Prove Theorem 4.8.9.

3. Give a "computational" proof of Theorem 4.8.8. (*Suggestion*: Letting $Q = P^{-1}$, deduce 4.8.4 from 4.8.3. Hence, $\mathbf{v}_j' \in [\mathbf{v}_1, \ldots, \mathbf{v}_n]$; etc.)

4. If

$$P = \begin{bmatrix} 1 & -2 \\ 2 & -4 \end{bmatrix}, \qquad A' = \begin{bmatrix} 3 & -6 \\ 1 & -2 \end{bmatrix}, \qquad A = \begin{bmatrix} -1 & 1 \\ 2 & 0 \end{bmatrix}$$

then $PA' = AP$. Why does this *not* imply that the matrices $A$ and $A'$ are similar? (ANS)

5. In each of the following, determine whether or not the given matrices $A$, $A'$ are similar over $\mathbf{R}$. (*Suggestions*: (1) The equation $PA' = AP$ is equivalent to a system of four linear equations in four unknowns. (2) A two-by-two matrix

$$\begin{bmatrix} a & b \\ c & d \end{bmatrix}$$

has an inverse if and only if $ad \neq bc$; cf. Sec. 2.4, Exercise 8.) (ANS)

(a) $A = \begin{bmatrix} 2 & 1 \\ 1 & 3 \end{bmatrix}$; $\quad A' = \begin{bmatrix} 1 & -1 \\ 1 & 4 \end{bmatrix}$.

(b) $A = \begin{bmatrix} 2 & -1 \\ 4 & -2 \end{bmatrix}$; $\quad A' = \begin{bmatrix} 2 & 6 \\ -1 & -3 \end{bmatrix}$.

(c) $A = \begin{bmatrix} 2 & -1 \\ 4 & -2 \end{bmatrix}$; $\quad A' = \begin{bmatrix} 6 & 9 \\ -4 & -6 \end{bmatrix}$.

(d) $A = \begin{bmatrix} 1 & 1 \\ 4 & 1 \end{bmatrix}$; $\quad A' = \begin{bmatrix} 2 & 0 \\ -1 & -1 \end{bmatrix}$.

(e) $A = \begin{bmatrix} 0 & -1 \\ 2 & 3 \end{bmatrix}$; $\quad A' = \begin{bmatrix} 1 & 4 \\ 0 & 2 \end{bmatrix}$.

6. (*Generalization of Theorem 4.8.5*) Let $V$, $W$ be vector spaces over $F$ having finite dimensions $m$, $n$, respectively, and let

$\mathbf{T} \in \mathrm{Lin}(V, W)$. Assume that $\mathbf{T}$ has matrix $A$ with respect to ordered bases $(\mathbf{v}_1, \ldots, \mathbf{v}_m)$ and $(\mathbf{w}_1, \ldots, \mathbf{w}_n)$ of $V$ and $W$, respectively; and that $\mathbf{T}$ has matrix $A'$ with respect to ordered bases $(\mathbf{v}'_1, \ldots, \mathbf{v}'_m)$ and $(\mathbf{w}'_1, \ldots, \mathbf{w}'_n)$ of $V$ and $W$, respectively. If the matrices $P, Q, R, S$ are defined by the equations

$$\mathbf{v}_i = \sum_{j=1}^{m} p_{ij}\mathbf{v}'_j \qquad (1 \leq i \leq m)$$

$$\mathbf{v}'_j = \sum_{k=1}^{m} q_{jk}\mathbf{v}_k \qquad (1 \leq j \leq m)$$

$$\mathbf{w}_\alpha = \sum_{\beta=1}^{n} r_{\alpha\beta}\,\mathbf{w}'_\beta \qquad (1 \leq \alpha \leq n)$$

$$\mathbf{w}'_\beta = \sum_{\gamma=1}^{n} s_{\beta\gamma}\,\mathbf{w}_\gamma \qquad (1 \leq \beta \leq n),$$

prove that $PA' = AR$. (*Note*: The equation $Q = P^{-1}$ is still valid, by the same argument as before, and similarly $S = R^{-1}$.)

## 4.9 Affine transformations

A linear transformation may be defined as a mapping which preserves linear combinations (cf. Sec. 4.1, Exercise 2 and Theorem 4.1.4(c)). Now as we saw in Section 3.6, the role played by linear combinations in algebra is somewhat analogous to the role played by *barycentric* combinations in geometry; thus, in studying vector spaces from a geometric standpoint, it is natural to turn our attention to mappings which preserve barycentric combinations. Not surprisingly, such mappings are closely related to L.T.'s, as the following theorem shows.

4.9.1 **Theorem.** Let $V$, $W$ be vector spaces over a field $F$ and let $\mathbf{M}$ be a mapping of $V \to W$. Then the following two statements are equivalent:

(a) Whenever $k$ is a positive integer and $a_1, \ldots, a_k$ are scalars such that $a_1 + \cdots + a_k = 1$, then

$$(a_1P_1 + \cdots + a_kP_k)\mathbf{M} = a_1(P_1\mathbf{M}) + \cdots + a_k(P_k\mathbf{M})$$

for all points $P_1, \ldots, P_k$ in $V$.

(b) $\mathbf{M}$ can be expressed as the product of a linear transformation and a translation; that is, $\mathbf{M} = \mathbf{S}\mathscr{T}_\mathbf{w}$ for some $\mathbf{S} \in \mathrm{Lin}(V, W)$ and some $\mathbf{w} \in W$.

In addition, if $F$ does not have characteristic 2, then statements (a) and (b) are also equivalent to the following statement:

(c) Whenever $a$, $b$ are scalars such that $a + b = 1$, then

$$(aP + bQ)\mathbf{M} = a(P\mathbf{M}) + b(Q\mathbf{M})$$

for all points $P$, $Q$ in $V$.

Condition (c) states that $\mathbf{M}$ preserves barycentric combinations of two points; condition (a) states that $\mathbf{M}$ preserves $b$-combinations of any number of points. (c) does *not* imply (a) if char $F = 2$ (see Exercise 13, this section), although (a) clearly implies (c) in all cases. A mapping $\mathbf{M}$ which satisfies condition (a) or (b) (equivalently, all three conditions) is called an *affine transformation* (abbreviated A.T.). In this section we shall discuss some of the basic algebraic and geometric properties of A.T.'s; in the next section we shall specialize our considerations to a particularly important class of A.T.'s, the *isometries*.

*Proof of Theorem 4.9.1.* The proof proceeds in several steps.

(1) Since a $b$-combination is a special case of a linear combination, it is clear that every L.T. satisfies condition (a).

(2) *Every translation satisfies* (*a*). Indeed, if $\mathbf{w} \in W$ and if $a_1 + \cdots + a_k = 1$, then for all points $P_1, \ldots, P_k$ in $W$, we have

$$\begin{aligned}\left(\sum_{i=1}^{k} a_i P_i\right)\mathcal{T}_{\mathbf{w}} &= \left(\sum_{i=1}^{k} a_i P_i\right) + \mathbf{w} \\ &= \sum_{i=1}^{k} a_i P_i + \left(\sum_{i=1}^{k} a_i\right)\mathbf{w} \qquad \left(\text{since } \sum_{i=1}^{k} a_i = 1\right) \\ &= \sum_{i=1}^{k} a_i (P_i + \mathbf{w}) = \sum_{i=1}^{k} a_i (P_i \mathcal{T}_{\mathbf{w}})\end{aligned}$$

as claimed.

(3) If $\mathbf{M}_1 : V \to W$ and $\mathbf{M}_2 : W \to Y$ both preserve $b$-combinations, *so does their product* $\mathbf{M}_1\mathbf{M}_2$. The proof is exactly the same as for linear combinations (Sec. 4.3); we leave it to you to verify that the argument remains valid. (This argument is really quite general; roughly speaking, *anything* which is preserved by each of two mappings is also preserved by their product.)

(4) By putting steps (1), (2), and (3) together, we see that (b) implies (a). Since (a) trivially implies (c), it remains only to show that (a) $\Rightarrow$ (b) and that if char $F \neq 2$, then (c) $\Rightarrow$ (b).

Assume either that (a) holds, or that (c) holds with char $F \neq 2$. Let $\mathbf{w} = \mathbf{0M}$, $\mathbf{S} = \mathbf{M}\mathscr{T}_{-\mathbf{w}}$. Then $\mathbf{w} \in W$, $\mathbf{S}$ is a mapping of $V \to W$, and $\mathbf{M} = \mathbf{S}\mathscr{T}_{\mathbf{w}}$; thus, we need only show that $\mathbf{S}$ is linear. Let $t$ be any scalar; then for all $\mathbf{u} \in V$, we have

$$\begin{aligned}(t\mathbf{u})\mathbf{S} &= (t\mathbf{u})\mathbf{M} - \mathbf{w} = (t\mathbf{u} + (1-t)\mathbf{0})\mathbf{M} - \mathbf{w}\\ &= t(\mathbf{uM}) + (1-t)\,(\mathbf{0M}) - \mathbf{w} \qquad \text{(since (a) or (c) holds)}\\ &= t(\mathbf{uS} + \mathbf{w}) + (1-t)\mathbf{w} - \mathbf{w} = t(\mathbf{uS})\end{aligned}$$

so that $\mathbf{S}$ preserves scalar multiplication. To show that $\mathbf{S}$ preserves addition, suppose first that char $F \neq 2$. Here division by 2 is allowed, so that for any $\mathbf{u}$, $\mathbf{v}$ in $V$, we have

$$\begin{aligned}\left(\frac{1}{2}\mathbf{u} + \frac{1}{2}\mathbf{v}\right)\mathbf{S} &= \left(\frac{1}{2}\mathbf{u} + \frac{1}{2}\mathbf{v}\right)\mathbf{M} - \mathbf{w}\\ &= \frac{1}{2}(\mathbf{uM}) + \frac{1}{2}(\mathbf{vM}) - \mathbf{w} \qquad \text{(since (a) or (c) holds)}\\ &= \frac{1}{2}(\mathbf{uM} - \mathbf{w}) + \frac{1}{2}(\mathbf{vM} - \mathbf{w})\\ &= \frac{1}{2}(\mathbf{uS}) + \frac{1}{2}(\mathbf{vS}).\end{aligned}$$

Since we have already shown that $\mathbf{S}$ preserves scalar multiplication (in particular, multiplication by 2), it follows that $(\mathbf{u} + \mathbf{v})\mathbf{S} = \mathbf{uS} + \mathbf{vS}$ as desired. On the other hand, if char $F = 2$, then $1_F + 1_F + 1_F = 1_F$, so that by applying 4.9.1(a) with $k = 3$, $a_1 = a_2 = a_3 = 1_F$, $P_3 = O$, we get

$$(P_1 + P_2)\mathbf{M} = P_1\mathbf{M} + P_2\mathbf{M} + O\mathbf{M} \qquad (\text{all } P_1, P_2 \in V);$$

that is,

$$\begin{aligned}(P_1 + P_2)\mathbf{S} + \mathbf{w} &= (P_1\mathbf{S} + \mathbf{w}) + (P_2\mathbf{S} + \mathbf{w}) + \mathbf{w}\\ &= P_1\mathbf{S} + P_2\mathbf{S} + \mathbf{w}\\ &\qquad (\text{since } \mathbf{w} + \mathbf{w} = (1_F + 1_F)\mathbf{w} = \mathbf{0})\end{aligned}$$

and by subtracting $\mathbf{w}$ we get $(P_1 + P_2)\mathbf{S} = P_1\mathbf{S} + P_2\mathbf{S}$ as desired. This completes the proof of Theorem 4.9.1. ∎

*Remark*: There are striking similarities between the proof that $\mathbf{S}$ is linear (above) and parts of the proof of Theorem 3.4.2. Can you find them? Can you think of a reason why such similarities should be present?

In the finite-dimensional case, we can write the equations of an affine transformation **M** in a form similar to 4.2.14. To do so, write $\mathbf{M} = \mathbf{S}\mathcal{T}_{\mathbf{w}}$ as in 4.9.1(b); then **S** has equations of the form 4.2.14, where $A = (a_{ij})$ is the matrix of **S**. If the components of the vector **w** are $c_1$, ..., $c_n$, it easily follows that the equations of **M** are

$$
\begin{aligned}
x_1' &= a_{11}x_1 + \cdots + a_{m1}x_m + c_1 \\
x_2' &= a_{12}x_1 + \cdots + a_{m2}x_m + c_2 \\
&\cdots \\
x_n' &= a_{1n}x_1 + \cdots + a_{mn}x_m + c_n.
\end{aligned}
\tag{4.9.2}
$$

It is not hard to show that the mapping **S** and the vector **w** in 4.9.1(b) are uniquely determined by **M** (see Exercise 3, this section). We shall call **S** the *linear part* of **M**, and **w** the *translation vector* of **M**. In this connection, the following result will be needed later.

**4.9.3 Theorem.** Let $\mathbf{M}_1 : V \to W$ and $\mathbf{M}_2 : W \to Y$ be affine transformations and let $\mathbf{S}_i$ be the linear part of $\mathbf{M}_i$ ($i = 1, 2$). Then $\mathbf{S}_1\mathbf{S}_2$ is the linear part of $\mathbf{M}_1\mathbf{M}_2$.

(Note that $\mathbf{M}_1\mathbf{M}_2$ is already known to be affine, by step (3) in the proof of 4.9.1.)

*Proof.* We have $\mathbf{M}_1 = \mathbf{S}_1\,\mathcal{T}_{\mathbf{w}}$, $\mathbf{M}_2 = \mathbf{S}_2\,\mathcal{T}_{\mathbf{y}}$ for some $\mathbf{w} \in W$, $\mathbf{y} \in Y$. Then for all $\mathbf{u} \in V$,

$$
\begin{aligned}
\mathbf{u}\mathbf{M}_1\mathbf{M}_2 &= (\mathbf{u}\mathbf{S}_1\mathcal{T}_{\mathbf{w}})\,(\mathbf{S}_2\mathcal{T}_{\mathbf{y}}) = (\mathbf{u}\mathbf{S}_1 + \mathbf{w})\mathbf{S}_2 + \mathbf{y} \\
&= \mathbf{u}\mathbf{S}_1\mathbf{S}_2 + \mathbf{w}\mathbf{S}_2 + \mathbf{y} \qquad \text{(since } \mathbf{S}_2 \text{ is linear)}
\end{aligned}
$$

and, hence,

$$\mathbf{M}_1\mathbf{M}_2 = (\mathbf{S}_1\mathbf{S}_2)\mathcal{T}_{\mathbf{w}\mathbf{S}_2+\mathbf{y}}$$

from which the theorem follows. ▮

Since flats are closed under *b*-combinations, conditions (a) and (c) of 4.9.1 make sense even if the domain of **M** is a flat rather than a vector space, and it is useful to extend the definition of A.T. to the latter situation. We do so as follows: if $S$, $S'$ are flats in the vector spaces $V$, $W$ (respectively), a mapping $\mathbf{M} : S \to S'$ will be called *affine* if it satisfies 4.9.1(a) with $V$ replaced by $S$. The following theorem is useful in this context.

**4.9.4 Theorem.** Let $V$, $W$ be vector spaces over $F$. Let $S$, $S'$ be flats in $V$, $W$ (respectively), with $\dim S = m < \infty$. Let $P_0, P_1, \ldots, P_m$ be $m + 1$ barycentrically independent points in $S$, and let

$P_0', P_1', \ldots, P_m'$ be arbitrary points in $S'$. Then:

(a) There is a unique affine mapping $\mathbf{M} : S \to S'$ which maps

$$P_0 \to P_0', P_1 \to P_1', \ldots, P_m \to P_m'.$$

In fact, $\mathbf{M}$ is given by

$$\mathbf{M} : \sum_{i=0}^{m} a_i P_i \to \sum_{i=0}^{m} a_i P_i'$$

(all scalars $a_0, \ldots, a_m$ such that $\sum_{i=0}^{m} a_i = 1$).

(b) If $\dim V < \infty$, then the mapping $\mathbf{M}$ of part (a) can be extended to an affine transformation $\mathbf{M}^*$ of $V \to W$.

Theorem 4.9.4(a) asserts that an affine mapping is uniquely determined by what it does to a maximal barycentrically independent set. This result is analogous to Theorem 4.2.2, which asserted that a *linear* transformation is uniquely determined by its effect on a maximal *linearly* independent set (i.e., a basis). The proof of 4.9.4(a) is like that of 4.2.2 (using Theorem 3.6.5 in place of 2.6.4); we omit the details. To prove 4.9.4(b), extend $\{P_0, \ldots, P_m\}$ to a maximal $b$-independent subset $\{P_0, \ldots, P_m, P_{m+1}, \ldots, P_n\}$ of $V$, where $n = \dim V$ (this is possible by Sec. 3.5, Exercise 5); also let $P_{m+1}', \ldots, P_n'$ be any $n - m$ points whatever in $W$. By 4.9.4(a), the mapping $\mathbf{M}^* : V \to W$ defined by

$$\text{(4.9.5)} \qquad \mathbf{M}^* : \sum_{i=0}^{n} a_i P_i \to \sum_{i=0}^{n} a_i P_i' \qquad (a_0 + \cdots + a_n = 1)$$

is affine; since $\mathbf{M}^* \restriction S = \mathbf{M}$, 4.9.4(b) follows. ∎

The *linear part* of the mapping $\mathbf{M}^*$ in 4.9.5 can be found explicitly. Indeed, let $\mathbf{u}_i = \mathbf{Ar}\, P_0P_i$, $\mathbf{v}_i = \mathbf{Ar}\, P_0'P_i'$; then $(\mathbf{u}_1, \ldots, \mathbf{u}_n)$ is an ordered basis of $V$ and, hence, there is a unique linear transformation $\mathbf{S}$: $V \to W$ which maps $\mathbf{u}_1 \to \mathbf{v}_1, \ldots, \mathbf{u}_n \to \mathbf{v}_n$. If we let $\mathbf{w} = -P_0\mathbf{S} + P_0'$, a straightforward calculation (which we leave to you!) shows that $P_i(\mathbf{S}\mathcal{T}_{\mathbf{w}}) = P_i'$ for all $i$ $(0 \leq i \leq n)$. Since $\mathbf{S}\mathcal{T}_{\mathbf{w}}$ is affine, the "uniqueness" assertion in 4.9.4 implies that $\mathbf{M}^* = \mathbf{S}\mathcal{T}_{\mathbf{w}}$. Changing notation, we have shown that if an affine transformation $\mathbf{M} : V \to W$ maps $P \to P'$ and $Q \to Q'$, then

(4.9.6)
(a) the linear part of $\mathbf{M}$ maps $Q - P \to Q' - P'$;
(b) the translation vector of $\mathbf{M}$ equals $P' - P\mathbf{S}$, where $\mathbf{S}$ is the linear part of $\mathbf{M}$.

4.9.7 **Example.** Let **M** be the affine transformation of $\mathbf{R}_2 \to \mathbf{R}_2$ which maps $(1, 1) \to (2, 3), (3, 2) \to (3, 8), (2, 3) \to (1, 7)$. Find the equations of **M** in the form 4.9.2.

*Solution.* Write $\mathbf{M} = \mathbf{S}\mathcal{T}_{\mathbf{w}}$, where **S** is linear. By 4.9.6, **S** maps $(2, 1) \to (1, 5)$ and $(1, 2) \to (-1, 4)$. We leave it to you to deduce from the latter (by linearity) that **S** maps $(1, 0) \to (1, 2)$ and $(0, 1) \to (-1, 1)$; hence, **S** maps

$$(x, y) \to (x - y, 2x + y).$$

Again, by 4.9.6, we have

$$\mathbf{w} = (2, 3) - (1, 1)\mathbf{S} = (2, 3) - (0, 3) = (2, 0).$$

Hence, **M** maps $(x, y) \to (x - y + 2, 2x + y)$, and the equations of **M** are

$$x' = x - y + 2$$
$$y' = 2x + y. \quad \blacksquare$$

Just as Theorem 4.9.4(a) was the affine analog of Theorem 4.2.2, so parts (a) and (c) of the next theorem are the affine analogs of Theorems 4.6.6 and 4.1.4(d).

4.9.8 **Theorem.** Let $V$, $W$ be vector spaces over $F$, let $S$ be a flat in $V$, and let $\mathbf{M} : S \to W$ be affine. Then

(a) For any points $P_0, ..., P_k$ in $S$, the image of the flat $\langle P_0, ..., P_k \rangle$ under **M** is the flat $\langle P_0\mathbf{M}, ..., P_k\mathbf{M} \rangle$.

(b) The image of any $k$-flat in $S$ under **M** is a flat of dimension $\leq k$.

(c) **M** preserves barycentric dependence; that is, if the points $P_0, ..., P_k$ in $S$ are $b$-dependent, then the points $P_0\mathbf{M}, ..., P_k\mathbf{M}$ are $b$-dependent.

(d) **M** preserves parallelism; that is, if $Y_1$, $Y_2$ are parallel $k$-flats in $S$, then their images $Y_1\mathbf{M}$, $Y_2\mathbf{M}$ are parallel flats in $W$ such that $\dim Y_1\mathbf{M} = \dim Y_2\mathbf{M}$.

*Proof.* Part (a) follows from Theorem 3.6.1 and the fact that **M** preserves $b$-combinations. For part (b), we note that by 3.5.5 any $k$-flat $Y$ in $S$ is determined by $k + 1$ points $P_0, ..., P_k$; by part (a), $Y\mathbf{M} = \langle P_0\mathbf{M}, ..., P_k\mathbf{M} \rangle$ which is a flat of dimension $\leq k$ by Theorem 3.5.2. Part (c) follows from (a) and (b) since $k + 1$ points are $b$-dependent if and only if they determine a flat of dimension $< k$.

To prove (d), assume first that $S$ is a vector space (i.e., a subspace of $V$). By 4.9.1(b), we can write $\mathbf{M} = \mathbf{M}_0 \cdot \mathcal{T}_\mathbf{w}$, where $\mathbf{M}_0$ is linear. If $Y_1$, $Y_2$ are parallel $k$-flats in $S$, then we may write $Y_2 = Y_1 + \mathbf{u}$ for some $\mathbf{u} \in S$ (Theorem 3.4.4(g)). Then

$$\begin{aligned} Y_2\mathbf{M} &= (Y_1 + \mathbf{u})\,(\mathbf{M}_0 \cdot \mathcal{T}_\mathbf{w}) \\ &= (Y_1 + \mathbf{u})\mathbf{M}_0 + \mathbf{w} = Y_1\mathbf{M}_0 + \mathbf{u}\mathbf{M}_0 + \mathbf{w} \\ &= (Y_1\mathbf{M}_0 + \mathbf{w}) + \mathbf{u}\mathbf{M}_0 = Y_1(\mathbf{M}_0 \cdot \mathcal{T}_\mathbf{w}) + \mathbf{u}\mathbf{M}_0 \\ &= Y_1\mathbf{M} + \mathbf{u}\mathbf{M}_0 \end{aligned}$$

which implies (d). The more general case where $S$ is a flat (so that 4.9.1(b) is not immediately applicable) can be handled as follows: let $Y = \langle Y_1, Y_2 \rangle$ and let $V_0$ be the (finite-dimensional!) subspace of $V$ spanned by $\langle Y_1, Y_2 \rangle$. By 4.9.4, $\mathbf{M} \restriction Y$ can be extended to an affine transformation $\mathbf{M}^*$ with domain $V_0$. Since $V_0$ is a subspace, (d) holds for $\mathbf{M}^*$; since $\mathbf{M}^*$ and $\mathbf{M}$ agree on $Y_1$ and $Y_2$, it follows that (d) holds for $\mathbf{M}$ as well. ▮

Theorem 4.9.8(b) shows, in particular, that the image of a line under an affine transformation is always a line or a point; that is, A.T.'s *preserve collinearity*. We might ask whether, conversely, mappings which preserve collinearity are necessarily affine. It is rather surprising that the answer to this question depends on the field of scalars; indeed, the following theorem, which partially answers the question for the field $\mathbf{R}$, would be false if we replaced $\mathbf{R}$ by $\mathbf{C}$.

**4.9.9 Theorem.** Let $n$ be an integer greater than 1. If $\mathbf{M}$ is a one-to-one mapping of $\mathbf{R}_n$ onto $\mathbf{R}_n$ which maps collinear points into collinear points, then $\mathbf{M}$ is affine. If in addition $\mathbf{M}$ maps $\mathbf{0} \to \mathbf{0}$, then $\mathbf{M}$ is linear.

The proof, which is long, can be found in Appendix D. For a counterexample over the field $\mathbf{C}$, see Section 4.1, Exercise 11. It would be interesting to know whether or not the "one-to-one and onto" hypotheses can be weakened; Exercises 12 and 14 (below) suggest that they cannot be entirely eliminated. Regarding the last sentence of Theorem 4.9.9, see Exercise 6 below.

**One-to-one A.T.'s**

Affine transformations, like other mappings, behave better if they are one-to-one. Some results in the one-to-one case are collected in the following theorem.

4.9.10 **Theorem.** Let $V$, $W$ be vector spaces over $F$, let $S$ be a flat in $V$, let $\mathbf{M}$ be a one-to-one affine mapping of $S \to W$, and let $S' = \operatorname{Ran} \mathbf{M}$. Then the following statements hold:

(a) The inverse mapping $\mathbf{M}^{-1} : S' \to S$ is affine.

(b) The image of any $k$-flat in $S$ under $\mathbf{M}$ is a $k$-flat in $S'$.

(c) $\mathbf{M}$ preserves barycentric independence; that is, if $P_0, \ldots, P_k$ are $b$-independent points in $S$, then the points $P_0\mathbf{M}, \ldots, P_k\mathbf{M}$ are $b$-independent.

(d) Two $k$-flats in $S$ are parallel if and only if their images under $\mathbf{M}$ are parallel.

*Proof.* We first observe that the mapping $\mathbf{M}^{-1}$ exists (since $\mathbf{M} : S \to S'$ is one-to-one and onto) and that $S'$ is a flat by 4.9.8(a). It is not hard to show that if $\mathbf{M}$ satisfies 4.9.1(a), so does $\mathbf{M}^{-1}$ (the argument is like the proof of Theorem 1.8.6); thus, 4.9.10(a) holds. Parts (b), (c), and (d) are all proved by applying the corresponding parts of Theorem 4.9.8 to $\mathbf{M}^{-1}$ as well as $\mathbf{M}$. For example, if $Y$ is a $k$-flat in $S$, then 4.9.8(b) tells us that $Y\mathbf{M}$ is an $h$-flat in $S'$, where $h \leqslant k$. Since $\mathbf{M}^{-1}$ is affine also, we similarly have $\dim(Y\mathbf{M})\mathbf{M}^{-1} \leqslant \dim Y\mathbf{M}$; that is, $\dim Y \leqslant \dim Y\mathbf{M}$; that is, $k \leqslant h$. Hence, $k = h$, proving 4.9.10(b). The proofs of parts (c) and (d) are similar; we leave the details as an exercise. ∎

*Remarks*: 1. If $S$, $S'$ are flats in $V$, $W$ (respectively) such that $\dim S \leqslant \dim S'$, a one-to-one affine transformation $\mathbf{M} : S \to S'$ always exists. Indeed, if $\dim S = m$, we need only choose $m + 1$ $b$-independent points $P_0, \ldots, P_m$ in $S$ and $m + 1$ $b$-independent points $P'_0, \ldots, P'_m$ in $S'$ and then define $\mathbf{M}$ as in 4.9.4(a) (cf. Exercise 10, this section).

2. All of the results which we have obtained for affine transformations hold in particular for linear transformations, since every L.T. is affine. (Among other things, this gives us a solution to Sec. 4.1, Exercise 10.)

### A Euclidean interpretation of A.T.'s

If $P$ is any point on a line $\langle A, B\rangle$, we can, of course, express $P$ as a barycentric combination $aA + bB$, where $a + b = 1$. In the space $\mathbf{R}_n$ or, more generally, in any real Euclidean space, the numbers $a$ and $b$ have a geometric interpretation; specifically, the ratio $|b/a|$ is the ratio between the distances $d(A, P)$ and $d(B, P)$ (see Sec. 3.2, Exercise 8). Since affine transformations are defined to be mappings which preserve barycentric combinations, it follows that in the real Euclidean case the A.T.'s are precisely those mappings which *preserve ratios of distances along lines*. That is, if $\mathbf{M}$ is affine and $A$, $P$, $B$ are collinear, then

$$\frac{d(A, P)}{d(B, P)} = \frac{d(A\mathbf{M}, P\mathbf{M})}{d(B\mathbf{M}, P\mathbf{M})}$$

provided that neither denominator is zero. The equation need *not* hold if the points are noncollinear. Figure 4.9.1 illustrates the situation where $\mathbf{M}$ is the mapping defined in Example 4.9.7; here $|\overline{BC}|/|\overline{CD}| = |\overline{B'C'}|/|\overline{C'D'}|$ since $B$, $C$, $D$ are collinear, but $|\overline{AB}|/|\overline{BC}| \neq |\overline{A'B'}|/|\overline{B'C'}|$.

**Exercises**

In the Exercises below, $V$ and $W$ denote vector spaces over a field $F$.

1. Let $\mathbf{M}$ be the affine transformation of $\mathbf{R}_2 \to \mathbf{R}_2$ which maps $(1, 1) \to (3, 2)$, $(2, -1) \to (8, -2)$, $(3, 2) \to (8, 9)$. Find the equations of $\mathbf{M}$. (ANS)

Figure 4.9.1

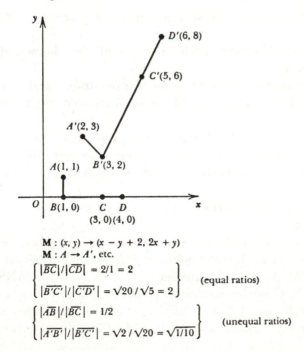

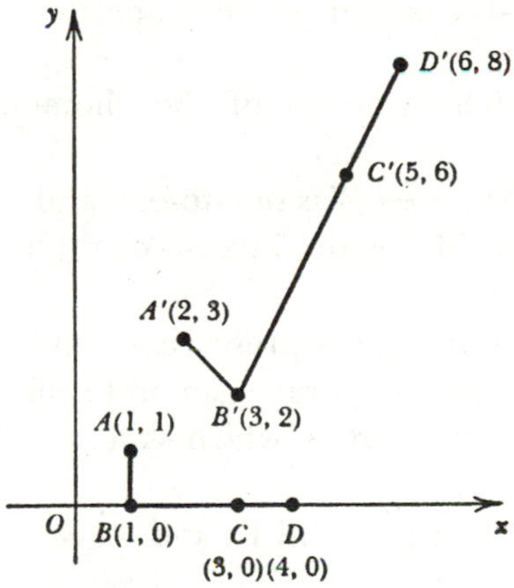


$\mathbf{M} : (x, y) \to (x - y + 2, 2x + y)$
$\mathbf{M} : A \to A'$, etc.

$$\left\{\begin{array}{l} |\overline{BC}|/|\overline{CD}| = 2/1 = 2 \\ |\overline{B'C'}|/|\overline{C'D'}| = \sqrt{20}/\sqrt{5} = 2 \end{array}\right\} \quad \text{(equal ratios)}$$

$$\left\{\begin{array}{l} |\overline{AB}|/|\overline{BC}| = 1/2 \\ |\overline{A'B'}|/|\overline{B'C'}| = \sqrt{2}/\sqrt{20} = \sqrt{1/10} \end{array}\right\} \quad \text{(unequal ratios)}$$

2. Same problem if **M** maps $(1, 2) \to (9, 1)$, $(-1, 1) \to (8, 0)$, $(2, 0) \to (2, -1)$. (ANS)

*3. Prove that the vector **w** and the mapping **S** in 4.9.1(b) are uniquely determined by **M**; that is, if also $\mathbf{M} = \mathbf{S}'\mathcal{T}_{\mathbf{y}}$, where $\mathbf{S}'$ is linear and $\mathbf{y} \in W$, then $\mathbf{w} = \mathbf{y}$ and $\mathbf{S} = \mathbf{S}'$. (*Hint*: Find **0M**.)

4. If $\mathbf{M} : V \to W$ is affine and $S$ is a flat in $V$, then by Theorem 4.9.8 the image of $S$ under **M** is a flat $S'$ in $W$. Prove that the linear part of **M** maps the direction space of $S$ onto the direction space of $S'$.

5. Let $\mathbf{M} : V \to W$ be affine. Although **M** can be expressed as an L.T. times a translation, **M** *cannot* necessarily be expressed as a translation times an L.T. (Multiplication of mappings is noncommutative!) Show that the following two statements are equivalent:
   (a) **M** can be expressed in the form $\mathcal{T}_{\mathbf{v}}\mathbf{S}$, where $\mathbf{v} \in V$ and $\mathbf{S} \in \text{Lin}(V, W)$.
   (b) $\mathbf{0} \in \text{Ran } \mathbf{M}$.

   Show further that if statement (a) holds, then the mapping **S** of statement (a) equals the linear part of **M** and the translation vector of **M** is **vS**.

6. Prove that an affine transformation which maps $\mathbf{0} \to \mathbf{0}$ is linear.

7. (a) Interpret Theorem 4.9.3 in terms of the "homomorphism" concept.
   (b) If the affine mapping $\mathbf{M} : V \to W$ is one-to-one and onto, show that the linear part of $\mathbf{M}^{-1}$ is the inverse of the linear part of **M**.

8. As we have seen in this section, A.T.'s preserve a number of geometric properties; for instance, parallelism and collinearity. Name some geometric properties which A.T.'s do *not* preserve.

9. Let $V$, $W$ be vector spaces over the field **R**, and let $\mathbf{M} : V \to W$ be affine.
   (a) Does **M** preserve "betweenness"? That is, if $P$ lies between $A$ and $B$ (where $P$, $A$, $B$ are points in $V$), must $P\mathbf{M}$ lie between $A\mathbf{M}$ and $B\mathbf{M}$? Justify your answer. (ANS)
   (b) If $A$, $B$, $C$ are noncollinear points in $V$ and $P$ lies "inside" triangle $ABC$ (cf. Sec. 3.6), must $P\mathbf{M}$ lie inside the triangle whose vertices are $A\mathbf{M}$, $B\mathbf{M}$, $C\mathbf{M}$? Justify your answer. (ANS)

*10. Prove: if the points $P'_0, \ldots, P'_m$ of Theorem 4.9.4(a) are $b$-independent, then the mapping **M** of that theorem is one-to-one.

11. Carry out the suggested proof of parts (c) and (d) of Theorem 4.9.10.

12. Show by means of counterexamples:
(a) That Theorem 4.9.9 is false if $n = 1$.
(b) That Theorem 4.9.9 is false if the word "onto" is replaced by "into". (*Suggestion*: It is known that the sets $\mathbf{R}$ and $\mathbf{R}_n$ have the same cardinality for all $n \in \mathbf{Z}^+$.)
(c) That Theorem 4.9.9 is false if $\mathbf{R}$ is replaced by $F$, where $F$ is the subfield of $\mathbf{R}$ consisting of all numbers of the form
$x + y\sqrt{2} \qquad (x, y \in \mathbf{Q})$.
(*Suggestion*: Imitate Sec. 4.1, Exercise 11.)

13. Show, by means of a counterexample, that 4.9.1(c) does not imply 4.9.1(a) if the field $F$ is allowed to have characteristic 2.

14. Exhibit a mapping $\mathbf{M} : \mathbf{R}_2 \to \mathbf{R}_2$ which maps collinear points into collinear points and whose range is more than a line, but which is not affine.

15. If **M** maps each point of $\mathbf{R}_2$ onto its perpendicular projection on the line $y = mx + b$ (see Fig. 4.9.2), show that **M** is affine and find the equations of **M**. (Compare with Sec. 4.2, Exercise 9.) (ANS)

16. Let $E$ be a real Euclidean space of finite dimension and let $S$ be a flat in $E$. For each point $P \in E$, let $PM$ be the perpendicular projection of $P$ on $S$. Show that **M** is affine. (Compare with Sec. 4.2, Exercise 10.) (SUG)

Figure 4.9.2

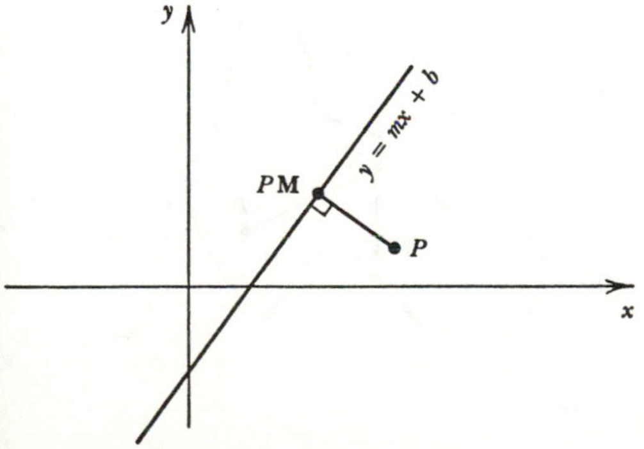

## 4.10 Isometries and congruence

Throughout this section, the letter $E$, either alone or with a subscript or superscript attached (e.g., $E_1$, $E'$, etc.), will denote a real Euclidean space.

Consider the two regions $S$ and $S'$ which are shown in Figure 4.10.1. (We include both boundary and interior as part of the region.) If one were to cut out $S'$ from the page, flip it over, rotate it slightly, and then paste it on top of $S$, the boundaries of $S$ and $S'$ would coincide; the point $P'$ would find itself on top of $P$, and similarly $Q'$ would be on top of $Q$, $R'$ on top of $R$. Roughly speaking, $S$ and $S'$ "look alike"; they have the "same size and shape". Two such sets $S$ and $S'$ are said to be *congruent*.

To make the concept of congruence a bit more formal, we observe that the act of "pasting $S'$ on top of $S$" really amounts to constructing a *one-to-one correspondence* between $S$ and $S'$ (under which, for instance, $P \to P'$ and $Q \to Q'$). The important thing about this correspondence is that it preserves all geometric properties; for example, the distance from $P$ to $Q$ equals the distance from $P'$ to $Q'$, the angles $\angle PQR$ and $\angle P'Q'R'$ are equal, and so on. Two sets are congruent if and only if such a correspondence between them exists.

Our description of "congruence" is still not completely rigorous; we have not specified precisely *which* geometric properties our one-to-one correspondence must preserve. We could conceivably make a long list of such properties; but to do so would be a waste of effort, for it turns out that preservation of *distance* is all we need; once this is specified, *the preservation of all other geometric properties follows automat-*

Figure 4.10.1

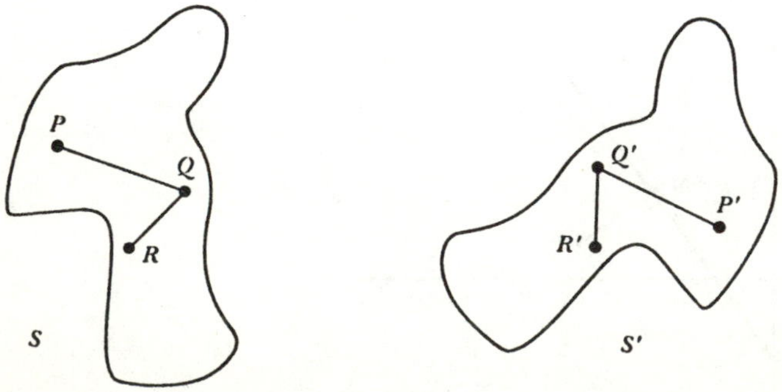

*ically*. For example, by solving equation 3.9.5 for cos(**u**, **v**) and letting **u** = **Ar** $BA$, **v** = **Ar** $BC$, we obtain

$$\cos ABC = \frac{d(A, B)^2 + d(B, C)^2 - d(A, C)^2}{2d(A, B)d(B, C)}$$

which implies that any mapping **M** which preserves distance must also preserve *the angle determined by three points*; that is, if **M** maps $A \to A'$, $B \to B'$, $C \to C'$, then

(4.10.1) $\angle ABC = \angle A'B'C'$.

Similarly, any mapping which preserves distance is affine (Theorem 4.10.6 below) and, hence, preserves collinearity, parallelism, "betweenness", and so on. In fact, it is no exaggeration to say that in real Euclidean spaces *all* geometric concepts can be described in terms of the "distance" concept. We are thus led to make the following formal definition:

4.10.2 **Definition.** If $S$ and $S'$ are subsets of the real Euclidean spaces $E$ and $E'$, respectively, a mapping $\mathbf{M} : S \to S'$ is called an *isometry*, or *Euclidean transformation*, if

(4.10.3) $d(P, Q) = d(P\mathbf{M}, Q\mathbf{M})$ (all $P$, $Q$ in $S$).

$S$ is said to be *congruent* to $S'$ if $S'$ is the image of $S$ under an isometry.

For convenience, we shall refer to an isometry of a set $S$ into *itself* as an *isometry of* $S$.

Note that Definition 4.10.2 appears to broaden our previous definition of "isometry" (Sec. 3.8) by allowing the domain of an isometry to be an arbitrary *subset* of $E$, not necessarily the whole space. However, the following theorem shows that, even under this broader definition, the assumption that the domain is an entire Euclidean space usually results in no loss of generality.

4.10.4 **Theorem.** Let $E$, $E'$ be real Euclidean spaces of finite dimension such that

$$\dim E \leqslant \dim E'$$

and let $S$ be a subset of $E$. Then any isometry **M** of $S$ into $E'$ can be extended to an isometry **M*** of $E$ into $E'$.

(We recall that **M*** "extends" **M** if the domain of **M*** includes that of **M** and if $x\mathbf{M}^* = x\mathbf{M}$ for all $x$ in the domain of **M**.) The proof of Theorem 4.10.4, which is quite long, may be found in Appendix A.

To illustrate how Theorem 4.10.4 can come in handy, we prove that *two triangular regions ABC and A'B'C' in E (each consisting of vertices, sides, and interior) are congruent if their corresponding sides are equal in length.* This statement corresponds to the "side-side-side" theorem of classical Euclidean geometry. Of course, 4.10.1 implies at once that corresponding angles are equal; however, although this fact alone would suffice for a proof according to Euclid, it is not sufficient for us since our definition of "congruent" is stronger (*how*?). We proceed instead as follows: let $S = \{A, B, C\}$ (a set of just three points) and define $\mathbf{M} : S \to E$ by

$$A\mathbf{M} = A', \qquad B\mathbf{M} = B', \qquad C\mathbf{M} = C'.$$

The assumption, "corresponding sides are equal in length", means precisely that $\mathbf{M}$ is an isometry. Hence, by 4.10.4 we can extend $\mathbf{M}$ to an isometry $\mathbf{M}^*$ of $E$. Since $\mathbf{M}^*$ preserves $b$-combinations (see 4.10.8 below), and since a triangular region consists precisely of all $b$-combinations of the vertices with nonnegative coefficients (Sec. 3.6), the result follows.∎

It should already be evident that isometries are quite important in geometry. In Chapters 6 and 9, we shall take up the problem of classifying isometries by type (for instance, every isometry of $\mathbf{R}_2$ is either a reflection, a rotation, a translation, or a reflection followed by a translation). In the present section we are more concerned with the general behavior of isometries; we shall prove that all isometries are affine, shall obtain a condition on a square matrix $A$ which is necessary and sufficient for $A$ to represent an isometry, and shall discuss the existence (or nonexistence) of isometries between certain types of sets. We begin with the following preliminary theorem.

4.10.5 **Theorem.** (a) Every isometry is one-to-one. (b) If $\mathbf{M}$ is an isometry of $S$ onto $S'$, then $\mathbf{M}^{-1} : S' \to S$ is also an isometry. (c) If $\mathbf{M} : S_1 \to S_2$ and $\mathbf{N} : S_2 \to S_3$ are isometries, then $\mathbf{MN} : S_1 \to S_3$ is an isometry. (d) Every translation of a real Euclidean space $E$ is an isometry of $E$.

*Proof.* If $P$, $Q$ are distinct points in the domain of an isometry $\mathbf{M}$, then $d(P, Q) \neq 0$; hence, $d(P\mathbf{M}, Q\mathbf{M}) \neq 0$; hence, $P\mathbf{M} \neq Q\mathbf{M}$; hence, $\mathbf{M}$ is one-to-one. It follows that if $\mathbf{M}$ is an isometry of $S$ *onto* $S'$, then the mapping $\mathbf{M}^{-1} : S' \to S$ exists, and it is easily shown that if 4.10.3 holds for $\mathbf{M}$, then it also holds for $\mathbf{M}^{-1}$. Similarly, if 4.10.3 holds for $\mathbf{M}$ and $\mathbf{N}$, then it holds for $\mathbf{MN}$ since

$$d(P(\mathbf{MN}), Q(\mathbf{MN})) = d((P\mathbf{M})\mathbf{N}, (Q\mathbf{M})\mathbf{N})$$
$$= d(P\mathbf{M}, Q\mathbf{M}) = d(P, Q).$$

Thus, parts (a), (b), and (c) of 4.10.5 are established. Part (d) is the same as Section 3.3, Exercise 3 (with $\mathbf{R}_n$ replaced by $E$; the argument is the same); do it now if you haven't done it already. ∎

*Remarks*: It follows from Theorem 4.10.5 that the isometries of $S$ onto $S$ (where $S$ is any subset of $E$) constitute a group under multiplication. It also follows that *congruence is an equivalence relation*; indeed, 4.10.5(b) and (c) show that congruence is symmetric and transitive, and reflexivity is trivial.

**4.10.6 Theorem.** If $E$ and $E'$ are real Euclidean spaces, then every isometry of $E \to E'$ can be expressed as a product **ST** where **S** is a *linear* isometry of $E \to E'$ and **T** is a translation of $E'$. In particular, every isometry of $E \to E'$ is an affine transformation.

This result was stated previously as Exercise 15 in Section 4.1. To prove it, let $\mathbf{M} : E \to E'$ be an isometry and let $\mathbf{u} = \mathbf{0M}$. By Theorem 4.10.5(d) the translation $\mathscr{T}_{-\mathbf{u}}$ is an isometry of $E'$; hence, by 4.10.5(c), the product $\mathbf{S} = \mathbf{M}\mathscr{T}_{-\mathbf{u}}$ is an isometry of $E \to E'$. Since **S** maps $\mathbf{0} \to \mathbf{0}$ (why?), Theorem 4.1.6 implies that **S** is linear; since $\mathbf{M} = \mathbf{S}\mathscr{T}_{\mathbf{u}}$ (why?), the theorem follows. ∎

**4.10.7 Corollary.** If $E$ and $E'$ have the same finite dimension, then every isometry of $E \to E'$ is one-to-one and onto.

*Proof.* "One-to-one" is already known (Theorem 4.10.5(a)). If $\mathbf{M} : E \to E'$ is an isometry, write $\mathbf{M} = \mathbf{S}\mathscr{T}_{\mathbf{u}}$ as in the proof of 4.10.6. Since **S** is one-to-one and linear, **S is** onto by Theorem 4.7.3; and $\mathscr{T}_{\mathbf{u}}$ is onto since it has an inverse (Theorem 3.3.5). Hence, the product $\mathbf{M} = \mathbf{S}\mathscr{T}_{\mathbf{u}}$ is onto. ∎

It follows from 4.10.6 and 4.10.5(a) that all of the results which we have obtained for affine transformations in Section 4.9, including the various parts of Theorem 4.9.10, are valid for isometries. In particular, every isometry of one real Euclidean space into another must

**(4.10.8)**
(a) preserve barycentric combinations;
(b) preserve *b*-dependence and *b*-independence;
(c) preserve parallelism;
(d) map *k*-flats onto *k*-flats.

It follows from Theorem 4.10.6 that in studying isometries it will often suffice to study only the *linear* isometries (also called *orthogonal transformations*). The next theorem lists several ways in which linear isometries can be characterized. (The equivalence of statements (a) and (c) was Exercise 16 in Sec. 4.1.)

**4.10.9 Theorem.** If $E$, $E'$ are real Euclidean spaces and $\mathbf{T}$ is a mapping of $E \to E'$, the following statements are equivalent:

(a) $\mathbf{T}$ is an isometry and $\mathbf{0T} = \mathbf{0}$.

(b) $\mathbf{T}$ is a linear isometry.

(c) $\mathbf{T}$ preserves inner products; that is,

**(4.10.10)** $p(\mathbf{u}, \mathbf{v}) = p(\mathbf{uT}, \mathbf{vT}) \qquad (\text{all } \mathbf{u}, \mathbf{v} \in E).$

(d) $\mathbf{T}$ is linear, and $\|\mathbf{u}\| = \|\mathbf{uT}\|$ for all $\mathbf{u} \in E$.

*Proof.* We shall show that (a) $\Rightarrow$ (b) $\Rightarrow$ (d) $\Rightarrow$ (c) $\Rightarrow$ (a). The implication (a) $\Rightarrow$ (b) is already known, by Theorem 4.1.6. If (b) holds, then

$$\|\mathbf{u}\| = d(\mathbf{u}, \mathbf{0}) = d(\mathbf{uT}, \mathbf{0T}) = d(\mathbf{uT}, \mathbf{0}) = \|\mathbf{uT}\|$$

proving (d). If (d) holds, then by Theorem 3.7.5(c) we have

$$\begin{aligned} 2p(\mathbf{u}, \mathbf{v}) &= \|\mathbf{u}\|^2 + \|\mathbf{v}\|^2 - \|\mathbf{u} - \mathbf{v}\|^2 \\ &= \|\mathbf{uT}\|^2 + \|\mathbf{vT}\|^2 - \|(\mathbf{u} - \mathbf{v})\mathbf{T}\|^2 \\ &= \|\mathbf{uT}\|^2 + \|\mathbf{vT}\|^2 - \|\mathbf{uT} - \mathbf{vT}\|^2 \quad (\text{since } \mathbf{T} \text{ is linear}) \\ &= 2p(\mathbf{uT}, \mathbf{vT}) \end{aligned}$$

proving (c). Finally, if (c) holds, then

$$0 = p(\mathbf{0}, \mathbf{0}) = p(\mathbf{0T}, \mathbf{0T})$$

so that $\mathbf{0T} = \mathbf{0}$ by 3.7.4(e); also, for any $\mathbf{u}, \mathbf{v} \in E$ we have

$$\begin{aligned} (d(\mathbf{u}, \mathbf{v}))^2 = \|\mathbf{u} - \mathbf{v}\|^2 &= \|\mathbf{u}\|^2 + \|\mathbf{v}\|^2 - 2p(\mathbf{u}, \mathbf{v}) \\ &= p(\mathbf{u}, \mathbf{u}) + p(\mathbf{v}, \mathbf{v}) - 2p(\mathbf{u}, \mathbf{v}) \end{aligned}$$

so that 4.10.10 implies $d(\mathbf{u}, \mathbf{v}) = d(\mathbf{uT}, \mathbf{vT})$ and, hence, $\mathbf{T}$ is an isometry, proving (a). ∎

*Remark*: Since $\cos(\mathbf{u}, \mathbf{v}) = p(\mathbf{u}, \mathbf{v})/\|\mathbf{u}\| \cdot \|\mathbf{v}\|$, it follows from conditions (c) and (d) of Theorem 4.10.9 that a linear isometry $\mathbf{T}$ preserves the *angle between two vectors*:

**(4.10.11)** $\angle(\mathbf{u}, \mathbf{v}) = \angle(\mathbf{uT}, \mathbf{vT}).$

As we have already seen, *all* isometries preserve the *angle determined by three points* (equation 4.10.1), but 4.10.1 does *not* imply 4.10.11; in order for an isometry to satisfy 4.10.11, it must be linear.

In the finite-dimensional case, the next theorem gives another useful characterization of linear isometries.

**4.10.12 Theorem.** Let $\mathbf{T} \in \mathrm{Lin}(E, E')$ and suppose $E$ has an (ordered) orthonormal basis $(\mathbf{u}_1, \ldots, \mathbf{u}_n)$. Then **T** is an isometry if and only if the vectors $\mathbf{u}_1\mathbf{T}, \ldots, \mathbf{u}_n\mathbf{T}$ are orthonormal.

*Proof.* If **T** is an isometry, then equation 4.10.10 holds. It follows that equation 3.8.1 for the vectors $\mathbf{u}_i$ implies the same equation for the vectors $\mathbf{u}_i\mathbf{T}$, so that the $\mathbf{u}_i\mathbf{T}$ are orthonormal. Conversely, suppose that $\mathbf{u}_1\mathbf{T}, \ldots, \mathbf{u}_n\mathbf{T}$ are orthonormal. Letting

$$\mathbf{v} = \sum_{i=1}^{n} a_i\mathbf{u}_i, \qquad \mathbf{w} = \sum_{i=1}^{n} b_i\mathbf{u}_i$$

be any vectors in $E$, we have

$$p(\mathbf{v}, \mathbf{w}) = \sum_{i=1}^{n} a_i b_i$$

by Theorem 3.8.17; and

$$\begin{aligned} p(\mathbf{vT}, \mathbf{wT}) &= p\left(\left(\sum_{i=1}^{n} a_i\mathbf{u}_i\right)\mathbf{T}, \left(\sum_{i=1}^{n} b_i\mathbf{u}_i\right)\mathbf{T}\right) \\ &= p\left(\sum_{i=1}^{n} a_i(\mathbf{u}_i\mathbf{T}), \sum_{i=1}^{n} b_i(\mathbf{u}_i\mathbf{T})\right) \\ &\qquad \text{(since } \mathbf{T} \text{ is assumed linear)} \\ &= \sum_{i=1}^{n} a_i b_i \\ &\qquad \text{(by 3.8.17, since the vectors } \mathbf{u}_i\mathbf{T} \text{ are orthonormal)} \end{aligned}$$

so that **T** satisfies 4.10.10; hence, **T** is an isometry. ∎

**The matrix of a linear isometry**

In Example 4.2.16, we showed that the matrix of a certain reflection in $\mathbf{R}_2$ (with respect to the canonical basis) was

$$\begin{bmatrix} \frac{3}{5} & \frac{4}{5} \\ \frac{4}{5} & -\frac{3}{5} \end{bmatrix}.$$

Observe that the two row vectors (3/5, 4/5) and (4/5, −3/5) are orthonormal: they each have length 1 and their dot product is zero. The same is true of the two column vectors. Similar statements hold regarding the matrix of a rotation (Example 4.2.20). The following theorem shows that this is no accident.

**4.10.13 Theorem.** Let $E$ be a finite-dimensional real Euclidean space, let $\mathbf{T} \in \text{Lin}(E, E)$, and let $A$ be the matrix of $\mathbf{T}$ with respect to some orthonormal basis of $E$. Then the following statements are equivalent:

(a) $\mathbf{T}$ is an isometry.
(b) The row vectors of $A$ are orthonormal.
(c) The column vectors of $A$ are orthonormal.
(d) $A^t = A^{-1}$.

(A matrix $A$ over $\mathbf{R}$ which satisfies these conditions is called an *orthogonal matrix.*)

*Proof.* Let $(\mathbf{u}_1, \ldots, \mathbf{u}_n)$ be the given orthonormal basis of $E$. Then

$$\mathbf{u}_i\mathbf{T} = \sum_{k=1}^{n} a_{ik}\mathbf{u}_k$$

and, hence, by 3.8.17, we have

$$\begin{aligned} p(\mathbf{u}_i\mathbf{T}, \mathbf{u}_j\mathbf{T}) &= (a_{i1}, \ldots, a_{in}) \cdot (a_{j1}, \ldots, a_{jn}) \\ &= (i\text{th row of } A) \cdot (j\text{th row of } A). \end{aligned}$$

Hence, the rows of $A$ are orthonormal if and only if the vectors $\mathbf{u}_i\mathbf{T}$ are, which by 4.10.12 occurs if and only if $\mathbf{T}$ is an isometry. Thus, statements (a) and (b) are equivalent. Moreover,

$$\begin{aligned} \text{rows of } A \text{ are orthonormal} &\Leftrightarrow (i\text{th row of } A) \cdot (j\text{th row of } A) = \delta_{ij} \\ &\Leftrightarrow (i\text{th row of } A) \cdot (j\text{th column of } A^t) = \delta_{ij} \\ &\Leftrightarrow (AA^t)_{ij} = \delta_{ij} \\ &\Leftrightarrow AA^t = I. \end{aligned}$$

Similar reasoning shows that the *columns* of $A$ are orthonormal if and only if $A^tA = I$. In view of Theorem 4.7.5, it follows that (b) $\Leftrightarrow$ (c) $\Leftrightarrow$ (d). ∎

*Remark*: By 4.10.13, a square matrix over $\mathbf{R}$ has orthonormal rows if and only if it has orthonormal columns. However, a similar statement with the word "orthonormal" replaced by "orthogonal" would be false; for example, the rows of the matrix

$$\begin{bmatrix} 2 & -1 & 3 \\ 1 & 2 & 0 \\ -6 & 3 & 5 \end{bmatrix}$$

are mutually orthogonal but the columns are not.

The preceding theorems have established some properties of isometries, but they have not tried to answer the question of precisely which subsets of a given real Euclidean space $E$ are isometric (congruent) to each other. For example, given a line $L$ in $E$, which subsets of $E$ are congruent to $L$? If your guess is "the lines in $E$", you have guessed right. Similarly, the subsets congruent to a given plane are the various planes in $E$; any two planes "look alike". More generally, the subsets congruent to a given $k$-flat are precisely the $k$-flats, as we now show.

**4.10.14 Theorem.** Let $E$ and $E'$ be real Euclidean spaces of finite dimension such that $\dim E \leqslant \dim E'$. Then

(a) If $S$ is a $k$-flat in $E$ and $S'$ is a $k$-flat in $E'$, then there exists an isometry $\mathbf{M} : E \to E'$ which maps $S$ onto $S'$. In addition, if the $k$-flats $S$ and $S'$ are subspaces, then the isometry $\mathbf{M}$ may be chosen to be linear.

(b) Conversely, if $S$ is a $k$-flat in $E$ and $\mathbf{M} : S \to E'$ is an isometry, then $S\mathbf{M}$ is a $k$-flat in $E'$.

*Proof.* We may write $S = W + \mathbf{s}$, $S' = W' + \mathbf{t}$, where $W$, $W'$ are $k$-dimensional subspaces of $E$, $E'$, respectively. By Theorem 3.8.15, we may choose orthonormal bases $(\mathbf{u}_1, ..., \mathbf{u}_k)$ of $W$, $(\mathbf{u}'_1, ..., \mathbf{u}'_k)$ of $W'$ and then extend these bases to orthonormal bases $(\mathbf{u}_1, ..., \mathbf{u}_m)$ of $E$, $(\mathbf{u}'_1, ..., \mathbf{u}'_n)$ of $E'$, where $m \leqslant n$ by hypothesis. By Theorem 4.2.2 there exists $\mathbf{T} \in \mathrm{Lin}(E, E')$ such that

$$\mathbf{u}_i\mathbf{T} = \mathbf{u}'_i \qquad (i = 1, ..., m).$$

By 4.10.12, $\mathbf{T}$ is an isometry; by 4.6.6, $\mathbf{T}$ maps $W$ onto $W'$. Hence, the mapping

$$\mathbf{M} = \mathscr{T}_{-\mathbf{s}}\mathbf{T}\mathscr{T}_{\mathbf{t}}$$

is an isometry of $E \to E'$ which maps $S$ onto $S'$. If $S$ and $S'$ are subspaces, then we may take $\mathbf{s}$ and $\mathbf{t}$ to be zero, so that $\mathbf{M} = \mathbf{T}$ is linear. This proves 4.10.14(a). To prove 4.10.14(b), extend the given isometry $\mathbf{M}$ to an isometry $\mathbf{M}^* : E \to E'$ (possible by 4.10.4) and then use 4.10.8(d) to deduce that $\mathbf{M}^*$ (and, hence, also $\mathbf{M}$) maps $S$ onto a $k$-flat. ∎

As shown earlier, the angle formed by three points is preserved under isometries. The next theorem states a similar result for half lines. (Refer back to Sec. 3.9 for the definition of "half line", etc.)

**4.10.15 Theorem.** Let $H_1$, $H_2$ be half lines in $E$ with the same initial point $A$. Let $\mathbf{M}$ be an isometry of $E \to E'$ and let $A'$, $H_1'$, $H_2'$ be the images of $A$, $H_1$, $H_2$ (respectively) under $\mathbf{M}$.

Then $H_1'$ and $H_2'$ are half lines having the initial point $A'$, and

$$\angle(H_1', H_2') = \angle(H_1, H_2).$$

*Proof.* Exercise.

**4.10.16 Theorem.** (Converse of 4.10.15) Assume that $E$, $E'$ have finite dimension with $\dim E \leqslant \dim E'$. Let $H_1$, $H_2$ be half lines in $E$ with the same initial point $A$; let $H_1'$, $H_2'$ be half lines in $E'$ with the same initial point $A'$; and assume that $\angle(H_1, H_2) = \angle(H_1', H_2')$.

Then there is an isometry of $E \to E'$ which maps $A$ onto $A'$, $H_1$ onto $H_1'$, $H_2$ onto $H_2'$.

(Interpretation of Theorem 4.10.16: *any two pairs of half lines which make the same angle are congruent.*)

*Proof.* Let $\alpha = \angle(H_1, H_2) = \angle(H_1', H_2')$. By Theorem 3.9.13(d) there are unique points $B_1$, $B_2$, $B_1'$, $B_2'$ on $H_1$, $H_2$, $H_1'$, $H_2'$, respectively, such that

(4.10.17) $d(A, B_1) = d(A, B_2) = d(A', B_1') = d(A', B_2') = 1$

(see Fig. 4.10.2). From equation 3.9.5, we then obtain

Figure 4.10.2

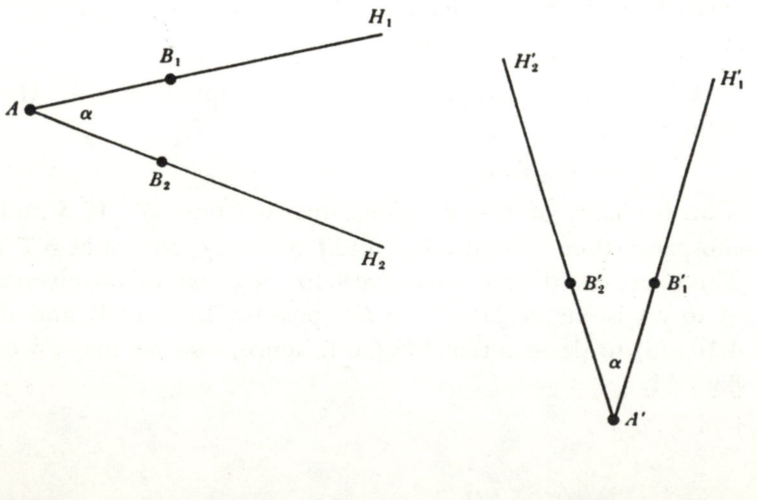

(4.10.18) $d(B_1, B_2) = (2 - 2\cos\alpha)^{1/2} = d(B'_1, B'_2)$.

By 4.10.17 and 4.10.18, the mapping $\mathbf{M}_0 : \{A, B_1, B_2\} \to \{A', B'_1, B'_2\}$ defined by

$$\mathbf{M}_0 : A \to A', B_1 \to B'_1, B_2 \to B'_2$$

is an isometry. (This is all right even if $B_1 = B_2$, since then 4.10.18 implies $B'_1 = B'_2$ and $\mathbf{M}_0$ is still well defined.) Hence, by Theorem 4.10.4, we can extend $\mathbf{M}_0$ to an isometry $\mathbf{M} : E \to E'$. Since $\mathbf{M}$ preserves barycentric combinations, $\mathbf{M}$ must map $H(A, B_1)$ onto $H(A', B'_1)$, that is, $H_1$ onto $H'_1$ (cf. Theorem 3.9.13(b)); similarly, $\mathbf{M}$ maps $H_2$ onto $H'_2$. ∎

You should compare the proof of 4.10.16 with the "side-side-side" proof given earlier in the section. *Question*: What standard theorem on "congruent triangles" follows from Theorem 4.10.16? How?

**Isometries and change of coordinates in $\mathbf{R}_n$**

In Section 3.10 we discussed orthogonal coordinate systems in $\mathbf{R}_n$. They are related to isometries via the following theorem:

**4.10.19 Theorem.** (a) If $K_1$, $K_2$ are any two orthogonal coordinate systems in $\mathbf{R}_n$, then the mapping $\mathbf{M} : \mathbf{R}_n \to \mathbf{R}_n$ defined by

(4.10.20) $\mathbf{M} : (x_1, \ldots, x_n)_{K_1} \to (x_1, \ldots, x_n)_{K_2}$ (all $x_i \in \mathbf{R}$)

is an isometry.

(b) Conversely, if $\mathbf{M} : \mathbf{R}_n \to \mathbf{R}_n$ is an isometry and $K_1$ is any orthogonal coordinate system in $\mathbf{R}_n$, then there exists an orthogonal coordinate system $K_2$ such that $\mathbf{M}$ is described by 4.10.20.

*Proof of (a).* Let $K_1 = (P; \mathbf{u}_1, \ldots, \mathbf{u}_n)$ and $K_2 = (Q; \mathbf{v}_1, \ldots, \mathbf{v}_n)$ (notations as in Sec. 3.10). By Theorem 3.10.3,

(4.10.21)
$$(x_1, \ldots, x_n)_{K_1} = P + \sum_{i=1}^{n} x_i\mathbf{u}_i$$
$$(x_1, \ldots, x_n)_{K_2} = Q + \sum_{i=1}^{n} x_i\mathbf{v}_i.$$

If $\mathbf{S} : \mathbf{R}_n \to \mathbf{R}_n$ is defined by

(4.10.22) $\mathbf{S} : \sum_{i=1}^{n} x_i\mathbf{u}_i \to \sum_{i=1}^{n} x_i\mathbf{v}_i$

then $\mathbf{S}$ is an isometry by Theorem 4.10.12, and it is easy to see from 4.10.21 that

$$\mathbf{M} = \mathcal{T}_{-P}\mathbf{S}\mathcal{T}_{Q}.$$

Hence, $\mathbf{M}$ is an isometry by Theorem 4.10.5.

*Proof of* (*b*). Given $K_1 = (P; \mathbf{u}_1, ..., \mathbf{u}_n)$ and given an isometry $\mathbf{M}$, we define

$$Q = P\mathbf{M}; \qquad \mathbf{S} = \mathcal{T}_P\mathbf{M}\mathcal{T}_{-Q}; \qquad \mathbf{v}_i = \mathbf{u}_i\mathbf{S} \qquad (1 \leq i \leq n).$$

By 4.10.5, $\mathbf{S}$ is an isometry; since $\mathbf{0S} = P\mathbf{M} - Q = \mathbf{0}$, $\mathbf{S}$ is linear. By 4.10.12, it follows that the vectors $\mathbf{v}_1, ..., \mathbf{v}_n$ are orthonormal, so that $K_2 = (Q; \mathbf{v}_1, ..., \mathbf{v}_n)$ is an orthogonal coordinate system and 4.10.21 holds. Since $\mathbf{S}$ is linear, 4.10.22 holds; hence, 4.10.21 implies that the mapping

$$\mathbf{M} = \mathcal{T}_{-P}\mathbf{S}\mathcal{T}_{Q}$$

satisfies 4.10.20. ∎

As an illustration, let $K_1 = (P; \mathbf{u}_1, \mathbf{u}_2)$ be the coordinate system in $\mathbf{R}_2$ whose axes are the lines $L_1$ and $L_2$ shown in Figure 4.10.3, and let $K_2 = (Q; \mathbf{v}_1, \mathbf{v}_2)$ be the system whose axes are $L_1'$ and $L_2'$. Since the point $A$ shown in the figure has $K_1$ coordinates $(1, 3)$ (that is, $A = (1, 3)_{K_1}$), and similarly $A' = (1, 3)_{K_2}$, the isometry $\mathbf{M}$ of Theorem 4.10.19(a) maps $A \to A'$. Similarly, $\mathbf{M}$ maps $B \to B'$, $C \to C'$. Note that $d(A, B) = d(A', B')$, and so forth, as must be the case for an isometry.

Theorem 4.10.19 has a useful corollary. Let $f$ be a function of $n$ real variables, and let $K$ be an orthogonal coordinate system in $\mathbf{R}_n$.

Figure 4.10.3

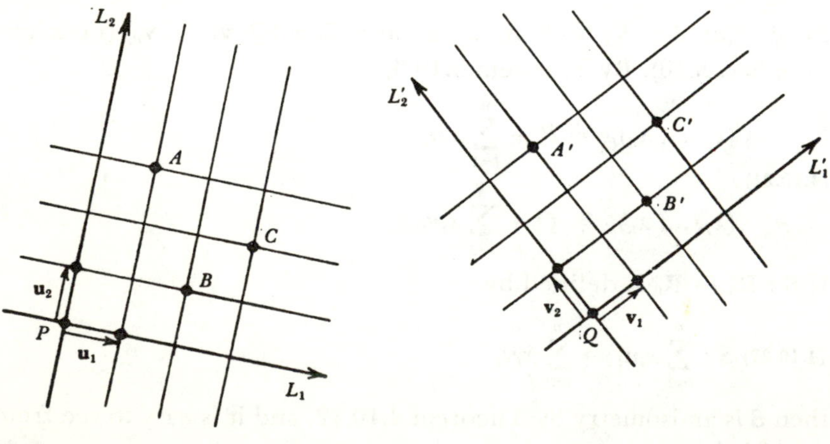

By the *graph of the equation* $f = 0$ *in the coordinate system* $K$, we mean the set of all points $(x_1, \ldots, x_n)_K$ such that $f(x_1, \ldots, x_n) = 0$. Our corollary may now be stated as follows:

**4.10.23 Theorem.** If $f$ is a function of $n$ real variables, then the graph of $f = 0$ in any given orthogonal coordinate system (in $\mathbf{R}_n$) is congruent to the graph of $f = 0$ in any other orthogonal coordinate system.

*Proof.* For $i = 1, 2$, let $\mathscr{S}_i$ be the graph of $f = 0$ in the coordinate system $K_i$, and let $\mathbf{M}$ be defined by 4.10.20. Then for any point $P = (x_1, \ldots, x_n)_{K_1}$,

$$\begin{aligned} P \in \mathscr{S}_1 &\Leftrightarrow (x_1, \ldots, x_n)_{K_1} \in \mathscr{S}_1 \Leftrightarrow f(x_1, \ldots, x_n) = 0 \\ &\Leftrightarrow (x_1, \ldots, x_n)_{K_2} \in \mathscr{S}_2 \Leftrightarrow P\mathbf{M} \in \mathscr{S}_2. \end{aligned}$$

Hence, $\mathscr{S}_1$ is mapped onto $\mathscr{S}_2$ by the isometry $\mathbf{M}$, so that $\mathscr{S}_1$ and $\mathscr{S}_2$ are congruent by definition. ▮

Much of what is done in elementary "analytic geometry" courses tacitly depends on either Theorem 4.10.19 or Theorem 4.10.23. *Example*: suppose we wish to obtain a geometric description of the curve $\mathscr{E} \subseteq \mathbf{R}_2$ whose equation (in "usual" $(x, y)$ coordinates) is

**(4.10.24)** $8x^2 + 17y^2 - 12xy - 44x - 42y + 73 = 0;$

how do we proceed? Answer: by a suitable change of coordinates (using a procedure to be developed in Sec. 7.1), we obtain new axes (say, an $x'$ axis and a $y'$ axis) such that equation 4.10.24 is equivalent to the $(x', y')$ equation

$$\frac{(x')^2}{20} + \frac{(y')^2}{5} = 1.$$

Hence, by Theorem 4.10.23, $\mathscr{E}$ is congruent to the graph of the equation

$$\frac{x^2}{20} + \frac{y^2}{5} = 1.$$

Since the latter represents an ellipse, $\mathscr{E}$ must be an ellipse as well (see Exercise 11 below).

Similarly, suppose we find that a certain locus $L'$ in $\mathbf{R}_2$ has the equation

$$y' = mx' + b$$

in $(x', y')$ coordinates. It follows from Theorem 4.10.23 that $L'$ is congruent to the locus $L$ whose equation is $y = mx + b$ in $(x, y)$ coordinates. Since $L$ is a line, and since anything congruent to a line is a line (Theorem 4.10.14(b)), it follows that $L'$ is a line.

In similar vein: given a plane $S$ in $\mathbf{R}_3$, there exists an orthogonal coordinate system $K$ such that, if $K$ coordinates are denoted $(x', y', z')$, then the equation of $S$ in the system $K$ is $z' = 0$. *Proof*: let $S_0$ be the plane $[\mathbf{e}_1, \mathbf{e}_2]$ whose equation is $z = 0$ in "usual" coordinates. By 4.10.14, there exists an isometry $\mathbf{M}$ which maps $S_0$ onto $S$; by 4.10.19(b), there exists a coordinate system $K$ such that

$$\mathbf{M} : (x, y, z) \rightarrow (x, y, z)_K \qquad \text{(all } x, y, z).$$

Hence,

$$\begin{aligned} z = 0 &\Leftrightarrow (x, y, z) \in S_0 \\ &\Leftrightarrow (x, y, z)_K \in S_0\mathbf{M} \Leftrightarrow (x, y, z)_K \in S \end{aligned}$$

or, changing notation,

$$z' = 0 \Leftrightarrow (x', y', z')_K \in S.$$

The result follows.

In a similar vein: a common type of problem in elementary analytic geometry courses is to "prove analytically" some geometric statement, for example, "the diagonals of a parallelogram bisect each other". We often start such a proof by making certain simplifying assumptions; for example, that one vertex is at the origin and that one side lies along the $x$ axis. Even though these assumptions hold only for certain special parallelograms, we lose no generality in making the assumptions; for Theorem 4.10.19 easily implies (*how*?) that *every parallelogram is congruent to one of the special parallelograms, and hence has the same geometric properties*. We could reach this same conclusion by reasoning as in our "side-side-side" proof, using 4.10.4 instead of 4.10.19. See Exercise 2 below.

**Exercises**

1. Compare Definition 4.10.2 with the standard high-school definition of "congruent triangles".
2. In the last paragraph of the section, it was asserted that every parallelogram is congruent to a parallelogram having one vertex at the origin and one side along the $x$ axis. Justify this assertion formally in two different ways: (a) using Theorem 4.10.19; (b) using Theorem 4.10.4.

3. (a) State and prove a formal theorem corresponding to the "side-angle-side" theorem of high school geometry. (SUG)
(b) State and prove a formal theorem corresponding to the "angle-side-angle" theorem of high school geometry. (SUG)

*4. Prove Theorem 4.10.15.

5. Let $P$, $P'$ be points in $\mathbf{R}_2$; let $L$, $L'$ be lines in $\mathbf{R}_2$. Show that there exists an isometry of $\mathbf{R}_2$ which maps $P \to P'$ and $L$ onto $L'$ *if and only if* the shortest distance from $P$ to $L$ equals the shortest distance from $P'$ to $L'$. (*Suggestion*: Consider the perpendicular projection of $P$ on $L$.) (SUG)

6. In each case, find the equations of an isometry $\mathbf{M} : \mathbf{R}_2 \to \mathbf{R}_2$ which satisfies the given conditions, or else show that $\mathbf{M}$ does not exist. (*Suggestion*: Use the preceding exercise and 4.9.5.)
(a) $\mathbf{M}$ maps $(2, 1) \to (3, 1)$ and maps the line $y = x + 2$ onto the line $y = 7x - 5$. (Two possible solutions.) (SOL)
(b) $\mathbf{M}$ maps $(2, 1) \to (3, 1)$ and the line $x = 4$ onto the line $x + y = 5$. (ANS)
(c) $\mathbf{M}$ maps $(2, 1) \to (3, 1)$ and the line $7y = x + 5$ onto the line $y = x - 2$. (Four possible solutions.) (SUG)
(d) $\mathbf{M}$ maps $(2, 1) \to (3, 1)$, $(9, 5) \to (4, 9)$, $(5, -1) \to (0, 3)$. (ANS)
(e) $\mathbf{M}$ maps $(2, 1) \to (3, 1)$ and the line $x = -3$ onto the line $3y = 4x + 16$. (ANS)

*7. If $m$, $n$ are integers with $m \leq n$, show that any real Euclidean space $E$ of dimension $m$ is a subspace of some real Euclidean space $E^*$ of dimension $n$. (SOL)

8. If $\mathbf{T} \in \text{Lin}(\mathbf{R}_2,\mathbf{R}_2)$ and $A$ is the matrix of $\mathbf{T}$ with respect to the canonical basis, deduce from Theorem 4.10.13 that $\mathbf{T}$ is an isometry if and only if $A$ has one of the two forms
$$\begin{bmatrix} a & b \\ -b & a \end{bmatrix}; \quad \begin{bmatrix} a & b \\ b & -a \end{bmatrix}$$
where $a$, $b$ are real numbers such that $a^2 + b^2 = 1$.

9. *Reflections* were introduced in Section 3.8. Let $W$ be a subspace of $\mathbf{R}_n$ and let $\mathcal{U}_W$ be the reflection of $\mathbf{R}_n$ through $W$. Since $\mathcal{U}_W$ is an isometry and maps $\mathbf{0} \to \mathbf{0}$, it follows that $\mathcal{U}_W$ is linear; hence, $\mathcal{U}_W$ corresponds to some matrix $A$ with respect to the canonical basis. Show that $A = A^t$. (A matrix with this property is said to be *symmetric*.) Compare this result with that of the preceding exercise.

10. If the rows of the matrix $Q$ of Theorem 4.8.9 are orthonormal, then Theorem 4.10.13 implies that $Q^t = Q^{-1} = P$. Using this fact, show that Theorem 3.10.4 may be regarded as a special case of Theorem 4.8.9.
11. An *ellipse* is defined to be the locus of points (in a plane), the sum of whose distances from two fixed points $F_1$, $F_2$ (in the plane) is a constant greater than $d(F_1, F_2)$. Prove: anything congruent to an ellipse must be an ellipse.
12. State and prove results, similar to that of the preceding exercise, involving "hyperbola" and "parabola".

# 5 Matrices and determinants

## 5.0 Introduction

Associated with every square matrix $A$ (over a field $F$) is a scalar which is called the "determinant" of $A$. Determinants crop up in a surprising number of different situations in algebra, geometry, and analysis. Simultaneous equations, maxima and minima, quadric surfaces, area and volume, orientation – these are only some of the topics to which determinants are relevant. This chapter is devoted to the definition and basic properties of determinants and a few of the applications.

There are basically two approaches to determinants. The "traditional", or "old-fashioned", approach is to define the determinant via an explicit formula and then derive from the formula various algebraic properties of determinants. The "modern" approach is to define a "determinant function" *axiomatically* by listing some of its properties (as one defines "field", for example), and *then* show that there is one and only one function having these properties. Each approach has its advantages and its adherents. The second approach can be more easily motivated (for instance, by interpreting the determinant as the signed volume of a parallelepiped of suitable dimension); on the other hand, the explicitness of the first approach enables us to avoid a complicated existence proof. It is for the latter reason that my own taste tends toward the first approach; thus, the definition here will be the old-fashioned one. Later in the chapter, however, one of the exercises asks you to show that the determinant is indeed the only function having certain specified properties; see Section 5.6, Exercise 5.

## 5.1 Permutations

The traditional approach to determinants depends on some knowledge of the properties of permutations. This section will supply what is needed in this respect, after which we can get down to the main business of the chapter.

By a *permutation* of a set $A$, we mean a one-to-one mapping of $A$ onto $A$. The result of Example 1.6.2 shows that the set of all permutations of $A$ is a group under multiplication. This group is called the *symmetric group on A* and is denoted Sym($A$), or $\mathbf{S}_A$. At present we are concerned primarily with permutations of the set

$$E_n = \{1, 2, ..., n\} = \{k \in \mathbf{Z} : 1 \leqslant k \leqslant n\}$$

where $n$ is a positive integer. The group Sym($E_n$) will be abbreviated $\mathbf{S}_n$. By Section 1.14, Exercise 6, $\mathbf{S}_n$ has exactly $n!$ elements.

Let $\pi$ be an element of $\mathbf{S}_n$. If $\pi$ maps $1 \to b_1$, $2 \to b_2$, ..., $n \to b_n$, we may write

$$\pi = \begin{pmatrix} 1 & 2 & \cdots & n \\ b_1 & b_2 & \cdots & b_n \end{pmatrix}.$$

As an illustration, the definition of multiplication of mappings in the group $\mathbf{S}_5$ gives

$$\begin{pmatrix} 1 & 2 & 3 & 4 & 5 \\ 3 & 1 & 4 & 5 & 2 \end{pmatrix}\begin{pmatrix} 1 & 2 & 3 & 4 & 5 \\ 2 & 1 & 4 & 3 & 5 \end{pmatrix} = \begin{pmatrix} 1 & 2 & 3 & 4 & 5 \\ 4 & 2 & 3 & 5 & 1 \end{pmatrix}.$$

In this notation for permutations, each element of $E_n$ is written directly above its image. However, our emphasis will be on an alternative notation which is based on the concept of *cycles*. A cycle is a permutation of a special type, which we now define.

**5.1.1 Definition.** Let $k$ be a positive integer not exceeding $n$, and let $a_1$, ..., $a_k$ be distinct elements of $E_n$. We define

**(5.1.2)** $(a_1 \quad a_2 \quad \ldots \quad a_k)$

to be the permutation $\sigma \in \mathbf{S}_n$ which maps $a_1 \to a_2$, $a_2 \to a_3$, ..., $a_{k-1} \to a_k$, $a_k \to a_1$, and $b \to b$ for all $b$ different from $a_1$, ..., $a_k$. The permutation $(a_1\ a_2 \cdots a_k)$ is called a *k-cycle*, or a *cycle of length k*.

Observe that in the notation 5.1.2 it does not matter which element $a_i$ is written first, since

$$\begin{aligned}(a_1 \quad a_2 \quad a_3 \quad \cdots \quad a_k) &= (a_2 \quad a_3 \quad \cdots \quad a_k \quad a_1)\\ &= (a_3 \quad \cdots \quad a_k \quad a_1 \quad a_2)\\ &= \text{(etc.)}.\end{aligned}$$

However, once we have decided which of the elements $a_1, \ldots, a_k$ to write first, the order of the remaining elements is determined.

**Example**
In the group $\mathbf{S}_6$,

$$(5\ 2\ 4\ 3) = (2\ 4\ 3\ 5) = \begin{pmatrix} 1 & 2 & 3 & 4 & 5 & 6 \\ 1 & 4 & 5 & 3 & 2 & 6 \end{pmatrix}$$

$$(3\ 1) = (1\ 3) = \begin{pmatrix} 1 & 2 & 3 & 4 & 5 & 6 \\ 3 & 2 & 1 & 4 & 5 & 6 \end{pmatrix}$$

$$(1\ 4\ 5) = (4\ 5\ 1) = (5\ 1\ 4) = \begin{pmatrix} 1 & 2 & 3 & 4 & 5 & 6 \\ 4 & 2 & 3 & 5 & 1 & 6 \end{pmatrix}.$$

If $k \geqslant 2$, the elements of $E_n$ which are moved by the cycle $(a_1\ a_2 \cdots a_k)$ are precisely the elements $a_1, \ldots, a_k$ themselves. (We say that $\pi$ *moves* $x$ if $x\pi \neq x$; $x$ is *fixed* by $\pi$ if $x\pi = x$.) However, if $k = 1$, Definition 5.1.1 is to be interpreted thus: $(a_1)$ is the permutation which maps $a_1 \to a_1$ and maps $b \to b$ for all $b \neq a_1$. In short, a cycle of length 1 fixes every element of $E_n$ and, hence, equals the identity mapping $\mathbf{I}$ (which of course is the identity of the group $\mathbf{S}_n$).

Two permutations $\pi$, $\sigma$ are said to be *disjoint* if no element of $E_n$ is moved by both $\pi$ and $\sigma$; that is, if the set of elements moved by $\pi$ is disjoint (in the sense of Sec. 1.1) from the set of elements moved by $\sigma$. (More generally, we say that permutations $\sigma_1, \ldots, \sigma_m$ are disjoint if every pair of them is disjoint.) According to this definition, the cycles $(a_1 \cdots a_k)$ and $(b_1 \cdots b_h)$ (where $h > 1$, $k > 1$) are disjoint if and only if no $a_i$ equals any $b_j$. A cycle of length 1, on the other hand, moves no elements and, hence, is disjoint from every permutation.

**5.1.3** **Theorem.** If the permutations $\sigma, \pi \in \mathbf{S}_n$ are disjoint, then $\pi\sigma = \sigma\pi$.

*Proof.* We must show that

$$x(\pi\sigma) = x(\sigma\pi) \tag{5.1.4}$$

for all $x \in E_n$. First, suppose $x$ is moved by $\pi$. Then $x$ and $x\pi$ are different and, hence, have different $\pi$-images; that is, $x\pi \neq (x\pi)\pi$, so that $x\pi$ is moved by $\pi$. Since $\pi$, $\sigma$ are disjoint, both $x$ and $x\pi$ are fixed by $\sigma$. Hence,

**(5.1.5)** $$x(\pi\sigma) = (x\pi)\sigma = x\pi = (x\sigma)\pi = x(\sigma\pi)$$

so that 5.1.4 holds. By similar reasoning, 5.1.4 holds if $x$ is moved by $\sigma$; while if $x$ is moved by neither $\pi$ nor $\sigma$, 5.1.4 is trivially true since both sides reduce to $x$. ▮

**5.1.6 Theorem.** Every element $\pi \in \mathbf{S}_n$ can be expressed as a product of disjoint cycles. This expression for $\pi$ is unique except for the order of the factors and the inclusion or exclusion of cycles of length 1.

Cycles of length 1 can, of course, be included or excluded at will, since they equal **I**; and the order of the factors may be changed at will by virtue of Theorem 5.1.3. As an example, we have

$$\begin{aligned}\begin{pmatrix}1&2&3&4&5&6&7&8&9\\4&2&1&3&7&8&5&9&6\end{pmatrix} &= (1\ 4\ 3)\,(5\ 7)\,(6\ 8\ 9)\\ &= (1\ 4\ 3)\,(6\ 8\ 9)\,(5\ 7)\\ &= (5\ 7)\,(6\ 8\ 9)\,(1\ 4\ 3)\\ &= (5\ 7)\,(6\ 8\ 9)\,(1\ 4\ 3)\,(2)\\ &\quad \text{(etc.)}.\end{aligned}$$

The expression for $\pi$ as a product of disjoint cycles is called the *cycle representation* of $\pi$. To compute it for a given $\pi$, begin by finding the image of any element $a_1 \in E_n$ under $\pi$; say the image is $a_2$. Then find the image of $a_2$ (say, $a_3$), then the image of $a_3$, and so on, until you reach an element $a_k$ whose image is the element $a_1$ that you started with. You now have a cycle $(a_1\ a_2\ \cdots\ a_k)$. If $k < n$ (so that there still remain $n - k$ elements of $E_n$ other than $a_1, \ldots, a_k$), start with any one of the remaining elements and repeat the process to obtain another cycle. Continue the process until all elements of $E_n$ have been taken care of.

Although performing the above process a few times should convince you of the truth of Theorem 5.1.6, it does not constitute a proof. A completely rigorous proof of Theorem 5.1.6 is fairly long; we give one in Appendix B.

A cycle $(a_1\ a_2)$ of length 2 is called a *transposition*; its effect is to interchange two elements of $E_n$ while leaving all other elements fixed. We shall need the fact that *every permutation* $\pi \in \mathbf{S}_n$ *can be expressed as a product of transpositions*. In view of Theorem 5.1.6, it will suffice to prove this when $\pi$ is a cycle.

**5.1.7 Lemma.** Every $k$-cycle can be expressed as a product of $k - 1$ transpositions.

*Proof.* The desired expression is

$$(a_1 \quad a_2 \quad \cdots \quad a_{k-1} \quad a_k) = (a_1 \quad a_k)\,(a_2 \quad a_k)\cdots(a_{k-1} \quad a_k). \tag{5.1.8}$$

To verify 5.1.8, simply show that both sides do the same thing to each $x \in E_n$. For example, take $x = a_2$. By the definition of multiplication of mappings, the effect of applying the right side of 5.1.8 to $a_2$ is to apply each of the mappings $(a_1\ a_k)$, ..., $(a_{k-1}\ a_k)$ in succession. The following diagram shows what happens:

$$a_2 \xrightarrow{(a_1\ a_k)} a_2 \xrightarrow{(a_2\ a_k)} a_k \xrightarrow{(a_3\ a_k)} a_3 \xrightarrow{(a_4\ a_k)} a_3 \ \ldots\ a_3 \xrightarrow{(a_{k-1}\ a_k)} a_3.$$

But the left side of 5.1.8 also maps $a_2 \to a_3$. We leave it to you to perform similar verifications for the other values of $x$ (see Exercise 2, this section). ∎

It should be noted that Lemma 5.1.7 holds even for $k = 1$, since by convention the empty product equals **I** (cf. Sec. 1.12).

Lemma 5.1.7 motivates the next definition (5.1.9) and implies the next theorem (5.1.10).

**5.1.9 Definition.** Let $\pi \in \mathbf{S}_n$, and let $\sigma_1, \sigma_2, \ldots, \sigma_m$ be the disjoint cycles such that $\pi = \sigma_1\sigma_2 \cdots \sigma_m$. If the length of the cycle $\sigma_i$ is $k_i$ $(i = 1, \ldots, m)$, we define the *index* of $\pi$, denoted "ind $\pi$", by

$$\text{ind } \pi = \sum_{i=1}^{m} (k_i - 1).$$

Note that although cycles of length 1 can be included or not in the cycle representation of $\pi$, their inclusion does not affect the index of $\pi$ (*why not*?); thus, the index depends only on $\pi$. In view of Lemma 5.1.7, the following theorem is now immediate.

**5.1.10 Theorem.** Every permutation $\pi \in \mathbf{S}_n$ is expressible as a product of exactly ind $\pi$ transpositions.

In general, ind $\pi$ is not the *only* number of transpositions whose product is $\pi$. For example,

$$\operatorname{ind}(1\ 4\ 2)(3\ 5) = (3 - 1) + (2 - 1) = 3$$
$$(1\ 4\ 2)(3\ 5) = (1\ 2)(4\ 2)(3\ 5)$$
$$= \text{product of 3 transpositions}$$

which agrees with Theorem 5.1.10; but we also have

$$(1\ 4\ 2)(3\ 5) = (1\ 3)(2\ 5)(3\ 4)(2\ 3)(1\ 5)$$
$$= \text{product of 5 transpositions.}$$

However, the number of transpositions is not totally arbitrary, as the next theorem shows.

**5.1.11 Theorem.** If the permutation $\pi \in \mathbf{S}_n$ can be expressed as a product of $t$ transpositions, then $t \equiv \operatorname{ind} \pi \pmod 2$; that is, $t$ is even if ind $\pi$ is even, odd if ind $\pi$ is odd.

The proof of Theorem 5.1.11 depends on the following lemma.

**5.1.12 Lemma.** Let $\pi \in \mathbf{S}_n$ and let $\tau = (c_1\ c_2)$ be a transposition in $\mathbf{S}_n$. Let $\pi = \sigma_1\sigma_2 \cdots \sigma_m$ be the representation of $\pi$ as a product of disjoint cycles. Then

(a) If $c_1$ and $c_2$ are moved by the same cycle $\sigma_i$, then $\operatorname{ind}(\pi\tau) = \operatorname{ind} \pi - 1$.

(b) If $c_1$ and $c_2$ are not moved by the same cycle $\sigma_i$, then $\operatorname{ind}(\pi\tau) = \operatorname{ind} \pi + 1$.

The proof of the lemma is left as an exercise. Try doing it by brute force; that is, write down the actual expression for each cycle $\sigma_i$, perform the multiplications, and so on.

Once Lemma 5.1.12 is established, Theorem 5.1.11 can be proved by induction on $t$. This, too, we leave as an exercise.

We define an *even* permutation to be one whose index is even; similarly, an *odd* permutation is one whose index is odd. Theorem 5.1.11, thus, says that any product of an even number of transpositions is always an even permutation, and similarly for "odd".

**5.1.13 Theorem.** (a) For all $\pi$, $\sigma$ in $\mathbf{S}_n$, $\operatorname{ind} \pi + \operatorname{ind} \sigma \equiv \operatorname{ind}(\pi\sigma) \pmod 2$. (b) For all $\pi \in \mathbf{S}_n$, $\operatorname{ind} \pi^{-1} = \operatorname{ind} \pi$.

*Proof of (a).* Let $p = \operatorname{ind} \pi$, $s = \operatorname{ind} \sigma$. By Theorem 5.1.10, $\pi$ equals a product of $p$ transpositions, say, $\pi = \tau_1\tau_2 \cdots \tau_p$. Similarly, $\sigma$ equals a product of $s$ transpositions, say, $\sigma = \xi_1\xi_2 \cdots \xi_s$. Then

$$\pi\sigma = \tau_1 \cdots \tau_p\xi_1 \cdots \xi_s$$

so that $\pi\sigma$ is a product of $p + s$ transpositions. Hence, by Theorem 5.1.11, $p + s \equiv \text{ind}\,(\pi\sigma) \pmod 2$ as asserted.

We leave the proof of part (b) as an exercise. (*Suggestion*: Compare the cycle representation of $\pi$ with that of $\pi^{-1}$.) ▮

We conclude this section with a result which has nothing to do with cycles or index but will be needed for definitions and proofs later on.

**5.1.14 Theorem.** (Generalized Commutative Law) Assume that "$*$" is a commutative and associative binary operation on a set $A$, $n$ is a positive integer, $a_1, \ldots, a_n$ are elements of $A$, and $\pi \in \mathbf{S}_n$. Then

$$\prod_{i=1}^{n} a_i = \prod_{i=1}^{n} a_{i\pi}.$$

In plain English, Theorem 5.1.14 asserts that if $*$ is associative and any two elements of $A$ commute, then any $n$ elements of $A$ commute; that is, their product in any order is the same as their product in any other order. In view of Theorem 5.1.10, it suffices to prove 5.1.14 in the case where $\pi$ is a transposition. In fact, since any transposition is the product of transpositions of the form $(k \quad k + 1)$ (see Exercise 3 below), we may further assume that $\pi = (k \quad k + 1)$ for some $k$. In this case,

$$\prod_{i=1}^{n} a_i = \left(\prod_{i=1}^{k-1} a_i\right) * a_k * a_{k+1} * \left(\prod_{i=k+2}^{n} a_i\right)$$

(cf. Sec. 1.12, Exercise 7)

$$= \left(\prod_{i=1}^{k-1} a_i\right) * a_{k+1} * a_k * \left(\prod_{i=k+2}^{n} a_i\right)$$

(since $*$ is commutative)

$$= \left(\prod_{i=1}^{k-1} a_{i\pi}\right) * a_{k\pi} * a_{(k+1)\pi} * \left(\prod_{i=k+2}^{n} a_{i\pi}\right)$$

$$= \prod_{i=1}^{n} a_{i\pi},$$

as desired. *Question*: Where was the associativity of $*$ used? ▮

A consequence of 5.1.14 is that if $*$ is an associative and commutative binary operation on $A$, and if $a_i$ ($i \in I$) are elements of $A$ where $I$ is any *finite* set of indices (not necessarily the set $\{1, 2, ..., n\}$), then the product

$$\prod_{i \in I} a_i$$

can be unambiguously defined. Indeed, since $I$ is finite the elements of $I$ can be denoted $\alpha_j$ ($1 \leqslant j \leqslant n$; see Sec. 1.14), and we then define

$$\prod_{i \in I} a_i = \prod_{j=1}^{n} a_{\alpha_j}.$$

It follows easily from 5.1.14 that the right side of the above equation is unaffected by the order in which we number the elements of $I$.

Of course, if the operation on $A$ is "addition", the preceding discussion applies to sums rather than products.

**Exercises**

1. Express each of the following permutations as a product of disjoint cycles, and find the index of each permutation. (ANS)

   (a) $\begin{pmatrix} 1 & 2 & 3 & 4 & 5 & 6 & 7 & 8 & 9 \\ 1 & 8 & 3 & 9 & 2 & 7 & 6 & 5 & 4 \end{pmatrix}$.

   (b) $(1\ \ 5\ \ 3\ \ 4)\,(2\ \ 4\ \ 3)\,(1\ \ 6\ \ 5)$.
   (This is already a product of cycles, but not of *disjoint* cycles.)

   (c) $\begin{pmatrix} 1 & 2 & 3 & 4 & 5 & 6 & 7 & 8 & 9 & 10 \\ 3 & 9 & 5 & 7 & 10 & 2 & 6 & 8 & 4 & 1 \end{pmatrix}$.

2. Verify that the left and right sides of 5.1.8 do the same thing to every $x \in E_n$.

*3. If $1 \leqslant a < b \leqslant n$, verify the following equation in $\mathbf{S}_n$: $(a\ \ b) = (a\ \ a+1)\,(a+1\ \ a+2) \cdots (b-2\ \ b-1)\,(b-1\ \ b)\,(b-2\ \ b-1) \cdots (a+1\ \ a+2)\,(a\ \ a+1)$. (For example, $(3\ \ 7) = (3\ \ 4)\,(4\ \ 5)\,(5\ \ 6)\,(6\ \ 7)\,(5\ \ 6)\,(4\ \ 5)\,(3\ \ 4)$.)

*4. Prove Lemma 5.1.12.

*5. Using Lemma 5.1.12, prove Theorem 5.1.11 by induction on $t$ (starting with the case $t = 0$). (SUG)

*6. Prove Theorem 5.1.13(b).

7. For any permutation

$$\pi = \begin{pmatrix} 1 & 2 & \cdots & n \\ b_1 & b_2 & \cdots & b_n \end{pmatrix},$$

let $N(\pi)$ be the number of ordered pairs $(i, j)$ such that

(5.1.15) $i < j, \qquad b_i > b_j.$

(For example, if

$$\pi = (5 \quad 2 \quad 4 \quad 3) = \begin{pmatrix} 1 & 2 & 3 & 4 & 5 & 6 \\ 1 & 4 & 5 & 3 & 2 & 6 \end{pmatrix}$$

then the ordered pairs $(i, j)$ satisfying 5.1.15 are (2, 4), (2, 5), (3, 4), (3, 5), (4, 5); hence, $N(\pi) = 5$.) Prove that $N(\pi) \equiv \text{ind } \pi \pmod 2$, and, hence, that $\pi$ is even (odd) if and only if the number $N(\pi)$ is even (odd). (SUG)

8. Prove: if $n \geqslant 2$, then exactly half of the elements of $\mathbf{S}_n$ are even and half are odd; thus, there are $(n!)/2$ permutations of each type. (*Suggestion*: Find a one-to-one correspondence between the set of all odd permutations and the set of all even permutations.)
9. Theorem 5.1.13(a) may be regarded as an assertion that a certain mapping is a homomorphism. What mapping, between what sets?

## 5.2 Definition of the determinant

Let $A = (a_{ij})$ be a square $(n \times n)$ matrix over a field $F$. We define the *determinant* of $A$ (denoted det $A$) to be the scalar

(5.2.1)
$$\det A = \sum_{\pi \in \mathbf{S}_n} (-1)^{\text{ind } \pi} a_{1,1\pi} a_{2,2\pi} \cdots a_{n,n\pi} = \sum_{\pi \in \mathbf{S}_n} (-1)^{\text{ind } \pi} \left( \prod_{i=1}^{n} a_{i,i\pi} \right).$$

The determinant of a nonsquare matrix is undefined. Many authors use the notation $|A|$ instead of det $A$, but this can sometimes lead to confusion (for example, when discussing the absolute value of a determinant); hence, we shall stick to "det $A$" in this book.

The sum in 5.2.1 is a sum of $n!$ terms, one for each permutation $\pi$ in $\mathbf{S}_n$. The order in which the terms are added does not matter (cf. the discussion at the end of Sec. 5.1). Each of the $n!$ summands is itself a product of $n$ entries from $A$, with a factor of $\pm 1$ attached. In any one of these products, the row subscripts of the $n$ factors are 1, 2, ..., $n$, so that there is one factor from each row of $A$. The column subscripts of the $n$ factors are $1\pi$, $2\pi$, ..., $n\pi$; since these indices are simply 1, 2, ..., $n$ rearranged (*why*?), there is one factor from each column. For example, in the case $n = 3$, the term in 5.2.1 which corresponds to the permutation

$$\pi = \begin{pmatrix} 1 & 2 & 3 \\ 3 & 2 & 1 \end{pmatrix} \tag{5.2.2}$$

is

$$-a_{13}a_{22}a_{31}$$

which is minus the product of the entries indicated below in boldface type:

$$A = \begin{bmatrix} a_{11} & a_{12} & \mathbf{a}_{13} \\ a_{21} & \mathbf{a}_{22} & a_{23} \\ \mathbf{a}_{31} & a_{32} & a_{33} \end{bmatrix}.$$

As you can see, one entry in each row is in boldface and one entry in each column is in boldface. The minus sign occurs because this particular permutation $\pi$ is odd; in fact, in cycle notation, equation 5.2.2 becomes

$$\pi = (1 \quad 3)$$

which has odd index $2 - 1 = 1$.

The term corresponding to each permutation can be computed in similar fashion. In the case $n = 3$, the complete list of permutations (and corresponding summands in 5.2.1) is given in Table 5.2.0.

**5.2.0. Table**

| | cycle representation of $\pi$ | ind $\pi$ | $(-1)^{\text{ind } \pi}$ | $a_{1,1\pi}a_{2,2\pi}a_{3,3\pi}$ |
|---|---|---|---|---|
| $\mathbf{I} = \begin{pmatrix} 1 & 2 & 3 \\ 1 & 2 & 3 \end{pmatrix}$ | (cycles of length 1) | 0 | 1 | $a_{11}a_{22}a_{33}$ |
| $\begin{pmatrix} 1 & 2 & 3 \\ 1 & 3 & 2 \end{pmatrix}$ | (2 3) | 1 | $-1$ | $a_{11}a_{23}a_{32}$ |
| $\begin{pmatrix} 1 & 2 & 3 \\ 2 & 1 & 3 \end{pmatrix}$ | (1 2) | 1 | $-1$ | $a_{12}a_{21}a_{33}$ |
| $\begin{pmatrix} 1 & 2 & 3 \\ 3 & 2 & 1 \end{pmatrix}$ | (1 3) | 1 | $-1$ | $a_{13}a_{22}a_{31}$ |
| $\begin{pmatrix} 1 & 2 & 3 \\ 2 & 3 & 1 \end{pmatrix}$ | (1 2 3) | 2 | 1 | $a_{12}a_{23}a_{31}$ |
| $\begin{pmatrix} 1 & 2 & 3 \\ 3 & 1 & 2 \end{pmatrix}$ | (1 3 2) | 2 | 1 | $a_{13}a_{21}a_{32}$ |

Hence, the definition of a $3 \times 3$ determinant is

$$\det\begin{bmatrix} a_{11} & a_{12} & a_{13} \\ a_{21} & a_{22} & a_{23} \\ a_{31} & a_{32} & a_{33} \end{bmatrix} = a_{11}a_{22}a_{33} - a_{11}a_{23}a_{32} - a_{12}a_{21}a_{33}$$
$$- a_{13}a_{22}a_{31} + a_{12}a_{23}a_{31} + a_{13}a_{21}a_{32}.$$

Observe that three of the six terms have minus signs attached; these are the terms corresponding to odd permutations. (If $n \geqslant 2$, exactly half of the permutations in $\mathbf{S}_n$ are odd; cf. Sec. 5.1, Exercise 8.)

The cases $n = 1$ and $n = 2$ can be treated similarly but more easily. We leave it to you to verify that

**(5.2.3)** $$\det\begin{bmatrix} a_{11} & a_{12} \\ a_{21} & a_{22} \end{bmatrix} = a_{11}a_{22} - a_{12}a_{21}$$

and that

**(5.2.4)** $$\det\,[a_{11}] = a_{11}.$$

In Section 4.5, we defined *triangular* matrices. The determinant of a triangular matrix is especially easy to compute, as the following theorem shows.

**5.2.5** **Theorem.** If the $n \times n$ matrix $A = (a_{ij})$ (over $F$) is triangular, then its determinant is the product of the entries on the main diagonal; that is,

$$\det A = a_{11}a_{22} \cdots a_{nn} = \prod_{i=1}^{n} a_{ii}.$$

For example, the lower triangular matrix

$$\begin{bmatrix} 1 & 0 & 0 & 0 \\ 2 & 5 & 0 & 0 \\ 3 & 6 & 8 & 0 \\ 4 & 7 & 9 & 10 \end{bmatrix}$$

has determinant 400 ($= 1 \cdot 5 \cdot 8 \cdot 10$). To prove Theorem 5.2.5, suppose first that $A$ is lower triangular; that is, $a_{ij} = 0$, whenever $i < j$. Equivalently, $a_{ij} \neq 0$ only if $j \leqslant i$. Hence, if any summand

**(5.2.6)** $$(-1)^{\text{ind}\,\pi}\, a_{1,1\pi}a_{2,2\pi} \cdots a_{n,n\pi}$$

in 5.2.1 is nonzero, we must have

$$1\pi \leqslant 1;\ 2\pi \leqslant 2;\ \ldots;\ n\pi \leqslant n.$$

But $1\pi \leqslant 1$ implies $1\pi = 1$. Hence, $2\pi \neq 1$, since $\pi$ is one-to-one.

Hence, $2\pi \leqslant 2$ implies $2\pi = 2$. Continuing this argument, we obtain $3\pi = 3, \ldots, n\pi = n$, so that $\pi = \mathbf{I}$ and 5.2.6 becomes

$$a_{11}a_{22} \cdots a_{nn}.$$

Since all other summands in 5.2.1 are zero, the theorem follows. The proof in the upper triangular case is similar.

**5.2.7 Corollary.** Det $I = 1$.

For a matrix which is not triangular, the formula 5.2.1 is normally too cumbersome for actual computation unless $n \leqslant 3$. (Even for $n = 4$ there are already $4! = 24$ summands.) Other methods for computing determinants must therefore be developed. We shall do so in the next two sections.

**Exercises**

*1. Derive 5.2.3 and 5.2.4 from Definition 5.2.1.

2. Show that

$$\det \begin{bmatrix} 1 & 2 & 3 \\ 4 & 5 & 6 \\ 7 & 8 & 9 \end{bmatrix} = 0.$$

3. Compute

$$\det \begin{bmatrix} 2 & 5 & -1 \\ 3 & 1 & 4 \\ 0 & 6 & 1 \end{bmatrix}.$$

(ANS)

4. In elementary courses, students are sometimes taught to compute $3 \times 3$ determinants as follows: adjoin to the given matrix a fourth and a fifth column equal to the first and the second column, respectively, thus:

$$\begin{bmatrix} a_{11} & a_{12} & a_{13} & a_{11} & a_{12} \\ a_{21} & a_{22} & a_{23} & a_{21} & a_{22} \\ a_{31} & a_{32} & a_{33} & a_{31} & a_{32} \end{bmatrix};$$

then take the sum of the three diagonal products indicated by solid arrows, minus the sum of the three diagonal products indicated by dotted arrows. Show that this method is valid for $3 \times 3$ determinants, *but that a similar method for $n \times n$ determinants is not valid*. Draw a moral from this.

5. What value should be assigned to the determinant of a $0 \times 0$ matrix (the empty matrix) so as to be consistent with 5.2.1 and with previous conventions? (This question will be an-

swered in Sec. 5.3; try to answer it yourself without looking ahead.)

6. If $A$ is a square matrix over $F$ and $A^*$ is defined as in Sec. 4.4, Exercise 16, show that $\det(A^*) = (\det A)^*$.

## 5.3 The expansion by minors

By a matrix of *degree n* we shall mean an $n \times n$ matrix; the determinant of an $n \times n$ matrix will similarly be called a determinant of degree $n$. In this section we shall derive an expression for the general determinant of degree $n$ in terms of determinants of degree $n - 1$.

First, a definition. Let $A = (a_{ij})$ be an $n \times n$ matrix. For any *fixed* indices $i$ and $j$, we can obtain a new matrix from $A$ by deleting the row and column which contain the entry $a_{ij}$; that is, by deleting the $i$th row and $j$th column of $A$. The new matrix, thus obtained, is a submatrix of $A$ of degree $n - 1$ which we call the *(i, j) minor of A*. For example, if

**(5.3.1)** $$A = \begin{bmatrix} 1 & 2 & 3 \\ 4 & 5 & 6 \\ 7 & 8 & 9 \end{bmatrix}$$

then the (2, 3) minor of $A$ is the submatrix $\begin{bmatrix} 1 & 2 \\ 7 & 8 \end{bmatrix}$ (see Fig. 5.3.1).

**5.3.2 Theorem.** Let $A = (a_{ij})$ be an $n \times n$ matrix over the field $F$, and let $M(i, j)$ denote the $(i, j)$ minor of $A$. Then for each $i$,

**(5.3.3)** $$\det A = \sum_{j=1}^{n} (-1)^{i+j} a_{ij} \det M(i, j);$$

and for each $j$,

Figure 5.3.1

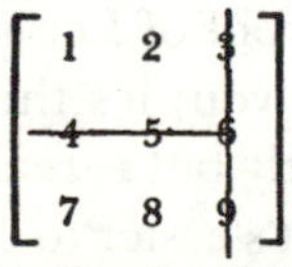


Finding the (2, 3) minor of a given matrix

**(5.3.4)** $$\det A = \sum_{i=1}^{n} (-1)^{i+j} a_{ij} \det M(i, j).$$

Formula 5.3.3 is called the *expansion of det A by minors of the ith row* (remember that $i$ is fixed in 5.3.3); similarly, 5.3.4 is the expansion of det $A$ by minors of the $j$th column. As an illustration, let us compute the determinant of the matrix 5.3.1 by using minors of the second column ($j = 2$ in equation 5.3.4):

$$\begin{aligned}\det\begin{bmatrix}1&2&3\\4&5&6\\7&8&9\end{bmatrix} &= -a_{12}\det M(1, 2) + a_{22}\det M(2, 2)\\ &\quad - a_{32}\det M(3, 2)\\ &= -2\det\begin{bmatrix}4&6\\7&9\end{bmatrix} + 5\det\begin{bmatrix}1&3\\7&9\end{bmatrix}\\ &\quad - 8\det\begin{bmatrix}1&3\\4&6\end{bmatrix}\\ &= -2(-6) + 5(-12) -8(-6)\\ &= 0.\end{aligned}$$

Let us now prove Theorem 5.3.2. Throughout the proof (until almost the end) we shall regard both $i$ and $j$ as fixed. The matrix $M(i, j)$ has degree $n - 1$ and we will denote its entries by $c_{rs}$ ($1 \leq r \leq n - 1$, $1 \leq s \leq n - 1$). Also, for each integer $k$ ($1 \leq k \leq n$), we define a permutation $\epsilon_k \in \mathbf{S}_n$ by

$$\epsilon_k = (k \quad k + 1 \quad k + 2 \quad \cdots \quad n).$$

**5.3.5** **Lemma.** For all $r$ and $s$ ($1 \leq r \leq n - 1$, $1 \leq s \leq n - 1$), we have

$$[M(i, j)]_{rs} = c_{rs} = a_{r\epsilon_i, s\epsilon_j}.$$

The lemma asserts that the entries in the $r$th row of $M(i, j)$ appear in the $(r\epsilon_i)$th row of $A$, and that the entries in the $s$th column of $M(i, j)$ appear in the $(s\epsilon_j)$th column of $A$. We leave the proof of Lemma 5.3.5 as an exercise. (Don't let this exercise intimidate you; it's the sort of thing that looks formidable when written abstractly but is really quite simple if one thinks about what it means. In fact, it's easier to convince oneself that Lemma 5.3.5 is true than to write down a formal proof of it. If you suspect that that's why the proof is omitted, you're right.)

We now consider the quantity

(5.3.6) $d_{ij} = (-1)^{i+j} a_{ij} \det M(i, j).$

Evaluating det $M(i, j)$ by means of Definition 5.2.1 and substituting the result into 5.3.6, we have

$$\text{(5.3.7)}\quad \begin{aligned} d_{ij} &= (-1)^{i+j} a_{ij} \sum_{\tau \in \mathbf{S}_{n-1}} (-1)^{\operatorname{ind}\tau} \left( \prod_{r=1}^{n-1} c_{r,r\tau} \right) \\ &= (-1)^{i+j} a_{ij} \sum_{\tau \in \mathbf{S}_{n-1}} (-1)^{\operatorname{ind}\tau} \left( \prod_{r=1}^{n-1} a_{r\epsilon_i, r\tau\epsilon_j} \right) \end{aligned}$$

by Lemma 5.3.5. Now, the elements $\tau \in \mathbf{S}_{n-1}$ may be regarded as elements of $\mathbf{S}_n$ which map $n \to n$; for all such $\tau$, we then have

$$n\epsilon_i = i; \qquad (n\tau)\epsilon_j = n\epsilon_j = j$$

so that $a_{ij} = a_{n\epsilon_i, n\tau\epsilon_j}$. Substituting this for $a_{ij}$ in 5.3.7, we obtain

$$\text{(5.3.8)}\quad d_{ij} = (-1)^{i+j} \sum_{\substack{\tau \in \mathbf{S}_n \\ \tau : n \to n}} (-1)^{\operatorname{ind}\tau} \left( \prod_{r=1}^{n} a_{r\epsilon_i, r\tau\epsilon_j} \right)$$

with the product inside parentheses now consisting of $n$ factors instead of $n - 1$, the factor for $r = n$ being $a_{ij}$. We now let

$$k = r\epsilon_i; \qquad \pi = \epsilon_i^{-1}\tau\epsilon_j.$$

As $r$ takes on the values $1, \ldots, n$ once each, so does $k = r\epsilon_i$ since $\epsilon_i$ is a permutation of $1, \ldots, n$; also, $r = k\epsilon_i^{-1}$ so that

$$r\tau\epsilon_j = k(\epsilon_i^{-1}\tau\epsilon_j) = k\pi.$$

It follows that

$$\text{(5.3.9)}\quad \prod_{r=1}^{n} a_{r\epsilon_i, r\tau\epsilon_j} = \prod_{k=1}^{n} a_{k,k\pi}$$

(the order of the factors in the product being immaterial since multiplication of scalars is commutative; cf. Theorem 5.1.14). Also, by Theorem 5.1.13, we have

$$\operatorname{ind}\pi = \operatorname{ind}(\epsilon_i^{-1}\tau\epsilon_j) \equiv \operatorname{ind}\epsilon_i + \operatorname{ind}\tau + \operatorname{ind}\epsilon_j \pmod 2$$

and since ind $\epsilon_k = n - k$ for all $k$, this gives

$$\begin{aligned} \operatorname{ind}\pi &\equiv \operatorname{ind}\tau + (n - i) + (n - j) \pmod 2 \\ &\equiv \operatorname{ind}\tau + i + j \pmod 2 \end{aligned}$$

so that

$$(5.3.10) \quad (-1)^{i+j}(-1)^{\text{ind}\,\tau} = (-1)^{\text{ind}\,\pi}.$$

Now distinct permutations $\tau$ correspond to distinct permutations $\pi$ (*why*?), and the condition that $\tau$ maps $n \to n$ is equivalent to the condition that $\pi : i \to j$ (proof: exercise). Hence, by substituting 5.3.9 and 5.3.10 into 5.3.8, we obtain

$$(5.3.11) \quad d_{ij} = \sum_{\substack{\pi \in \mathbf{S}_n \\ \pi : i \to j}} (-1)^{\text{ind}\,\pi} \left( \prod_{k=1}^{n} a_{k,k\pi} \right)$$

which is precisely the right side of 5.2.1 except for the restriction that $\pi : i \to j$. If we now keep $i$ fixed but allow $j$ to vary, every $\pi$ in $\mathbf{S}_n$ maps $i$ into *some* integer $j$ from 1 to $n$; hence, by summing 5.3.11 over all $j$ ($j = 1, \ldots, n$), we obtain the entire right side of 5.2.1. In short,

$$\det A = \sum_{j=1}^{n} d_{ij}$$

for each fixed $i$. Similarly, if we fix $j$ but allow $i$ to vary, each $\pi \in \mathbf{S}_n$ maps some integer $i$ into $j$, so that by summing 5.3.11 over all $i$, we get

$$\det A = \sum_{i=1}^{n} d_{ij}$$

for each fixed $j$. Substituting 5.3.6 for $d_{ij}$, we obtain Theorem 5.3.2.

*Remarks*: 1. In the case $n = 1$, equation 5.3.3 (or 5.3.4) reduces to

$$\det A = a_{11} \det M(1, 1)$$

and $M(1, 1)$ is a matrix of degree 0, the *empty matrix*. By 5.2.4, on the other hand, $\det A = a_{11}$. This suggests that in order to make our results consistent we should adopt the convention that *the determinant of the empty matrix is 1*. Fortunately, this convention agrees with 5.2.1. We see this as follows: when $n = 0$, $\mathbf{S}_n$ is the set of all one-to-one mappings of $\varnothing$ onto $\varnothing$. There is exactly one such mapping, namely, the *empty mapping* (think about that!). Since the empty mapping has no cycles, its index (Definition 5.1.9) is the empty sum, which by prior convention is zero. Hence, by 5.2.1, the empty determinant equals $(-1)^0$ times the empty product. But the empty product equals 1 by prior convention, and the desired result follows!

2. In numerical computations using the expansion by minors, we

usually look for a row or column having as many zeros as possible. For example, if we expand the determinant of the matrix

$$A = \begin{bmatrix} 4 & -3 & 1 & 2 \\ 3 & 1 & 0 & 5 \\ -1 & -1 & 2 & 1 \\ 6 & -2 & 0 & 1 \end{bmatrix}$$

by minors of the first row, we have four $3 \times 3$ determinants to compute; but if we expand by minors of the third column, we have only two:

$$\det A = 1 \det \begin{bmatrix} 3 & 1 & 5 \\ -1 & -1 & 1 \\ 6 & -2 & 1 \end{bmatrix} - 0 + 2 \det \begin{bmatrix} 4 & -3 & 2 \\ 3 & 1 & 5 \\ 6 & -2 & 1 \end{bmatrix} - 0.$$

Ideally, we hope to find a row or a column in which all but one entry is zero. If no such row or column is present, the method of minors may be cumbersome unless combined with the results to be obtained in the next section. So read on . . .

**Exercises**

*1. Prove Lemma 5.3.5.

*2. If $\pi = \epsilon_i^{-1}\tau\epsilon_j$, where $\tau \in \mathbf{S}_n$ and $\epsilon_i$, $\epsilon_j$ are as defined in Lemma 5.3.5, prove that $\tau : n \to n$ if and only if $\pi : i \to j$. (This fact was used in the proof of Theorem 5.3.2.)

3. Compute

$$\det \begin{bmatrix} 0 & 1 & 0 & 3 \\ 0 & 3 & 0 & 5 \\ 2 & 0 & -2 & 4 \\ 2 & 1 & 1 & 1 \end{bmatrix}$$

by minors of the second row; also by minors of the third column. (Your answers should agree!) (ANS)

## 5.4 Properties of determinants

In this section we shall prove a number of general theorems about determinants, and shall indicate how some of these theorems can be used to make it easier to compute determinants. *Throughout this section, A will denote an $n \times n$ matrix over the field F.* Also, for convenience in notation, the $i$th row of $A$ will be denoted $A_{(i)}$. When we speak of "adding" one row to another, or of multiplying a row by a scalar, it will be understood that the rows are being regarded as *row vectors* (elements of $F_n$). Similarly, by the "zero row" we mean the row ($n$-tuple) consisting of all zeros, since (0, ..., 0) is the zero element of $F_n$.

**5.4.1 Theorem.** If $A'$ is the matrix obtained from $A$ by multiplying one row $A_{(i)}$ by the scalar $c$ (that is, by replacing $A_{(i)}$ by $cA_{(i)}$), then $\det A' = c \det A$.

**5.4.2 Theorem.** If any one row of $A$ is the zero row, then $\det A = 0$.

Both 5.4.1 and 5.4.2 are proved by expanding the determinants by minors of the row in question. For Theorem 5.4.1, expand by minors of the $i$th row (equation 5.3.3); the effect of replacing $A_{(i)}$ by $cA_{(i)}$ is to replace each term $a_{ij}$ in 5.3.3 by $ca_{ij}$. (The terms $M(i, j)$ are unaffected since the row $A_{(i)}$ is excluded from the matrix $M(i, j)$.) The constant $c$ can then be factored out, giving

$$\begin{aligned}\det A' &= \sum_{j=1}^{n} (-1)^{i+j} c a_{ij} \det M(i, j) \\ &= c\sum_{j=1}^{n} (-1)^{i+j} a_{ij} \det M(i, j) \\ &= c \det A.\end{aligned}$$

For Theorem 5.4.2, choose $i$ such that the $i$th row of $A$ is the zero row; then all terms $a_{ij}$ in 5.3.3 are zero and, hence, $\det A = 0$. ▮

**5.4.3 Theorem.** (a) If $A'$ is the matrix obtained from $A$ by interchanging two rows of $A$ (that is, by replacing $A_{(r)}$ by $A_{(s)}$ and $A_{(s)}$ by $A_{(r)}$, where $r \neq s$), then $\det A' = -\det A$.

(b) If two rows of $A$ are equal, then $\det A = 0$.

*Proof of* (*a*). In the case $n = 2$ (the smallest value of $n$ under consideration), we can write

$$A = \begin{bmatrix} a & b \\ c & d \end{bmatrix}; \qquad \det A = ad - bc$$

by 5.2.3. We then have

$$\begin{aligned} A' = \begin{bmatrix} c & d \\ a & b \end{bmatrix}; \qquad \det A' &= cb - da \\ &= -(ad - bc) = -\det A \end{aligned}$$

as desired. Completing the proof by induction, we may assume that Theorem 5.4.3(a) holds for matrices of degree $k$, and that $A$ is a matrix of degree $n = k + 1$. Since $k \geq 2$, $k + 1$ is at least 3, and hence for given $r$, $s$, we may choose a row $A_{(i)}$ such that $i \neq r$, $i \neq s$. If we now interchange $A_{(r)}$ and $A_{(s)}$, the effect on the right side of 5.3.3 is to leave the terms $a_{ij}$ unchanged, but to interchange two rows of each minor $M(i, j)$. But each $M(i, j)$ has degree $k$ (one less than the degree of $A$), and *we have assumed that 5.4.3(a) is true for matrices of degree k*; hence, the interchange of $A_{(r)}$ with $A_{(s)}$ results in multiplying each of the terms det $M(i, j)$ in equation 5.3.3 by $-1$. Hence,

$$\begin{aligned}\det A' &= \sum_{j=1}^{n} (-1)^{i+j} a_{ij} (-\det M(i,j)) \\ &= -\sum_{j=1}^{n} (-1)^{i+j} a_{ij} \det M(i,j) \\ &= -\det A\end{aligned}$$

as desired.

The proof of Theorem 5.4.3(b) is similar, and we leave it to you. ▮

(*Remark*: The following is a "proof" of 5.4.3(b) which is not totally valid: if the two equal rows are interchanged, then $A$ itself does *not* change and, hence, by 5.4.3(a), we have $\det A = -\det A$, or $2 \det A = 0$; dividing by 2, $\det A = 0$. The trouble with this argument is that if $F$ happens to be a field of characteristic 2 (see the end of Sec. 3.4), then we cannot "divide by 2" since $2_F = 0_F$. The induction proof, on the other hand, works no matter what the field of scalars may be.)

The next theorem strengthens Theorem 5.4.1. As in Section 5.1, $E_n$ denotes the set $\{1, 2, ..., n\}$.

**5.4.4 Theorem.** Let $i$ be a fixed integer in $E_n$. Let $A$, $B$, $C$ be three $n \times n$ matrices over $F$ which differ only in the $i$th row (that is, $A_{(h)} = B_{(h)} = C_{(h)}$, for all $h \neq i$), and suppose that $C_{(i)}$ is a linear combination of $A_{(i)}$ and $B_{(i)}$: $C_{(i)} = aA_{(i)} + bB_{(i)}$, where $a$, $b$ are scalars. Then

$$\det C = a \det A + b \det B.$$

As an illustration,

$$2 \det \begin{bmatrix} 1 & 2 & 3 \\ 4 & 5 & 6 \\ 7 & 8 & 9 \end{bmatrix} - \det \begin{bmatrix} 1 & 2 & 3 \\ -2 & 6 & 1 \\ 7 & 8 & 9 \end{bmatrix} = \det \begin{bmatrix} 1 & 2 & 3 \\ 10 & 4 & 11 \\ 7 & 8 & 9 \end{bmatrix}.$$

Theorem 5.4.4 is proved by expanding the determinant by minors of the $i$th row; by now you should be able to supply the details yourself.

**5.4.5 Theorem.** Let $c$ be a scalar, and let $i$, $k$ be *distinct* integers in $E_n$. If $A'$ is the matrix obtained from $A$ by replacing $A_{(i)}$ by $A_{(i)} + cA_{(k)}$, then $\det A' = \det A$. In other words, a determinant remains unchanged when we add to one row a scalar multiple of another row.

*Proof.* If $B$ is the matrix obtained from $A$ by replacing $A_{(i)}$ by $A_{(k)}$, then $B_{(i)} = A_{(k)} = B_{(k)}$; that is, two rows of $B$ are equal, so that $\det B = 0$ by Theorem 5.4.3. In addition, the matrices $A$, $B$, $A'$ differ only in the $i$th row, and by definition, we have

$$A'_{(i)} = A_{(i)} + cA_{(k)} = A_{(i)} + cB_{(i)}.$$

Hence, by Theorem 5.4.4,

$$\det A' = \det A + c \det B = \det A + c(0) = \det A. \blacksquare$$

Theorem 5.4.5 is probably the single most important theorem about determinants, both for theoretical reasons and for its use in computation. To illustrate the latter, suppose we wish to find $\det A$, where

$$A = \begin{bmatrix} 4 & -3 & 1 & 2 \\ 3 & 1 & 0 & 5 \\ -1 & -1 & 2 & 1 \\ 6 & -2 & 0 & 1 \end{bmatrix}.$$

As in Section 5.3, we look for a row or column with as many zeros as possible; here, we see that the third column has two zeros. It would be desirable if this column had *three* zeros; thus, we look for a way to replace $a_{33}$ by 0. One way to do so is to replace the row $A_{(3)}$ by $A_{(3)} - 2A_{(1)}$. By Theorem 5.4.5, this leaves the determinant unchanged; hence,

$$\det A = \det \begin{bmatrix} 4 & -3 & 1 & 2 \\ 3 & 1 & 0 & 5 \\ \mathbf{-9} & \mathbf{5} & \mathbf{0} & \mathbf{-3} \\ 6 & -2 & 0 & 1 \end{bmatrix}.$$

Now, expanding by minors of the third column, we get

$$\det A = +1 \cdot \det \begin{bmatrix} 3 & 1 & 5 \\ -9 & 5 & -3 \\ 6 & -2 & 1 \end{bmatrix}.$$

We have thus reduced our $4 \times 4$ determinant to a $3 \times 3$ determinant.

We could now reduce this $3 \times 3$ determinant to a $2 \times 2$ determinant by the same method, if we wished. Actually, we shall show in Section 5.6 that *any* square matrix can be reduced to a triangular matrix by means of operations of the type described in Theorem 5.4.5; and once the matrix is triangular, Theorem 5.2.5 gives us the determinant immediately.

The following theoretical application of Theorem 5.4.5 is of particular importance since it connects determinants to the topic of linear dependence (Chapter 2). Suppose that $A$ is an $n \times n$ matrix whose row vectors are linearly dependent. Then one row of $A$, say, $A_{(i)}$, is a linear combination of the other rows:

**(5.4.6)** $$A_{(i)} = c_1 A_{(1)} + \cdots + c_{i-1}A_{(i-1)} + c_{i+1}A_{(i+1)} + \cdots + c_n A_{(n)}.$$

By applying Theorem 5.4.5 $n - 1$ times in succession, we see that the determinant of $A$ is unchanged if the row $A_{(i)}$ is replaced by

**(5.4.7)** $$A_{(i)} - \sum_{k \neq i} c_k A_{(k)}.$$

But by 5.4.6, the row 5.4.7 is zero; hence, by 5.4.2, the determinant is zero also. We have proved:

**5.4.8** **Theorem.** If the row vectors of a square matrix $A$ are linearly dependent, then $\det A = 0$.

In Section 5.6 we shall prove the converse: if the rows are independent, then $\det A \neq 0$. This means that problems like Example 2.4.3 can be solved simply by computing a determinant, without the need to solve equations; just let $A$ be the matrix whose rows are the given vectors and then determine whether $\det A$ is zero or nonzero. (Problems like Sec. 2.4, Exercise 2(c) cannot be done quite so simply, since a nonsquare matrix does not have a determinant.)

Everything that we have said about rows in this section applies to columns also; specifically, we have

**5.4.9** **Theorem.** Theorems 5.4.1 through 5.4.5 and 5.4.8 are valid for columns in place of rows.

The proofs are the same as before, except that equation 5.3.4 is used instead of 5.3.3.

**5.4.10** **Theorem.** For any square matrix $A$ over $F$, $\det A = \det A^t$.

Theorem 5.4.10 implies that all properties of determinants

are symmetric with respect to "rows" and "columns" (this provides an alternate proof of 5.4.9). Our proof of 5.4.10 is by induction on $n$, the degree of $A$. The case $n = 1$ is trivial. Assume that the theorem holds for $k \times k$ matrices, and let $A$ have degree $n = k + 1$. For convenience in notation we let $A = (a_{ij})$, $B = (b_{ij}) = A^t$, and we let

$M(i, j)$ = the $(i, j)$ minor of $A$
$N(i, j)$ = the $(i, j)$ minor of $B$.

By definition of the transpose, $b_{ij} = a_{ji}$; similarly,

(5.4.11) $N(i, j) = (M(j, i))^t$

(*why*?). Hence, by equation 5.3.3, with $i = 1$,

$$\begin{aligned}\det B &= \sum_{j=1}^{n} (-1)^{1+j} b_{1j} \det N(1, j)\\ &= \sum_{j=1}^{n} (-1)^{1+j} a_{j1} \det (M(j, 1))^t\\ &= \sum_{j=1}^{n} (-1)^{j+1} a_{j1} \det M(j, 1)\\ &\qquad\qquad \text{(since } M(j, 1) \text{ is of degree } k\text{)}\\ &= \sum_{i=1}^{n} (-1)^{i+1} a_{i1} \det M(i, 1)\\ &\qquad\qquad \text{(changing notation)}\\ &= \det A \qquad \text{(equation 5.3.4 with } j = 1\text{).} \ \blacksquare\end{aligned}$$

(For an alternate proof of Theorem 5.4.10, see Exercise 5 below.)

**Exercises**

*1. Prove Theorem 5.4.3(b) by induction.

*2. Prove Theorem 5.4.4 by the method suggested.

3. Compute the determinant of each of the following matrices. In each case, begin by using Theorem 5.4.5 to simplify the matrix. (ANS)

$$\text{(a)} \begin{bmatrix} 1 & 2 & 3 & 4 \\ 1 & 0 & 1 & -3 \\ -1 & 0 & -2 & 1 \\ 2 & 1 & 2 & 1 \end{bmatrix}. \quad \text{(b)} \begin{bmatrix} 3 & 1 & 0 & 1 \\ -2 & 1 & 1 & 2 \\ 0 & 1 & -1 & 3 \\ 1 & 1 & 3 & 4 \end{bmatrix}.$$

4. Justify equation 5.4.11.

5. Prove Theorem 5.4.10 directly from Definition 5.2.1 without using the expansion by minors. (*Hint*: Let $\tau = \pi^{-1}$.)

6. Do Example 2.4.3, using determinants.

7. If $\mathbf{u} = (a_1, a_2, a_3)$ and $\mathbf{v} = (b_1, b_2, b_3)$ are vectors in $F_3$, we define a vector $\mathbf{u} \times \mathbf{v}$ in $F_3$, called the *cross product* of $\mathbf{u}$ and $\mathbf{v}$, by
$$\mathbf{u} \times \mathbf{v} = \left(\det \begin{bmatrix} a_2 & a_3 \\ b_2 & b_3 \end{bmatrix}, -\det \begin{bmatrix} a_1 & a_3 \\ b_1 & b_3 \end{bmatrix}, \det \begin{bmatrix} a_1 & a_2 \\ b_1 & b_2 \end{bmatrix}\right).$$
Use theorems on determinants to prove that $(\mathbf{u} \times \mathbf{v}) \cdot \mathbf{u} = (\mathbf{u} \times \mathbf{v}) \cdot \mathbf{v} = 0$. Interpret this result geometrically when $F = R$. (SUG)
8. Let $\mathbf{u}, \mathbf{v}, \mathbf{w}$ be vectors in $F_3$. Let $A$ be the $3 \times 3$ matrix whose row vectors are $\mathbf{u}, \mathbf{v}, \mathbf{w}$; let $B$ be the $3 \times 3$ matrix whose row vectors are $\mathbf{v} \times \mathbf{w}, \mathbf{w} \times \mathbf{u}, \mathbf{u} \times \mathbf{v}$ (cf. the preceding exercise). Discover (by trial and error, if necessary) a relationship between $\det A$ and $\det B$. Can you prove it?
9. Show (by example) that the cross product in $F_3$ is *not* associative; that is, the vectors $\mathbf{u} \times (\mathbf{v} \times \mathbf{w})$ and $(\mathbf{u} \times \mathbf{v}) \times \mathbf{w}$ are not always equal.
10. Let $A$ be a square matrix of degree $n \geq 3$. Suppose that for each $i$ $(1 \leq i \leq n)$, the sequence of entries in the row $A_{(i)}$ is an arithmetic progression. Prove that $\det A = 0$. (This generalizes Sec. 5.2, Exercise 2.)

## 5.5 Row and column operations on matrices

In Theorems 5.4.1, 5.4.3(a), and 5.4.5, certain operations are described, by means of which we may obtain a new matrix from an old one. Since these operations have a number of uses, it is worth paying some attention to them. In this section *matrices are not necessarily square* unless it is so stated; as usual, it is understood that all matrix entries lie in a field $F$. In addition, we shall use the notations

$A_{(i)} = i$th row (vector) of $A$
$A^{(j)} = j$th column (vector) of $A$

throughout this section *and for the remainder of the book*.

We begin by listing the three types of *elementary row operations* (e.r.o.'s) that can be performed on an arbitrary $m \times n$ matrix $A$:

*Type* 1. Interchange two rows of $A$.

*Type* 2. Multiply one row of $A$ by a *nonzero* scalar $c$.

*Type* 3. Add to one row $A_{(i)}$ a scalar multiple of another row $A_{(k)}$; that is, replace $A_{(i)}$ by $A_{(i)} + bA_{(k)}$. (Of course, $k \neq i$.)

By a *row operation* we shall mean a finite number of *elementary* row operations performed in a definite order. For example, the sentence

(5.5.1) "Multiply the second row by 2, then add three times the second row to the third row"

describes a row operation which is not elementary; it consists of an e.r.o. of Type 2 followed by an e.r.o. of Type 3. (The order in which the operations are performed matters; cf. Exercise 1 below.)

If $A'$ is the matrix obtained from $A$ by means of the row operation $\mathcal{R}$, we shall write

(5.5.2) $\mathcal{R}(A) = A'$

(note the use of left-hand notation!), or

$$A \xrightarrow{\mathcal{R}} A'.$$

The following comments are now in order:

1. Elementary row operations of Types 1 and 3 obviously can be performed only on matrices having at least two rows. (Operations of Type 2, of course, may be performed even on one-rowed matrices.)

2. Each elementary row operation is reversible. To be precise, if $\mathcal{R}$ is an e.r.o. of any given Type (1, 2, or 3), then there exists an elementary row operation $\mathcal{R}'$ of the *same type*, such that whenever $\mathcal{R}(A) = A'$, then $\mathcal{R}'(A') = A$. (See Exercise 2 below.) Note that $\mathcal{R}'$ depends only on $\mathcal{R}$, not on $A$.

3. For each elementary row operation $\mathcal{R}$, there exists a scalar $k = k(\mathcal{R})$, *depending only on* $\mathcal{R}$, such that

(5.5.3) $\det \mathcal{R}(A) = k \det A$

for all square matrices $A$ to which $\mathcal{R}$ can be applied. (For example, if $\mathcal{R}$ is the operation, "multiply the fifth row by $-2$", then $k = -2$ and 5.5.3 holds for all square matrices of degree at least 5.) To be specific, the theorems of Section 5.4 imply that $k = -1$ if $\mathcal{R}$ is of Type 1, $k = 1$ if $\mathcal{R}$ is of Type 3, and $k = c$ if $\mathcal{R}$ multiplies one row by $c$ as in Type 2.

4. The scalar $k = k(\mathcal{R})$ which we have just described is always nonzero. (We always have $1 \neq 0$ in a field, and we assumed $c \neq 0$ in the definition of Type 2.) It follows immediately that if $\mathcal{R}$ is any e.r.o., then

(5.5.4)
$$\begin{aligned} \det A = 0 &\Leftrightarrow \det \mathcal{R}(A) = 0 \\ \det A \neq 0 &\Leftrightarrow \det \mathcal{R}(A) \neq 0 \end{aligned}$$

for all square matrices $A$ to which $\mathcal{R}$ can be applied.

5. The fact that equations 5.5.3 and 5.5.4 hold for *elementary* row operations easily implies that these equations also hold for *all* row operations $\mathcal{R}$, not necessarily elementary.

6. Operations of Type 1 are actually superfluous: any interchange of two rows may also be accomplished by performing a finite sequence of operations of Types 2 and 3. (See Exercise 3 below.)

7. *Column* operations are defined similarly to row operations, and similar comments about them are valid. Note, however, a difference in notation: if $A'$ is obtained from $A$ by means of the *column* operation $\mathscr{C}$, we shall write

$$A\mathscr{C} = A'$$

using *right-hand notation* (the operation written to the right of the matrix) instead of the left-hand notation 5.5.2 which we use for row operations. The following theorem may suggest to you a reason for our choice of notation.

5.5.5 **Theorem.** Let $A$, $B$ be matrices such that the product $AB$ is defined. Then for all row operations $\mathscr{R}$,

$$(\mathscr{R}(A))B = \mathscr{R}(AB);$$

and for all column operations $\mathscr{C}$,

$$A(B\mathscr{C}) = (AB)\mathscr{C}.$$

(*Note*: It is understood, of course, that we consider only row operations $\mathscr{R}$ which are *applicable* to $A$, and column operations which are applicable to $B$. For example, the operation 5.5.1 applies only to matrices with at least three rows.)

In informal English, Theorem 5.5.5 asserts that the effect of applying a row operation (column operation) to a product is the same as if the operation were applied only to the first factor (second factor). The proof must be divided into several cases, only one of which shall be worked out in detail here. Suppose $\mathscr{R}$ is an elementary row operation of Type 3: "Add $c$ times the $k$th row to the $h$th row." Then $\mathscr{R}$ affects only the $h$th row of the matrices to which it is applied; hence, for all $i \neq h$ and all $j$, we have, by 4.4.11,

$$\begin{aligned}[(\mathscr{R}(A))B]_{ij} &= (\mathscr{R}(A))_{(i)} \cdot B^{(j)} \\ &= A_{(i)} \cdot B^{(j)} \\ &= (AB)_{ij} = [\mathscr{R}(AB)]_{ij}\end{aligned}$$

while for $i = h$, we have

$$\begin{aligned}[(\mathscr{R}(A))B]_{hj} &= (\mathscr{R}(A))_{(h)} \cdot B^{(j)} \\ &= A_{(h)} + cA_{(k)} \cdot B^{(j)} \\ &= A_{(h)} \cdot B^{(j)} + c(A_{(k)} \cdot B^{(j)}) \quad \text{(Theorem 3.7.2)} \\ &= (AB)_{hj} + c(AB)_{kj} = [\mathscr{R}(AB)]_{hj}.\end{aligned}$$

Hence, $[\mathscr{R}(A)]B = \mathscr{R}(AB)$ as desired. The proof for e.r.o.'s of Types 1 and 2, and for the three types of elementary column operations, are similar. In the nonelementary case, where $\mathscr{R}$ (or $\mathscr{C}$) is a sequence of, say, $r$ elementary operations, the theorem can be proved by induction on $r$; we leave the details to you. ∎

In order to state our next theorem on row and column operations, we need a definition. Let $A$ be an $m \times n$ matrix over $F$; then the row vectors of $A$ are elements of the vector space $F_n$. The subspace of $F_n$ that is spanned by these row vectors is called the *row space of A*. Similarly, the *column space of A* is the subspace of $F_m$ that is spanned by the column vectors of $A$. In symbols,

$$\text{row space of } A = [A_{(1)}, \ldots, A_{(m)}] \subseteq F_n$$
$$\text{column space of } A = [A^{(1)}, \ldots, A^{(n)}] \subseteq F_m.$$

5.5.6 **Theorem.** If $\mathscr{R}$ is a row operation such that $A \xrightarrow{\mathscr{R}} A'$, then the matrices $A$ and $A'$ have the same row space. In particular, the rows of $A$ are linearly independent if and only if the rows of $A'$ are linearly independent.

Similarly, if $\mathscr{C}$ is a column operation such that $B \xrightarrow{\mathscr{C}} B'$, then $B$ and $B'$ have the same column space; the columns of $B$ are independent if and only if the columns of $B'$ are independent.

*Proof.* It suffices to prove the assertions in the elementary case (*why*?). If $\mathscr{R}$ is elementary of Type 1, it is trivial that $A$ and $A'$ have the same row space; the same assertion when $\mathscr{R}$ is elementary of Type 2 or 3 is an immediate consequence of Theorem 2.3.16. Moreover, since the *dimension* of the row space equals the maximum number of independent rows (Theorem 2.6.9), it follows that $A$ and $A'$ have the same number of independent rows. In particular, all $m$ rows of $A$ are independent if and only if all $m$ rows of $A'$ are independent. The proof for elementary column operations is similar. ∎

**Exercises**

1. If $\mathscr{R}_1$ is the operation, "add twice the first row to the second row", and if $\mathscr{R}_2$ is the operation, "subtract the second row from the first row", show (by means of an example) that $\mathscr{R}_1$ followed by $\mathscr{R}_2$ is not the same as $\mathscr{R}_2$ followed by $\mathscr{R}_1$. (ANS)

*2. For each elementary row operation $\mathscr{R}$, find an elementary row operation $\mathscr{R}'$, of the same "Type" as $\mathscr{R}$, such that $\mathscr{R}'$ reverses the effect of $\mathcal{R}$; that is, whenever $\mathcal{R}(A) = A'$, then $\mathcal{R}'(A') = A$. (Consider the three possible types separately.)

3. Show that any elementary operation of Type 1 is equivalent to (i.e., has the same effect as) a finite sequence of operations of Types 2 and 3.
4. If

$$A = \begin{bmatrix} 1 & 1 & 5 & 0 \\ 1 & -1 & 1 & 2 \\ 0 & 2 & 1 & -5 \end{bmatrix}; \qquad B = \begin{bmatrix} 2 & -3 \\ 1 & 0 \\ -1 & 1 \\ 1 & 2 \end{bmatrix}$$

and $\mathcal{R}$ is the operation 5.5.1, compute $[\mathcal{R}(A)]B$ and $\mathcal{R}(AB)$ separately and verify that they are equal. (ANS)

*5. Fill in the missing parts of the proof of Theorem 5.5.5.

6. If $A$ is the matrix

$$\begin{bmatrix} 1 & 3 \\ 2 & 6 \end{bmatrix}$$

over **R**, draw a picture which exhibits (as geometric objects) the row space of $A$ and the column space of $A$. What property do these two spaces appear to have in common?

7. Let $n$ be a positive integer, let $A$ and $B$ be matrices over $F$ each having $n$ columns, and let $S$ be any subset of $\{1, 2, ..., n\}$. Also, let $A'$ be the submatrix of $A$ consisting of only those columns $A^{(j)}$ such that $j \in S$; similarly, let $B'$ be the submatrix of $B$ consisting of those columns $B^{(j)}$ such that $j \in S$.

   Prove: if the row spaces of $A$ and $B$ are equal, then the row spaces of $A'$ and $B'$ are equal. (*Hint*: Use Theorem 4.6.6.) (SOL)

## 5.6 Echelon form of a matrix

Since row and column operations preserve certain properties of a matrix, one way of investigating such properties is to reduce the matrix, by means of such operations, to a matrix whose form is particularly simple or easy with which to work. (For instance, in studying the determinant we might try to reduce the matrix to triangular form and then apply Theorem 5.2.5.) In this section, we shall develop a procedure by which any matrix can be reduced to *echelon form*, which for square matrices turns out to be a special case of triangular form. The reduction can even be accomplished using row operations only, without column operations. As usual, we assume that all matrices have their entries in a field $F$.

**5.6.1 Definition.** Let $A$ be an $m \times n$ matrix over $F$. The matrix $A$ is said to be an *echelon matrix*, or to be in *echelon form*, if it has the following three properties:

(a) For some integer $r$ $(0 \leq r \leq m)$, the first $r$ rows of $A$ are nonzero and the remaining $m - r$ rows are zero.

(b) For each positive integer $k < r$, the first nonzero entry in the $k$th row lies strictly to the left of the first nonzero entry in the $(k + 1)$th row.

(c) For each positive integer $k \leq r$, the first nonzero entry in the $k$th row is the only nonzero entry in its column.

An echelon matrix thus looks like

$$\left[\begin{array}{ccccccccccccccccccc}
\overbrace{0 & \cdots & 0} & c_1 & \overbrace{* & \cdots & *} & 0 & \overbrace{* & \cdots & *} & 0 & \cdots & 0 & \overbrace{* & \cdots & *} \\
0 & \cdots & 0 & 0 & 0 & \cdots & 0 & c_2 & * & \cdots & * & 0 & \cdots & 0 & * & \cdots & * \\
0 & \cdots & 0 & 0 & 0 & \cdots & 0 & 0 & 0 & \cdots & 0 & c_3 & \cdots & 0 & * & \cdots & * \\
\vdots & & \vdots & \vdots & \vdots & & \vdots & \vdots & \vdots & & \vdots & \vdots & \text{(etc.)} & \vdots & \vdots & & \vdots \\
0 & \cdots & 0 & 0 & 0 & \cdots & 0 & 0 & 0 & \cdots & 0 & 0 & \cdots & c_r & * & \cdots & * \\
0 & \cdots & 0 & 0 & 0 & \cdots & 0 & 0 & 0 & \cdots & 0 & 0 & \cdots & 0 & 0 & \cdots & 0 \\
\vdots & & \vdots & \vdots & \vdots & & \vdots & \vdots & \vdots & & \vdots & \vdots & & \vdots & \vdots & & \vdots \\
0 & \cdots & 0 & 0 & 0 & \cdots & 0 & 0 & 0 & \cdots & 0 & 0 & \cdots & 0 & 0 & \cdots & 0
\end{array}\right]
\begin{array}{l}
\left.\vphantom{\begin{array}{c}0\\0\\0\\0\\0\end{array}}\right\} r \text{ nonzero rows} \\
\left.\vphantom{\begin{array}{c}0\\0\\0\end{array}}\right\} m - r \text{ zero rows}
\end{array} \tag{5.6.2}$$

with the understanding that (1) the entries $c_1, \ldots, c_r$ are nonzero; (2) the bracketed columns may or may not appear; and (3) entries denoted by asterisks are entries whose values are either unknown to us or not of interest to us; they may or may not be zero. To illustrate, the matrix

$$\begin{bmatrix}
0 & 2 & 0 & 3 & 0 & 0 & 5 & 0 & 0 \\
0 & 0 & 1 & 3 & 0 & 0 & 3 & 2 & 0 \\
0 & 0 & 0 & 0 & 5 & 0 & -1 & 0 & 0 \\
0 & 0 & 0 & 0 & 0 & 4 & 2 & 0 & 0 \\
0 & 0 & 0 & 0 & 0 & 0 & 0 & 0 & -2 \\
0 & 0 & 0 & 0 & 0 & 0 & 0 & 0 & 0
\end{bmatrix}$$

has the form 5.6.2 with $m = 6$, $r = 5$.

**5.6.3 Definition.** A *reduced echelon matrix* is an echelon matrix in which the first nonzero entry in each nonzero row equals 1. (That is, $c_1 = \cdots = c_r = 1$ in 5.6.2.)

**5.6.4 Theorem.** (a) Every matrix can be transformed into echelon form by means of a finite sequence of elementary row operations of Type 3. (b) Every matrix can be transformed into reduced echelon form by means of a finite sequence of elementary row operations of Types 2 and 3.

Our proof of Theorem 5.6.4(a) is constructive: it will describe in detail "how to do it". The method is to reduce each column in succession to the proper form, starting from the left. For simplicity in notation, the successive matrices to which the given matrix $A$ is reduced shall also be denoted $A$; although this is technically incorrect, we hope it will be understandable. You may find it helpful to write down a numerical example and work through the example step-by-step as you follow the proof.

If $A$ is the zero matrix, then $A$ is already in echelon form (with $r = 0$) and there is nothing to prove. Hence, we may assume that $A \neq 0$. Consider the leftmost nonzero column of $A$, say, the $j$th column. Our first object is to obtain a nonzero entry in the $(1, j)$ position (the top position in the $j$th column). If $a_{1j} \neq 0$, this is already done; if $a_{1j} = 0$, our object is accomplished by choosing a row index $h$ such that $a_{hj} \neq 0$ (possible by assumption) and then adding the $h$th row to the first row (a row operation of Type 3). Observe that this row operation does not affect the columns preceding the $j$th column: adding zero to zero still gives zero. *We now have an entry* $c_1 \neq 0$ *in the* $(1, j)$ *position*. For each $i \neq 1$ $(2 \leq i \leq m)$, we now replace the row $A_{(i)}$ by $A_{(i)} - c_1^{-1}a_{ij}A_{(1)}$ (still using only operations of Type 3). This again has no effect on the columns preceding the $j$th column, but the effect on the $j$th column is to replace $a_{ij}$ by $a_{ij} - c_1^{-1}a_{ij}c_1 = 0$. Thus, *all entries in the* $j$*th column below* $c_1$ *are now zero*, so that the columns up to and including the $j$th column look the way they should (see 5.6.2).

Next, we look for a column having a nonzero entry *below the first row*. If there is no such column, then we already have echelon form, q.e.d. If there is such a column, choose the leftmost such column, say, the $k$th column. Evidently, $k > j$, since the first $j$ columns have all zeros below the first row. The columns preceding the $k$th column also have all zeros below the first row, by choice of $k$. If the entry $a_{2k}$ in the $(2, k)$ position is zero, we make it nonzero by choosing a nonzero entry $a_{hk}$ for which $h \neq 1$ (possible by choice of $k$) and then adding $A_{(h)}$ to $A_{(2)}$. The columns preceding the $k$th column are unaffected by this operation since the $h$th row has only zeros in those columns. *We now have an entry* $c_2 \neq 0$ *in the* $(2, k)$ *position*. We then make the rest of the $k$th column zero by replacing each row $A_{(i)}$ $(i \neq 2)$ by $A_{(i)} - c_2^{-1}a_{ik}A_{(2)}$. Once again, this does not affect the columns preceding the $k$th, since

$A_{(2)}$ has only zeros in those columns. Thus, the columns up to and including the $k$th column (the column containing the entry $c_2$) all look the way they should.

It should be evident now how to continue the process. If we do not yet have echelon form, choose the leftmost column (say, the $s$th column) which has a nonzero entry *below the second row*, make the entry in the $(3, s)$ position nonzero (if necessary, by adding $A_{(h)}$ to $A_{(3)}$, where $h \neq 1, 2$), and so on. Since $A$ has only finitely many rows, the process ends after finitely many steps, and Theorem 5.6.4(a) is proved. Moreover, once we have a matrix of the form 5.6.2, we can transform it into *reduced* echelon form simply by multiplying the first row by $c_1^{-1}$, ..., the $r$th row by $c_r^{-1}$. Since the latter operations are of Type 2, this proves Theorem 5.6.4(b). ∎

Any (reduced) echelon matrix into which $A$ can be transformed (as in Theorem 5.6.4) is called a *(reduced) echelon form of $A$*.

Reduction of a matrix to its reduced echelon form provides a systematic approach to certain types of algebraic problems. The following is an illustration.

5.6.5 **Example.** In $\mathbf{R}_3$, show that $[(1, 2, 3), (0, 4, 1)] = [(1, 6, 4), (1, -2, 2)]$.

*Solution.* These two subspaces are the row spaces of the matrices

$$\begin{bmatrix} 1 & 2 & 3 \\ 0 & 4 & 1 \end{bmatrix}, \quad \begin{bmatrix} 1 & 6 & 4 \\ 1 & -2 & 2 \end{bmatrix},$$

respectively. Since row operations preserve the row space of a matrix, it will suffice to show that both of the above matrices can be reduced by row operations to the same reduced echelon matrix. Following the procedure described above, we have

$$\begin{bmatrix} 1 & 2 & 3 \\ 0 & 4 & 1 \end{bmatrix} \xrightarrow[\text{to } A_{(1)}]{\text{add } -\frac{1}{2}A_{(2)}} \begin{bmatrix} 1 & 0 & 5/2 \\ 0 & 4 & 1 \end{bmatrix} \xrightarrow[\text{by } \frac{1}{4}]{\text{multiply second row}} \begin{bmatrix} 1 & 0 & 5/2 \\ 0 & 1 & 1/4 \end{bmatrix};$$

$$\begin{bmatrix} 1 & 6 & 4 \\ 1 & -2 & 2 \end{bmatrix} \xrightarrow[\text{to } A_{(2)}]{\text{add } -A_{(1)}} \begin{bmatrix} 1 & 6 & 4 \\ 0 & -8 & -2 \end{bmatrix}$$

$$\xrightarrow[\text{to } A_{(1)}]{\text{add } \frac{3}{4}A_{(2)}} \begin{bmatrix} 1 & 0 & 5/2 \\ 0 & -8 & -2 \end{bmatrix} \xrightarrow[\text{row by } -\frac{1}{8}]{\text{multiply second}} \begin{bmatrix} 1 & 0 & 5/2 \\ 0 & 1 & 1/4 \end{bmatrix}$$

and the desired result follows. (Conversely, if the two reduced echelon matrices had *not* turned out to be the same, then the two given subspaces of $\mathbf{R}_3$ could not have been equal; cf. Exercise 6(a), this section.) Note that this example coincides with Example 2.3.14; our present method is essentially the same as the second of the two methods described in Section 2.3 (cf. item 2.3.19), except that now the sequence of steps is systematic: we reduce each column in turn to the proper form. ∎

Another application of the echelon form is in solving systems of simultaneous linear equations (see Sec. 5.10). In addition, the echelon form can be used to obtain simple proofs of certain theorems about matrices and determinants. In doing so, we shall find the following result useful.

**5.6.6** **Theorem.** Let $A^*$ be a reduced echelon form of $A$. Then the number of nonzero rows of $A^*$ equals the dimension of the row space of $A$. If this number is $r$, then the $r \times r$ identity matrix is a submatrix of $A^*$.

*Proof.* $A^*$ has the form 5.6.2 with all $c_i = 1$. We shall show that the $r$ nonzero rows of $A^*$ are linearly independent. Suppose $b_1, \ldots, b_r$ are scalars such that

$$b_1 A^*_{(1)} + \cdots + b_r A^*_{(r)} = \mathbf{0}. \tag{5.6.7}$$

For each $h$ $(1 \leq h \leq r)$ let the column containing the entry $c_h$ be the $(k_h)$th column. Since the $n$-tuple 5.6.7 is zero, its $(k_h)$th coordinate is zero; that is,

$$\begin{aligned} 0 &= b_1 a^*_{1k_h} + \cdots + b_r a^*_{rk_h} \\ &= b_1 0 + \cdots + b_h c_h + \cdots + b_r 0 = b_h c_h = b_h. \end{aligned}$$

Thus, all the $b$'s are zero and, hence, the row vectors $A^*_{(1)}, \ldots, A^*_{(r)}$ are independent as claimed. Since the remaining rows of $A^*$ are zero, it is clear that $r$ is the dimension of the row space of $A^*$, hence, also of $A$, by Theorem 5.5.6. Moreover, since all $c_i = 1$, we obtain the $r \times r$ identity matrix from the first $r$ rows of $A^*$ and the $(k_1)$th, ..., $(k_r)$th columns. ∎

In Section 4.7, we exhibited several conditions on a linear transformation $\mathbf{T} : V \to V$ (and on the corresponding square matrix $A$) which are equivalent to the assertion that $\mathbf{T}$ is one-to-one and onto. Using Theorems 5.6.4 and 5.6.6, we can now add two additional conditions to the list.

**5.6.8 Theorem.** If $A$ is a square $(n \times n)$ matrix over the field $F$, then the following statements are equivalent:

(a) The row vectors of $A$ are linearly independent.

(b) The column vectors of $A$ are linearly independent.

(c) $\det A \neq 0$.

(d) There exists a (not necessarily elementary) row operation $\mathcal{R}$ such that $\mathcal{R}(A) = I$.

(e) The matrix $A$ has an inverse $A^{-1}$ such that $AA^{-1} = A^{-1}A = I$.

Moreover, if $V$ is an $n$-dimensional vector space over $F$ and $\mathbf{T}$ is a linear transformation of $V \to V$ whose matrix (with respect to some basis of $V$) is $A$, then each of the following conditions on $\mathbf{T}$ is equivalent to statements (a)–(e):

(f) $\mathbf{T}$ is one-to-one.

(g) $\mathbf{T}$ is onto.

(h) $\mathbf{T}$ has an inverse in $\text{Lin}(V, V)$.

(i) $\text{Nul}\, \mathbf{T} = \{\mathbf{0}\}$.

The two new conditions are, of course, (c) and (d); equivalence of all the others was established in Section 4.7. Hence, to prove Theorem 5.6.8, it suffices to show that statements (a), (c), and (d) are equivalent. If the rows of $A$ are independent, then the row space has dimension $n$, so that by Theorem 5.6.6 the matrix $I_n$ is a submatrix of $A^*$, where $A^* = \mathcal{R}(A)$ is the reduced echelon form of $A$. Since $I_n$ and $A^*$ are both $n \times n$ matrices, it follows that $I_n = A^* = \mathcal{R}(A)$. This proves that (a) $\Rightarrow$ (d). Since $\det I = 1$, (d) and equations 5.5.4 imply that $\det A \neq 0$, so that (d) $\Rightarrow$ (c). Finally, (c) $\Rightarrow$ (a) by the contrapositive of Theorem 5.4.8. ∎

**5.6.9 Definition.** (a) A square matrix $A$ over $F$ is said to be *nonsingular* if it satisfies any one of conditions (a)–(e) of Theorem 5.6.8 (equivalently, if it satisfies all five conditions); otherwise, $A$ is *singular*. (The terms "singular", "nonsingular" are applied only to square matrices.)

(b) If $\mathbf{T} \in \text{Lin}(V, V)$, where $V$ is finite dimensional, $\mathbf{T}$ is *nonsingular* if it satisfies any one of conditions (f)–(i) of Theorem 5.6.8 (equivalently, if it satisfies all of these conditions); otherwise $\mathbf{T}$ is *singular*.

Note that according to 5.6.8 and 5.6.9, if $\mathbf{T} \in \text{Lin}(V, V)$ has matrix $A$, then $\mathbf{T}$ is nonsingular if and only if $A$ is.

**5.6.10 Theorem.** (a) If $A$ and $B$ are $n \times n$ matrices over $F$, then $\det AB = (\det A)(\det B)$. (The determinant of a product is the product of the determinants.) (b) If $A$ is nonsingular, then det $A^{-1} = (\det A)^{-1}$.

*Proof.* If $A$ is nonsingular, then by Theorem 5.6.8(d) we have $\mathscr{R}(A) = I$ for some $\mathscr{R}$. As we saw in Section 5.5, there is a nonzero scalar $k = k(\mathscr{R})$ depending only on $\mathscr{R}$, such that $\mathscr{R}$ has the effect of multiplying all determinants by $k$. Thus, we have

$$\begin{aligned} k \det A &= \det \mathscr{R}(A) = \det I = 1 \\ k \det AB &= \det \mathscr{R}(AB) = \det[(\mathscr{R}(A))B] \qquad \text{(by Theorem 5.5.5)} \\ &= \det IB = \det B = 1 \cdot \det B = (k \det A)(\det B). \end{aligned}$$

Dividing by $k$, we obtain det $AB$ = (det $A$) (det $B$) as desired. On the other hand, if $A$ is singular, then Theorem 5.6.8 implies that the rows of $A$ are dependent. Hence, the rows of $AB$ are dependent (Sec. 4.4, Exercise 12), and by Theorem 5.6.8, we obtain

$$\det AB = 0 = 0 (\det B) = (\det A)(\det B).$$

Thus, 5.6.10(a) is proved in all cases.

Theorem 5.6.10(b) is an immediate corollary of 5.6.10(a). Indeed, by Theorem 1.9.6, the nonsingular $n \times n$ matrices over $F$ form a group under multiplication; by 5.6.10(a), "det" is a homomorphism of this group into the multiplicative group $F^*$ of nonzero scalars. Hence, 5.6.10(b) follows from Theorem 1.8.4. ∎

An interesting consequence of 5.6.10 is

**5.6.11** **Theorem.** Let $V$ be a finite-dimensional vector space over $F$ and let $\mathbf{T} \in \operatorname{Lin}(V, V)$. If $\mathbf{T}$ has a matrix $A$ with respect to one basis of $V$ and matrix $A'$ with respect to another basis of $V$, then det $A$ = det $A'$.

*Proof.* As shown in Section 4.8, $A' = P^{-1}AP$ for some square matrix $P$. By 5.6.10, we then have

$$\begin{aligned} \det A' = \det P^{-1}AP &= (\det P^{-1})(\det A)(\det P) \\ &= (\det P)^{-1}(\det A)(\det P) \\ &= \det A \end{aligned}$$

since multiplication of scalars is commutative even though multiplication of matrices is not! ∎

Theorem 5.6.11 shows that if $\mathbf{T} \in \operatorname{Lin}(V, V)$, where dim $V < \infty$, we may unambiguously define the *determinant of* $\mathbf{T}$ to be the determinant of the matrix of $\mathbf{T}$ with respect to any basis of $V$. This definition may even be extended to affine transformations (see Sec. 4.9): if $\mathbf{T} : V \to V$ is affine, we define the "determinant of $\mathbf{T}$" to be the determinant of the linear part of $\mathbf{T}$. By Theorem 5.6.10(a), we have

$$\det \mathbf{T}_1\mathbf{T}_2 = (\det \mathbf{T}_1)\,(\det \mathbf{T}_2)$$

whenever the mappings $\mathbf{T}_i : V \to V$ are linear; the same formula must then hold for all *affine* mappings of $V \to V$ by Theorem 4.9.3.

In the case $V = \mathbf{R}_n$, the determinant of an affine transformation $\mathbf{T}$ has a simple geometric interpretation: *if $\mathcal{S}$ is any region in $\mathbf{R}_n$ and $\mathcal{S}'$ is its image under $\mathbf{T}$, then the n-dimensional volume of $\mathcal{S}'$ equals the n-dimensional volume of $\mathcal{S}$ multiplied by $|\det \mathbf{T}|$.* (A proof will be sketched in the next section.) As a special case, suppose $\mathbf{T}$ is an isometry and let $\mathcal{S}$ be the interior of the unit "sphere":

$$\mathcal{S} = \{(x_1, \ldots, x_n) \in \mathbf{R}_n : \sum_{i=1}^{n} x_i^2 < 1\} = \{\mathbf{u} \in \mathbf{R}_n : |\mathbf{u}| < 1\}.$$

Then the linear part of $\mathbf{T}$ maps $\mathcal{S}$ onto itself (*why*?), and, hence, $|\det \mathbf{T}| = 1$. Thus, we have

**5.6.12 Theorem.** If $\mathbf{T}$ is an isometry of $\mathbf{R}_n$, then $\det \mathbf{T} = \pm 1$.

Although the argument leading to Theorem 5.6.12 depended on facts about volume, we can give an alternate proof of 5.6.12 which avoids the subject of volume entirely. The proof is as follows: by definition of $\det \mathbf{T}$, we lose no generality in assuming $\mathbf{T}$ is linear (note that the linear part of an isometry is an isometry, cf. Theorem 4.10.6). Then by Theorem 4.10.13, $\mathbf{T}$ has a matrix $A$ such that $A^t = A^{-1}$. Thus,

$$\begin{aligned} AA^t &= AA^{-1} = I \\ 1 &= \det I = (\det A)\,(\det A^t) \qquad \text{(by 5.6.10)} \\ &= (\det A)^2 \qquad \text{(by 5.4.10)} \\ &= (\det \mathbf{T})^2 \end{aligned}$$

which implies the result. ▮

The following special case of 5.6.12 is of some interest:

**5.6.13 Theorem.** If $S$ is a $k$-flat in $\mathbf{R}_n$ and $\mathcal{M}_S : \mathbf{R}_n \to \mathbf{R}_n$ is the reflection through $S$, then $\det \mathcal{M}_S = (-1)^{n-k}$.

*Proof.* By Theorem 3.8.30(b), the linear part of $\mathcal{M}_S$ is the reflection through the direction space of $S$. Hence, we may assume without loss of generality that $S$ itself is a subspace. By Theorem 3.8.18, $\dim S^{\perp} = n - k$ and $\mathbf{R}_n = S \oplus S^{\perp}$. From the latter, $\mathbf{R}_n$ has a basis $\{\mathbf{u}_1, ..., \mathbf{u}_k, \mathbf{v}_{k+1}, ..., \mathbf{v}_n\}$ such that $\{\mathbf{u}_1, ..., \mathbf{u}_k\}$ is a basis of $S$ and $\{\mathbf{v}_{k+1}, ..., \mathbf{v}_n\}$ is a basis of $S^{\perp}$. By Section 3.8, Exercise 13, $\mathcal{M}_S$ maps each $\mathbf{u}_i$ into itself and each $\mathbf{v}_j$ into $-\mathbf{v}_j$. Hence, with respect to the ordered basis $(\mathbf{u}_1, ..., \mathbf{u}_k, \mathbf{v}_{k+1}, ..., \mathbf{v}_n)$, the matrix of $\mathcal{M}_S$ is

$$\begin{bmatrix} 1 & & & & & \\ & \ddots & & & & \\ & & 1 & & & \\ & & & -1 & & \\ & & & & \ddots & \\ & & & & & -1 \end{bmatrix}$$

($k$ entries positive, $n - k$ entries negative), whose determinant is clearly $(-1)^{n-k}$. ∎

*Comment*: It will be shown later (cf. Sec. 9.4, Exercise 3) that every isometry of $\mathbf{R}_n$ is a product of reflections. Combined with Theorem 5.6.13, this provides yet another proof of Theorem 5.6.12.

**Exercises**

*1. Prove formally that in the case $m = n$, the general echelon form 5.6.2 is upper triangular.

2. (a) Reduce each of the following matrices to echelon form *using only row operations of Type 3*:

$$A = \begin{bmatrix} 2 & 6 & 0 & -2 & 2 \\ 0 & 0 & 0 & 4 & -2 \\ -1 & 0 & 1 & -1 & 0 \\ 2 & 0 & -2 & 0 & 1 \end{bmatrix}; \quad B = \begin{bmatrix} 0 & 1 & 0 & -1 \\ 2 & -2 & -1 & -1 \\ 4 & 0 & 1 & -5 \\ 4 & -1 & 0 & -2 \end{bmatrix};$$

$$C = \begin{bmatrix} 2 & 1 & -2 \\ -2 & 1 & 5 \\ 0 & 2 & 3 \\ 4 & 8 & 5 \end{bmatrix}.$$

(b) Use the echelon form obtained in part (a) to evaluate det $B$. (ANS)

(c) Find the *reduced* echelon form of each of the matrices $A$, $B$, $C$. (ANS)

(*Note*: Answers to part (a) are not unique, but answers to part (c) are unique. Cf. Exercise 6.)

3. (a) Use the reduced echelon form to determine whether the two subspaces

$W_1 = [(2, 3, 1), (1, 2, 3)]; \quad W_2 = [(1, 0, -7), (-1, 1, 12)]$

of $\mathbf{R}_3$ are equal. (ANS)

(b) Same problem if $W_1 = [(2, 3, 4), (0, 1, 0)]$, $W_2 = [(1, 2, 0), (-2, -4, 5)]$. (ANS)

*4. (a) Suppose that the $n \times n$ matrix $A$ has the block form

$$A = \begin{bmatrix} B & * \\ 0 & C \end{bmatrix}$$

where $B$ and $C$ are square submatrices of degrees $k$ and $n - k$ (respectively) and the $(n - k) \times k$ submatrix in the lower left corner is the zero matrix. (The $k \times (n - k)$ submatrix in the upper right corner may be arbitrary.) Prove that

$\det A = (\det B)(\det C)$.

(*Suggested method*: Use Theorems 5.6.4(a) and 5.2.5. *Alternate method*: Expand det $A$ by minors of the first column and use induction on $k$.)

(b) If

$$A = \begin{bmatrix} B & 0 \\ * & C \end{bmatrix}$$

where $B$ and $C$ are square submatrices of $A$, prove that det $A = (\det B)(\det C)$.

5. Prove that the determinant, regarded as a mapping of Mat($n$, $n$; $F$) $\to F$ (that is, $A \to \det A$ for each $n \times n$ matrix $A$), is the only mapping of Mat($n$, $n$; $F$) $\to F$ which has the properties described in Theorems 5.4.1 and 5.4.5 and assigns to the identity matrix the value 1. (Some authors *define* a "determinant" to be a mapping having these three properties and then prove formula 5.2.1 as a theorem.) (SOL)

6. (a) Prove: if $A$ and $B$ are $m \times n$ reduced echelon matrices having the same row space, then $A = B$. (This is not an easy exercise. You may find the result of Sec. 5.5, Exercise 7 useful.) (SOL)

(b) Deduce from (a) that the reduced echelon form of any given matrix is unique.

7. Let $A$ be a $3 \times 4$ matrix over $F$. If exactly three of the four $3 \times 3$ submatrices of $A$ have zero determinant, show that one column of $A$ is the zero column. Can this result be generalized?

8. Let $V$ be a finite-dimensional vector space, let $\mathbf{T} \in \text{Lin}(V, V)$

and let $k$ be a positive integer. Prove: **T** is singular if and only if $\mathbf{T}^k$ is singular. (This shows that the positive powers of **T** are either all singular or all nonsingular.) (SUG)

## 5.7 Elementary matrices, areas, and volumes

By an *elementary matrix* we mean any matrix of the form $\mathcal{R}(I)$ where $\mathcal{R}$ is an *elementary* row operation. (The lack of symmetry between rows and columns in this definition is only apparent, not actual; see Exercise 3, this section.) An elementary matrix $\mathcal{R}(I)$ is said to be of Type 1, 2, or 3 according to whether $\mathcal{R}$ is of Type 1, 2, or 3. For example, the elementary 2 × 2 matrices are the following:

$$\textit{Type 1:} \begin{bmatrix} 0 & 1 \\ 1 & 0 \end{bmatrix}.$$

$$\textit{Type 2:} \begin{bmatrix} k & 0 \\ 0 & 1 \end{bmatrix} \quad \text{or} \quad \begin{bmatrix} 1 & 0 \\ 0 & k \end{bmatrix} \qquad (k \neq 0).$$

$$\textit{Type 3:} \begin{bmatrix} 1 & k \\ 0 & 1 \end{bmatrix} \quad \text{or} \quad \begin{bmatrix} 1 & 0 \\ k & 1 \end{bmatrix}.$$

You should try writing down the elementary 3 × 3 matrices yourself (or, if you are more ambitious, the elementary $n \times n$ matrices).

The significance of elementary matrices lies in the following theorem:

**5.7.1 Theorem.** Every nonsingular matrix is a product of elementary matrices.

*Proof.* For any row operation $\mathcal{R}$, Theorem 5.5.5 implies that

$$\mathcal{R}(A) = \mathcal{R}(IA) = \mathcal{R}(I) \cdot A;$$

in other words, applying a row operation to a given matrix has the same effect as *multiplying the matrix (on the left) by the corresponding elementary matrix*. If $A$ is nonsingular, then $A \to I$ via a finite sequence of elementary row operations. Since every row operation is reversible (Sec. 5.5), it is also true that $I$ can be transformed into $A$, say by performing the elementary row operations $\mathcal{R}_1, \mathcal{R}_2, \ldots, \mathcal{R}_k$ in that order. Then

$$I \xrightarrow{\mathcal{R}_1} \mathcal{R}_1(I) \xrightarrow{\mathcal{R}_2} \mathcal{R}_2(I)\,\mathcal{R}_1(I) \xrightarrow{\mathcal{R}_3} \cdots \xrightarrow{\mathcal{R}_k} [\mathcal{R}_k(I) \cdots \mathcal{R}_2(I)\,\mathcal{R}_1(I)] = A$$

which is the desired result. ∎

In the preceding section it was asserted that if **T** is an affine transformation of $\mathbf{R}_n \to \mathbf{R}_n$, then the $n$-dimensional volume of the region $\mathscr{S}\mathbf{T}$ equals $|\det \mathbf{T}|$ times the volume of $\mathscr{S}$. (A formal statement of this, for readers who know some measure theory, is that $\mu(\mathscr{S}\mathbf{T}) = |\det \mathbf{T}| \cdot \mu(\mathscr{S})$ for all Borel sets $\mathscr{S}$, where $\mu$ is Lebesgue measure in $\mathbf{R}_n$.) Since this is not the place to do measure theory or to try to define "area" and "volume" rigorously, we cannot give a completely formal proof of the result; however, we shall sketch an informal proof in the case $n = 2$. (The proof for arbitrary $n$ is not essentially different.) The idea is to prove the result first when **T** is linear and its matrix is elementary, and then use Theorem 5.7.1 to extend the result to the general nonsingular case. (The singular case is treated separately.) Once we have the result when **T** is linear, its validity when **T** is affine follows from the fact that translations preserve area.

To fix notation, assume that **T** is linear and has matrix $A$ with respect to the canonical basis of $\mathbf{R}_2$, let $\mathscr{S}$ be a region in $\mathbf{R}_2$, and let $\mu(\mathscr{S})$ denote the area of $\mathscr{S}$. Suppose first that $A$ is elementary and that $\mathscr{S}$ is the rectangular region bounded by the vertical lines $x = a$ and $x = b$ and the horizontal lines $y = c$ and $y = d$ (see Fig. 5.7.1(a)). That is,

**(5.7.2)** $$\mathscr{S} = \{(x, y) : a \leqslant x \leqslant b,\ c \leqslant y \leqslant d\}.$$

If $A$ is the Type 2 matrix:

$$\begin{bmatrix} k & 0 \\ 0 & 1 \end{bmatrix}$$

with $k > 0$, it is not hard to show (using 4.2.14) that the image of $\mathscr{S}$ under **T** is the rectangle

$$\mathscr{S}\mathbf{T} = \{(x, y) : ak \leqslant x \leqslant bk,\ c \leqslant y \leqslant d\}$$

(see Fig. 5.7.1(b)). We then have $|\det A| = k$, $\mu(\mathscr{S}) = (b - a)(d - c)$,

$$\begin{aligned} \mu(\mathscr{S}\mathbf{T}) &= (bk - ak)(d - c) \\ &= k(b - a)(d - c) = |\det A|\ \mu(\mathscr{S}) \end{aligned}$$

as desired. If $k < 0$, similar reasoning gives $\mu(\mathscr{S}\mathbf{T}) = -k(b - a)(d - c) = |\det A|\ \mu(\mathscr{S})$. If

$$A = \begin{bmatrix} 1 & 0 \\ 0 & k \end{bmatrix}$$

the argument is similar. If, instead, $A$ is of Type 3, say,

$$A = \begin{bmatrix} 1 & 0 \\ k & 1 \end{bmatrix}$$

then $\mathscr{S}\mathbf{T}$ is the region bounded by $y = c$, $y = d$, $x = a + ky$, $x = b + ky$ (see Fig. 5.7.1(c)); this is a parallelogram with the same height and base (and, hence, the same area) as $\mathscr{S}$. Since $\det A = 1$ in this case, the desired result holds in this case too. A similar argument works when

$$A = \begin{bmatrix} 1 & k \\ 0 & 1 \end{bmatrix}$$

or when $A$ is of Type 1 (actually, though, Type 1 is superfluous; cf. Theorem 5.6.4(b) and/or Exercise 3, Sec. 5.5). Moreover, once the result is known when $\mathscr{S}$ has the special form 5.7.2, its validity for arbitrary regions follows from the fact that any region can be approximated arbitrarily closely by a union of rectangles of the form 5.7.2. (The preceding sentence can be made rigorous, but we cannot do so here; however, it is hoped that the argument seems intuitively reasonable.) Hence, our informal proof is complete in the case where $A$ is elementary.

In the more general case, where $A$ is nonsingular, we have $A = E_1E_2 \cdots E_k$, where the $E$'s are elementary. It follows that $\mathbf{T} = \mathbf{T}_1\mathbf{T}_2 \cdots \mathbf{T}_k$, where $\mathbf{T}_i$ is the L.T. having matrix $E_i$. By the result already proved in the elementary case,

$$\begin{aligned}
&\mu(\mathscr{S}\mathbf{T}_1) = |\det E_1|\ \mu(\mathscr{S}) \\
&\mu(\mathscr{S}\mathbf{T}_1\mathbf{T}_2) = |\det E_2|\ \mu(\mathscr{S}\mathbf{T}_1) \\
&\quad\vdots \\
&\mu(\mathscr{S}\mathbf{T}_1\mathbf{T}_2 \cdots \mathbf{T}_{k-1}\mathbf{T}_k) = |\det E_k|\ \mu(\mathscr{S}\mathbf{T}_1\mathbf{T}_2 \cdots \mathbf{T}_{k-1}).
\end{aligned}$$

By substituting the first of these equations into the second, then the second into the third, ..., the $(k-1)$th into the $k$th, we obtain

$$\begin{aligned}
\mu(\mathscr{S}\mathbf{T}) = \mu(\mathscr{S}\mathbf{T}_1 \cdots \mathbf{T}_k) &= |\det E_k| \cdots |\det E_1|\ \mu(\mathscr{S}) \\
&= |\det (E_1 \cdots E_k)|\ \mu(\mathscr{S}) \\
&\qquad\qquad \text{(by 5.6.10 and commutativity in } \mathbf{R}) \\
&= |\det A|\ \mu(\mathscr{S})
\end{aligned}$$

as desired.

Figure 5.7.1

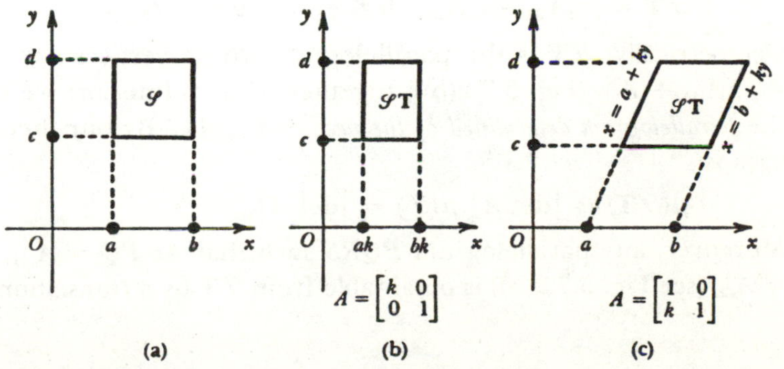

If instead $A$ is singular, then $\mathbf{T} : \mathbf{R}_2 \to \mathbf{R}_2$ is not onto, so that Ran $\mathbf{T}$ has dimension $\leq 1$. Hence, $\mathscr{S}\mathbf{T}$ is a subset of a line and has area 0. Since also det $A = 0$ in this case, the theorem reduces to $0 = 0$, q.e.d. This completes the proof when $\mathbf{T}$ is linear. In the more general case, where $\mathbf{T}$ is affine, write $\mathbf{T} = \mathbf{S}\mathscr{T}_{\mathbf{u}}$, where $\mathbf{S}$ is linear; then

$$\mu(\mathscr{S}\mathbf{T}) = \mu[(\mathscr{S}\mathbf{S})\mathscr{T}_{\mathbf{u}}] = \mu(\mathscr{S}\mathbf{S})$$

(since translations preserve area)

$$= |\det \mathbf{S}|\ \mu(\mathscr{S})$$

(by the linear case already proved)

$$= |\det \mathbf{T}|\ \mu(\mathscr{S})$$

(by definition of det $\mathbf{T}$).

The proof (for $n = 2$) is now complete in all cases. ▮

*Comment*: In Section 1.12, we called attention to the technique of proving a general theorem by reducing it to a special case. This technique was used no fewer than three times in the proof above: (1) we reduced the affine case to the linear case; (2) we reduced the general nonsingular case to the elementary case; (3) in treating the elementary case, we reduced the case of a general region to that of a rectangle.

We obtain an interesting application of our theorem on areas by taking the region $\mathscr{S}$ to be the unit square (in $\mathbf{R}_2$) and its interior, bounded by the lines $x = 0$, $x = 1$, $y = 0$, $y = 1$. It is clear that

$$\mathscr{S} = \{c_1\mathbf{e}_1 + c_2\mathbf{e}_2 : 0 \leq c_1 \leq 1, 0 \leq c_2 \leq 1\}$$

and that $\mu(\mathscr{S}) = 1$ (see Fig. 5.7.2(a)). If $\mathbf{T} \in \text{Lin}(\mathbf{R}_2, \mathbf{R}_2)$ is arbitrary and

$$A = \begin{bmatrix} a & b \\ c & d \end{bmatrix}$$

is its matrix, then $\mathbf{T}$ maps $\mathbf{e}_1$ into the vector

$$a\mathbf{e}_1 + b\mathbf{e}_2 = (a, b) = A_{(1)}$$

and $\mathbf{e}_2$ into $(c, d) = A_{(2)}$. Hence,

$$\mathscr{S}\mathbf{T} = \{c_1A_{(1)} + c_2A_{(2)} : 0 \leq c_1 \leq 1, 0 \leq c_2 \leq 1\}.$$

Geometrically, $\mathscr{S}\mathbf{T}$ is the parallelogram whose vertices are $O$, $A_{(1)}$, $A_{(2)}$, $A_{(1)} + A_{(2)}$ (Fig. 5.7.2(b)), together with its interior; we call this the *parallelogram determined by the vectors* $A_{(1)}$, $A_{(2)}$. By our theorem its area is

$$\mu(\mathscr{S}\mathbf{T}) = |\det A|\ \mu(\mathscr{S}) = |\det A|.$$

Moreover, any parallelogram $PQRS$ such that $\mathbf{Ar}\ PQ = A_{(1)}$, $\mathbf{Ar}\ PS = A_{(2)}$ (see Fig. 5.7.2(b)) is obtainable from $\mathscr{S}\mathbf{T}$ by a translation (*why*?)

and, thus, also has area equal to $|\det A|$. This gives us the area of any parallelogram as a determinant once the vertices are known. For example, if

$$P = (2, 4); \qquad Q = (4, 7); \qquad R = (8, 8); \qquad S = (6, 5)$$

(check that this is a parallelogram!), then $\mathbf{Ar}\, PQ = (2, 3)$, $\mathbf{Ar}\, PS = (4, 1)$, and the area of $PQRS$ is

$$\left|\det \begin{bmatrix} 2 & 3 \\ 4 & 1 \end{bmatrix}\right| = |2 - 12| = 10.$$

Similarly, in $\mathbf{R}_n$, the volume of the $n$-dimensional "parallelepiped" determined by $n$ given vectors is $|\det A|$, where $A$ is the $n \times n$ matrix whose rows are the given vectors.

The latter result can be generalized even further: suppose $\mathscr{P}$ is the $m$-dimensional parallelepiped determined by $m$ vectors $\mathbf{u}_1, \ldots, \mathbf{u}_m$ in $\mathbf{R}_n$. Formally,

$$\mathscr{P} = \left\{ \sum_{i=1}^{m} c_i \mathbf{u}_i : 0 \leqslant c_i \leqslant 1, \text{ all } i \right\}.$$

(For example, if $m = 2$ and $n = 3$, then $\mathscr{P}$ is a parallelogram in

Figure 5.7.2

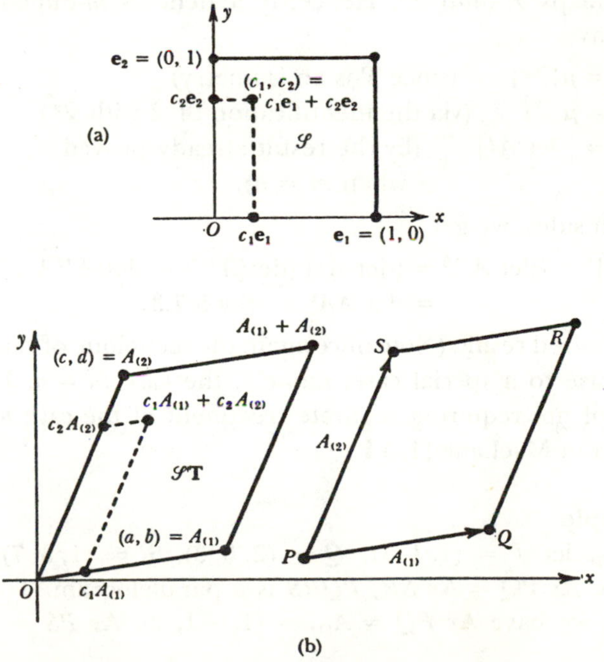

3-space.) It can be shown that the *m-dimensional volume* of $\mathcal{P}$ is equal to

$$(\det AA^t)^{1/2}$$

where $A$ is the $m \times n$ matrix whose rows are $\mathbf{u}_1, \ldots, \mathbf{u}_m$. (Note that $AA^t$ is a *square* matrix and thus has a determinant!) The proof is as follows: for each vector $\mathbf{v} = (c_1, \ldots, c_m)$ in $\mathbf{R}_m$, we define a vector $\mathbf{v}^* \in \mathbf{R}_n$ by

$$\mathbf{v}^* = (c_1, \ldots, c_m, 0, \ldots, 0).$$

By identifying $\mathbf{v}$ with $\mathbf{v}^*$, we identify any region $\mathcal{S} \subseteq \mathbf{R}_m$ with a corresponding region $\mathcal{S}^* \subseteq \mathbf{R}_n$; the $m$-dimensional volumes of $\mathcal{S}$ and $\mathcal{S}^*$ are the same. Now, by Theorem 4.10.14 there exists a linear isometry $\mathbf{T} : \mathbf{R}_n \to \mathbf{R}_n$ which maps the subspace $[\mathbf{u}_1, \ldots, \mathbf{u}_m]$ into $\mathbf{R}_m^*$; thus, there are vectors $\mathbf{v}_1, \ldots, \mathbf{v}_m \in \mathbf{R}_m$ such that $\mathbf{u}_i\mathbf{T} = \mathbf{v}_i^*$. If $M$ is the matrix of $\mathbf{T}$, then by Section 4.4, Exercise 10, the matrix $AM$ is the $m \times n$ matrix whose rows are $\mathbf{v}_1^*, \ldots, \mathbf{v}_m^*$. If $A'$ is the $m \times m$ matrix whose rows are $\mathbf{v}_1, \ldots, \mathbf{v}_m$, then $AM$ is obtained from $A'$ by adding $n - m$ columns of zeros. It easily follows that $(AM)(AM)^t = A'(A')^t$. But $MM^t = I$ by Theorem 4.10.13; hence,

$$\begin{aligned} A'(A')^t &= (AM)(AM)^t = AM(M^tA^t) \\ &= A(MM^t)A^t = AIA^t = AA^t. \end{aligned} \tag{5.7.3}$$

Now, if $\mathcal{Q}$ is the parallelepiped in $\mathbf{R}_m$ determined by $\mathbf{v}_1, \ldots, \mathbf{v}_m$, it is clear that $\mathbf{T}$ maps $\mathcal{P}$ onto $\mathcal{Q}^*$. Hence, if $\mu$ denotes $m$-dimensional volume, we have

$$\begin{aligned} \mu(\mathcal{P}) &= \mu(\mathcal{Q}^*) && \text{(since } \mathbf{T} \text{ is an isometry)} \\ &= \mu(\mathcal{Q}) && \text{(via the identification of } \mathcal{Q} \text{ with } \mathcal{Q}^*) \\ &= |\det A'| && \text{(by the result already proved when } m = n). \end{aligned}$$

Squaring both sides, we get

$$\begin{aligned} [\mu(\mathcal{P})]^2 = (\det A')^2 = (\det A')\,(\det(A')^t) &= \det[A'(A')^t] \\ &= \det AA^t \qquad \text{(by 5.7.3)} \end{aligned}$$

which is the desired result. (Note once again the technique of reducing the general case to a special case, namely, the case $m = n$. For an alternate proof not requiring separate treatment of the case $m = n$, see Birkhoff and MacLane [1].) ∎

**Example**

In $\mathbf{R}_3$, let $P = (1, 1, 1)$, $Q = (2, 0, 3)$, $R = (1, 3, 7)$, $S = (0, 4, 5)$. Since $\mathbf{Ar}\ PQ = \mathbf{Ar}\ SR$, $PQRS$ is a parallelogram. To compute its area, we have $\mathbf{Ar}\ PQ = A_{(1)} = (1, -1, 2)$, $\mathbf{Ar}\ PS = A_{(2)} = (-1, 3, 4)$,

$$AA^t = \begin{bmatrix} 1 & -1 & 2 \\ -1 & 3 & 4 \end{bmatrix} \begin{bmatrix} 1 & -1 \\ -1 & 3 \\ 2 & 4 \end{bmatrix} = \begin{bmatrix} 6 & 4 \\ 4 & 26 \end{bmatrix}$$

$$\det AA^t = 140$$

and, thus, $PQRS$ has area $\sqrt{140}$.

The following result is of some interest in connection with the preceding discussion:

5.7.4 **Theorem.** For any $m \times n$ matrix $A$ over any field $F$,

$$\det AA^t = \sum_B (\det B)^2$$

where the sum is taken over all $m \times m$ submatrices $B$ of $A$.

A proof of 5.7.4 is outlined in Appendix C (with some details omitted); here we only point out that in the special case $m = 2$, $n = 3$, $F = \mathbf{R}$, Theorem 5.7.4 reduces to something fairly familiar. Indeed, letting $\mathbf{u} = (a_1, a_2, a_3)$ and $\mathbf{v} = (b_1, b_2, b_3)$ be the two row vectors of $A$ in this case, the definition of matrix multiplication (equation 4.4.11) implies that

$$AA^t = \begin{bmatrix} \mathbf{u}\cdot\mathbf{u} & \mathbf{u}\cdot\mathbf{v} \\ \mathbf{v}\cdot\mathbf{u} & \mathbf{v}\cdot\mathbf{v} \end{bmatrix}$$

so that

$$\begin{aligned} \det AA^t &= |\mathbf{u}|^2\,|\mathbf{v}|^2 - (\mathbf{u}\cdot\mathbf{v})^2 \\ &= (|\mathbf{u}|\cdot|\mathbf{v}|)^2 - (|\mathbf{u}|\cdot|\mathbf{v}|\cos(\mathbf{u},\mathbf{v}))^2 \qquad (5.7.5) \\ &= (|\mathbf{u}|\cdot|\mathbf{v}|\sin(\mathbf{u},\mathbf{v}))^2. \end{aligned}$$

On the other hand, $A$ has exactly three $2 \times 2$ submatrices $B$, and the determinants of these submatrices are (up to sign) the components of the vector $\mathbf{u} \times \mathbf{v}$ (Sec. 5.4, Exercise 7); hence,

$$\sum_B (\det B)^2 = |\mathbf{u} \times \mathbf{v}|^2. \qquad (5.7.6)$$

Comparing 5.7.5 with 5.7.6, we see that in this case Theorem 5.7.4 reduces to

$$|\mathbf{u} \times \mathbf{v}| = |\mathbf{u}|\,|\mathbf{v}|\sin(\mathbf{u},\mathbf{v}), \qquad (5.7.7)$$

a well-known result. It is also well known, in fact geometrically obvious, that the right side of 5.7.7 is the area of the parallelogram determined by $\mathbf{u}$ and $\mathbf{v}$; see Figure 5.7.3. This agrees with our previous result that the area equals the square root of $\det AA^t$.

**Exercises**

1. (a) Write down the elementary $3 \times 3$ matrices.
   (b) Write down the elementary $n \times n$ matrices.
2. Express the nonsingular matrix
   $$\begin{bmatrix} -1 & 3 & 2 \\ 1 & 1 & -3 \\ -1 & 5 & 1 \end{bmatrix}$$
   explicitly as a product of elementary matrices.
3. Show that the set of all matrices of the form $I\mathscr{C}$, where $\mathscr{C}$ is an elementary *column* operation, is identical to the set of all matrices of the form $\mathscr{R}(I)$, where $\mathscr{R}$ is an elementary row operation.
4. Given that the area inside the circle $x^2 + y^2 = 1$ is $\pi$, find the area inside the ellipse $x^2/a^2 + y^2/b^2 = 1$ by using a result from this section. (SOL)
5. Given any point $P$ on the ellipse $x^2/a^2 + y^2/b^2 = 1$, let $f(P)$ be the maximum area of a triangle inscribed in this ellipse and having one vertex at $P$. Show that $f(P)$ is independent of $P$. (This exercise may be done "informally".) (SUG)
6. Find the area of the triangle with vertices $(0, -2, 4)$, $(2, -2, 5)$, $(6, 6, -1)$ by the method of this section. (Regard a triangle as half a parallelogram.) (ANS)
7. The quantity exhibited in Theorem 5.7.4 has been interpreted in this section as the square of an $m$-dimensional volume. Justify calling this result the "generalized Pythagorean Theorem".

Figure 5.7.3

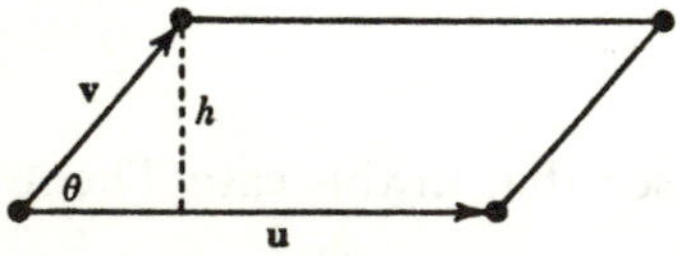


Base = $|\mathbf{u}|$
Height = $h$ = $|\mathbf{v}| \sin\theta$
Area = base × height = $|\mathbf{u}|\,|\mathbf{v}| \sin\theta$

## 5.8 The computation of $A^{-1}$

If $A$ is an $n \times n$ matrix over the field $F$, then $A^{-1}$ exists if and only if $\mathscr{R}(A) = I$ for some sequence $\mathscr{R}$ of elementary operations (Theorem 5.6.8). By Theorem 5.5.5, we then have

$$\mathscr{R}(I) = \mathscr{R}(AA^{-1}) = \mathscr{R}(A) \cdot A^{-1} = IA^{-1} = A^{-1}.$$

In short, *the same sequence $\mathscr{R}$ of row operations which transforms $A$ into $I$ also transforms $I$ into $A^{-1}$*. Since $I$ is the reduced echelon form of $A$, we can find $\mathscr{R}$ by the standard procedure described in the proof of Theorem 5.6.4; this gives us a practical method for finding $A^{-1}$. This method can easily be programmed for computers.

**5.8.1** **Example.** If

$$A = \begin{bmatrix} 0 & 2 & 0 \\ 1 & 1 & -1 \\ 2 & 1 & -1 \end{bmatrix},$$

compute $A^{-1}$.

*Solution.* By the result above, if we apply row operations to the $3 \times 6$ matrix $B = [A \mid I]$ in such a way as to reduce $A$ to $I$, we will end up with the $3 \times 6$ matrix $[I \mid A^{-1}]$. The computations follow:

$$\left[\begin{array}{ccc|ccc} 0 & 2 & 0 & 1 & 0 & 0 \\ 1 & 1 & -1 & 0 & 1 & 0 \\ 2 & 1 & -1 & 0 & 0 & 1 \end{array}\right] \xrightarrow[B_{(1)},\, B_{(2)}]{\text{interchange}} \left[\begin{array}{ccc|ccc} 1 & 1 & -1 & 0 & 1 & 0 \\ 0 & 2 & 0 & 1 & 0 & 0 \\ 2 & 1 & -1 & 0 & 0 & 1 \end{array}\right]$$

$$\xrightarrow[\text{to } B_{(3)}]{\text{add } -2B_{(1)}} \left[\begin{array}{ccc|ccc} 1 & 1 & -1 & 0 & 1 & 0 \\ 0 & 2 & 0 & 1 & 0 & 0 \\ 0 & -1 & 1 & 0 & -2 & 1 \end{array}\right] \xrightarrow[B_{(2)} \text{ by } \frac{1}{2}]{\text{multiply}} \left[\begin{array}{ccc|ccc} 1 & 1 & -1 & 0 & 1 & 0 \\ 0 & 1 & 0 & \frac{1}{2} & 0 & 0 \\ 0 & -1 & 1 & 0 & -2 & 1 \end{array}\right]$$

$$\xrightarrow[\text{add } B_{(2)} \text{ to } B_{(3)}]{\text{add } -B_{(2)} \text{ to } B_{(1)};} \left[\begin{array}{ccc|ccc} 1 & 0 & -1 & -\frac{1}{2} & 1 & 0 \\ 0 & 1 & 0 & \frac{1}{2} & 0 & 0 \\ 0 & 0 & 1 & \frac{1}{2} & -2 & 1 \end{array}\right]$$

$$\xrightarrow[\text{to } B_{(1)}]{\text{add } B_{(3)}} \left[\begin{array}{ccc|ccc} 1 & 0 & 0 & 0 & -1 & 1 \\ 0 & 1 & 0 & \frac{1}{2} & 0 & 0 \\ 0 & 0 & 1 & \frac{1}{2} & -2 & 1 \end{array}\right].$$

Hence,

$$A^{-1} = \begin{bmatrix} 0 & -1 & 1 \\ \frac{1}{2} & 0 & 0 \\ \frac{1}{2} & -2 & 1 \end{bmatrix}.$$

(This should be checked: multiply $A$ by $A^{-1}$ and see if you get $I$.) Note that in our first step we obtained a "1" in the (1, 1) position by an interchange of rows rather than by an operation of Type 3, thus, saving one step. Precise adherence to the steps described in the proof of Theorem 5.6.4 is not always the quickest method, though it has the virtue of being systematic. ∎

*Remark*: Instead of using row operations, we could reduce

$$\left[\frac{A}{I}\right] \text{ to } \left[\frac{I}{A^{-1}}\right]$$

by means of column operations. However, *row and column operations cannot be mixed* when computing $A^{-1}$ by this method; stay with one or the other. The reason why we cannot mix the two lies in the difference in the form of the two equations which appear in Theorem 5.5.5.

Although the method above is practical for computation, it does not give a general formula for the entries in $A^{-1}$ in terms of those in $A$. The following theorem provides such a formula.

**5.8.2** **Theorem.** If $A$ is a nonsingular $n \times n$ matrix over $F$ and if $M(i, j)$ denotes the $(i, j)$ minor of $A$, then

**(5.8.3)** $$(A^{-1})_{ij} = \frac{(-1)^{i+j} \det M(j, i)}{\det A} \qquad \text{(all } i, j).$$

Note that the minor which appears in 5.8.3 is $M(j, i)$, *not* $M(i, j)$ (the indices are reversed). Note also that we are not dividing by zero; $\det A$ is different from zero, since $A$ is nonsingular.

*Proof of Theorem 5.8.2.* Let $c_{ij}$ be the right side of 5.8.3, and let $C = (c_{ij})$; we must show that $C = A^{-1}$. By Theorem 4.7.5, it suffices to prove that $AC = I$. But

**(5.8.4)** $$\begin{aligned}(AC)_{ik} &= \sum_{j=1}^{n} a_{ij}c_{jk} = \sum_{j=1}^{n} a_{ij}\left(\frac{(-1)^{j+k} \det M(k, j)}{\det A}\right) \\ &= \frac{1}{\det A} \sum_{j=1}^{n} (-1)^{k+j} a_{ij} \det M(k, j).\end{aligned}$$

If $k \neq i$, let $B$ be the matrix obtained from $A$ by replacing the $k$th row $A_{(k)}$ by $A_{(i)}$. Then $a_{ij} = b_{kj}$ so that the right side of 5.8.4 reduces to $(1/\det A)$ $(\det B)$ by 5.3.3. (Note that $B$ has the same $(k, j)$ minor as $A$.) Since the $i$th and $k$th rows of $B$ are the same, $\det B = 0$ and, hence, $(AC)_{ik} = 0$. On the other hand, if $k = i$, then the right side of 5.8.4

reduces to (1/det $A$) (det $A$) so that $(AC)_{ii} = 1$. Hence, for *all* $i$ and $k$, $(AC)_{ik} = \delta_{ik}$ and, hence, $AC = I$. ▮

As an illustration, if $A$ is the matrix of Example 5.8.1, then

$$(A^{-1})_{32} = \frac{(-1)^5 \det M(2,3)}{\det A} = \frac{-\det \begin{bmatrix} 0 & 2 \\ 2 & 1 \end{bmatrix}}{\det A} = \frac{4}{-2} = -2$$

which agrees with our previous result.

Theorem 5.8.2 is not normally practical for computation unless $n \leqslant 3$; when $n > 3$ the method of Example 5.8.1 is usually quicker. However, Theorem 5.8.2 is sometimes useful as a theoretical tool; for example, we shall use it to derive Cramer's Rule in Section 5.10.

**Exercises**

1. Prove that if the $2 \times 2$ matrix

$$A = \begin{bmatrix} a & b \\ c & d \end{bmatrix}$$

is nonsingular, then its inverse is

$$\frac{1}{D}\begin{bmatrix} d & -b \\ -c & a \end{bmatrix}$$

where $D = \det A$. Use this result to compute the inverse of

$$\begin{bmatrix} 5 & 3 \\ 6 & 4 \end{bmatrix}.$$

2. Compute the inverse of the matrix

$$A = \begin{bmatrix} 0 & 1 & -1 \\ 2 & 5 & 4 \\ 1 & 2 & 3 \end{bmatrix}$$

by each of the two methods described in this section. Check that your answers agree. (ANS)

3. Prove: if $A$ is a square matrix whose entries are integers, then the entries in $A^{-1}$ are integers if and only if $\det A = \pm 1$.

4. Suppose that the $n \times n$ matrix $A$ has the block form

$$A = \begin{bmatrix} B & D \\ 0 & C \end{bmatrix}$$

where $B$, $C$ are square submatrices of degrees $k$, $n - k$, respectively. Prove: if $B$ and $C$ are nonsingular, then $A$ is nonsingular and $A^{-1}$ has the form

$$A^{-1} = \begin{bmatrix} B^{-1} & E \\ 0 & C^{-1} \end{bmatrix}.$$

5. Deduce from the preceding exercise: if $A$ is a triangular nonsingular matrix with diagonal entries $a_{11}, a_{22}, \ldots, a_{nn}$, then $A^{-1}$ is triangular and has diagonal entries $a_{11}{}^{-1}, a_{22}{}^{-1}, \ldots, a_{nn}{}^{-1}$.
6. Prove: if $A$ is a $2 \times 2$ matrix such that $A^2 = I$ and $\det A = 1$, then $A = \pm I$. Assume that char $F \neq 2$.

## 5.9 The rank of a matrix

Theorem 5.6.8 asserted that the rows (or columns) of a square matrix $A$ are linearly independent if and only if $\det A \neq 0$. This result is actually a special case of a much more general theorem which we now state.

**5.9.1 Theorem.** Let $A$ be any $m \times n$ matrix over the field $F$. Then the following three numbers are equal:

(a) The dimension of the row space of $A$.

(b) The dimension of the column space of $A$.

(c) The greatest integer $d$ such that some $d \times d$ submatrix of $A$ has nonzero determinant.

To prove 5.9.1, let $r$ be the dimension of the row space; let $s$ be the dimension of the column space. By hypothesis, $A$ has a $d \times d$ submatrix $B$ such that $\det B \neq 0$. By 5.6.8, the $d$ rows of $B$ are independent. Hence, the corresponding $d$ rows of $A$ are independent (Sec. 4.5, Exercise 5), so that the row space of $A$ has dimension $\geq d$; that is,

$$r \geq d.$$

Conversely, by definition of $r$, we may choose $r$ independent rows of $A$; let $C$ be the $r \times n$ submatrix of $A$ consisting of these $r$ rows. Then the row space of $C$ also has dimension $r$. By Theorem 5.6.6, $I_r$ is a submatrix of $\mathcal{R}(C)$ for some row operation $\mathcal{R}$; since $I_r$ involves *all* of the rows of $C$, it follows that $I_r = \mathcal{R}(B)$ for some $r \times r$ submatrix $B$ of $C$. Since $\det I_r = 1$, $\det B \neq 0$ so that $C$ (and, hence, $A$) has an $r \times r$ submatrix with nonzero determinant. Hence, $d \geq r$. Since we have already proved $r \geq d$, it follows that $d = r$. Applying this result to the matrix $A^t$ in place of $A$ (and using Theorem 5.4.10), we obtain $d = s$. ∎

We define the *rank* of a matrix $A$ to be the dimension of its row space. By Theorem 5.9.1, of course, the rank could also be defined as the dimension of the column space, or as the integer $d$ in 5.9.1(c). By Theorem 5.6.6, the rank also equals the number of nonzero rows in the reduced echelon form of $A$. The concept of rank is applicable to such diverse topics as systems of simultaneous equations (Sec. 5.10) and singular points of quadric surfaces (Sec. 7.6).

*Remark*: It follows from Theorem 5.5.6 that the rank of a matrix is unchanged under row or column operations. Also, since the row space of $A$ is the column space of $A^t$, it is clear that $A$ and $A^t$ have the same rank.

Since matrices correspond to linear transformations, it is natural to ask whether the rank of a matrix has an interpretation in terms of L.T.'s. The following theorem answers this question in the affirmative.

**5.9.2 Theorem.** Let $V$, $W$ be finite-dimensional vector spaces over $F$; let $\mathbf{T} \in \text{Lin}(V, W)$ and let $A$ be the matrix of $\mathbf{T}$ with respect to ordered bases $(\mathbf{v}_1, ..., \mathbf{v}_m)$ of $V$, $(\mathbf{w}_1, ...\mathbf{w}_n)$ of $W$. Then the rank of $A$ equals the dimension of the range of $\mathbf{T}$.

*Proof.* As in the proof of Theorem 4.7.1, the mapping

$$\mathbf{S} : \sum_{j=1}^{n} c_j \mathbf{w}_j \to (c_1, \ldots, c_n)$$

is an isomorphism of $W \to F_n$ which maps $\mathbf{v}_i\mathbf{T} \to A_{(i)}$. Hence, the spaces $[\mathbf{v}_1\mathbf{T}, ..., \mathbf{v}_m\mathbf{T}]$ (the range of $\mathbf{T}$) and $[A_{(1)}, \ldots, A_{(m)}]$ (the row space of $A$) are isomorphic and, thus, have the same dimension. ∎

We define the *rank of* $\mathbf{T}$ to be the rank of the corresponding matrix $A$; thus, by Theorem 5.9.2,

$$\text{rank } \mathbf{T} = \dim(\text{Ran } \mathbf{T}).$$

In particular, the rank of $\mathbf{T}$ depends only on $\mathbf{T}$, not on the choice of basis.

It follows from Theorem 5.9.2 that "similar" matrices (see Sec. 4.8) have the same rank. Indeed, if $A' = P^{-1}AP$, then the results of Section 4.8 show that $A$, $A'$ represent the same linear transformation $\mathbf{T}$ with respect to different bases. Hence, rank $A$ = rank $\mathbf{T}$ = rank $A'$.

An interesting consequence of Theorem 5.9.1 which we will need later is

**5.9.3 Theorem.** Let $F$, $L$ be fields such that $F$ is a subfield of $L$; let $n$ be a positive integer. If $\mathbf{u}_1, ..., \mathbf{u}_k$ are linearly independent vectors in $F_n$, then these vectors remain independent when regarded as elements of $L_n$. (Note that $F_n \subseteq L_n$, since $F \subseteq L$.)

*Proof.* Let $A$ be the $k \times n$ matrix whose row vectors are $\mathbf{u}_1, \ldots, \mathbf{u}_k$. Then the $\mathbf{u}_i$ are independent if and only if the row space of $A$ has dimension $k$; that is (by Theorem 5.9.1), if and only if some $k \times k$ submatrix of $A$ has nonzero determinant. But the value of the determinant is the same whether regarded as an element of $F$ or of $L$. ∎

**5.9.4** **Corollary.** If $F$ is a subfield of $L$ and $W'$ is a subspace of $L_n$, then $W = W' \cap F_n$ is a subspace of $F_n$ and

(5.9.5) $$\dim_F W \leqslant \dim_L W'.$$

*Proof.* That $W$ is a subspace of $F_n$ is an easy consequence of Theorem 2.2.4. By Theorem 5.9.3, any basis of $W$ over $F$ remains independent when regarded as a subset of $W'$; hence, 5.9.5 holds. ∎

**Exercises**

1. In $\mathbf{R}_5$, find the dimension of the subspace
$$W = [(0, 2, 1, 0, 3), (2, 0, 1, 1, 1), (4, 2, 3, 2, 5), (-2, 4, 1, -1, 5)]$$
by using the fact that the rank of a matrix is unchanged by elementary operations. (ANS)
2. In the proof of Theorem 5.9.1, the proof of the inequality $d \geqslant r$ depended on Theorem 5.6.6. Construct an alternate proof of this inequality using Theorems 2.6.15 and 5.6.8 without referring to 5.6.6. (Of course, since 5.6.8 itself depends on 5.6.6, we don't really gain much by the alternate approach, except variety.)
3. Prove: if $A$, $B$ are matrices over $F$ such that the product $AB$ exists, then (a) rank $AB \leqslant$ rank $B$; (b) rank $AB \leqslant$ rank $A$. (Part (a) is an immediate consequence of an exercise in Sec. 4.6.) (SOL)
4. Prove: if $A$ is any $m \times n$ matrix over $F$ and $B$ is any $n \times p$ matrix over $F$, then
$$\text{rank } AB \geqslant \text{rank } A + \text{rank } B - n.$$
(This exercise is not easy. *Suggestion*: Let the matrices correspond to L.T.'s and use Theorem 4.6.7.) Note that this provides a *lower* bound for the rank of $AB$, in contrast to Exercise 3 which provides *upper* bounds. (SOL)

5. If $AB = 0$, Exercise 4 implies that rank $A$ + rank $B \leq n$. Show that the latter inequality could also be deduced from a theorem in Sec. 3.8, at least when $F = \mathbf{R}$. (SUG)

*6. Deduce from previous exercises that if $A$ is square and nonsingular, then (a) rank $AB$ = rank $B$ whenever $AB$ is defined, and (b) rank $BA$ = rank $B$ whenever $BA$ is defined. (SUG)

7. (a) Let $\mathbf{S}, \mathbf{T} \in \text{Lin}(V, V)$, where dim $V$ is finite. Prove: if $\mathbf{ST} = \mathbf{TS}$, then rank $\mathbf{S}$ + rank $\mathbf{ST}^2 \geq 2$ rank $\mathbf{ST}$.
(*Suggestion*: Apply Theorem 4.6.7 to the spaces Ran $\mathbf{S}$, Ran $\mathbf{ST}$ in place of $V$.) Compare with Exercise 4.
(b) Show, by means of a counterexample, that the result of part (a) is false if the hypothesis $\mathbf{ST} = \mathbf{TS}$ is omitted.

8. (a) If $\mathbf{u}, \mathbf{v} \in F_3$, use a theorem of this section to prove that the cross product $\mathbf{u} \times \mathbf{v}$ (as defined in Sec. 5.4, Exercise 7) is zero if and only if the vectors $\mathbf{u}, \mathbf{v}$ are linearly dependent. Interpret geometrically in the case $F = \mathbf{R}$. (This result will be generalized in Sec. 5.14.)
(b) If $\mathbf{u}, \mathbf{v}$ are independent vectors in $\mathbf{R}_3$ and $W = [\mathbf{u}, \mathbf{v}]$, deduce from part (a) that the vector $\mathbf{u} \times \mathbf{v}$ constitutes a basis of the subspace $W^\perp$ of $\mathbf{R}_3$.

9. If $a \to a^*$ is a ring isomorphism of $F \to F$, and if $A = (a_{ij})$ is any matrix over $F$, prove that the matrix $A^* = (a_{ij}^*)$ has the same rank as $A$.

10. For any field $F$, let $F_\infty$ be the vector space of all infinite sequences $(c_1, c_2, c_3, \ldots)$, such that the $c_i$ lie in $F$. (The vector space operations are defined in the usual way.) Prove that Theorem 5.9.3 remains valid if $n$ is replaced by $\infty$. (SOL)

## 5.10 Systems of linear equations

We consider in this section the general system of $m$ simultaneous linear equations in $n$ unknowns:

$$\begin{aligned} a_{11}x_1 + a_{12}x_2 + \cdots + a_{1n}x_n &= c_1 \\ a_{21}x_1 + a_{22}x_2 + \cdots + a_{2n}x_n &= c_2 \\ &\cdots \\ a_{m1}x_1 + a_{m2}x_2 + \cdots + a_{mn}x_n &= c_m \end{aligned} \tag{5.10.1}$$

where the scalars $a_{ij}$ and $c_i$ belong to an arbitrary field $F$. Our object is (1) to show how to find the solutions $(x_1, \ldots, x_n)$ of 5.10.1, and (2) to discuss the nature of the set of all solutions.

We begin by interpreting 5.10.1 as a matrix equation. Let $A$ be the $m \times n$ matrix $(a_{ij})$ and let $X$ and $C$ be the column matrices

$$X = \begin{bmatrix} x_1 \\ \vdots \\ x_n \end{bmatrix}; \quad C = \begin{bmatrix} c_1 \\ \vdots \\ c_m \end{bmatrix}.$$

Then the $i$th equation in 5.10.1 may be written $A_{(i)} \cdot X^{(1)} = c_i$; that is, $(AX)_{i1} = C_{i1}$. Hence, the entire system 5.10.1 is the same as the single matrix equation

(5.10.2) $AX = C.$

The standard method of solving 5.10.2 (usually called *Gaussian elimination*) is based on the use of elementary row operations. By Theorem 5.6.4(b), there exists a finite sequence $\mathcal{R}$ of elementary row operations such that the matrix $A^* = \mathcal{R}(A)$ is a reduced echelon matrix. Let $C^* = \mathcal{R}(C)$. By applying $\mathcal{R}$ to both sides of 5.10.2, we get

(5.10.3) $\mathcal{R}(AX) = \mathcal{R}(C) = C^*.$

But by Theorem 5.5.5, $\mathcal{R}(AX) = [\mathcal{R}(A)]X = A^*X$, so that 5.10.3 is the same as

(5.10.4) $A^*X = C^*$

which has the same form as 5.10.2 except that $A$, $C$ have been replaced by $\mathcal{R}(A)$, $\mathcal{R}(C)$, respectively; equivalently, *the* $m \times (n + 1)$ *matrix* $B = [A \mid C]$ *has been replaced by* $\mathcal{R}(B)$. Moreover, since every row operation is reversible, we can go the other way and obtain 5.10.2 from 5.10.4. Hence, 5.10.2 and 5.10.4 are equivalent; a column vector $X$ will satisfy 5.10.2 if and only if it satisfies 5.10.4. The advantage of 5.10.4 is that the form of $A^*$ is particularly simple.

**5.10.5 Example.** Solve the system

(5.10.6)
$$\begin{aligned} x_1 + 2x_2 - 2x_3 - x_4 \phantom{{}- 7x_5} &= 3 \\ 2x_1 + 3x_2 - 5x_3 + x_4 - 7x_5 &= 4 \\ x_2 + \phantom{5}x_3 - x_4 + \phantom{7}x_5 &= 2 \\ -x_1 + \phantom{3}x_2 + 5x_3 - x_4 \phantom{{}- 7x_5} &= 3. \end{aligned}$$

*Solution.* Here we have

$$[A \mid C] = \left[\begin{array}{rrrrr|r} 1 & 2 & -2 & -1 & 0 & 3 \\ 2 & 3 & -5 & 1 & -7 & 4 \\ 0 & 1 & 1 & -1 & 1 & 2 \\ -1 & 1 & 5 & -1 & 0 & 3 \end{array}\right].$$

By applying row operations, we reduce $[A \mid C]$ to

(5.10.7) $$[A^* \mid C^*] = \left[\begin{array}{ccccc|c} 1 & 0 & -4 & 0 & 1 & -1 \\ 0 & 1 & 1 & 0 & -2 & 2 \\ 0 & 0 & 0 & 1 & -3 & 0 \\ 0 & 0 & 0 & 0 & 0 & 0 \end{array}\right].$$

(We have omitted the detailed computations; you should fill them in. The procedure is the standard one described in Sec. 5.6.) Hence, the system 5.10.6 is equivalent to the system

(5.10.8) $$\begin{aligned} x_1 \quad - 4x_3 \quad + x_5 &= -1 \\ x_2 + x_3 \quad - 2x_5 &= 2 \\ x_4 - 3x_5 &= 0 \\ 0 &= 0. \end{aligned}$$

The equation $0 = 0$ is clearly superfluous and can be omitted. (Note that if the last entry in the column $C^*$ had been something other than zero, say, $c_4^* \neq 0$, then the last equation 5.10.8 would have been $0 = c_4^*$, impossible. In such a case, 5.10.8 and, hence, also 5.10.6, would have had no solutions.) We may thus work only with the three nonzero equations 5.10.8; they correspond to the three nonzero rows of $A^*$. By Theorem 5.6.6, we know that the $3 \times 3$ identity matrix will be a submatrix of these three rows; we locate it (by inspection of 5.10.7) in the first, second, and fourth columns, which correspond to the unknowns $x_1, x_2, x_4$. Our next step is to transfer the terms involving the *other* unknowns ($x_3$ and $x_5$) to the right side of 5.10.8. Doing so, we obtain

$$\begin{aligned} x_1 &= 4x_3 - x_5 - 1 \\ x_2 &= -x_3 + 2x_5 + 2 \\ x_4 &= 3x_5 \end{aligned}$$

which expresses $x_1$, $x_2$, and $x_4$ in terms of $x_3$ and $x_5$. It is clear that if $x_3$ and $x_5$ are given arbitrary values, $x_1$, $x_2$, and $x_4$ are then uniquely determined. Conversely, any values of the $x_j$ which satisfy these three equations must also satisfy 5.10.8 and, hence, 5.10.6. Thus, the solution to 5.10.6 is finally given by

$$\begin{aligned} x_3 &= \text{any value} \\ x_5 &= \text{any value} \\ x_1 &= 4x_3 - x_5 - 1 \\ x_2 &= -x_3 + 2x_5 + 2 \\ x_4 &= 3x_5. \end{aligned}$$

This solution may be written compactly in the vector form

(5.10.9) $$X = (4x_3 - x_5 - 1, -x_3 + 2x_5 + 2, x_3, 3x_5, x_5).$$ ∎

The procedure used in the above example is perfectly general. To solve any system 5.10.1, first reduce $[A \mid C]$ to $[A^* \mid C^*]$ by row operations, where $A^*$ is a reduced echelon matrix. (*Note: Column operations cannot be used. Why not?*) If $A$ has rank $r$, then the nonzero rows of $A^*$ will be the first $r$ rows; the last $m - r$ rows of $A^*$ are zero. If any of the last $m - r$ entries in $C^*$ are nonzero, then the system has no solutions. If the last $m - r$ entries in $C^*$ are all zero, we can omit the zero rows and work with the $r$ nonzero rows, which give us a system of $r$ equations in $n$ unknowns (in our example, these were the first three equations 5.10.8). We next find (by inspection) $r$ columns of $A^*$ which contain the $r \times r$ identity submatrix; these columns exist by Theorem 5.6.6. The $n - r$ unknowns $x_j$ corresponding to the *other* columns are then transferred to the right side of our $r$ equations, enabling us to express $r$ of the unknowns in terms of the other $n - r$ unknowns. The latter $n - r$ unknowns may take on any values.

Since the steps in the above procedure are the same in every example, it is clear that this method of solving equations can easily be programmed for computers.

Returning for a moment to Example 5.10.5, we remark that by using the definitions of addition and scalar multiplication of $n$-tuples, we can express the solution vector 5.10.9 in the form

$$X = x_3\,(4, -1, 1, 0, 0) + x_5\,(-1, 2, 0, 3, 1) + (-1, 2, 0, 0, 0) \tag{5.10.9'}$$

which shows that the set of all solutions of the system 5.10.6 is in fact the plane through the point $(-1, 2, 0, 0, 0)$ which has direction space $[(4, -1, 1, 0, 0,), (-1, 2, 0, 3, 1)]$. More generally, the set of solutions of any system $AX = C$, if nonempty, is a flat (see Theorem 5.10.18 below).

Some further comments:

1. *Condition for existence of a solution.* If we think of the column $C$ and the columns $A^{(1)}, \ldots, A^{(n)}$ of $A$ as vectors in $F_m$, the equations 5.10.1 are the same as the vector equation

$$x_1A^{(1)} + \cdots + x_nA^{(n)} = C. \tag{5.10.10}$$

Hence, 5.10.1 has a solution if and only if there exist scalars $x_j$ satisfying 5.10.10; that is, if and only if $C$ is a linear combination of the columns of $A$.

2. *Condition for uniqueness of a solution.* Assuming that a solution exists, we have seen that the general solution assigns arbitrary values

to $n - r$ of the unknowns, where $r$ is the rank of $A$. Hence, the solution is unique if and only if $n - r = 0$, that is, $r = n$. Since $r$ is the number of independent columns of $A$ and $n$ is the total number of columns, $r = n$ means that the columns of $A$ are independent. Combining this result with the preceding paragraph, we see that an arbitrary system 5.10.1 (equivalently, 5.10.2) has a unique solution if and only if *the columns of A are independent and C is a linear combination of them.*

In the special case where $A$ is square ($m = n$), then by definition the columns of $A$ are independent if and only if $A$ is nonsingular. In this case, the $n$ independent columns belong to $F_n$, and by Corollary 2.6.14(b), they span $F_n$ so that $C$ is automatically a linear combination of them. We conclude:

**5.10.11 Theorem.** If the matrix $A$ is square ($m = n$), then 5.10.1 has a unique solution if and only if $A$ is nonsingular.

Note the following corollary of 5.10.11: if $A$ is square and 5.10.2 has a unique solution for some particular $C$, then 5.10.2 has a unique solution for *every* $C$.

In the nonsingular case, the unique solution can in fact be found explicitly in terms of the entries in $A$, as follows:

**5.10.12 Theorem.** ("Cramer's Rule") If $A$ is square and nonsingular and if $B(j)$ is the matrix obtained from $A$ by replacing the $j$th column of $A$ by the column $C$, then the unique solution of 5.10.1 is given by

$$x_j = \frac{\det B(j)}{\det A} \qquad (j = 1, \ldots, n).$$

*Proof.* Since $A$ is nonsingular, $A^{-1}$ exists. Multiplying 5.10.2 by $A^{-1}$ on the left, we obtain $X = A^{-1}C$. Using the definition of matrix multiplication, this gives

$$\begin{aligned} x_j = X_{j1} &= \sum_{i=1}^{n} (A^{-1})_{ji} C_{i1} \\ &= \sum_{i=1}^{n} (A^{-1})_{ji} c_i \\ &= \frac{1}{\det A} \sum_{i=1}^{n} (-1)^{i+j} c_i \det M(i,j) \end{aligned}$$

by Theorem 5.8.2, where $M(i, j)$ is the $(i, j)$ minor of $A$. The matrix

$B(j)$ has the same $(i, j)$ minor as $A$ (*why*?), and $c_i = (B(j))_{ij}$; hence,

$$x_j = \frac{1}{\det A} \sum_{i=1}^{n} (-1)^{i+j} (B(j))_{ij} \det M(i, j)$$
$$= \frac{1}{\det A} (\det B(j))$$

by Theorem 5.3.2. ▮

Even if $A$ is not square, we can still use Cramer's Rule if the rows of $A$ are independent. Indeed, if the $m$ rows are independent, then $m$ of the columns of $A$ are independent by Theorem 5.9.1. If we transfer to the right side the terms in 5.10.1 which correspond to the other $n - m$ columns, we are left with a nonsingular $m \times m$ matrix on the left side and we may apply Cramer's Rule to it, regarding the right side as a new $C$. For example, given the system

$$\begin{aligned} 2x - \phantom{2}y + 3z &= 4 \\ x + 2y - \phantom{3}z &= 7, \end{aligned}$$

we may transfer the $z$ terms to the right:

$$\begin{aligned} 2x - \phantom{2}y &= 4 - 3z \\ x + 2y &= 7 + \phantom{3}z. \end{aligned}$$

By Theorem 5.10.12, the solution is

$$x = \frac{\det \begin{bmatrix} 4 - 3z & -1 \\ 7 + z & 2 \end{bmatrix}}{\det \begin{bmatrix} 2 & -1 \\ 1 & 2 \end{bmatrix}} = \frac{15 - 5z}{5} = 3 - z$$

$$y = \frac{\det \begin{bmatrix} 2 & 4 - 3z \\ 1 & 7 + z \end{bmatrix}}{\det \begin{bmatrix} 2 & -1 \\ 1 & 2 \end{bmatrix}} = \frac{10 + 5z}{5} = 2 + z$$

$z =$ any value.

Of course, we would have obtained the same solution by the method of reduction by row operations.

*Remark*: The above argument shows that *if the rows of $A$ are independent, then the system $AX = C$ always has a solution*. A second proof of this fact is as follows: if the $m$ rows are independent, then the column space of $A$ has dimension $m$; hence, it is all of $F_m$; hence, it contains $C$; hence, the previously stated "Condition for existence of a solution" is satisfied. Still a third proof (using linear transformations!) goes as follows: let $\mathbf{T} \in \mathrm{Lin}(F_n, F_m)$ be the mapping whose matrix with respect to the canonical bases of $F_n$ and $F_m$ is $A^t$. Then for any column vector $X$, $X^t$ is a row vector and by Theorem 4.4.14, we have

$$\mathbf{T} : X^t \to X^t A^t = (AX)^t.$$

Hence, if $S$ is the set of solutions of 5.10.2 in $F_n$, we have

$$\begin{aligned} S &= \{X \in F_n : AX = C\} = \{X \in F_n : (AX)^t = C^t\} \\ &= \{X \in F_n : X^t\mathbf{T} = C^t\} \qquad \text{(5.10.13)} \\ &= \{X \in F_n : X\mathbf{T} = C\} \end{aligned}$$

(since a column matrix and its transpose represent the same *vector* even though they are different as *matrices*!). That is, $S$ consists of the pre-images of $C$ under $\mathbf{T}$. If the rows of $A$ are independent, then the columns of $A^t$ are independent; hence, $\mathbf{T}$ is onto (Theorem 4.7.2); hence, $C$ has at least one pre-image under $\mathbf{T}$, that is, 5.10.2 has at least one solution! (A similar argument using 4.7.1 instead of 4.7.2 shows that if the *columns* of $A$ are independent, then 5.10.2 has *at most* one solution. Thes arguments beautifully illustrate the power of the "linear transformation" concept in linear algebra.)

**Homogeneous equations**

The system 5.10.1 (or 5.10.2) is called *homogeneous* if $C = 0$. In this case, it is obvious that $X = \mathbf{0}$ is a solution of 5.10.2. As shown above, the solution is unique if and only if the columns of $A$ are independent; thus, if the columns are dependent, the system $AX = 0$ has at least one nonzero solution. In the case where $A$ is square, $AX = 0$ has a nonzero solution if and only if $A$ is singular.

The following theorem gives more precise information about the set of solutions of a homogeneous system:

**5.10.14 Theorem.** If $A$ is an $m \times n$ matrix (over $F$) of rank $r$, and if $S$ is the set of all solutions $X$ ( in $F_n$) of the system $AX = 0$, then $S$ is a subspace of $F_n$ of dimension $n - r$.

*Proof.* As shown above, if $\mathbf{T} \in \mathrm{Lin}(F_n, F_m)$ has matrix $A^t$, then $S = \{X \in F_n : X\mathbf{T} = \mathbf{0}\}$ (cf. 5.10.13); that is, $S = \mathrm{Nul}\,\mathbf{T}$, which is a subspace of dimension $n - r$ by Theorems 4.6.7 and 5.9.2. ∎

*Alternate proof.* As we have seen earlier, arbitrary values may be assigned to $n - r$ of the unknowns $x_j$ (say, $x_{h_1}, \ldots, x_{h_{n-r}}$) and then the values of the remaining $x$'s are uniquely determined. Hence, the (linear) mapping $\mathbf{M} : S \to F_{n-r}$ defined by

$$(x_1, \ldots, x_n) \to (x_{h_1}, \ldots, x_{h_{n-r}})$$

is one-to-one and onto; that is, $\mathbf{M}$ is an isomorphism. If $\mathbf{u}_i = (x_1, \ldots, x_n)$ is the unique solution (element of $S$) such that

$$x_{h_i} = 1; \qquad x_{h_j} = 0 \qquad (j \neq i, 1 \leqslant j \leqslant n - r)$$

then $\mathbf{M}$ maps $\mathbf{u}_i \to \mathbf{e}_i$; since $\{\mathbf{e}_1, \ldots, \mathbf{e}_{n-r}\}$ is a basis of $F_{n-r}$, $\{\mathbf{u}_1, \ldots, \mathbf{u}_{n-r}\}$ is a basis of $S$, and Theorem 5.10.14 follows. (Note that this proof exhibits a specific basis.) ▮

Theorem 5.10.14 has a geometric interpretation which we have come across in an earlier chapter: let $W$ be any $r$-dimensional subspace of $F_n$, let $\{\mathbf{w}_1, \ldots, \mathbf{w}_r\}$ be a basis of $W$, and let $A$ be the $r \times n$ matrix whose row vectors are $\mathbf{w}_1, \ldots, \mathbf{w}_r$. (The matrix $A$ then has rank $r$, since its $r$ rows are independent.) The $i$th equation 5.10.1 is $A_{(i)} \cdot X = c_i$; if $C = \mathbf{0}$, this becomes $A_{(i)} \cdot X = \mathbf{w}_i \cdot X = 0$ for all $i$. In other words, the elements of $S$ (the solutions of $AX = 0$) are precisely *the vectors which are perpendicular to all of the* $\mathbf{w}_i$. (Just as in the case $F = \mathbf{R}$, we say that two vectors $\mathbf{u}, \mathbf{v}$ in $F_n$ are "perpendicular" if $\mathbf{u} \cdot \mathbf{v} = 0$.) If we define

**(5.10.15)** $W^\perp = \{\mathbf{v} \in F_n : \mathbf{v} \perp \mathbf{w} \text{ for all } \mathbf{w} \in W\}$

(this agrees with our previous definition of $W^\perp$ when $F = \mathbf{R}$), it follows that $S = W^\perp$. Thus, Theorem 5.10.14 implies that *if $W$ is an $r$-dimensional subspace of $F_n$, then $W^\perp$ is an $(n - r)$-dimensional subspace of $F_n$.* In short, the first assertion of Theorem 3.8.18(b) remains true if we replace $E$ by $F_n$. Theorem 3.8.18(d) remains true also; indeed, the definition 5.10.15 implies immediately that $W \subseteq (W^\perp)^\perp$, and we have

$$\dim (W^\perp)^\perp = n - \dim W^\perp = n - (n - \dim W) = \dim W;$$

hence, $(W^\perp)^\perp = W$. It is of some interest that parts (a) and (c) of 3.8.18, as well as the assertion about bases in part (b), do *not* remain true when $E$ is replaced by $F_n$; see Exercise 12, this section.

Theorem 5.10.14 shows that the solutions of any homogeneous system form a subspace. Conversely, we have

**5.10.16 Theorem.** Every subspace of $F_n$ is the set of solutions of some homogeneous system. Specifically, if $S$ is a $k$-dimensional subspace of $F_n$, then there exists an $m \times n$ matrix $A$ over $F$, where $m$

$= n - k$, such that (a) the rows of $A$ are independent, and (b) $S$ is the set of all solutions $X \in F_n$ of the system $AX = 0$.

*Proof.* Let $W = S^{\perp}$, and then choose $A$ as in the preceding discussion so that the set of solutions of $AX = 0$ is $W^{\perp}$. But $W^{\perp} = (S^{\perp})^{\perp} = S$; moreover, $A$ was chosen to have independent rows, and the number of rows of $A$ was $r = \dim W = n - \dim S = n - k$. ∎

As an illustration, take $k = 2$, $n = 3$, $m = n - k = 1$. Theorem 5.10.16 then asserts that every plane through the origin in $F_3$ is the set of solutions of some system $AX = 0$, where $A$ is a $1 \times 3$ matrix whose single row is linearly independent (i.e., nonzero). That is, every such plane has an equation of the form

**(5.10.17)** $a_1x_1 + a_2x_2 + a_3x_3 = 0$

where $a_1, a_2, a_3$ are not all zero. (Does this look familiar?) Conversely, every such equation represents a plane through the origin, by 5.10.14.

The set of solutions of a *nonhomogeneous* system may also be interpreted geometrically, as the following theorem shows.

**5.10.18 Theorem.** Let $A$ be an $m \times n$ matrix (over $F$) of rank $r$, and let $S$ be the set of all solutions $X \in F_n$ of the system $AX = C$. If $S$ is nonempty, then $S$ is an $(n - r)$-dimensional flat whose direction space is the set of all solutions of $AX = 0$.

*Proof.* Let $S_0$ be the set of solutions of $AX = 0$, and assume $S \neq \varnothing$ so that $S$ contains at least one vector $X_0$. Then $AX_0 = C$. Hence, for any $X \in F_n$,

$$X \in S \Leftrightarrow AX = C \Leftrightarrow AX = AX_0 \Leftrightarrow A(X - X_0) = 0$$
$$\Leftrightarrow X - X_0 \in S_0 \Leftrightarrow X \in S_0 + X_0.$$

Hence, $S = S_0 + X_0$. Since $S_0$ is a subspace of dimension $n - r$, $S_0 + X_0$ is a flat of dimension $n - r$ whose direction space is $S_0$. ∎

Conversely, we have

**5.10.19 Theorem.** If $S$ is any $k$-flat in $F_n$, then there exists an $(n - k) \times n$ matrix $A$ and an $(n - k) \times 1$ matrix $C$ such that (a) the rows of $A$ are independent, and (b) $S$ is the set of all solutions $X \in F_n$ of the system $AX = C$.

*Proof.* Exercise.

It follows from Theorem 5.10.18 (taking $n = 3$, $m = r = 1$, $n - r = 2$) that any equation of the form

**(5.10.20)** $a_1x_1 + a_2x_2 + a_3x_3 = c \qquad (a_1, a_2, a_3 \text{ not all zero})$

represents a plane in $F_3$; in fact, the plane 5.10.17 is the direction space of the plane 5.10.20. (It follows that the planes 5.10.17 and 5.10.20 are parallel.) Conversely, every plane in $F_3$ has an equation of the form 5.10.20, by Theorem 5.10.19 (take $k = 2$, $n = 3$, $n - k = 1$).

Similarly, the case $k = 1$, $n = 3$ of Theorem 5.10.19 asserts that every line in $F_3$ is the set of solutions of some pair of equations of the form

$$\begin{aligned} a_1x_1 + a_2x_2 + a_3x_3 &= c_1 \\ b_1x_1 + b_2x_2 + b_3x_3 &= c_2. \end{aligned}$$

Since each of these equations separately represents a plane, it follows that every line in $F_3$ is the intersection of two planes.

### Exercises

1. Solve each of the following systems over the field **R**, using the method of Gaussian elimination. In each case, express the set of all solutions as a specific flat, as was done for the solution to Example 5.10.5 in the text. (ANS)

(a)
$$\begin{aligned} 2x_1 - 3x_2 + 7x_3 + 7x_4 &= -4 \\ x_1 + x_2 - 4x_3 + x_4 &= 3 \\ x_2 - 3x_3 - x_4 &= 2. \end{aligned}$$

(b)
$$\begin{aligned} x_1 - x_2 + 5x_3 - x_4 &= 3 \\ -2x_1 + x_2 + x_3 - x_4 &= -5 \\ 2x_2 - x_3 - x_4 &= -2. \end{aligned}$$

(c)
$$\begin{aligned} x_1 - x_2 + x_3 &= 1 \\ 2x_1 + x_2 &= 3 \\ x_1 - 4x_2 + 3x_3 &= 6. \end{aligned}$$

(d)
$$\begin{aligned} x_1 - x_2 + 2x_3 &= 5 \\ x_1 + x_2 + x_3 &= 5 \\ 3x_1 + 2x_2 - x_3 &= -3. \end{aligned}$$

2. Use Cramer's Rule to solve each of the following systems over **R**:

(a)
$$\begin{aligned} x + 3y - z + w &= 5 \\ 2x + 6y + 3z - w &= 9. \end{aligned}$$

(b)
$$\begin{aligned} x - y + z &= -3 \\ 2x + y + z &= 4 \\ x + 3y - 2z &= 7. \end{aligned}$$
(ANS)

3. In $\mathbf{R}_3$, let $W_1$ be the subspace $[(1, 1, 1), (2, 0, 1)]$ and let $W_2$ be the subspace $[(0, 3, 1), (1, 0, 1)]$. Show that $\dim(W_1 \cap W_2) = 1$ and find a vector which spans $W_1 \cap W_2$. (This problem involves solving a system of three equations in four unknowns.) (ANS)
4. Prove Theorem 5.10.19.
5. Using arguments like those in the section, prove that the general equation of a line in $F_2$ has the form $a_1x + a_2y = c$ ($a_1$, $a_2$ not both 0).
6. Let $S$ be the 2-dimensional subspace $[(1, 2, -3, 0), (4, 2, -2, 1)] \subseteq \mathbf{R}_4$. Find a system $AX = 0$ of two homogeneous equations in four unknowns, the set of solutions of which is precisely $S$. (SUG)
7. Let $S$ be the plane in $\mathbf{R}_4$ determined by the points $(1, 1, -1, 2)$, $(2, 3, -4, 2)$, $(5, 3, -3, 3)$. Find a system $AX = C$ of two equations in four unknowns, the set of solutions of which is $S$.
8. Prove: if $\mathbf{u} = (a_1, a_2, a_3)$ and $\mathbf{v} = (b_1, b_2, b_3)$ are *independent* vectors in $F_3$, then the solutions $(x, y, z)$ of the system
$$\begin{aligned} a_1x + a_2y + a_3z &= 0 \\ b_1x + b_2y + b_3z &= 0 \end{aligned}$$
are precisely the scalar multiples of $\mathbf{u} \times \mathbf{v}$, where $\mathbf{u} \times \mathbf{v}$ is as defined in Section 5.4, Exercise 7. (This is usually the simplest way of solving two homogeneous equations in three unknowns. Compare with Sec. 5.9, Exercise 8.) (SOL)
9. Solve the system
$$\begin{aligned} 2x - y + 5z &= 0 \\ 3x + 2y - 2z &= 0 \end{aligned}$$
by using the method of the preceding exercise.
10. Let $c_0, \ldots, c_n$ be $n + 1$ *distinct* elements of a field $F$. Prove that the "Vandermonde matrix"
$$A = \begin{bmatrix} 1 & c_0 & c_0^2 & \cdots & c_0^n \\ 1 & c_1 & c_1^2 & \cdots & c_1^n \\ & & & \cdots & \\ 1 & c_n & c_n^2 & \cdots & c_n^n \end{bmatrix}$$
is nonsingular. (*Suggestion*: Show that the existence of a non-

zero solution of the system $AX = 0$ would contradict a certain theorem on polynomials. If you're lazy, see Mostow [11], Sec. 11.3 for an alternate proof.)

11. Let $c_0, \ldots, c_n$ be $n + 1$ distinct elements of $F$, and let $y_0, \ldots, y_n$ be arbitrary elements of $F$ (not necessarily distinct). Prove: there exists a unique polynomial $f \in F[x]$ such that

    (a) $f(c_i) = y_i \quad (i = 0, 1, \ldots, n)$; and
    (b) $\deg f \leqslant n$ or $f = 0$. (SUG)

12. The statement, "If $W$ is a subspace of $F_n$, then the sum $W + W^{\perp}$ is direct", although true when $F = \mathbf{R}$, is not true for arbitrary fields $F$. Find a counterexample: (a) when $F = \mathbf{C}$ (the complex numbers); (b) when $F = \mathbf{Z}_{(5)}$.

13. Find the fallacy in the following "proof" of the (false) assertion that the set of solutions of $AX = C$ is always a subspace.

    "Proof": In the "Alternate proof" of Theorem 5.10.14, the argument showing that the mapping $\mathbf{M} : S \to F_{n-r}$ was one-to-one and onto did not depend on the fact that the system $AX = 0$ is homogeneous; it remains valid for the system $AX = C$. Moreover, $\mathbf{M}$ is linear by Section 4.1, Exercise 4. Hence, $S$ is still isomorphic to $F_{n-r}$ and must therefore be a vector space.

## 5.11 Eigenvectors and eigenvalues

Suppose we know that a given linear transformation $\mathbf{T} : \mathbf{R}_3 \to \mathbf{R}_3$ is a rotation about some axis, but we do not know the axis. In order to find it, we observe that a vector $\mathbf{u}$ lies on the axis if and only if it is fixed (mapped into itself) by $\mathbf{T}$; thus, it suffices to find the vectors $\mathbf{u}$ such that $\mathbf{uT} = \mathbf{u}$ (see Fig. 5.11.1). More generally, given a linear

Figure 5.11.1

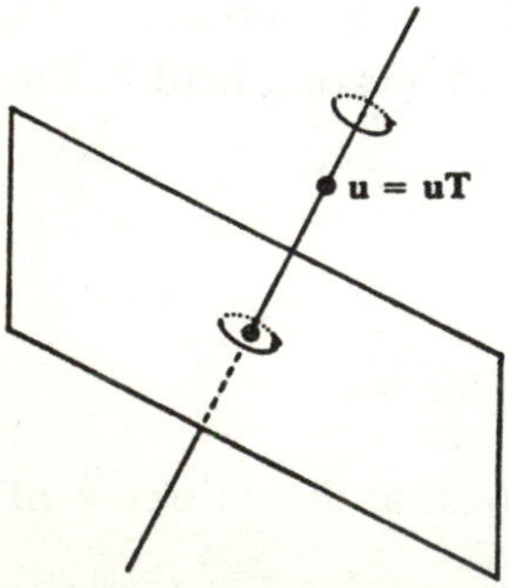

transformation $\mathbf{T} : V \to V$, we sometimes wish to find the vectors $\mathbf{u}$ such that

$$\mathbf{uT} = \lambda\mathbf{u}$$

for some scalar $\lambda$; such vectors $\mathbf{u}$ are called *eigenvectors* of $\mathbf{T}$. In this section we shall show how to find all eigenvectors of a given linear transformation. (Geometrically this is equivalent to finding the lines through the origin which are mapped into themselves by $\mathbf{T}$; see Fig. 5.11.2.) Throughout the section, $V$ will denote a *finite-dimensional* vector space over a field $F$.

**5.11.1 Definition.** Let $\mathbf{T} \in \text{Lin}(V, V)$.

(a) A scalar $\lambda \in F$ is called an *eigenvalue* of $\mathbf{T}$, if there exists a *nonzero* vector $\mathbf{u} \in V$ such that $\mathbf{uT} = \lambda\mathbf{u}$.

(b) If $\lambda \in F$, all vectors $\mathbf{u} \in V$ such that $\mathbf{uT} = \lambda\mathbf{u}$ are called $\lambda$-*eigenvectors* of $\mathbf{T}$. By an *eigenvector* of $\mathbf{T}$ we mean a $\lambda$-eigenvector of $\mathbf{T}$ for some $\lambda$.

The reason we require a *nonzero* vector $\mathbf{u}$ in Definition 5.11.1(a) is that when $\mathbf{u} = \mathbf{0}$ the equation $\mathbf{uT} = \lambda\mathbf{u}$ is trivially satisfied by all scalars $\lambda$. That is, for every $\lambda$ the vector $\mathbf{0}$ is a $\lambda$-eigenvector. We call $\lambda$ an eigenvalue only if there is some other $\lambda$-eigenvector besides $\mathbf{0}$.

The next two theorems will provide the solution to the "eigenvector problem": Theorem 5.11.2 tells us what the $\lambda$-eigenvectors are for a given $\lambda$, while 5.11.3 tells us how to find the eigenvalues $\lambda$.

**5.11.2 Theorem.** Let $\mathbf{T} \in \text{Lin}(V, V)$. If $\lambda \in F$, then the set of all $\lambda$-eigenvectors of $\mathbf{T}$ is equal to the subspace $\text{Nul}(\mathbf{T} - \lambda\mathbf{I})$. (We shall refer to this subspace as the $(\mathbf{T}, \lambda)$-*eigenspace*.)

Figure 5.11.2

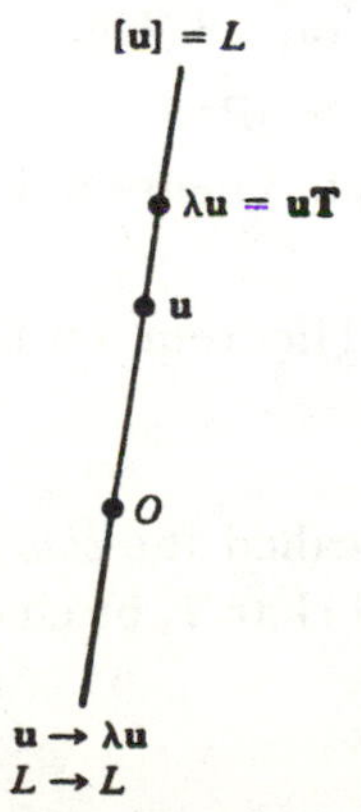

*Proof.* $\mathbf{u}$ is a $\lambda$-eigenvector $\Leftrightarrow \mathbf{uT} = \lambda\mathbf{u}$

$$\Leftrightarrow \mathbf{uT} - \lambda\mathbf{u} = \mathbf{0}$$
$$\Leftrightarrow \mathbf{uT} - \mathbf{u}(\lambda\mathbf{I}) = \mathbf{0}$$
$$\Leftrightarrow \mathbf{u}(\mathbf{T} - \lambda\mathbf{I}) = \mathbf{0}$$
$$\Leftrightarrow \mathbf{u} \in \mathrm{Nul}(\mathbf{T} - \lambda\mathbf{I}).$$ ∎

(*Question*: What idea used in this proof has also arisen in several previous proofs?)

**5.11.3 Theorem.** Let $n = \dim V$, let $\mathbf{T} \in \mathrm{Lin}(V, V)$, and let $x$ be an indeterminate over $F$. If $A$ is the matrix of $\mathbf{T}$ with respect to any basis of $V$, then $\det(A - xI)$ is a polynomial of degree $n$ over $F$; that is,

(5.11.4) $$\det(A - xI) = c_0 + c_1x + \cdots + c_nx^n,$$

where the $c_i$ belong to $F$ and $c_n \neq 0$. The eigenvalues of $\mathbf{T}$ are precisely the roots of the polynomial 5.11.4. Moreover, this polynomial depends only on $\mathbf{T}$, not on the choice of basis; that is, if $A'$ is the matrix of $\mathbf{T}$ with respect to some other basis of $V$, then

(5.11.5) $$\det(A - xI) = \det(A' - xI).$$

The proof that $\det(A - xI)$ is a polynomial of degree $n$ will be given in Section 5.12, where we shall actually obtain an explicit formula for the coefficients $c_i$. The rest of Theorem 5.11.3 is proved as follows: By Definition 5.11.1,

$\lambda$ is an eigenvalue of $\mathbf{T} \Leftrightarrow$ there exists a nonzero $\lambda$-eigenvector

$\Leftrightarrow \mathrm{Nul}(\mathbf{T} - \lambda\mathbf{I}) \neq \{\mathbf{0}\}$ (by 5.11.2)

$\Leftrightarrow \det(A - \lambda I) = 0$ (by 5.6.8)

$\Leftrightarrow \lambda$ is a root of $\det(A - xI)$.

Also, if $A'$ is the matrix of $\mathbf{T}$ with respect to some other basis of $V$, then $A' = P^{-1}AP$ for some $P$ as shown in Section 4.8. Hence,

$$A' - xI = P^{-1}AP - xI = P^{-1}AP - P^{-1}(xI)P$$
(cf. Theorem 4.4.13(c))
$$= P^{-1}(A - xI)P$$

so that the same argument as in the proof of Theorem 5.6.12 shows that $\det(A' - xI) = \det(A - xI)$. ∎

**5.11.6 Definition.** The polynomial 5.11.4 is called the *characteristic polynomial of* $\mathbf{T}$ (or of $A$) and is denoted char $\mathbf{T}$, or char($\mathbf{T}$; $x$). (It is also denoted char $A$ or char($A$; $x$).)

*Remarks*: 1. Although determinants were defined only for matrices with entries in a field (Def. 5.2.1), it is still all right to work with det $(A - xI)$ even though $x$ is an indeterminate and does not belong to the field $F$. The easiest way to justify this is to observe that since $F[x]$ is an integral domain (Sec. 1.16), it is contained in some field (Sec. 1.15); hence, there does exist a field simultaneously containing both $x$ and the entries in $A$.

2. The last part of Theorem 5.11.3 shows that similar matrices $A$, $A'$ have the same characteristic polynomial. Hence, *similar matrices have the same eigenvalues*, where by an "eigenvalue of $A$" we mean a root of char$(A; x)$ (equivalently, an eigenvalue of some transformation $\mathbf{T} \in$ Lin$(V, V)$ which has $A$ as its matrix).

3. Over the field $\mathbf{R}$, every polynomial of odd degree has at least one root (Sec. 1.16); hence, *if $n$ is odd, every linear transformation of $\mathbf{R}_n \to \mathbf{R}_n$ has an eigenvalue*. In particular, this is true when $n = 3$: every L.T. of real 3-space maps some line into itself.

**5.11.7 Example.** If $\mathbf{T} : \mathbf{R}_3 \to \mathbf{R}_3$ is defined by

$$\mathbf{T} : (x, y, z) \to (2x - 2z, 2x, 2x - 2y),$$

find all eigenvalues and eigenvectors of $\mathbf{T}$.

*Solution.* The matrix of $\mathbf{T}$ with respect to the canonical basis is

$$A = \begin{bmatrix} 2 & 2 & 2 \\ 0 & 0 & -2 \\ -2 & 0 & 0 \end{bmatrix}.$$

Since the matrix $I$ has 1's along the main diagonal and zeros elsewhere, the matrix $A - xI$ is obtained from $A$ by subtracting $x$ from all entries along the main diagonal. Thus,

$$\begin{aligned} \text{char}\,(\mathbf{T}; x) = \det(A - xI) &= \det \begin{bmatrix} 2-x & 2 & 2 \\ 0 & -x & -2 \\ -2 & 0 & -x \end{bmatrix} \\ &= (2 - x) \det \begin{bmatrix} -x & -2 \\ 0 & -x \end{bmatrix} - 2 \det \begin{bmatrix} 2 & 2 \\ -x & -2 \end{bmatrix} \\ &= (2 - x)(x^2) - 2(-4 + 2x) \\ &= -x^3 + 2x^2 - 4x + 8 = -(x^2 + 4)(x - 2). \end{aligned}$$

(Note that this is a polynomial of degree 3, as expected!) Since $x^2 + 4$ has no real roots, $\lambda = 2$ is the only real root of char $\mathbf{T}$ and, hence, the only eigenvalue of $\mathbf{T}$. The corresponding eigenvectors $\mathbf{u} = (x, y, z)$

are the elements of Nul(**T** − 2**I**) and can be found by solving the equation **u**(**T** − 2**I**) = **0**, which by Theorem 4.4.14 is equivalent to the matrix equation $\mathbf{u}(A - 2I) = 0$; that is,

$$[x \quad y \quad z]\begin{bmatrix} 0 & 2 & 2 \\ 0 & -2 & -2 \\ -2 & 0 & -2 \end{bmatrix} = [0 \quad 0 \quad 0].$$

Carrying out the matrix multiplication, we obtain a system of three linear equations:

$$\begin{aligned} -2z &= 0 \\ 2x - 2y \phantom{-2z} &= 0 \\ 2x - 2y - 2z &= 0, \end{aligned}$$

whose general solution is

$$x = y = \text{any value}$$
$$z = 0.$$

Hence, the eigenvectors of **T** are the vectors of the form $(k, k, 0)$, $k \in \mathbf{R}$. Equivalently, the eigenspace is the line $[(1, 1, 0)]$. This line is mapped into itself by **T**, since each eigenvector **u** is mapped into 2**u**. ▮

Note that in the above example, **T** has only one eigenvalue. If there were more than one eigenvalue, we would compute the eigenvectors for each eigenvalue separately.

An eigenvalue $\lambda$ of **T** is called an eigenvalue of *multiplicity m*, if it is a root of char **T** of multiplicity $m$; that is, if

$$\text{char}(\mathbf{T}; x) = (x - \lambda)^m g(x),$$

where $g(x)$ is not divisible by $(x - \lambda)$. It will be shown in Chapter 8 that if $\lambda$ is an eigenvalue of multiplicity $m$, then the dimension of the corresponding $(\mathbf{T}, \lambda)$-eigenspace (the space Nul($\mathbf{T} - \lambda\mathbf{I}$)) is less than or equal to $m$. In particular, if $m = 1$, then the eigenspace must be one dimensional (clearly, not 0 dimensional, since for every eigenvalue there is a *nonzero* eigenvector). This was the case in Example 5.11.7, in which $\lambda = 2$ was a root of char **T** of multiplicity 1.

### Exercises

1. In computing the characteristic polynomial of a matrix $A$, is it permissible to first reduce $A$ to a simpler form by means of elementary row operations? Why or why not? (ANS)
2. In each of the following, $\mathbf{T} \in \text{Lin}(F_n, F_n)$ has the given matrix $A$ with respect to the canonical basis. Find all eigenvalues and eigenvectors of **T**.

(a) $F = \mathbf{R}$; $n = 3$; $A = \begin{bmatrix} -2 & -3 & -3 \\ 1 & 2 & 1 \\ 1 & 1 & 2 \end{bmatrix}$. (ANS)

(b) $F = \mathbf{R}$; $n = 2$; $A = \begin{bmatrix} 3 & -2 \\ 4 & -3 \end{bmatrix}$.

(c) $F = \mathbf{C}$; $n = 2$; $A = \begin{bmatrix} 3 & -2 \\ 5 & -3 \end{bmatrix}$.

(d) $F = \mathbf{R}$; $n = 3$; $A = \begin{bmatrix} 0 & 1 & 3 \\ 1 & -1 & 2 \\ -1 & 0 & -3 \end{bmatrix}$. (ANS)

3. It was asserted in this section that if $\lambda$ is an eigenvalue of $\mathbf{T}$ of multiplicity $m$, then the $(\mathbf{T}, \lambda)$-eigenspace has dimension $\leq m$. Are the results of Exercise 2 consistent with this assertion?
4. (a) Interpret the mapping $\mathbf{T}$ of Exercise 2(a) geometrically. Draw a picture.
   (b) Do the same for the mapping of Exercise 2(b).
5. In Example 5.11.7, the cross product of the first two columns of the matrix $A - 2I$ is

   $(0, 0, -2) \times (2, -2, 0) = (-4, -4, 0)$.

   We notice that this vector spans the eigenspace corresponding to the eigenvalue 2. Discuss the validity of this method for finding eigenvectors. What hypotheses are needed? (ANS)
6. If $E$ is a real Euclidean space and $\mathbf{T} \in \mathrm{Lin}(E, E)$ is an isometry,
   (a) Show that any eigenvalue of $\mathbf{T}$ must equal $\pm 1$.
   (b) Prove that if $\mathbf{u}$, $\mathbf{v}$ are eigenvectors of $\mathbf{T}$ corresponding to the eigenvalues $1, -1$, respectively, then $\mathbf{u} \perp \mathbf{v}$.
7. By Section 4.10, Exercise 8, the matrix of a linear isometry of $\mathbf{R}_2$ (with respect to the canonical basis) must have one of the two forms

   (a) $\begin{bmatrix} a & b \\ -b & a \end{bmatrix}$; (b) $\begin{bmatrix} a & b \\ b & -a \end{bmatrix}$,

   where $a^2 + b^2 = 1$. By computing the characteristic polynomial of each matrix, show that matrix (a) has no eigenvalues if $b \neq 0$, but matrix (b) has the eigenvalues $1, -1$. Using part (b) of the preceding exercise, deduce that matrix (b) represents a reflection through a one-dimensional subspace of $\mathbf{R}_2$. What sort of mapping does matrix (a) represent?

*8. Let $\mathbf{T} \in \text{Lin}(\mathbf{R}_n, \mathbf{R}_n)$, let $A$ be the matrix of $\mathbf{T}$ with respect to the canonical basis, and suppose that $A = A^t$. Prove: eigenvectors corresponding to *different* eigenvalues of $\mathbf{T}$ are orthogonal. (*Hint*: See Sec. 4.5, Exercise 3.) (SOL)

*9. Let $\mathbf{T} \in \text{Lin}(V, V)$ and let $c$ be a scalar. Prove: if the eigenvalues of $\mathbf{T}$ are $\lambda_1, \ldots, \lambda_n$, then the eigenvalues of $c\mathbf{T}$ are $c\lambda_1, \ldots, c\lambda_n$.

10. Find a condition on the polynomial char($\mathbf{T}$; $x$) which is equivalent to the statement, "$\mathbf{T}$ is singular". (ANS)

11. Eigenvalues and eigenvectors are useful not only in linear algebra and geometry but in other areas as diverse as physics and economics. Look up some of these other applications in the literature. (See, e.g., [15], Secs. 5.3 and 5.4.)

## 5.12 The coefficients of the characteristic polynomial

In this section we finish the proof of Theorem 5.11.3 by finding scalars $c_0, \ldots, c_n$ (with $c_n \neq 0$) which satisfy equation 5.11.4. For convenience in stating the result, we introduce the following notations:

1. For all integers $n \geq 0$, $E_n = \{j \in \mathbf{Z} : 1 \leq j \leq n\}$. (We used this notation in Secs. 1.14 and 5.1.)

2. For any finite set $S$, $\#S$ is the number of elements in $S$.

3. Let $A = (a_{ij})$ be any $n \times n$ matrix and let $S$ be any subset of $E_n$. Then $A_S$ denotes the submatrix of $A$ consisting of only those entries $a_{ij}$ such that $i \in S, j \in S$. For example, if

$$A = (a_{ij}) = \begin{bmatrix} 1 & 2 & 3 \\ 4 & 5 & 6 \\ 7 & 8 & 9 \end{bmatrix},$$

then

$$A_{\{2,3\}} = \begin{bmatrix} a_{22} & a_{23} \\ a_{32} & a_{33} \end{bmatrix} = \begin{bmatrix} 5 & 6 \\ 8 & 9 \end{bmatrix}.$$

We can now state the formula for the $c_i$.

**5.12.1 Theorem.** Let $A$ be any $n \times n$ matrix over the field $F$, and let

$$c_i = (-1)^i \sum_{\substack{S \subseteq E_n \\ \#S = n-i}} \det A_S \qquad (i = 0, 1, \ldots, n) \tag{5.12.2}$$

(the sum runs over all subsets $S$ of $E_n$ which have exactly $n - i$ elements). Then equation 5.11.4 holds. In particular, $c_n = (-1)^n \neq 0$, and $c_0 = \det A$.

The special value $c_n$ is obtained from the general formula as follows: when $i = n$, then $n - i = 0$ and the only set $S$ appearing in 5.12.2 is $S = \varnothing$. Since $A_S$ is then the empty matrix which has determinant 1 (cf. Sec. 5.3), we obtain

$$c_n = (-1)^n \cdot 1 = (-1)^n.$$

The value $c_0 = \det A$ is derived similarly.

To illustrate, let

$$n = 3; \qquad A = \begin{bmatrix} a_{11} & a_{12} & a_{13} \\ a_{21} & a_{22} & a_{23} \\ a_{31} & a_{32} & a_{33} \end{bmatrix}.$$

Then

$$c_0 = \det A.$$

$$c_1 = -(\det A_{\{1,2\}} + \det A_{\{1,3\}} + \det A_{\{2,3\}})$$

$$= -\left(\det \begin{bmatrix} a_{11} & a_{12} \\ a_{21} & a_{22} \end{bmatrix} + \det \begin{bmatrix} a_{11} & a_{13} \\ a_{31} & a_{33} \end{bmatrix} + \det \begin{bmatrix} a_{22} & a_{23} \\ a_{32} & a_{33} \end{bmatrix}\right).$$

$$c_2 = \det A_{\{1\}} + \det A_{\{2\}} + \det A_{\{3\}} = a_{11} + a_{22} + a_{33}.$$

$$c_3 = (-1)^3 = -1.$$

(Try computing char **T** this way for the mapping **T** of Example 5.11.7, and see if you get the same result as before.)

The proof of Theorem 5.12.1 requires the following algebraic lemma.

**5.12.3 Lemma.** Let $n \geqslant 0$ be a fixed integer and let $a_i, b_i$ $(i = 1, 2, \dots, n)$ be elements of a fixed commutative ring with unity. Then

$$\text{(5.12.4)} \quad \prod_{i=1}^{n} (a_i + b_i) = \sum_{S \subseteq E_n} \left[ \left( \prod_{i \in S} a_i \right) \left( \prod_{j \in S'} b_j \right) \right]$$

where the sum is taken over all subsets $S$ of $E_n$, and where $S'$ denotes the relative complement of $S$ in $E_n$; that is,

$$S' = \{ j \in E_n : j \notin S \}.$$

(Note that the products on the right side of 5.12.4 were defined at the end of Sec. 5.1. The same definition applies to the sum on the right side, since $E_n$ has only finitely many subsets $S$.)

*Proof.* When $n = 0$, $E_n = \varnothing$ (the empty set). The only subset of $E_n$ is $S = \varnothing$, and we also have $S' = \varnothing$. Since the empty product equals the unity $e$ of the given ring (Example 1.12.14), 5.12.4 reduces to $e = e \cdot e$ which is obviously true. Thus, 5.12.4 holds when $n = 0$. Continuing by induction on $n$, we assume that 5.12.4 holds for $n = k$; that is,

$$\textbf{(5.12.5)}\quad \prod_{i=1}^{k} (a_i + b_i) = \sum_{T \subseteq E_k} \left[ \left( \prod_{i \in T} a_i \right) \left( \prod_{j \in T^*} b_j \right) \right]$$

where $T^*$ denotes the relative complement of $T$ in $E_k$. Under this assumption, we must show that 5.12.4 holds for $n = k + 1$. Let $S$ be any subset of $E_{k+1}$ and let $S' = \{j \in E_{k+1} : j \notin S\}$. We consider two cases, according to whether $k + 1$ belongs to $S$ or to $S'$.

*Case* 1. $k + 1 \in S$. Let $T = S \cap E_k$; clearly, $T \subseteq E_k$ and

$$\textbf{(5.12.6)}\quad S = T \cup \{k + 1\}; \qquad S' = T^*.$$

Hence, we have

$$\textbf{(5.12.7)}\quad \prod_{i \in S} a_i = \left( \prod_{i \in T} a_i \right) a_{k+1}; \qquad \prod_{j \in S'} b_j = \prod_{j \in T^*} b_j.$$

Moreover, 5.12.6 describes a one-to-one correspondence between the sets $S \subseteq E_{k+1}$ which contain $k + 1$ and the sets $T \subseteq E_k$. Hence, from 5.12.7, we get

$$\textbf{(5.12.8)}\quad \begin{aligned} \sum_{\substack{S \subseteq E_{k+1} \\ k+1 \in S}} \left[ \left( \prod_{i \in S} a_i \right) \left( \prod_{j \in S'} b_j \right) \right] &= \sum_{T \subseteq E_k} \left[ \left( \prod_{i \in T} a_i \right) a_{k+1} \left( \prod_{j \in T^*} b_j \right) \right] \\ &= \left( \prod_{i=1}^{k} (a_i + b_i) \right) a_{k+1} \qquad \text{(by 5.12.5).} \end{aligned}$$

*Case* 2. $k + 1 \in S'$. Letting $T = S$, we have $T \subseteq E_k$ and $S = T$, $S' = T^* \cup \{k + 1\}$. Hence, instead of 5.12.7, we have

$$\prod_{i \in S} a_i = \prod_{i \in T} a_i; \qquad \prod_{j \in S'} b_j = \left( \prod_{j \in T^*} b_j \right) b_{k+1}$$

and, thus,

$$\textbf{(5.12.9)}\quad \begin{aligned} \sum_{\substack{S \subseteq E_{k+1} \\ k+1 \in S'}} \left[ \left( \prod_{i \in S} a_i \right) \left( \prod_{j \in S'} b_j \right) \right] &= \sum_{T \subseteq E_k} \left[ \left( \prod_{i \in T} a_i \right) \left( \prod_{j \in T^*} b_j \right) b_{k+1} \right] \\ &= \left( \prod_{i=1}^{k} (a_i + b_i) \right) b_{k+1} \qquad \text{(by 5.12.5).} \end{aligned}$$

Now for every $S \subseteq E_{k+1}$, either $k + 1 \in S$ or $k + 1 \in S'$. Hence, by adding equations 5.12.8 and 5.12.9, we obtain

$$\sum_{S\subseteq E_{k+1}}\left[\left(\prod_{i\in S}a_i\right)\left(\prod_{j\in S'}b_j\right)\right]=\left(\prod_{i=1}^{k}(a_i+b_i)\right)a_{k+1}+\left(\prod_{i=1}^{k}(a_i+b_i)\right)b_{k+1}$$
$$=\left(\prod_{i=1}^{k}(a_i+b_i)\right)(a_{k+1}+b_{k+1})=\prod_{i=1}^{k+1}(a_i+b_i)$$

so that 5.12.4 holds for $n = k + 1$, as desired.∎

Now that Lemma 5.12.3 is established, we turn to the proof of Theorem 5.12.1. By Definition 5.2.1, we have

$$\det(A-xI)=\sum_{\pi\in S_n}(-1)^{\operatorname{ind}\pi}\left(\prod_{i=1}^{n}(A-xI)_{i,i\pi}\right)$$
$$=\sum_{\pi\in S_n}(-1)^{\operatorname{ind}\pi}\left(\prod_{i=1}^{n}(a_{i,i\pi}-x\delta_{i,i\pi})\right)$$
$$(5.12.10)\qquad =\sum_{\pi\in S_n}(-1)^{\operatorname{ind}\pi}\left[\sum_{S\subseteq E_n}\left(\prod_{i\in S}a_{i,i\pi}\right)\left(\prod_{j\in S'}(-x)\delta_{j,j\pi}\right)\right]$$

(by Lemma 5.12.3)

$$=\sum_{S\subseteq E_n}\sum_{\pi\in S_n}(-1)^{\operatorname{ind}\pi}\left(\prod_{i\in S}a_{i,i\pi}\right)\left(\prod_{j\in S'}(-x)\delta_{j,j\pi}\right).$$

Consider the factor $y_\pi^{(S)}=\prod_{j\in S'}(-x)\delta_{j,j\pi}$. Since $\delta_{rs}=0$ if $r\neq s$, it follows that $y_\pi^{(S)}\neq 0$ only if $j=j\pi$ for all $j\in S'$; that is, only if the integers moved by $\pi$ all belong to $S$. In this case, $\pi$ can be regarded as an element of Sym($S$), and each $\delta_{j,j\pi}=1$ so that

$$y_\pi^{(S)}=\prod_{j\in S'}(-x)=(-x)^{\#S'}=(-x)^{n-\#S}.$$

Hence, 5.12.10 reduces to

$$(5.12.11)\qquad \det(A-xI)=\sum_{S\subseteq E_n}(-x)^{n-\#S}\sum_{\pi\in\operatorname{Sym}(S)}(-1)^{\operatorname{ind}\pi}\left(\prod_{i\in S}a_{i,i\pi}\right).$$

But it is clear from Definition 5.2.1 that the inner sum

$$\sum_{\pi\in\operatorname{Sym}(S)}(-1)^{\operatorname{ind}\pi}\left(\prod_{i\in S}a_{i,i\pi}\right)$$

is the determinant of the submatrix $A_S$. Hence, 5.12.11 reduces to

$$\det(A-xI)=\sum_{S\subseteq E_n}(-x)^{n-\#S}\det A_S$$
$$=\sum_{i=0}^{n}\left(\sum_{\#S=n-i}\det A_S\right)(-x)^i$$

(the last line is obtained from the preceding line by setting $i = n - \#S$). Clearly, this is a polynomial of degree $n$ in $x$, and the coefficient of $x^i$ in this polynomial is equal to

$$c_i = (-1)^i \sum_{\#S=n-i} \det A_S$$

as desired. This completes the proof of Theorem 5.12.1.∎

We recall from Section 1.16 that a polynomial is said to *split* if it can be expressed as a product of first-degree polynomials. Using this terminology, we have the following corollary of Theorem 5.12.1.

**5.12.12 Theorem.** Let $A$ be an $n \times n$ matrix over the field $F$, and suppose that the characteristic polynomial of $A$ splits; that is,

$$\text{char}(A; x) = (\lambda_1 - x)(\lambda_2 - x) \cdots (\lambda_n - x).$$

For each integer $k \in E_n$, let

$$\sigma_k(A) = \sum_{\substack{S \subseteq E_n \\ \#S = k}} \det A_S.$$

Then for all $k$ $(1 \leqslant k \leqslant n)$, we have $\sigma_k(A) = \sigma_k(A')$, where $A'$ is the diagonal matrix

$$A' = \begin{bmatrix} \lambda_1 & & & \\ & \lambda_2 & & \\ & & \ddots & \\ & & & \lambda_n \end{bmatrix}.$$

(The notation indicates that $A'$ has zeros everywhere except on the main diagonal.)

*Proof.* By Theorem 5.12.1, $(-1)^{n-k}\sigma_k(A)$ equals the coefficient of $x^{n-k}$ in char$(A; x)$. But

$$\begin{aligned} \text{char}(A'; x) &= \det(A' - xI) \\ &= \prod_{i=1}^{n} (\lambda_i - x) = \text{char}(A; x) \end{aligned}$$

and the result follows.∎

Theorem 5.12.12 is more general than it looks. Indeed, let $A$ be *any* $n \times n$ matrix over $F$. Even if the polynomial char$(A; x)$ does not split over $F$, it must split over some larger field $L$ which contains $F$ (cf. Sec. 1.16); we may then regard $A$ as a matrix over $L$, and Theorem 5.12.12 still applies. For example, suppose $A = (a_{ij})$ is a $3 \times 3$ matrix with real coefficients whose characteristic polynomial is

$$\operatorname{char}(A; x) = -x^3 + 2x^2 - x + 2.$$

This polynomial does not split over **R**; but over **C** (the complex numbers) we have

$$\operatorname{char}(A; x) = -(x - 2)(x + i)(x - i);$$

$$A' = \begin{bmatrix} i & & \\ & -i & \\ & & 2 \end{bmatrix}.$$

Hence, by Theorem 5.12.12, we have

$$a_{11} + a_{22} + a_{33} = \sigma_1(A) = \sigma_1(A') = i - i + 2 = 2$$

$$\det\begin{bmatrix} a_{11} & a_{12} \\ a_{21} & a_{22} \end{bmatrix} + \det\begin{bmatrix} a_{11} & a_{13} \\ a_{31} & a_{33} \end{bmatrix} + \det\begin{bmatrix} a_{22} & a_{23} \\ a_{32} & a_{33} \end{bmatrix}$$
$$= \sigma_2(A) = \sigma_2(A') = i(-i) + 2i + 2(-i) = 1$$

$$\det A = \sigma_3(A) = \sigma_3(A') = \det A' = 2.$$

These results, of course, could also have been obtained directly from Theorem 5.12.1.

**Exercises**

1. (a) Find a matrix $A$ with real entries whose characteristic polynomial is $x^2 + 2x + 5$.
   (b) Find a matrix $A$ with real entries whose characteristic polynomial is $-x^3 + x^2 - x + 1$.
2. Prove: if $a, b, c, d$ are real numbers such that $b$ and $c$ have the same sign (both positive or both negative), then the matrix
   $$\begin{bmatrix} a & b \\ c & d \end{bmatrix}$$
   has two distinct real eigenvalues.
3. Obtain a condition on the real numbers $a, b, c, d$ which is equivalent to the assertion that the matrix
   $$\begin{bmatrix} a & b \\ c & d \end{bmatrix}$$
   has an eigenvalue of multiplicity 2.

## 5.13 Some special cases

If a linear transformation **T** (or its matrix) satisfies certain special conditions, it may follow that the characteristic polynomial of **T** can be factored (partially or completely). Some results along such lines will be obtained in this section.

**5.13.1** **Theorem.** Let $A$ be an $n \times n$ matrix over $F$, and suppose that $A$ can be written in the block form

$$A = \begin{bmatrix} B & 0 \\ * & C \end{bmatrix}$$

where $B$ is square of degree $n_1$, $C$ is square of degree $n_2$, and $n_1 + n_2 = n$. Then

$$\text{char}(A; x) = \text{char}(B; x) \cdot \text{char}(C; x).$$

*Proof.* Since $B$ and $C$ are square matrices, the main diagonal of $A$ consists of the main diagonal of $B$ together with the main diagonal of $C$. Hence,

$$A - xI = \begin{bmatrix} B - xI & 0 \\ * & C - xI \end{bmatrix}.$$

It follows from Section 5.6, Exercise 4 that

$$\det(A - xI) = \det(B - xI) \cdot \det(C - xI)$$

which is the desired result. ∎

**5.13.2** **Corollary.** If the matrix $A$ can be written in the block form

$$\begin{bmatrix} B_1 & 0 & \cdots & 0 \\ * & B_2 & \cdots & 0 \\ \vdots & \vdots & \ddots & \vdots \\ * & * & \cdots & B_r \end{bmatrix}$$

where the matrices $B_1, B_2, \ldots, B_r$ are square and all entries above the blocks $B_i$ are zero, then

$$\text{char}(A; x) = \text{char}(B_1; x) \cdot \text{char}(B_2; x) \cdots \text{char}(B_r; x).$$

This corollary easily follows from Theorem 5.13.1 by induction on $r$. Clearly, the same result holds for matrices of the form

$$A = \begin{bmatrix} B_1 & * & \cdots & * \\ 0 & B_2 & \cdots & * \\ \vdots & \vdots & \ddots & \vdots \\ 0 & 0 & \cdots & B_r \end{bmatrix}$$

in which all entries *below* the blocks $B_i$ are zero.

As a special case, suppose all of the $B_i$ have degree 1. Then each $B_i$ has the single entry $a_{ii}$, so that

$$\text{char}(B_i; x) = a_{ii} - x.$$

Hence, we obtain the following corollary of 5.13.2:

5.13.3 **Theorem.** Suppose $A = (a_{ij})$ is a triangular matrix of degree $n$ over $F$. Then

$$\operatorname{char}(A; x) = (a_{11} - x)(a_{22} - x) \cdots (a_{nn} - x) = \prod_{i=1}^{n} (a_{ii} - x)$$

and the eigenvalues of $A$ are precisely $a_{11}, a_{22}, \ldots, a_{nn}$ (the entries along the main diagonal of $A$).

For the purpose of stating further results, the following definition will be convenient.

5.13.4 **Definition.** Let $V$ be a vector space, let $W$ be a subspace of $V$, and let $\mathbf{T} \in \operatorname{Lin}(V, V)$. The subspace $W$ is said to be **T**-*stable* (or **T**-*invariant*), if $W\mathbf{T} \subseteq W$; that is, if **T** maps every element of $W$ into an element of $W$.

For example, a nonzero vector $\mathbf{u} \in V$ is an eigenvector of **T** if and only if the one-dimensional subspace $[\mathbf{u}]$ is **T**-stable (why?)

If $W$ is a **T**-stable subspace, then $\mathbf{T} \restriction W$ is an element of $\operatorname{Lin}(W, W)$ and, hence, has a characteristic polynomial of its own. In this situation, we shall show that $\operatorname{char}(\mathbf{T} \restriction W)$ is a factor of char **T**; more precisely, that

$$\operatorname{char}(\mathbf{T}; x) = \operatorname{char}(\mathbf{T} \restriction W; x) \cdot \operatorname{char}(\mathbf{T}_*; x) \tag{5.13.5}$$

where $\mathbf{T}_*$ is a certain linear mapping of the quotient space $V/W$ into itself. The mapping $\mathbf{T}_*$ is defined as follows: letting $\mathbf{v}_* = \mathbf{v} + W$ be the coset of $W$ which contains $\mathbf{v}$ (as in Sec. 2.7), we let

$$\mathbf{v}_*\mathbf{T}_* = (\mathbf{v}\mathbf{T})_* = \mathbf{v}\mathbf{T} + W \qquad (\text{all } \mathbf{v} \in V). \tag{5.13.6}$$

5.13.7 **Theorem.** Let $V$ be a finite-dimensional vector space over $F$, let $\mathbf{T} \in \operatorname{Lin}(V, V)$, and let $W$ be a **T**-stable subspace of $V$. Then the mapping $\mathbf{T}_*$ defined by 5.13.6 is a well-defined element of Lin $(V/W, V/W)$, and equation 5.13.5 holds.

*Proof.* The fact that $\mathbf{T}_*$ is well defined and linear is left as an exercise (below). To prove 5.13.5, choose an ordered basis $(\mathbf{u}_1, \ldots, \mathbf{u}_k)$ of $W$ and extend it to an ordered basis $(\mathbf{u}_1, \ldots, \mathbf{u}_k, \mathbf{u}_{k+1}, \ldots, \mathbf{u}_n)$ of $V$. By Theorem 4.6.11, $((\mathbf{u}_{k+1})_*, \ldots, (\mathbf{u}_n)_*)$ is then an ordered basis of $V/W$.

Let

$A$ = matrix of $\mathbf{T}$ with respect to the basis $(\mathbf{u}_1, ..., \mathbf{u}_n)$ of $V$;
$B$ = matrix of $\mathbf{T} \restriction W$ with respect to the basis $(\mathbf{u}_1, ..., \mathbf{u}_k)$ of $W$;
$C$ = matrix of $\mathbf{T}_*$ with respect to the basis $((\mathbf{u}_{k+1})_*, ..., (\mathbf{u}_n)_*)$ of $V/W$.

It will suffice to prove that

$$(5.13.8) \quad A = \begin{bmatrix} B & 0 \\ * & C \end{bmatrix};$$

indeed, 5.13.8 clearly implies 5.13.5 in view of Theorem 5.13.1.

Since $A$ is the matrix of $\mathbf{T}$ with respect to the basis $(\mathbf{u}_1, ..., \mathbf{u}_n)$, we have

$$(5.13.9) \quad \mathbf{u}_i\mathbf{T} = a_{i1}\mathbf{u}_1 + \cdots + a_{ik}\mathbf{u}_k + \cdots + a_{in}\mathbf{u}_n \qquad (1 \leqslant i \leqslant n).$$

If $i \leqslant k$, then $\mathbf{u}_i \in W$, so that $\mathbf{u}_i\mathbf{T} \in W$ (since $W$ is $\mathbf{T}$-stable); that is, $\mathbf{u}_i\mathbf{T} \in [\mathbf{u}_1, ..., \mathbf{u}_k]$. Hence, for $i \leqslant k$ the coefficients of $\mathbf{u}_{k+1}, ..., \mathbf{u}_n$ in 5.13.9 are zero:

$$a_{ij} = 0 \qquad (1 \leqslant i \leqslant k, k+1 \leqslant j \leqslant n).$$

This gives us the zero block in equation 5.13.8; it also shows that for $i \leqslant k$, equation 5.13.9 reduces to

$$(5.13.10) \quad \mathbf{u}_i\mathbf{T} = a_{i1}\mathbf{u}_1 + \cdots + a_{ik}\mathbf{u}_k \qquad (1 \leqslant i \leqslant k).$$

But equations 5.13.10 express the images of the *basis vectors of W* as linear combinations of those basis vectors; hence, the coefficients $a_{ij}$ in 5.13.10 $(1 \leqslant i \leqslant k, 1 \leqslant j \leqslant k)$ constitute the matrix of $\mathbf{T} \restriction W$. That is, the matrix $B$ consists of the entries in just the first $k$ rows and columns of $A$, as indicated by 5.13.8.

We must still show that the matrix $C$ coincides with the lower right corner of $A$. To do so, we observe that by 5.13.6 we have for all $i$

$$(5.13.11) \quad \begin{aligned} (\mathbf{u}_i)_*\mathbf{T}_* &= (\mathbf{u}_i\mathbf{T})_* = (a_{i1}\mathbf{u}_1 + \cdots + a_{in}\mathbf{u}_n)_* \qquad \text{(by 5.13.9)} \\ &= a_{i1}(\mathbf{u}_1)_* + \cdots + a_{ik}(\mathbf{u}_k)_* + \cdots + a_{in}(\mathbf{u}_n)_*, \end{aligned}$$

the last line being obtained from the definitions of addition and scalar multiplication in $V/W$. Now, for $i \leqslant k$, we have $\mathbf{u}_i \in W$ and, hence, $(\mathbf{u}_i)_* = W = \mathbf{0}_*$ (the zero element of $V/W$). Hence, the terms involving $(\mathbf{u}_1)_*, ..., (\mathbf{u}_k)_*$ drop out of equation 5.13.11, and we obtain

$$(5.13.12) \quad (\mathbf{u}_i)_*\mathbf{T}_* = a_{i,k+1}(\mathbf{u}_{k+1})_* + \cdots + a_{in}(\mathbf{u}_n)_*.$$

The equations 5.13.12 for $k + 1 \leqslant i \leqslant n$ (ignoring other values of $i$) express the images under $\mathbf{T}_*$ of the basis vectors of $V/W$ as linear combinations of those basis vectors. Hence, the coefficients $a_{ij}$ in these equations ($k + 1 \leq i \leq n$, $k + 1 \leq j \leq n$) constitute the matrix of $\mathbf{T}_*$. That is, the matrix $C$ consists of the entries in just the last $n - k$ rows and columns of $A$, which agrees with 5.13.8. ∎

An argument exactly like the preceding one can be used to prove that if $W_1$, $W_2$ are *two* **T**-stable subspaces of $V$ such that $V = W_1 \oplus W_2$, then

$$\operatorname{char}(\mathbf{T}; x) = \operatorname{char}(\mathbf{T} \restriction W_1; x) \cdot \operatorname{char}(\mathbf{T} \restriction W_2; x).$$

In this situation, the matrix equation analogous to 5.13.8 is

$$A = \begin{bmatrix} B_1 & 0 \\ 0 & B_2 \end{bmatrix}$$

where $A$, $B_1$, $B_2$ are the matrices of $\mathbf{T}$, $\mathbf{T} \restriction W_1$, $\mathbf{T} \restriction W_2$, respectively. (In the proof, choose bases as in Theorem 2.8.2(e).) This result can be extended by induction to a direct sum of $r$ subspaces, giving us the following theorem:

**5.13.13 Theorem.** Suppose $W_1, \ldots, W_r$ are **T**-stable subspaces of $V$ such that $V = W_1 \oplus \cdots \oplus W_r$. Then

$$\operatorname{char}(\mathbf{T}; x) = \operatorname{char}(\mathbf{T} \restriction W_1; x) \cdot \operatorname{char}(\mathbf{T} \restriction W_2; x) \cdots \operatorname{char}(\mathbf{T} \restriction W_r; x),$$

and with respect to bases chosen as in Theorem 2.8.2(e), we have

$$A = \begin{bmatrix} B_1 & 0 & \cdots & 0 \\ 0 & B_2 & \cdots & 0 \\ \vdots & \vdots & \ddots & \vdots \\ 0 & 0 & \cdots & B_r \end{bmatrix}$$

where $A$ is the matrix of $\mathbf{T}$ and $B_i$ is the matrix of $\mathbf{T} \restriction W_i$ ($1 \leq i \leq r$).

Theorem 5.13.13 shows that any way of expressing $V$ as a direct sum of **T**-stable subspaces produces a factorization of the polynomial char**T**. We might ask whether the converse holds: if

$$\operatorname{char} \mathbf{T} = f_1 f_2 \cdots f_r$$

where the $f_i$ are monic polynomials over $F$, need it follow that there exist **T**-stable subspaces $W_1, \ldots, W_r$ of $V$ such that

$$V = W_1 \oplus \cdots \oplus W_r,$$
$$\operatorname{char}(\mathbf{T} \restriction W_i) = \pm f_i \qquad (\text{each } i)? \tag{5.13.14}$$

The answer is "yes" under certain additional assumptions; the assumptions which suffice are (1) that the polynomials $f_i$ are *relatively prime* (i.e., no two $f_i$ have any nonconstant common factor), and (2) that the field $F$ is a "perfect field". For the definition of "perfect field", see more advanced texts on field theory; suffice it to say here that any field containing $\mathbf{Q}$ is perfect, and so is any finite field. In particular, the fields $\mathbf{Q}$, $\mathbf{R}$, $\mathbf{C}$, $\mathbf{Z}_{(p)}$ ($p$ prime) are all perfect. We shall not prove 5.13.14 here (the result will not be needed, and the proof is beyond the scope of this book), but in Section 8.4 there is a numerical example which illustrates how to compute the spaces $W_i$ when the polynomials $f_i$ are not too complicated.

**Exercise**

*If $\mathbf{T} \in \text{Lin}(V, V)$, $W$ is a $\mathbf{T}$-stable subspace of $V$, and $\mathbf{T}_* : V/W \to V/W$ is defined by 5.13.6,

(a) Prove that $\mathbf{T}_*$ is well defined; that is, $\mathbf{v}_*\mathbf{T}_*$ depends only on $\mathbf{v}_*$, not on $\mathbf{v}$.

(b) Prove that $\mathbf{T}_*$ is linear.

## 5.14 The generalized cross product

Let $\mathbf{u} = (a_1, a_2, a_3)$ and $\mathbf{v} = (b_1, b_2, b_3)$ be vectors in $F_3$. In Section 5.4, Exercise 7, we defined the *cross product* of $\mathbf{u}$ and $\mathbf{v}$ by the equation

**(5.14.1)** $$\mathbf{u} \times \mathbf{v} = \left(\det \begin{bmatrix} a_2 & a_3 \\ b_2 & b_3 \end{bmatrix}, -\det \begin{bmatrix} a_1 & a_3 \\ b_1 & b_3 \end{bmatrix}, \det \begin{bmatrix} a_1 & a_2 \\ b_1 & b_2 \end{bmatrix}\right).$$

In the case $F = \mathbf{R}$, both the length and the direction of $\mathbf{u} \times \mathbf{v}$ are geometrically significant: $\mathbf{u} \times \mathbf{v}$ is perpendicular to $\mathbf{u}$ and $\mathbf{v}$ (Sec. 5.4, Exercise 7), and the length of $\mathbf{u} \times \mathbf{v}$ equals the area of the parallelogram determined by $\mathbf{u}$ and $\mathbf{v}$ (Sec. 5.7).

In this section, we generalize our considerations to $F_n$; specifically, we define the *cross product of $n - 1$ vectors in $F_n$*. For convenience in stating both the definition and subsequent results, we adopt the following notation: whenever $\mathbf{u}_1, \ldots, \mathbf{u}_m$ are vectors in $F_n$, then

$$R(\mathbf{u}_1, \ldots, \mathbf{u}_m)$$

denotes the $m \times n$ matrix whose row vectors (in order) are $\mathbf{u}_1, \ldots, \mathbf{u}_m$.

**5.14.2 Definition.** Let $n \geq 2$ and let $\mathbf{u}_1, \ldots, \mathbf{u}_{n-1}$ be $n - 1$ vectors in $F_n$ (not necessarily distinct). Let $B_j$ be the $(n - 1) \times (n - 1)$ matrix obtained from $R(\mathbf{u}_1, \ldots, \mathbf{u}_{n-1})$ by deleting the $j$th column. Then the *(generalized) cross product* of $\mathbf{u}_1, \ldots, \mathbf{u}_{n-1}$ is the vector

$$\mathbf{u}_1 \times \cdots \times \mathbf{u}_{n-1} = ((-1)^{n-1} \det B_1, (-1)^{n-2} \det B_2, \ldots, (-1)^{n-n} \det B_n).$$

The vectors $\mathbf{u}_1, \ldots, \mathbf{u}_{n-1}$ will be called *factors* of $\mathbf{u}_1 \times \cdots \times \mathbf{u}_{n-1}$.

Note that the generalized cross product, unlike most other forms of "multiplication" which we encounter in mathematics, is *not* a binary operation; it is an "$(n-1)$-ary" operation. We cannot define $\mathbf{u}_1 \times \mathbf{u}_2$ in $F_n$; we cannot break up the expression $\mathbf{u}_1 \times \cdots \times \mathbf{u}_{n-1}$ by inserting parentheses.

It is clear that Definition 5.14.2 reduces to 5.14.1 when $n = 3$. Moreover, in the case $F = \mathbf{R}$, we shall show that the generalized cross product has the same geometric significance for arbitrary $n$ as it has when $n = 3$. To begin with, *the generalized cross product is perpendicular to each of its factors*. To establish this fact, it will be useful to prove the following result which is slightly more general:

**5.14.3 Theorem.** Let $\mathbf{u}_1, \ldots, \mathbf{u}_{n-1}, \mathbf{w}$ be any vectors in $F_n$ and let $\mathbf{v} = \mathbf{u}_1 \times \cdots \times \mathbf{u}_{n-1}$. If $D = R(\mathbf{u}_1, \ldots, \mathbf{u}_{n-1}, \mathbf{w})$, then
$\det D = \mathbf{v} \cdot \mathbf{w}$.

*Proof.* It is clear that the $(n,j)$ minor of $D$ is the matrix $B_j$ of Definition 5.14.2. Hence, if $\mathbf{w} = (c_1, \ldots, c_n)$ and we expand $\det D$ by minors of the last row, we get

$$\begin{aligned}\det D &= \sum_{j=1}^{n} (-1)^{n+j} c_j \det B_j \\ &= \sum_{j=1}^{n} (-1)^{n-j} c_j \det B_j \qquad \text{(why?)} \\ &= ((-1)^{n-1} \det B_1, \ldots, (-1)^{n-n} \det B_n) \cdot (c_1, \ldots, c_n) \\ &= \mathbf{v} \cdot \mathbf{w}. \ \blacksquare\end{aligned}$$

**5.14.4 Corollary.** If $\mathbf{u}_1, \ldots, \mathbf{u}_{n-1} \in F_n$, then $\mathbf{u}_1 \times \cdots \times \mathbf{u}_{n-1}$ is perpendicular to each $\mathbf{u}_i$. ∎

*Proof.* If we let $\mathbf{w} = \mathbf{u}_i$ in Theorem 5.14.3, then $D$ has two equal rows so that $\det D = 0$. Hence, $\mathbf{v} \cdot \mathbf{u}_i = 0$ by 5.14.3, so that $\mathbf{v} \perp \mathbf{u}_i$. ∎

The geometric significance of the *length* of the cross product also generalizes to $n$ dimensions: *if* $\mathbf{u}_1, \ldots, \mathbf{u}_{n-1}$ *are elements of* $\mathbf{R}_n$, *then* $|\mathbf{u}_1 \times \cdots \times \mathbf{u}_{n-1}|$ *equals the* $(n-1)$*-dimensional volume of the parallelepiped in* $\mathbf{R}_n$ *determined by the vectors* $\mathbf{u}_1, \ldots, \mathbf{u}_{n-1}$. It was shown in Section 5.7 that this volume equals

$$(\det AA^t)^{1/2}$$

where $A = R(\mathbf{u}_1, ..., \mathbf{u}_{n-1})$; hence, what we really wish to show is that

(5.14.5) $$|\mathbf{u}_1 \times \cdots \times \mathbf{u}_{n-1}| = (\det AA^t)^{1/2}.$$

To prove 5.14.5, we observe that by Theorem 5.7.4,

(5.14.6) $$\det AA^t = \sum_B (\det B)^2$$

where $B$ runs over all $(n - 1) \times (n - 1)$ submatrices of $A$. But these submatrices are precisely the $B_j$ of Definition 5.14.2; their determinants are (up to sign) the coordinates of $\mathbf{u}_1 \times \cdots \times \mathbf{u}_{n-1}$. Since the sum of the squares of these coordinates is $|\mathbf{u}_1 \times \cdots \times \mathbf{u}_{n-1}|^2$ by definition of length, 5.14.6 reduces to

$$\det AA^t = |\mathbf{u}_1 \times \cdots \times \mathbf{u}_{n-1}|^2$$

and 5.14.5 then follows by taking square roots.

Two other expressions for the length of the cross product are given by the following theorem.

**5.14.7 Theorem.** Let $\mathbf{u}_1, ..., \mathbf{u}_{n-1}$ be vectors in $\mathbf{R}_n$ and let $\mathbf{v} = \mathbf{u}_1 \times \cdots \times \mathbf{u}_{n-1}$; also, let $C = (c_{ij})$ be the $(n - 1) \times (n - 1)$ matrix defined by

$$c_{ij} = \cos(\mathbf{u}_i, \mathbf{u}_j).$$

(If $\mathbf{u}_i$ or $\mathbf{u}_j$ is zero, $c_{ij}$ may be assigned any arbitrary real value.) Then

(5.14.8) $$|\mathbf{u}_1 \times \cdots \times \mathbf{u}_{n-1}| = (\det R(\mathbf{u}_1, ..., \mathbf{u}_{n-1}, \mathbf{v}))^{1/2};$$

and

(5.14.9) $$|\mathbf{u}_1 \times \cdots \times \mathbf{u}_{n-1}| = |\mathbf{u}_1||\mathbf{u}_2| \cdots |\mathbf{u}_{n-1}| (\det C)^{1/2}.$$

*Proof.* By Theorem 5.14.3, $\det R(\mathbf{u}_1, ..., \mathbf{u}_{n-1}, \mathbf{v}) = \mathbf{v} \cdot \mathbf{v} = |\mathbf{v}|^2$; hence, $|\mathbf{v}| = (\det R(\mathbf{u}_1, ..., \mathbf{u}_{n-1}, \mathbf{v}))^{1/2}$, proving 5.14.8. To prove 5.14.9, let $A = R(\mathbf{u}_1, ..., \mathbf{u}_{n-1})$; then

$$\begin{aligned}(AA^t)_{ij} &= A_{(i)} \cdot (A^t)^{(j)} = A_{(i)} \cdot A_{(j)} \\ &= \mathbf{u}_i \cdot \mathbf{u}_j = |\mathbf{u}_i||\mathbf{u}_j| c_{ij}.\end{aligned}$$

In other words, $AA^t$ is obtained from $C$ by multiplying the $i$th row by $|\mathbf{u}_i|$ for all $i$ $(1 \leq i \leq n-1)$ and then multiplying the $j$th column by $|\mathbf{u}_j|$ for all $j$ $(1 \leq j \leq n - 1)$. Hence, by Theorem 5.4.1 (applied $2n - 2$ times in succession), we have

$$\det AA^t = (|\mathbf{u}_1||\mathbf{u}_2| \cdots |\mathbf{u}_{n-1}|)^2 \det C.$$

Taking square roots and using 5.14.5, we obtain 5.14.9. ∎

*Remarks*: 1. If the vectors $\mathbf{u}_i$ are mutually orthogonal, then cos $(\mathbf{u}_i, \mathbf{u}_j) = \delta_{ij}$ so that $C = I$. Hence, in this case, 5.14.9 implies that $|\mathbf{u}_1 \times \cdots \times \mathbf{u}_{n-1}| = |\mathbf{u}_1||\mathbf{u}_2| \cdots |\mathbf{u}_{n-1}|$. Geometrically, this means that the volume of an $(n-1)$-dimensional "rectangular solid" is the product of the lengths of its $n-1$ distinct (nonparallel) edges, a not surprising result. In the nonrectangular case, there is an additional factor $(\det C)^{1/2}$. (*Question*: What does this imply about the magnitude of det $C$?)

2. Since the entries in $C$ are the numbers

$$\cos(\mathbf{u}_i, \mathbf{u}_j) = \frac{\mathbf{u}_i \cdot \mathbf{u}_j}{|\mathbf{u}_i||\mathbf{u}_j|},$$

the quantity $|\mathbf{u}_1||\mathbf{u}_2| \cdots |\mathbf{u}_{n-1}| (\det C)^{1/2}$ depends only on the lengths of the $\mathbf{u}_i$ and the dot products of pairs of the $\mathbf{u}_i$. Since both lengths and dot products are preserved by linear isometries (Theorem 4.10.9), it follows that if $\mathbf{T}$ is a linear isometry, then the quantity 5.14.9 and, hence, also the quantity 5.14.8, are unchanged if each $\mathbf{u}_i$ is replaced by $\mathbf{u}_i\mathbf{T}$. That is, whenever $\mathbf{T}$ is a linear isometry of $\mathbf{R}_n$, we have

**(5.14.10)** $$|\mathbf{u}_1 \times \cdots \times \mathbf{u}_{n-1}| = |\mathbf{u}_1\mathbf{T} \times \cdots \times \mathbf{u}_{n-1}\mathbf{T}|$$

and

**(5.14.11)** $$\det R(\mathbf{u}_1, \ldots, \mathbf{u}_{n-1}, \mathbf{u}_1 \times \cdots \times \mathbf{u}_{n-1}) = \det R(\mathbf{u}_1\mathbf{T}, \ldots, \mathbf{u}_{n-1}\mathbf{T}, \mathbf{u}_1\mathbf{T} \times \cdots \times \mathbf{u}_{n-1}\mathbf{T})$$

for all $\mathbf{u}_1, \ldots, \mathbf{u}_{n-1}$ in $\mathbf{R}_n$.

By 5.14.10, a linear isometry $\mathbf{T}$ preserves the length of the cross product. We might ask whether $\mathbf{T}$ preserves the cross product itself; that is, whether the image of the cross product equals the cross product of the images. This is not quite true, but it is "almost" true: it is true except for a possible minus sign. A precise statement of this fact (which we shall need for our discussion of orientation in Sec. 6.7) is as follows:

**5.14.12 Theorem.** Let $\mathbf{u}_1, \ldots, \mathbf{u}_{n-1} \in \mathbf{R}_n$ and let $\mathbf{T}$ be a linear isometry of $\mathbf{R}_n$. Then

**(5.14.13)** $$(\mathbf{u}_1\mathbf{T}) \times (\mathbf{u}_2\mathbf{T}) \times \cdots \times (\mathbf{u}_{n-1}\mathbf{T}) = \epsilon((\mathbf{u}_1 \times \cdots \times \mathbf{u}_{n-1})\mathbf{T})$$

where $\epsilon = \det \mathbf{T} = \pm 1$.

*Proof.* We shall show below (Theorem 5.14.15) that $\mathbf{u}_1 \times \cdots \times \mathbf{u}_{n-1} = \mathbf{0}$ if and only if the $\mathbf{u}_i$ are linearly dependent. If $\mathbf{u}_1, \ldots, \mathbf{u}_{n-1}$ are dependent, then linearity of $\mathbf{T}$ implies that $\mathbf{u}_1\mathbf{T}, \ldots, \mathbf{u}_{n-1}\mathbf{T}$ are dependent so that both sides of 5.14.13 reduce to zero. Hence, we may assume for the remainder of the proof that the vectors $\mathbf{u}_1, \ldots, \mathbf{u}_{n-1}$ are linearly independent and that $\mathbf{u}_1 \times \cdots \times \mathbf{u}_{n-1} \neq \mathbf{0}$. Since $\mathbf{T}$ is one-to-one and linear, the vectors $\mathbf{u}_i\mathbf{T}$ are independent also and, hence, the subspace $W = [\mathbf{u}_1\mathbf{T}, \ldots, \mathbf{u}_{n-1}\mathbf{T}]^{\perp}$ is one-dimensional (Theorem 3.8.18(b)). By Corollary 5.14.4, $(\mathbf{u}_1\mathbf{T}) \times \cdots \times (\mathbf{u}_{n-1}\mathbf{T})$ is orthogonal to each vector $\mathbf{u}_i\mathbf{T}$ and, hence, belongs to $W$. Similarly, $\mathbf{u}_1 \times \cdots \times \mathbf{u}_{n-1}$ is orthogonal to each $\mathbf{u}_i$; since $\mathbf{T}$ preserves dot products, it follows that $(\mathbf{u}_1 \times \cdots \times \mathbf{u}_{n-1})\mathbf{T}$ is orthogonal to each $\mathbf{u}_i\mathbf{T}$ and, hence, belongs to $W$. Moreover, $(\mathbf{u}_1 \times \cdots \times \mathbf{u}_{n-1})\mathbf{T} \neq \mathbf{0}$ since $\mathbf{T}$ is one-to-one. It follows that the vector $(\mathbf{u}_1 \times \cdots \times \mathbf{u}_{n-1})\mathbf{T}$ spans $W$, so that 5.14.13 must hold for some scalar $\epsilon$. To find the value of $\epsilon$, we use the result of Section 4.4, Exercise 10, which tells us that

$$R(\mathbf{u}_1\mathbf{T}, \ldots, \mathbf{u}_{n-1}\mathbf{T}, (\mathbf{u}_i \times \cdots \times \mathbf{u}_{n-1})\mathbf{T})$$
$$= [R(\mathbf{u}_1, \ldots, \mathbf{u}_{n-1}, \mathbf{u}_1 \times \cdots \times \mathbf{u}_{n-1})] \cdot A$$

where $A$ is the matrix of $\mathbf{T}$. Taking the determinant of both sides, we obtain

$$\begin{aligned}
&\det R(\mathbf{u}_1\mathbf{T}, \ldots, \mathbf{u}_{n-1}\mathbf{T}, (\mathbf{u}_1 \times \cdots \times \mathbf{u}_{n-1})\mathbf{T}) \\
&\quad = \det R(\mathbf{u}_1, \ldots, \mathbf{u}_{n-1}, \mathbf{u}_1 \times \cdots \times \mathbf{u}_{n-1}) \det A \qquad \text{(by 5.6.10)} \\
&\quad = \det R(\mathbf{u}_1\mathbf{T}, \ldots, \mathbf{u}_{n-1}\mathbf{T}, \mathbf{u}_1\mathbf{T} \times \cdots \times \mathbf{u}_{n-1}\mathbf{T}) \det A \qquad \text{(by 5.14.11)} \\
&\quad = \det R(\mathbf{u}_1\mathbf{T}, \ldots, \mathbf{u}_{n-1}\mathbf{T}, \epsilon(\mathbf{u}_1 \times \cdots \times \mathbf{u}_{n-1})\mathbf{T}) \det A \qquad \text{(by 5.14.13)} \\
&\quad = \epsilon \det A \det R(\mathbf{u}_1\mathbf{T}, \ldots, \mathbf{u}_{n-1}\mathbf{T}, (\mathbf{u}_1 \times \cdots \times \mathbf{u}_{n-1})\mathbf{T}) \qquad \text{(by 5.4.1).}
\end{aligned} \tag{5.14.14}$$

Since $(\mathbf{u}_1 \times \cdots \times \mathbf{u}_{n-1})\mathbf{T}$ is a nonzero element of $W$, the vectors $\mathbf{u}_1\mathbf{T}, \ldots, \mathbf{u}_{n-1}\mathbf{T}, (\mathbf{u}_1 \times \cdots \times \mathbf{u}_{n-1})\mathbf{T}$ are independent. It follows that the factor $\det R(\mathbf{u}_1\mathbf{T}, \ldots, \mathbf{u}_{n-1}\mathbf{T}, (\mathbf{u}_1 \times \cdots \times \mathbf{u}_{n-1})\mathbf{T})$ in equation 5.14.14 is nonzero and may be cancelled from the equation. This gives

$$\epsilon \det A = 1.$$

Since $\det A = \pm 1$ by Theorem 5.6.12, it follows that $\epsilon = \det A = \det \mathbf{T} = \pm 1$ as desired. ∎

Let us now prove the theorem which we referred to at the beginning of the preceding proof.

**5.14.15 Theorem.** If $\mathbf{u}_1, \ldots, \mathbf{u}_{n-1} \in \mathbf{R}_n$, then $\mathbf{u}_1 \times \cdots \times \mathbf{u}_{n-1} = \mathbf{0}$ if and only if the vectors $\mathbf{u}_1, \ldots, \mathbf{u}_{n-1}$ are linearly dependent.

*Proof.* If $A = R(\mathbf{u}_1, \ldots, \mathbf{u}_{n-1})$, the coordinates of $\mathbf{u}_1 \times \cdots \times \mathbf{u}_{n-1}$ are (up to sign) the determinants of the $(n-1) \times (n-1)$ submatrices of $A$. Thus, $\mathbf{u}_1 \times \cdots \times \mathbf{u}_{n-1} = \mathbf{0}$ if and only if all such determinants are zero; that is, if and only if the number $d$ in Theorem 5.9.1(c) is less than $n - 1$. But by Theorem 5.9.1, $d$ equals the dimension of the row space of $A$; hence, $d < n - 1$ if and only if the $n - 1$ rows of $A$ are dependent, that is, if and only if the vectors $\mathbf{u}_i$ are dependent. ∎

**Exercises**

1. If $\mathbf{u}_1 = (1, 0, 1, 1)$, $\mathbf{u}_2 = (2, 1, 0, 3)$, $\mathbf{u}_3 = (-1, 2, 1, 0)$ (in $\mathbf{R}_4$), compute $\mathbf{u}_1 \times \mathbf{u}_2 \times \mathbf{u}_3$ using Definition 5.14.2. Verify that your answer is orthogonal to each $\mathbf{u}_i$. (ANS)
2. For each $i$, show that the generalized cross product in $F_n$ is a linear function of its $i$th factor. That is,
$$\begin{aligned}\mathbf{u}_1 \times \cdots \times \mathbf{u}_{i-1} \times (a\mathbf{v} + b\mathbf{w}) \times \mathbf{u}_{i+1} \times \cdots \times \mathbf{u}_{n-1} \\ = a(\mathbf{u}_1 \times \cdots \times \mathbf{u}_{i-1} \times \mathbf{v} \times \mathbf{u}_{i+1} \times \cdots \times \mathbf{u}_{n-1}) \\ + b(\mathbf{u}_1 \times \cdots \times \mathbf{u}_{i-1} \times \mathbf{w} \times \mathbf{u}_{i+1} \times \cdots \times \mathbf{u}_{n-1}).\end{aligned}$$
(SUG)
3. If $\mathbf{u}_1, \ldots, \mathbf{u}_{n-1} \in F_n$ and $\pi \in \mathbf{S}_{n-1}$, show that
$$\mathbf{u}_{1\pi} \times \cdots \times \mathbf{u}_{(n-1)\pi} = (-1)^{\text{ind } \pi} (\mathbf{u}_1 \times \cdots \times \mathbf{u}_{n-1}).$$
4. Let $\mathbf{u}_1, \ldots, \mathbf{u}_n$ be $n$ vectors in $F_n$, where $n \geq 2$. For each $i$ ($1 \leq i \leq n$), let $\mathbf{v}_i$ be the generalized cross product of all the $\mathbf{u}$'s except $\mathbf{u}_i$. (For example, $\mathbf{v}_3 = \mathbf{u}_1 \times \mathbf{u}_2 \times \mathbf{u}_4 \times \cdots \times \mathbf{u}_n$.)
   (a) If some vector $\mathbf{u}_i$ is replaced by $\mathbf{u}_i + k\mathbf{u}_j$, where $j$ is an index different from $i$, how are $\mathbf{v}_1, \ldots, \mathbf{v}_n$ affected? (ANS)
   (b) If some $\mathbf{u}_i$ is replaced by $c\mathbf{u}_i$ ($c \in F$), how are $\mathbf{v}_1, \ldots, \mathbf{v}_n$ affected? (ANS)
5. With notations as in the preceding exercise, prove that if $\mathbf{u}_1, \ldots, \mathbf{u}_n$ are linearly dependent, then $\mathbf{v}_1, \ldots, \mathbf{v}_n$ are linearly dependent.
6. With notations as in Exercise 4, let $U = R(\mathbf{u}_1, \ldots, \mathbf{u}_n)$, $V = R(\mathbf{v}_1, \ldots, \mathbf{v}_n)$. Prove that $\det V = (-1)^{n(n-1)/2}(\det U)^{n-1}$.

   (This generalizes Sec. 5.4, Exercise 8. *Suggestion*: Use the results of Exercise 4 above.)

# 6 Further study of isometries

## 6.1 Rotations in $\mathbf{R}_2$

In this section we will formally define "rotations" in the real plane $\mathbf{R}_2$. In so doing, we shall need the concept of the *angle between two half lines*, which we defined in Section 3.9. As before, $H(A, P)$ will denote the half line (ray) with initial point $A$ which passes through $P$ ($P \neq A$). The following notation will also be useful: if $A$ is a fixed point in the real Euclidean space $E$, then the set of all half lines in $E$ having initial point $A$ will be denoted

$$\text{Ha}_E(A)$$

or simply $\text{Ha}(A)$, if it is clear which space $E$ we are discussing. If $E = \mathbf{R}_n$, $\text{Ha}_E(A)$ will be denoted $\text{Ha}_n(A)$.

To motivate our definition, we begin informally. Let $A$ be a point in $\mathbf{R}_2$, and imagine the plane being physically rotated about $A$ through an angle $\alpha$, counterclockwise. Figure 6.1.1 exhibits the effect of this motion: the points $P$ and $Q$ are moved (mapped!) into $P'$ and $Q'$, respectively, the half line $H_1 = H(A, P)$ is mapped into $H_1'$, and so forth. Clearly,

**(6.1.1)** $\angle(H_1, H_1') = \alpha = \angle(H_2, H_2')$.

Also, since triangles $PAQ$ and $P'AQ'$ are congruent (side-angle-side), we have $|\overline{PQ}| = |\overline{P'Q'}|$; that is,

**(6.1.2)** $d(P, Q) = d(P', Q')$.

Equation 6.1.2 is, of course, the condition which defines an isometry.

Conversely, suppose $\mathbf{T} : \mathbf{R}_2 \to \mathbf{R}_2$ is a mapping which fixes the point $A$, satisfies 6.1.2 for all $P$ and $Q$, and satisfies 6.1.1 for all $H_1$, $H_2$ in $\text{Ha}(A)$. (We assume $P'$, $H_1'$, etc. denote the images of $P$, $H_1$, etc.) We claim that $\mathbf{T}$ must then be a (clockwise or counterclockwise) rotation about $A$ through the angle $\alpha$. Indeed, for any point $P \neq A$, let $H_1 = H(A, P)$; then $P\mathbf{T} = P'$ lies on $H_1\mathbf{T} = H_1'$ and, hence, $\angle PAP' = \angle(H_1, H_1') = \alpha$. Since also $d(A, P) = d(A', P') = d(A, P')$, it follows at once that $P'$ is obtained by rotating $P$ about $A$ through the angle $\alpha$. The only difficulty is in showing that the direction of rotation (clockwise or counterclockwise) is the same for all points $P$. However, if the direction is not uniform, then some point $P$ is rotated clockwise, some point $Q$ is rotated counterclockwise (as in Fig. 6.1.2) and, moreover, we may choose $P$ and $Q$ in such a way that $d(P, Q)$ is arbitrarily small. (That is, it is less than an arbitrary number $\epsilon > 0$. Readers with some acquaintance with topology should try proving this using the connectedness of the plane.) It is then not hard to deduce that, for suitably chosen $P$ and $Q$, $d(P, Q) \neq d(P', Q')$, contrary to equation 6.1.2.

The preceding discussion motivates the following formal definition:

**6.1.3** **Definition.** Let $A$ be a point in $\mathbf{R}_2$. A mapping $\mathbf{T}: \mathbf{R}_2 \to \mathbf{R}_2$ is called a *rotation about* $A$ if it has the following three properties:

(a) $\mathbf{T}$ maps $A$ into $A$.
(b) $\mathbf{T}$ is an isometry.

Figure 6.1.1

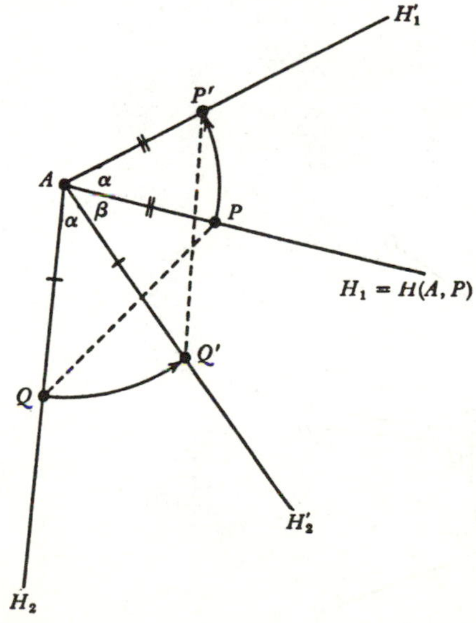

(c) There exists an angle $\alpha$ such that

$$\angle(H, H\mathbf{T}) = \alpha \qquad (\text{all } H \in \text{Ha}(A)).$$

The set of all rotations about $A$ will be denoted Rot($A$).

In connection with 6.1.3(c), note that if $\mathbf{T}$ is an isometry which fixes $A$ and if $H \in \text{Ha}(A)$, then also $H\mathbf{T} \in \text{Ha}(A)$ (Theorem 4.10.15) so that $\angle(H, H\mathbf{T})$ is defined; 6.1.3(c) then asserts that this angle is the same for all $H \in \text{Ha}(A)$. We shall call this angle $\alpha$ the *angle of* $\mathbf{T}$, and shall call $\mathbf{T}$ a *rotation through* $\alpha$. Note also that if $P \neq A$ is a point on $H$, then $P' = P\mathbf{T}$ lies on $H\mathbf{T}$ so that $H = H(A, P)$, $H\mathbf{T} = H(A, P')$, $\angle(H, H\mathbf{T}) = \angle(\mathbf{Ar}\, AP, \mathbf{Ar}\, AP')$. Hence, 6.1.3(c) may be rewritten in the form

$$\angle(\mathbf{Ar}\, AP, \mathbf{Ar}\, AP') = \alpha \qquad (\text{all } P \neq A, \text{ where } P' \text{ denotes the image of } P \text{ under } \mathbf{T}).$$

In particular, if $A = O$ (the origin), then 6.1.3(c) reduces to

$$\angle(\mathbf{u}, \mathbf{uT}) = \alpha \qquad (\text{all } \mathbf{u} \neq \mathbf{0} \text{ in } \mathbf{R}_2).$$

The case $A = O$ deserves special attention since rotations about $O$ are linear (Theorem 4.1.6). Let $\mathbf{T} \in \text{Rot}(O)$, and let

$$\begin{bmatrix} a & b \\ c & d \end{bmatrix}$$

be the matrix of $\mathbf{T}$ with respect to the canonical basis. By Theorem 4.10.13, the rows of this matrix are orthonormal, so that

$$\begin{aligned} a^2 + b^2 &= 1 \\ c^2 + d^2 &= 1 \\ ac + bd &= 0. \end{aligned}$$

Figure 6.1.2

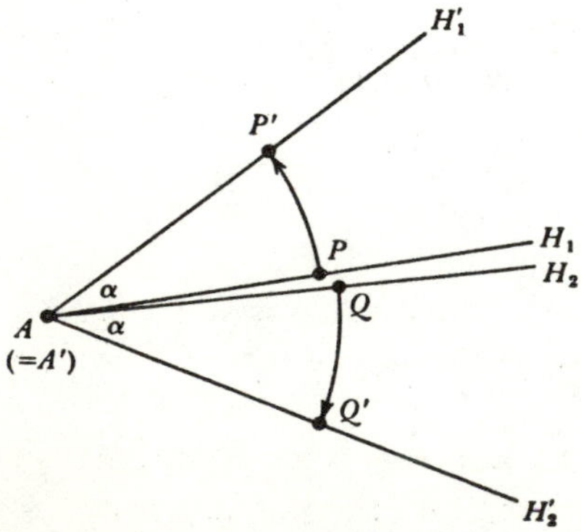

For any vector $(x, y)$, $\mathbf{T}$ maps $(x, y) \to (ax + cy, bx + dy)$; in particular, $\mathbf{T}$ maps $(1, 0) \to (a, b)$ and, hence,

$$\angle((1, 0), (a, b)) = \alpha$$

$$\begin{aligned}\cos\alpha &= \cos((1, 0), (a, b)) \\ &= \frac{(1, 0) \cdot (a, b)}{|(1, 0)|\;|(a, b)|} \\ &= \frac{a}{1 \cdot (a^2 + b^2)^{1/2}} = a.\end{aligned}$$

Similarly, $\mathbf{T} : (0, 1) \to (c, d)$ and, hence, $\cos\alpha = \cos((0, 1), (c, d)) = d$. Thus, $a = d = \cos\alpha$. Since $0 = ac + bd = a(b + c)$, either $a = 0$ or $c = -b$. If $a = 0$, then $b^2 = a^2 + b^2 = 1$ which implies $b = \pm 1$; and, similarly, $c = \pm 1$. *We claim that $c = -b$ in this case too*; for if not, then $c = b = \pm 1$ and we would have

$$\mathbf{T} : (1, 1) \to (a + c, b + d) = (b, b)$$

$$\cos\alpha = \frac{(1, 1) \cdot (b, b)}{|(1, 1)|\;|(b, b)|} = \frac{2b}{2} = b \neq 0$$

contradicting the fact that $\cos\alpha = a = 0$. Thus, $c = -b$ *in all cases*. Moreover, since $a^2 + b^2 = 1$ and $a = \cos\alpha$, we must have $b = \pm\sin\alpha$. Hence, the matrix of $\mathbf{T}$ has one of the two forms

**(6.1.4)** (a) $\begin{bmatrix} \cos\alpha & \sin\alpha \\ -\sin\alpha & \cos\alpha \end{bmatrix}$; (b) $\begin{bmatrix} \cos\alpha & -\sin\alpha \\ \sin\alpha & \cos\alpha \end{bmatrix}$.

Conversely, if $\mathbf{T} \in \mathrm{Lin}(\mathbf{R}_2, \mathbf{R}_2)$ has the matrix 6.1.4(a), then $\mathbf{T}$ is an isometry by Theorem 4.10.13, $\mathbf{T}$ obviously maps $O \to O$, and for any nonzero vector $\mathbf{u} = (x, y)$, we have

$$\cos(\mathbf{u}, \mathbf{u}\mathbf{T}) = \frac{(x, y) \cdot (x, y)\mathbf{T}}{|(x, y)|\;|(x, y)\mathbf{T}|} = \frac{(x, y) \cdot (x, y)\mathbf{T}}{|(x, y)|^2}$$

(cf. 4.10.9(d))

$$= \frac{(x, y) \cdot (x\cos\alpha - y\sin\alpha, x\sin\alpha + y\cos\alpha)}{|(x, y)|^2}$$

$$= \frac{x^2\cos\alpha + y^2\cos\alpha}{x^2 + y^2} = \cos\alpha.$$

Hence, $\mathbf{T}$ is a rotation about $O$ through the angle $\alpha$. A similar argument shows that the matrix 6.1.4(b) represents a rotation about $O$ through the angle $\alpha$. We have thus proved:

**6.1.5** **Theorem.** Let $\alpha$ be an angle and let $\mathbf{T}$ be a mapping of $\mathbf{R}_2 \to \mathbf{R}_2$. Then the following two statements are equivalent:

(a) $\mathbf{T}$ is a rotation about the origin through the angle $\alpha$.

(b) $\mathbf{T}$ is linear, and its matrix (with respect to the canonical basis) is one of the two matrices exhibited in 6.1.4.

By Theorem 3.9.8, a pair $(a, b)$ of real numbers represents the cosine and sine of an angle if and only if $a^2 + b^2 = 1$ and $b \geqslant 0$. Hence, from Theorem 6.1.5, we immediately deduce the following corollary:

**6.1.6 Theorem.** In $\mathbf{R}_2$, the rotations about the origin are precisely the linear transformations whose matrices (with respect to the canonical basis) have the form

$$\begin{bmatrix} a & b \\ -b & a \end{bmatrix} \qquad (a^2 + b^2 = 1). \tag{6.1.7}$$

Matrices of the form 6.1.7 are called *rotation matrices*. Note that Theorem 6.1.6 is precisely the result which we obtained informally in Section 4.2. In fact, in Example 4.2.20, we obtained, for the *counterclockwise* rotation through $O$, the matrix 6.1.4(a); a similar argument for the clockwise rotation would have led to the matrix 6.1.4(b). Our next step (as you can probably guess by now!) is to let our informal results motivate a formal definition.

**6.1.8 Definition.** The rotation $\mathbf{T}$ through the angle $\alpha$ about $O$ is called *counterclockwise* if its matrix is the matrix 6.1.4(a), *clockwise* if its matrix is the matrix 6.1.4(b). (*Remark*: If $\mathbf{T}$ has the matrix 6.1.7, it is clear that $\mathbf{T}$ is clockwise if $b < 0$, counterclockwise if $b > 0$.)

If $\sin \alpha = 0$ (equivalently, $b = 0$ in 6.1.7), then the two matrices 6.1.4 are the same. In this case, either $\cos \alpha = 1$ and $\mathbf{T} = \mathbf{I}$ (the rotation through 0°), or $\cos \alpha = -1$ and $\mathbf{T} = -\mathbf{I}$ (the rotation through 180°). In all other cases, the clockwise and counterclockwise rotations through $\alpha$ about $O$ are different.

The product of two rotations about $O$ is again a rotation about $O$, as is suggested by Figure 6.1.3. (In the figure, $\mathbf{T}$, $\mathbf{S}$, and $\mathbf{TS}$ are the counterclockwise rotations through angles $\alpha$, $\beta$, $\alpha + \beta$, respectively.) That is, *the set* Rot($O$) *is closed under multiplication*. This can be proved formally; in fact, we can say more:

**6.1.9 Theorem.** (a) Rot($O$) is an abelian group under the operation of multiplication of mappings. (b) For any angle $\alpha$, the clockwise and counterclockwise rotations through $\alpha$ (about $O$) are inverses of each other in the group Rot($O$).

*Proof.* If $a^2 + b^2 = c^2 + d^2 = 1$, then

$$\begin{bmatrix} a & b \\ -b & a \end{bmatrix} \begin{bmatrix} c & d \\ -d & c \end{bmatrix} = \begin{bmatrix} ac - bd & ad + bc \\ -(ad + bc) & ac - bd \end{bmatrix}; \tag{6.1.10}$$

$$(ac - bd)^2 + (ad + bc)^2 = (a^2 + b^2)(c^2 + d^2) = 1.$$

Hence, the product of any two rotation matrices is a rotation matrix. Also, it easily follows from 6.1.10 that multiplication of rotation matrices is commutative; and matrix multiplication is always associative (Theorem 4.4.13(d)). By taking $a = 1$, $b = 0$ we see that the matrix $I$ has the form 6.1.7, and it is easily seen that the product of the two matrices 6.1.4(a) and 6.1.4(b) is $I$, so that these two matrices are inverses of each other. (Note that they are also transposes of each other, which agrees with Theorem 4.10.13.) Since the set of all rotation matrices is isomorphic to Rot($O$), Theorem 6.1.9 follows. ▮

*Remark*: We have already seen other results like Theorem 6.1.9 in previous chapters: in Section 3.3 it was shown that the *translations* of a vector space $V$ form an abelian group, and in Section 4.10 we saw that the *isometries* of a real Euclidean space $E$ form a group (*not* abelian). In the case $E = V = \mathbf{R}_2$, it is clear that the group of translations and the group of rotations about $O$ are both subgroups of the group of isometries.

**6.1.11 Theorem.** (a) If $\mathbf{u}, \mathbf{v} \in \mathbf{R}_2$ and $|\mathbf{u}| = |\mathbf{v}| \neq 0$, then there is a unique rotation about $O$ which maps $\mathbf{u} \to \mathbf{v}$. (b) If $H, H' \in$ Ha($O$), there is a unique rotation about $O$ which maps $H$ onto $H'$.

Figure 6.1.3

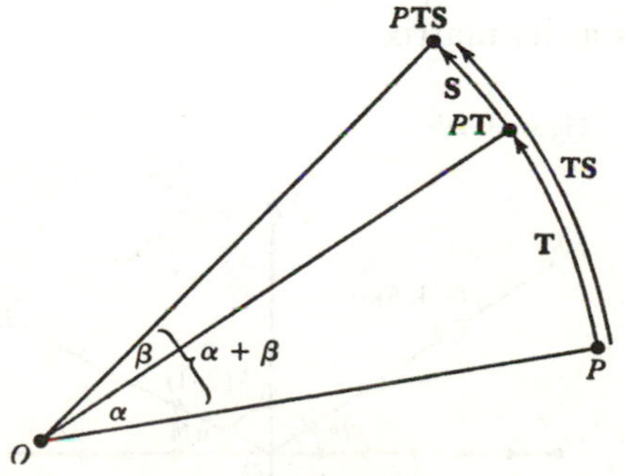

*Proof.* Since any ray with initial point $O$ contains a unique point $\mathbf{u} \in \mathbf{R}_2$ having any given positive absolute value (cf. Theorem 3.9.13(d)), it is clear that (b) is a consequence of (a); hence, it will suffice to prove (a). Let $|\mathbf{u}| = |\mathbf{v}| = k$; then $\mathbf{u}/k = (a, b)$, $\mathbf{v}/k = (c, d)$, where

$$a^2 + b^2 = c^2 + d^2 = 1.$$

By Theorem 6.1.6, there are rotations $\mathbf{T}_1$, $\mathbf{T}_2$ whose matrices are

$$\begin{bmatrix} a & b \\ -b & a \end{bmatrix}; \quad \begin{bmatrix} c & d \\ -d & c \end{bmatrix}$$

respectively. Clearly, $\mathbf{T}_1$ maps $(k, 0) \to (ka, kb) = \mathbf{u}$, $\mathbf{T}_2$ maps $(k; 0) \to (kc, kd) = \mathbf{v}$. Hence, $\mathbf{T}_1^{-1}\mathbf{T}_2$ maps $\mathbf{u} \to \mathbf{v}$, and, by Theorem 6.1.9, $\mathbf{T}_1^{-1}\mathbf{T}_2$ is a rotation about $O$. If $\mathbf{S}$ were any other rotation about $O$ which maps $\mathbf{u} \to \mathbf{v}$, then $\mathbf{T}_1\mathbf{S}$ maps $(1, 0) \to \mathbf{v}/k = (c, d)$. Hence, the matrix of $\mathbf{T}_1\mathbf{S}$ has $(c, d)$ as its first row and, thus, must have $(-d, c)$ as its second row by Theorem 6.1.6. It follows that $\mathbf{T}_1\mathbf{S} = \mathbf{T}_2$, so that $\mathbf{S} = \mathbf{T}_1^{-1}\mathbf{T}_2$ proving uniqueness. Figure 6.1.4 exhibits the various rotations mentioned in the proof. ∎

Note that in the above proof, $\mathbf{T}_1^{-1}$ has the matrix

$$\begin{bmatrix} a & -b \\ b & a \end{bmatrix}$$

(*why*?) so that $\mathbf{S} = \mathbf{T}_1^{-1}\mathbf{T}_2$ has the matrix

$$\begin{bmatrix} a & -b \\ b & a \end{bmatrix} \begin{bmatrix} c & d \\ -d & c \end{bmatrix}$$

which can be computed explicitly. For example, if $H = H(O, (2, 1))$ and $H' = H(O, (-4, 3))$ (see Fig. 6.1.5), then

$$(a, b) = \mathbf{u}/k = (2, 1)/\sqrt{5}; \qquad (c, d) = \mathbf{v}/k = (-4, 3)/5$$

(the denominators chosen to make $|\mathbf{u}|/k = |\mathbf{v}|/k = 1$), and the rotation $\mathbf{S}$ which maps $H \to H'$ has as its matrix

Figure 6.1.4

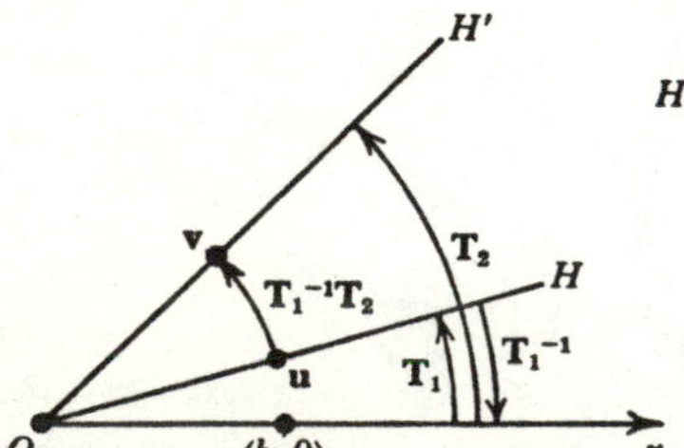


Figure 6.1.5

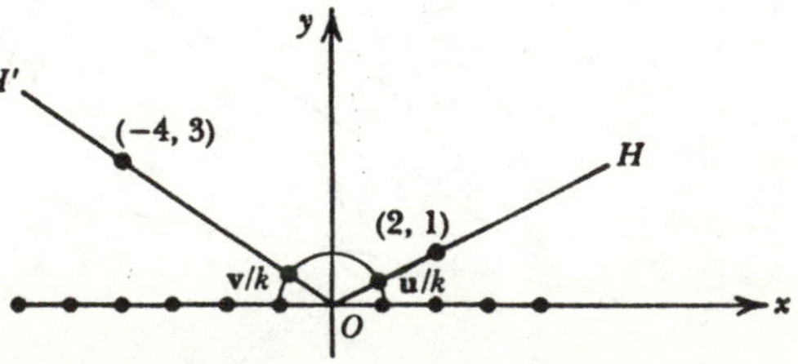

$$(1/\sqrt{5})\,(1/5)\begin{bmatrix}2 & -1\\ 1 & 2\end{bmatrix}\begin{bmatrix}-4 & 3\\ -3 & -4\end{bmatrix} = (1/5\sqrt{5})\begin{bmatrix}-5 & 10\\ -10 & -5\end{bmatrix}$$

$$= \begin{bmatrix}-1/\sqrt{5} & 2/\sqrt{5}\\ -2/\sqrt{5} & -1/\sqrt{5}\end{bmatrix}.$$

As a check, we have

$$[2\ \ 1]\begin{bmatrix}-1/\sqrt{5} & 2/\sqrt{5}\\ -2/\sqrt{5} & -1/\sqrt{5}\end{bmatrix} = [-4/\sqrt{5}\ \ 3/\sqrt{5}] = [-4\ \ 3]/\sqrt{5}$$

so that $H\mathbf{S} = H'$ as expected.

**Exercises**

1. Find the matrix of the rotation about $O$ which maps $(1, 2) \to (2, 1)$. (ANS)
2. If a rotation about $O$ maps $H(O, (7, 24)) \to H(O, (4, 3))$, then it maps $(3, 1)$ into what point? Draw a figure to illustrate. (ANS)
3. Using Definition 6.1.8, show that a rotation about $O$ is "clockwise" or "counterclockwise" according to whether the image of the point $(1, 0)$ lies in the lower or upper half plane, respectively.
4. Prove: if $A$ is the matrix of a clockwise rotation, then $-A$ is the matrix of a counterclockwise rotation, and vice versa.

*5. Let $H_1, H'_1, H_2, H'_2 \in \mathrm{Ha}_2(O)$ and assume that $\angle(H_1, H'_1) = \angle(H_2, H'_2)$. Let $\mathbf{T}$ be the rotation about $O$ which maps $H_1$ onto $H'_1$ (cf. Theorem 6.1.11(b)). Prove that $\mathbf{T}$ either maps $H_2$ onto $H'_2$ or maps $H'_2$ onto $H_2$. (The two possibilities are shown in Figure 6.2.1 at the beginning of the next section.)

## 6.2 Oriented angles in $\mathbf{R}_2$

Let $H_1, H'_1, H_2, H'_2$ be half lines in $\mathbf{R}_2$ with initial point $O$ (the origin), and assume that $\angle\,(H_1, H'_1) = \alpha$. By Theorem 6.1.11 there is a rotation $\mathbf{T}$ which maps $H_1$ onto $H'_1$. If $\mathbf{T}$ also maps $H_2 \to H'_2$, then $\angle(H_2, H'_2) = \alpha$. However, the converse is not true: if $\angle(H_2, H'_2) = \alpha$, $\mathbf{T}$ need not map $H_2 \to H'_2$; it may instead map $H'_2 \to H_2$ (see Fig. 6.2.1). These are, in fact, the only two possibilities (Sec. 6.1, Exercise 5); which one of them occurs depends on whether the angle $\alpha$ from $H_2$ to $H'_2$ has the same orientation (clockwise or counterclockwise) as that from $H_1$ to $H'_1$, or whether it has opposite orientation. This suggests the desirability of introducing the concept of "oriented angle" as a means of distinguishing between, say, the angle from $H_1$ to $H'_1$ and the angle from $H'_1$ to $H_1$.

One way of doing this formally is as follows: if $H_1, H'_1, H_2, H'_2$ are half lines in $\mathbf{R}_2$ with initial point at $O$, call the ordered pairs $(H_1, H'_1)$ and $(H_2, H'_2)$ *equivalent* if there exists a rotation $\mathbf{T}$ about $O$ which simultaneously maps $H_1 \to H'_1$ and $H_2 \to H'_2$. "Equivalence" is indeed an equivalence relation, as we now show. The symmetric property is trivial: if $\mathbf{T}$ maps $H_1 \to H'_1$ and $H_2 \to H'_2$, then obviously, $\mathbf{T}$ maps $H_2 \to H'_2$ and $H_1 \to H'_1$. Reflexivity (i.e., $(H, H') \sim (H, H')$) means that there exists $\mathbf{T} \in \mathrm{Rot}(O)$ which maps $H \to H'$ and $H \to H'$, which is precisely the assertion of Theorem 6.1.11(b). Finally, if $\mathbf{T}, \mathbf{S}$ are rotations about $O$ such that

$$\mathbf{T}: H_1 \to H'_1, H_2 \to H'_2 \qquad (\text{i.e., } (H_1, H'_1) \sim (H_2, H'_2));$$
$$\mathbf{S}: H_2 \to H'_2, H_3 \to H'_3 \qquad (\text{i.e., } (H_2, H'_2) \sim (H_3, H'_3)),$$

then the "uniqueness" part of 6.1.11(b) (applied to $H_2 \to H'_2$) implies that $\mathbf{T} = \mathbf{S}$, so that $\mathbf{T}$ maps $H_1 \to H'_1$ and $H_3 \to H'_3$. Thus, $(H_1, H'_1) \sim (H_3, H'_3)$, proving transitivity. Hence, we do have an equivalence relation on the set of all ordered pairs of half lines with initial point $O$. Now, define the *oriented angle* from $H$ to $H'$ (*notation*: Or $\angle(H, H')$) to be the equivalence class to which the pair $(H, H')$ belongs. Thus, for example, in Figure 6.2.1(a), we have

$$\text{Or } \angle(H_1, H'_1) = \text{Or } \angle(H_2, H'_2)$$

since $\mathbf{T}$ maps $H_1 \to H'_1$ and $H_2 \to H'_2$; but in Figure 6.2.1(b), we have

Figure 6.2.1

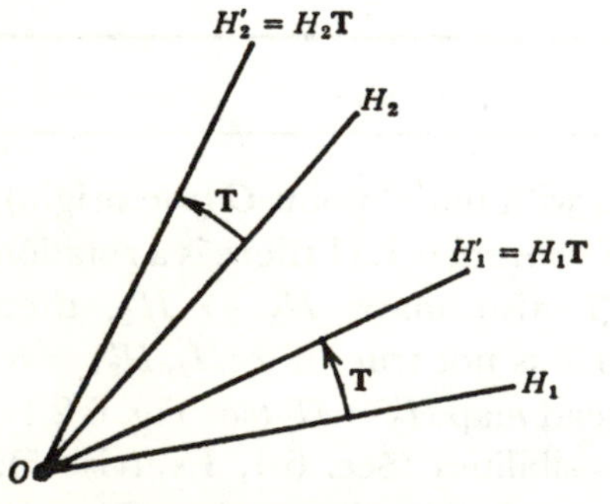


angles from $H_1$ to $H'_1$, $H_2$ to $H'_2$ have same orientation

(a)

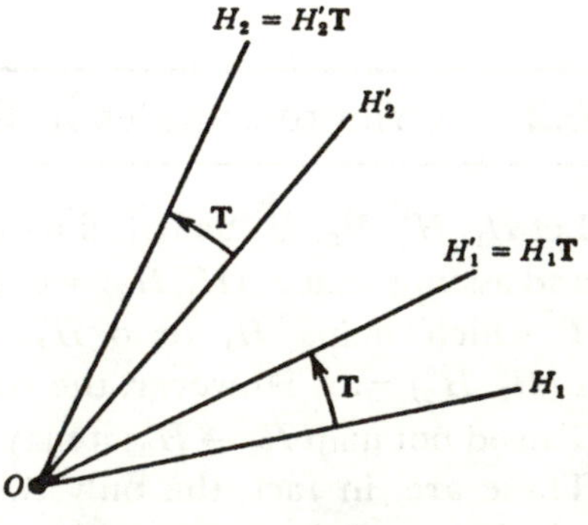


angles from $H_1$ to $H'_1$, $H_2$ to $H'_2$ have opposite orientation

(b)

$$\text{Or } \angle(H_1, H'_1) = \text{Or } \angle(H'_2, H_2) \neq \text{Or } \angle(H_2, H'_2).$$

We shall often denote oriented angles by Greek letters such as $\theta$, $\psi$, and others.

*Notation:* The set of all oriented angles will be denoted **Or**.

**6.2.1** **Theorem.** The mapping

$$\mathbf{T} \to \text{Or } \angle(H, H\mathbf{T}) \qquad (\mathbf{T} \in \text{Rot}(O), H \in \text{Ha}_2(O)) \tag{6.2.2}$$

is well defined and is a one-to-one mapping of Rot($O$) onto **Or**.

By "well defined" we mean that Or $\angle(H, H\mathbf{T})$ depends only on **T**, not on $H$, a fact which is immediate from the definition of "oriented angle". To show that the mapping 6.2.2 is one-to-one, suppose Or $\angle(H, H\mathbf{T})$ = Or $\angle(H, H\mathbf{S})$; then by definition, some rotation simultaneously maps $H \to H\mathbf{T}$ and $H \to H\mathbf{S}$. Hence, $H\mathbf{T} = H\mathbf{S}$, so that $\mathbf{T} = \mathbf{S}$ by the uniqueness assertion of Theorem 6.1.11(b). To show that the mapping 6.2.2 is onto, let $\theta$ be any oriented angle; then $\theta$ = Or $\angle(H, H')$ for some $H, H' \in \text{Ha}(O)$. By 6.1.11, there exists $\mathbf{T} \in$ Rot($O$) which maps $H \to H'$, so that

$$\theta = \text{Or } \angle(H, H') = \text{Or } \angle(H, H\mathbf{T}),$$

that is, $\theta$ is the image of **T** under the mapping 6.2.2.

**6.2.3.** **Definition.** If $\theta$ is the oriented angle corresponding to **T** via the mapping 6.2.2, we call $\theta$ the *oriented angle of* **T**, and we write

$$\mathbf{T} = \mathcal{R}_{O,\theta}$$

to indicate that **T** is the rotation about $O$ whose oriented angle is $\theta$. We call $\theta$ a *clockwise* or *counterclockwise* oriented angle according to whether **T** is a clockwise or counterclockwise rotation.

*Remark*: As shown in Section 6.1, each (*non*oriented) angle $\alpha$ corresponds to two (not necessarily distinct) rotations through $\alpha$, one clockwise and one counterclockwise. Hence, by Theorem 6.2.1, each angle $\alpha$ corresponds to two oriented angles, one clockwise and one counterclockwise. In fact, if $\alpha = \angle(H_1, H_2)$, then the oriented angles corresponding to $\alpha$ are

$$\text{Or } \angle(H_1, H_2); \qquad \text{Or } \angle(H_2, H_1). \tag{6.2.4}$$

Indeed, by 6.1.11, there exists $\mathbf{T} \in$ Rot($O$) such that $\mathbf{T} : H_1 \to H_2$; by definition, **T** is a rotation through $\alpha$. Hence, the two rotations about

$O$ through $\alpha$ are precisely $\mathbf{T}$ and $\mathbf{T}^{-1}$ (Theorem 6.1.9(b)), and since $\mathbf{T} : H_1 \to H_2$, we clearly have $\mathbf{T}^{-1} : H_2 \to H_1$, from which our assertion follows. (Compare with Sec. 6.1, Exercise 5. *Question*: When will the two oriented angles 6.2.4 be equal?)

As we remarked in Section 3.9, our definition of "angle" had the effect of restricting all angles to the interval from 0° to 180°. When assigning a number of degrees (or radians) to *oriented* angles, it is usual to identify a *counterclockwise* oriented angle with the corresponding angle, and a clockwise oriented angle with "minus" the angle; thus, in Figure 6.2.2, we have

$$\text{Or } \angle(H_1, H_2) = 45°$$
$$\text{Or } \angle(H_2, H_1) = -45°.$$

Thus, oriented angles range from $-180°$ to $+180°$ ( a full 360° range).

Although we will not define "degrees" or "radians" formally (except for some special angles), it is possible, even without them, to define *addition of oriented angles*. Let $\theta$ and $\psi$ be any oriented angles, and let $H_1$, $H_2$ be half lines with initial point $O$ such that

**(6.2.5)** $\theta = \text{Or } \angle(H_1, H_2)$.

We can then find $H_3 \in \text{Ha}(O)$ such that

**(6.2.6)** $\psi = \text{Or } \angle(H_2, H_3)$

(indeed, if $\mathbf{S}$ is the rotation corresponding to $\psi$, then by Theorem 6.2.1 we may take $H_3$ to be $H_2\mathbf{S}$). We then define

**(6.2.7)** $\theta + \psi = \text{Or } \angle(H_1, H_2) + \text{Or } \angle(H_2, H_3) = \text{Or } \angle(H_1, H_3)$.

Figure 6.2.3(a) illustrates this definition when $\theta$ and $\psi$ have the same orientation, Figure 6.2.3(b) when they have opposite orientation. Actually, for our definition to be valid, it must be shown that the right side of 6.2.7 depends only on $\theta$ and $\psi$, not on the particular choice of half lines. To show this, let $\theta$ and $\psi$ be given and suppose $H_1$, $H_2$, $H_3$ satisfy 6.2.5 and 6.2.6. If $\mathbf{T}$, $\mathbf{S}$ are, respectively, the rotations which map $H_1 \to H_2$ and $H_2 \to H_3$ (see Theorem 6.1.11(b)), then by 6.2.5

Figure 6.2.2

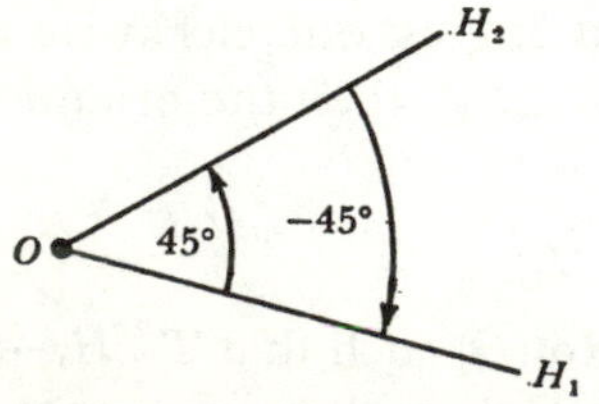

and 6.2.6 we see that **T** and **S** correspond to $\theta$ and $\psi$ (respectively) under the mapping 6.2.2; similarly, since **TS** maps $H_1 \to H_3$ (why?), **TS** corresponds to Or $\angle(H_1, H_3)$ under the mapping 6.2.2. Thus, the right side of 6.2.7 is uniquely determined by **TS**, which in turn is uniquely determined by **T** and **S**, which in turn are uniquely determined by $\theta$ and $\psi$. This proves that addition of oriented angles is well defined, as we wished to show. We have also proved more: we have shown that if **T**, **S** are the rotations corresponding to $\theta$, $\psi$ (respectively), then **TS** corresponds to $\theta + \psi$. That is,

(6.2.8) $\mathscr{R}_{O,\theta+\psi} = \mathbf{TS} = \mathscr{R}_{O,\theta}\mathscr{R}_{O,\psi}$.

By this time you should recognize the form of equation 6.2.8 as that of a homomorphism equation. Specifically, 6.2.8 tells us that the mapping $\theta \to \mathscr{R}_{O,\theta}$ is a homomorphism. This homomorphism is clearly the inverse of the mapping 6.2.2 and, hence, is one-to-one and onto; that is, it is an *isomorphism*. Thus, the set of all oriented angles (under addition) is isomorphic to the set of all rotations about $O$ (under multiplication). We can, thus, translate Theorem 6.1.9 into the following result:

**6.2.9 Theorem.** The set **Or** of all oriented angles is an abelian group under addition, and the mapping 6.2.2 is an isomorphism of Rot($O$) onto **Or**. In the group **Or**, the clockwise and counterclockwise oriented angles corresponding to any given angle $\alpha$ are additive inverses of each other.

Figure 6.2.3

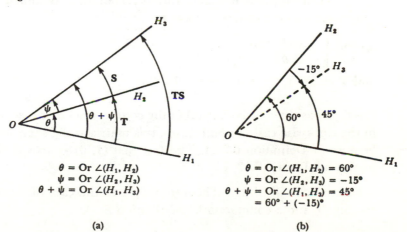

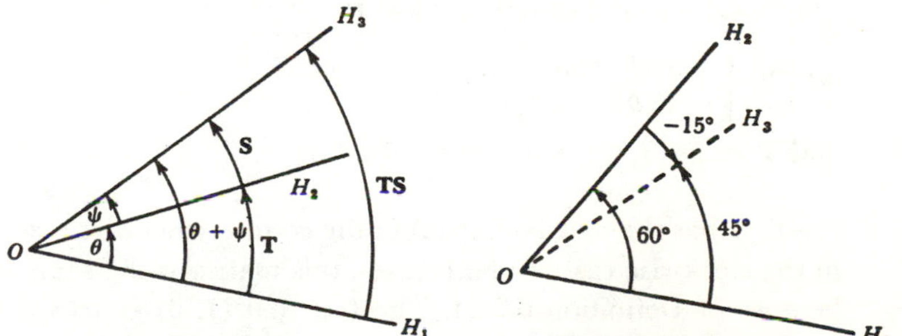


$\theta$ = Or $\angle(H_1, H_2)$
$\psi$ = Or $\angle(H_2, H_3)$
$\theta + \psi$ = Or $\angle(H_1, H_3)$

(a)

$\theta$ = Or $\angle(H_1, H_2) = 60°$
$\psi$ = Or $\angle(H_2, H_3) = -15°$
$\theta + \psi$ = Or $\angle(H_1, H_3) = 45°$
$= 60° + (-15)°$

(b)

As we have already shown, if $\alpha = \angle(H_1, H_2)$, then the two oriented angles corresponding to $\alpha$ are Or $\angle(H_1, H_2)$ and Or $\angle(H_2, H_1)$ (see 6.2.4). Hence, the last part of Theorem 6.2.9 asserts that

$$\text{(6.2.10)}\quad \text{Or } \angle(H_1, H_2) = -\text{Or } \angle(H_2, H_1) \qquad (\text{all } H_1, H_2 \in \text{Ha}(O)).$$

This shows that "minus" an oriented angle (as defined formally in the additive group **Or**) has the same meaning as in the informal discussion above (see text between equations 6.2.4 and 6.2.5).

### Sines and cosines of oriented angles

As stated earlier, it is customary to identify a counterclockwise oriented angle with the corresponding angle (between 0° and 180°) and a clockwise oriented angle with "minus" the angle (between 0° and −180°). This suggests defining the *sine* and *cosine* of an oriented angle as follows:

**6.2.11 Definition.** Let $\alpha$ be an angle, and let $\theta$ be the corresponding *counterclockwise* oriented angle (so that $-\theta$ is the corresponding clockwise oriented angle). Then

$$\cos\theta = \cos\alpha; \qquad \sin\theta = \sin\alpha$$
$$\cos(-\theta) = \cos\alpha; \qquad \sin(-\theta) = -\sin\alpha.$$

In view of 3.9.7 and 3.9.8, it follows that *two real numbers represent the cosine and sine of an oriented angle if and only if the sum of their squares is* 1.

**6.2.12 Theorem.** For any oriented angle $\theta$ (clockwise or counterclockwise), the matrix of the corresponding rotation $\mathbf{T} = \mathcal{R}_{O,\theta}$ with respect to the canonical basis is

$$\text{(6.2.13)}\quad \begin{bmatrix} \cos\theta & \sin\theta \\ -\sin\theta & \cos\theta \end{bmatrix}$$

and $\mathbf{T}$ maps $(1, 0) \to (\cos\theta, \sin\theta)$.

*Proof.* $\mathbf{T}$ has the matrix 6.1.4(a) in the counterclockwise case, 6.1.4(b) in the clockwise case. In both cases, this matrix is the same as 6.2.13 because of Definition 6.2.11. The fact that $(1, 0) \to (\cos\theta, \sin\theta)$ is immediate from 6.2.13. ▮

Figure 6.2.4 illustrates Theorem 6.2.12 in the case of a clockwise rotation $\mathbf{T}$ and a counterclockwise rotation $\mathbf{S}$.

**6.2.14 Theorem.** (The addition formulas) For all oriented angles $\theta$, $\psi$,

$$\cos(\theta + \psi) = \cos\theta\cos\psi - \sin\theta\sin\psi;$$
$$\sin(\theta + \psi) = \cos\theta\sin\psi + \sin\theta\cos\psi.$$

*Proof.* By Theorem 6.2.12, $\theta$ and $\psi$ correspond, respectively, to the matrices

$$\text{(6.2.15)}\quad \begin{bmatrix} \cos\theta & \sin\theta \\ -\sin\theta & \cos\theta \end{bmatrix}; \quad \begin{bmatrix} \cos\psi & \sin\psi \\ -\sin\psi & \cos\psi \end{bmatrix}.$$

Since *addition* of oriented angles corresponds to *multiplication* of rotations (and, hence, to multiplication of matrices), it follows that the matrix

$$\begin{bmatrix} \cos(\theta + \psi) & \sin(\theta + \psi) \\ -\sin(\theta + \psi) & \cos(\theta + \psi) \end{bmatrix}$$

which corresponds to $\theta + \psi$ must equal the product of the two matrices 6.2.15. But this product, by direct computation, is

$$\begin{bmatrix} \cos\theta\cos\psi - \sin\theta\sin\psi & \cos\theta\sin\psi + \sin\theta\cos\psi \\ -(\cos\theta\sin\psi + \sin\theta\cos\psi) & \cos\theta\cos\psi - \sin\theta\sin\psi \end{bmatrix}$$

and the theorem follows. ∎

**6.2.16 Theorem.** No two distinct oriented angles have the same sine and cosine.

Figure 6.2.4

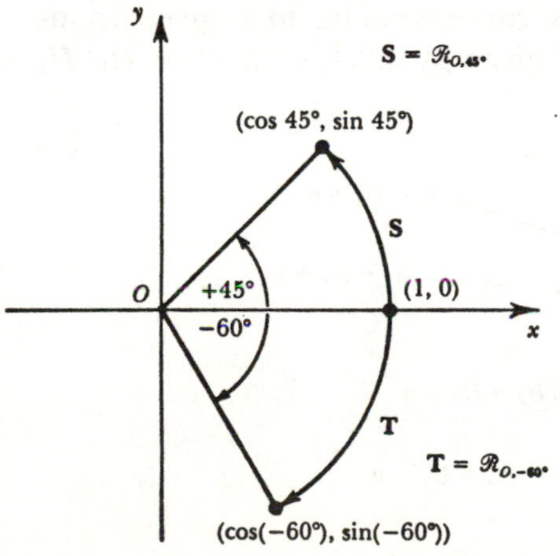

*Proof.* If they had the same sine and cosine, then they would correspond to the same matrix and, hence, to the same rotation, contradicting Theorem 6.2.1. (Of course, two oriented angles may have the same sine *or* the same cosine; e.g., sin 60° = sin 120°.) ∎

**Oriented angles and rotations at a point other than $O$**

So far we have defined Or $\angle(H_1, H_2)$ only when the common initial point of $H_1$, $H_2$ is $O$. Suppose instead that $H_1$, $H_2$ are half lines in $\mathbf{R}_2$ which have a common initial point $\mathbf{u} \neq \mathbf{0}$ (we identify vectors with points, as in Sec. 3.1 and elsewhere). Figure 6.2.5 suggests how the oriented angle may be defined in this case. We proceed as follows: the images of $H_1$, $H_2$ under the translation $\mathcal{T}_{-\mathbf{u}}$ are two half lines $H_1'$, $H_2'$ whose common initial point is $O$ (see Theorem 4.10.15), and we define

$$\text{Or } \angle(H_1, H_2) = \text{Or } \angle(H_1', H_2').$$

It is easy to deduce from this definition the more general fact that if $H_1, H_2 \in \text{Ha}(A)$ for any fixed point $A$ and if $\mathbf{v}$ is any element of $\mathbf{R}_2$, then

**(6.2.17)** $\text{Or } \angle(H_1, H_2) = \text{Or } \angle(H_1\mathcal{T}_{\mathbf{v}}, H_2\mathcal{T}_{\mathbf{v}}).$

(*Proof*: By definition, Or $\angle(H_1, H_2) = $ Or $\angle(H_1\mathcal{T}_{-A}, H_2\mathcal{T}_{-A})$; similarly, since $H_1\mathcal{T}_{\mathbf{v}}$ and $H_2\mathcal{T}_{\mathbf{v}}$ have common initial point $A + \mathbf{v}$, we have Or $\angle(H_1\mathcal{T}_{\mathbf{v}}, H_2\mathcal{T}_{\mathbf{v}}) = $ Or $\angle(H_1\mathcal{T}_{\mathbf{v}}\mathcal{T}_{-(A+\mathbf{v})}, H_2\mathcal{T}_{\mathbf{v}}\mathcal{T}_{-(A+\mathbf{v})})$. But $\mathcal{T}_{\mathbf{v}}\mathcal{T}_{-(A+\mathbf{v})}$ equals $\mathcal{T}_{-A}$, and 6.2.17 follows.) In short, we have generalized the definition of oriented angle in such a way that *translations preserve oriented angles*. Since translations also preserve angles (Theorem 4.10.15), the two oriented angles corresponding to a given (nonoriented) angle $\angle(H_1, H_2)$ are still given by 6.2.4, even when $H_1$, $H_2$

Figure 6.2.5

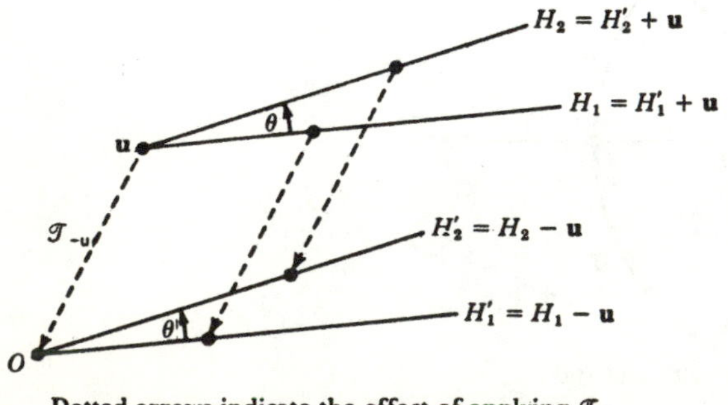


Dotted arrows indicate the effect of applying $\mathcal{T}_{-\mathbf{u}}$

belong to Ha($A$) rather than Ha($O$). Equation 6.2.7 also remains valid at points other than $O$; that is, if $A$ is any point in $\mathbf{R}_2$, then

(6.2.18) $$\text{Or } \angle(H_1, H_2) + \text{Or } \angle(H_2, H_3) = \text{Or } \angle(H_1, H_3)$$

*for all half lines* $H_1$, $H_2$, $H_3$ *in* Ha($A$). Indeed, if $H_i' = H_i - A$ ($i = 1, 2, 3$), then $H_i' \in$ Ha($O$) and Or $\angle(H_i, H_j) =$ Or $\angle(H_i', H_j')$ by definition, so that 6.2.18 follows from 6.2.7. (A similar argument shows that equation 6.2.10 remains valid at points other than $O$.)

Pairs of half lines in Ha($A$) correspond to rotations about $A$ in the same way that pairs of half lines in Ha($O$) correspond to rotations about $O$. Specifically, we have the following analog to Theorem 6.1.11(b):

**6.2.19 Theorem.** If $A$ is any point in $\mathbf{R}_2$ and if $H_1, H_2 \in$ Ha($A$), then there is a unique rotation $\mathbf{T}$ about $A$ which maps $H_1$ onto $H_2$. In fact, if

$$\theta = \text{Or } \angle(H_1, H_2),$$

then $\mathbf{T}$ is given by

$$\mathbf{T} = \mathcal{T}_{-A}\mathcal{R}_{O,\theta}\mathcal{T}_A.$$

*Proof.* If we let $H_i' = H_i\mathcal{T}_{-A} = H_i - A$ as in Figure 6.2.6, then $H_1', H_2' \in$ Ha($O$) and

$$\text{Or } \angle(H_1', H_2') = \text{Or } \angle(H_1, H_2) = \theta.$$

Figure 6.2.6

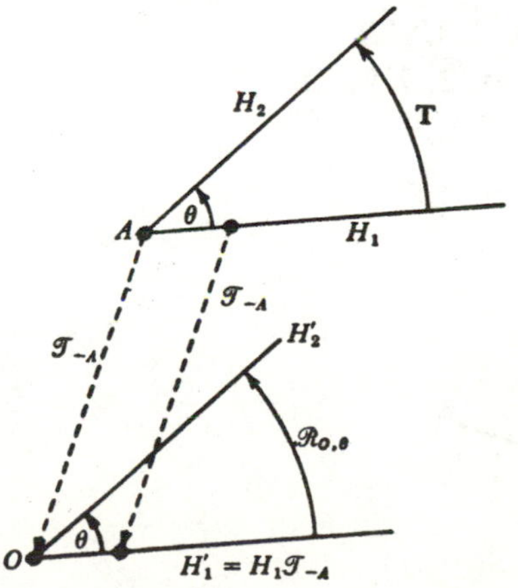

Hence, $\mathcal{R}_{O,\theta}$ maps $H_1'$ onto $H_2'$. Hence, we have

$$H_1 \xrightarrow{\mathcal{T}_{-A}} H_1' \xrightarrow{\mathcal{R}_{O,\theta}} H_2' \xrightarrow{\mathcal{T}_A} H_2$$

so that the mapping $\mathbf{T} = \mathcal{T}_{-A}\mathcal{R}_{O,\theta}\mathcal{T}_A$ maps $H_1$ onto $H_2$. To show that $\mathbf{T} \in \mathrm{Rot}(A)$, we verify the three parts of 6.1.3: (a) $\mathbf{T} : A \to A$ by direct computation; (b) $\mathbf{T}$ is an isometry since it is the product of isometries; and (c) for any $H \in \mathrm{Ha}(A)$, we have

$$\begin{aligned} \angle(H, H\mathbf{T}) &= \angle(H\mathcal{T}_{-A}, H\mathbf{T}\mathcal{T}_{-A}) \qquad \text{(by Theorem 4.10.15)} \\ &= \angle(H\mathcal{T}_{-A}, H\mathcal{T}_{-A}\mathcal{R}_{O,\theta}) \end{aligned}$$

which is the same for all $H$, since $\mathcal{R}_{O,\theta}$ is a rotation about $O$.

Finally, to prove that $\mathbf{T}$ is unique, suppose $\mathbf{S} \in \mathrm{Rot}(A)$ also maps $H_1 \to H_2$. Then by an argument like the above, $\mathcal{T}_A\mathbf{S}\mathcal{T}_{-A}$ is a rotation about $O$ which maps $H_1' \to H_2'$. Hence,

$$\mathcal{T}_A\mathbf{S}\mathcal{T}_{-A} = \mathcal{R}_{O,\theta};$$

multiplying by $\mathcal{T}_{-A}$ on the left and $\mathcal{T}_A$ on the right, we get $\mathbf{S} = \mathcal{T}_{-A}\mathcal{R}_{O,\theta}\mathcal{T}_A = \mathbf{T}$. ▮

**6.2.20 Example.** Let $H_1, H_2$ be the half lines with initial point (2, 3) which pass through the points (3, 4) and (3, 10), respectively (see Figure 6.2.7). Find the equations of the rotation $\mathbf{T}$ about (2, 3) which maps $H_1$ onto $H_2$.

*Solution.* Here $A = (2, 3)$, $H_i' = H_i - (2, 3)$. We first find the rotation $\mathcal{R}_{O,\theta}$ which maps $H_1' \to H_2'$. This can be done by the method of the example at the very end of Section 6.1: since $(1, 1) \in H_1'$ and $(1, 7) \in H_2'$, we have (using the notations of that example)

$$\mathbf{u}/k = (1, 1)/\sqrt{2} = (1/\sqrt{2}, 1/\sqrt{2})$$
$$\mathbf{v}/k = (1, 7)/\sqrt{50} = (1/5\sqrt{2}, 7/5\sqrt{2})$$

so that

Figure 6.2.7

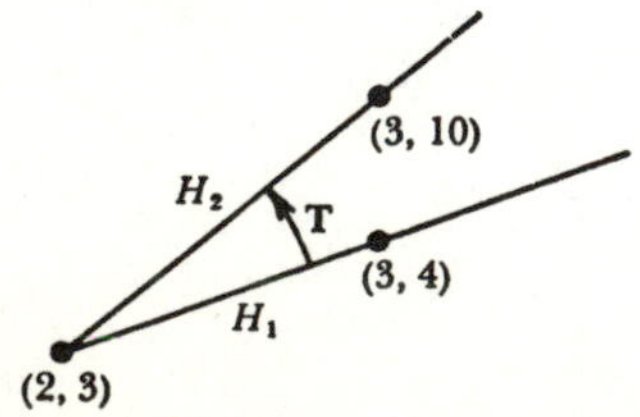

$$\text{matrix of } \mathcal{R}_{O,\theta} = \begin{bmatrix} 1/\sqrt{2} & -1/\sqrt{2} \\ 1/\sqrt{2} & 1/\sqrt{2} \end{bmatrix} \begin{bmatrix} 1/5\sqrt{2} & 7/5\sqrt{2} \\ -7/5\sqrt{2} & 1/5\sqrt{2} \end{bmatrix}$$

$$= \begin{bmatrix} 4/5 & 3/5 \\ -3/5 & 4/5 \end{bmatrix}$$

$$\mathcal{R}_{O,\theta} : (x, y) \to \left(\frac{4}{5}x - \frac{3}{5}y, \frac{3}{5}x + \frac{4}{5}y\right)$$

$$(x, y) \xrightarrow{\mathcal{T}_{-A}} (x - 2, y - 3)$$

$$\xrightarrow{\mathcal{R}_{O,\theta}} \left(\frac{4}{5}(x-2) - \frac{3}{5}(y-3), \frac{3}{5}(x-2) + \frac{4}{5}(y-3)\right)$$

$$\xrightarrow{\mathcal{T}_{A}} \left(\frac{4}{5}(x-2) - \frac{3}{5}(y-3) + 2, \frac{3}{5}(x-2) + \frac{4}{5}(y-3) + 3\right)$$

$$= \left(\frac{4}{5}x - \frac{3}{5}y + \frac{11}{5}, \frac{3}{5}x + \frac{4}{5}y - \frac{3}{5}\right).$$

Hence, the equations of **T** are

$$x' = \frac{4}{5}x - \frac{3}{5}y + \frac{11}{5}$$

$$y' = \frac{3}{5}x + \frac{4}{5}y - \frac{3}{5}$$

where $(x', y')$ denotes the image of $(x, y)$ under **T**. ▮

*Remarks*: 1. It is clear from the above equations that $(x', y')$ is obtained from $(x, y)$ by first performing the rotation $\mathcal{R}_{O,\theta}$ and then adding the vector $\mathbf{v} = (11/5, -3/5)$. In other words, $\mathbf{T} = \mathcal{R}_{O,\theta}\,\mathcal{T}_{\mathbf{v}}$. A generalization of this argument shows that *any rotation in* $\mathbf{R}_2$ *equals a rotation about O followed by a translation*. In the next section, we shall prove the converse: any (nonidentity) rotation about $O$, followed by a translation, equals a rotation about some point.

2. The rotation described in Theorem 6.2.19 is called the *rotation about A through the oriented angle* $\theta$ and is denoted $\mathcal{R}_{A,\theta}$. (This reduces to previous definitions when $A = O$.) It is clear from Theorem 6.2.19 that *every* rotation about a given point $A$ is of the form

$$\mathcal{R}_{A,\theta} = \mathcal{T}_{-A}\,\mathcal{R}_{O,\theta}\,\mathcal{T}_{A}$$

for some $\theta$. (Note the similarity between this result and Theorem 3.8.32.) It follows that the mapping

$$\mathcal{R}_{O,\theta} \to \mathcal{R}_{A,\theta}$$

is an isomorphism of Rot($O$) → Rot($A$) (cf. Sec. 1.8, Exercise 7); in particular, Rot($A$) is an abelian group. Since we have seen that the mapping $\theta \to \mathcal{R}_{O,\theta}$ is an isomorphism of **Or** → Rot($O$), it also follows that the mapping

$$\theta \to \mathcal{R}_{A,\theta}$$

is an isomorphism of **Or** onto Rot($A$). For all $\theta \in \mathbf{Or}$ and all rays $H \in \mathrm{Ha}(A)$, we have

$$\mathrm{Or}\ \angle(H, H\mathcal{R}_{A,\theta}) = \theta;$$

indeed, if we have Or $\angle(H, H\mathcal{R}_{A,\theta}) = \psi$, then, by Theorem 6.2.19, we have $\mathcal{R}_{A,\theta} = \mathcal{R}_{A,\psi}$ and, hence, $\theta = \psi$.

**6.2.21 Theorem.** Rotations preserve oriented angles. That is, if $\mathbf{T} : \mathbf{R}_2 \to \mathbf{R}_2$ is any rotation, $P$ is any point in $\mathbf{R}_2$, and $H_1, H_2 \in \mathrm{Ha}(P)$, then

$$\mathrm{Or}\ \angle(H_1, H_2) = \mathrm{Or}\ \angle(H_1\mathbf{T}, H_2\mathbf{T}).$$

*Proof.* We first assume that $\mathbf{T}$ is a rotation about $O$ and that $H_1, H_2$ belong to Ha($O$). Then by definition of oriented angle,

$$\mathrm{Or}\ \angle(H_1, H_1\mathbf{T}) = \mathrm{Or}\ \angle(H_2, H_2\mathbf{T}).$$

Hence, we have

$$\begin{aligned}
\mathrm{Or}\ \angle(H_1, H_2) &= \mathrm{Or}\ \angle(H_1, H_1\mathbf{T}) + \mathrm{Or}\ \angle(H_1\mathbf{T}, H_2) && \text{(by 6.2.7)}\\
&= \mathrm{Or}\ \angle(H_2, H_2\mathbf{T}) + \mathrm{Or}\ \angle(H_1\mathbf{T}, H_2) && \text{(substitution)}\\
&= \mathrm{Or}\ \angle(H_1\mathbf{T}, H_2) + \mathrm{Or}\ \angle(H_2, H_2\mathbf{T}) && \text{(since the group } \mathbf{Or} \text{ is abelian!)}\\
&= \mathrm{Or}\ \angle(H_1\mathbf{T}, H_2\mathbf{T})
\end{aligned}$$

as desired.

In the more general case where $H_1, H_2 \in \mathrm{Ha}(P)$, $P \neq O$ (but still assuming $\mathbf{T} \in \mathrm{Rot}(O)$), let $H_i' = H_i - P$; then $H_i' \in \mathrm{Ha}(O)$ and

$$H_i\mathbf{T} = (H_i' + P)\mathbf{T} = H_i'\mathbf{T} + P\mathbf{T}$$

since $\mathbf{T}$ is linear. Hence,

$$\begin{aligned}
\mathrm{Or}\ \angle(H_1, H_2) &= \mathrm{Or}\ \angle(H_1', H_2') && \text{(by 6.2.17 with } \mathbf{v} = -P)\\
&= \mathrm{Or}\ \angle(H_1'\mathbf{T}, H_2'\mathbf{T}) && \text{(by the case already proved)}\\
&= \mathrm{Or}\ \angle(H_1\mathbf{T}, H_2\mathbf{T}) && \text{(by 6.2.17 with } \mathbf{v} = P\mathbf{T}).
\end{aligned}$$

Finally, let **T** be an arbitrary rotation, not necessarily about $O$. Then by Theorem 6.2.19, we may write $\mathbf{T} = \mathcal{T}_{-A}\,\mathcal{R}_{O,\theta}\,\mathcal{T}_A$ for some $A$, $\theta$. But $\mathcal{R}_{O,\theta}$ preserves oriented angles by the case just proved, and translations preserve oriented angles by 6.2.17. Hence, **T** preserves oriented angles, q.e.d. ∎

(Note again the technique, mentioned in Secs. 1.12 and 5.7, of using a special case to prove the general case.)

**Exercises**

*1. Let 0° denote the zero element (additive identity) of the group **Or**. Prove that $0° = \text{Or } \angle(H, H)$ for all rays $H$ in $\mathbf{R}_2$. Also prove that $\cos 0° = 1$, $\sin 0° = 0$. (SUG)

*2. Prove that for all $\theta \in \mathbf{Or}$,

$$\cos 2\theta = \cos^2\theta - \sin^2\theta$$
$$\sin 2\theta = 2\sin\theta\cos\theta.$$

(Here "$2\theta$" means "$\theta + \theta$", as in Sec. 1.12.) (SUG)

*3. Prove: For any $\psi \in \mathbf{Or}$, there are exactly two oriented angles $\theta$ such that $2\theta = \psi$. (*Suggestion*: Use the preceding exercise.) Also, prove that if $\psi \neq 0°$, these two oriented angles $\theta$ have opposite orientation: one is clockwise and the other is counterclockwise. (SOL)

*4. Let $P_\theta$ be the point $(\cos\theta, \sin\theta)$. Prove: the rotation $\mathcal{R}_{O,\psi-\theta}$ maps $P_\theta \to P_\psi$. Deduce that the oriented angle from $H(O, P_\theta)$ to $H(O, P_\psi)$ is $\psi - \theta$. Draw a picture to illustrate. (SUG)

*5. If **u**, **v** are nonzero vectors in $\mathbf{R}_2$, then we can write $\mathbf{u} = \mathbf{Ar}\ AB$, $\mathbf{v} = \mathbf{Ar}\ AC$ for suitable points $A$, $B$, $C$; we then define Or $\angle(\mathbf{u}, \mathbf{v})$ (the oriented angle from **u** to **v**) to be the oriented angle from $H(A, B)$ to $H(A, C)$. Draw a figure to illustrate this. Prove: (1) this definition of Or $\angle(\mathbf{u}, \mathbf{v})$ depends only on **u** and **v**, not on the particular choice of arrows to represent **u** and **v**; (2) Or $\angle(\mathbf{u}, \mathbf{v}) = -$ Or $\angle(\mathbf{v}, \mathbf{u})$; (3) for all nonzero vectors **u**, **v**, **w** in $\mathbf{R}_2$, Or $\angle(\mathbf{u},\mathbf{v})$ + Or $\angle(\mathbf{v}, \mathbf{w})$ = Or $\angle(\mathbf{u}, \mathbf{w})$; (4) if $t_1$, $t_2$ are *positive* real numbers, then Or $\angle(\mathbf{u}, \mathbf{v})$ = Or $\angle(t_1\mathbf{u}, t_2\mathbf{v})$.

6. Find the equations of the rotation which maps $H((2, 1), (7, 2))$ onto $H((2, 1), (3, 6))$. (ANS)

7. Show how to extend the concepts of "clockwise rotation" and "counterclockwise rotation" to rotations about an arbitrary point $A \in \mathbf{R}_2$.

8. Show that Theorem 6.1.9(b) remains valid if we replace $O$ by an arbitrary point $A$.

*9. As seen earlier, $-\mathbf{I}$ is a rotation about $O$; the oriented angle corresponding to this rotation (via the isomorphism 6.2.2) is denoted 180°. Let $L$ be any line in $\mathbf{R}_2$ and let $A$ be any point on $L$. If we denote by $H_1$ and $H_2$ the two rays into which $L$ is divided by $A$ (see Fig. 6.2.8; also see Theorem 3.9.14), prove that Or $\angle(H_1, H_2) = 180°$. Also prove that $\cos 180° = -1$, $\sin 180° = 0$, and that

$180° + 180° = 0°$. (SOL)

*10. If $\mathbf{u}, \mathbf{v}$ are nonzero vectors in $\mathbf{R}_2$, prove that
(a) Or $\angle(\mathbf{u}, -\mathbf{u}) = 180°$
(b) Or $\angle(-\mathbf{u}, \mathbf{v}) = \text{Or } \angle(\mathbf{u}, \mathbf{v}) + 180°$.
(See Exercise 5.) Draw pictures to illustrate.

11. If $\theta$ is the unique counterclockwise oriented angle such that $2\theta = 180°$ (see Exercise 3), then $\theta$ is denoted 90°. Prove that $\cos 90° = 0$, $\sin 90° = 1$, and that Or $\angle(\mathbf{e}_1, \mathbf{e}_2) = 90°$.

12. If $\theta$ is the unique counterclockwise oriented angle such that $2\theta = 90°$ (cf. the preceding exercise), then $\theta$ is denoted 45°. Prove that $\cos 45° = \sin 45° = \frac{1}{2}\sqrt{2}$.

13. If $\theta, \psi$ are the unique oriented angles such that

$\cos\theta = \frac{1}{2}\sqrt{3}, \qquad \sin\theta = \frac{1}{2}$

$\cos\psi = \frac{1}{2}, \qquad \sin\psi = \frac{1}{2}\sqrt{3}$

then $\theta$ is denoted 30° and $\psi$ is denoted 60°. Justify this notation by showing that $2\theta = \psi$ and $\theta + \psi = 90°$.

14. Prove: If $\mathbf{u}, \mathbf{v}$ are nonzero vectors in $\mathbf{R}_2$, then $\mathbf{u} \perp \mathbf{v}$ if and only if Or $\angle(\mathbf{u}, \mathbf{v}) = \pm 90°$. (SUG)

15. (Generalizing Exercise 13.) If $\theta, \psi$ are oriented angles, show that the following two statements are equivalent:
(a) $\cos\theta = \sin\psi$ and $\cos\psi = \sin\theta$.
(b) $\theta + \psi = 90°$.

Figure 6.2.8

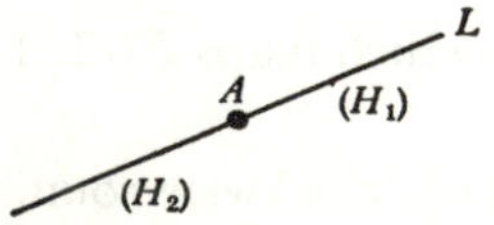

## 6.3 The isometries of $\mathbf{R}_2$

The object of this section is to completely classify the isometries of $\mathbf{R}_2$ according to geometric type; our final result is Theorem 6.3.18 below. Along the way, we derive a new characterization of "rotation" (Theorem 6.3.6) as well as other results of interest.

We begin with the *linear* isometries. It turns out that every linear isometry of $\mathbf{R}_2$ is either (1) a rotation about $O$, or (2) a reflection through some line containing $O$. We shall prove this by showing that the matrices of these two types of mappings are the only $2 \times 2$ matrices which satisfy Theorem 4.10.13. We have already found the matrices of rotations about $O$; the next theorem tells us which matrices represent reflections. As in Section 3.8, $\mathcal{M}_L$ will denote the reflection through $L$; also, it will be convenient to let $L(\theta)$ denote the line through the points $(0, 0)$ and $(\cos\theta, \sin\theta)$.

**6.3.1 Theorem.** For any oriented angle $\theta$, the reflection $\mathcal{M}_{L(\theta)} : \mathbf{R}_2 \to \mathbf{R}_2$ is a linear isometry and its matrix (with respect to the canonical basis) is

$$\begin{bmatrix} \cos 2\theta & \sin 2\theta \\ \sin 2\theta & -\cos 2\theta \end{bmatrix}. \tag{6.3.2}$$

*Proof.* By Section 3.8, Exercise 11, $\mathcal{M}_{L(\theta)}$ is an isometry; since $L(\theta)$ passes through $O$, $\mathcal{M}_{L(\theta)}$ maps $O \to O$ and is, hence, linear. Let

$$\mathbf{u} = (a, b) = (\cos\theta, \sin\theta)$$
$$\mathbf{v} = (-b, a) = (-\sin\theta, \cos\theta);$$

then $\mathbf{u} \in L(\theta)$ and $\mathbf{v} \perp \mathbf{u}$ (see Fig. 6.3.1). By Section 3.8, Exercise 13, we have

$$\mathcal{M}_{L(\theta)} : \mathbf{u} \to \mathbf{u}; \qquad \mathcal{M}_{L(\theta)} : \mathbf{v} \to -\mathbf{v}.$$

Also, by Section 6.2, Exercise 2 (or by Theorem 6.2.14, if you prefer),

$$\cos 2\theta = a^2 - b^2$$
$$\sin 2\theta = 2ab.$$

Now, if $\mathbf{S}$ is the L.T. which has the matrix 6.3.2, then Theorem 4.4.14 gives

$$\mathbf{uS} = [a \ \ b]\begin{bmatrix} \cos 2\theta & \sin 2\theta \\ \sin 2\theta & -\cos 2\theta \end{bmatrix}$$

$$= [a \;\; b]\begin{bmatrix} a^2 - b^2 & 2ab \\ 2ab & -(a^2 - b^2)\end{bmatrix}$$
$$= (a(a^2 + b^2), b(a^2 + b^2))$$
$$= (a, b) \qquad (\text{since } a^2 + b^2 = \cos^2\theta + \sin^2\theta = 1)$$
$$= \mathbf{u}$$

and a similar computation shows that $\mathbf{vS} = -\mathbf{v}$. Hence, $\mathbf{S}$ and $\mathcal{M}_{L(\theta)}$ have the same effect on $\mathbf{u}$ and $\mathbf{v}$. But $\{\mathbf{u}, \mathbf{v}\}$ is a basis of $\mathbf{R}_2$ (why?), and both $\mathbf{S}$ $\mathcal{M}_{L(\theta)}$ are linear. Hence, $\mathbf{S} = \mathcal{M}_{L(\theta)}$ by Theorem 4.2.2. ▌

Figure 6.3.2 illustrates the effect of the reflection $\mathcal{M}_{L(\theta)}$ on the canonical basis. The oriented angle from the positive $x$ axis to the line $L(O, \mathbf{e}_1\mathcal{M}_{L(\theta)})$ is $2\theta$ (see figure), so that $\mathbf{e}_1\mathcal{M}_{L(\theta)}$ has coordinates $(\cos 2\theta, \sin 2\theta)$ which is precisely the first row of the matrix 6.3.2. Similarly, the oriented angle from the $x$ axis to $L(O, \mathbf{e}_2\mathcal{M}_{L(\theta)})$ is $2\theta - 90°$ (why?), so that $\mathbf{e}_2\mathcal{M}_{L(\theta)} = (\cos(2\theta - 90°), \sin(2\theta - 90°)) = (\sin 2\theta, -\cos 2\theta)$ which again agrees with 6.3.2.

It is clear that the matrix 6.3.2 has the form

**(6.3.3)** $$\begin{bmatrix} a & b \\ b & -a\end{bmatrix} \qquad (a^2 + b^2 = 1).$$

Conversely, for any $a$, $b$ such that $a^2 + b^2 = 1$, we have shown previously that $a = \cos\psi$, $b = \sin\psi$ for some oriented angle $\psi$. By Section 6.2, Exercise 3, there exists an oriented angle $\theta$ such that $2\theta = \psi$; and for this oriented angle $\theta$, the matrix 6.3.3 is the same as 6.3.2. Since every line through $O$ has the form $L(\theta)$ for some $\theta \in \mathbf{Or}$ (*why*?), it

Figure 6.3.1

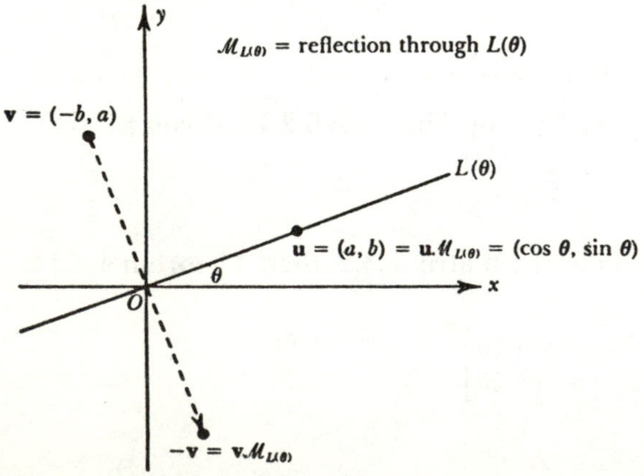

follows that the reflections through lines through $O$ (in $\mathbf{R}_2$) are precisely the L.T.'s whose matrix has the form 6.3.3.

Consider now *any* linear isometry $\mathbf{T}$ of $\mathbf{R}_2$; let

$$A = \begin{bmatrix} a & b \\ c & d \end{bmatrix}$$

be the matrix of $\mathbf{T}$. By Theorem 4.10.13, the row vectors of $A$ are orthonormal; thus, $a^2 + b^2 = 1$, $c^2 + d^2 = 1$, $ac + bd = 0$. Similarly, the columns of $A$ are orthonormal, so that $a^2 + c^2 = 1$. Hence, $c^2 = 1 - a^2 = b^2$, $c = \pm b$. Similarly, $d = \pm a$, so that

$$A = \begin{bmatrix} a & b \\ \pm b & \pm a \end{bmatrix}. \tag{6.3.4}$$

If the two $\pm$ signs in 6.3.4 are the same, then $0 = ac + bd = \pm 2ab$ so that either $a$ or $b$ equals 0; in either case, since $+0 = -0$, we may assume the two $\pm$ signs in 6.3.4 are opposite. But then $A$ has either the form 6.3.3 (a reflection through a line through $O$) or the form 6.1.7 (a rotation about $O$). We have thus proved:

**6.3.5 Theorem.** The linear isometries of $\mathbf{R}_2$ are precisely (a) the rotations about $O$, and (b) the reflections $\mathcal{M}_L$, where $L$ is a line through $O$.

Figure 6.3.2

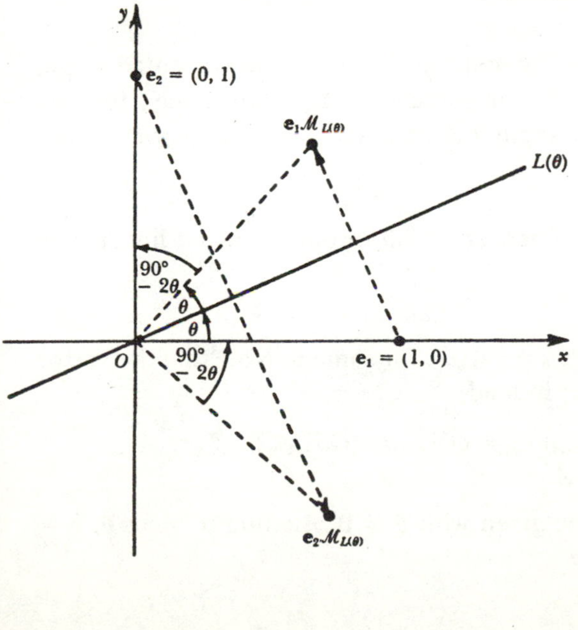

In treating the more general case of isometries which are not necessarily linear, the following theorem will be needed; this theorem is also of interest in itself, since it allows us to simplify the definition of "rotation" by removing all reference to angles.

**6.3.6** **Theorem.** If **T** is an isometry of $\mathbf{R}_2$ which leaves exactly one point fixed, then **T** is a rotation about that point. Conversely, every rotation except **I** leaves exactly one point fixed.

*Proof.* For every half line $H_1 \in \text{Ha}(A)$, the rotation $\mathcal{R}_{A,\theta}$ maps $H_1 \to H_2$, where Or $\angle(H_1, H_2) = \theta$, $H_2 \in \text{Ha}(A)$. If $\theta \neq 0°$, then $H_1 \neq H_2$ (Sec. 6.2, Exercise 1) and, hence, $H_1$, $H_2$ have no point in common except $A$ (Theorem 3.9.13(c)), from which it follows that $\mathcal{R}_{A,\theta}$ fixes only the point $A$. If instead $\theta = 0°$, then $\mathcal{R}_{A,\theta} = \mathbf{I}$ (cf. Theorem 1.8.4(a)). The "conversely" assertion follows. Now, let $\mathbf{T} : \mathbf{R}_2 \to \mathbf{R}_2$ be an isometry having exactly one fixed point $A$. The mapping

$$\mathbf{S} = \mathcal{T}_A \mathbf{T} \mathcal{T}_{-A} \tag{6.3.7}$$

is an isometry (*why*?) and maps $O \to O$ (*why*?), so that **S** is linear. If **S** were a reflection $\mathcal{M}_L$, then **S** would fix every point **u** on $L$, so that **T** would fix every point $\mathbf{u} + A$ on $L + A$ (*why*?), R.A.A. Hence, **S** must be a rotation about $O$, say $\mathbf{S} = \mathcal{R}_{O,\theta}$. But then we have

$$\begin{aligned} \mathbf{T} &= \mathcal{T}_{-A}(\mathcal{T}_A \mathbf{T} \mathcal{T}_{-A})\mathcal{T}_A = \mathcal{T}_{-A}\mathbf{S}\mathcal{T}_A \\ &= \mathcal{T}_{-A}\mathcal{R}_{O,\theta}\mathcal{T}_A = \mathcal{R}_{A,\theta} \end{aligned}$$

so that **T** is a rotation about $A$. ∎

**6.3.8** **Theorem.** If $A$ is any point in $\mathbf{R}_2$ and $\theta$ is any oriented angle, then $\mathcal{R}_{A,\theta} = \mathcal{R}_{O,\theta}\mathcal{T}_\mathbf{u}$ for some $\mathbf{u} \in \mathbf{R}_2$. Conversely, for any $\mathbf{u} \in \mathbf{R}_2$ and any oriented angle $\theta \neq 0°$, $\mathcal{R}_{O,\theta}\mathcal{T}_\mathbf{u} = \mathcal{R}_{A,\theta}$ for some point $A$.

*Proof.* For any $A$ and $\theta$, Theorem 4.9.3 implies that the linear part of

$$\mathcal{R}_{A,\theta} = \mathcal{T}_{-A}\mathcal{R}_{O,\theta}\mathcal{T}_A$$

is $\mathbf{I}\mathcal{R}_{O,\theta}\mathbf{I} = \mathcal{R}_{O,\theta}$; hence, $\mathcal{R}_{A,\theta} = \mathcal{R}_{O,\theta}\mathcal{T}_\mathbf{u}$ for some **u**. Note that the vector **u** can be found explicitly; indeed,

$$\begin{aligned} \mathbf{u} &= O\mathcal{T}_\mathbf{u} = (O\mathcal{R}_{O,\theta})\mathcal{T}_\mathbf{u} = O\mathcal{R}_{A,\theta} = (O\mathcal{T}_{-A})\mathcal{R}_{O,\theta}\mathcal{T}_A \\ &= (-A)\mathcal{R}_{O,\theta} + A. \end{aligned} \tag{6.3.9}$$

Conversely, let **u** and $\theta$ be given with $\theta \neq 0°$. Letting $a = \cos\theta$, $b = \sin\theta$, $\mathbf{u} = (c, d)$, we have

(6.3.10) $(x, y) \xrightarrow{\mathcal{R}_{O,\theta}} (ax - by, bx + ay) \xrightarrow{\mathcal{T}_{\mathbf{u}}} (ax - by + c, bx + ay + d).$

Hence, $(x, y)$ is a fixed point of $\mathcal{R}_{O,\theta}\mathcal{T}_{\mathbf{u}}$ if and only if $x = ax - by + c$, $y = bx + ay + d$; that is, if and only if $(x, y)$ is a solution of the system

(6.3.11)
$$\begin{aligned}(a - 1)x - by &= -c\\ bx + (a - 1)y &= -d.\end{aligned}$$

But the system 6.3.11 has a unique solution, since the determinant

$$\begin{bmatrix} a - 1 & -b \\ b & a - 1 \end{bmatrix} = (a - 1)^2 + b^2$$

is nonzero (if it were zero, then we would have $a = 1$, $b = 0$ and, hence, $\theta = 0°$). Thus, $\mathcal{R}_{O,\theta}\mathcal{T}_{\mathbf{u}}$ has a unique fixed point and, hence, is a rotation by Theorem 6.3.6; say,

$$\mathcal{R}_{O,\theta}\mathcal{T}_{\mathbf{u}} = \mathcal{R}_{A,\psi}.$$

But by the first half of the theorem (which we have already proved), the linear part of $\mathcal{R}_{A,\psi}$ is $\mathcal{R}_{O,\psi}$. Hence, $\mathcal{R}_{O,\theta} = \mathcal{R}_{O,\psi}$, so that $\theta = \psi$ and, thus, $\mathcal{R}_{O,\theta}\mathcal{T}_{\mathbf{u}} = \mathcal{R}_{A,\theta}$. ▮

Observe that equations 6.3.11 enable us to find the fixed point $A$ if $\theta$ and $\mathbf{u}$ are known; conversely, equation 6.3.9 determines $\mathbf{u}$ if $\theta$ and $A$ are known. Perhaps some numerical examples would be appropriate at this point.

**6.3.12. Example.** If **T** is the mapping of $\mathbf{R}_2 \to \mathbf{R}_2$ whose equations are

$$\begin{aligned} x' &= \tfrac{1}{2}x - \tfrac{1}{2}\sqrt{3}\,y + 4\\ y' &= \tfrac{1}{2}\sqrt{3}\,x + \tfrac{1}{2}y - 2,\end{aligned}$$

describe **T** geometrically. (As usual, $(x', y')$ denotes the image of $(x, y)$ under **T**.)

*Solution.* From the form of the given equations, it is clear that **T** is the L.T. whose matrix is

(6.3.13)
$$\begin{bmatrix} 1/2 & \sqrt{3}/2 \\ -\sqrt{3}/2 & 1/2 \end{bmatrix}$$

followed by the translation $\mathcal{T}_{(4,-2)}$. (Remember that the entries in the matrix occur in transposed position when writing $x'$, $y'$ in terms of $x$, $y$; cf. Sec. 4.2.) Clearly, 6.3.13 is the matrix of the rotation $\mathcal{R}_{O,\theta}$ where $\theta = 60°$ (see Sec. 6.2, Exercise 13), so that $\mathbf{T} = \mathcal{R}_{O,\theta}\mathcal{T}_{\mathbf{u}}$ where $\mathbf{u} = (4, -2)$. The system 6.3.11 is now

$$-\tfrac{1}{2}x - \tfrac{1}{2}\sqrt{3}y = -4$$
$$\tfrac{1}{2}\sqrt{3}\,x - \tfrac{1}{2}y = 2;$$

solving, we get $x = 2 + \sqrt{3}$, $y = -1 + 2\sqrt{3}$. Hence, $\mathbf{T}$ is a counterclockwise rotation through 60° about the point $A = (2 + \sqrt{3}, -1 + 2\sqrt{3})$. ∎

**6.3.14 Example.** Find the equations of **T**, if **T** is the counterclockwise rotation through 45° about the point (2, 3).

*Solution.* Here $A = (2, 3)$, $\theta = 45°$ (if **T** had been clockwise, we would have $\theta = -45°$). The matrix of $\mathcal{R}_{O,\theta}$ is

$$\begin{bmatrix} \cos 45° & \sin 45° \\ -\sin 45° & \cos 45° \end{bmatrix} = \begin{bmatrix} 1/\sqrt{2} & 1/\sqrt{2} \\ -1/\sqrt{2} & 1/\sqrt{2} \end{bmatrix}.$$

Hence, by 6.3.9,

$$\mathbf{u} = (2, 3) + [-2 \quad -3]\begin{bmatrix} 1/\sqrt{2} & 1/\sqrt{2} \\ -1/\sqrt{2} & 1/\sqrt{2} \end{bmatrix}$$
$$= (2, 3) + (1/\sqrt{2}, -5/\sqrt{2})$$
$$= (2 + 1/\sqrt{2}, 3 - 5/\sqrt{2}).$$

Since $\mathbf{T} = \mathcal{R}_{O,\theta}\mathcal{T}_{\mathbf{u}}$, it follows that the equations of **T** are

$$x' = x/\sqrt{2} - y/\sqrt{2} + (2 + 1/\sqrt{2})$$
$$y' = x/\sqrt{2} + y/\sqrt{2} + (3 - 5/\sqrt{2}).$$ ∎

**6.3.15 Example.** Describe geometrically the effect of the translation $\mathcal{T}_{(1,1)}$ followed by the counterclockwise rotation $\mathcal{R}$ through 45° about the point (0, 2).

*Solution.* By the same method as in Example 6.3.14, we obtain the equations of the given rotation $\mathcal{R}$ as

$$x' = x/\sqrt{2} - y/\sqrt{2} + \sqrt{2}$$
$$y' = x/\sqrt{2} + y/\sqrt{2} + (2 - \sqrt{2}).$$

Hence, the effect on any point $(x, y)$ of the translation $\mathcal{T}_{(1,1)}$ followed by the rotation is

$$(x, y) \xrightarrow{\mathcal{T}_{(1,1)}} (x + 1, y + 1) \xrightarrow{\mathcal{R}} ((x + 1)/\sqrt{2} - (y + 1)/\sqrt{2} + \sqrt{2},$$
$$(x + 1)/\sqrt{2} + (y + 1)/\sqrt{2} + 2 - \sqrt{2})$$
$$= (x/\sqrt{2} - y/\sqrt{2} + \sqrt{2}, x/\sqrt{2} + y/\sqrt{2} + 2).$$

This gives us the equations of $\mathbf{T} = \mathcal{T}_{(1,1)}\mathcal{R}$; we can then use the method of Example 6.3.12 to determine **T** geometrically. If you carry this out in detail, you should be able to show that **T** is the 45° rotation about the point $(-1 - 1/\sqrt{2}, 2 + 1/\sqrt{2})$. ∎

The following theorem shows that in general the product of two rotations **S**, **T** is a rotation, even if **S** and **T** are rotations about different points.

**6.3.16 Theorem.** Let $\theta$ and $\psi$ be oriented angles such that $\theta + \psi \neq 0°$, and let $A$, $B$ be any points in $\mathbf{R}_2$. Then $\mathcal{R}_{A,\theta}\mathcal{R}_{B,\psi} = \mathcal{R}_{C,\theta+\psi}$ for some point $C$.

*Proof.* By Theorem 6.3.8, we have $\mathcal{R}_{A,\theta} = \mathcal{R}_{O,\theta}\mathcal{T}_{\mathbf{u}}$, $\mathcal{R}_{B,\psi} = \mathcal{R}_{O,\psi}\mathcal{T}_{\mathbf{v}}$ for some $\mathbf{u}, \mathbf{v} \in \mathbf{R}_2$. Hence,

$$\begin{aligned}
\mathcal{R}_{A,\theta}\mathcal{R}_{B,\psi} &= \mathcal{R}_{O,\theta}\mathcal{T}_{\mathbf{u}}\mathcal{R}_{O,\psi}\mathcal{T}_{\mathbf{v}} \\
&= \mathcal{R}_{O,\theta}(\mathcal{T}_{\mathbf{u}}\mathcal{R}_{O,\psi}\mathcal{T}_{-\mathbf{u}})\mathcal{T}_{\mathbf{u}+\mathbf{v}} \\
&= \mathcal{R}_{O,\theta}\mathcal{R}_{-\mathbf{u},\psi}\mathcal{T}_{\mathbf{u}+\mathbf{v}} \\
&= \mathcal{R}_{O,\theta}(\mathcal{R}_{O,\psi}\mathcal{T}_{\mathbf{w}})\mathcal{T}_{\mathbf{u}+\mathbf{v}} \qquad \text{(Theorem 6.3.8)} \\
&= \mathcal{R}_{O,\theta+\psi}\mathcal{T}_{\mathbf{w}+\mathbf{u}+\mathbf{v}} \\
&= \mathcal{R}_{C,\theta+\psi} \quad \text{for some } C \qquad \text{(Theorem 6.3.8)},
\end{aligned}$$

as desired. ∎

Note that the point $C$ in Theorem 6.3.16 can be computed explicitly Using the notations of the proof, 6.3.9 gives

$$\mathbf{u} = A + (-A)\mathcal{R}_{O,\theta}; \qquad \mathbf{v} = B + (-B)\mathcal{R}_{O,\psi};$$
$$\mathbf{w} = -\mathbf{u} + \mathbf{u}\mathcal{R}_{O,\psi};$$
$$\mathbf{u} + \mathbf{v} + \mathbf{w} = \mathbf{v} + \mathbf{u}\mathcal{R}_{O,\psi} = B + (A - B)\mathcal{R}_{O,\psi} - A\mathcal{R}_{O,\theta+\psi}.$$

The fixed point $C = (x, y)$ can then be obtained from equations 6.3.11, replacing the vector $(c, d)$ in those equations by the vector $\mathbf{u} + \mathbf{v} + \mathbf{w}$ and the pair $(a, b)$ by $(\cos(\theta + \psi), \sin(\theta + \psi))$.

Theorems 6.2.19 and 6.3.8 relate rotations to translations. Similar results relating reflections to translations have already been proved in Section 3.8 (see Theorems 3.8.30(b) and 3.8.32 and Exercise 14). We shall need one more such result, which we now state. For the purpose of subsequent application to higher dimensions, we state the result in an arbitrary real Euclidean space of finite dimension.

**6.3.17** **Theorem.** Let $E$ be a finite-dimensional real Euclidean space. Let $S$ be a flat in $E$ having direction space $W$; let $\mathbf{u}$ be any vector in $E$. Then there exists a flat $S_1$ parallel to $S$, and a vector $\mathbf{w} \in W$, such that

$$\mathcal{M}_S\mathcal{T}_\mathbf{u} = \mathcal{M}_{S_1}\mathcal{T}_\mathbf{w}.$$

In other words, a reflection $\mathcal{M}_S$ times a translation is equal to a reflection through a flat times a translation *in the direction of the flat*. (Formally, we say the translation $\mathcal{T}_\mathbf{w}$ is "in the direction of $S$", if $\mathbf{w}$ belongs to the direction space of $S$. In Figure 6.3.3, $\mathcal{T}_\mathbf{w}$ is in the direction of both $S$ and $S_1$.)

To prove Theorem 6.3.17, we use Theorem 3.8.18(c) to write $\mathbf{u} = \mathbf{w} + \mathbf{v}$ where $\mathbf{w} \in W$, $\mathbf{v} \in W^\perp$. The flat $S_1 = S + \frac{1}{2}\mathbf{v}$ is parallel to $S$, and since $\frac{1}{2}\mathbf{v}$ is in $W^\perp$ it follows from Section 3.8, Exercise 14 that

$$\mathcal{M}_{S_1} = \mathcal{M}_S\mathcal{T}_\mathbf{v}.$$

Multiplying by $\mathcal{T}_\mathbf{w}$ on the right, we get

$$\mathcal{M}_S\mathcal{T}_\mathbf{u} = \mathcal{M}_S\mathcal{T}_{\mathbf{v}+\mathbf{w}} = \mathcal{M}_S\mathcal{T}_\mathbf{v}\mathcal{T}_\mathbf{w} = \mathcal{M}_{S_1}\mathcal{T}_\mathbf{w}$$

as desired. ▮

(Note that this proof shows us how to find $S_1$ and $\mathbf{w}$ explicitly: $\mathbf{w}$ is the projection of $\mathbf{u}$ on $W$, and $S_1 = S + \frac{1}{2}\mathbf{v} = S + \frac{1}{2}(\mathbf{u} - \mathbf{w})$.) Figure 6.3.3 illustrates Theorem 6.3.17 in the case where $S$ is a line.

Figure 6.3.3

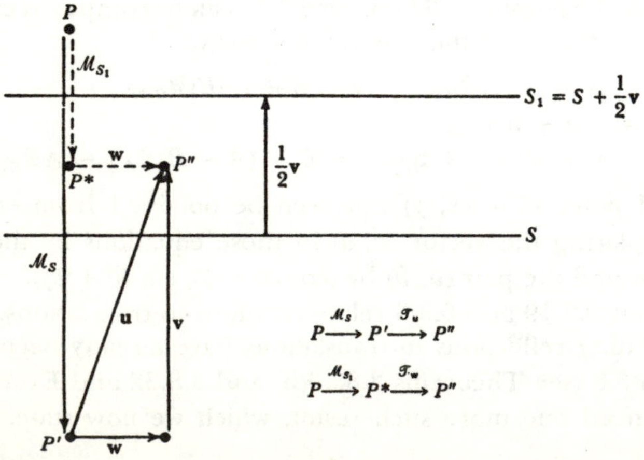

**6.3.18 Theorem.** Every isometry of $\mathbf{R}_2$ is of one of the following four types:

(a) A rotation $\mathcal{R}_{A,\theta}$.

(b) A nontrivial translation $\mathcal{T}_{\mathbf{u}}$. (We say $\mathcal{T}_{\mathbf{u}}$ is "nontrivial" if $\mathbf{u} \neq \mathbf{0}$.)

(c) A reflection $\mathcal{M}_L$ through some line $L$.

(d) A reflection $\mathcal{M}_L$ followed by a nontrivial translation in the direction of $L$.

Moreover, these four types are mutually exclusive. Isometries of types (a) and (b) preserve oriented angles, while isometries of types (c) and (d) reverse oriented angles. That is, if $\mathbf{T}$ is of type (a) or (b), then for all points $A \in \mathbf{R}_2$, we have

**(6.3.19)** $$\text{Or } \angle(H_1, H_2) = \text{Or } \angle(H_1\mathbf{T}, H_2\mathbf{T}) \qquad (\text{all } H_i \in \text{Ha}(A))$$

while if $\mathbf{T}$ is of type (c) or (d), then for all points $A \in \mathbf{R}_2$, we have

**(6.3.20)** $$\text{Or } \angle(H_1, H_2) = \text{Or } \angle(H_2\mathbf{T}, H_1\mathbf{T}) \qquad (\text{all } H_i \in \text{Ha}(A)).$$

*Proof.* By Theorem 6.3.5 the only *linear* isometries of $\mathbf{R}_2$ are of types (a) and (c). If the isometry $\mathbf{T}$ is nonlinear, then by Theorem 4.10.6 $\mathbf{T}$ equals a linear isometry followed by a translation; that is, $\mathbf{T} = \mathcal{R}_{O,\theta}\mathcal{T}_{\mathbf{u}}$ or $\mathcal{M}_L\mathcal{T}_{\mathbf{u}}$. If $\mathbf{T} = \mathcal{R}_{O,\theta}\mathcal{T}_{\mathbf{u}}$, then $\mathbf{T}$ is a rotation if $\theta \neq 0°$ (Theorem 6.3.8), while if $\theta = 0°$, then $\mathcal{R}_{O,\theta} = \mathbf{I}$, $\mathbf{T} = \mathcal{T}_{\mathbf{u}}$ which is of type 6.3.18(b). (We may assume $\mathbf{u} \neq \mathbf{0}$ in type (b) since the case $\mathcal{T}_{\mathbf{u}} = \mathcal{T}_O = \mathbf{I}$ is covered by type (a).) If instead $\mathbf{T} = \mathcal{M}_L\mathcal{T}_{\mathbf{u}}$, then, by Theorem 6.3.17, we can write $\mathbf{T} = \mathcal{M}_{L_1}\mathcal{T}_{\mathbf{w}}$, where $\mathbf{w}$ is in the direction of $L_1$; hence, $\mathbf{T}$ is of type 6.3.18(d) if $\mathbf{w} \neq \mathbf{0}$, type (c) if $\mathbf{w} = \mathbf{0}$. Thus, in every case $\mathbf{T}$ is one of the four types listed.

It was shown in Section 6.2 that isometries of types 6.3.18(a) and (b) satisfy 6.3.19. To show that an isometry $\mathcal{M}_L$ of type 6.3.18(c) satisfies 6.3.20, consider first the special case where $L$ passes through $O$ and $H_1, H_2 \in \text{Ha}(O)$. Here, $\mathcal{M}_L$ is linear and has the matrix 6.3.2 for some $\theta$. Letting $P_\psi$ denote the point $(\cos\psi, \sin\psi)$, a straightforward computation (which we leave to you) shows that

**(6.3.21)** $$\mathcal{M}_L : P_\psi \to P_{2\theta-\psi} \qquad (\text{all } \psi).$$

Now, each of the half lines $H_i$ $(i = 1, 2)$ contains a point $P_{\psi_i}$; by 6.3.21, $H_i\mathcal{M}_L$ then contains $P_{2\theta-\psi_i}$. Hence, Section 6.2, Exercise 4 implies that

$$\begin{aligned}
\text{Or } \angle(H_1, H_2) &= \psi_2 - \psi_1 \\
\text{Or } \angle(H_1\mathcal{M}_L, H_2\mathcal{M}_L) &= (2\theta - \psi_2) - (2\theta - \psi_1) \\
&= -(\psi_2 - \psi_1) \\
&= -\text{Or } \angle(H_1, H_2) \\
&= \text{Or } \angle(H_2, H_1)
\end{aligned}$$

which is the same as 6.3.20. The more general case (where $L$ need not contain $O$ and the initial point $A$ of $H_1$, $H_2$ is arbitrary) can now be deduced from the special case just as in the proof of Theorem 6.2.21. Moreover, since reflections reverse oriented angles and translations preserve them, it follows that isometries of the form $\mathcal{M}_L\mathcal{T}_{\mathbf{u}}$ must reverse them, so that 6.3.20 also holds when $\mathbf{T}$ is of type (d).

Since there exist oriented angles $\theta$ such that $\theta \neq -\theta$ (*proof*?), no isometry $\mathbf{T}$ can satisfy *both* 6.3.19 and 6.3.20; hence, isometries of types 6.3.18 (a) or (b) are never equal to isometries of types (c) or (d). An isometry $\mathcal{R}_{A,\theta}$ of type (a) cannot equal a translation $\mathcal{T}_{\mathbf{u}}$ of type (b), since $\mathcal{R}_{A,\theta}$ maps $A \to A$ while $\mathcal{T}_{\mathbf{u}}$ maps $A \to A + \mathbf{u} \neq A$. Finally, an isometry $\mathcal{M}_L$ of type (c) cannot equal an isometry of type (d), since the former fixes every point on $L$ while the latter leaves no point fixed (Exercise 6 below). This completes the proof of Theorem 6.3.18. ∎

**Exercises**

*1. Verify equation 6.3.21, and draw a picture to illustrate it.

2. Draw a picture illustrating equation 6.3.20 in the case where $\mathbf{T}$ is a reflection $\mathcal{M}_L$ and the point $A$ does not lie on $L$.

3. Prove that for all $\theta, \psi \in \mathbf{Or}$,
   (a) $\mathcal{R}_{O,2\theta}\mathcal{M}_{L(\psi)} = \mathcal{M}_{L(\psi-\theta)}$;
   (b) $\mathcal{R}_{O,2\theta}\mathcal{M}_{[\mathbf{e}_1]} = \mathcal{M}_{L(-\theta)}$. (SUG)

4. Show that $\mathcal{M}_{L(\psi)}\mathcal{M}_{L(\psi+\theta)} = \mathcal{R}_{O,2\theta}$. Deduce that every rotation about $O$ in $\mathbf{R}_2$ is the product of two reflections. Draw a picture to illustrate.

5. Generalize Exercise 4 as follows: every rotation about $A$ in $\mathbf{R}_2$ is the product of two reflections $\mathcal{M}_{L_1}$, $\mathcal{M}_{L_2}$ such that the lines $L_i$ pass through $A$. Conversely, if $L_1$ and $L_2$ pass through $A$, then $\mathcal{M}_{L_1}\mathcal{M}_{L_2}$ is a rotation about $A$. (SOL)

*6. Theorem 6.3.18 lists four types of isometries of $\mathbf{R}_2$. Show that an isometry of Type 6.3.18(d) has no fixed points. (SUG)

7. (a) Using results from this section and Section 3.8 (theorems and previous exercises), prove that every isometry of $\mathbf{R}_2$ is a product of at most three reflections through lines. Show that the number three in the preceding sentence is "best possible"; that is, there exists an isometry which is not expressible as a product of fewer than three reflections.
   (b) Suppose an isometry $\mathbf{T}$ of $\mathbf{R}_2$ is expressible as a product of reflections in more than one way, for instance, $\mathbf{T} = \mathcal{M}_{L_1}\mathcal{M}_{L_2} \cdots \mathcal{M}_{L_k} = \mathcal{M}_{N_1}\mathcal{M}_{N_2} \cdots \mathcal{M}_{N_h}$, where the $L_i$, $N_j$ are lines.

Prove that $h$ and $j$ are both odd or both even. (More concisely, $k \equiv h \pmod 2$.) (SUG)

8. If $\mathbf{T} = \mathcal{M}_L\mathcal{T}_{\mathbf{u}}$, where $\mathcal{T}_{\mathbf{u}}$ is in the direction of the line $L$ (i.e., $\mathbf{T}$ is an isometry of type 6.3.18(d)), show that $L$ and $\mathbf{u}$ are uniquely determined by $\mathbf{T}$.
9. Let $\mathbf{T}$ be the rotation through $\theta$ about (2, 2), where $\cos\theta = 3/5$ and $\sin\theta = 4/5$. Then $\mathbf{T}$ is expressible in the form $\mathcal{R}_{O,\theta}\mathcal{T}_{\mathbf{u}}$. Find $\mathbf{u}$. Also find the image of (0, 3) under $\mathbf{T}$. (ANS)
10. For any $\mathbf{u} \in \mathbf{R}_2$, $A \in \mathbf{R}_2$, $\theta \in \mathbf{Or}$, prove that $\mathcal{T}_{\mathbf{u}}\mathcal{R}_{A,\theta} = \mathcal{R}_{O,\theta}\mathcal{T}_{\mathbf{v}}$, where
    $\mathbf{v} = A + (\mathbf{u} - A)\,\mathcal{R}_{O,\theta}$.
11. In Theorem 6.3.16 it is assumed that $\theta + \psi \neq 0°$. Take care of the case $\theta + \psi = 0°$ by showing that $\mathcal{R}_{A,\theta}\mathcal{R}_{B,-\theta}$ is a translation $\mathcal{T}_{\mathbf{u}}$. Find $\mathbf{u}$ explicitly in terms of $A$, $B$, $\theta$. (ANS)
12. If $\mathbf{T}$ is the counterclockwise rotation through 30° about (1, 2) and $\mathbf{S}$ is the counterclockwise rotation through 60° about (−5, 0), find the fixed point of the rotation $\mathbf{TS}$. (ANS)
13. Describe geometrically (i.e., express explicitly in one of the four forms specified by Theorem 6.3.18) each of the following mappings $\mathbf{T} : \mathbf{R}_2 \to \mathbf{R}_2$. (In each case, $\mathbf{T} : (x, y) \to (x', y')$.)

(a)
$$x' = \frac{3}{5}x + \frac{4}{5}y + 3$$
$$y' = \frac{4}{5}x - \frac{3}{5}y + 4.$$
(SOL)

(b)
$$x' = \frac{5}{13}x - \frac{12}{13}y + \frac{64}{13}$$
$$y' = \frac{12}{13}x + \frac{5}{13}y - \frac{44}{13}.$$
(ANS)

(c)
$$x' = \frac{3}{5}x + \frac{4}{5}y - 3$$
$$y' = \frac{4}{5}x - \frac{3}{5}y + 6.$$
(ANS)

(d)
$$x' = \frac{1}{2}x + \frac{1}{2}\sqrt{3}y$$
$$y' = \frac{1}{2}\sqrt{3}x - \frac{1}{2}y + 8.$$
(ANS)

(e)
$$x' = -\frac{1}{2}x + \frac{1}{2}\sqrt{3}y + \sqrt{3}$$
$$y' = -\frac{1}{2}\sqrt{3}x - \frac{1}{2}y + 3.$$
(ANS)

14. The rotation $\mathcal{R}_{P,\theta}$ maps $(0, -2) \to (-6, -2)$ and $(4, 1) \to (-2, -5)$. Find $\theta$ and $P$. (*Hint*: Use 4.9.6 and 6.3.8.) (SOL)
15. The rotation $\mathcal{R}_{P,\theta}$ maps $(5, 4) \to (2, 5)$ and $(6, 6) \to (0, 6)$. Find $\theta$ and $P$. (ANS)
16. If $L$ is the line through (0, 1) and (2, 3), find the equations of the reflection $\mathcal{M}_L$. (SOL)
17. If $L$ is the line through (2, 1) and (1, 3), find the equations of $\mathcal{M}_L$. (ANS)
18. For each of the mappings **T** of Exercise 13, find the equations of $\mathbf{T}^{-1}$ and describe $\mathbf{T}^{-1}$ geometrically. (*Suggestion*: When **T** is of type 6.3.18(d), use Sec. 3.8, Exercise 16(b).) (SOL/ANS)
19. An isometry $\mathbf{T}: \mathbf{R}_2 \to \mathbf{R}_2$ maps $(1, 2) \to (1, 2)$ and $(4, 6) \to (1, 7)$. Show that there are only two possibilities for **T**, and find the equations of **T** in each case. (SOL)
20. Using the notations of Theorem 6.3.16, and assuming that $\theta, \psi, A, B$ are given, show how to construct $C$ with straightedge and compass. (Don't worry about being too formal in this exercise.) (SOL)
21. Let $G$ be the group of all isometries of $\mathbf{R}_2$; let $S$ be the subset of $G$ consisting of all isometries of types 6.3.18(a) and (b) (that is, all rotations and translations). Show that $S$ is a subgroup of $G$. Is $S$ abelian?
22. (For readers familiar with elementary group theory) (a) Show that the subgroup $S$ of the preceding exercise has index 2 in $G$, and that $S$ is a *normal* subgroup of $G$.

    (b) Let $T$ be the group of all translations of $\mathbf{R}_2$. Show that $T$ is normal in both $S$ and $G$. What are the cosets of $T$ in $S$?

    (c) Show that the quotient group $S/T$ is isomorphic to the group **Or**.
23. Find the determinant of each of the isometries listed in Theorem 6.3.18. Compare with Theorems 5.6.12 and 5.6.13.
24. Let $M$ be the matrix 6.2.13 (the matrix of the rotation $\mathcal{R}_{O,\theta}$ with respect to the canonical basis); let $M'$ be the matrix of $\mathcal{R}_{O,\theta}$ with respect to some other *orthonormal* basis $(\mathbf{u}_1, \mathbf{u}_2)$ of $\mathbf{R}_2$. Prove that either $M' = M$ or $M' = M^t$. Prove that the first case ($M' = M$) occurs if and only if Or $\angle(\mathbf{u}_1, \mathbf{u}_2) = 90°$, while the second case occurs if and only if Or $\angle(\mathbf{u}_1, \mathbf{u}_2) = -90°$.
25. If $\mathbf{T}: \mathbf{R}_2 \to \mathbf{R}_2$ is a linear isometry whose characteristic polynomial is $x^2 - x + 1$, show that there are only two possibilities for **T**, and find them. (ANS)

## 6.4 Rotations in higher dimensions

If $W$ is any 2-dimensional subspace of $\mathbf{R}_n$, then by Theorem 4.10.14 there exists a linear isometry $\mathbf{M}$ of $\mathbf{R}_2$ onto $W$. Hence, $W$ is algebraically and geometrically isomorphic to $\mathbf{R}_2$, and for every definition or theorem in $\mathbf{R}_2$ there is an analogous definition or theorem in $W$. Specifically, let us define a mapping $\mathbf{T}: W \to W$ to be a *rotation about A* (where $A$ is a fixed point in $W$) if $\mathbf{T}$ satisfies conditions (a), (b), and (c) of Definition 6.1.3. (The same definition would be used if $W$ were a 2-flat not necessarily passing through the origin.) The angle in 6.1.3(c) will still be called the "angle of $\mathbf{T}$", etc. One slight problem does arise: how to distinguish between the *clockwise* and *counterclockwise* rotations through a given angle about a given point. The nature of the difficulty is suggested in Figure 6.4.2: if $W$ is a plane in $\mathbf{R}_3$ (say) and $P$ and $Q$ are points on opposite sides of this plane, then the circular arrow will be seen as counterclockwise by an observer at $P$ while an observer at $Q$ will see the same arrow as clockwise. (Try this yourself: draw a heavy dark circular arrow on a piece of transparent paper and then hold the paper between yourself and a friend.) In $\mathbf{R}_2$ we succeeded in making the necessary distinction by looking at the matrix of $\mathbf{T}$ with respect to the canonical basis (Def. 6.1.8), but in an arbitrary 2-dimensional subspace $W$ no one basis is "canonical"; the best we can do is to call $\mathbf{T}$ "clockwise" or "counterclockwise" *with respect to a particular (orthonormal) basis of W*. Similarly, the *oriented angle* of $\mathbf{T}$ must be defined with respect to a particular basis, although the

Figure 6.4.1

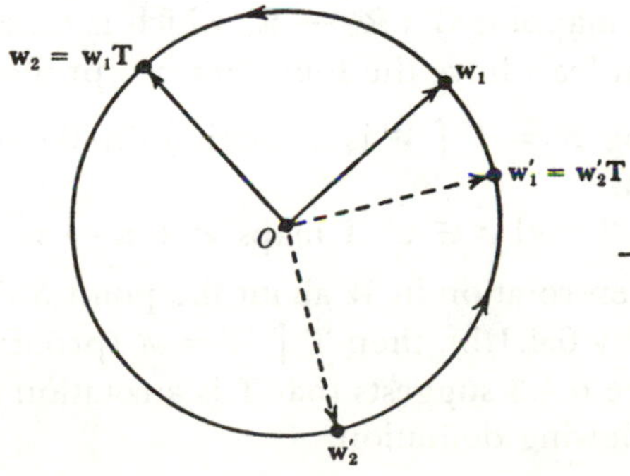


Figure 6.4.2

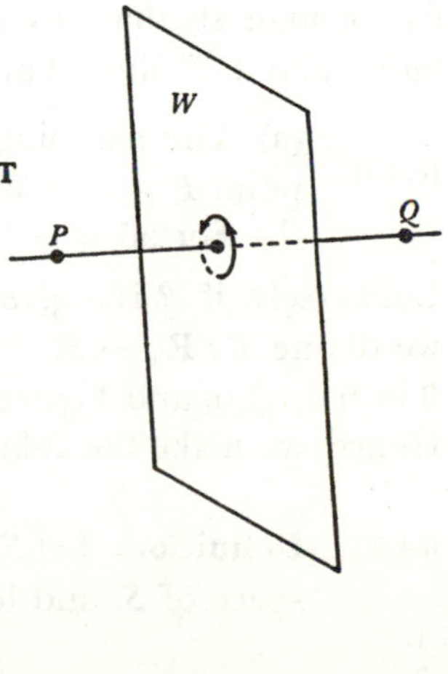

"angle" of **T** is independent of the basis chosen. (For example, the rotation **T** shown in Figure 6.4.1 maps $\mathbf{w}_1 \to \mathbf{w}_2, \mathbf{w}_2' \to \mathbf{w}_1'$. With respect to the basis $(\mathbf{w}_1, \mathbf{w}_2)$, **T** is counterclockwise and its oriented angle is 90°. With respect to the basis $(\mathbf{w}_1', \mathbf{w}_2')$, **T** is clockwise and its oriented angle is $-90°$.) Formally, if $W$ is a 2-dimensional subspace of $\mathbf{R}_n$ and $\mathbf{T} : W \to W$ is a rotation about $O$, we define the oriented angle of **T** (with respect to a given orthonormal basis of $W$) to be the unique oriented angle $\theta$ such that **T** has the matrix 6.2.13 with respect to the given basis. Oriented angles at a point other than $O$ can then be defined in the same manner as in $\mathbf{R}_2$ (cf. Fig. 6.2.5). It is still true that every linear isometry of $W$ is a rotation or reflection and that the matrices of these isometries are given by 6.2.13 and 6.3.2; similarly, Theorem 6.3.18 holds for $W$ in place of $\mathbf{R}_2$. These assertions and others may be proved for $W$ by imitating the proofs for $\mathbf{R}_2$; for an alternate method of proof, see Exercise 1 below.

Up to now our discussion of "rotations" has been limited to rotations of a *plane* about a fixed point. There also exist rotations of higher-dimensional spaces; the earth, for example (a 3-dimensional object), is said to *rotate about its axis*, the "axis" being an imaginary line through the North and South Poles. Note that for each type of rotation the *set of fixed points* (a single point in a plane, an axis in 3-space) has dimension exactly 2 less than that of the space being rotated. This suggests that a "rotation in $\mathbf{R}_n$" (however we might define it) ought to be, in some sense, a rotation "about" an $(n - 2)$-flat. To treat this formally, we fix notation as follows: let $S$ be an $(n - 2)$-flat in $\mathbf{R}_n$, let $U$ be the direction space of $S$, and let $W = U^\perp$. Then by Theorem 3.8.18 we have $U = W^\perp$, $\dim W = 2$, and every element of $\mathbf{R}_n$ is uniquely expressible in the form $\mathbf{w} + \mathbf{u}$ ($\mathbf{w} \in W$, $\mathbf{u} \in U$). Figure 6.4.3 illustrates this situation in the case $n = 3$ ($\dim U = 1$). A glance at the figure suggests that any mapping $\mathbf{T} : \mathbf{R}_n \to \mathbf{R}_n$ which is called a "rotation about $S$" should at least have the following two properties:

**(6.4.1)**
(a) The mapping $\mathscr{R} = \mathbf{T} \restriction W$ is a rotation (in $W$) about the point $P = S \cap W$.
(b) For all $\mathbf{w} \in W$ and $\mathbf{u} \in U$, **T** maps $\mathbf{w} + \mathbf{u} \to \mathbf{w}\mathscr{R} + \mathbf{u}$.

Conversely, if $\mathscr{R}$ is a given rotation in $W$ about the point $S \cap W$ and we define $\mathbf{T} : \mathbf{R}_n \to \mathbf{R}_n$ by 6.4.1(b), then $\mathbf{T} \restriction W = \mathscr{R}$ (proof: let $\mathbf{u} = \mathbf{0}$ in 6.4.1(b)), and Figure 6.4.3 suggests that **T** is a rotation about $S$. Hence, we make the following definition:

**6.4.2** **Definition.** Let $S$ be an $(n - 2)$-flat in $\mathbf{R}_n$, let $U$ be the direction space of $S$, and let $W = U^\perp$. By a *rotation about* $S$ (in $\mathbf{R}_n$), we mean a mapping $\mathbf{T} : \mathbf{R}_n \to \mathbf{R}_n$ which satisfies the two conditions 6.4.1.

The flat $S$ in Definition 6.4.2 is called the *axis* of the rotation $\mathbf{T}$. If $\mathbf{T} \neq \mathbf{I}$, this terminology is unambiguous since $S$ is uniquely determined by $\mathbf{T}$. See Exercise 3 below.

6.4.3 **Remark.** By Remark 3.8.31, $S \cap W$ is indeed a single point, as we have tacitly assumed in 6.4.1(a); in fact, $S \cap W'$ is a single point whenever $W'$ is a coset of $W$. (See Fig. 6.4.3.)

The following theorem generalizes Theorem 6.3.8 to $n$ dimensions.

6.4.4 **Theorem.** Let $U$ be an $(n - 2)$-dimensional subspace of $\mathbf{R}_n$ and let $W = U^{\perp}$.

Figure 6.4.3

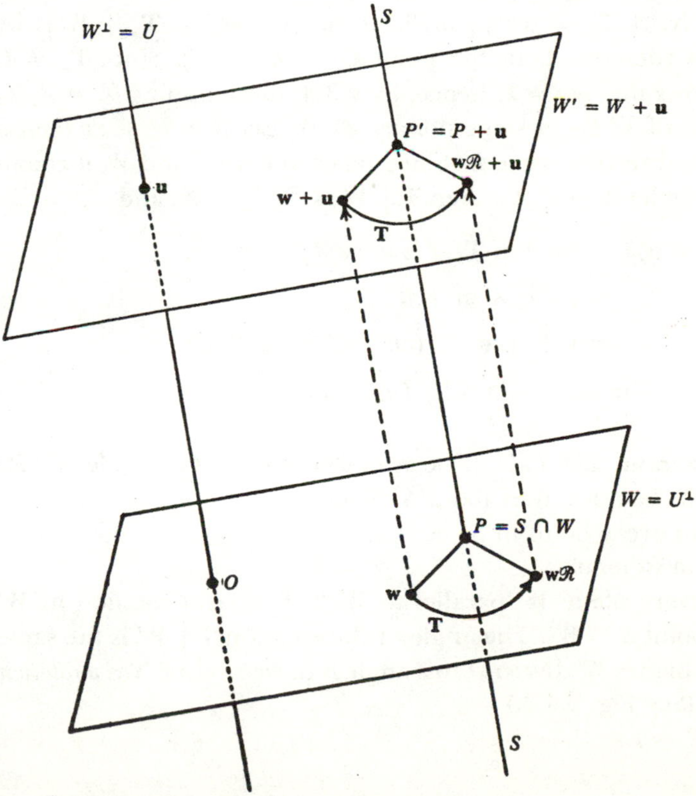

(a) If $S$ is a flat in $\mathbf{R}_n$ with direction space $U$, and if $\mathbf{T}: \mathbf{R}_n \to \mathbf{R}_n$ is a rotation about $S$, then $\mathbf{T} = \mathbf{T}_*\mathcal{T}_\mathbf{z}$, where $\mathbf{T}_*$ is a rotation about $U$ and $\mathbf{z}$ is some element of $W$.

(b) Conversely, if $\mathbf{T}_*$ is any rotation about $U$ such that $\mathbf{T}_* \neq \mathbf{I}$, and if $\mathbf{z} \in W$, then $\mathbf{T}_*\mathcal{T}_\mathbf{z}$ is a rotation about some flat $S$ whose direction space is $U$.

*Proof.* Let $S$, $\mathbf{T}$ be as in 6.4.4(a). By definition, the mapping $\mathcal{R} = \mathbf{T} \restriction W$ is a rotation in $W$. Since dim $W = 2$, Theorem 6.3.8 may be applied to $W$ in place of $\mathbf{R}_2$; hence, $\mathcal{R} = \mathcal{R}_*\mathcal{T}_\mathbf{z}$, where $\mathcal{R}_*: W \to W$ is a rotation about $O$ and $\mathbf{z}$ is some element of $W$. Define $\mathbf{T}_*: \mathbf{R}_n \to \mathbf{R}_n$ by

$$\mathbf{T}_*: \mathbf{w} + \mathbf{u} \to \mathbf{w}\mathcal{R}_* + \mathbf{u} \qquad (\text{all } \mathbf{w} \in W, \mathbf{u} \in U).$$

$\mathbf{T}_*$ is then a rotation about $U$ by Definition 6.4.2; and since

$$(\mathbf{w} + \mathbf{u})\mathbf{T} = \mathbf{w}\mathcal{R} + \mathbf{u} = (\mathbf{w}\mathcal{R}_* + \mathbf{z}) + \mathbf{u} = (\mathbf{w}\mathcal{R}_* + \mathbf{u}) + \mathbf{z}$$
$$= (\mathbf{w} + \mathbf{u})\mathbf{T}_* + \mathbf{z} \qquad (\text{all } \mathbf{w} \in W, \mathbf{u} \in U)$$

it follows that $\mathbf{T} = \mathbf{T}_*\mathcal{T}_\mathbf{z}$, proving 6.4.4(a).

Conversely, if $\mathbf{T}_*$, $\mathbf{z}$ are as in 6.4.4(b), then $\mathcal{R}_* = \mathbf{T}_* \restriction W$ is by definition a rotation about the point $O$ $(= U \cap W)$. Since $\mathbf{T}_* \neq \mathbf{I}$, 6.4.1(b) shows that $\mathcal{R}_* \neq \mathbf{I}$; hence, by 6.3.8, the mapping $\mathcal{R} = \mathcal{R}_*\mathcal{T}_\mathbf{z}$ is a rotation of $W$ about some point $P \in W$. Let $S = U + P$; then $S$ is a flat with direction space $U$. Since $P$ lies in both $S$ and $W$, it equals $S \cap W$. If we let $\mathbf{T} = \mathbf{T}_*\mathcal{T}_\mathbf{z}$, then $\mathbf{T} \restriction W = \mathcal{R}_*\mathcal{T}_\mathbf{z} = \mathcal{R}$, and

$$(\mathbf{w} + \mathbf{u})\mathbf{T} = (\mathbf{w} + \mathbf{u})\mathbf{T}_* + \mathbf{z} = (\mathbf{w}\mathcal{R}_* + \mathbf{u}) + \mathbf{z}$$
$$= (\mathbf{w}\mathcal{R}_* + \mathbf{z}) + \mathbf{u}$$
$$= \mathbf{w}\mathcal{R} + \mathbf{u} \qquad (\text{all } \mathbf{w} \in W, \mathbf{u} \in U).$$

Hence, $\mathbf{T}$ is a rotation about $S$ by Definition 6.4.2. ∎

**6.4.5** **Theorem.** Let $S$, $U$, $W$ be as in Definition 6.4.2 and let $\mathbf{T}: \mathbf{R}_n \to \mathbf{R}_n$ be a rotation about $S$. Then:

(a) $\mathbf{T}$ fixes every point in $S$.

(b) $\mathbf{T}$ is an isometry.

(c) For every plane $W'$ parallel to $W$, $\mathbf{T} \restriction W'$ is a rotation in $W'$ about the point $S \cap W'$. The angle of the rotation $\mathbf{T} \restriction W'$ is the same for all such planes $W'$. (*Note*: If this angle is $\alpha$, we say that $\mathbf{T}$ is a *rotation through* $\alpha$.) (See Fig. 6.4.3.)

(d) det $\mathbf{T} = 1$.

*Proof.* (a) Let $P = S \cap W$. Since $P \in S$ and $U$ is the direction space of $S$, $S = U + P$. Since $\mathscr{R} = \mathbf{T} \restriction W$ is a rotation about $P$ by 6.4.1(a), $P\mathscr{R} = P$ and, hence,

$$(P + \mathbf{u})\mathbf{T} = P\mathscr{R} + \mathbf{u} = P + \mathbf{u} \qquad (\text{all } \mathbf{u} \in U)$$

by 6.4.1(b). Hence, $\mathbf{T}$ fixes every point of $S$.

(b) To prove that $\mathbf{T}$ is an isometry, let $\mathbf{v}_1 = \mathbf{w}_1 + \mathbf{u}_1$ and $\mathbf{v}_2 = \mathbf{w}_2 + \mathbf{u}_2$ be arbitrary elements of $\mathbf{R}_n$ ($\mathbf{w}_i \in W$, $\mathbf{u}_i \in U$). Then

$$\begin{aligned} d(\mathbf{v}_1, \mathbf{v}_2)^2 = d(\mathbf{w}_1 + \mathbf{u}_1, \mathbf{w}_2 + \mathbf{u}_2)^2 &= |(\mathbf{w}_1 + \mathbf{u}_1) - (\mathbf{w}_2 + \mathbf{u}_2)|^2 \\ &= |(\mathbf{w}_1 - \mathbf{w}_2) + (\mathbf{u}_1 - \mathbf{u}_2)|^2 \\ &= |\mathbf{w}_1 - \mathbf{w}_2|^2 + |\mathbf{u}_1 - \mathbf{u}_2|^2 \qquad (\text{Theorem 3.7.15}); \end{aligned}$$

$$\begin{aligned} d(\mathbf{v}_1\mathbf{T}, \mathbf{v}_2\mathbf{T})^2 &= d(\mathbf{w}_1\mathscr{R} + \mathbf{u}_1, \mathbf{w}_2\mathscr{R} + \mathbf{u}_2)^2 \\ &= |(\mathbf{w}_1\mathscr{R} + \mathbf{u}_1) - (\mathbf{w}_2\mathscr{R} + \mathbf{u}_2)|^2 \\ &= |(\mathbf{w}_1\mathscr{R} - \mathbf{w}_2\mathscr{R}) + (\mathbf{u}_1 - \mathbf{u}_2)|^2 \\ &= |\mathbf{w}_1\mathscr{R} - \mathbf{w}_2\mathscr{R}|^2 + |\mathbf{u}_1 - \mathbf{u}_2|^2 \\ &\qquad\qquad (\text{Theorem 3.7.15}) \\ &= |\mathbf{w}_1 - \mathbf{w}_2|^2 + |\mathbf{u}_1 - \mathbf{u}_2|^2 \\ &\qquad\qquad (\text{since } \mathscr{R} \text{ is an isometry}) \\ &= d(\mathbf{v}_1, \mathbf{v}_2)^2. \end{aligned}$$

(c) Any plane $W'$ parallel to $W$ has the form $W' = W + \mathbf{v}_0$ for some $\mathbf{v}_0 \in \mathbf{R}_n$. Writing $\mathbf{v}_0 = \mathbf{w}_0 + \mathbf{u}$ ($\mathbf{w}_0 \in W$, $\mathbf{u} \in U$), we have $W' = W + \mathbf{v}_0 = (W + \mathbf{w}_0) + \mathbf{u} = W + \mathbf{u}$ (Theorem 2.7.7(b)). Since

$$(\mathbf{w} + \mathbf{u})\mathbf{T} = \mathbf{w}\mathscr{R} + \mathbf{u} \in W + \mathbf{u} \qquad (\text{all } \mathbf{w} \in W),$$

$\mathbf{T}$ maps $W'$ into itself. Also, $S \cap W'$ is a single point $P'$ as we have seen; in fact, we must have $P' = P + \mathbf{u}$ since $P + \mathbf{u}$ belongs to both $S$ and $W'$. By 6.4.5(a), $\mathbf{T}$ fixes $P'$; hence, to prove 6.4.5(c) it suffices to show that for all $\mathbf{w}' = \mathbf{w} + \mathbf{u}$ in $W'$ (corresponding to all $\mathbf{w} \in W$),

**(6.4.6)** $$\angle(\mathbf{w}' - P', \mathbf{w}'\mathbf{T} - P') = \angle(\mathbf{w} - P, \mathbf{w}\mathbf{T} - P).$$

(Indeed, the right side of 6.4.6 is the angle of the rotation $\mathscr{R}$, which is thus independent of $\mathbf{w}$.) But 6.4.6 is easy, since $\mathbf{w}' - P' = \mathbf{w} - P$ and $\mathbf{w}'\mathbf{T} - P' = \mathbf{w}\mathbf{T} - P$.

(d) If $\mathbf{T}_*$ is the linear part of $\mathbf{T}$, then $\mathbf{T}_*$ is a rotation about $U$ by Theorem 6.4.4. If $\mathscr{R}_* = \mathbf{T}_* \restriction W$, then $\mathscr{R}_*$ is a rotation of $W$ about $O$ and thus has a matrix $M$ of the form 6.2.13 with respect to some orthonormal basis $(\mathbf{w}_1, \mathbf{w}_2)$ of $W$. Let $\{\mathbf{u}_1, ..., \mathbf{u}_{n-2}\}$ be any basis of $U$; then

$$\mathbf{w}_i\mathbf{T}_* = \mathbf{w}_i\mathscr{R}_* \qquad (i = 1, 2)$$

$$\mathbf{u}_j\mathbf{T}_* = \mathbf{u}_j \qquad (j = 1, ..., n - 2).$$

Hence, the matrix of $\mathbf{T}_*$ with respect to the ordered basis ($\mathbf{w}_1$, $\mathbf{w}_2$, $\mathbf{u}_1$, ..., $\mathbf{u}_{n-2}$) of $\mathbf{R}_n$ has the block form

$$\begin{bmatrix} M & 0 \\ 0 & I_{n-2} \end{bmatrix}$$

from which $\det \mathbf{T}_* = \det M = \cos^2\theta + \sin^2\theta = 1$. But $\det \mathbf{T} = \det \mathbf{T}_*$ by definition, proving 6.4.5(d). ▮

*Remark*: It can be shown, conversely, that any mapping $\mathbf{T} : \mathbf{R}_n \to \mathbf{R}_n$ which satisfies 6.4.5(b) and 6.4.5(c) is a rotation about $S$ (Exercise 2 below). This provides a purely geometric characterization of rotations in $\mathbf{R}_n$, in contrast to 6.4.1 which is partly algebraic.

**Exercises**

1. Let $W$ be a 2-dimensional subspace of $\mathbf{R}_n$. Justify the assertions of this section (regarding such subspaces) by the following method:
   (a) Let $\mathbf{M}$ be a linear isometry of $\mathbf{R}_2$ onto $W$. (We know $\mathbf{M}$ exists by Theorem 4.10.14.) Prove that every isometry of $W$ has the form $\mathbf{M}^{-1}\mathbf{S}\mathbf{M}$, where $\mathbf{S}$ is an isometry of $\mathbf{R}_2$. Show conversely that every such mapping $\mathbf{M}^{-1}\mathbf{S}\mathbf{M}$ is an isometry of $W$.
   (b) Show that if $\mathbf{S} : \mathbf{R}_2 \to \mathbf{R}_2$ is a rotation about $A$, then $\mathbf{M}^{-1}\mathbf{S}\mathbf{M}$ is a rotation about the point $A\mathbf{M}$; also, show that if $A = O$ and $\mathbf{S}$ has the matrix 6.2.13 with respect to the basis ($\mathbf{e}_1$, $\mathbf{e}_2$) of $\mathbf{R}_2$, then $\mathbf{M}^{-1}\mathbf{S}\mathbf{M}$ has the matrix 6.2.13 with respect to the basis ($\mathbf{e}_1\mathbf{M}$, $\mathbf{e}_2\mathbf{M}$) of $W$.
   (c) If $L$ is a line in $\mathbf{R}_2$ and $\mathbf{S} : \mathbf{R}_2 \to \mathbf{R}_2$ is the reflection through $L$, show that $\mathbf{M}^{-1}\mathbf{S}\mathbf{M} : W \to W$ is the reflection of $W$ through the line $L' = L\mathbf{M}$.
   (d) Deduce from (a), (b), and (c) that Theorem 6.3.5 holds for $W$ in place of $\mathbf{R}_2$.
   (e) Justify Theorem 6.3.18 for $W$ similarly.
   (f) Justify the analog of Theorem 6.3.8 for $W$ similarly.
2. Let $S$ be an $(n-2)$-flat in $\mathbf{R}_n$ with direction space $U$; let $W = U^\perp$. Prove: if $\mathbf{T} : \mathbf{R}_n \to \mathbf{R}_n$ is a mapping which satisfies 6.4.5(b) and 6.4.5(c), then $\mathbf{T}$ is a rotation about $S$. (This is the converse of Theorem 6.4.5.)
3. If $\mathbf{T} : \mathbf{R}_n \to \mathbf{R}_n$ is a rotation about an $(n-2)$-flat $S$ and if $\mathbf{T} \neq \mathbf{I}$, prove that the set of fixed points of $\mathbf{T}$ is precisely $S$. (This strengthens 6.4.5(a) and shows that $S$ is uniquely determined by $\mathbf{T}$.) (SUG)

4. If $\mathbf{T}: \mathbf{R}_n \to \mathbf{R}_n$ is a rotation about the $(n-2)$-flat $S$ and if $\mathbf{v}$ is any element of $\mathbf{R}_n$, prove that the mapping $\mathbf{M} = \mathcal{T}_{-\mathbf{v}}\mathbf{T}\mathcal{T}_{\mathbf{v}}$ is a rotation about the flat $S + \mathbf{v}$. (*Hint*: Use Theorems 6.4.4 and 4.9.3.) (SOL)
5. (Converse of the preceding exercise.) If $S$ is an $(n-2)$-flat in $\mathbf{R}_n$ and if $\mathbf{v} \in \mathbf{R}_n$, then every rotation $\mathbf{M}$ about $S + \mathbf{v}$ is of the form $\mathcal{T}_{-\mathbf{v}}\mathbf{T}\mathcal{T}_{\mathbf{v}}$, where $\mathbf{T}$ is a rotation about $S$. (This result generalizes Theorem 6.2.19.)

## 6.5 The isometries of $\mathbf{R}_3$

Once the isometries of $\mathbf{R}_2$ are known, it is not hard to find all isometries of $\mathbf{R}_3$. The method is to show that any linear isometry $\mathbf{T}$ of $\mathbf{R}_3$ must map some 2-dimensional subspace of $\mathbf{R}_3$ into itself. (Such a subspace was called **T**-*stable* in Sec. 5.13.) To be specific, we prove the following:

6.5.1 **Theorem.** Let $\mathbf{T}: \mathbf{R}_3 \to \mathbf{R}_3$ be a linear isometry. Then:
(a) There exists a nonzero vector $\mathbf{u} \in \mathbf{R}_3$ such that $\mathbf{uT} = \pm\mathbf{u}$.
(b) For any such $\mathbf{u}$, the 2-dimensional subspace $[\mathbf{u}]^\perp$ of $\mathbf{R}_3$ is **T**-stable.

*Proof.* Since 3 is odd, the mapping $\mathbf{T} \in \text{Lin}(\mathbf{R}_3, \mathbf{R}_3)$ has at least one eigenvalue $\lambda$ (cf. third remark following Def. 5.11.6). This means, by definition, that $\mathbf{uT} = \lambda\mathbf{u}$ for some $\mathbf{u} \neq \mathbf{0}$. Since $\mathbf{T}$ preserves length, $|\mathbf{u}| = |\mathbf{uT}| = |\lambda\mathbf{u}|$ and, hence, $\lambda = \pm 1$, proving (a). Since $\mathbf{T}$ is linear and maps $\mathbf{u}$ into $\pm\mathbf{u}$, it must map $[\mathbf{u}]$ onto $[\mathbf{u}]$; since $\mathbf{T}$ preserves orthogonality (Theorem 4.10.9(c)), it follows that $\mathbf{T}$ must map $[\mathbf{u}]^\perp$ onto $[\mathbf{u}]^\perp$, so that $[\mathbf{u}]^\perp$ is a **T**-stable subspace. We have $\dim [\mathbf{u}]^\perp = 2$ by 3.8.18(b). ▮

Now, let $\mathbf{T}$ be any linear isometry of $\mathbf{R}_3$, choose $\mathbf{u}$ as in Theorem 6.5.1(a), and let $W = [\mathbf{u}]^\perp$. Since $W$ is **T**-stable, $\mathbf{T} \restriction W$ is a linear isometry of $W$; hence, $\mathbf{T} \restriction W$ is a rotation about $O$ or a reflection through some line containing $O$ (Secs. 6.3–6.4). By choice of $\mathbf{u}$, $\mathbf{uT} = \pm\mathbf{u}$; moreover, the effect of $\mathbf{T}$ on $W$ and $\mathbf{u}$ determines $\mathbf{T}$ completely since $\mathbf{T}$ is linear and $[W, \mathbf{u}] = \mathbf{R}_3$. Hence, there are only the following four possibilities for $\mathbf{T}$:

*Case* 1. $\mathbf{uT} = \mathbf{u}$; $\mathcal{R} = \mathbf{T} \restriction W$ *is a rotation about* $O$. By linearity,

$$(\mathbf{w} + k\mathbf{u})\mathbf{T} = \mathbf{wT} + k(\mathbf{uT}) = \mathbf{w}\mathcal{R} + k\mathbf{u} \qquad (\text{all } \mathbf{w} \in W, k \in \mathbf{R}).$$

Since $W \cap [\mathbf{u}] = \{O\}$, it follows that $\mathbf{T}$ satisfies 6.4.1 if we let $S = U = [\mathbf{u}]$. Hence, $\mathbf{T}$ is a *rotation about the axis* $[\mathbf{u}]$.

*Case* 2. $\mathbf{uT} = -\mathbf{u}$; $\mathbf{T} \restriction W$ *is a rotation about O*. The reflection $\mathcal{M}_W$ fixes all elements of $W$ and maps $\mathbf{u} \to -\mathbf{u}$. The product $\mathbf{T}\mathcal{M}_W$ (which is an isometry by 4.10.5(c)) thus has the same effect as $\mathbf{T}$ on the subspace $W$ but maps $\mathbf{u} \to \mathbf{u}$. Hence, $\mathbf{T}\mathcal{M}_W$ satisfies the hypotheses of Case 1 (above) and, thus, is a rotation about $[\mathbf{u}]$. Since $\mathcal{M}_W\mathcal{M}_W = \mathbf{I}$, we have $\mathbf{T} = (\mathbf{T}\mathcal{M}_W)\mathcal{M}_W$ so that $\mathbf{T}$ is a *rotation about* $[\mathbf{u}]$ *followed by a reflection through the plane* $W = [\mathbf{u}]^{\perp}$. The effect of $\mathbf{T}$ is shown in Figure 6.5.1.

*Case* 3. $\mathbf{uT} = \mathbf{u}$; $\mathbf{T} \restriction W = \mathcal{M}_L$. Here, $L$ is a one-dimensional subspace of $W$, say, $L = [\mathbf{v}_1]$. Choose $\mathbf{v}_2$ such that $\{\mathbf{v}_1, \mathbf{v}_2\}$ is an orthogonal basis of $W$; then $\mathbf{T}$ maps $\mathbf{u} \to \mathbf{u}$, $\mathbf{v}_1 \to \mathbf{v}_1$, $\mathbf{v}_2 \to -\mathbf{v}_2$. Hence, by linearity,

**(6.5.2)** $$\mathbf{T} : a\mathbf{u} + b\mathbf{v}_1 + c\mathbf{v}_2 \to a\mathbf{u} + b\mathbf{v}_1 - c\mathbf{v}_2.$$

Since $\mathbf{v}_2 \perp [\mathbf{u}, \mathbf{v}_1]$ and $[\mathbf{u}, \mathbf{v}_1, \mathbf{v}_2] = \mathbf{R}_3$, it follows from 6.5.2 that $\mathbf{T}$ is the *reflection through the plane* $[\mathbf{u}, \mathbf{v}_1]$. The effect of $\mathbf{T}$ is shown in Figure 6.5.2.

*Case* 4. $\mathbf{uT} = -\mathbf{u}$; $\mathbf{T} \restriction W = \mathcal{M}_L$. Letting $\mathbf{v}_1$, $\mathbf{v}_2$ have the same meaning as above, we have

$$\mathbf{T} : \mathbf{v}_1 \to \mathbf{v}_1,\ \mathbf{v}_2 \to -\mathbf{v}_2,\ \mathbf{u} \to -\mathbf{u}.$$

Hence, $\mathbf{T} \restriction [\mathbf{u}, \mathbf{v}_2]$ is the rotation $-\mathbf{I}$ (the rotation through 180°). Since $[\mathbf{u}, \mathbf{v}_2] = [\mathbf{v}_1]^{\perp}$ and $\mathbf{T}$ maps $\mathbf{v}_1 \to \mathbf{v}_1$, this is simply a special case of Case 1 which has already been treated. We have thus established the following result:

**6.5.3 Theorem.** Every linear isometry of $\mathbf{R}_3 \to \mathbf{R}_3$ is of one of the following three types:

(a) A rotation about a line through the origin.

Figure 6.5.1

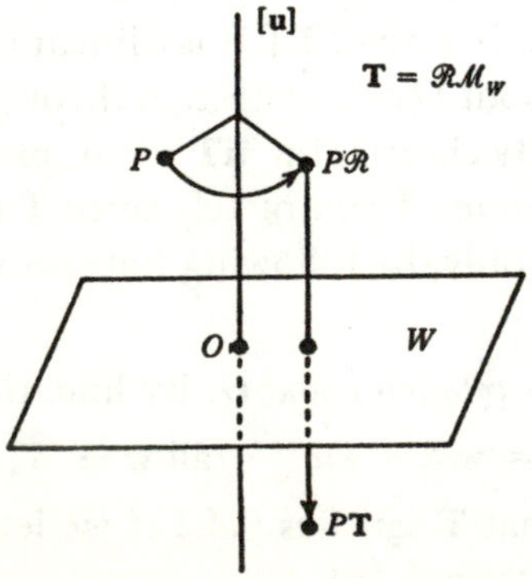

(b) A reflection through a plane containing the origin.

(c) A rotation through a nonzero angle about an axis [**u**], followed by a reflection through the plane $[\mathbf{u}]^{\perp}$.

Note that in type 6.5.3(c) the angle may be assumed nonzero since the case $\alpha = 0°$ is identical with type 6.5.3(b).

In $\mathbf{R}_2$ there were only two types of linear isometries (rotation, reflection) and the matrices of these two types were easy to recognize (and tell apart) by inspection. In $\mathbf{R}_3$ it is still easy to tell by inspection whether a given linear transformation **T** is an isometry (by Theorem 4.10.13, we need only observe whether the rows of the matrix of **T** are orthonormal), and we can determine which of the three "types" **T** represents by computing the characteristic polynomial of **T**, as we see from Table 6.5.4.

(To determine such things as the axis of rotation or the plane of reflection, we must compute eigenvectors also; we shall give examples later in this section.)

Let us verify the results given in Table 6.5.4 for the case where **T** is a rotation about the axis [**u**] through an angle $\alpha$ different from 0°,

Figure 6.5.2

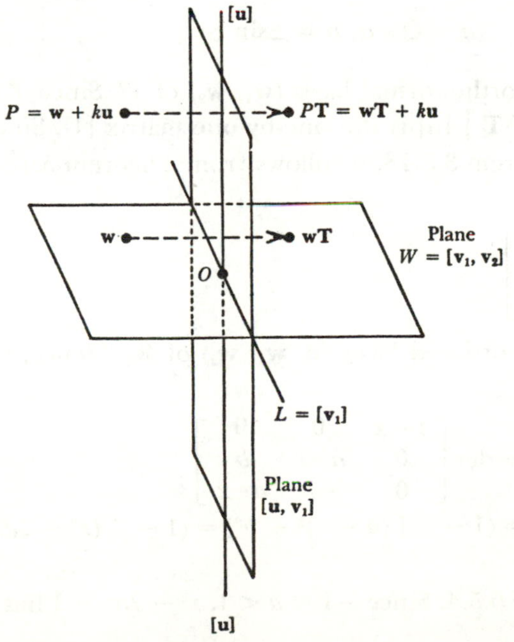

### 6.5.4 Table of linear isometries of $\mathbf{R}_3$

| Type | Description | Real Eigenvalues | Characteristic polynomial | Determinant |
|---|---|---|---|---|
| (a) | $\mathbf{I}$ | 1, 1, 1 | $-(x-1)^3$ | 1 |
| (a) | Rotation through 180° about axis | 1, −1, −1 | $-(x-1)(x+1)^2$ | 1 |
| (a) | Rotation through $\alpha$ about axis ($\alpha \neq 0°, 180°$) | 1 | $-(x-1)(x^2-2ax+1)$ where $a = \cos\alpha$, $a \neq \pm 1$ | 1 |
| (b) | Reflection through plane | 1, 1, −1 | $-(x-1)^2(x+1)$ | −1 |
| (c) | Type (c) with angle of rotation 180° | −1, −1, −1 | $-(x+1)^3$ | −1 |
| (c) | Type (c) with angle of rotation $\alpha$ ($\alpha \neq 0°, 180°$) | −1 | $-(x+1)(x^2-2ax+1)$ where $a = \cos\alpha$, $a \neq \pm 1$ | −1 |

(*Note*: Each eigenvalue is counted as many times as its multiplicity.)

180°. Since the restriction of $\mathbf{T}$ to $W = [\mathbf{u}]^\perp$ is a rotation through $\alpha$, $\mathbf{T} \restriction W$ has the matrix

$$\begin{bmatrix} a & b \\ -b & a \end{bmatrix} \qquad (a = \cos\alpha,\ b = \pm\sin\alpha)$$

with respect to an orthonormal basis $(\mathbf{w}_1, \mathbf{w}_2)$ of $W$. Since $\mathbf{T}$ maps $\mathbf{u} \to \mathbf{u}$, the matrix of $\mathbf{T} \restriction [\mathbf{u}]$ is the one-by-one matrix [1]. Since $\mathbf{R}_3 = [\mathbf{u}] \oplus [\mathbf{u}]^\perp$ by Theorem 3.8.18, it follows from Theorem 5.13.13 that $\mathbf{T}$ has the matrix

$$\textbf{(6.5.5)} \qquad \begin{bmatrix} 1 & 0 & 0 \\ 0 & a & b \\ 0 & -b & a \end{bmatrix}$$

with respect to the ordered basis $(\mathbf{u}, \mathbf{w}_1, \mathbf{w}_2)$ of $\mathbf{R}_3$. Hence, $\det \mathbf{T} = a^2 + b^2 = 1$, and

$$\operatorname{char}(\mathbf{T}; x) = \det \begin{bmatrix} 1-x & 0 & 0 \\ 0 & a-x & b \\ 0 & -b & a-x \end{bmatrix}$$
$$= (1-x)\,[\,(a-x)^2 + b^2] = (1-x)(x^2 - 2ax + 1)$$

as asserted in Table 6.5.4. Since $-1 < a < 1$, $x^2 - 2ax + 1$ has no real roots and, hence, the only eigenvalue of **T** is $\lambda = 1$ (with multiplicity 1).

Similarly, let us verify the results of the last line of Table 6.5.4 (Type (c) with $\alpha \neq 0°, 180°$). Here, $\mathbf{T} = \mathbf{S}_1\mathbf{S}_2$, where $\mathbf{S}_1$ is a rotation about the axis $[\mathbf{u}]$ and $\mathbf{S}_2$ is a reflection through the plane $W = [\mathbf{u}]^\perp$. With respect to the basis $(\mathbf{u}, \mathbf{w}_1, \mathbf{w}_2)$ described above, $\mathbf{S}_1$ has the matrix 6.5.5. With respect to the same basis, $\mathbf{S}_2$ has the matrix

$$\begin{bmatrix} -1 & 0 & 0 \\ 0 & 1 & 0 \\ 0 & 0 & 1 \end{bmatrix}$$

since $\mathbf{S}_2$ maps $\mathbf{u} \to -\mathbf{u}$, $\mathbf{w}_1 \to \mathbf{w}_1$, $\mathbf{w}_2 \to \mathbf{w}_2$. Hence, **T** has the matrix

$$\begin{bmatrix} 1 & 0 & 0 \\ 0 & a & b \\ 0 & -b & a \end{bmatrix}\begin{bmatrix} -1 & 0 & 0 \\ 0 & 1 & 0 \\ 0 & 0 & 1 \end{bmatrix} = \begin{bmatrix} -1 & 0 & 0 \\ 0 & a & b \\ 0 & -b & a \end{bmatrix}$$

and the results of Table 6.5.4 follow. We leave it to you to verify the remaining lines of Table 6.5.4 in the same manner.

6.5.6 **Example.** Let $\mathbf{T} : (x, y, z) \to (x', y', z')$ be the mapping of $\mathbf{R}_3 \to \mathbf{R}_3$ whose equations are

$$x' = \frac{7}{9}x - \frac{4}{45}y + \frac{28}{45}z$$

$$y' = \frac{4}{9}x + \frac{7}{9}y - \frac{4}{9}z$$

$$z' = -\frac{4}{9}x + \frac{28}{45}y + \frac{29}{45}z.$$

Show that **T** is a rotation about an axis, and find the axis and the angle of rotation.

*Solution.* The matrix of **T** with respect to the canonical basis is

$$A = \begin{bmatrix} \frac{7}{9} & \frac{4}{9} & -\frac{4}{9} \\ -\frac{4}{45} & \frac{7}{9} & \frac{28}{45} \\ \frac{28}{45} & -\frac{4}{9} & \frac{29}{45} \end{bmatrix}.$$

By direct computation, the rows of $A$ are orthonormal (so that **T** is an isometry), and

$$\text{char } A = -x^3 + \frac{11}{5}x^2 - \frac{11}{5}x + 1$$
$$= -(x-1)\left(x^2 - \frac{6}{5}x + 1\right).$$

Hence, **T** is a rotation about an axis through the (nonoriented) angle $\alpha$ such that $\cos \alpha = 3/5$. (Use of a table gives $\alpha = 53°$ to the nearest degree.) The axis of rotation consists of the vectors fixed by **T**, that is, the eigenvectors of **T** corresponding to the eigenvalue $\lambda = 1$. These eigenvectors can be found by the method of Example 5.11.7 (or by the method of Sec. 5.11, Exercise 5); they turn out to be the vectors of the form $(2k, 2k, k)$. Thus, the axis of rotation is the line $[(2, 2, 1)]$.There is still some ambiguity due to the fact that there are *two* rotations about this axis through the given angle $\alpha$, having opposite orientations. The ambiguity may be removed as follows: choose an eigenvector $\mathbf{u} \neq \mathbf{0}$, choose a vector $\mathbf{v}_1 \neq \mathbf{0}$ orthogonal to **u** (this may be done by inspection), and let $\mathbf{v}_2 = \mathbf{u} \times \mathbf{v}_1$. The vectors $\mathbf{v}_1$, $\mathbf{v}_2$ then constitute an orthogonal basis of $[\mathbf{u}]^\perp$ (*why*?); letting $\mathbf{w}_i = \mathbf{v}_i/|\mathbf{v}_i|$ $(i = 1, 2)$, it follows that $\{\mathbf{w}_1, \mathbf{w}_2\}$ is an orthonormal basis of $[\mathbf{u}]^\perp$. (This is the easiest way to actually compute an orthonormal basis of $[\mathbf{u}]^\perp$. In our present example, we could choose $\mathbf{u} = (2, 2, 1)$ and $\mathbf{v}_1 = (1, -2, 2)$, from which $\mathbf{v}_2 = \mathbf{u} \times \mathbf{v}_1 = (6, -3, -6)$, $\mathbf{w}_1 = \frac{1}{3}(1, -2, 2)$, $\mathbf{w}_2 = \frac{1}{9}(6, -3, -6) = \frac{1}{3}(2, -1, -2)$). Since $\mathbf{T} \restriction [\mathbf{u}]^\perp$ is a rotation through $\alpha$, we have

**(6.5.7)** $\quad \mathbf{w}_1\mathbf{T} = (\cos \alpha)\mathbf{w}_1 \pm (\sin \alpha)\mathbf{w}_2$

and the $\pm$ sign can be determined by directly computing $\mathbf{w}_1\mathbf{T}$ from the original equations of **T**. (In this example,

$$\mathbf{w}_1\mathbf{T} = \frac{1}{3}\left(\frac{11}{5}, -2, -\frac{2}{5}\right) = \frac{3}{5}\mathbf{w}_1 + \frac{4}{5}\mathbf{w}_2$$

so that the sign is plus.) If the sign is plus (and if we are using a right-handed coordinate system), then an observer (say, Jones) stationed at the point **u** and looking at the plane $[\mathbf{u}]^\perp$ will see the rotation as counterclockwise, while another observer (Smith) looking at the same plane from the point $-\mathbf{u}$ will see it as clockwise. If, however, the $\pm$ sign in 6.5.7 is a minus sign, then Jones will see the rotation as clockwise and Smith will see it as counterclockwise. In a left-handed coordinate system the words "clockwise" and "counterclockwise" in the two preceding sentences must be interchanged. For justification of these assertions, see Section 6.7. (The justification is only semi-formal; obviously total rigor is impossible when using words like "observer" and "see".) ∎

**6.5.8 Example.** Let $\mathbf{T} : \mathbf{R}_3 \to \mathbf{R}_3$ be the rotation through 30° about the axis [(1, 2, 2)]. Find the equations of $\mathbf{T}$. (Assume that we are using a right-handed coordinate system and that an observer at (1, 2, 2) looking towards the origin would see the rotation as counterclockwise.)

*Solution.* The problem is equivalent to finding the matrix $A$ of $\mathbf{T}$ with respect to the canonical basis. We already know the matrix of $\mathbf{T}$ with respect to another basis; indeed, if $\mathbf{u}$ is an eigenvector and $\{\mathbf{w}_1, \mathbf{w}_2\}$ is an orthonormal basis of $[\mathbf{u}]^\perp$, then the matrix $A'$ of $\mathbf{T}$ with respect to the ordered basis $(\mathbf{u}, \mathbf{w}_1, \mathbf{w}_2)$ is the matrix 6.5.5, where

(6.5.9) $\quad a = \cos 30° = \frac{1}{2}\sqrt{3}; \qquad b = \pm \sin 30° = \pm \frac{1}{2}.$

We choose $\mathbf{u}$ to be (1/3, 2/3, 2/3) rather than (1, 2, 2), so that $|\mathbf{u}| = 1$ and the basis $(\mathbf{u}, \mathbf{w}_1, \mathbf{w}_2)$ will be orthonormal. One possible choice of the vectors $\mathbf{w}_1, \mathbf{w}_2$ is

$$\mathbf{w}_1 = (0, 1/\sqrt{2}, -1/\sqrt{2});$$
$$\mathbf{w}_2 = \mathbf{u} \times \mathbf{w}_1 = (-4/3\sqrt{2}, 1/3\sqrt{2}, 1/3\sqrt{2}).$$

The discussion of orientation in the preceding example implies that the $\pm$ sign in 6.5.9 is a plus sign, so that

$$A' = \begin{bmatrix} 1 & 0 & 0 \\ 0 & \frac{1}{2}\sqrt{3} & \frac{1}{2} \\ 0 & -\frac{1}{2} & \frac{1}{2}\sqrt{3} \end{bmatrix}$$

To find $A$, we use the results of Section 4.8: $A'Q = QA$, $A = Q^{-1}A'Q$, where $Q$ is the matrix whose entries are the coefficients of $\mathbf{e}_1, \mathbf{e}_2, \mathbf{e}_3$ in the equations for $\mathbf{u}, \mathbf{w}_1, \mathbf{w}_2$ (cf. equation 4.8.4). That is, *the rows of $Q$ are the vectors* $\mathbf{u}, \mathbf{w}_1, \mathbf{w}_2$. But $\mathbf{u}, \mathbf{w}_1, \mathbf{w}_2$ are orthonormal, hence, the rows of $Q$ are orthonormal, so that Theorem 4.10.13 implies that $Q^{-1} = Q^t$. Thus, we have $A = Q^tA'Q$

$$= \begin{bmatrix} \frac{1}{3} & 0 & -\frac{4}{3\sqrt{2}} \\ \frac{2}{3} & \frac{1}{\sqrt{2}} & \frac{1}{3\sqrt{2}} \\ \frac{2}{3} & \frac{-1}{\sqrt{2}} & \frac{1}{3\sqrt{2}} \end{bmatrix} \begin{bmatrix} 1 & 0 & 0 \\ 0 & \frac{\sqrt{3}}{2} & \frac{1}{2} \\ 0 & -\frac{1}{2} & \frac{\sqrt{3}}{2} \end{bmatrix} \begin{bmatrix} \frac{1}{3} & \frac{2}{3} & \frac{2}{3} \\ 0 & \frac{1}{\sqrt{2}} & -\frac{1}{\sqrt{2}} \\ -\frac{4}{3\sqrt{2}} & \frac{1}{3\sqrt{2}} & \frac{1}{3\sqrt{2}} \end{bmatrix}$$

$$= \frac{1}{18}\begin{bmatrix} 2+8\sqrt{3} & 10-2\sqrt{3} & -2-2\sqrt{3} \\ -2-2\sqrt{3} & 8+5\sqrt{3} & 11-4\sqrt{3} \\ 10-2\sqrt{3} & 5-4\sqrt{3} & 8+5\sqrt{3} \end{bmatrix}$$

Hence, the equations of **T** are $x' = (1/18)[(2 + 8\sqrt{3})x - (2 + 2\sqrt{3})y + (10 - 2\sqrt{3})z]$, etc. ▮

**6.5.10 Example.** If $\mathbf{T}: \mathbf{R}_3 \to \mathbf{R}_3$ is the rotation of Example 6.5.8 followed by the reflection through the plane $[(1, 2, 2)]^{\perp}$, find the equations of **T**.

*Solution.* The method is the same as for Example 6.5.8, except that in $A'$ the entry in the (1, 1) position is $-1$ instead of 1 (why?). ▮

Theorem 6.5.3 classifies only the *linear* isometries of $\mathbf{R}_3$. However, since every isometry of $\mathbf{R}_3$ equals a linear isometry followed by a translation, we know the nonlinear isometries as well. For more details in the nonlinear case, see the exercises. In Section 9.4 we shall extend our considerations to the space $\mathbf{R}_n$.

**Exercises**

1. Two of the six rows in Table 6.5.4 were verified in the text. Verify the others.
2. Find the equations of the rotation through 45° about the axis $[(1, -1, \sqrt{2})]$. Assume that the coordinate system is right-handed and that an observer at $(1, -1, \sqrt{2})$ looking towards the origin sees the rotation as clockwise. (ANS)
3. Find the equations of the reflection through the plane $W = [(1, -1, 0), (3, 1, 1)]$. (SOL)
4. Let **T** be a rotation through a nonzero angle about the line [**u**] and let **w** be a vector orthogonal to **u**. Prove that the product $\mathbf{T}\mathcal{T}_{\mathbf{w}}$ is a rotation about a line parallel to [**u**]. (Compare with Theorem 6.3.8.)
5. Let **M** be an isometry of $\mathbf{R}_3$ and let **T** be the linear part of **M**. Prove:

   (a) If **T** is of type 6.5.3(a), then **M** is a rotation about a line $L$ followed by a translation *in the direction of* $L$. (*Suggestion*: Use the result of Exercise 4.)

(b) If **T** is of type 6.5.3(b), then **M** is a reflection through a plane $S$ followed by a translation *in the direction of* $S$. (*Suggestion*: Use Theorem 3.8.30.)

(c) If **T** is of type 6.5.3(c), then **M** is a rotation about a line $L$ followed by a reflection through a plane perpendicular to $L$. (*Suggestion*: Use the results of previous exercises.) (SOL)

6. (a) Prove that every rotation about an axis (in $\mathbf{R}_3$) is the product of two reflections through planes. (Compare with Sec. 6.3, Exercises 4 and 5.)

   (b) Prove that every isometry of $\mathbf{R}_3$ is the product of at most four reflections through planes. (Compare with Sec. 6.3, Exercise 7(a)). (SOL)

7. Prove: If $\mathbf{T}_1$, $\mathbf{T}_2$ are rotations (in $\mathbf{R}_3$) about different axes which both pass through the origin, then $\mathbf{T}_1\mathbf{T}_2$ is a rotation about some axis through the origin.

8. Show that the result of the preceding exercise remains true if "the origin" is replaced by some other fixed point. (*Suggestion*: Use the results of Sec. 6.4, Exercises 4 and 5.)

9. Prove: if $\mathbf{T}_1$, $\mathbf{T}_2$ are rotations about *parallel* axes in $\mathbf{R}_3$, then $\mathbf{T}_1\mathbf{T}_2$ is either (a) a rotation about an axis parallel to the others, or (b) a translation. (SUG)

10. Two lines in $\mathbf{R}_3$ are *skew* if they are neither parallel nor intersecting. Let $\mathbf{T}_1$, $\mathbf{T}_2$ be rotations about skew axes $L_1$, $L_2$ in $\mathbf{R}_3$, with $\mathbf{T}_1 \neq \mathbf{I}$, $\mathbf{T}_2 \neq \mathbf{I}$.

    (a) Prove that $\mathbf{T}_1\mathbf{T}_2$ has no fixed points and, hence, cannot be a rotation about an axis. (*Suggestion*: Let $P$ be fixed by $\mathbf{T}_1\mathbf{T}_2$ and consider the perpendicular bisector of $\overline{PP'}$, where $P' = P\mathbf{T}_1$.)

    (b) Prove that $\mathbf{T}_1\mathbf{T}_2$ is not a translation. (*Suggestion*: Consider linear parts.)

    (c) Deduce that $\mathbf{T}_1\mathbf{T}_2$ is a nontrivial rotation followed by a nontrivial translation. (SOL)

11. Let $\mathbf{T} : (x, y, z) \to (x', y', z')$ be the mapping of $\mathbf{R}_3 \to \mathbf{R}_3$ defined by the equations $x' = -z$, $5y' = 3x - 4y$, $5z' = 4x + 3y$. Show that **T** is an isometry of type 6.5.3(c) with axis $[(1, -3, 1)]$, and find the angle of rotation.

## 6.6 Rigid motions and rigid mappings

In this section we change our point of view from the geometric to the physical in order to motivate the mathematical definition of a "rigid motion" in $\mathbf{R}_n$. Since none of what we do in this section will be needed for formal proofs in subsequent chapters, we will allow ourselves the luxury of assuming some knowledge beyond the scope of this book. Specifically, in this section and the next, we shall assume familiarity with (1) the basic facts about continuous functions in $\mathbf{R}_n$, and (2) radian measure.

We start with an informal illustration. Imagine an infinite roulette wheel with center at the origin. By introducing an $x$ axis, a $y$ axis, and a unit of length, we can assign, to each point $P$ on the wheel, a pair of coordinates $(x, y)$. This gives us a one-to-one correspondence between the wheel (a set of points) and $\mathbf{R}_2$ (a set of ordered pairs). Now, let us spin the wheel; as we do so, let us imagine that the points $P$ move with the wheel but that the $x$ and $y$ axes do not; the axes remain in a fixed stationary position. Thus, as a point $P$ is carried along by the revolving wheel, *its coordinates change*. In other words, the coordinates of a given point $P$ on the wheel are a function of time. We can indicate this in symbols by letting $P(t)$ denote the coordinates of $P$ at time $t$. For example, if the wheel revolves 90° counterclockwise between the time $t = 0$ and the time $t = 1$ (say), and if $P$ is the point whose coordinates at time $t = 0$ are $(a, b)$, then the coordinates of $P$ one time-unit later will be $(-b, a)$ (see Fig. 6.6.1). That is, $P(0) = (a, b)$ and $P(1) = (-b, a)$. (Note the distinction in notation: $P$ belongs to the wheel but $P(0)$, $P(1)$ belong to $\mathbf{R}_2$.) As $t$ goes from 0 to 1, $P$

Figure 6.6.1

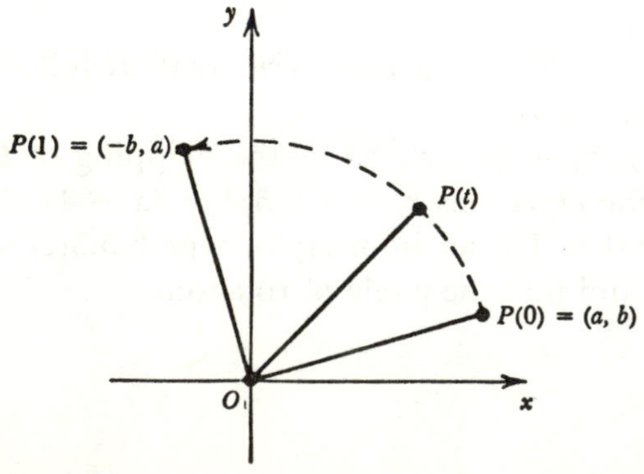

moves along a circular arc from $(a, b)$ to $(-b, a)$. The actual expression for $P(t)$ will depend on how the rate of revolution varies with time. However, regardless of what the rate is, the following two observations can be made about the motion of the wheel:

1. The motion is *continuous*: each point $P$ traces out a continuous path as $t$ goes from 0 to 1. Formally: for each $P$, $P(t)$ is a continuous function of $t$.

2. The motion is *rigid*: for each pair of points $P$ and $Q$, the segment $\overline{PQ}$ is of constant length throughout the motion. (You can convince yourself of this by using an actual wheel.) In Figure 6.6.2, for instance, the segments $\overline{P(0)Q(0)}$ and $\overline{P(t)Q(t)}$ have the same length. Saying the same thing in terms of distance,

**(6.6.1)** $d(P(t), Q(t)) = d(P(0), Q(0))$.

Conditions 1 and 2 can be restated in a different form. Suppose that for each value of $t$, we let $\mathbf{f}_t$ be the mapping of $\mathbf{R}_2 \to \mathbf{R}_2$ defined by

$$\mathbf{f}_t : P(0) \to P(t).$$

Then $P(t) = P(0)\mathbf{f}_t$, and 6.6.1 becomes

**(6.6.2)** $d(P(0)\mathbf{f}_t, Q(0)\mathbf{f}_t) = d(P(0), Q(0))$.

Equation 6.6.2 simply says that $\mathbf{f}_t$ is an isometry. Moreover, $\mathbf{f}_0$ maps $P(0) \to P(0)$ so that $\mathbf{f}_0 = \mathbf{I}$; and $\mathbf{f}_1$ maps $P(0) \to P(1)$ so that $\mathbf{f}_1$ is (in our example) the rotation through 90°. That is, $\mathbf{f}_1$ is the "end result" of our motion. A mapping which is the end result of a rigid motion will be called a "rigid mapping" (this terminology is unconventional but convenient). The distinction between motion and mapping can be expressed as follows: the motion is essentially the *process* by which the point $P$ moves from $(a, b)$ to $(-b, a)$; the mapping is the *rule* which assigns to $(a, b)$ the image $(-b, a)$.

Figure 6.6.2

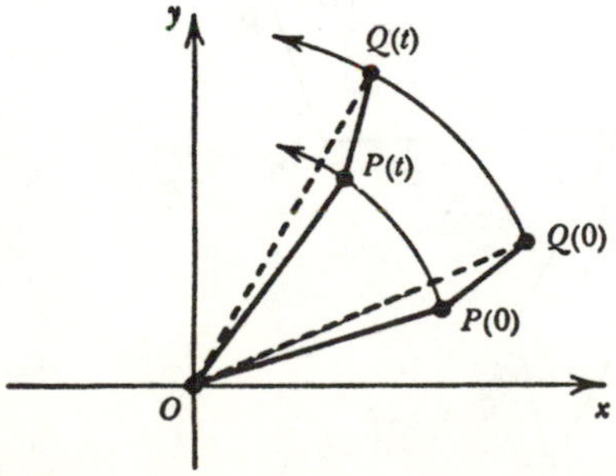

The preceding discussion leads us to make the following formal definition:

**6.6.3 Definition.** A *rigid motion* in $\mathbf{R}_n$ is a set of mappings $\mathbf{f}_t : \mathbf{R}_n \to \mathbf{R}_n$ such that

(a) $\mathbf{f}_t$ is defined for all real numbers $t$ such that $0 \leqslant t \leqslant 1$.

(b) For each $P \in \mathbf{R}_n$, $P\mathbf{f}_t$ is a continuous function of $t$ on the interval $0 \leqslant t \leqslant 1$.

(c) For each $t$ $(0 \leqslant t \leqslant 1)$, $\mathbf{f}_t$ is an isometry.

(d) $\mathbf{f}_0 = \mathbf{I}$.

Under the above conditions, the mapping $\mathbf{f}_1$ is called the *rigid mapping* (in $\mathbf{R}_n$) corresponding to the given rigid motion.

Note that in 6.6.3(b) we have changed our notation and written $P$ instead of $P(0)$.

It is immediate from Definition 6.6.3 that every rigid mapping is an isometry. However, not every isometry of $\mathbf{R}_n$ is a rigid mapping in $\mathbf{R}_n$. For example, let $L$ be a line in $\mathbf{R}_2$ and let $\mathcal{M}_L$ be the reflection through $L$ in $\mathbf{R}_2$. If we regard $\mathbf{R}_2$ as a subset of $\mathbf{R}_3$, it is possible to realize $\mathcal{M}_L$ by means of a rigid motion in $\mathbf{R}_3$ (namely, the 180° rotation about $L$; see Fig. 6.6.3). However, it can be proven that $\mathcal{M}_L$ does not

Figure 6.6.3

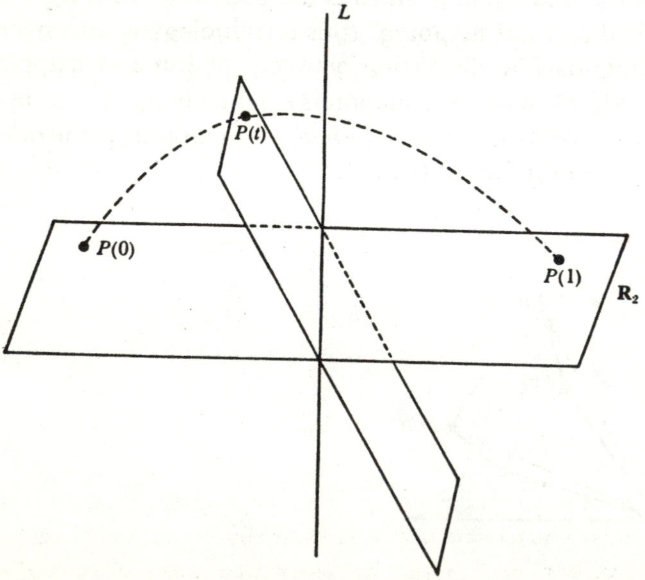

correspond to any rigid motion that stays entirely within $\mathbf{R}_2$; hence, $\mathcal{M}_L$ is not a rigid mapping according to our definition.

Which isometries *are* rigid mappings? The answer is quite simple: *an isometry of* $\mathbf{R}_n$ *is a rigid mapping in* $\mathbf{R}_n$ *if and only if its determinant is 1*. A complete proof of this result would not be practical here, but an outline of the proof can be given. Readers with some background in analysis are invited to try to fill in the missing details.

1. We show first that any rigid mapping $\mathbf{T} : \mathbf{R}_n \to \mathbf{R}_n$ must have determinant 1. As in Definition 6.6.3, $\mathbf{T}$ corresponds to a set of functions $\mathbf{f}_t$, where $\mathbf{f}_1 = \mathbf{T}$. Each $\mathbf{f}_t$ is an isometry and, hence, is the product of a linear isometry and a translation. In symbols,

$$\mathbf{vf}_t = \mathbf{vS}_t + \mathbf{u}_t \qquad (\text{all } \mathbf{v} \in \mathbf{R}_n)$$

where $\mathbf{S}_t$ is a linear isometry depending on $t$ and $\mathbf{u}_t$ is an element of $\mathbf{R}_n$ depending on $t$. For each vector $\mathbf{v} \in \mathbf{R}_n$, the vectors $\mathbf{vf}_t$ and $\mathbf{0f}_t = \mathbf{u}_t$ are continuous functions of $t$ by 6.6.3(b) and, hence, $\mathbf{vS}_t$ is a continuous function of $t$; in particular, the vectors $\mathbf{e}_1\mathbf{S}_t, \ldots, \mathbf{e}_n\mathbf{S}_t$ (which are the row vectors of the matrix of $\mathbf{S}_t$) are continuous functions of $t$. Since the determinant of a matrix is a continuous function of the rows of the matrix (*why*?), it follows that $\det \mathbf{f}_t = \det \mathbf{S}_t$ is a continuous function of $t$. But for all $t$, $\det \mathbf{f}_t = \pm 1$ by Theorem 5.6.12. By continuity, the function $\det \mathbf{f}_t$ cannot take on *both* values 1 and $-1$ (otherwise, it would take on all values between 1 and $-1$, R.A.A.); hence, $\det \mathbf{f}_t$ is constant throughout the interval $0 \leqslant t \leqslant 1$. Since $\det \mathbf{f}_0 = \det \mathbf{I} = 1$, it follows that $\det \mathbf{f}_1 = 1$, that is, $\det \mathbf{T} = 1$.

2. *Any translation* $\mathcal{T}_\mathbf{u}$ *is a rigid mapping*. Proof: for any $t$, let $\mathbf{f}_t = \mathcal{T}_{t\mathbf{u}}$. These functions $\mathbf{f}_t$ satisfy Definition 6.6.3, and $\mathbf{f}_1 = \mathcal{T}_\mathbf{u}$.

3. *Any rotation is a rigid mapping*. Proof: let $\mathbf{T}$ be a rotation of $\mathbf{R}_n$ about the $(n-2)$-flat $S$; let $U$ be the direction space of $S$ and let $W = U^\perp$. By definition, $\mathcal{R} = \mathbf{T} \restriction W$ is a rotation of $W$ about the point $P = S \cap W$. Hence, $\mathcal{R}$ corresponds to some oriented angle $\theta$ with respect to some fixed orthonormal basis $(\mathbf{w}_1, \mathbf{w}_2)$ of $W$. For each $t$ $(0 \leqslant t \leqslant 1)$, let $\mathcal{R}_t$ be the rotation of $W$ about $P$ whose oriented angle with respect to the given basis is $t\theta$. (This definition, of course, depends upon the existence of radian measure for angles.) Now, define $\mathbf{f}_t : \mathbf{R}_n \to \mathbf{R}_n$ by

$$\mathbf{f}_t : \mathbf{w} + \mathbf{u} \to \mathbf{w}\mathcal{R}_t + \mathbf{u} \qquad (\text{all } \mathbf{w} \in W, \mathbf{u} \in U).$$

Then the functions $\mathbf{f}_t$ $(0 \leqslant t \leqslant 1)$ satisfy Definition 6.6.3, and $\mathbf{f}_1 = \mathbf{T}$.

4. *The product of rigid mappings is a rigid mapping*. Intuitively, this is obvious; how would you prove it formally? (See Exercise 1 below.)

5. *Every isometry of* $\mathbf{R}_n$ *which has determinant 1 is a product of rotations*

*and translations*. This will be proved for arbitrary $n$ in Section 9.4 (see Theorem 9.4.20). It follows at once, using steps 2, 3, and 4, that every isometry of determinant 1 is a rigid mapping. Since the converse was shown in step 1, the proof is complete.

*Remark*: By Theorem 6.3.18, it follows that an isometry of $\mathbf{R}_2$ is a rigid mapping if and only if it preserves oriented angles. More generally, rigid mappings in $\mathbf{R}_n$ preserve "orientation"; we shall make this idea precise in the next section.

**Exercises**

1. Prove formally: (a) the product of two rigid mappings in $\mathbf{R}_n$ is a rigid mapping; (b) the inverse of a rigid mapping is a rigid mapping.
2. Deduce from the results of this section (and from a suitable result from Sec. 5.14) that a rigid mapping preserves generalized cross products. That is, if $\mathbf{u}_1, \ldots, \mathbf{u}_{n-1}$ are $n - 1$ vectors in $\mathbf{R}_n$ and $\mathbf{T}$ is a rigid mapping in $\mathbf{R}_n$, then $(\mathbf{u}_1 \times \cdots \times \mathbf{u}_{n-1})\mathbf{T} = (\mathbf{u}_1\mathbf{T}) \times \cdots \times (\mathbf{u}_{n-1}\mathbf{T})$.
3. Show that if we identify $\mathbf{R}_n$ with a subset of $\mathbf{R}_{n+1}$ (for instance, the subset $\{(x_1, \ldots, x_n, 0) : x_1 \in \mathbf{R}, \ldots, x_n \in \mathbf{R}\}$), then any isometry $\mathbf{M}$ of $\mathbf{R}_n$, even if not a rigid mapping in $\mathbf{R}_n$, can be regarded as the restriction to $\mathbf{R}_n$ of a rigid mapping in $\mathbf{R}_{n+1}$. (Figure 6.6.3 illustrates this assertion when $n = 2$ and $\mathbf{M}$ is a reflection $\mathcal{M}_L$.)

## 6.7 Orientation

In Figure 6.7.1, $\mathbf{u}_1$ and $\mathbf{u}_2$ are orthonormal vectors (represented as arrows) in $\mathbf{R}_2$; $\mathbf{v}_1$ and $\mathbf{v}_2$ are likewise orthonormal. The two pairs $(\mathbf{u}_1, \mathbf{u}_2)$ and $(\mathbf{v}_1, \mathbf{v}_2)$ are *oppositely oriented*: the 90° rotation which maps $\mathbf{u}_1 \to \mathbf{u}_2$ is clockwise whereas the 90° rotation which maps $\mathbf{v}_1 \to \mathbf{v}_2$ is counterclockwise. We shall say that an orthonormal pair such as $(\mathbf{u}_1, \mathbf{u}_2)$ is *left-handed* (since $\mathbf{u}_1$ lies, in a sense, to the left of $\mathbf{u}_2$) while the pair $(\mathbf{v}_1, \mathbf{v}_2)$ is *right-handed*.

To distinguish between a right-handed and a left-handed pair in Figure 6.7.1, all we need is our eyesight. However, let us consider a higher-dimensional situation: two orthonormal vectors in a 2-dimensional subspace $W$ of $\mathbf{R}_3$. Here, eyesight is not enough; as you can see from Figure 6.7.2, the orientation of the pair of vectors depends also on the *location of the observer*. The pair $(\mathbf{u}_1, \mathbf{u}_2)$ appears left-handed to Smith (located on one side of $W$) but right-handed to Jones (located on the other side of $W$); the pair $(\mathbf{v}_1, \mathbf{v}_2)$ appears right-handed to Smith but left-handed to Jones.

Analogous considerations arise with respect to *three* orthonormal vectors. In $\mathbf{R}_3$, the orthonormal triple $(\mathbf{u}_1, \mathbf{u}_2, \mathbf{u}_3)$ is usually called a *right-handed triple* if, when we represent $\mathbf{u}_1, \mathbf{u}_2, \mathbf{u}_3$ as arrows with a common initial point $P$, an observer located at the terminal point of $\mathbf{u}_1$ sees the pair $(\mathbf{u}_2, \mathbf{u}_3)$ as right-handed. (In this situation, we also say that the orthogonal coordinate system $(P; \mathbf{u}_1, \mathbf{u}_2, \mathbf{u}_3)$ is a *right-handed system*.) "Left-handed triple" and "left-handed coordinate system" are defined similarly. In Figure 6.7.3, the triple $(\mathbf{u}_1, \mathbf{u}_2, \mathbf{u}_3)$ is right-handed while the triple $(\mathbf{v}_1, \mathbf{v}_2, \mathbf{v}_3)$ is left-handed. (Perhaps you have seen an equivalent and more popular definition of the words "right-handed" and "left-handed" in terms of the motion of a screw: if the triple $(\mathbf{u}_1, \mathbf{u}_2, \mathbf{u}_3)$ is right-handed and the head of the screw is rotated in the "$\mathbf{u}_1$ to $\mathbf{u}_2$" direction, the screw moves in the direction of $\mathbf{u}_3$. See, for example, [11, page 480].) In the familiar 3-dimensional physical universe in which we live, every orthonormal triple is either right-handed or left-handed and we have no trouble telling which is which. If, however, we imagine this universe of ours to be a mere hyperplane in some larger 4-dimensional space, two 4-dimensional observers (named Bnx and Qjz, say) on opposite sides of our hyperplane will not agree on what they see: the triples within our hyperplane which appear right-handed to Bnx will appear left-handed to Qjz and vice versa. (To make intuitive sense out of this – which is not easy – try thinking of "time" as the fourth dimension. One observer looks forward in time,

Figure 6.7.1

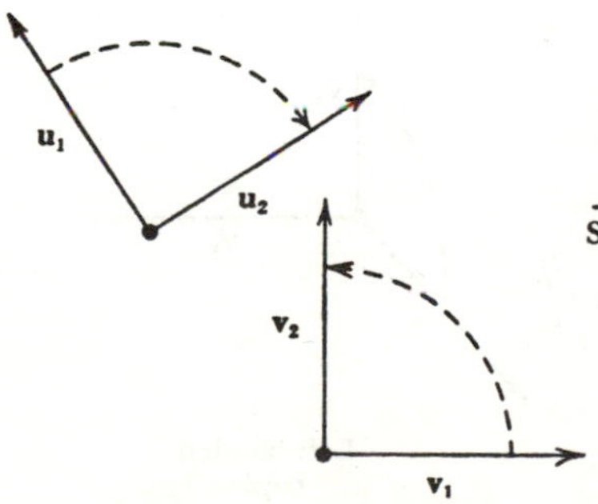


Figure 6.7.2

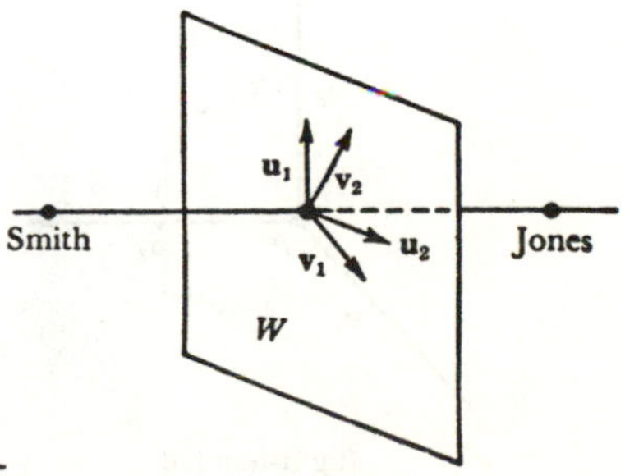

the other looks backward; thus, while one sees a screw being tightened, the other sees a screw being loosened.).

Out of all this seemingly hopeless ambiguity, what can be salvaged? Returning to Figure 6.7.2, we ask: do Smith's view of the vectors $\mathbf{u}_i$, $\mathbf{v}_i$ and Jones's view of these vectors have anything in common? Admittedly Smith and Jones disagree on which one of the pairs $(\mathbf{u}_1, \mathbf{u}_2)$, $(\mathbf{v}_1, \mathbf{v}_2)$ is right-handed and which one is left-handed. However, they do agree that the orientations of these two pairs are *different*. Similarly, our mythical 4-dimensional observers Bnx and Qjz will agree that the two triples exhibited in Figure 6.7.3 are oppositely oriented, even if they apply different labels to the orientations.

It thus seems reasonable to hope that the concepts of "same orientation" and "opposite orientation" can be formalized without appealing to physical intuition, pictures, or the like. Such a hope is justified, and the means of fulfilling it are quite close at hand – as close as the preceding section, in fact. Imagine for a moment that the three arrows exhibited in Figure 6.7.3(a) are actually three mutually perpendicular metal rods; imagine further that you pick up this configuration of three rods in your hand and physically move it through space. If the motion is rigid, the rods will remain mutually perpendicular at every instant throughout the motion. A little thought (better yet, actual experimentation) should convince you that the orientation of the configuration remains the same throughout the motion. For example, you will find it impossible, no matter how hard you try, to transform the three rods rigidly from the position shown in Figure 6.7.3(a) to the (oppositely oriented) position shown in Figure 6.7.3(b).

Figure 6.7.3

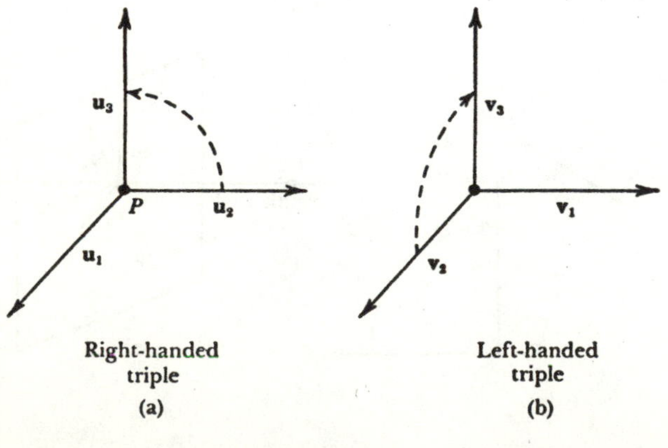

Likewise, the pair $(\mathbf{u}_1, \mathbf{u}_2)$ in Figure 6.7.1 cannot be mapped onto $(\mathbf{v}_1, \mathbf{v}_2)$ by any rigid motion that stays within the plane, but it can be mapped onto the pair $(\mathbf{v}_2, \mathbf{v}_1)$ (*how*?), which has the same orientation as $(\mathbf{u}_1, \mathbf{u}_2)$. All of this seems to suggest a definition: two orthonormal $n$-tuples in $\mathbf{R}_n$ have the "same orientation" if one can be transformed into the other by a rigid motion. In formalizing such a definition we must be somewhat careful since in our informal remarks above we have been regarding vectors as arrows (not points). Thus, we proceed as follows:

Let $\mathbf{u}_1, \ldots, \mathbf{u}_n$ be orthonormal vectors in $\mathbf{R}_n$ and let $P$ be a point in $\mathbf{R}_n$; let $P_i = P + \mathbf{u}_i$ $(i = 1, \ldots, n)$ so that $\mathbf{u}_i = \mathbf{Ar}\ PP_i$. Likewise, let $\mathbf{v}_1, \ldots, \mathbf{v}_n$ be orthonormal vectors in $\mathbf{R}_n$ with $\mathbf{v}_i = \mathbf{Ar}\ QQ_i$ for suitable points $Q, Q_1, \ldots, Q_n$. (See Fig. 6.7.4 for the case $n = 2$.) We say that the $(n + 1)$-tuple $(P; \mathbf{u}_1, \ldots, \mathbf{u}_n)$ is *rigidly equivalent* to the $(n + 1)$-tuple $(Q; \mathbf{v}_1, \ldots, \mathbf{v}_n)$, if there exists a rigid mapping in $\mathbf{R}_n$ which maps $P \to Q$, $P_1 \to Q_1, \ldots, P_n \to Q_n$. It follows easily from Exercise 1 of the preceding section that "rigid equivalence" is an equivalence relation. Moreover, for any two points $P, P'$, the $(n + 1)$-tuples $(P; \mathbf{u}_1, \ldots, \mathbf{u}_n)$ and $(P'; \mathbf{u}_1, \ldots, \mathbf{u}_n)$ are rigidly equivalent (proof: the translation which maps $P \to P'$ is a rigid mapping). Hence, if we define "rigid equivalence" of $(\mathbf{u}_1, \ldots, \mathbf{u}_n)$, $(\mathbf{v}_1, \ldots, \mathbf{v}_n)$ to mean rigid equivalence of $(P; \mathbf{u}_1, \ldots, \mathbf{u}_n)$ and $(Q; \mathbf{v}_1, \ldots, \mathbf{v}_n)$, it follows that *rigid equivalence of orthonormal*

Figure 6.7.4

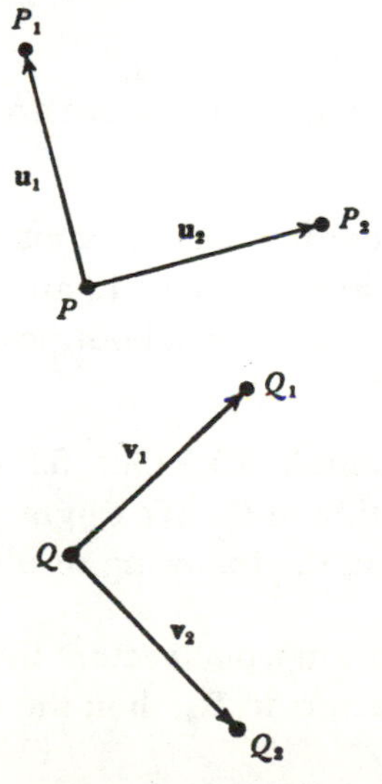

*n-tuples is a well-defined equivalence relation.* The equivalence classes of this relation will be called *orientations in* $\mathbf{R}_n$. Thus, to say that two orthonormal $n$-tuples have the same orientation is to say that they belong to the same equivalence class; that is, they are rigidly equivalent.

The determination of whether two given orthonormal $n$-tuples do, in fact, have the same orientation is (at least theoretically) a simple matter; we need only compute two $n$-by-$n$ determinants. For convenience in stating the relevant theorem, we shall (as in Sec. 5.14) denote by $R(\mathbf{u}_1, ..., \mathbf{u}_n)$ the $n$-rowed matrix whose $n$ rows are the vectors $\mathbf{u}_1$, ..., $\mathbf{u}_n$.

**6.7.1 Theorem.** Two orthonormal $n$-tuples $(\mathbf{u}_1, ..., \mathbf{u}_n)$ and $(\mathbf{v}_1, ..., \mathbf{v}_n)$ in $\mathbf{R}_n$ have the same orientation if and only if the $n \times n$ matrices $R(\mathbf{u}_1, ..., \mathbf{u}_n)$ and $R(\mathbf{v}_1, ..., \mathbf{v}_n)$ have the same determinant.

*Proof.* By definition the two $n$-tuples have the same orientation if and only if there is a rigid mapping $\mathbf{T} : \mathbf{R}_n \to \mathbf{R}_n$ such that

**(6.7.2)** $\mathbf{T} : \mathbf{0} \to \mathbf{0}, \mathbf{u}_1 \to \mathbf{v}_1, ..., \mathbf{u}_n \to \mathbf{v}_n.$

Any rigid mapping which maps $\mathbf{0} \to \mathbf{0}$ is a linear isometry, and a linear isometry is a rigid mapping if and only if it has determinant 1 (cf. Sec. 6.6). Now, by Theorems 4.2.2 and 4.10.12, there is a unique linear isometry $\mathbf{T}$ satisfying 6.7.2; by the result of Section 4.4, Exercise 10, if $A$ is the matrix of $\mathbf{T}$, then

$$R(\mathbf{u}_1, ..., \mathbf{u}_n) \cdot A = R(\mathbf{v}_1, ..., \mathbf{v}_n).$$

Hence, $\det \mathbf{T} = 1$ if and only if $\det R(\mathbf{u}_1, ..., \mathbf{u}_n) = \det R(\mathbf{v}_1, ..., \mathbf{v}_n)$, and the theorem follows. ▮

**6.7.3 Corollary.** If $n$ is any positive integer, then there are exactly two orientations in $\mathbf{R}_n$.

*Proof.* There are exactly two possible determinants for a matrix with orthonormal rows, namely, 1 and $-1$ (Theorems 4.10.13 and 5.6.12). Since $\det R(-\mathbf{u}_1, \mathbf{u}_2, ..., \mathbf{u}_n) = -\det R(\mathbf{u}_1, \mathbf{u}_2, ..., \mathbf{u}_n)$, both possibilities actually occur. ▮

In connection with the solution to Example 6.5.6 (Sec. 6.5), certain assertions were made regarding orientation in $\mathbf{R}_3$. We can now justify those assertions. In so doing, we shall find the following result useful:

**6.7.4 Theorem.** If $\mathbf{u}_1, ..., \mathbf{u}_{n-1}$ are orthonormal vectors in $\mathbf{R}_n$ and if $\mathbf{v}_1, ..., \mathbf{v}_{n-1}$ are orthonormal vectors in $\mathbf{R}_n$, then the $n$-tuples

$$(\mathbf{u}_1, \ldots, \mathbf{u}_{n-1}, \mathbf{u}_1 \times \cdots \times \mathbf{u}_{n-1}); \quad (\mathbf{v}_1, \ldots, \mathbf{v}_{n-1}, \mathbf{v}_1 \times \cdots \times \mathbf{v}_{n-1})$$

have the same orientation.

*Proof.* Both of these $n$-tuples are orthonormal by the results of Section 5.14; hence, there is a linear isometry $\mathbf{T}$ of $\mathbf{R}_n$ which maps

$$\mathbf{u}_1 \to \mathbf{v}_1, \ldots, \mathbf{u}_{n-1} \to \mathbf{v}_{n-1},$$
$$\mathbf{u}_1 \times \cdots \times \mathbf{u}_{n-1} \to \mathbf{v}_1 \times \cdots \times \mathbf{v}_{n-1}.$$

By Theorem 5.14.12,

$$\begin{aligned}\mathbf{v}_1 \times \cdots \times \mathbf{v}_{n-1} &= (\det \mathbf{T})[(\mathbf{u}_1 \times \cdots \times \mathbf{u}_{n-1})\mathbf{T}] \\ &= (\det \mathbf{T})\,(\mathbf{v}_1 \times \cdots \times \mathbf{v}_{n-1});\end{aligned}$$

hence, $\det \mathbf{T} = 1$ so that $\mathbf{T}$ is a rigid mapping. The result follows. ▮

We now turn to the situation described in Example 6.5.6; that is, we consider a (linear) rotation $\mathbf{T}$ in $\mathbf{R}_3$. The axis of rotation is a one-dimensional subspace $[\mathbf{u}]$; we lose no generality in assuming that $|\mathbf{u}| = 1$. Let $\mathbf{w}_1$ be a vector of length 1 orthogonal to $\mathbf{u}$ and let $\mathbf{w}_2 = \mathbf{u} \times \mathbf{w}_1$; then $\{\mathbf{w}_1, \mathbf{w}_2\}$ is an orthonormal basis of $[\mathbf{u}]^\perp$. If $\alpha$ is the angle of rotation, then the matrix of $\mathbf{T} \restriction [\mathbf{u}]^\perp$ with respect to the ordered basis $(\mathbf{w}_1, \mathbf{w}_2)$ is

$$\begin{bmatrix} \cos\alpha & \pm\sin\alpha \\ \mp\sin\alpha & \cos\alpha \end{bmatrix};$$

hence, in particular, we have

$$\mathbf{w}_1\mathbf{T} = (\cos\alpha)\mathbf{w}_1 \pm (\sin\alpha)\mathbf{w}_2 \tag{6.7.5}$$

(equation 6.7.5 is the same as 6.5.7). In Section 6.5 we asserted that if our coordinate system is right-handed (i.e., the triple $(\mathbf{e}_1, \mathbf{e}_2, \mathbf{e}_3)$ is right-handed), then an observer at the point $\mathbf{u}$ will see the rotation $\mathbf{T} \restriction [\mathbf{u}]^\perp$ as counterclockwise if the $\pm$ sign in 6.7.5 is plus, clockwise if the sign is minus. To see that this assertion is valid, we argue as follows:

Since the triple $(\mathbf{e}_1, \mathbf{e}_2, \mathbf{e}_3)$ is assumed right-handed, and since

$$\mathbf{e}_1 \times \mathbf{e}_2 = \mathbf{e}_3; \qquad \mathbf{u} \times \mathbf{w}_1 = \mathbf{w}_2,$$

Theorem 6.7.4 implies that the triple $(\mathbf{u}, \mathbf{w}_1, \mathbf{w}_2)$ is right-handed. Hence, (by definition) our observer at $\mathbf{u}$ (imagine yourself as this observer) will see the pair $(\mathbf{w}_1, \mathbf{w}_2)$ as right-handed (see Fig. 6.7.5). Hence, you, the observer, will see a rotation $\mathbf{T}$ of the plane $[\mathbf{w}_1, \mathbf{w}_2]$ as counterclockwise if and only if the image of $\mathbf{w}_1$ under this rotation lies on the same side of the line $[\mathbf{w}_1]$ as the point $\mathbf{w}_2$ (again, see Fig. 6.7.5); that is, if and only if the coefficient of $\mathbf{w}_2$ in the expression for

$\mathbf{w}_1\mathbf{T}$ is positive (why?); that is, if and only if the sign in equation 6.7.5 is a plus sign – which is precisely the desired assertion.

In a left-handed coordinate system the words "clockwise" and "counterclockwise" must be interchanged in our description of how the observer at $\mathbf{u}$ sees the rotation $\mathbf{T} \upharpoonright [\mathbf{u}]^{\perp}$. You should be able to supply the argument, which is like the one just given.

**Exercise**

Given any ordered basis $(\mathbf{u}_1, ..., \mathbf{u}_n)$ of $\mathbf{R}_n$, the Gram-Schmidt process (Sec. 3.8) can be applied to give us an *orthogonal* basis $(\mathbf{u}'_1, ..., \mathbf{u}'_n)$. (Specifically, $\mathbf{u}'_1 = \mathbf{u}_1$; and after $\mathbf{u}'_1, ..., \mathbf{u}'_k$ have been found, $\mathbf{u}'_{k+1}$ is obtained by subtracting from $\mathbf{u}_{k+1}$ its projections on $\mathbf{u}'_1, ..., \mathbf{u}'_k$.) If we then divide each $\mathbf{u}_i$ by its length, we obtain an *orthonormal* basis $(\mathbf{u}''_1, ..., \mathbf{u}''_n)$ of $\mathbf{R}_n$. We shall define the *orientation* of the $n$-tuple $(\mathbf{u}_1, ..., \mathbf{u}_n)$ to be the orientation of $(\mathbf{u}''_1, ..., \mathbf{u}''_n)$. Thus, every linearly independent $n$-tuple of vectors in $\mathbf{R}_n$ is assigned an orientation.

(a) Prove: two linearly independent $n$-tuples $(\mathbf{u}_1, ..., \mathbf{u}_n)$, $(\mathbf{v}_1, ..., \mathbf{v}_n)$ have the same orientation if and only if the determinants

$$\det R(\mathbf{u}_1, ..., \mathbf{u}_n); \qquad \det R(\mathbf{v}_1, ..., \mathbf{v}_n)$$

have the same sign (i.e., are both positive or both negative).

(b) Interpret geometrically (with justification) the two different orientations in $\mathbf{R}_2$ (as defined in this exercise). Do likewise in $\mathbf{R}_3$.

Figure 6.7.5

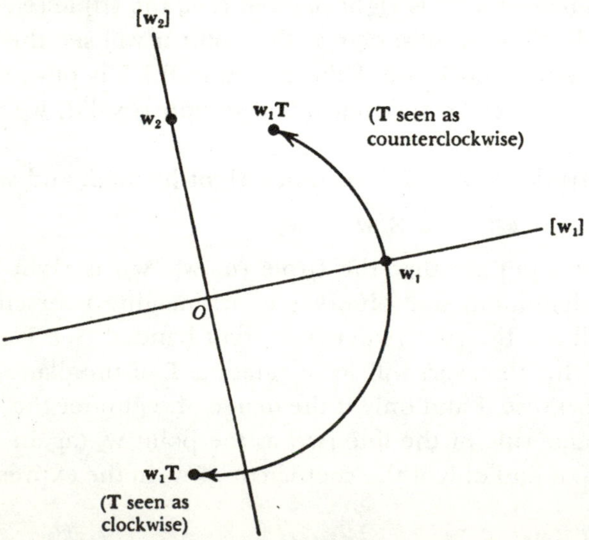

# 7 Conics and quadrics in $\mathbf{R}_n$

## 7.0 Introduction

The equation

(7.0.1) $8x^2 + 17y^2 - 12xy - 44x - 42y + 73 = 0$

can be transformed, by an orthogonal change of coordinates in $\mathbf{R}_2$, into the equation

(7.0.2) $$\frac{x^2}{20} + \frac{y^2}{5} = 1.$$

That is, there exists an orthogonal coordinate system $K$ in $\mathbf{R}_2$ such that a point $P = (x, y)$ in $\mathbf{R}_2$ satisfies 7.0.1 if and only if the $K$ coordinates of $P$ satisfy 7.0.2. As we noted in Section 4.10, it follows that 7.0.1 represents the same type of geometric locus as 7.0.2, namely, an ellipse.

In Section 7.1, we shall develop a systematic procedure by which equations like 7.0.1 can be transformed, by an orthogonal change of coordinates, into simpler forms like 7.0.2. The procedure is not limited to 2-space; it works for any second-degree equation in $n$ real variables. In Sections 7.3 and 7.4, we apply the method to the cases $n = 2$ and $n = 3$ to obtain a complete geometric classification of all second-degree equations in $\mathbf{R}_2$ and $\mathbf{R}_3$. In the remaining sections of the chapter, we discuss certain other aspects of the algebra and geometry of real second-degree equations, some of which eventually lead us into a brief introduction to projective geometry.

As the preceding paragraphs suggest, most of what we do in this chapter will take place in the space $\mathbf{R}_n$; this is due to the fact that the proofs of some key theorems depend strongly on properties of *real* scalars and would be invalid over other fields. However, in Sections 7.6 and 7.7, we do generalize *some* of our considerations to arbitrary vector spaces.

## 7.1 Equations of the second degree

If you have studied elementary analytic geometry, you know that the equation of an ellipse, a hyperbola, or a parabola in $\mathbf{R}_2$ is an equation of the second degree (examples: $x^2/a^2 + y^2/b^2 = 1$, $4x^2 - 3y^2 = 12$, $y = 8(x - 2)^2 + 2$). The form of the equation is simplest if the axes of the curve are parallel to the coordinate axes; but even if they are not, the equation is always of second degree — that is, it has the form

$$Ax^2 + Bxy + Cy^2 + Dx + Ey + F = 0 \tag{7.1.1}$$

where $A, ..., F$ are constants. Similarly, any *quadric surface* (ellipsoid, hyperboloid, paraboloid, cone, elliptic cylinder, etc.) in $\mathbf{R}_3$ always has an equation of the form

$$Ax^2 + By^2 + Cz^2 + Dxy + Exz + Fyz + Gx + Hy + Jz + K = 0 \tag{7.1.2}$$

where $A, ..., K$ are constants.

The generalization of the foregoing to $n$-space leads us to consider the set of solutions $(x_1, ..., x_n)$ (in $\mathbf{R}_n$) of an equation of the form

$$\sum_{i=1}^{n}\sum_{j=1}^{n} a_{ij}x_ix_j + \sum_{i=1}^{n} b_ix_i + c = 0. \tag{7.1.3}$$

Equation 7.1.3 is the *general second-degree equation in n variables*. It is easy to see that equations 7.1.1 and 7.1.2 are the cases $n = 2$, $n = 3$ of equation 7.1.3. For example, if $n = 2$ and we replace $(x_1, x_2)$ by $(x, y)$, 7.1.3 becomes

$$(a_{11}x^2 + a_{12}xy + a_{21}yx + a_{22}y^2) + (b_1x + b_2y) + c = 0. \tag{7.1.4}$$

Since $xy = yx$, 7.1.4 can be rewritten as

$$a_{11}x^2 + (a_{12} + a_{21})xy + a_{22}y^2 + b_1x + b_2y + c = 0 \tag{7.1.5}$$

which clearly has the form 7.1.1. Conversely, to put any equation 7.1.1 into the form 7.1.3 (equivalently, the form 7.1.5), simply let

$$a_{11} = A,\ a_{12} = a_{21} = \tfrac{1}{2}B,\ a_{22} = C,\ b_1 = D,\ b_2 = E,\ c = F. \tag{7.1.6}$$

Similarly, if $n = 3$ and we replace $(x_1, x_2, x_3)$ by $(x, y, z)$, 7.1.3 becomes

$$a_{11}x^2 + a_{22}y^2 + a_{33}z^2 + (a_{12} + a_{21})xy + (a_{13} + a_{31})xz + (a_{23} + a_{32})yz + b_1x + b_2y + b_3z + c = 0 \tag{7.1.7}$$

which has the form 7.1.2. Conversely, any equation 7.1.2 can be put into the form 7.1.7 by letting $a_{11} = A$, $a_{22} = B$, $a_{33} = C$,

**(7.1.8)** $a_{12} = a_{21} = \frac{1}{2}D, \quad a_{13} = a_{31} = \frac{1}{2}E, \quad a_{23} = a_{32} = \frac{1}{2}F$

and so on.

Observe that in 7.1.6 we could have let $a_{12}$, $a_{21}$ be any two scalars whose sum was $B$. Similarly, in 7.1.8 we could have let $a_{13}$, $a_{31}$ be any two scalars whose sum was $E$; and so on. The reason we chose $a_{12}$, $a_{21}$ to be *equal* in 7.1.6 is that the $2 \times 2$ matrix $(a_{ij})$ thereby becomes symmetric. (A square matrix $M = (m_{ij})$ is said to be *symmetric* if $m_{ij} = m_{ji}$ for all $i, j$; equivalently, if $M = M^t$.) Likewise, our choices in 7.1.8 make the $3 \times 3$ matrix $(a_{ij})$ symmetric. A similar argument in $\mathbf{R}_n$ shows that *any equation of the form 7.1.3 is equivalent to an equation of the same form such that the $n \times n$ matrix $(a_{ij})$ is symmetric*. We shall see below why symmetry of the matrix is useful.

Consider an equation of the form 7.1.3, where $x_1, \ldots, x_n$ represent coordinates in some fixed orthogonal coordinate system in $\mathbf{R}_n$. For the sake of simplicity, we shall assume our coordinate system is the "usual" system $K_0 = (O; \mathbf{e}_1, \ldots, \mathbf{e}_n)$. We further assume that the matrix $M = (a_{ij})$ is symmetric; $M$ is called the *matrix of the equation*. (We use the letter $M$ rather than $A$ in order to avoid confusion with the $A$ in equations 7.1.1, 7.1.2, 7.1.6.) $M$ is, of course, a matrix of degree $n$. We shall also have occasion to consider the matrix $M^* = (a^*_{ij})$ of degree $n + 1$ which we define as follows:

**(7.1.9)**
$$\begin{aligned} a^*_{ij} &= a_{ij} \qquad (1 \leq i \leq n, 1 \leq j \leq n) \\ a^*_{n+1,j} &= \frac{1}{2}b_j \qquad (1 \leq j \leq n) \\ a^*_{i,n+1} &= \frac{1}{2}b_i \qquad (1 \leq i \leq n) \\ a^*_{n+1,n+1} &= c. \end{aligned}$$

$M^*$ is called the *extended matrix* of equation 7.1.3; it too is symmetric. It is easy to see that if we define $x_{n+1}$ to be equal to 1, then equation 7.1.3 is the same as the equation

**(7.1.10)** $$\sum_{i=1}^{n+1} \sum_{j=1}^{n+1} a^*_{ij} x_i x_j = 0.$$

The advantage of 7.1.10 over 7.1.3 is that 7.1.10 is homogeneous: *all* terms are now of the second degree in the $x$'s, whereas in 7.1.3 there are first-degree terms $b_i x_i$ and a constant term $c$. The disadvantage of 7.1.10 is the special role played by $x_{n+1}$, which denotes a constant 1 rather than a coordinate of a point.

Let $\mathcal{G}$ be the graph of equation 7.1.3 in $\mathbf{R}_n$; that is, $\mathcal{G}$ is the set of all points $(x_1, \ldots, x_n) \in \mathbf{R}_n$ which satisfy 7.1.3. *We now show how to find a coordinate system in which the equation of $\mathcal{G}$ has a particularly simple form.* Our first step is to anticipate the future by borrowing a theorem from Chapter 9. The theorem in question is Theorem 9.5.7, and it tells us (among other things) that *if* $\mathbf{T}$ *is a linear transformation of* $\mathbf{R}_n \to \mathbf{R}_n$ *whose matrix (with respect to any given orthonormal basis) is symmetric, then* $\mathbf{R}_n$ *has an orthonormal basis whose elements are eigenvectors of* $\mathbf{T}$. In particular, let $M = (a_{ij})$ be the (symmetric) matrix of equation 7.1.3, and let $\mathbf{T}$ be the linear transformation which has $M$ as its matrix with respect to the canonical basis $(\mathbf{e}_1, \ldots, \mathbf{e}_n)$. By the theorem, there exist orthonormal eigenvectors $\mathbf{u}_1, \ldots, \mathbf{u}_n$ of $\mathbf{T}$. Thus,

$$\mathbf{u}_i\mathbf{T} = \lambda_i\mathbf{u}_i \qquad (i = 1, \ldots, n)$$

for certain scalars $\lambda_1, \ldots, \lambda_n$. Clearly, the matrix of $\mathbf{T}$ with respect to the new basis $(\mathbf{u}_1, \ldots, \mathbf{u}_n)$ is the diagonal matrix

$$\textbf{(7.1.11)} \quad M' = \begin{bmatrix} \lambda_1 & & & \\ & \lambda_2 & & \\ & & \ddots & \\ & & & \lambda_n \end{bmatrix}.$$

Hence, $M$ is similar to $M'$ (Sec. 4.8),

$$\operatorname{char}(M; x) = \operatorname{char}(\mathbf{T}; x) = (\lambda_1 - x) \cdots (\lambda_n - x),$$

and the $\lambda_i$ are the eigenvalues of $\mathbf{T}$ and, hence, of $M$.

Since $\mathbf{u}_1, \ldots, \mathbf{u}_n$ are orthonormal, the $(n + 1)$-tuple $K_1 = (O; \mathbf{u}_1, \ldots, \mathbf{u}_n)$ is an orthogonal coordinate system in $\mathbf{R}_n$ (cf. Sec. 3.10). If $P = (x_1, \ldots, x_n)$ is any element of $\mathbf{R}_n$, we shall denote by $(x'_1, \ldots, x'_n)$ the coordinates of $P$ in the coordinate system $K_1$; that is,

$$P = \sum_{i=1}^{n} x'_i\mathbf{u}_i = \sum_{i=1}^{n} x_i\mathbf{e}_i.$$

By Theorem 3.10.4,

$$\textbf{(7.1.12)} \quad x'_i = (\text{Ar } OP) \cdot \mathbf{u}_i.$$

If we let $Q = (q_{ij})$ be the $n \times n$ matrix whose $i$th row vector is $\mathbf{u}_i$, then

$$\textbf{(7.1.13)} \quad \mathbf{u}_i = (q_{i1}, \ldots, q_{in}) = \sum_{j=1}^{n} q_{ij}\mathbf{e}_j.$$

If we replace $\mathbf{v}_i$, $\mathbf{v}'_i$ by $\mathbf{e}_i$, $\mathbf{u}_i$ in equation 4.8.4, then 4.8.4 is the same as 7.1.13. Moreover, the rows of $Q$ (the vectors $\mathbf{u}_i$) are orthonormal and, hence, $Q^{-1} = Q^t$ by Theorem 4.10.13. Theorems 4.8.5 and 4.8.9 thus give

**(7.1.14)** $X' = XQ^{-1} = XQ^t; \qquad X = X'Q$

where $X$, $X'$ are the row matrices $[x_1 \cdots x_n]$, $[x'_1 \cdots x'_n]$, respectively. By taking the transpose of both sides of the equation $X' = XQ^t$, we obtain $(X')^t = QX^t$; that is,

**(7.1.15)** $$x'_i = (X')^t_{i1} = \sum_{j=1}^{n} q_{ij}(X^t)_{j1} = \sum_{j=1}^{n} q_{ij}x_j.$$

Note that 7.1.12 and 7.1.15 agree, since **Ar** $OP = P = (x_1, \ldots, x_n)$ and $\mathbf{u}_i = (q_{i1}, \ldots, q_{in})$. (Our faith in the consistency of mathematics is reaffirmed!) Similarly, by taking the transpose of the last equation 7.1.14, we get $X^t = Q^t(X')^t$ or

**(7.1.16)** $$x_j = (X^t)_{j1} = \sum_{i=1}^{n} (Q^t)_{ji}\,(X')^t_{i1} = \sum_{i=1}^{n} q_{ij}x'_i.$$

We next observe that $XMX^t$ is a $1 \times 1$ matrix whose sole entry is

**(7.1.17)** $$\begin{aligned}(XMX^t)_{11} &= \sum_{i=1}^{n}\sum_{j=1}^{n} X_{1i}M_{ij}\,(X^t)_{j1}\\ &= \sum_{i=1}^{n}\sum_{j=1}^{n} x_i a_{ij} x_j\end{aligned}$$

and that similarly $X'M'(X')^t$ is a $1 \times 1$ matrix whose sole entry is

**(7.1.18)** $$\begin{aligned}(X'M'(X')^t)_{11} &= \sum_{i=1}^{n}\sum_{j=1}^{n} x'_i a'_{ij} x'_j\\ &= \sum_{i=1}^{n} \lambda_i x'_i x'_i \qquad \text{(by 7.1.11).}\end{aligned}$$

But $Q^{-1}M' = MQ^{-1}$ by Theorem 4.8.5; hence, by 7.1.14, we have

$$\begin{aligned}X'M'(X')^t &= (XQ^{-1})M'(XQ^t)^t = X(Q^{-1}M')\,(QX^t)\\ &= X(MQ^{-1})\,(QX^t)\\ &= XMX^t\end{aligned}$$

so that 7.1.17 and 7.1.18 give

**(7.1.19)** $$\sum_{i=1}^{n}\sum_{j=1}^{n} a_{ij}x_i x_j = \sum_{i=1}^{n} \lambda_i (x'_i)^2.$$

This shows that the equation of $\mathscr{G}$ in the coordinate system $K_1$ has no "cross terms"; the only terms of second degree are those involving $(x'_i)^2$.

We next consider how the change of coordinates affects the first-degree terms. Let $B$ be the *column* matrix ($n \times 1$ matrix) whose entries are $b_1, \ldots, b_n$, and let $B' = QB$. That is,

(7.1.20) $$b'_i = \sum_{j=1}^{n} q_{ij} b_j = \mathbf{u}_i \cdot (b_1, \ldots, b_n),$$

where the scalars $b'_i$ are the entries in the column matrix $B'$. Since $X$ is a $1 \times n$ matrix and $B$ is an $n \times 1$ matrix, $XB$ is a $1 \times 1$ matrix whose sole entry is

$$(XB)_{11} = \sum_{i=1}^{n} X_{1i} B_{i1} = \sum_{i=1}^{n} x_i b_i.$$

Similarly, the sole entry in $X'B'$ is $\Sigma\, x'_i b'_i$. But 7.1.14 implies that

$$X'B' = (XQ^{-1})\,(QB) = XB;$$

that is,

(7.1.21) $$\sum_{i=1}^{n} x'_i b'_i = \sum_{i=1}^{n} x_i b_i.$$

Thus, the first-degree terms in the equation of $\mathcal{G}$ remain of first degree in the new coordinate system. Combining 7.1.19 and 7.1.21, we see that equation 7.1.3 is the same as

(7.1.22) $$\sum_{i=1}^{n} \lambda_i (x'_i)^2 + \sum_{i=1}^{n} b'_i x'_i + c = 0$$

which is the equation of $\mathcal{G}$ in the coordinate system $K_1 = (O; \mathbf{u}_1, \ldots, \mathbf{u}_n)$.

Further simplification of the equation can be effected by a translation of axes. Our procedure is based on the observation that for each *nonzero* eigenvalue $\lambda_i$, the term involving $x'_i$ in 7.1.22 can be absorbed into the $(x'_i)^2$ term by completing the square. Specifically, we have

(7.1.23) $$\lambda_i (x'_i + b'_i/2\lambda_i)^2 = \lambda_i (x'_i)^2 + b'_i x'_i + (b'_i)^2/4\lambda_i.$$

Let $P_0$ be the point whose coordinates $(y_1, \ldots, y_n)$ *in the system* $K_1$ are defined by

(7.1.24) $$y_i = \begin{Bmatrix} -b'_i/2\lambda_i & \text{if } \lambda_i \neq 0 \\ 0 & \text{if } \lambda_i = 0 \end{Bmatrix},$$

and let $K_2$ be the coordinate system $(P_0; \mathbf{u}_1, \ldots, \mathbf{u}_n)$. $K_2$ is obtained from $K_1$ by a translation of axes; the axes of $K_2$ have the same directions as those of $K_1$ but pass through $P_0$ rather than $O$. It is easy to see (using Theorem 3.10.3) that the coordinates $(x''_1, \ldots, x''_n)$ of any point $P$ in the system $K_2$ are related to the coordinates $(x'_1, \ldots, x'_n)$ of $P$ in the system $K_1$ by the equations

(7.1.25) $x_i'' = x_i' - y_i.$

It then follows from 7.1.23 that

$$\lambda_i(x_i')^2 + b_i'x_i' = \lambda_i(x_i'')^2 - (b_i')^2/4\lambda_i$$
$$(\text{if } \lambda_i \neq 0).$$

If $S$ is the subset of $E_n = \{1, 2, ..., n\}$ consisting of those integers $i$ such that $\lambda_i \neq 0$ and if $S'$ is the relative complement of $S$ in $E_n$, it follows that equation 7.1.22 is the same as the equation

$$\sum_{i\in S} \lambda_i(x_i'')^2 + \sum_{i\in S'} b_i'x_i'' + k = 0 \qquad (7.1.26)$$

where

$$k = c - \sum_{i\in S} (b_i')^2/4\lambda_i. \qquad (7.1.27)$$

Equation 7.1.26 is the equation of $\mathscr{G}$ in the coordinate system $K_2 = (P_0; \mathbf{u}_1, ..., \mathbf{u}_n)$. Its advantage over 7.1.22 is that we now have only one term corresponding to each coordinate $x_i''$.

We now make a further simplification whose effect is to replace the sum of the constant term $k$ and the first-degree terms $b_i'x_i''$ in 7.1.26 by a *single* first-degree term. Assume at least one number $b_i'$ $(i \in S')$ is nonzero, and let $i_1, ..., i_m$ be the integers belonging to $S'$. Since the number

$$\beta = \left(\sum_{i\in S'} (b_i')^2\right)^{1/2} \qquad (7.1.28)$$

is nonzero, the vector

$$\hat{\mathbf{u}}_{i_1} = \frac{1}{\beta}\sum_{i\in S'} b_i'\mathbf{u}_i = \frac{1}{\beta}\sum_{r=1}^{m} b_{i_r}'\mathbf{u}_{i_r} \qquad (7.1.29)$$

is well defined; it has length 1 and belongs to the subspace $W = [\mathbf{u}_{i_1}, ..., \mathbf{u}_{i_m}]$. Hence, the set $\{\hat{\mathbf{u}}_{i_1}\}$ can be extended to an orthonormal basis $\{\hat{\mathbf{u}}_{i_1}, ..., \hat{\mathbf{u}}_{i_m}\}$ of $W$; say,

$$\hat{\mathbf{u}}_{i_j} = \sum_{r=1}^{m} z_{jr}\mathbf{u}_{i_r} \qquad (1 \leq j \leq m). \qquad (7.1.30)$$

If any element of the subspace $W$ is written

$$\mathbf{w} = \sum_{j=1}^{m} x_{i_j}''\mathbf{u}_{i_j} = \sum_{j=1}^{m} \hat{x}_{i_j}\hat{\mathbf{u}}_{i_j}, \qquad (7.1.31)$$

the same argument which led to 7.1.14 shows that $\hat{X} = X''Z^t$, where $\hat{X}$, $X''$ are the obvious $1 \times m$ row matrices and $Z = (z_{jr})$ is of degree $m$. In particular,

$$\text{(7.1.32)} \quad \hat{x}_{i_1} = \hat{X}_{11} = \sum_{r=1}^{m} X''_{1r} Z^t_{r1} = \sum_{r=1}^{m} x''_{i_r} z_{1r}.$$

But comparison of 7.1.29 with 7.1.30 shows that $z_{1r} = b'_{i_r}/\beta$. Hence, 7.1.32 gives

$$\beta \hat{x}_{i_1} = \sum_{i \in S'} b'_i x''_i$$

so that

$$\text{(7.1.33)} \quad \beta(\hat{x}_{i_1} + k/\beta) = \sum_{i \in S'} b'_i x''_i + k$$

where $k$ is as in 7.1.26 and 7.1.27. We now let

$$\bar{\mathbf{u}}_i = \begin{Bmatrix} \mathbf{u}_i & (i \in S) \\ \hat{\mathbf{u}}_i & (i \in S') \end{Bmatrix}$$

$$\text{(7.1.34)} \quad \bar{x}_i = \begin{Bmatrix} x''_i & (i \in S) \\ \hat{x}_{i_1} + k/\beta & (i = i_1) \\ \hat{x}_i & (i \in S', i \neq i_1) \end{Bmatrix}$$

$$\bar{P} = P_0 - \frac{k}{\beta} \hat{\mathbf{u}}_{i_1}.$$

Clearly, the $\bar{\mathbf{u}}_i$ are still orthonormal. Referring to 7.1.31 and 7.1.34, it easily follows from Theorem 3.10.3 that if $\mathcal{H}$ is the coordinate system defined by

$$\mathcal{H} = (\bar{P}; \bar{\mathbf{u}}_1, \ldots, \bar{\mathbf{u}}_n),$$

then the point having coordinates $x''_i$ in the system $K_2$ will have coordinates $\bar{x}_i$ in the system $\mathcal{H}$. By 7.1.33, equation 7.1.26 now becomes

$$\text{(7.1.35)} \quad \sum_{i \in S} \lambda_i (\bar{x}_i)^2 + \beta \bar{x}_{i_1} = 0 \qquad (\beta \neq 0)$$

which is the equation of $\mathcal{G}$ in the coordinate system $\mathcal{H}$. Remember that for 7.1.35 to be valid, at least one of the scalars $b'_i$ (for which $i \in S'$) must be nonzero. If instead all such numbers $b'_i$ are zero, 7.1.26 reduces to

$$\text{(7.1.36)} \quad \sum_{i \in S} \lambda_i (x''_i)^2 + k = 0.$$

Thus, *in all cases*, we have managed to find an orthogonal coordinate system in which the equation of $\mathcal{G}$ has either the form 7.1.35 or the form 7.1.36. Observe that in 7.1.35 we have $i_1 \in S'$, that is, $i_1 \notin S$, so that no two terms of 7.1.35 involve the same coordinate.

*Remarks*: 1. If the matrix $M$ is not the zero matrix, then $M' \neq 0$ also and, hence, at least one $\lambda_i$ is nonzero. It easily follows that there exists at least one $n$-tuple $(x_1'', \ldots, x_n'')$ which does *not* satisfy equation 7.1.36. (For example, if $\lambda_h \neq 0$, we may let $x_h''$ be any number different from $\pm\sqrt{-k/\lambda_h}$ and $x_i'' = 0$ for all $i \neq h$.) Similarly, there exists at least one $n$-tuple $(x_1, \ldots, x_n)$ not satisfying 7.1.35. Hence, *if* $M \neq 0$, *the locus* $\mathcal{G}$ *cannot be the whole space* $\mathbf{R}_n$.

2. The computation of the orthonormal eigenvectors $\mathbf{u}_i$ of $\mathbf{T}$ is a straightforward matter once the eigenvalues $\lambda_i$ are known. If $\lambda^{(1)}$, ..., $\lambda^{(r)}$ are the *distinct* eigenvalues, we first use the method of Section 5.11 to compute, for each $i$, a basis for the $(\mathbf{T}, \lambda^{(i)})$-eigenspace $V^{(i)}$. (*Note*: The number of elements in this basis will equal the multiplicity of $\lambda^{(i)}$; see Theorem 8.4.33 in the next chapter.) Now, replace the basis of $V^{(i)}$ by an *orthonormal* basis $B_i$ of $V^{(i)}$ (using the Gram-Schmidt process if necessary). The set $B_1 \cup \ldots \cup B_r$ is still an orthonormal set (Sec. 5.11, Exercise 8) and has the right number of elements; hence, it is an orthonormal basis of $\mathbf{R}_n$ consisting of eigenvectors of $\mathbf{T}$.

We now illustrate the preceding discussion by means of some numerical examples. In these examples, we will need to assume as known the standard geometric descriptions of second-degree equations in $\mathbf{R}_2$ and $\mathbf{R}_3$ which have one of the simple forms 7.1.35 or 7.1.36; for instance, that $x^2/a^2 + y^2/b^2 = 1$ represents an ellipse in $\mathbf{R}_2$ and an elliptic cylinder in $\mathbf{R}_3$, that $x^2/a^2 + y^2/b^2 - z^2/c^2 = 1$ represents a hyperboloid of one sheet, and so forth. For a brief development of this material, see Appendix F.

**7.1.37 Example.** Describe geometrically the surface in $\mathbf{R}_3$ whose equation is $x^2 - yz - 2x + 3y - 5 = 0$. (*Note*: It is understood that our coordinate system is $K_0 = (O; \mathbf{e}_1, \ldots, \mathbf{e}_n)$ unless some other coordinate system is specified. It is also understood that the letters $x$, $y$, $z$ represent the coordinates of the "general" point in $\mathbf{R}_3$ in the system $K_0$.)

*Solution.* Here we have

$$M = \begin{bmatrix} 1 & 0 & 0 \\ 0 & 0 & -1/2 \\ 0 & -1/2 & 0 \end{bmatrix};$$

the eigenvalues and eigenvectors can be computed as in Section 5.11 (or 5.12). We find that $\operatorname{char}(M; x) = -(x - 1)(x - 1/2)(x + 1/2)$, so that

$$\lambda_1 = 1, \qquad \lambda_2 = \tfrac{1}{2}, \qquad \lambda_3 = -\tfrac{1}{2}.$$

Since each distinct eigenvalue has multiplicity 1, each eigenspace has dimension 1; hence, to find the $\mathbf{u}_i$, we need only find an eigenvector of length 1 corresponding to each $\lambda_i$. By computation, we obtain

$$\mathbf{u}_1 = (1, 0, 0); \qquad \mathbf{u}_2 = (0, 1/\sqrt{2}, -1/\sqrt{2});$$
$$\mathbf{u}_3 = (0, 1/\sqrt{2}, 1/\sqrt{2}).$$

(Any $\mathbf{u}_i$ could be replaced by $-\mathbf{u}_i$.) Equation 7.1.20 gives

$$b_1' = \mathbf{u}_1 \cdot (-2, 3, 0) = -2$$
$$b_2' = \mathbf{u}_2 \cdot (-2, 3, 0) = 3/\sqrt{2}$$
$$b_3' = \mathbf{u}_3 \cdot (-2, 3, 0) = 3/\sqrt{2}.$$

Equation 7.1.24 then gives

$$y_1 = 1; \qquad y_2 = -3/\sqrt{2}; \qquad y_3 = 3/\sqrt{2}$$
$$P_0 = y_1\mathbf{u}_1 + y_2\mathbf{u}_2 + y_3\mathbf{u}_3 = (1, 0, 3).$$

From 7.1.27, $k = -5 - (1 + 9/4 - 9/4) = -6$. Since all $\lambda_i \neq 0$, the set $S'$ in 7.1.26 is empty. Hence, in this case, 7.1.26 becomes

$$(x'')^2 + \frac{1}{2}(y'')^2 - \frac{1}{2}(z'')^2 - 6 = 0$$

which we recognize as the equation of a hyperboloid of one sheet whose principal axis is the $z''$ axis, as shown in Figure 7.1.1. The center of the hyperboloid is the "origin" of the $(x'', y'', z'')$ coordinate system, namely, the point $P_0 = (1, 0, 3)$. The principal axis ($z''$ axis) is the line $P_0 + [\mathbf{u}_3]$, whose equations in $(x, y, z)$ coordinates are

$$x = 1, \qquad z = y + 3. \blacksquare$$

Figure 7.1.1

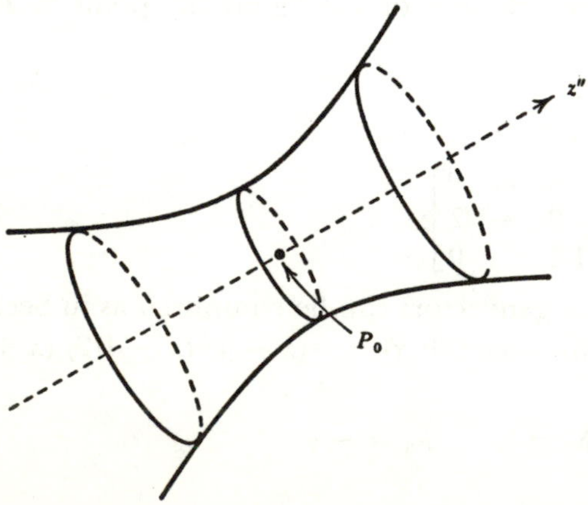

**7.1.38 Example.** Describe the surface in $\mathbf{R}_3$ whose equation is

(7.1.39) $\frac{1}{4}x^2 + y^2 + z^2 + xy + xz + 2yz + 3x + 1 = 0.$

*Solution.* We have

$$M = \begin{bmatrix} \frac{1}{4} & \frac{1}{2} & \frac{1}{2} \\ \frac{1}{2} & 1 & 1 \\ \frac{1}{2} & 1 & 1 \end{bmatrix}; \text{char}(M; x) = -x^3 + \frac{9}{4}x^2 = -x^2\left(x - \frac{9}{4}\right)$$

$$\lambda_1 = \frac{9}{4}; \qquad \lambda_2 = \lambda_3 = 0.$$

The $\lambda_1$-eigenspace, by computation, consists of vectors of the form $(a, 2a, 2a)$; hence, we may take

$$\mathbf{u}_1 = \left(\frac{1}{3}, \frac{2}{3}, \frac{2}{3}\right)$$

as the $\lambda_1$-eigenvector of length 1. The $\lambda_2$-eigenspace is $[\mathbf{u}_1]^\perp$ (*why*?) and thus $\{\mathbf{u}_2, \mathbf{u}_3\}$ may be chosen to be any orthonormal basis of $[\mathbf{u}_1]^\perp$; for example, we may choose

$$\mathbf{u}_2 = \left(\frac{2}{3}, \frac{1}{3}, -\frac{2}{3}\right); \quad \mathbf{u}_3 = \mathbf{u}_1 \times \mathbf{u}_2 = \left(-\frac{2}{3}, \frac{2}{3}, -\frac{1}{3}\right)$$

(cf. the solution to Example 6.5.6). Then by 7.1.20,

$$b_1' = \mathbf{u}_1 \cdot (3, 0, 0) = 1$$
$$b_2' = \mathbf{u}_2 \cdot (3, 0, 0) = 2$$
$$b_3' = \mathbf{u}_3 \cdot (3, 0, 0) = -2$$

and by 7.1.24 and 7.1.27 we get

$$y_1 = -\frac{2}{9}; \qquad y_2 = y_3 = 0$$

$$P_0 = \sum_{i=1}^{3} y_i\mathbf{u}_i = -\frac{2}{9}\mathbf{u}_1 = \left(-\frac{2}{27}, -\frac{4}{27}, -\frac{4}{27}\right)$$

$$k = 1 - \frac{1}{9} = \frac{8}{9}.$$

In equation 7.1.26, $S' = \{2, 3\}$; since $b_2', b_3'$ are not both zero, we keep going. By 7.1.28 and 7.1.29,

$$\beta = ((b_2')^2 + (b_3')^2)^{1/2} = \sqrt{8} = 2\sqrt{2}$$

$$\hat{\mathbf{u}}_2 = \frac{1}{\beta}(b_2'\mathbf{u}_2 + b_3'\mathbf{u}_3) = (4, -1, -1)/3\sqrt{2}.$$

We must find $\hat{\mathbf{u}}_3$ such that $\{\hat{\mathbf{u}}_2, \hat{\mathbf{u}}_3\}$ is an orthonormal basis of $[\mathbf{u}_2, \mathbf{u}_3]$. This can be done by the Gram-Schmidt process: the vector

$$\mathbf{u}_3 - \left(\frac{\mathbf{u}_3 \cdot \hat{\mathbf{u}}_2}{\hat{\mathbf{u}}_2 \cdot \hat{\mathbf{u}}_2}\right)\hat{\mathbf{u}}_2 = \left(0, \frac{1}{2}, -\frac{1}{2}\right)$$

is orthogonal to $\hat{\mathbf{u}}_2$ and lies in $[\mathbf{u}_2, \mathbf{u}_3]$, and we then divide this vector by its length to obtain

$$\hat{\mathbf{u}}_3 = (0, 1/\sqrt{2}, -1/\sqrt{2}).$$

From 7.1.34,

$$\tilde{P} = P_0 - (k/\beta)\hat{\mathbf{u}}_2 = (-10/27, -2/27, -2/27)$$

$$\tilde{\mathbf{u}}_1 = \left(\frac{1}{3}, \frac{2}{3}, \frac{2}{3}\right); \qquad \tilde{\mathbf{u}}_2 = (4, -1, -1)/3\sqrt{2};$$

$$\tilde{\mathbf{u}}_3 = (0, 1/\sqrt{2}, -1/\sqrt{2})$$

and by 7.1.35, the equation of the given surface in the coordinate system $(\tilde{P}; \tilde{\mathbf{u}}_1, \tilde{\mathbf{u}}_2, \tilde{\mathbf{u}}_3)$ is

**(7.1.40)** $$\frac{9}{4}\tilde{x}^2 + 2\sqrt{2}\,\tilde{y} = 0.$$

We recognize this as the equation of a parabolic cylinder whose axis is the $\tilde{z}$ axis, that is, the line $\tilde{P} + [\tilde{\mathbf{u}}_3]$. Figure 7.1.2 shows the approximate shape of the graph. ▮

**7.1.41** **Example.** Show that the curve $x^2 + 4xy + y^2 - 10x - 2y - 11 = 0$ (in $\mathbf{R}_2$) is a hyperbola, and find its center, foci, axes, intercepts on transverse axis, and asymptotes.

*Solution.* By the same procedure as in previous examples, we find that the curve has the equation

**(7.1.42)** $$3(x'')^2 - (y'')^2 - 9 = 0$$

with respect to a coordinate system $(P_0; \mathbf{u}_1, \mathbf{u}_2)$, where

Figure 7.1.2

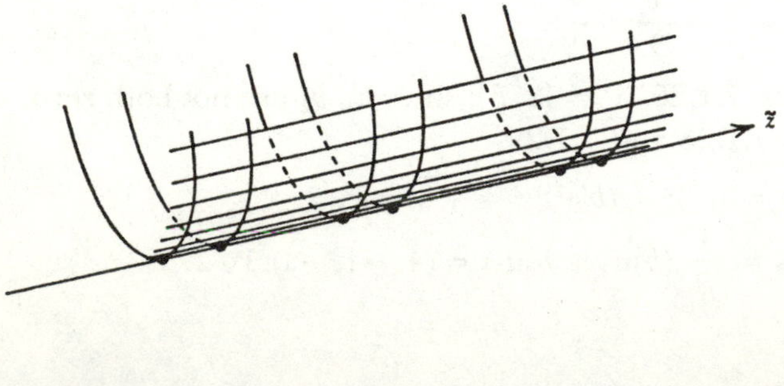

$$\mathbf{u}_1 = (1/\sqrt{2}, 1/\sqrt{2});$$
$$\mathbf{u}_2 = (1/\sqrt{2}, -1/\sqrt{2}); \qquad P_0 = (-1, 3)$$

(we omit details). Dividing equation 7.1.42 by 9, we obtain

$$\frac{(x'')^2}{3} - \frac{(y'')^2}{9} = 1.$$

We recognize this as the equation of a hyperbola having $x''$ intercepts at $x'' = \pm\sqrt{3}$ and foci at the points $x'' = \pm(3 + 9)^{1/2} = \pm\sqrt{12}$ on the $x''$ axis. In $(x, y)$ coordinates, the foci are at the points $F_1$, $F_2$, where

$$F_1 = P_0 + \sqrt{12}\mathbf{u}_1 = (\sqrt{6} - 1, \sqrt{6} + 3)$$
$$F_2 = P_0 - \sqrt{12}\mathbf{u}_1 = (-1 - \sqrt{6}, 3 - \sqrt{6}).$$

The center is, of course, the point $P_0 = (-1, 3)$. The $x''$ axis (transverse axis) is the line $P_0 + [\mathbf{u}_1]$ whose equation in $(x, y)$ coordinates is $y = x + 4$ (*why*?); the $y''$ axis $P_0 + [\mathbf{u}_2]$ has equation $x + y = 2$. The $x''$ intercepts are the points

$$P_0 \pm \sqrt{3}\,\mathbf{u}_1 = (-1 \pm \sqrt{3}/\sqrt{2}, 3 \pm \sqrt{3}/\sqrt{2}).$$

The asymptotes are the lines which satisfy the equation

$$\frac{(x'')^2}{3} - \frac{(y'')^2}{9} = 0;$$

thus, they are the lines $y'' = \pm\sqrt{3}\,x''$. The $(x, y)$ equations of these lines may be found using Theorem 3.10.4, which gives

$$\text{(7.1.43)} \quad \begin{aligned} x'' &= (x + 1, y - 3) \cdot \mathbf{u}_1 = (x + y - 2)/\sqrt{2} \\ y'' &= (x + 1, y - 3) \cdot \mathbf{u}_2 = (x - y + 4)/\sqrt{2}. \end{aligned}$$

(Equations 7.1.43 could also be obtained from 7.1.15, 7.1.24, and 7.1.25.) Substituting 7.1.43 into the two equations $y'' = \pm\sqrt{3}x''$, we obtain the equations of the two asymptotes in $(x, y)$ coordinates. The graph is sketched in Figure 7.1.3. ▮

7.1.44 **Example.** Find the equation of the ellipse whose vertices (on the major and minor axes) are the points (1, 1), (9, 5), (4, 5), (6, 1). (See Fig. 7.1.4.)

*Solution.* We label the vertices $P$, $Q$, $R$, $S$ as in Figure 7.1.4. Clearly, the center $P_0$ is the point (5, 3) (the common midpoint of $\overline{QS}$ and $\overline{PR}$). Taking the $x''$ and $y''$ axes to be the axes of the ellipse (as shown in the figure), the corresponding vectors $\mathbf{u}_1$, $\mathbf{u}_2$ are found by dividing **Ar** $P_0R$ and **Ar** $P_0Q$ by their respective lengths; thus,

$$\mathbf{u}_1 = (\mathbf{Ar}\ P_0R)/|\mathbf{Ar}\ P_0R| = (4, 2)/\sqrt{20} = (2/\sqrt{5}, 1/\sqrt{5})$$
$$\mathbf{u}_2 = (\mathbf{Ar}\ P_0Q)/|\mathbf{Ar}\ P_0Q| = (-1, 2)/\sqrt{5} = (-1/\sqrt{5}, 2/\sqrt{5}).$$

From elementary analytic geometry we know that the equation of the ellipse in $(x'', y'')$ coordinates is

**(7.1.45)** $(x'')^2/20 + (y'')^2/5 = 1$

since the semiaxes of the ellipse have lengths $\sqrt{20}$ and $\sqrt{5}$. Now, $P_0 = (5, 3)$ so that Theorem 3.10.4 gives

Figure 7.1.3

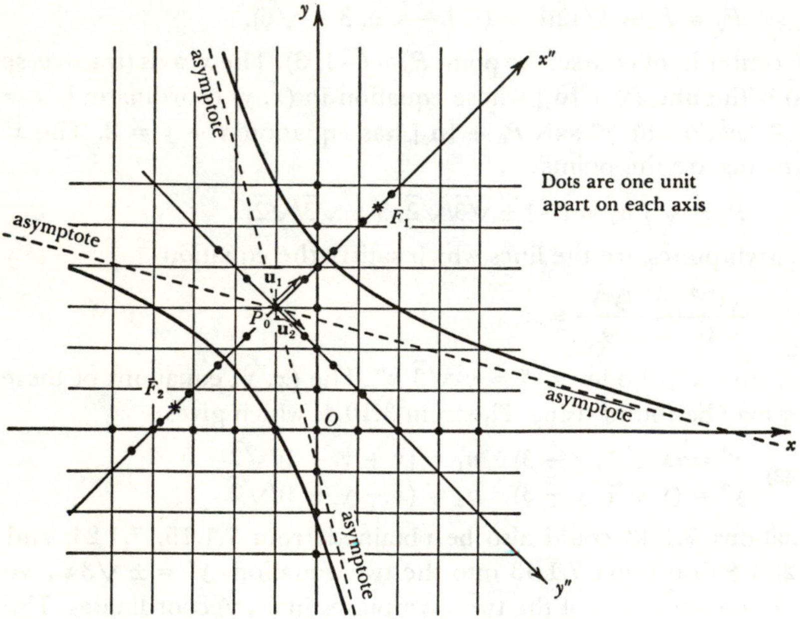


Figure 7.1.4

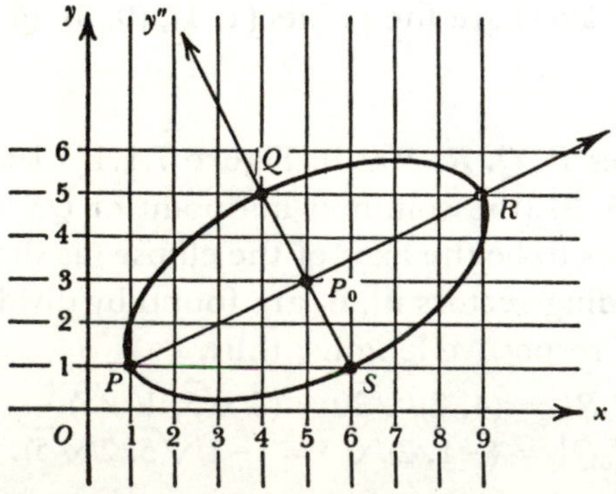

$$x'' = (x - 5, y - 3) \cdot \mathbf{u}_1 = (2x + y - 13)/\sqrt{5}$$
$$y'' = (x - 5, y - 3) \cdot \mathbf{u}_2 = (-x + 2y - 1)/\sqrt{5}. \tag{7.1.46}$$

Now substitute 7.1.46 into 7.1.45. The result (after expanding and clearing of fractions) is

$$8x^2 + 17y^2 - 12xy - 44x - 42y + 73 = 0. \blacksquare$$

**Exercises**

1. Classify each of the following loci in $\mathbf{R}_2$ geometrically. Whenever applicable, locate center, foci, vertices, aysmptotes. Draw graphs. (ANS)
   (a) $x^2 + 4xy + 4y^2 + 2x + 6y + 11 = 0$.
   (b) $15x^2 + 8xy - 34x - 8y + 15 = 0$.
   (c) $5x^2 + 8y^2 + 4xy + (14x - 52y)/\sqrt{5} - 11 = 0$.
   (d) $11x^2 + 24xy + 4y^2 + 26x + 32y + 15 = 0$.
   (e) $x^2 - 2xy + y^2 - 6x - 10y + 9 = 0$.
   (f) $x^2 + xy + y^2 = 3$.
2. Find the equation for each of the following curves in $\mathbf{R}_2$:
   (a) The parabola with vertex (1, 2) and focus (5, 5).
   (b) The ellipse with foci at (2, 0) and (8, 8) and vertices (on the major axis) at (−1, −4) and (11, 12). (ANS)
3. Describe geometrically the following loci in $\mathbf{R}_3$. Locate axes, center, vertices when applicable. (ANS)
   (a) $2xy + 2xz + 2yz + 2x = 1$.
   (b) $x^2 + y^2 + z^2 + 2xy + 4y - 5 = 0$.
   (c) $7(x^2 + y^2) - 5z^2 - 32xy + 16xz - 16yz + 2x + 34y + 10z - 23 = 0$.

*4. (a) Let $K = (P_0; \mathbf{u}_1, \ldots, \mathbf{u}_n)$ be any orthogonal coordinate system in $\mathbf{R}_n$. Let $Q = (q_{ij})$ be the $n \times n$ matrix whose $i$th row is $\mathbf{u}_i$; let $P_0 = (h_1, \ldots, h_n)$, and let $x'_1, \ldots, x'_n$ denote the coordinates of the point $(x_1, \ldots, x_n)$ in the system $K$. Prove that

$$x'_i = \sum_{j=1}^{n} q_{ij}(x_j - h_j) \qquad \text{(all } i\text{);} \tag{7.1.47}$$

$$x_j = \sum_{i=1}^{n} q_{ij}x'_i + h_j \qquad \text{(all } j\text{).} \tag{7.1.48}$$

(This generalizes equations 7.1.15–7.1.16 to the case where $P_0$ is not necessarily $O$.)

(b) Deduce from (a) that if a locus in $\mathbf{R}_n$ has an equation of second degree in a given orthogonal coordinate system, then it also has an equation of second degree in any other orthogonal coordinate system.

## 7.2 The rank of the extended matrix

The *matrix* and *extended* matrix of a second-degree equation were defined in Section 7.1. These definitions can be generalized, as follows: let $K$ be an orthogonal coordinate system in $\mathbf{R}_n$ and consider an equation of the form

$$\text{(7.2.1)} \quad \sum_{i=1}^{n}\sum_{j=1}^{n} a'_{ij}x'_ix'_j + \sum_{i=1}^{n} b'_ix'_i + c' = 0$$

where $x'_1, \ldots, x'_n$ denote coordinates in the system $K$. If the matrix $M_1 = (a'_{ij})$ is symmetric, $M_1$ is called the *matrix of equation 7.2.1 (with respect to the system K)*. The *extended matrix* $M_1^*$ of this equation (with respect to $K$) is defined by

$$\text{(7.2.2)} \quad M_1^* = \left[\begin{array}{ccc|c} & & & b'_1/2 \\ & M_1 & & \vdots \\ & & & b'_n/2 \\ \hline b'_1/2 & \cdots & b'_n/2 & c' \end{array}\right].$$

Normally, the phrase "with respect to $K$" may be omitted since the system $K$ will be understood from the context. For example, the extended matrix $M^*$ of equation 7.1.39 and the extended matrix $\tilde{M}^*$ of equation 7.1.40 are (respectively)

$$M^* = \begin{bmatrix} \frac{1}{4} & \frac{1}{2} & \frac{1}{2} & \frac{3}{2} \\ \frac{1}{2} & 1 & 1 & 0 \\ \frac{1}{2} & 1 & 1 & 0 \\ \frac{3}{2} & 0 & 0 & 1 \end{bmatrix}; \quad \tilde{M}^* = \begin{bmatrix} \frac{9}{4} & 0 & 0 & 0 \\ 0 & 0 & 0 & \sqrt{2} \\ 0 & 0 & 0 & 0 \\ 0 & \sqrt{2} & 0 & 0 \end{bmatrix}.$$

Both $M^*$ and $\tilde{M}^*$ have rank 3. It is no accident that the ranks of $M^*$ and $\tilde{M}^*$ are equal. More generally, we shall show in this section that *any* orthogonal change of coordinates leaves unchanged both the rank and the determinant of the extended matrix of a second-degree equation. The precise statement of the result is as follows:

**7.2.3 Theorem.** Assume that the matrix $M = (a_{ij})$ of equation 7.1.3 is symmetric. Let $K = (P_0; \mathbf{u}_1, \ldots, \mathbf{u}_n)$ be an orthogonal coordinate system in $\mathbf{R}_n$; for any point $P = (x_1, \ldots, x_n)$ in $\mathbf{R}_n$, let $x'_1, \ldots, x'_n$ denote the coordinates of $P$ in the system $K$; that is,

$$(x_1, \ldots, x_n) = (x'_1, \ldots, x'_n)_K.$$

Then:

(a) If the expressions for $x_1, \ldots, x_n$ in terms of $x'_1, \ldots, x'_n$ (equations 7.1.48) are substituted for $x_1, \ldots, x_n$ in equation 7.1.3, the resulting equation has the form

$$\sum_{i=1}^{n}\sum_{j=1}^{n} a'_{ij}x'_ix'_j + \sum_{i=1}^{n} b'_ix'_i + c' = 0 \tag{7.2.4}$$

where the matrix $M_1 = (a'_{ij})$ is still symmetric.

(b) The matrices $M$ and $M_1$ are similar.

(c) The extended matrices $M^*$ and $M_1^*$ of equations 7.1.3 and 7.2.4 (respectively) have the same rank and determinant.

(d) If $P_0 = O$, the matrices $M^*$ and $M_1^*$ are similar.

Theorem 7.2.3 will be used in Sections 7.3 and 7.4 to help us classify second-degree equations by geometric type. In Section 7.6 the rank of the extended matrix will be given a nice geometric interpretation.

*Proof of Theorem 7.2.3.* Let $h_1, \ldots, h_n$, $Q$ be as defined in Section 7.1, Exercise 4(a), and let $B$, $B'$, $H$ be the *column* matrices whose entries are respectively $(b_1, \ldots, b_n)$, $(b'_1, \ldots, b'_n)$, $(h_1, \ldots, h_n)$. By direct computation, we find that substitution of 7.1.48 into 7.1.3 yields 7.2.4, where

$$\begin{aligned}
&\text{(a)}\quad a'_{ij} = \sum_{k=1}^{n}\sum_{s=1}^{n} q_{ik}a_{ks}q_{js}\\
&\text{(b)}\quad b'_i = 2\sum_{k=1}^{n}\sum_{s=1}^{n} q_{ik}a_{ks}h_s\\
&\qquad\qquad + \sum_{k=1}^{n} q_{ik}b_k\\
&\text{(c)}\quad c' = \sum_{i=1}^{n}\sum_{j=1}^{n} h_ia_{ij}h_j\\
&\qquad\qquad + \sum_{i=1}^{n} b_ih_i + c.
\end{aligned} \tag{7.2.5}$$

By 7.2.5(a), we have

$$M_1 = QMQ^t = QMQ^{-1} \tag{7.2.6}$$

so that $M_1$ and $M$ are similar; also,

$$M_1^t = (QMQ^t)^t = Q^{tt}M^tQ^t = QMQ^t = M_1$$

so that $M_1$ is still symmetric. Thus, parts (a) and (b) of 7.2.3 are proved.

Also, by parts (b) and (c) of 7.2.5, we have

(7.2.7) $$\begin{aligned} B' &= 2QMH + QB \\ c' &= H^tMH + H^tB + c. \end{aligned}$$

Using 7.2.2, 7.2.6, 7.2.7, and the results from Section 4.5 on "block multiplication," we obtain

(7.2.8) $$M_1^* = Q^*M^*(Q^*)^t$$

where $Q^*$ is the $(n + 1) \times (n + 1)$ matrix having the block form

(7.2.9) $$Q^* = \begin{bmatrix} Q & 0 \\ H^t & 1 \end{bmatrix}.$$

Since $Q^t = Q^{-1}$, it follows from the results of Section 5.6 (Exercise 4 and Theorems 5.6.10 and 5.6.12) that $Q^*$ is nonsingular and det $Q^*$ = det $Q = \pm 1$, so that det $M_1^*$ = det $M^*$; also, since multiplication by a nonsingular matrix leaves rank unchanged (Sec. 5.9, Exercise 6), rank $M_1^*$ = rank $M^*$, proving 7.2.3(c). Finally, if $P_0 = O$, then $H = 0$, so that from 7.2.9 we get $(Q^*)^t = (Q^*)^{-1}$; hence, $M_1^*$ and $M^*$ are similar, by 7.2.8. ▮

**Exercises**

1. In each of the following, verify by direct computation that the extended matrices of equations 7.1.3, 7.1.22, 7.1.26, and (if applicable) 7.1.35 have the same rank and determinant. (ANS)
   (a) Example 7.1.38.
   (b) Example 7.1.41.
   (c) Section 7.1, Exercise 1(c).
   (d) Section 7.1, Exercise 1(d).
   (e) Section 7.1, Exercise 3(a).
   (f) Section 7.1, Exercise 3(c).
2. Let $\mathscr{G}$ be the graph of an equation 7.1.3 in $\mathbf{R}_n$, let $M$, $M^*$ be the matrix and extended matrix of this equation, and let $\mathscr{H}$ be a subset of $\mathbf{R}_n$ such that $\mathscr{H}$ is congruent to $\mathscr{G}$ (cf. Sec. 4.10). Prove: $\mathscr{H}$ is the graph of a second-degree equation whose matrix is similar to $M$ and whose extended matrix has the same rank and determinant as $M^*$. (SUG)

## 7.3 Classification of second-degree equations in $\mathbf{R}_2$

Consider the curve

**(7.3.1)** $x^2 + 4xy + y^2 - 10x - 2y - 11 = 0$

which was shown to be a hyperbola in Example 7.1.41. The matrix and extended matrix of equation 7.3.1 are, respectively,

$$M = \begin{bmatrix} 1 & 2 \\ 2 & 1 \end{bmatrix}; \qquad M^* = \begin{bmatrix} 1 & 2 & -5 \\ 2 & 1 & -1 \\ -5 & -1 & -11 \end{bmatrix}.$$

If we are interested *only* in the fact that the curve is a hyperbola (without worrying about the location of vertices, axes, etc.), there is no need to compute eigenvectors $\mathbf{u}_i$, or even to compute eigenvalues $\lambda_i$; the desired result is already implied merely by the fact that det $M < 0$ and rank $M^* = 3$. In this section, we shall show that the geometric type of *any* second-degree equation in $\mathbf{R}_2$ which has a nonempty graph is uniquely determined by the rank of $M^*$ and the sign of det $M$.

We begin with the observation that the matrix $M$ of any equation 7.1.3 has the same determinant as the matrix $M'$ in 7.1.11, since they both represent the same linear transformation. That is,

$$\det M = \lambda_1\lambda_2 \dots \lambda_n.$$

In the case $n = 2$, this reduces to

**(7.3.2)** $\det M = \lambda_1\lambda_2.$

(Note that 7.3.2 is a special case of Theorem 5.12.12.) For the remainder of this section, $\mathcal{G}$ will denote the graph in $\mathbf{R}_2$ of an equation 7.1.3 (with $n = 2$) and $M$, $M^*$ will denote the matrix and extended matrix of the equation, respectively; the eigenvalues of $M$ will be denoted $\lambda_1$, $\lambda_2$. The case where $M = 0$ (i.e., where the equation is only of first degree) is trivial and will not be treated here; hence, we assume from now on that $M \neq 0$. It follows that $M' \neq 0$ so that $\lambda_1$, $\lambda_2$ are not both zero. There are three possibilities:

*Case* 1. $\lambda_1, \lambda_2$ *are both nonzero and have the same sign*. Here 7.3.2 implies det $M > 0$. We may write $\lambda_1 = \epsilon/a^2$, $\lambda_2 = \epsilon/b^2$, where $\epsilon = \pm 1$. Since $S' = \emptyset$, equation 7.1.26 reduces to

**(7.3.3)** $\epsilon[(x'')^2/a^2 + (y'')^2/b^2] + k = 0.$

The extended matrix of 7.3.3 is

**(7.3.4)** $$\begin{bmatrix} \epsilon/a^2 & 0 & 0 \\ 0 & \epsilon/b^2 & 0 \\ 0 & 0 & k \end{bmatrix}$$

which clearly has rank 3 if $k \neq 0$, rank 2 if $k = 0$. If $k = 0$, then 7.3.3 is satisfied only by $x'' = y'' = 0$ and $\mathcal{G}$ is thus a single point. If $k$ has the same sign as $\epsilon$, then 7.3.3 has no solution and $\mathcal{G}$ is empty; if $k$ and $\epsilon$ have opposite signs, then 7.3.3 is the equation of an ellipse.

*Case* 2. *$\lambda_1$, $\lambda_2$ are nonzero and have opposite signs*. In this case, det $M < 0$. Letting $\lambda_1 = \epsilon/a^2$, $\lambda_2 = -\epsilon/b^2$ (where $\epsilon = \pm 1$), 7.1.26 becomes

$$(x'')^2/a^2 - (y'')^2/b^2 = -\epsilon k$$

which is the equation of a hyperbola if $k \neq 0$ and represents the two lines $x''/a = \pm y''/b$ if $k = 0$. In the latter case, the two lines intersect at the point $x'' = y'' = 0$ (i.e., the point $P_0$, using the notation of Sec. 7.1). Once again, the extended matrix has rank 3 if $k \neq 0$, rank 2 if $k = 0$.

*Case* 3. *One of $\lambda_1$, $\lambda_2$ is zero*. Here, det $M = 0$. We lose no generality in assuming $\lambda_1 \neq 0$, $\lambda_2 = 0$. Equation 7.1.26 reduces to

$$\lambda_1(x'')^2 + b_2' y'' + k = 0$$

whose extended matrix is

$$\text{(7.3.5)} \qquad \begin{bmatrix} \lambda_1 & 0 & 0 \\ 0 & 0 & \frac{1}{2}b_2' \\ 0 & \frac{1}{2}b_2' & k \end{bmatrix}.$$

If $b_2' \neq 0$, this matrix has rank 3 and we may use equation 7.1.35:

$$\lambda_1 \bar{x}^2 + \beta \bar{y} = 0 \qquad (\beta \neq 0)$$

which clearly represents a parabola. If instead $b_2' = 0$, then equation 7.1.26 becomes

$$\lambda_1(x'')^2 + k = 0.$$

In this case, the graph is empty if $k$ has the same sign as $\lambda_1$; it represents the pair of parallel lines

$$x'' = \pm\,(-k/\lambda_1)^{1/2}$$

if $k$, $\lambda_1$ have opposite signs; it consists of the single line $x'' = 0$ if $k = 0$. Evidently, the matrix 7.3.5 has rank 1 if $b_2' = k = 0$, rank 2 if $b_2' = 0$ but $k \neq 0$.

We, hence, obtain the classification in Table 7.3.6.

**7.3.6 Table of the graphs of Equation 7.1.3 when $n = 2$.**
($M$, $M^*$ denote the matrix and extended matrix of the equation, respectively. We assume $M \neq 0$.)

| Sign of det $M$ | Graph if rank $M^* = 3$ | Graph if rank $M^* = 2$ | Graph if rank $M^* = 1$ |
|---|---|---|---|
| $>0$ | ellipse *or* empty | point | — |
| $<0$ | hyperbola | 2 intersecting lines | — |
| $=0$ | parabola | 2 parallel lines *or* empty | line |

**7.3.7 Example.** Describe the geometric type of the equation

(7.3.8) $$x^2 + 4xy + 4y^2 + 2x + 4y - 3 = 0$$

without computing eigenvalues or eigenvectors.

*Solution.* Here we have

$$M = \begin{bmatrix} 1 & 2 \\ 2 & 4 \end{bmatrix}; \qquad M^* = \begin{bmatrix} 1 & 2 & 1 \\ 2 & 4 & 2 \\ 1 & 2 & -3 \end{bmatrix}$$

$$\det M = 0; \qquad \text{rank } M^* = 2.$$

Hence, the graph $\mathcal{G}$ of 7.3.8 is either a pair of parallel lines or is empty. One simple way of showing that $\mathcal{G}$ is nonempty is to observe that the left side of 7.3.8 is negative when $x = y = 0$ and is positive when $x = y = 1$; hence, by the intermediate value theorem there must be some point at which the left side of 7.3.8 is zero. Thus, $\mathcal{G}$ is not empty; hence $\mathcal{G}$ is a pair of parallel lines.

*Remark concerning the preceding example.* Once we know that the graph consists of one or more lines, the actual lines may be found without computing eigenvectors. The method is to factor the left side of the equation. Thus, in the above example, we would factor the left side of 7.3.8:

(7.3.9) $$x^2 + 4xy + 4y^2 + 2x + 4y - 3 = (x + 2y + d_1)(x + 2y + d_2).$$

(The coefficients of $x$ and $y$ in each factor are obtained by factoring $x^2 + 4xy + 4y^2$.) If we expand 7.3.9, we see that $d_1$ and $d_2$ must satisfy the equations

$$d_1 + d_2 = 2; \qquad d_1 d_2 = -3$$

from which we get $d_1 = -1$, $d_2 = 3$ (or vice versa). Hence, $\mathcal{G}$ consists of the two lines

$$x + 2y - 1 = 0; \qquad x + 2y + 3 = 0.$$

It is easily proved (*how*?) that this method always works when det $M$

$\leq 0$ and rank $M^* < 3$. In the case where the graph is empty, the solutions of the equations for $d_1$ and $d_2$ will be imaginary. (This provides a second way to determine whether the graph is nonempty. This method, unlike the trick used in Example 7.3.7, is guaranteed to work.)

As shown by Table 7.3.6, the classification of second-degree equations in $\mathbf{R}_2$ according to the sign of det $M$ and the rank of $M^*$ produces seven cases. In two of these cases, the classification is insufficient to tell us whether the graph $\mathcal{G}$ is empty or nonempty. One of these two cases (det $M = 0$, rank $M^* = 2$) can be treated by the method described in the preceding paragraph. The other case (det $M > 0$, rank $M^* = 3$) is most easily taken care of via the following theorem:

**7.3.10 Theorem.** Let $M$, $M^*$ be the matrix and extended matrix (respectively) of the equation

(7.3.11) $$Ax^2 + Bxy + Cy^2 + Dx + Ey + F = 0$$

(in $\mathbf{R}_2$), and assume det $M > 0$. Then the graph of 7.3.11 is empty if and only if $A$ det $M^* > 0$.

For example, the matrix and extended matrix of the equation

(7.3.12) $$8x^2 + 17y^2 - 12xy + 44x + 42y + 73 = 0$$

are

$$M = \begin{bmatrix} 8 & -6 \\ -6 & 17 \end{bmatrix}; \qquad M^* = \begin{bmatrix} 8 & -6 & 22 \\ -6 & 17 & 21 \\ 22 & 21 & 73 \end{bmatrix}$$

We have det $M = 100 > 0$ and det $M^* = -10{,}000 < 0$, $A = 8$, $A$ det $M^* < 0$. Hence, the graph of 7.3.12 is nonempty. Since det $M^* \neq 0$, $M^*$ has rank 3 and, thus, equation 7.3.12 represents an ellipse.

To prove Theorem 7.3.10, we need the following lemma:

**7.3.13 Lemma.** Let $N = (n_{ij})$ be a real symmetric $3 \times 3$ matrix, and assume that

(7.3.14) $$\det \begin{bmatrix} n_{11} & n_{12} \\ n_{21} & n_{22} \end{bmatrix} > 0.$$

Then the following statements are equivalent:

(a) $n_{11}$ det $N > 0$.

(b) All eigenvalues of $N$ have the same sign (i.e., they are either all positive or all negative).

To prove the lemma, we again borrow from Chapter 9: Theorem

9.7.4 (parts (a) and (f)) tells us that the eigenvalues of $N$ are all *positive* if and only if

(7.3.15) $$n_{11} > 0; \qquad \det \begin{bmatrix} n_{11} & n_{12} \\ n_{21} & n_{22} \end{bmatrix} > 0; \qquad \det N > 0.$$

Now the eigenvalues of $N$ are all *negative* if and only if those of $-N$ are all positive (Sec. 5.11, Exercise 9); that is, if and only if

$$-n_{11} > 0; \qquad \det \begin{bmatrix} -n_{11} & -n_{12} \\ -n_{21} & -n_{22} \end{bmatrix} > 0; \qquad \det(-N) > 0,$$

that is, if and only if

(7.3.16) $$n_{11} < 0; \qquad \det \begin{bmatrix} n_{11} & n_{12} \\ n_{21} & n_{22} \end{bmatrix} > 0; \qquad \det N < 0.$$

Combining 7.3.15 and 7.3.16 (and remembering 7.3.14), the lemma follows.

We now prove Theorem 7.3.10. Since $\det M > 0$, we have the situation of Case 1 of the discussion preceding Table 7.3.6. Let $M^*$, $M_1^*$, $M_2^*$ be the extended matrices of (respectively) equations 7.1.3, 7.1.22, 7.1.26; thus, in our present situation, $M_2^*$ is the matrix 7.3.4 and

$$M_1^* = \begin{bmatrix} \lambda_1 & 0 & \frac{1}{2}b_1' \\ 0 & \lambda_2 & \frac{1}{2}b_2' \\ \frac{1}{2}b_1' & \frac{1}{2}b_2' & c \end{bmatrix}.$$

Let $\epsilon$, $k$ be as in equations 7.3.3 and 7.3.4. Our proof will consist of showing that each of the following statements is equivalent to the next one.

(1) The graph $\mathscr{G}$ of equation 7.3.11 is empty.
(2) $\epsilon$ and $k$ have the same sign.
(3) $\epsilon k > 0$.
(4) $\epsilon \det M_2^* > 0$.
(5) $\epsilon \det M_1^* > 0$.
(6) $\lambda_1 \det M_1^* > 0$.
(7) All eigenvalues of $M_1^*$ have the same sign.
(8) All eigenvalues of $M^*$ have the same sign.
(9) $A \det M^* > 0$.

Equivalence of (1) and (2) was shown in the prior discussion of Case 1. (2) $\Leftrightarrow$ (3) is obvious. Since $\det M_2^* = \lambda_1\lambda_2 k$ and $\lambda_1\lambda_2 = \det M > 0$, it follows that (3) $\Leftrightarrow$ (4). By Theorem 7.2.3, (4) $\Leftrightarrow$ (5). Since $\lambda_1 = \epsilon/a^2$, (5) $\Leftrightarrow$ (6). By Lemma 7.3.13, (6) $\Leftrightarrow$ (7) and (8) $\Leftrightarrow$ (9). Finally, $M_1^*$ and $M^*$ are similar (Theorem 7.2.3(d)) and, hence, have the same eigenvalues (cf. Sec. 5.11) so that (7) $\Leftrightarrow$ (8). Thus, each of statements (1) through (9) is indeed equivalent to the next one. Hence, (1) $\Leftrightarrow$ (9), which is the desired assertion. ▮

In the case $\det M = 0$, rank $M^* < 3$, the empty and nonempty cases may be distinguished by the following theorem analogous to Theorem 7.3.10.

**7.3.17 Theorem.** Let $M$, $M^*$ be the matrix and extended matrix (respectively) of equation 7.3.11 in $\mathbf{R}_2$, and assume $\det M = 0$, rank $M^* < 3$. Then the graph of 7.3.11 is empty if and only if

$$4(A + C)F > D^2 + E^2.$$

The proof depends on the following lemma, which takes the place of Lemma 7.3.13.

**7.3.18 Lemma.** If $N$ is a real symmetric $n \times n$ matrix of rank $< 3$, then $N$ has two nonzero eigenvalues of the same sign (counting multiplicities, see Note below) if and only if $\sigma_2(N) > 0$, where $\sigma_k(N)$ is defined as in Theorem 5.12.12.

(*Note*: The phrase about "two eigenvalues counting multiplicities" means that if $\operatorname{char}(N; x) = (\lambda_1 - x)(\lambda_2 - x) \cdots (\lambda_n - x)$, there are two distinct indices $i$ and $j$ such that $\lambda_i$, $\lambda_j$ have the same sign. In the case where $\lambda_i = \lambda_j$, the "two" eigenvalues would really be one eigenvalue of multiplicity 2.)

*Proof of 7.3.18.* Let $\operatorname{char}(N; x) = (\lambda_1 - x)(\lambda_1 - x) \cdots (\lambda_n - x)$. Since $N$ is symmetric, the discussion in Section 7.1 shows that $N$ is similar to the diagonal matrix $N'$ whose diagonal entries are $\lambda_1, \ldots, \lambda_n$. It easily follows that the rank of $N$ equals the number of indices $i$ such that $\lambda_i \neq 0$. Since it is assumed that rank $N < 3$, we may assume that $\lambda_3 = \cdots = \lambda_n = 0$. Hence, Theorem 5.12.12 implies that $\sigma_2(N) = \lambda_1\lambda_2$, and Lemma 7.3.18 follows.

*Proof of 7.3.17.* Here we have the situation of Case 3 of the discussion at the beginning of the section, with $b_2' = 0$. The matrices $M_1^*$, $M_2^*$ (notation as above) are given by

$$M_1^* = \begin{bmatrix} \lambda_1 & 0 & \frac{1}{2}b_1' \\ 0 & 0 & 0 \\ \frac{1}{2}b_1' & 0 & c \end{bmatrix}; \qquad M_2^* = \begin{bmatrix} \lambda_1 & 0 & 0 \\ 0 & 0 & 0 \\ 0 & 0 & k \end{bmatrix}.$$

Each of the following statements may now be shown to be equivalent to the next one:

(1) The graph $\mathscr{G}$ of equation 7.3.11 is empty.
(2) $\lambda_1$, $k$ have the same sign.
(3) $\lambda_1 k > 0$.
(4) $\sigma_2(M_1^*) > 0$.
(5) $M_1^*$ has two nonzero eigenvalues of the same sign.
(6) $M^*$ has two nonzero eigenvalues of the same sign.
(7) $\sigma_2(M^*) > 0$.

(Using 7.1.27, a computation shows that $\sigma_2(M_1^*) = \lambda_1 k$ and, hence, (3) $\Leftrightarrow$ (4); you should be able to fill in the other reasons yourself.) Since $\det M = 0$, a straightforward computation gives

$$\sigma_2(M^*) = \tfrac{1}{4}[4(A + C)F - D^2 - E^2]$$

and Theorem 7.3.17 follows. ∎

## Exercises

1. Determine the geometric type of each of the following loci in $\mathbf{R}_2$. *without* computing eigenvalues or eigenvectors. (ANS)
   (a) $x^2 + 2y^2 + 2xy + 2x + 2y = 2$.
   (b) $x^2 + y^2 + 4xy + 2x + 6y = 0$.
   (c) $x^2 + y^2 + 4xy + 2x + 6y + 2/3 = 0$.
   (d) $2x^2 + 5y^2 - 2xy + 4x - 8y + 4 = 0$.
   (e) $4x^2 + 9y^2 - 12xy + 4x - 6y + 1 = 0$.
   (f) $x^2 + 2y^2 - 2xy - 4y + 5 = 0$.
2. (a) In each part of Exercise 1 for which the graph consists of one or more lines, find the lines by factoring the equation. (ANS)
   (b) Show that this "factoring" method will always work for second-degree equations in $\mathbf{R}_2$ when $\det M \leqslant 0$ and rank $M^* < 3$.
3. Prove: If the graph of a second-degree equation in $\mathbf{R}_2$ is a pair of parallel lines or a single line, the first *two* rows of the cor-

responding extended matrix $M^*$ are linearly dependent; that is, they are scalar multiples of each other or one of them is the zero row.

*4. Prove: The rank of any real symmetric $n \times n$ matrix $N$ equals the largest integer $r$ such that $\sigma_r(N) \neq 0$. (*Hint*: See the proof of 7.3.18.)

## 7.4 Quadric surfaces in $\mathbf{R}_3$

The graph of a second-degree equation in $\mathbf{R}_3$ is called a *quadric*, or *quadric surface*. (More generally, the graph of a second-degree equation in $\mathbf{R}_n$ may be called an *n-quadric*.) The classification of second-degree equations in $\mathbf{R}_3$, like that in $\mathbf{R}_2$, is based on the signs of the eigenvalues; and these signs are again determined uniquely if one knows the signs of certain determinants. To be precise, let $M = (a_{ij})$ be the (symmetric) matrix of equation 7.1.3 where $n = 3$, and let $\sigma_k(M)$ be defined as in Theorem 5.12.12. That is,

$$\sigma_1(M) = a_1 + a_2 + a_3$$

$$\sigma_2(M) = \det\begin{bmatrix} a_{11} & a_{12} \\ a_{21} & a_{22} \end{bmatrix} + \det\begin{bmatrix} a_{11} & a_{13} \\ a_{31} & a_{33} \end{bmatrix} + \det\begin{bmatrix} a_{22} & a_{23} \\ a_{32} & a_{33} \end{bmatrix}$$

$$\sigma_3(M) = \det M.$$

Using Theorem 5.12.12, it is not hard to show that the signs of $\sigma_1(M)$, $\sigma_2(M)$, $\sigma_3(M)$ uniquely determine the signs of the eigenvalues of $M$; we leave the details as an exercise (the actual results appear in Table 7.4.1). Further distinctions can be made according to the rank of the extended matrix $M^*$, just as in the preceding section. Leaving the details to you, the end result is the classification in Table 7.4.1 (cf. Appendix F).

**7.4.1** **Table of the graphs of Equation 7.1.3 when $n = 3$** (see facing page). ($M$, $M^*$ denote the matrix and extended matrix of the equation, respectively; we assume $M \neq 0$. The eigenvalues of $M$ are denoted $\lambda_1$, $\lambda_2$, $\lambda_3$, each eigenvalue being repeated as many times as its multiplicity.)

**7.4.2** **Example.** Describe the surface in $\mathbf{R}_3$ whose equation is

$$x^2 + 4y^2 + 4xy + 2xz + 4yz + 2x + 4y + 1 = 0.$$

**7.4.1 Table of the graphs of Equation 7.1.3 when $n = 3$**

| Signs of $\sigma_1(M)$, $\sigma_2(M)$, $\sigma_3(M)$ | Signs of $\lambda_1$, $\lambda_2$, $\lambda_3$ | Graph if rank $M^* = 4$ | Graph if rank $M^* = 3$ | Graph if rank $M^* = 2$ | Graph if rank $M^* = 1$ |
|---|---|---|---|---|---|
| $\sigma_2 > 0$; $\sigma_3 \neq 0$; $\sigma_1$, $\sigma_3$ have same sign | all the same and nonzero | ellipsoid *or* empty | point | — | — |
| $\sigma_3 \neq 0$; either $\sigma_2 < 0$ or $\sigma_1$, $\sigma_3$ have opposite sign | all nonzero, not all the same | hyperboloid of one *or* two sheets | elliptic cone | — | — |
| $\sigma_2 > 0$, $\sigma_3 = 0$ | $\lambda_1$, $\lambda_2$ nonzero and of same sign; $\lambda_3 = 0$ | elliptic paraboloid | elliptic cylinder *or* empty | line | — |
| $\sigma_2 < 0$, $\sigma_3 = 0$ | $\lambda_1$, $\lambda_2$ nonzero and of opposite signs; $\lambda_3 = 0$ | hyperbolic paraboloid | hyperbolic cylinder | two intersecting planes | — |
| $\sigma_2 = \sigma_3 = 0$ | $\lambda_1 \neq 0$, $\lambda_2 = \lambda_3 = 0$ | — | parabolic cylinder | 2 parallel planes *or* empty | plane |

*Solution.* The matrix and extended matrix of this equation are

$$M = \begin{bmatrix} 1 & 2 & 1 \\ 2 & 4 & 2 \\ 1 & 2 & 0 \end{bmatrix}; \quad M^* = \begin{bmatrix} 1 & 2 & 1 & 1 \\ 2 & 4 & 2 & 2 \\ 1 & 2 & 0 & 0 \\ 1 & 2 & 0 & 1 \end{bmatrix}.$$

Then $\sigma_1(M) = 5$, $\sigma_2(M) = -5$, $\sigma_3(M) = \det M = 0$, rank $M^* = 3$. Hence, the surface is a hyperbolic cylinder. ▮

Four of the thirteen cases listed in Table 7.4.1 are ambiguous: in three cases, the graph may be empty or nonempty, while in one case the graph is a hyperboloid which may have either one or two sheets. Let us consider these cases one by one. In what follows, $M_1^*$, $M_2^*$ denote the extended matrices of equations 7.1.22 and 7.1.26; $\lambda_1, \lambda_2, \lambda_3$ denote the eigenvalues of $M$; and $\mathscr{G}$ denotes the graph of equation 7.1.3. We use the letters $x, y, z$ in place of $x_1, x_2, x_3$; and so forth.

*Case* 1. *$\lambda_1, \lambda_2, \lambda_3$ are nonzero and of the same sign; rank $M^*$* $= 4$. This is the case in which equation 7.1.26 reduces to

$$\lambda_1(x'')^2 + \lambda_2(y'')^2 + \lambda_3(z'')^2 + k = 0$$

with $k \neq 0$; the matrix $M_2^*$ is the diagonal matrix whose diagonal entries are $\lambda_1, \lambda_2, \lambda_3, k$. Clearly,

$$\begin{aligned} \mathscr{G} \text{ is empty} &\Leftrightarrow k \text{ has the same sign as the } \lambda_i \\ &\Leftrightarrow \lambda_1\lambda_2\lambda_3 k > 0 \\ &\Leftrightarrow \det M_2^* > 0 \\ &\Leftrightarrow \det M^* > 0 \qquad \text{(by Theorem 7.2.3).} \end{aligned}$$

*Case* 2. *Two of $\lambda_1, \lambda_2, \lambda_3$ are zero, and rank $M^*$* $= 2$. This is the case where equation 7.1.26 reduces to

$$\lambda_1(x'')^2 + k = 0$$

with $k \neq 0$. In this case, an argument exactly like the proof of 7.3.17 shows that

$$\begin{aligned} \mathscr{G} \text{ is empty} &\Leftrightarrow \sigma_2(M^*) > 0 \\ &\Leftrightarrow 4(a_{11} + a_{22} + a_{33})c > b_1^2 + b_2^2 + b_3^2 \end{aligned}$$

where $a_{ij}, b_i, c$ are as in 7.1.3.

*Case* 3. *$\lambda_1, \lambda_2$ are nonzero and of the same sign, $\lambda_3$* $= 0$, *and rank $M^*$* $= 3$. Here equation 7.1.26 is

$$\lambda_1(x'')^2 + \lambda_2(y'')^2 + k = 0$$

with $k \neq 0$, and

(7.4.3) $$M_1^* = \begin{bmatrix} \lambda_1 & 0 & 0 & \frac{1}{2}b_1' \\ 0 & \lambda_2 & 0 & \frac{1}{2}b_2' \\ 0 & 0 & 0 & 0 \\ \frac{1}{2}b_1' & \frac{1}{2}b_2' & 0 & c \end{bmatrix}; \qquad M_2^* = \begin{bmatrix} \lambda_1 & & & \\ & \lambda_2 & & \\ & & 0 & \\ & & & k \end{bmatrix}.$$

We claim that the following statements are equivalent:

(a) $\mathcal{G}$ is empty.
(b) $k$ has the same sign as $\lambda_1$ and $\lambda_2$.
(c) $\lambda_1\sigma_3(M_1^*) > 0$.
(7.4.4) (d) $M_1^*$ has three nonzero eigenvalues of the same sign.
(e) $M^*$ has three nonzero eigenvalues of the same sign.
(f) $\sigma_1(M^*)$ and $\sigma_3(M^*)$ have the same sign and are nonzero, and $\sigma_2(M^*) > 0$.

(It is understood that when we say "three nonzero eigenvalues" we are counting multiplicities, as we did in stating Lemma 7.3.18.) It is clear that (a) $\Leftrightarrow$ (b). Since 7.1.27 and 7.4.3 imply that

$$\sigma_3(M_1^*) = \sigma_3(M_2^*) = \lambda_1\lambda_2 k,$$

it easily follows that (b) $\Leftrightarrow$ (c). We obtain (c) $\Leftrightarrow$ (d) by applying Lemma 7.3.13 to the (3, 3) minor of $M_1^*$; (d) and (e) are equivalent for the usual reason (what?); and (e) $\Leftrightarrow$ (f) can be shown by means of Theorem 5.12.12 (remembering that one eigenvalue of $M^*$ must be zero since rank $M^* < 4$). Thus, statements (a) through (f) are equivalent; in particular, $\mathcal{G}$ is empty if and only if (f) holds.

*Case* 4. $\lambda_1, \lambda_2, \lambda_3$ *are nonzero and not all of the same sign; rank* $M^* =$ 4. Here equation 7.1.26 is

$$\lambda_1(x'')^2 + \lambda_2(y'')^2 + \lambda_3(z'')^2 + k = 0 \qquad (k \neq 0)$$

where, say, $\lambda_1$ and $\lambda_2$ are of one sign and $\lambda_3$ is of the opposite sign. It is easy to see that the hyperboloid is of *one sheet* if and only if $k$ has the same sign as $\lambda_3$, which in turn is equivalent (by an argument like that in Case 1 above) to the inequality

$$\det M^* > 0.$$

*Conclusion*: *For* $n = 3$, *the geometric type of any equation* 7.1.3 *is determined by the signs of the numbers* $\sigma_i(M)$ $(i = 1, 2, 3)$ *and* $\sigma_j(M^*)$ $(j = 1, 2, 3, 4)$. (Note that not even the rank of $M^*$ is needed here, in view of Sec. 7.3, Exercise 4.)

**Example**

Describe the surface in $\mathbf{R}_3$ whose equation is

$2x^2 + y^2 + z^2 + 2xy + 2xz - x - y = 0.$

*Solution.* We have

$$M = \begin{bmatrix} 2 & 1 & 1 \\ 1 & 1 & 0 \\ 1 & 0 & 1 \end{bmatrix}; \qquad M^* = \begin{bmatrix} 2 & 1 & 1 & -\frac{1}{2} \\ 1 & 1 & 0 & -\frac{1}{2} \\ 1 & 0 & 1 & 0 \\ -\frac{1}{2} & -\frac{1}{2} & 0 & 0 \end{bmatrix}$$

$\sigma_3(M) = \det M = 0$

$\sigma_2(M) = \det\begin{bmatrix} 2 & 1 \\ 1 & 1 \end{bmatrix} + \det\begin{bmatrix} 2 & 1 \\ 1 & 1 \end{bmatrix} + \det\begin{bmatrix} 1 & 0 \\ 0 & 1 \end{bmatrix} > 0$

$\sigma_4(M^*) = \det M^* = 0$

$\sigma_3(M^*) =$ (the sum of four $3 \times 3$ determinants) $= -3/4 < 0$

$\sigma_2(M^*) =$ (the sum of six $2 \times 2$ determinants) $= 5/2 > 0$

$\sigma_1(M^*) = 4 > 0.$

Since $\sigma_3(M^*) \neq 0$ and $\sigma_4(M^*) = 0$, we have rank $M^* = 3$, so that the surface is either empty or an elliptic cylinder. Since 7.4.4(f) does not hold, it follows that the surface is an elliptic cylinder. ∎

**Exercises**

1. Determine the geometric type of each of the following loci in $\mathbf{R}_3$ without computing eigenvalues or eigenvectors. (ANS)
   (a) $xy + z^2 = 6.$
   (b) $xy + z^2 = -6.$
   (c) $xy + z^2 = 0.$
   (d) $2x^2 - y^2 + 4yz + x - y + 1 = 0.$
   (e) $2x^2 + y^2 + z^2 + yz = 5.$
   (f) $2x^2 + y^2 + z^2 + 3yz = 5.$
   (g) $2x^2 + y^2 + z^2 + 2yz = 5.$
   (h) $(x + y + z)^2 + x - 6 = 0.$
   (i) $x^2 - 2y^2 + xy + xz + 2yz + 2x + 4y = 0.$
   (j) $2x^2 + 3y^2 + z^2 - 2xy + 4x - 4y + 4 = 0.$
   (k) $x^2 + y^2 + 2xy + 4xz + 4yz - x = 0.$
2. Prove: the coefficients of $x^2$, $y^2$, $z^2$ in the equation of an ellipsoid must all have the same sign. (SUG)

## 7.5 Conic sections

The intersection of a quadric surface in $\mathbf{R}_3$ with a plane in $\mathbf{R}_3$ is always congruent to a 2-quadric; that is, the intersection (if not empty) is necessarily an ellipse, hyperbola, parabola, point, line, or pair of lines. The obvious generalization of this result to higher dimensions is true; we state it as a theorem.

7.5.1 **Theorem.** If $\mathscr{G}$ is any $n$-quadric (i.e., graph of an equation 7.1.3 in $\mathbf{R}_n$) and if $S$ is any $k$-flat in $\mathbf{R}_n$, then $\mathscr{G} \cap S$ is congruent to a $k$-quadric.

*Proof.* We may write $S = W + P_0$, where $W$ is a subspace of dimension $k$. Choose an orthonormal basis $\{\mathbf{u}_1, ..., \mathbf{u}_k\}$ for $W$, extend it to an orthonormal basis $\{\mathbf{u}_1, ..., \mathbf{u}_n\}$ of $\mathbf{R}_n$, let $K$ be the coordinate system $(P_0; \mathbf{u}_1, ..., \mathbf{u}_n)$, and let $x_1', ..., x_n'$ denote coordinates in the system $K$. By Section 7.1, Exercise 4, the equation of $\mathscr{G}$ in the system $K$ is still of second degree, say

$$\sum_{i=1}^{n}\sum_{j=1}^{n} a_{ij}'x_i'x_j' + \sum_{i=1}^{n} b_i'x_i' + c' = 0.$$

The equations of $S$ in the system $K$ are $x_{k+1}' = \cdots = x_n' = 0$ (*why*?); hence, those of $\mathscr{G} \cap S$ are

$$\text{(a)}\ \sum_{i=1}^{k}\sum_{j=1}^{k} a_{ij}'x_i'x_j' + \sum_{i=1}^{k} b_i'x_i' + c' = 0;$$

$$\text{(b)}\ x_{k+1}' = \cdots = x_n' = 0.$$

It follows from Theorem 4.10.23 that $\mathscr{G} \cap S$ is congruent to the graph $\mathscr{H}$ (in $\mathbf{R}_n$) of

$$\text{(a)}\ \sum_{i=1}^{k}\sum_{j=1}^{k} a_{ij}'x_ix_j + \sum_{i=1}^{k} b_i'x_i + c' = 0$$

(7.5.2)

$$\text{(b)}\ x_{k+1} = \cdots = x_n = 0;$$

and $\mathscr{H}$ is in turn congruent to the $k$-quadric $\mathscr{K}$ in $\mathbf{R}_k$ whose equation is 7.5.2(a), since the isometry of $\mathbf{R}_k \to \mathbf{R}_n$ defined by

$$(x_1, ..., x_k) \to (x_1, ..., x_k, 0, ..., 0)$$

maps $\mathscr{K}$ onto $\mathscr{H}$. Hence, $\mathscr{G} \cap S$ is congruent to $\mathscr{K}$, as desired. ∎

7.5.3 **Example.** Describe the intersection of the cone $x^2 + y^2 = z^2$ (in $\mathbf{R}_3$) with the plane $2z = x + 4$.

*Solution.* The given plane $S$ has the equation $x - 2z + 4 = 0$; its direction space $W$ is the plane $x - 2z = 0$ (Theorem 5.10.18) which consists of all elements of $\mathbf{R}_3$ which are orthogonal to $(1, 0, -2)$. By inspection, one element of $W$ is $(0, 1, 0)$; and then the vector

$$(1, 0, -2) \times (0, 1, 0) = (2, 0, 1)$$

lies in $W$ and is orthogonal to $(0, 1, 0)$. Hence, we may let

$$\mathbf{u}_1 = (2/\sqrt{5}, 0, 1/\sqrt{5})$$
$$\mathbf{u}_2 = (0, 1, 0)$$
$$\mathbf{u}_3 = (1/\sqrt{5}, 0, -2/\sqrt{5})$$

and $\{\mathbf{u}_1, \mathbf{u}_2\}$ is an orthonormal basis of $W$ while $\{\mathbf{u}_1, \mathbf{u}_2, \mathbf{u}_3\}$ is an orthonormal basis of $\mathbf{R}_3$. By inspection, the point $P_0 = (0, 0, 2)$ lies in $S$ and, hence, $S = W + P_0$. If $x', y', z'$ represent the coordinates of the point $(x, y, z)$ in the system $K = (P_0; \mathbf{u}_1, \mathbf{u}_2, \mathbf{u}_3)$, then by applying equations 7.1.48 we find that the equation $x^2 + y^2 = z^2$ is equivalent to

$$\textbf{(7.5.4)} \qquad [(2/\sqrt{5})x' + (1/\sqrt{5})z']^2 + (y')^2 = [(1/\sqrt{5})x' - (2/\sqrt{5})z' + 2]^2.$$

Setting $z' = 0$ (the equation of $S$) and expanding, equation 7.5.4 becomes

$$(3x')^2 + 5(y')^2 - 4\sqrt{5}\,x' - 20 = 0$$

which can be seen to be the equation of an ellipse using the results of Section 7.3. ▌

*Alternate solution.* It is evident from Figure 7.5.1 that the curve of intersection is bounded but is not just a single point. Since the only bounded 2-quadrics containing more than one point are the ellipses, it follows that the curve is an ellipse. (To make this solution rigorous, you must show analytically that the curve is indeed bounded and contains two distinct points; you might wish to try this as an exercise. The idea is to do it without a lot of computation.)

The converse of Theorem 7.5.1 is trivial: any $k$-quadric $\mathscr{K}$ is congruent to the intersection of an $n$-quadric and a $k$-flat. (Indeed, if $\mathscr{K}$ has the equation 7.5.2(a) then $\mathscr{K}$ is congruent to the intersection of the $k$-flat $x_{k+1} = \cdots = x_n = 0$ with the $n$-quadric whose equation in $\mathbf{R}_n$ is 7.5.2(a).) To make the converse of 7.5.1 somewhat less trivial, we may impose restrictions on the $n$-quadric; for example, we may require the $n$-quadric to be of a particular geometric type. One well known result of this nature is the following: *every nonempty 2-quadric except a pair of parallel lines can be obtained as the intersection of a plane and a right circular cone in* $\mathbf{R}_3$. We shall sketch a proof of this result; because of it, 2-quadrics are often called *conic sections*, or *conics*.

The proof is based on the fact that any nonempty 2-quadric $\mathscr{K}$ except a pair of parallel lines is congruent to the graph of one of the following equations:

(7.5.5)
(a) $x^2 + by^2 + k = 0 \qquad (0 < b \leqslant 1, k \leqslant 0).$
(b) $x^2 + by^2 + k = 0 \qquad (b < 0, k \geqslant 0).$
(c) $x^2 + cy = 0.$

(This is easily seen by examining the various cases discussed in Sec. 7.3; note that we may interchange the roles of $x$ and $y$ if necessary, and that we may also divide an equation through by a constant.) Now let $\mathscr{G}$ be the right circular cone

$$x^2 + y^2 = Az^2$$

where we require $A$ to satisfy the inequalities

$$A > 0; \qquad A > -b;$$

and let the coordinate system $K = (P_0; \mathbf{u}_1, \mathbf{u}_2, \mathbf{u}_3)$ be defined by

(7.5.6)
$$\mathbf{u}_1 = (1, 0, 0)$$
$$\mathbf{u}_2 = (0, ((A + b)/(1 + A))^{1/2}, ((1 - b)/(1 + A))^{1/2})$$
(where $b = 0$ in case (c))
$$\mathbf{u}_3 = (0, -((1 - b)/(1 + A))^{1/2}, ((A + b)/(1 + A))^{1/2})$$
(where $b = 0$ in case (c))

Figure 7.5.1

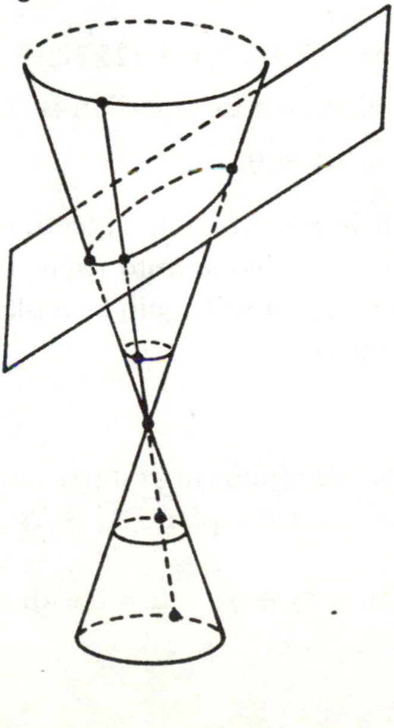

and

$$P_0 = \begin{cases} (0, (-Ak(1-b)/b(A+1))^{1/2}, (-k(A+b)/bA(A+1))^{1/2}) & \text{(in cases (a), (b)).} \\ (0, \tfrac{1}{4}c((1+A)/A)^{1/2}, -\tfrac{1}{4}(c/A)(1+A)^{1/2}) & \text{(in case (c)).} \end{cases} \tag{7.5.7}$$

If $x'$, $y'$, $z'$ denote $K$ coordinates, it is then a tedious but straightforward computation to verify that if we set $z' = 0$ in the $(x', y', z')$ equation of the cone $\mathscr{G}$, we obtain

$$(x')^2 + b(y')^2 + k = 0 \quad \text{(in cases (a), (b))}$$
$$(x')^2 + cy' = 0 \quad \text{(in case (c));}$$

hence, our original 2-quadric $\mathscr{K}$ is congruent to the intersection of $\mathscr{G}$ with the plane $z' = 0$. ∎

**7.5.8 Example.** Find a plane in $\mathbf{R}_3$ whose intersection with the cone $x^2 + y^2 = z^2$ is an ellipse with semiaxes of lengths 4 and 3.

*Solution.* Such an ellipse is congruent to the ellipse

$$(x^2/9) + (y^2/16) = 1$$

which is equivalent to equation 7.5.5(a) with $b = 9/16$, $k = -9$. Using formulas 7.5.6 and 7.5.7 (with $A = 1$) and 7.1.47, we obtain

$$\mathbf{u}_3 = (0, -(7/32)^{1/2}, (25/32)^{1/2})$$
$$P_0 = (0, (7/2)^{1/2}, (25/2)^{1/2})$$
$$z' = \mathbf{u}_3 \cdot ((x, y, z) - P_0) = -(7/32)^{1/2} y + (25/32)^{1/2} z - 9/4.$$

Hence, the required plane is the plane whose equation is

$$-(7/32)^{1/2} y + (25/32)^{1/2} z - 9/4 = 0. \tag{7.5.9}$$

(This answer to Example 7.5.8 is not unique; indeed, since any rotation about the $z$ axis maps the given cone onto itself, the image of the plane 7.5.9 under any such rotation will again be a plane which satisfies the conditions of the problem.)

### Exercises

1. Describe (by geometric type) the following intersections in $\mathbf{R}_3$:
   (a) The cone $x^2 + 2y^2 = z^2$ with the plane $2x + 2y - z = 3$. (ANS)
   (b) The cone $2xy + z^2 + 2x - 2y + 4z + 2 = 0$ with the plane $-x + 9y + 3\sqrt{2}z = 0$. (ANS)

2. Find a plane whose intersection with the cone $x^2 + y^2 = 3z^2$ is a parabola whose focus $F$ and vertex $V$ satisfy $d(F, V) = 2$. (ANS)

## 7.6 Singular points and the rank of $M^*$

Figure 7.6.1 is a sketch of a circular cone in $\mathbf{R}_3$. The vertex of the cone (point $V$ in the figure) has the following property: for any point $P \neq V$ on the cone, the entire line $L(P, V)$ lies on the cone. It is not hard to see that $V$ is the only point on the cone that has this property.

Figure 7.6.2 is a sketch of the 3-quadric which consists of two intersecting planes. If $V$ is any point on the line of intersection of the two planes, then $V$ has the same property as was described in the preceding paragraph. That is, the line joining $V$ to any other point $P$ on the quadric lies entirely on the quadric.

Points $V$ having this property with respect to a given quadric in $\mathbf{R}_n$ are called *singular points* of the quadric. Thus, a cone has one singular point, namely, its vertex; a pair of intersecting planes in $\mathbf{R}_3$ contains a line of singular points, namely, the line of intersection of the two planes. Some other examples:

(a) A pair of intersecting lines in $\mathbf{R}_2$ is a 2-quadric with one singular point, namely, the point of intersection.

Figure 7.6.1

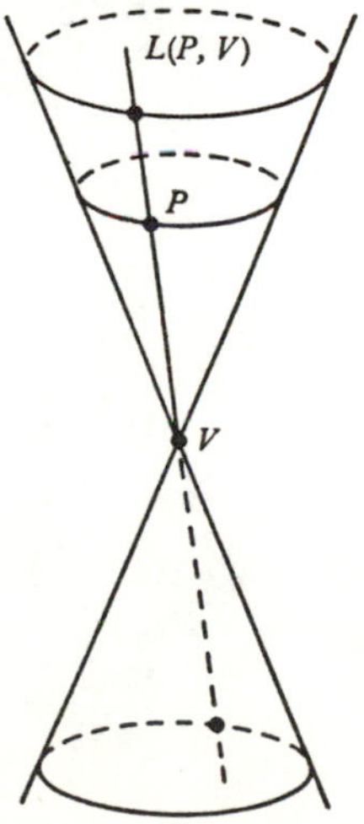


Figure 7.6.2

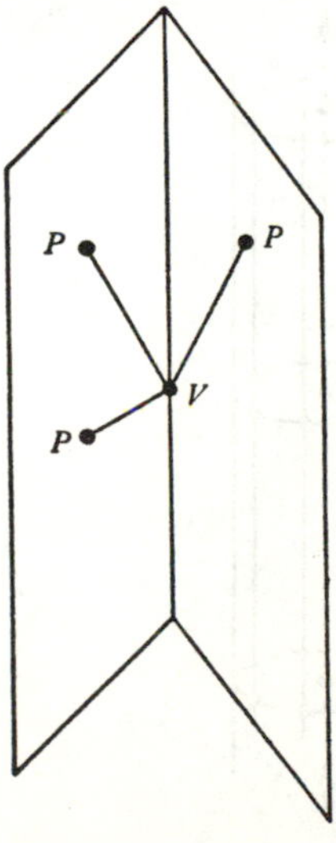

(b) A parabola has no singular points.

(c) If a quadric $\mathcal{G}$ consists of *just one point* $V$, that one point is a singular point. Indeed, it is vacuously true that $L(P, V) \subseteq \mathcal{G}$ "for all points $P \neq V$ on $\mathcal{G}$", since there are no such points $P$.

(d) Let $\mathcal{G}$ be a circular cylinder in $\mathbf{R}_3$ (Fig. 7.6.3). At first glance it looks as if $\mathcal{G}$ has no singular points. However, suppose we adopt the convention that *parallel lines meet at infinity*. Then the various lines on the cylinder which are parallel to the axis of the cylinder all meet at a "point at infinity", and *that point is a singular point*. Indeed, any finite point $P$ on the cylinder can be joined to the "point at infinity" by some line parallel to the axis, and that line lies entirely on the cylinder.

(e) In similar vein, if we adopt the convention that parallel planes meet in a "line at infinity", then the 3-quadric consisting of two parallel planes contains a line of singular points, namely, the "line at infinity" in which the two planes meet. (Close your eyes and try to visualize this.)

The reason for introducing objects at infinity in the present context is that they enable us to state a remarkably elegant theorem which relates the set of singular points of a quadric to the rank of the extended matrix of that quadric. To be precise, let $\mathcal{G}$ be a nonempty $n$-quadric and let $M^*$ be the extended matrix of the equation of $\mathcal{G}$. Our theorem then says the following: the set of singular points of $\mathcal{G}$

Figure 7.6.3

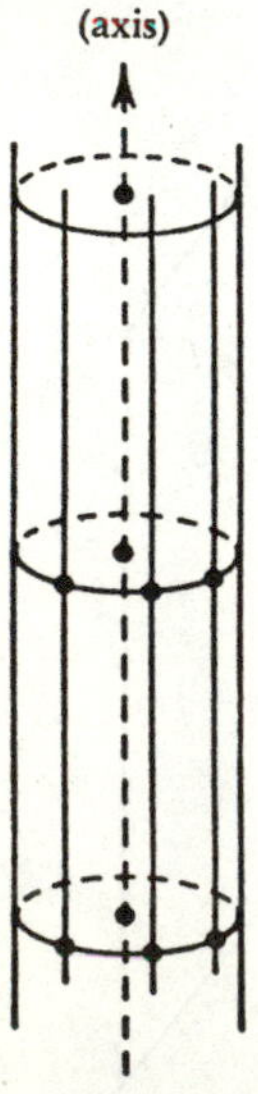

is a flat whose dimension equals $n$ − rank $M^*$, with the convention that the empty set is regarded as a flat of dimension −1. In particular, the set of singular points is empty if and only if rank $M^* = n + 1$. In Table 7.4.1, for example, the ellipsoids, hyperboloids, and paraboloids (for which rank $M^* = 4 = n + 1$) have no singular points, but each nonempty quadric for which rank $M^* = 3$ has exactly one singular point (possibly infinite!), each nonempty quadric for which rank $M^* = 2$ has precisely a line of singular points (possibly infinite!), and so on. Similar observations apply to Table 7.3.6 (verify this!).

In order to make the foregoing mathematically rigorous, we must formalize our ideas about "points at infinity", and so forth. For the sake of generality, we consider an arbitrary vector space $V$ over a field $F$. We proceed as follows:

(a) As we saw in Section 3.2, parallelism is an equivalence relation on the set of all lines in $V$. Let $\mathscr{P}$ be the set of all equivalence classes; thus, each element $\mathscr{L}$ of $\mathscr{P}$ is a class of parallel lines, and two lines belong to the same class $\mathscr{L}$ if and only if they are parallel.

(b) We define a *point at infinity* (or *infinite point*) to be an element of $\mathscr{P}$; that is, a point at infinity is (formally) a class of parallel lines. (Points in $V$ may be called *finite* points to distinguish them from points at infinity.) If the line $L$ belongs to the class $\mathscr{L}$, then we call $\mathscr{L}$ the point at infinity *corresponding* to $L$. We let $L^*$ be the set consisting of all points of $L$ together with the point at infinity corresponding to $L$; thus, $L^*$ has one more element than $L$. We call $L^*$ an *extended line*. It follows at once that if the lines $L_1$, $L_2$ are parallel then the extended lines $L_1^*$, $L_2^*$ have one element in common, namely, their common point at infinity.

(c) More generally, let $S$ be any $k$-flat in $V$. If we let

$$\mathscr{P}_S = \{\mathscr{L} \in \mathscr{P} : \text{Some line belonging to } \mathscr{L} \text{ lies in } S\},$$

the set $\mathscr{P}_S$ is called the $(k-1)$*-flat at infinity* (or *infinite* $(k-1)$*-flat*) *corresponding to* $S$. By definition, then, $\mathscr{P}_S$ consists of precisely those points at infinity which correspond to lines in $S$. We let

$$S^* = S \cup \mathscr{P}_S;$$

thus, $S^*$ consists of the (finite) $k$-flat $S$ together with the infinite $(k-1)$-flat corresponding to $S$. The set $S^*$ is called an *extended k-flat*. It is easy to show that if the $k$-flats $S_1$, $S_2$ are parallel, then the sets $\mathscr{P}_{S_1}$ and $\mathscr{P}_{S_2}$ are the same; thus, the extended $k$-flats $S_1^*$, $S_2^*$ intersect in their common $(k-1)$-flat at infinity.

(d) In particular, if $S$ is a 0-dimensional flat (a single point), $\mathscr{P}_S$ is clearly empty. (Conversely, $\mathscr{P}_S$ can be empty only if $S$ is a single point.)

Thus, according to our definition in the preceding paragraph, *the empty set is a* $(-1)$*-flat at infinity*.

(e) If $S$ is a line, then $\mathcal{P}_S$ consists of just one element, namely, the class $\mathcal{L}$ to which $S$ belongs (i.e., the point at infinity corresponding to $S$). Hence, a point at infinity is a 0-flat at infinity, and we see that paragraph (b) is a special case of paragraph (c).

(f) By a *generalized k-flat* (or *generalized flat of dimension k*) we shall mean *either* an extended $k$-flat *or* a $k$-flat at infinity. In particular, we see (in view of paragraphs (d) and (e)) that the "generalized points" (generalized 0-flats) are precisely the finite points and the points at infinity.

(g) Generalizing the notation in paragraph (c): if $T$ is *any* subset of $V$ (not necessarily a flat), then we write

$$\mathcal{P}_T = \{\mathcal{L} \in \mathcal{P} : \text{some line belonging to } \mathcal{L} \text{ is contained in } T\};$$
$$T^* = T \cup \mathcal{P}_T.$$

Thus, $T^*$ is obtained by adding to $T$ those points at infinity which correspond to lines in $T$. For example, if $T$ is the cylinder of Figure 7.6.3, then $T^*$ consists of $T$ together with a single point at infinity, namely, the point corresponding to the parallel lines which lie on the cylinder in the direction of the axis. In general, it is easy to see that for any subsets $T_1$, $T_2$ of $V$, we have

**(7.6.1)** $$T_1^* \cap T_2^* = (T_1 \cap T_2) \cup (\mathcal{P}_{T_1} \cap \mathcal{P}_{T_2});$$

we leave the proof to you.

(h) As a special case of (g), $V^* = V \cup \mathcal{P}$. The set $V^*$ (together with its structure, as defined in the preceding paragraphs) is called a *projective space* over $F$.

(i) Many statements about flats can be extended (or are analogous) to statements about generalized flats. One example of the latter is

**7.6.2** **Theorem.** Any two distinct generalized points $P_1, P_2 \in V^*$ lie in a unique generalized line.

*Proof.* Exercise. (For now, you will have to consider three cases separately: both points finite, both infinite, one finite and one infinite. However, in the next section we will give a proof that does not require division into cases.)

In connection with Theorem 7.6.2, note that if at least one of $P_1$, $P_2$ is finite, then the generalized line through $P_1$ and $P_2$ must be an extended line (not a line at infinity); we denote this extended line by $L^*(P_1, P_2)$.

We can now make precise the theorem about singular points which we have already stated informally.

**7.6.3 Definition.** Let $Q$ be any subset of $V$. By a *singular point of* $Q$, we mean a generalized (finite or infinite) point $J \in Q^*$ which has the property that, for every point $P \neq J$ in $Q$, the extended line $L^*(P, J)$ is entirely contained in $Q^*$.

**7.6.4 Theorem.** Let $Q$ be the graph of equation 7.1.3 in $\mathbf{R}_n$, and let $M$, $M^*$ be, respectively, the matrix and extended matrix of this equation. (It is assumed, as usual, that $M$ is symmetric.) Assume further that $M \neq 0$ and that $Q$ is nonempty.

Then the set of all singular points of $Q$ (finite and infinite) is a generalized flat of dimension $n$ − rank $M^*$.

Note that Theorem 7.6.4 specifies $\mathbf{R}$ as the field of scalars; our proof uses special properties of the real numbers. (Cf. Lemma A below).

*Note*: As far as we can determine, no other proof of 7.6.4 has been published, even though an analogous theorem about *projective* quadrics (Theorem 7.7.10 in the next section) is well known. In fact, this writer's query as to whether 7.6.4 might hold over a wider class of fields, published in the *Notices* of the American Mathematical Society (June 1973), has drawn no responses to date.

Before proving Theorem 7.6.4, we shall need the following three preliminary theorems; for the sake of generality, we state these theorems for an arbitrary vector space.

**7.6.5 Theorem.** Let $Q$ be any nonempty subset of the vector space $V$, and let $J$ be a point in $V^*$. Then $J$ is a singular point of $Q$ if and only if one of the following two statements holds:

(a) $J \in Q$; for every point $P \neq J$ in $Q$, the line $L(P, J)$ is contained in $Q$.

(b) $J$ is an element $\mathscr{L}$ of $\mathscr{P}$; for every point $P$ in $Q$, the line through $P$ belonging to the class $\mathscr{L}$ lies entirely in $Q$.

**7.6.6 Theorem.** Let $S$, $T$ be finite-dimensional flats in $V$. Then the following two statements are equivalent:

(a) $\mathscr{P}_S \subseteq \mathscr{P}_T$.

(b) $S$ is parallel to $T$.

(Cf. Sec. 3.4, Exercise 1 for the definition of "parallel".)

**7.6.7 Theorem.** If $W$ is any $(k + 1)$-dimensional subspace of $V$, and if $H$ is a hyperplane in $V$ such that $W \not\subseteq H$, then $W^* \cap H^*$ is a generalized $k$-flat.

*Proofs.* Theorem 7.6.5 is immediate from the definition of "singular point". To prove 7.6.6, suppose first that $\mathscr{P}_S \subseteq \mathscr{P}_T$. If $L$ is any line contained in $S$, and if $\mathscr{L}$ is the class of parallel lines to which $L$ belongs, then $\mathscr{L} \in \mathscr{P}_S$ by definition; hence, $\mathscr{L} \in \mathscr{P}_T$; hence, some line belonging to $\mathscr{L}$ lies in $T$; hence, $L$ is parallel to some line which lies in $T$. It follows from Section 3.4, Exercise 4, that $S$ is parallel to $T$, so that 7.6.6(a) implies 7.6.6(b). Conversely, suppose $S$ is parallel to $T$; then $S$ is parallel to some flat $S' \subseteq T$, where $\dim S' = \dim S$. By 3.4.4(g), $S'$ is the image of $S$ under a translation $\mathscr{T}_\mathbf{u}$. If $\mathscr{L}$ is any element of $\mathscr{P}_S$, then some line $L \in \mathscr{L}$ lies in $S$. The line $L\mathscr{T}_\mathbf{u}$ is parallel to $L$ and thus belongs to $\mathscr{L}$; also, since $L \subseteq S$, we have $L\mathscr{T}_\mathbf{u} \subseteq S\mathscr{T}_\mathbf{u} = S'$ and, hence, $L\mathscr{T}_\mathbf{u} \subseteq T$. Since we have found an element of $\mathscr{L}$ which lies in $T$, it follows that $\mathscr{L} \in \mathscr{P}_T$. Hence, 7.6.6(b) implies 7.6.6(a).

To prove 7.6.7, we consider separately the two possible cases:
(a) $W \cap H = \varnothing$;
(b) $W \cap H \neq \varnothing$.

In case (a), 7.6.1 implies that $W^* \cap H^* = \mathscr{P}_W \cap \mathscr{P}_H$. Also, $W \parallel H$ by Sec. 3.4, Exercise 3(a); hence, 7.6.6 implies $\mathscr{P}_W \subseteq \mathscr{P}_H$, so that $W^* \cap H^* = \mathscr{P}_W \cap \mathscr{P}_H = \mathscr{P}_W$ which by definition is a $k$-flat at infinity, q.e.d. In case (b), $W \cap H$ is a $k$-flat (Sec. 3.4, Exercise 3(b)) so that it will suffice to show that

$$W^* \cap H^* = (W \cap H)^*. \tag{7.6.8}$$

In view of 7.6.1, equation 7.6.8 is equivalent to the equation

$$\mathscr{P}_W \cap \mathscr{P}_H = \mathscr{P}_{W \cap H}. \tag{7.6.9}$$

It is trivial that the right side of 7.6.9 is contained in the left side. Conversely, suppose $\mathscr{L} \in \mathscr{P}_W \cap \mathscr{P}_H$. Then some line $L_1 \in \mathscr{L}$ lies in $W$, and some line $L_2 \in \mathscr{L}$ lies in $H$. Since $W \cap H \neq \varnothing$, we may choose a point $P \in W \cap H$. The line $L$ through $P$ belonging to the class $\mathscr{L}$ is parallel to both $L_1$ and $L_2$; hence, $L$ lies in both $W$ and $H$ by 3.4.1(b). Hence, $L \subseteq W \cap H$; hence, $\mathscr{L} \in \mathscr{P}_{W \cap H}$. This implies 7.6.9, hence, 7.6.8, hence, 7.6.7. ∎

We now turn to the proof of Theorem 7.6.4; the proof is quite long and will take up the remainder of this section. As in Section 7.1, the entries in $M$ and $M^*$ will be denoted $a_{ij}$ and $a_{ij}^*$, respectively. We recall from Section 7.1 that if $x_{n+1}$ is defined to be 1, equation 7.1.3 is equivalent to equation 7.1.10. Equation 7.1.10 is homogeneous in $x_1$, ..., $x_{n+1}$; 7.1.10 is also easier to write in matrix form than 7.1.3. Because of these considerations, it turns out that in proving Theorem 7.6.4 it is desirable to work in the vector space $\mathbf{R}_{n+1}$ instead of $\mathbf{R}_n$. We thus introduce the following notations, which will be used throughout the proof:

$\mathcal{Q}$ denotes the graph of the equation

$$\sum_{i=1}^{n+1}\sum_{j=1}^{n+1} a_{ij}^* x_i x_j = 0 \tag{7.6.10}$$

in $\mathbf{R}_{n+1}$; that is, $\mathcal{Q}$ is the set of all points $(x_1, ..., x_{n+1}) \in \mathbf{R}_{n+1}$ such that 7.6.10 holds. If $X = (x_1, ..., x_{n+1})$ is thought of as a row matrix, then the left side of 7.6.10 is the sole entry in the matrix $XM^*X^t$, so that 7.6.10 may be written as

$$XM^*X^t = 0. \tag{7.6.10'}$$

We let

$$H_0 = [\mathbf{e}_1, ..., \mathbf{e}_n] \subseteq \mathbf{R}_{n+1};$$
$$H = H_0 + \mathbf{e}_{n+1} \subseteq \mathbf{R}_{n+1};$$

thus, $H_0$ is an $n$-dimensional subspace of $\mathbf{R}_{n+1}$ and $H$ is an $n$-flat whose direction space is $H_0$. It is clear that $H_0$ consists precisely of all $(n + 1)$-tuples of the form $(x_1, ..., x_n, 0)$ and that $H$ consists of all $(n + 1)$-tuples of the form $(x_1, ..., x_n, 1)$. Hence, the equation of $H$ is $x_{n+1} = 1$. It follows that if we let

$$\bar{Q} = \mathcal{Q} \cap H,$$

then the mapping $(x_1, ..., x_n) \to (x_1, ..., x_n, 1)$ of $\mathbf{R}_n \to H$ maps $Q$ onto $\bar{Q}$. Since this mapping is an isometry, $Q$ and $\bar{Q}$ are congruent. Hence, it will suffice to work with $\bar{Q}$ instead of $Q$ for most of the proof.

We begin the proof with one nonalgebraic observation: if we define $f: \mathbf{R}_{n+1} \to \mathbf{R}$ by

$$f(x_1, ..., x_{n+1}) = \sum_{i=1}^{n+1}\sum_{j=1}^{n+1} a_{ij}^* x_i x_j,$$

then $f$ is a polynomial and, hence, a *continuous* function of $x_1$, ..., $x_{n+1}$.

This implies that if $P_1, P_2, \ldots$ is an infinite sequence of points in $\mathbf{R}_{n+1}$ such that

$$\lim_{i\to\infty} P_i = P \in \mathbf{R}_{n+1},$$

and if $f(P_i) = 0$ for all $i$, then $f(P) = 0$. Equivalently, if $P_i \in \mathcal{Q}$ (all $i$), then $P \in \mathcal{Q}$, since by definition $\mathcal{Q}$ consists of all points $P$ such that $f(P) = 0$. Let us state this as a lemma.

*Lemma A.* $\mathcal{Q}$ is topologically closed; that is, if $P_i \in \mathcal{Q}$ ($i = 1, 2, 3, \ldots$) and $\lim_{i\to\infty} P_i = P$, then $P \in \mathcal{Q}$.

Lemma A is the only part of the proof in which concepts external to this book are used. We now state and prove several other lemmas which we will need.

*Lemma B.* There exists a basis $\{\mathbf{u}_1, \ldots, \mathbf{u}_n\}$ of $H_0$ such that none of the vectors $\mathbf{u}_1, \ldots, \mathbf{u}_n$ belong to $\mathcal{Q}$.

*Proof.* Since $M \neq 0$, there must be at least one element $(x_1^{(0)}, \ldots, x_n^{(0)})$ in $\mathbf{R}_n$ which does *not* satisfy the equation

$$\sum_{i=1}^{n}\sum_{j=1}^{n} a_{ij}x_ix_j = 0$$

(see Sec. 7.1, first remark following equation 7.1.36). The corresponding element $P_0 = (x_1^{(0)}, \ldots, x_n^{(0)}, 0)$ of $H_0$ then does not satisfy equation 7.6.10 and, hence, is not in $\mathcal{Q}$. By Lemma A, it follows that all points sufficiently close to $P_0$ must also fail to lie in $\mathcal{Q}$. In particular, for some number $\epsilon > 0$, we have

$$P_0 + \delta\mathbf{e}_i \notin \mathcal{Q} \qquad (i = 1, \ldots, n)$$

for all numbers $\delta$ such that $0 \leq \delta < \epsilon$. Let $D$ be the $n \times n$ matrix each row of which is $(x_1^{(0)}, \ldots, x_n^{(0)})$. By Theorem 5.11.3, the expression

$$\det(D + \delta I_n) = \operatorname{char}(D; -\delta)$$

is a polynomial in $\delta$ of degree $n$; hence, it equals zero for at most $n$ values of $\delta$. Hence, for some $\delta$ ($0 < \delta < \epsilon$) we must have $\det(D + \delta I_n) \neq 0$; for this value of $\delta$, the rows of $D + \delta I_n$ are independent and, thus, the vectors

$$\mathbf{u}_i = P_0 + \delta\mathbf{e}_i \qquad (1 \leq i \leq n)$$

are independent. Hence, the set $\{\mathbf{u}_1, \ldots, \mathbf{u}_n\}$ has the desired properties.

*Lemma C.* $\mathcal{Q}$ is closed under scalar multiplication.

The proof of Lemma C is trivial; we leave it to you.

*Lemma D.* For all row vectors $X, Y \in \mathbf{R}_{n+1}$ and all scalars $a$, $b$,

$$(aX + bY)M^*(aX + bY)^t = a^2(XM^*X^t) + b^2(YM^*Y^t) + 2ab(XM^*Y^t).$$

*Proof.* Since a one-by-one matrix equals its own transpose, $XM^*Y^t = YM^*X^t$ (remember that $M^*$ is symmetric). The lemma easily follows (expand the left side via the distributive law, etc.).

*Lemma E.* If $J \in \mathcal{Q}$ and $Y \notin \mathcal{Q}$, then $J + aY \in \mathcal{Q}$, where

**(7.6.11)** $a = -2(JM^*Y^t)/(YM^*Y^t)$.

(*Note*: The vectors $J$ and $Y$ are considered to be row matrices for purposes of matrix multiplication. Note that both numerator and denominator of 7.6.11 are one-by-one matrices and, thus, may be regarded as scalars; the denominator is nonzero since $Y \notin \mathcal{Q}$. Regarding the identification of one-by-one matrices with scalars, cf. Sec. 4.4, Exercise 9.)

*Proof.* Since $J \in \mathcal{Q}$, $JM^*J^t = 0$. Hence, Lemma D gives

**(7.6.12)** $(J + aY)M^*(J + aY)^t = a^2(YM^*Y^t) + 2a(JM^*Y^t)$.

If we substitute 7.6.11 into 7.6.12, the right side of 7.6.12 becomes zero; hence, $J + aY \in \mathcal{Q}$.

*Lemma F.* If $\mathbf{u}$ is any nonzero vector and $P$ is any point such that the line $P + [\mathbf{u}]$ lies in $\tilde{Q}$, then the subspace $[P, \mathbf{u}]$ is contained in $\mathcal{Q}$.

*Proof.* $\mathcal{Q}$ contains (by hypothesis) everything of the form $P + c\mathbf{u}$, hence, also (by Lemma C) everything of the form

**(7.6.13)** $tP + tc\mathbf{u} \qquad (t \in \mathbf{R}, c \in \mathbf{R})$.

Any linear combination $aP + b\mathbf{u}$ can be expressed in the form 7.6.13 if $a \neq 0$; and anything of the form $b\mathbf{u}$ is a limit of points of the form $aP + b\mathbf{u}$ ($a \neq 0$) and, hence, lies in $\mathcal{Q}$ by Lemma A. Thus, everything in $[P, \mathbf{u}]$ belongs to $\mathcal{Q}$.

*Lemma G.* Let $S$ be the set of all vectors (row matrices) $X \in \mathbf{R}_{n+1}$ such that $XM^* = 0$. Then

(a) $S$ is a subspace of $\mathbf{R}_{n+1}$ of dimension $(n + 1) -$ rank $M^*$.
(b) Whenever $X \in S$ and $Y \in \mathcal{Q}$, then $[X, Y] \subseteq \mathcal{Q}$.
(c) $S \subseteq \mathcal{Q}$.

*Proof.* Part (a) follows from Theorem 5.10.14 (note that the transpose of $XM^*$ is $M^*X^t$). Part (c) is obvious since $XM^* = 0$ implies $XM^*X^t = 0X^t = 0$. To prove (b), let $X \in S$ and $Y \in \mathcal{Q}$. Then $XM^* = 0$ and $YM^*Y^t = 0$. Hence, for any scalars $a$, $b$, Lemma D gives

$$(aX + bY)M^*(aX + bY)^t = 0 + 0 + 0 = 0$$

so that $aX + bY \in \mathcal{Q}$. Since this holds for all $a$, $b$, $[X, Y] \subseteq \mathcal{Q}$ as desired.

*Lemma H.* The set of all singular points of $\bar{Q}$ is precisely $S^* \cap H^*$.

This Lemma is the heart of the whole proof. The proof of Lemma H is in four steps.

*Step* 1. *If $J$ is a finite point in $S^* \cap H^*$, then $J$ is a singular point of $\bar{Q}$.* Proof: Since $J$ is a finite point, $J \in S \cap H$; hence, by part (c) of Lemma G, $J \in \mathcal{Q} \cap H = \bar{Q}$. For any point $P \neq J$ in $\bar{Q}$, we have

$$L(J, P) \subseteq [J, P] \subseteq \mathcal{Q} \qquad \text{(by Lemma G, part (b))}$$
$$L(J, P) \subseteq H \qquad \text{(since } J \in H \text{ and } P \in H\text{)}$$

and, hence, $L(J, P) \subseteq \bar{Q}$. Thus, $J$ is a finite singular point of $\bar{Q}$ by Theorem 7.6.5(a).

*Step* 2. *If $\mathcal{L}$ is a point at infinity in $S^* \cap H^*$, then $\mathcal{L}$ is a singular point of $\bar{Q}$.* Proof: Since $\mathcal{L}$ is an element of $S^*$ and of $H^*$, some line $L_1 \in \mathcal{L}$ lies in $S$ and some line $L_2 \in \mathcal{L}$ lies in $H$. The common direction space of the parallel lines $L_1$, $L_2$ is a one-dimensional subspace $[\mathbf{u}]$; since $L_1 \subseteq S$ and $L_2 \subseteq H$, Theorem 2.7.7(d) implies that $[\mathbf{u}] \subseteq S$ and $[\mathbf{u}] \subseteq H_0$. Choose any point $P \in \bar{Q}$; then $P \in \mathcal{Q} \cap H$, and we have

$$P + [\mathbf{u}] \subseteq [P, \mathbf{u}] \subseteq \mathcal{Q} \qquad \text{(by Lemma G, part (b));}$$
$$P + [\mathbf{u}] \subseteq P + H_0 = H \qquad \text{(by Theorem 3.4.4(a))}$$

so that $P + [\mathbf{u}] \subseteq \bar{Q}$. Hence, $\mathcal{L}$ is a singular point of $\bar{Q}$ by 7.6.5(b).

*Step* 3. *If $J$ is a finite singular point of $\bar{Q}$, then $J \in S \cap H$.* Proof: Since $\bar{Q} \subseteq H$, certainly $J \in H$; thus, we need only show that $J \in S$. By Lemma B, there is a basis $\{\mathbf{u}_1, ..., \mathbf{u}_n\}$ of $H_0$ such that $\mathbf{u}_i \notin \mathcal{Q}$ $(1 \leq i \leq n)$. Suppose $J + k\mathbf{u}_i \in \mathcal{Q}$ for some $i$ and some scalar $k \neq 0$. Since also

$$J + k\mathbf{u}_i \in J + [\mathbf{u}_i] \subseteq J + H_0 = H,$$

it follows that $J + k\mathbf{u}_i \in \tilde{Q}$, so that the assumption that $J$ is a singular point implies that the line

$$L(J, J + k\mathbf{u}_i) = J + [\mathbf{u}_i]$$

lies in $\tilde{Q}$. Hence, Lemma F implies $\mathbf{u}_i \in \mathcal{Q}$, R.A.A. We have thus shown that

$$J + k\mathbf{u}_i \notin \mathcal{Q} \qquad (1 \leqslant i \leqslant n;\text{ all scalars } k \neq 0). \tag{7.6.14}$$

In particular, $J + \mathbf{u}_i \notin \mathcal{Q}$; hence, by Lemma E, $\mathcal{Q}$ contains the points $P_i = J + a_i(J + \mathbf{u}_i)$, where

$$a_i = -\,2JM^*(J^t + \mathbf{u}_i^t)/(J + \mathbf{u}_i)M^*(J^t + \mathbf{u}_i^t) \qquad (1 \leqslant i \leqslant n). \tag{7.6.15}$$

Remembering that $JM^*J^t = 0$, some simple algebra (using Lemma D) gives

$$1 + a_i = \mathbf{u}_i M^* \mathbf{u}_i^t/(J + \mathbf{u}_i)M^*(J + \mathbf{u}_i)^t.$$

Since $\mathbf{u}_i \notin \mathcal{Q}$, $\mathbf{u}_i M^* \mathbf{u}_i^t \neq 0$ and, hence, $1 + a_i \neq 0$. The point

$$(1 + a_i)^{-1}P_i = J + a_i\mathbf{u}_i/(1 + a_i)$$

lies in $\mathcal{Q}$ by Lemma C; hence, by 7.6.14, we must have $a_i = 0$, so that by 7.6.15, we have

$$JM^*\mathbf{u}_i^t = 0 \qquad (1 \leqslant i \leqslant n). \tag{7.6.16}$$

Since the $\mathbf{u}_i$ span the $n$-dimensional subspace $H_0$, and since $J \notin H_0$, it follows that the subspace $[\mathbf{u}_1, \ldots, \mathbf{u}_n, J\,]$ is all of $\mathbf{R}_{n+1}$. Hence, from the relation $JM^*J^t = 0$ together with 7.6.16, we obtain

$$JM^*X^t = 0 \qquad (\text{all } X \in \mathbf{R}_{n+1}).$$

But this means that the vector $JM^*$ is orthogonal to every vector $X$ in $\mathbf{R}_{n+1}$. Hence, $JM^* = 0$ (*why*?), so that $J \in S$.

*Step* 4. *If $\mathcal{L}$ is an infinite singular point of $\tilde{Q}$, then* $\mathcal{L} \in S^* \cap H^*$. Proof: Let $[\mathbf{u}]$ be the common direction space of the parallel lines in the class $\mathcal{L}$. Since

$$\mathcal{L} \in \tilde{Q}^* \subseteq H^*,$$

arguments like those in Step 2 show that $[\mathbf{u}] \subseteq H_0$ and, hence,

$$P \in H \Rightarrow P + [\mathbf{u}] \subseteq H. \tag{7.6.17}$$

Also, for any point $P$, the line through $P$ belonging to the class $\mathcal{L}$ is clearly $P + [\mathbf{u}]$; hence, by 7.6.5(b),

$$P \in \tilde{Q} \Rightarrow P + [\mathbf{u}] \subseteq \tilde{Q} \tag{7.6.18}$$

and by Lemma F, we then have

$$P \in \tilde{Q} \Rightarrow [P, \mathbf{u}] \subseteq \mathcal{Q}. \tag{7.6.19}$$

We next show that

**(7.6.20)** $\mathbf{u}M^*P^t = 0$ (all $P \in H$).

Indeed, if $P \in \tilde{Q}$, then $P$, $\mathbf{u}$, $P + \mathbf{u}$ all lie in $\mathcal{Q}$ by 7.6.19, and 7.6.20 then follows from Lemma D. On the other hand, if $P \in H$ but $P \notin \tilde{Q}$, then $P \notin \mathcal{Q}$; noting that $\mathbf{u} \in \mathcal{Q}$ by 7.6.19 since $\tilde{Q} \neq \varnothing$, Lemma E thus gives

$$\mathbf{u} + aP \in \mathcal{Q}; \qquad a = -2\mathbf{u}M^*P^t/PM^*P^t.$$

If $a \neq 0$, then $P + a^{-1}\mathbf{u} \in \mathcal{Q}$ by Lemma C; hence, $P + a^{-1}\mathbf{u} \in \tilde{Q}$ by 7.6.17; hence $P \in \tilde{Q}$ by 7.6.18, R.A.A. Hence, $a = 0$, from which 7.6.20 follows. Now, the subspace spanned by $H$ is all of $\mathbf{R}_{n+1}$ (see, for example, Sec. 3.4, Exercise 8); hence, it follows from 7.6.20, just as in the proof of Step 3, that $\mathbf{u}M^* = 0$ and $\mathbf{u} \in S$. Since $[\mathbf{u}]$ is a line belonging to the class $\mathcal{L}$, it follows that $\mathcal{L} \in S^*$ and, hence, $\mathcal{L} \in S^* \cap H^*$. This completes the proof of Step 4 and hence also the proof of Lemma H.

Since $O \notin H$, $S \not\subseteq H$; thus, Theorem 7.6.7 and Lemma G imply that $S^* \cap H^*$ is a generalized flat of dimension $n$ − rank $M^*$. Since $S^* \cap H^*$ is the set of singular points of $\tilde{Q}$ (by Lemma H), we have shown that *Theorem 7.6.4 holds for $\tilde{Q}$ in place of $Q$*. To prove the same result for $Q$, we use the fact that $Q$ and $\tilde{Q}$ are congruent. Specifically, we have seen that the isometry $\mathbf{T} : (x_1, ..., x_n, 1) \to (x_1, ..., x_n)$ of $H \to \mathbf{R}_n$ maps $\tilde{Q}$ onto $Q$. By Exercise 4 below, we can then define a mapping $\mathbf{T}^*$ of $H^* \to \mathbf{R}_n^*$ which maps generalized $k$-flats onto generalized $k$-flats and maps the singular points of $\tilde{Q}$ onto the singular points of $Q$. It follows at once that Theorem 7.6.4 holds for $Q$ as well as $\tilde{Q}$, and our proof is complete. ∎

### Exercises

*1. Prove equation 7.6.1 formally. More generally, if $T_1, ..., T_k$ are subsets of $V$, show that $T_1^* \cap \cdots \cap T_k^* = (T_1 \cap \cdots \cap T_k) \cup (\mathcal{P}_{T_1} \cap \cdots \cap \mathcal{P}_{T_k})$.

*2. Prove Theorem 7.6.2. (SUG)

3. (a) Deduce from Theorem 7.6.6 that if $S$ and $T$ are flats such that $\mathcal{P}_S = \mathcal{P}_T$, then dim $S$ = dim $T$.
   (b) Deduce from part (a) that the dimension of a "generalized flat" (as defined in this section) is uniquely determined; that is, the same set cannot be both a generalized $k$-flat and a generalized $h$-flat where $h \neq k$.

*4. Let $n$ be a positive integer; let $V$, $V'$ be vector spaces over a field $F$, each having dimension $\geqslant n$. Let $H_1$, $H_2$ be $n$-flats in $V$, $V'$, respectively, and let $\mathbf{T} : H_1 \to H_2$ be a one-to-one affine transformation. (The existence of such a mapping $\mathbf{T}$ was shown in Sec. 4.9.) If $\mathscr{L} \in \mathscr{P}_{H_1}$, then $L \subseteq H_1$ for some line $L \in \mathscr{L}$. Let $L'$ be the image of $L$ under $\mathbf{T}$; then $L'$ is a line in $H_2$ (why?), so that $L'$ belongs to some class $\mathscr{L}'$ of parallel lines in $V'$ such that $\mathscr{L}' \in \mathscr{P}_{H_2}$.

(a) Show that $\mathscr{L}'$ is uniquely determined by $\mathscr{L}$; that is, $\mathscr{L}'$ does not depend on which line $L$ (satisfying the conditions $L \subseteq H_1$, $L \in \mathscr{L}$) was chosen. (SUG)

(b) Deduce: if we define a mapping $\mathbf{T}^*$ with domain $H_1^*$ by

$PT^* = PT$ (all $P \in H_1$)

$\mathscr{L}\mathbf{T}^* = \mathscr{L}'$ (all $\mathscr{L} \in \mathscr{P}_{H_1}$,

where $\mathscr{L}'$ is determined by $\mathscr{L}$ as above),

then $\mathbf{T}^*$ is a well-defined mapping of $H_1^*$ into $H_2^*$.

(c) Prove: $\mathbf{T}^*$ is one-to-one and onto. (SUG)

(d) Prove: if $S_1$ is any subset of $H_1$, and if $S_1\mathbf{T} = S_2$, then $\mathscr{P}_{S_1}\mathbf{T}^* = \mathscr{P}_{S_2}$ and, hence, $S_1^*\mathbf{T}^* = S_2^*$.

(e) Deduce from (d) that the image under $\mathbf{T}^*$ of any generalized $k$-flat ($k \leqslant n$) is a generalized $k$-flat; and conversely, every generalized $k$-flat in $H_2^*$ is the image of some generalized $k$-flat in $H_1^*$.

(f) Prove: if $Q$ is any nonempty subset of $H_1$ and if $Q' = Q\mathbf{T}$, then the set of all singular points of $Q$ is mapped by $\mathbf{T}^*$ onto the set of all singular points of $Q'$.

5. Inspection of Tables 7.3.6 and 7.4.1 leads us to guess the following "theorem": *if a nonempty quadric $Q$ in $\mathbf{R}_n$ is entirely contained in a hyperplane, then $Q$ is a flat*. The italicized statement is in fact true. Prove it. (This exercise is quite difficult. The result of Sec. 3.4, Exercise 12 will come in handy; so will Lemma A of this section.)

6. If the field $\mathbf{R}$ is replaced by an arbitrary field of scalars $F$, the theorem of the preceding exercise may fail; even the requirement that $F$ have characteristic $\neq 2$ is not enough to imply the result. Find a counterexample over some field $F$ which *is not of characteristic* 2. That is, find an equation of the form 7.1.3 (with $a_{ij} \in F$, $b_i \in F$, $c \in F$ for $1 \leqslant i \leqslant n$, $1 \leqslant j \leqslant n$) whose graph in $F_n$ is nonempty and is contained in a hyperplane but is not a flat.

## 7.7 An interpretation of projective spaces within vector spaces

In the preceding section we defined a "point at infinity" to be a class of parallel lines in a vector space $V$; we then defined the "projective space" $V^*$ to be the union of $V$ with the set of all "points at infinity". Although this definition served our purposes at the time, it has an esthetic drawback: it makes a point at infinity a different kind of geometric object from a finite point. That is, *not all (generalized) points look alike*. (The drawback is not merely esthetic; it also complicates proofs involving generalized points, since finite points and points at infinity must be treated separately.) Similarly, a line at infinity is a different kind of object from a finite line, or even from an extended line; not all generalized lines "look alike". And so on.

Our object now is to rectify this situation. We consider an arbitrary vector space $V$ of dimension $n$ over $F$. What we shall do in this section is to construct, within a vector space $V'$ of dimension $n + 1$ over $F$ (a "finite" vector space of *one higher dimension*) a system $S$ which is formally isomorphic to $V^*$ but has the additional (nice) property that all of the "points" in $S$ "look alike" in the sense of being the same kind of geometric object; similarly, all of the "lines" in $S$ will look alike, and likewise for higher-dimensional "flats" in $S$. The system $S$ (or $V^*$, or any isomorphic system) is called a *projective space of dimension n* over $F$.

The construction of $S$ is quite close at hand. In fact, to give the secret away, we define $S$ to be *the set of all one-dimensional subspaces of* $V'$. In other words, an element ("point") of the projective $n$-space $S$ is a line through the origin in the $(n + 1)$-dimensional vector space $V'$. In order to motivate this definition of $S$, consider a hyperplane ($n$-flat) $H$ in $V'$ which does not pass through $O$. For any line $L$ through $O$ (in $V'$), either $L$ intersects $H$ in a point $\mathbf{u}$ (Fig. 7.7.1(a)) or $L$ is

Figure 7.7.1

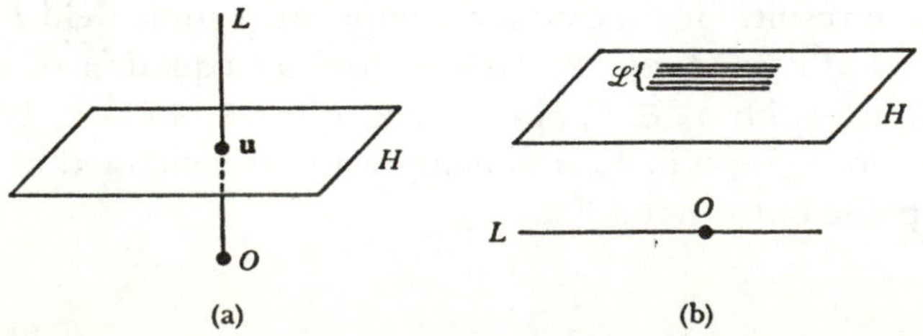

parallel to $H$ (Fig. 7.7.1(b)); in the latter case, if $\mathscr{L}$ is the class of parallel lines to which $L$ belongs, some of the lines in $\mathscr{L}$ lie in $H$ and, hence, $\mathscr{L} \in H^*$. Thus, *in either case*, $L$ corresponds to some generalized point in $H^*$, either the finite point $\mathbf{u}$ or the "point at infinity" $\mathscr{L}$. Conversely, every generalized point in $H^*$ corresponds to some one-dimensional subspace $L$. Thus, we have a one-to-one correspondence between $S$ and $H^*$. By Section 7.6, Exercise 4, we also have a 1–1 correspondence between $H^*$ and $V^*$; this gives us a correspondence between $S$ and $V^*$ as desired. Moreover, our previous assertion that all the elements of $S$ would "look alike" is clearly true by the definition of $S$.

Observe that in *both* Figures 7.7.1(a) and 7.7.1(b), the element of $H^*$ which corresponds to $L$ is the unique generalized point which belongs to both $L^*$ and $H^*$. Thus, the 1–1 mapping of $S \to H^*$ may be described elementwise by

$$L \to L^* \cap H^*.$$

More generally, if $k$ is any nonnegative integer (up to $n$), the $(k + 1)$-dimensional subspaces $W$ of $V'$ are in 1–1 correspondence with the generalized $k$-flats in $H^*$ via the correspondence $W \to W^* \cap H^*$. In Figure 7.7.2, for instance, $W$ is a 2-dimensional subspace of $V'$; $W^* \cap H^*$ is an extended line $M^*$ in $H^*$. Moreover, examination of Figure 7.7.2 should persuade you (at least informally!) that a one-dimensional subspace $L$ lies in $W$ if and only if the generalized point corresponding to $L$ (either $\mathbf{u}$ or $\mathscr{L}$, as the case may be) lies in the generalized line $M^*$ corresponding to $W$. Similar remarks apply to higher-dimensional cases. Let us now state all of this formally.

**7.7.1** **Theorem.** Let $V'$ be a vector space of dimension $n + 1$ over a field $F$, and let $H$ be an $n$-flat (hyperplane) in $V'$ such that $O \notin H$. For each subspace $W$ of $V'$, let $W\psi = W^* \cap H^*$. Then:

Figure 7.7.2

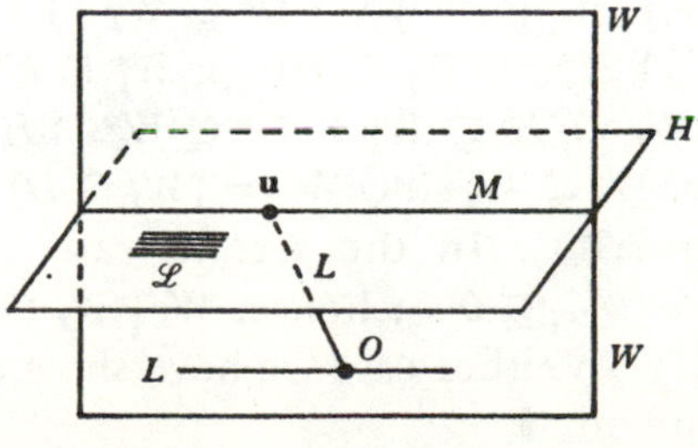

(a) For each fixed integer $k$ ($0 \leqslant k \leqslant n$), the correspondence $W \rightarrow W\psi$ is a one-to-one mapping of the set of all $(k + 1)$-dimensional subspaces of $V'$ onto the set of all generalized $k$-flats in $H^*$.

(b) If $W_1$, $W_2$ are arbitrary subspaces of $V'$ (not necessarily of the same dimension), then $W_1 \subseteq W_2$ if and only if $W_1\psi \subseteq W_2\psi$.

*Proof.* (1) If $W$ is a $(k + 1)$-dimensional subspace and $W \cap H = \varnothing$, then the proof of 7.6.7 implies that $W \parallel H$ and that $W^* \cap H^* = \mathscr{P}_W$ is a $k$-flat at infinity in $H^*$. If $W_1$, $W_2$ are two such subspaces and if $\mathscr{P}_{W_1} = \mathscr{P}_{W_2}$, then $W_1$ and $W_2$ are parallel (Theorem 7.6.6) and, hence, equal (why?); thus, distinct subspaces, both of which are disjoint from $H$, have distinct images under $\psi$. Conversely, any $k$-flat at infinity in $H^*$ has the form $\mathscr{P}_S$ for some $(k + 1)$-flat $S$ in $V'$; since $\mathscr{P}_S \subseteq H^*$, $\mathscr{P}_S \subseteq \mathscr{P}_H$ and, hence, $S$ is parallel to $H$ by Theorem 7.6.6. If $W$ is the direction space of $S$, then $W \parallel H$, so that $W \cap H = \varnothing$ and, hence, $W^* \cap H^* = \mathscr{P}_W = \mathscr{P}_S$; thus, $\mathscr{P}_S$ is the image of $W$ under $\psi$. It follows that under $\psi$ we have a 1–1 correspondence between *the* $(k + 1)$*-dimensional subspaces* $W$ *which do not intersect* $H$ and *the $k$-flats at infinity in* $H^*$.

(2) If instead $W \cap H \neq \varnothing$, then the proof of 7.6.7 implies that $W \cap H$ is a $k$-flat and $W\psi = W^* \cap H^* = (W \cap H)^*$ is an extended $k$-flat. Since $O \notin H$, the subspace spanned by $W \cap H$ is $(k + 1)$-dimensional (Sec. 3.4, Exercise 8) and is contained in $W$; hence, $W = [W \cap H]$. It follows that if $W_1$, $W_2$ are two (distinct) such subspaces, then $W_1 \cap H \neq W_2 \cap H$; this in turn implies $(W_1 \cap H)^* \neq (W_2 \cap H)^*$ so that $W_1$, $W_2$ have distinct images under $\psi$. Conversely, any extended $k$-flat in $H^*$ has the form $A^*$ for some $k$-flat $A$ contained in $H$. The subspace $W = [A]$ has dimension $k + 1$, and $W \cap H$ contains $A$ and is a $k$-flat (Sec. 3.4, Exercise 3(b)) so that $W \cap H = A$. Hence, $W^* \cap H^* = (W \cap H)^* = A^*$, so that $A^*$ is the image of $W$ under $\psi$. Thus, under $\psi$, we have a 1–1 correspondence between *the* $(k + 1)$*-dimensional subspaces* $W$ *which do intersect* $H$ and *the extended $k$-flats in* $H^*$.

(3) The results of paragraphs (1) and (2) clearly imply part (a) of Theorem 7.7.1. The "only if" assertion in part (b) is trivial: if $W_1 \subseteq W_2$, then $\mathscr{P}_{W_1} \subseteq \mathscr{P}_{W_2}$; hence, $W_1^* \subseteq W_2^*$; hence, $W_1^* \cap H^* \subseteq W_2^* \cap H^*$; hence, $W_1\psi \subseteq W_2\psi$. Conversely, suppose $W_1^* \cap H^* \subseteq W_2^* \cap H^*$. As we have seen above, either $W_1^* \cap H^* = (W_1 \cap H)^*$ or $W_1^* \cap H^* = \mathscr{P}_{W_1}$. In the first case, we have $W_1 \cap H \subseteq W_1^* \cap H^* \subseteq W_2^* \cap H^* \subseteq W_2^*$; hence, $W_1 \cap H \subseteq W_2$; hence, $W_1 \subseteq W_2$ since $W_1 = [W_1 \cap H]$ (cf. part (2) of proof for Theorem 7.7.1). In the second case, we have $\mathscr{P}_{W_1} \subseteq W_2^* \cap H^* \subseteq W_2^*$; hence, $\mathscr{P}_{W_1} \subseteq \mathscr{P}_{W_2}$; hence, $W_1 \parallel W_2$ (Theorem 7.6.6); hence, $W_1 \subseteq W_2$ (*why*?). In either case we have shown that $W_1 \subseteq W_2$, and our proof is complete. ▮

As a corollary of Theorem 7.7.1, we have

7.7.2 **Theorem.** Let $V$, $V'$ be vector spaces of dimensions $n$, $n + 1$ (respectively) over a field $F$. Then there exists a one-to-one mapping $\phi$ of the set of all subspaces of $V'$ onto the set of all generalized flats in $V^*$, such that:

(a) For each subspace $W$ of $V'$, the dimension of the generalized flat $W\phi \subseteq V^*$ is one less than the dimension of $W$.

(b) For any two subspaces $W_1$, $W_2$ of $V'$, $W_1 \subseteq W_2$ if and only if $W_1\phi \subseteq W_2\phi$.

*Proof.* Choose a hyperplane $H$ in $V'$ such that $O \notin H$. Define $\psi$ as in Theorem 7.7.1, and choose any one-to-one affine mapping $\mathbf{T}$ of $H \to V$. If we define $\mathbf{T}^* : H^* \to V^*$ as in Section 7.6, Exercise 4, then part (e) of that exercise implies that the mapping $\phi = \psi\mathbf{T}^*$ has the desired properties.

(*Remark*: The definition of $\psi$ in Theorem 7.7.1 implies that $\{\mathbf{0}\}\psi = \varnothing$. Hence, also $\{\mathbf{0}\}\phi = \varnothing$. Since, as we have seen, the empty set is considered to be a "flat at infinity" of dimension $-1$, Theorem 7.7.2 is still valid for the subspace $W = \{\mathbf{0}\}$.) ∎

It's time for some examples to show how the use of Theorem 7.7.2 simplifies proofs. In what follows we shall assume that the vector space $V$ is finite dimensional.

7.7.3 **Example.** Prove that any two distinct generalized points in $V^*$ lie in a unique generalized line in $V^*$. (This was Theorem 7.6.2.)

*Solution.* The proof of this assertion would formerly have required consideration of three separate cases, depending on whether the points were finite or infinite. Now, however, we can argue as follows (at least in the finite-dimensional case): let $V'$ be a vector space of dimension one more than $V$. Via the mapping $\phi$ of Theorem 7.7.2, generalized points and lines in $V^*$ correspond (respectively) to one-dimensional and two-dimensional subspaces of $V'$. Hence, the statement of Example 7.7.3 is equivalent to the assertion that *any two distinct one-dimensional subspaces of $V'$ lie in a unique two-dimensional subspace.* But the results of Chapter 2 easily imply (how?) that if $[\mathbf{u}]$ and $[\mathbf{v}]$ are distinct one-dimensional subspaces of $V'$, then $[\mathbf{u}, \mathbf{v}]$ is the unique two-dimensional subspace containing both $[\mathbf{u}]$ and $[\mathbf{v}]$. This is all we need. ∎

**7.7.4 Example.** The intersection of any collection of generalized flats in $V^*$ is a generalized flat.

*Solution.* Since Theorem 7.7.2 easily implies that

$$\bigcap_\alpha (W_\alpha\phi) = \left(\bigcap_\alpha W_\alpha\right)\phi \tag{7.7.5}$$

for any collection $(W_\alpha)$ of subspaces (cf. Exercise 1 below), the desired assertion is equivalent to the statement that the intersection of subspaces of $V'$ is a subspace; and the latter is already known by Section 2.2, Exercise 2. ∎

(Note again how short this proof is. Try constructing a proof directly from the definitions in Section 7.6 and see how long it takes!)

**7.7.6 Example.** Prove: if two distinct generalized $k$-flats $S_1$, $S_2$ in $V^*$ lie in the same generalized $(k+1)$-flat, then their intersection is a generalized $(k-1)$-flat.

*Solution.* By 7.7.2, we have $S_1 = W_1\phi$, $S_2 = W_2\phi$, where $W_1$, $W_2$ are $(k+1)$-dimensional subspaces of $V'$. Similarly, if $S$ is a generalized $(k+1)$-flat containing $S_1$ and $S_2$, we have $S = W\phi$, where $\dim W = k+2$. Since $S_1, S_2 \subseteq S$, 7.7.2(b) implies that $W_1, W_2 \subseteq W$. Since $W_1$, $W_2$ are distinct, their intersection has dimension strictly less than $k+1$ (why?). Thus,

$$\begin{aligned} k \geq \dim(W_1 \cap W_2) &= \dim W_1 + \dim W_2 - \dim(W_1 + W_2) \\ &\qquad \text{(by Theorem 2.6.17)} \\ &\geq \dim W_1 + \dim W_2 - \dim W \qquad (\text{since } W_1 + W_2 \subseteq W) \\ &= (k+1) + (k+1) - (k+2) = k. \end{aligned}$$

Hence, $\dim(W_1 \cap W_2) = k$, so that the generalized flat

$$S_1 \cap S_2 = (W_1\phi) \cap (W_2\phi) = (W_1 \cap W_2)\phi \qquad (\text{cf. } 7.7.5)$$

has dimension $k - 1$. ∎

### Collineations

Again let $V$, $V'$ be vector spaces of dimensions $n$, $n+1$ over $F$, and let $S_1$, $S_2$ be generalized flats in $V^*$. A mapping $\mathfrak{T} : S_1 \to S_2$ is called a *collineation* if (a) $\mathfrak{T}$ is one-to-one, and (b) the image under $\mathfrak{T}$ of every generalized line in $S_1$ is a generalized line in $S_2$.

One way of obtaining collineations is as follows: let $\mathbf{T}$ be a *one-to-one* linear transformation of $V' \to V'$. If $P$ is any element of $V^*$ and $L$

is the corresponding one-dimensional subspace of $V'$ (via the correspondence $\phi$ of Theorem 7.7.2), then $L\mathbf{T}$ is a one-dimensional subspace (why?) and, hence, $(L\mathbf{T})\phi$ is an element of $V^*$. If we define

$$P\mathfrak{T} = (L\mathbf{T})\phi,$$

then $\mathfrak{T}$ is a mapping of $V^* \rightarrow V^*$ and

$$(L\phi)\mathfrak{T} = (L\mathbf{T})\phi$$

for every one-dimensional subspace $L$ of $V'$. Since $\mathbf{T}$ maps 2-dimensional subspaces onto 2-dimensional subspaces, $\mathfrak{T}$ maps generalized lines onto generalized lines and is a collineation. We call $\mathfrak{T}$ the *collineation induced by* $\mathbf{T}$.

The existence of induced collineations has some interesting consequences. For example, consider the following theorem (Figure 7.7.3):

**7.7.7 Theorem.** Let $L_1$, $L_2$, $L_3$ be distinct lines in $V$ which intersect at a point $P$. For each $i$ (= 1, 2, 3), let $A_i$, $B_i$ be distinct points different from $P$ on $L_i$. If $\mathbf{Ar}\ A_1A_2 \parallel \mathbf{Ar}\ B_1B_2$ and $\mathbf{Ar}\ A_2A_3 \parallel \mathbf{Ar}\ B_2B_3$, then $\mathbf{Ar}\ A_1A_3 \parallel \mathbf{Ar}\ B_1B_3$.

This theorem was Exercise 17, Section 3.2. When $V = \mathbf{R}_n$ the theorem can be proved using similar triangles (formally, using the result and proof of Example 3.2.17); a more general solution appears in Appendix G. Let us now *interpret the theorem in terms of the projective space* $V^*$. Since the extended lines $L^*(A_1, A_2)$ and $L^*(B_1, B_2)$ both lie in the same extended plane (*why*?), they must have a point of intersection $C_3$ in that plane (cf. Example 7.7.6); similarly, $L^*(A_1, A_3)$ and $L^*(B_1, B_3)$ meet in a point $C_2$, and $L^*(A_2, A_3)$, $L^*(B_2, B_3)$ meet in a point $C_1$. Clearly, $\mathbf{Ar}\ A_1A_2 \parallel \mathbf{Ar}\ B_1B_2$ if and only if $C_3$ is a point at

Figure 7.7.3

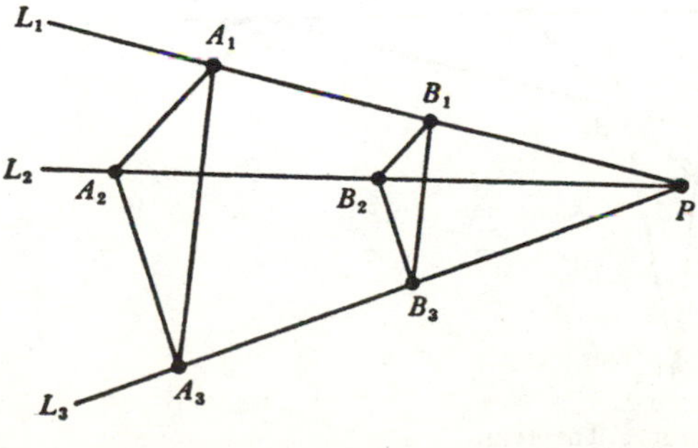

infinity; and so on. Thus, the last sentence of 7.7.7 may be restated: *if two of the points $C_1$, $C_2$, $C_3$ are points at infinity, then so is the third.* Now, the $A_i$ are all contained in a plane $S$, so that the $C_i$ are all in $S^*$ and, hence, are points at infinity if and only if they lie on the "line at infinity" $\mathcal{P}_S$. Thus, Theorem 7.7.7 may be further restated: *if two of the $C_i$ lie on $\mathcal{P}_S$, then so does the third.* Equivalently,

**(7.7.7′)** *If two of the $C_i$ lie on $\mathcal{P}_S$, then the $C_i$ are collinear.*

("Collinear" now means "on the same *generalized* line".)

We now come to the point of this whole discussion: that is, the point where induced collineations enter the picture. The idea, roughly speaking, is that the line $\mathcal{P}_S$ can be mapped onto any other given generalized line $L$ by some collineation and, hence, 7.7.7′ remains true if $\mathcal{P}_S$ is replaced by $L$. That is, the "if" clause in 7.7.7′ is superfluous; *the $C_i$ are collinear in any case.* Figure 7.7.4 illustrates the case where the $C_i$ are all finite points. Let us state and prove this result formally so as to show how to make the argument rigorous.

**7.7.8** **Theorem.** (Desargues' Theorem) Let $L_1$, $L_2$, $L_3$ be distinct lines in $V$ which intersect at a point $P$. For each $i$ (= 1, 2, 3), let $A_i$, $B_i$ be distinct points different from $P$ on $L_i$. Let $C_1$ be the point of intersection (in $V^*$) of the extended lines $L^*(A_2, A_3)$ and $L^*(B_2, B_3)$ (this point exists by 7.7.6); similarly, let $C_2$ be the intersection of $L^*(A_1, A_3)$ and $L^*(B_1, B_3)$, and let $C_3$ be the intersection of $L^*(A_1, A_2)$ and $L^*(B_1, B_2)$.

Figure 7.7.4

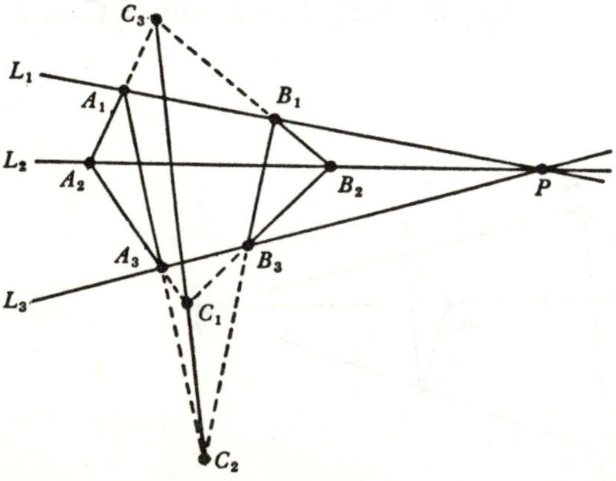


Desargues' Theorem

Then the points $C_1$, $C_2$, $C_3$ are collinear; that is, they lie on the same generalized line in $V^*$.

*Proof.* Assume first that the three lines $L_i$ all lie in the same plane $S$. We may, of course, assume the $C_i$ are distinct; let $L$ be the generalized line through $C_1$ and $C_3$. Via Theorem 7.7.2, the extended plane $S^*$ corresponds to a 3-dimensional subspace $W$ of $V'$ and the generalized lines $L$, $\mathcal{P}_S$ correspond respectively to 2-dimensional subspaces $M_1$, $M_2$ of $V'$. The results of earlier chapters easily imply the existence of a one-to-one linear transformation $\mathbf{T} : V' \to V'$ which maps $W$ onto $W$ and $M_1$ onto $M_2$ (cf. Theorems 4.2.2 and 2.6.16); the induced collineation $\mathfrak{T}$ maps $S^* \to S^*$ and $L$ onto $\mathcal{P}_S$. Write $A_i\mathfrak{T} = A_i'$, $B_i\mathfrak{T} = B_i'$, $C_i\mathfrak{T} = C_i'$, $P\mathfrak{T} = P'$. If $A_1$, $A_2$, $A_3$ are collinear, the proof is trivial; if $A_1$, $A_2$, $A_3$ are not collinear, it is easy to show that none of the $A_i$ lie on $L$. Hence, none of the $A_i'$ lie on $\mathcal{P}_S$; that is, the $A_i'$ are all finite points in $S$. Similarly, we may assume that the $B_i'$ are all finite points. On the other hand, $C_1$ and $C_3$ are on $L$, so that $C_1'$, $C_3'$ are points at infinity. If $P'$ is finite, it follows from Theorem 7.7.7 that $C_2'$ is a point at infinity, so that $C_2 \in L$, q.e.d. (If $P'$ is infinite, use Sec. 3.2, Exercise 18 in place of Theorem 7.7.7.)

If instead the lines $L_i$ are not in the same plane, then they determine a 3-flat which contains the distinct planes $S_1 = \langle A_1, A_2, A_3\rangle$, $S_2 = \langle B_1, B_2, B_3\rangle$. The intersection of the extended planes $S_1^*$, $S_2^*$ is a generalized line (Example 7.7.6); since the points $C_i$ lie in both $S_1^*$ and $S_2^*$, they lie in the line of intersection and are, hence, collinear. ∎

In connection with Theorem 7.7.8, it is rather striking that even in the case where everything lies in a single plane $S$, the proof requires the introduction of a 3-dimensional space (the subspace $W$ corresponding to $S^*$). In fact, there does not exist a purely "geometric" proof of Desargues' Theorem in two dimensions alone; all proofs which stay within the plane require the introduction of coordinates. (For a proof using *barycentric* coordinates, see Hausner [7].)

Desargues' Theorem is just one of many interesting results which, though possessing nonprojective interpretations, can be proved most neatly by "projective" methods. For further examples, see Exercises 4, 5, and 6 at the end of this section. The reader interested in pursuing this subject further is urged to consult one of the good books on projective geometry, such as Pedoe [12].

**Homogeneous coordinates and an analog of Theorem 7.6.4**

Let $V$, $V'$ be as above and let $(\mathbf{v}_1', \ldots, \mathbf{v}_{n+1}')$ be a fixed ordered basis of $V'$. By Theorem 7.7.2, any generalized point $P \in V^*$ corresponds to some one-dimensional subspace $W = [\mathbf{u}]$ of $V'$. If the expression for $\mathbf{u}$ in terms of the basis $(\mathbf{v}'_1, \ldots, \mathbf{v}'_{n+1})$ is

$$\mathbf{u} = c_1\mathbf{v}'_1 + \ldots + c_{n+1}\mathbf{v}'_{n+1}$$

then any other nonzero vector in $W$ has the form

$$\mathbf{w} = b\mathbf{u} = bc_1\mathbf{v}'_1 + \ldots + bc_{n+1}\mathbf{v}'_{n+1}$$

for some scalar $b \neq 0$. The components $(bc_1, \ldots, bc_{n+1})$ of $\mathbf{w}$ are called *homogeneous coordinates of the point P* with respect to the given basis of $V'$. In this manner, each generalized point $P$ in $V^*$ is assigned an $(n + 1)$-tuple of "homogeneous coordinates" which, though *not* unique, are unique *up to a nonzero constant factor*. (For example, if a point in $\mathbf{R}_2^*$ has homogeneous coordinates $(2, -1, 5)$, then the coordinates $(4, -2, 10)$ represent the same point.) Note that the $(n + 1)$-tuple $(0, \ldots, 0)$ does not represent any generalized point, since the subspace of $V'$ spanned by $\mathbf{0}$ is not one-dimensional.

By a *quadric in* $V^*$, we shall mean the graph in $V^*$ of an equation of the form

$$\sum_{i=1}^{n+1}\sum_{j=1}^{n+1} a_{ij}x_ix_j = 0; \tag{7.7.9}$$

that is, the set of all generalized points in $V^*$ whose homogeneous coordinates $(x_1, \ldots, x_{n+1})$ satisfy the equation. Note that if $(x_1, \ldots, x_{n+1})$ satisfies 7.7.9, then so does any scalar multiple of $(x_1, \ldots, x_{n+1})$; thus, with respect to a fixed basis of $V'$, the truth of the statement, "The point $P$ satisfies 7.7.9", depends only on $P$, not on which set of homogeneous coordinates of $P$ is chosen.

If $Q^*$ is the graph of 7.7.9 in $V^*$, we define a *singular point of* $Q^*$ to be a generalized point $J \in Q^*$ such that, for any (generalized) point $P \neq J$ in $Q^*$, the generalized line through $P$ and $J$ is contained in $Q^*$. The following result is known concerning singular points:

**7.7.10 Theorem.** Let $Q^*$ be the graph of 7.7.9 in $V^*$ and assume that the $(n + 1) \times (n + 1)$ matrix $M^* = (a_{ij})$ is symmetric. If the field of scalars does not have characteristic 2, then the set of all singular points of $Q^*$ is a generalized flat in $V^*$ of dimension $n - \operatorname{rank} M^*$.

Theorem 7.7.10 is the projective analog of Theorem 7.6.4. The proof of 7.7.10 is considerably simpler than that of 7.6.4; in particular, special properties of the field $\mathbf{R}$ are no longer needed. The extra difficulty in proving 7.6.4 had to do with the different way "singular points" were defined; specifically, with the fact that in Definition 7.6.3 we required $L^*(P, J) \subseteq Q^*$ only for points $P \in Q$, not for *generalized* points $P \in Q^*$. In other words, Definition 7.6.3 made it easier for a point $J$ to be a singular point, and hence harder to prove that the set of singular points was no larger than we wanted it to be.

To prove 7.7.10, let $Q'$ be the graph of 7.7.9 in $V'$ (that is, the set of all vectors in $V'$ whose components $x_1, \ldots, x_{n+1}$ with respect to the basis $(\mathbf{v}_1', \ldots, \mathbf{v}_{n+1}')$ satisfy the equation). Consider the system of linear equations

$$\text{(7.7.11)} \quad \sum_{i=1}^{n+1} x_i a_{ij} = 0 \qquad (j = 1, \ldots, n + 1).$$

The set $\mathscr{S}'$ of solutions of 7.7.11 in $V'$ is a subspace of dimension $(n + 1) - \text{rank } M^*$ (Theorem 5.10.14) and, hence, the set $\mathscr{S}$ of solutions of 7.7.11 in $V^*$ is a generalized flat of dimension $n - \text{rank } M^*$. It suffices to show that $\mathscr{S}$ is the set of singular points of $Q^*$. Equivalently (in view of the correspondence $\phi$ of Theorem 7.7.2), it suffices to show that $\mathscr{S}'$ is the set of all vectors $\mathbf{x} \in Q'$ having the property that

$$\text{(7.7.12)} \quad [\mathbf{x}, \mathbf{y}] \subseteq Q' \qquad (\text{all } \mathbf{y} \in Q').$$

To simplify notation, $X = [x_1\ x_2 \ldots x_{n+1}]$ will denote the row matrix whose entries are the components of the vector

$$\mathbf{x} = x_1\mathbf{v}_1' + \ldots x_{n+1}\mathbf{v}_{n+1}'$$

with respect to the given basis of $V'$. We remark at the outset that since $XM^* = 0 \Rightarrow XM^*X^t = 0$, any solution of 7.7.11 also satisfies 7.7.9; that is, $\mathscr{S}' \subseteq Q'$.

If $\mathbf{x} \in \mathscr{S}'$ and $\mathbf{y} \in Q'$, then $XM^* = 0 = YM^*Y^t$, from which we easily have

$$(aX + bY)M^*(aX^t + bY^t) = 0 \qquad (\text{all scalars } a, b)$$

and hence 7.7.12 holds. Conversely, *suppose that some* $\mathbf{x} \in Q'$ *satisfies 7.7.12 but not 7.7.11*; we shall complete the proof by deriving a contradiction from this assumption.

Let $Z = XM^*$. By 7.7.12, for any $\mathbf{y} \in Q'$, we have $\mathbf{x} + \mathbf{y} \in Q'$ so that

$$XM^*X^t = YM^*Y^t = (X + Y)M^*(X + Y)^t = 0$$

from which $2XM^*Y^t = 0$; hence, $XM^*Y^t = 0$ (division by 2 being permissible in our field!); hence, $ZY^t = 0$; hence, $\mathbf{y} \perp \mathbf{z}$. Thus,

$$Q' \subseteq [\mathbf{z}]^\perp.$$

Since we have assumed that 7.7.11 does not hold, $\mathbf{z} \neq \mathbf{0}$ and, hence,

$\dim[\mathbf{z}]^\perp = \dim V' - 1$ (cf. Theorem 5.10.14). In particular, $[\mathbf{z}]^\perp$ is not all of $V'$ and we can choose a vector $\mathbf{w}$ which is not in $[\mathbf{z}]^\perp$. (Hence, also $\mathbf{w} \notin Q'$.) Since $\mathbf{x} \in Q' \subseteq [\mathbf{z}]^\perp$, it follows by a dimension argument that

$$[\mathbf{x}, \mathbf{w}] \cap [\mathbf{z}]^\perp = [\mathbf{x}]. \tag{7.7.13}$$

Also, since $\mathbf{x} \in Q'$ and $\mathbf{w} \notin Q'$, the proof of Lemma E of Section 7.6 shows that if we let

$$a = -\frac{2XM^*W^t}{WM^*W^t}$$

then $\mathbf{x} + a\mathbf{w}$ belongs to $Q'$, hence, to $[\mathbf{x}]$ by 7.7.13. Since $\mathbf{x}$, $\mathbf{w}$ are linearly independent (*why*?), it follows that $a = 0$; hence, $XM^*W^t = 0$; hence, $\mathbf{w} \perp \mathbf{z}$, contradicting the choice of $\mathbf{w}$. This is the desired contradiction. ∎

**Exercises**

In the exercises below, $V$ and $V'$ denote vector spaces (over a field $F$) such that $\dim V' = 1 + \dim V$.

*1. Deduce equation 7.7.5 from Theorem 7.7.2. (SOL)

2. Let $V = \mathbf{R}_2$, $V' = \mathbf{R}_3$. Suppose that in the proof of Theorem 7.7.2 we choose $H$ to be the plane $z = 1$ (in $\mathbf{R}_3$) and $\mathbf{T}$ to be the mapping of $H \to \mathbf{R}_2$ defined by $(x, y, 1)\mathbf{T} = (x, y)$. A mapping $\phi$ is thereby determined, under which the one-dimensional subspaces of $\mathbf{R}_3$ correspond to the elements of $\mathbf{R}_2^\ddagger$. This in turn determines homogeneous coordinates $(x, y, z)$ (with respect to the canonical basis of $\mathbf{R}_3$) for the elements $P$ of $\mathbf{R}_2^\ddagger$.

(a) Show that if $(x, y, z)$ are homogeneous coordinates of $P$, then $P$ is a "finite" point (i.e., an element of $\mathbf{R}_2$) if and only if $z \neq 0$; and that in this case $P$ is the point $(x/z, y/z)$ of $\mathbf{R}_2$.

(b) By part (a), the homogeneous coordinates of a "point at infinity" $\mathscr{L} \in \mathbf{R}_2^\ddagger$ have the form $(a, b, 0)$. Here, $\mathscr{L}$ is by definition (Sec. 7.6) a class of parallel lines in $\mathbf{R}_2$. Show that the lines in this class have direction space $[(a, b)]$.

(c) Using (a) and (b), show that if $L$ is the line in $\mathbf{R}_2$ whose equation is $2x - y + 3 = 0$, and if $L^*$ is the corresponding "extended line" in $\mathbf{R}_2^\ddagger$, then the equation of $L^*$ in homogeneous coordinates is

$$2x - y + 3z = 0. \tag{7.7.14}$$

that is, $L^*$ consists of all generalized points in $\mathbf{R}_2^\ddagger$ whose homogeneous coordinates $(x, y, z)$ satisfy 7.7.14. (Equivalently, $L^*$ is the image under $\phi$ of the 2-dimensional subspace of $\mathbf{R}_3$ whose equation is 7.7.14.)

(d) Generalize part (c).

3. Generalize Exercise 2 to higher dimensions.
4. Deduce from Example 7.7.6 that if two distinct lines in $V$ lie in the same plane and have empty intersection, then they are parallel. (This is part of the case $k = 1$ of Theorem 3.4.9.)
5. Give a new proof of Theorem 3.4.8 using the ideas of this section. (*Suggestion*: Write $S_i^* = U_i\phi$ as in Theorem 7.7.2, then apply Theorem 3.4.5 to the $U_i$.) (SOL)
6. Deduce the result of Section 3.5, Exercise 10, from that of Example 3.5.7 in the same way that Desargues' Theorem was derived from Theorem 7.7.7.

# 8 The structure of linear transformations

## 8.0 Introduction

If $V$ is an $n$-dimensional vector space over a field $F$, the linear transformations of $V \to V$ correspond to $n \times n$ matrices over $F$. As we have seen, the precise correspondence depends on our choice of a basis of $V$. For example, if $V = \mathbf{R}_2$ and $\mathbf{T}$ is defined by

$$(x, y)\mathbf{T} = (-x + 4y, x - y),$$

then $\mathbf{T}$ has the matrix

$$A = \begin{bmatrix} -1 & 1 \\ 4 & -1 \end{bmatrix}$$

with respect to the canonical basis; but with respect to the ordered basis ((2, 1), (2, −1)) the matrix of $\mathbf{T}$ is

$$A' = \begin{bmatrix} 1 & 0 \\ 0 & -3 \end{bmatrix}$$

(verify this!). Now $A'$ is considerably simpler than $A$; by looking at $A'$, we can immediately see some things about $\mathbf{T}$ (for instance, that $\mathbf{T}$ has a line of fixed points) that are not obvious from a glance at $A$. In short, some matrices are nicer to work with than others. (Diagonal matrices are especially nice.) It is clearly desirable to be able to write the matrix of a given linear transformation in as simple a form as possible.

In this chapter, we shall show that *if the characteristic polynomial of* $\mathbf{T}$ *splits* (cf. Sec. 1.16), then there is a basis of $V$ with respect to which the matrix of $\mathbf{T}$ has zero entries everywhere except on the main diagonal and the "superdiagonal" (the diagonal immediately above the main diagonal); moreover, the superdiagonal consists entirely of 1's and 0's. This form of the matrix of $\mathbf{T}$ is called the *Jordan canonical form*, or simply the *Jordan form* (J.F.). To demonstrate the existence of the J.F., we first obtain it for a special class of L.T.'s called *nilpotent* mappings (in this case the J.F. is even simpler: it has all zeros on the diagonal). We then show that corresponding to each eigenvalue $\lambda_i$ of **T** there is a subspace $N^{(i)}$ of $V$ (called the (**T**, $\lambda_i$)-*characteristic subspace*), such that (1) $V$ is the direct sum of the subspaces $N^{(i)}$, and (2) the restriction of **T** to $N^{(i)}$ equals a nilpotent mapping plus $\lambda_i\mathbf{I}$. By putting these results together, we obtain the Jordan form for **T**.

As indicated above, the Jordan form exists only if char **T** splits. However, the consequent loss of generality in our results is not as serious as it might seem. This is due to the fact that a polynomial over a given field will always split over some larger field; thus, in certain (nonsplitting) situations we can still obtain useful results by *extending the field of scalars*. Applications of this idea occur both in this chapter and in Chapter 9.

## 8.1 Further theorems about the range and nullspace

This section is essentially preliminary; in it, we collect a few theorems which we will need later concerning the range and nullspace of a linear transformation. Throughout the section, $F$ will denote a field and $V$, $V'$, $V''$ will denote vector spaces over $F$.

**8.1.1 Theorem.** If $\mathbf{T} \in \text{Lin}\,(V, V')$ and $\mathbf{S} \in \text{Lin}\,(V', V'')$, then
(a) Ran $\mathbf{TS} \subseteq$ Ran $\mathbf{S}$
(b) Nul $\mathbf{TS} \supseteq$ Nul $\mathbf{T}$.

This theorem was stated earlier as Exercise 7, Section 4.6. To prove (a), we have

$$\text{Ran}\,\mathbf{TS} = V(\mathbf{TS}) = (V\mathbf{T})\mathbf{S} \subseteq V'\mathbf{S} = \text{Ran}\,\mathbf{S}.$$

To prove (b), let $\mathbf{v} \in$ Nul $\mathbf{T}$; then $\mathbf{vT} = \mathbf{0}$, so that

$$\mathbf{v}(\mathbf{TS}) = (\mathbf{vT})\mathbf{S} = \mathbf{0S} = \mathbf{0}$$

and, hence, $\mathbf{v} \in$ Nul $\mathbf{TS}$. ∎

As a special case of 8.1.1, let $V = V' = V''$, and let $k$ be any nonnegative integer. Letting $\mathbf{S} = \mathbf{T}^k$ in 8.1.1(a), we obtain Ran $\mathbf{T}^{k+1} \subseteq$ Ran $\mathbf{T}^k$; letting $\mathbf{T} = \mathbf{S}^k$ in 8.1.1(b), we obtain Nul $\mathbf{S}^{k+1} \supseteq$ Nul $\mathbf{S}^k$. Hence, the inclusions

$$V \supseteq \text{Ran}\,\mathbf{T} \supseteq \text{Ran}\,\mathbf{T}^2 \supseteq \text{Ran}\,\mathbf{T}^3 \supseteq \cdots \supseteq \text{Ran}\,\mathbf{T}^k \supseteq \cdots \tag{8.1.2}$$

$$\{\mathbf{0}\} \subseteq \text{Nul}\,\mathbf{T} \subseteq \text{Nul}\,\mathbf{T}^2 \subseteq \text{Nul}\,\mathbf{T}^3 \subseteq \cdots \subseteq \text{Nul}\,\mathbf{T}^k \subseteq \cdots \tag{8.1.3}$$

hold for all $\mathbf{T} \in \text{Lin}(V, V)$.

**8.1.4** **Theorem.** If $\mathbf{T} \in \text{Lin}(V, V)$ and $k$ is any positive integer, then the subspaces Ran $\mathbf{T}^k$ and Nul $\mathbf{T}^k$ are **T**-stable. (Cf. Def. 5.13.4.)

*Proof.* Since

$$(\text{Ran } \mathbf{T}^k)\mathbf{T} = (V\mathbf{T}^k)\mathbf{T} = V\mathbf{T}^{k+1}$$
$$= \text{Ran } \mathbf{T}^{k+1} \subseteq \text{Ran } \mathbf{T}^k \qquad \text{(by 8.1.2)}$$

it follows that Ran $\mathbf{T}^k$ is **T**-stable. If $\mathbf{v} \in \text{Nul } \mathbf{T}^k$, then $\mathbf{v} \in \text{Nul } \mathbf{T}^{k+1}$ so that $\mathbf{0} = \mathbf{v}\mathbf{T}^{k+1} = (\mathbf{v}\mathbf{T})\mathbf{T}^k$. Hence, $\mathbf{v}\mathbf{T} \in \text{Nul } \mathbf{T}^k$. It follows that Nul $\mathbf{T}^k$ is **T**-stable. ∎

**8.1.5** **Theorem.** Let $\mathbf{T} \in \text{Lin}(V, V)$ and suppose that Nul **T** = Nul $\mathbf{T}^2$. If $V$ is finite dimensional, then $V = (\text{Nul } \mathbf{T}) \oplus (\text{Ran } \mathbf{T})$.

*Proof.* We first show that

**(8.1.6)** $(\text{Nul } \mathbf{T}) \cap (\text{Ran } \mathbf{T}) = \{\mathbf{0}\}$.

Let $\mathbf{u} \in (\text{Nul } \mathbf{T}) \cap (\text{Ran } \mathbf{T})$; then

$$\mathbf{u} \in \text{Ran } \mathbf{T} \Rightarrow \mathbf{u} = \mathbf{v}\mathbf{T} \qquad \text{for some } \mathbf{v};$$
$$\mathbf{u} \in \text{Nul } \mathbf{T} \Rightarrow \mathbf{u}\mathbf{T} = \mathbf{0}$$
$$\Rightarrow (\mathbf{v}\mathbf{T})\mathbf{T} = \mathbf{0}$$
$$\Rightarrow \mathbf{v}(\mathbf{T}^2) = \mathbf{0}$$
$$\Rightarrow \mathbf{v} \in \text{Nul } \mathbf{T}^2$$
$$\Rightarrow \mathbf{v} \in \text{Nul } \mathbf{T} \qquad (\text{since Nul } \mathbf{T} = \text{Nul } \mathbf{T}^2)$$
$$\Rightarrow \mathbf{v}\mathbf{T} = \mathbf{0}$$
$$\Rightarrow \mathbf{u} = \mathbf{0}$$

which proves 8.1.6. It then follows from Theorem 2.6.17 that

$$\dim(\text{Nul } \mathbf{T} + \text{Ran } \mathbf{T}) = \dim \text{Nul } \mathbf{T} + \dim \text{Ran } \mathbf{T}$$
$$= \dim V \qquad \text{(by Theorem 4.6.7)}$$

so that

**(8.1.7)** $V = \text{Nul } \mathbf{T} + \text{Ran } \mathbf{T}$

by Theorem 2.6.16(b). The sum 8.1.7 is direct since 8.1.6 implies that condition 2.8.3(c) holds. ∎

*Note*: Theorem 8.1.5 is false without the hypothesis that dim $V$ is finite; cf. Exercise 1 below.

**Exercises**

1. Give an example to show that Theorem 8.1.5 is false if $V$ is infinite dimensional. (SUG)

2. Let $\mathbf{T} \in \text{Lin}(V, V)$; then by 8.1.2 we have $\text{Ran } \mathbf{T}^k \supseteq \text{Ran } \mathbf{T}^{k+1}$ for all $k$. Prove: If $\text{Ran } \mathbf{T}^k = \text{Ran } \mathbf{T}^{k+1}$ for some fixed $k$, then $\text{Ran } \mathbf{T}^k = \text{Ran } \mathbf{T}^{k+j}$ for all $j \in \mathbf{Z}^+$.
3. State and prove a theorem about nullspaces analogous to the result of the preceding exercise.
4. Let $\mathbf{T} \in \text{Lin}(V, V)$. We recall that if $W$ is a subspace of $V$, then $\mathbf{T} \restriction W$ denotes the restriction of $\mathbf{T}$ to $W$. Prove that for any integer $k \geqslant 0$, the nullspace of $\mathbf{T} \restriction (\text{Ran } \mathbf{T}^k)$ is the image under $\mathbf{T}^k$ of $\text{Nul } \mathbf{T}^{k+1}$.

## 8.2 Nilpotent transformations

Throughout this section, $V$ will denote a vector space (usually finite dimensional) over a field $F$.

8.2.1 **Definition.** If $\mathbf{T} \in \text{Lin}(V, V)$, we say that $\mathbf{T}$ is *nilpotent* if there exists an integer $m \geqslant 0$ such that $\mathbf{T}^m = \mathbf{O}$. The least such $m$ is called the *index of nilpotence* (or simply the *index*) of $\mathbf{T}$. Similarly, a square matrix $A$ is nilpotent if $A^m = 0$ for some integer $m \geqslant 0$; the least such $m$ is the index of nilpotence of $A$.

*Remark*: Although $m = 0$ is allowed by Definition 8.2.1, the case $m = 0$ can arise only if $V$ is the trivial vector space $\{\mathbf{0}\}$. Indeed, $\mathbf{T}^0 = \mathbf{I}$ so that if $\mathbf{T}^0 = \mathbf{O}$ then $\mathbf{I} = \mathbf{O}$, $\mathbf{v} = \mathbf{vI} = \mathbf{vO} = \mathbf{0}$ for all $\mathbf{v} \in V$, $V = \{\mathbf{0}\}$. Conversely, if $V = \{\mathbf{0}\}$, then the only element of $\text{Lin}(V, V)$ is $\mathbf{O}$ ($= \mathbf{I}$) and this mapping is trivially nilpotent of index 0. In all other cases, the index of a nilpotent L.T. must be positive.

A special type of nilpotent mapping is the following: suppose that $\mathbf{T} \in \text{Lin}(V, V)$ and that $V$ has a finite ordered basis $(\mathbf{v}_1, ..., \mathbf{v}_k)$ ($k > 0$) such that

(8.2.2) $$\mathbf{T} : \mathbf{v}_1 \to \mathbf{v}_2 \to \mathbf{v}_3 \to \cdots \to \mathbf{v}_k \to \mathbf{0}.$$

That is, $\mathbf{v}_i\mathbf{T} = \mathbf{v}_{i+1}$ for $1 \leqslant i < k$, and $\mathbf{v}_k\mathbf{T} = \mathbf{0}$. In this case, $\mathbf{T}$ is called *nilcyclic* and the ordered basis $(\mathbf{v}_1, ..., \mathbf{v}_k)$ is called a $\mathbf{T}$-*cyclic basis* of $V$. It is easy to see that any nilcyclic L.T. is nilpotent; indeed, 8.2.2 implies that every $\mathbf{v}_i$ is mapped into $\mathbf{0}$ by at most $k$ successive applications of $\mathbf{T}$, so that $\mathbf{v}_i\mathbf{T}^k = \mathbf{0}$ (all $i$) and, hence, $\mathbf{T}^k = \mathbf{O}$. The really interesting thing is that a partial converse holds: although not every nilpotent L.T. is nilcyclic, it is possible to express every nilpotent L.T. as a "sum" (in a certain sense) of nilcyclic mappings. Our next theorem states this result precisely; the theorem is quite important and much of our later work will depend on it.

**8.2.3 Theorem.** Let $V$ be finite dimensional and let $\mathbf{T} \in \text{Lin}(V, V)$ be nilpotent. Then there exist nonzero subspaces $W_1, \ldots, W_r$ of $V$ such that:

(a) Each subspace $W_i$ is $\mathbf{T}$-stable.
(b) $V = W_1 \oplus \cdots \oplus W_r$.
(c) For all $i$ $(1 \leq i \leq r)$, $\mathbf{T} \restriction W_i$ is nilcyclic.

Before proving Theorem 8.2.3, let us see what it tells us about the matrix of $\mathbf{T}$. Assume $\mathbf{T}$ is nilpotent and let $W_1, \ldots, W_r$ be as in the theorem. Since $\mathbf{T} \restriction W_i$ is nilcyclic, $W_i$ has an ordered basis $(\mathbf{w}_1^{(i)}, \ldots, \mathbf{w}_{k_i}^{(i)})$ such that

$$\begin{aligned}
\mathbf{w}_1^{(i)}\mathbf{T} &= \mathbf{w}_2^{(i)} = 0\mathbf{w}_1^{(i)} + 1\mathbf{w}_2^{(i)} + 0\mathbf{w}_3^{(i)} + \cdots + 0\mathbf{w}_{k_i}^{(i)}\\
\mathbf{w}_2^{(i)}\mathbf{T} &= \mathbf{w}_3^{(i)} = 0\mathbf{w}_1^{(i)} + 0\mathbf{w}_2^{(i)} + 1\mathbf{w}_3^{(i)} + \cdots + 0\mathbf{w}_{k_i}^{(i)}\\
&\cdots\\
\mathbf{w}_{k_i-1}^{(i)}\mathbf{T} &= \mathbf{w}_{k_i}^{(i)} = 0\mathbf{w}_1^{(i)} + 0\mathbf{w}_2^{(i)} + 0\mathbf{w}_3^{(i)} + \cdots + 1\mathbf{w}_{k_i}^{(i)}\\
\mathbf{w}_{k_i}^{(i)}\mathbf{T} &= \mathbf{0} \quad = 0\mathbf{w}_1^{(i)} + 0\mathbf{w}_2^{(i)} + 0\mathbf{w}_3^{(i)} + \cdots + 0\mathbf{w}_{k_i}^{(i)}.
\end{aligned}$$

Hence, the matrix of $\mathbf{T} \restriction W_i$ has the form

$$(8.2.4) \qquad B_i = \begin{bmatrix} 0 & 1 & 0 & \ldots & 0\\ 0 & 0 & 1 & \ldots & 0\\ \vdots & \vdots & \vdots & & \vdots\\ 0 & 0 & 0 & \ldots & 1\\ 0 & 0 & 0 & \ldots & 0 \end{bmatrix}$$

with respect to a $\mathbf{T}$-cyclic basis of $W_i$. The matrix 8.2.4 has 1's on the superdiagonal and zeros everywhere else. (The *superdiagonal* of a square matrix $A$ is the diagonal immediately above the main diagonal; it consists of the entries $a_{ij}$ for which $j = i + 1$.) Now the subspaces $W_i$ are $\mathbf{T}$-stable and $V$ is their direct sum. Hence, by Theorem 5.13.13, the matrix of $\mathbf{T}$ (with respect to the basis consisting of the vectors $w_j^{(i)}$, all $i$ and $j$) is

$$(8.2.5) \qquad \begin{bmatrix} B_1 & 0 & \ldots & 0\\ 0 & B_2 & \ldots & 0\\ \vdots & \vdots & \ddots & \vdots\\ 0 & 0 & \ldots & B_r \end{bmatrix} \qquad \text{(zeros everywhere except in the blocks } B_1, B_2, \ldots, B_r\text{)}$$

where the $B_i$ are as in 8.2.4.

As an illustration, if $r = 4$ and the subspaces $W_i$ have dimensions 3, 1, 4, and 2, respectively, then $\mathbf{T}$ has the matrix

$$\left[\begin{array}{ccc|c|cccc|cc}
0 & \mathbf{1} & 0 & & & & & & & \\
0 & 0 & \mathbf{1} & & & & & & & \\
0 & 0 & 0 & \mathbf{0} & & & & & & \\
\hline
 & & & 0 & \mathbf{0} & & & & & \\
\hline
 & & & & 0 & \mathbf{1} & 0 & 0 & & \\
 & & & & 0 & 0 & \mathbf{1} & 0 & & \\
 & & & & 0 & 0 & 0 & \mathbf{1} & & \\
 & & & & 0 & 0 & 0 & 0 & \mathbf{0} & \\
\hline
 & & & & & & & & 0 & \mathbf{1} \\
 & & & & & & & & 0 & 0
\end{array}\right]. \tag{8.2.6}$$

This matrix has 1's and 0's on the superdiagonal (indicated in boldface), zeros elsewhere. The parts of the superdiagonal which lie inside the blocks $B_i$ consist of 1's, but the $r - 1$ superdiagonal entries which lie outside these blocks are zero. It is clear that the general case follows the same pattern. Thus, we have

**8.2.7 Theorem.** If $V$ has finite dimension and $\mathbf{T} \in \text{Lin}(V, V)$ is nilpotent, then there exists a basis of $V$ with respect to which the matrix of $\mathbf{T}$ has 1's and 0's on the superdiagonal and zeros everywhere else. Specifically, the matrix of $\mathbf{T}$ has the form 8.2.5 where the $B_i$ are matrices of the form 8.2.4.

It should be strongly emphasized that the basis with respect to which $\mathbf{T}$ has the matrix 8.2.5 is *not*, in general, the canonical basis, though in particular cases it may be.

Let us now return to the proof of Theorem 8.2.3. It will suffice to prove parts (b) and (c) of 8.2.3, since (a) is an immediate consequence of (c) (*why*?). The proof will be by induction on $m$, where $m$ is the index of nilpotence of $\mathbf{T}$. If $m = 0$, then the theorem is vacuously true ($r = 0$ and $V = \{\mathbf{0}\}$ = the empty direct sum). We may thus assume by induction that $m > 0$ and that the theorem holds for all nilpotent mappings of smaller index than $m$.

Let $\mathbf{T}^*$ be the restriction of $\mathbf{T}$ to Ran $\mathbf{T}$. By Theorem 8.1.4, Ran $\mathbf{T}$ is $\mathbf{T}$-stable and, thus, $\mathbf{T}^*$ belongs to Lin(Ran $\mathbf{T}$, Ran $\mathbf{T}$). Moreover, for any element $\mathbf{u} \in$ Ran $\mathbf{T}$, we have $\mathbf{u} = \mathbf{vT}$ for some $\mathbf{v} \in V$, which implies that

$$\mathbf{u}(\mathbf{T}^*)^{m-1} = \mathbf{u}\mathbf{T}^{m-1} = (\mathbf{vT})\mathbf{T}^{m-1} = \mathbf{v}\mathbf{T}^m = \mathbf{v}\mathbf{O} = \mathbf{0}.$$

Hence, $\mathbf{T}^*$ is nilpotent of smaller index than $m$. By induction, it follows that Theorem 8.2.3 holds for $\mathbf{T}^*$; that is, there exist subspaces $U_1$, ..., $U_s$ of Ran $\mathbf{T}$ such that

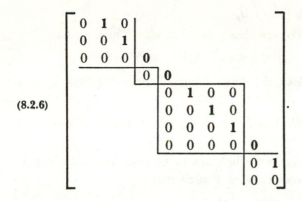

(8.2.8) $\text{Ran } \mathbf{T} = U_1 \oplus \cdots \oplus U_s$

where each $U_i$ has a $\mathbf{T}^*$-cyclic basis $(\mathbf{u}_1^{(i)}, \ldots, \mathbf{u}_{n_i}^{(i)})$. Thus, we have

(8.2.9) $\mathbf{u}_j^{(i)}\mathbf{T} = \mathbf{u}_{j+1}^{(i)} \qquad (1 \leqslant j \leqslant n_i - 1); \qquad \mathbf{u}_{n_i}^{(i)}\mathbf{T} = \mathbf{0}$

for all $i$ $(1 \leqslant i \leqslant s)$. Also, it follows from Theorem 2.8.3(e), in view of 8.2.8, that

(8.2.10) The vectors $\mathbf{u}_j^{(i)}$ $(1 \leqslant i \leqslant s,\ 1 \leqslant j \leqslant n_i)$ constitute a basis of Ran **T**.

Now, for each $i$, the vector $\mathbf{u}_1^{(i)}$ belongs to $U_i$ and, hence, to Ran **T**. Thus, we can choose vectors $\mathbf{w}_i \in V$ such that

(8.2.11) $\mathbf{w}_i\mathbf{T} = \mathbf{u}_1^{(i)} \qquad (i = 1, \ldots, s).$

We claim that

(8.2.12) The vectors $\mathbf{w}_1, \ldots, \mathbf{w}_s$ are independent;

(8.2.13) $[\mathbf{w}_1, \ldots, \mathbf{w}_s] \cap \text{Ran } \mathbf{T} = \{\mathbf{0}\}.$

The proof of 8.2.12 is easy: if the $\mathbf{w}_i$ were dependent, then their images $\mathbf{u}_1^{(i)}$ would be dependent (Theorem 4.1.4(d)), contradicting 8.2.10. To prove 8.2.13, let $\mathbf{v}$ belong to the intersection; then

(8.2.14) $$\mathbf{v} = \sum_{i=1}^{s} a_i\mathbf{w}_i = \sum_{i=1}^{s}\sum_{j=1}^{n_i} b_j^{(i)}\,\mathbf{u}_j^{(i)}$$

for some scalars $a_i$, $b_j^{(i)}$. If we apply **T** to both sides of 8.2.14, then by 8.2.9 and 8.2.11 we get

(8.2.15) $$\sum_{i=1}^{s} a_i\mathbf{u}_1^{(i)} = \sum_{i=1}^{s}\sum_{j=1}^{n_i-1} b_j^{(i)}\,\mathbf{u}_{j+1}^{(i)}.$$

Since the subscripts $j + 1$ on the right side of 8.2.15 are all $> 1$, no vector appears in both the right side and the left side of 8.2.15. Since all of the vectors in 8.2.15 are independent (by 8.2.10), this is possible only if all scalars in 8.2.15 are zero. In particular, all $a_i = 0$ and, hence, $\mathbf{v} = \mathbf{0}$, proving 8.2.13.

It follows from 8.2.10, 8.2.12, and 8.2.13 that all of the vectors

$$\mathbf{u}_j^{(i)} \qquad (\text{all } i, j); \qquad \mathbf{w}_1, \ldots, \mathbf{w}_s$$

are independent. Hence, this set of vectors can be extended to a basis of $V$ by adding certain vectors $\mathbf{v}_{s+1}, \ldots, \mathbf{v}_r$. For each index $h$ $(s + 1 \leqslant h \leqslant r)$, $\mathbf{v}_h\mathbf{T} \in \text{Ran } \mathbf{T}$; hence, there exist scalars $c_j^{(h,i)}$ such that

(8.2.16) $$\mathbf{v}_h\mathbf{T} = \sum_{i=1}^{s}\sum_{j=1}^{n_i} c_j^{(h,i)}\mathbf{u}_j^{(i)} \qquad (s + 1 \leqslant h \leqslant r).$$

Let

(8.2.17) $$\mathbf{w}_h = \mathbf{v}_h - \sum_{i=1}^{s}\sum_{j=1}^{n_i} c_j^{(h,i)}\mathbf{u}_{j-1}^{(i)} \qquad (s+1 \leqslant h \leqslant r)$$

where we define $\mathbf{u}_0^{(i)}$ to be $\mathbf{w}_i$ $(1 \leqslant i \leqslant s)$. Since $\mathbf{T}$ maps $\mathbf{u}_{j-1}^{(i)} \to \mathbf{u}_j^{(i)}$ for all $i, j$, we obtain

$$\mathbf{w}_h\mathbf{T} = \mathbf{v}_h\mathbf{T} - \sum_{i=1}^{s}\sum_{j=1}^{n_i} c_j^{(h,i)}\mathbf{u}_j^{(i)} = \mathbf{0} \qquad (s+1 \leqslant h \leqslant r).$$

Since the vectors $\mathbf{u}_j^{(i)}$, $\mathbf{w}_1, \ldots, \mathbf{w}_s, \mathbf{v}_{s+1}, \ldots, \mathbf{v}_r$ form a basis of $V$, it easily follows from Theorem 2.3.16(a) (using equation 8.2.17) that the vectors $\mathbf{u}_j^{(i)}$, $\mathbf{w}_1, \ldots, \mathbf{w}_s, \mathbf{w}_{s+1}, \ldots, \mathbf{w}_r$ also span $V$; in fact, they must form a basis of $V$ since there are the right number of them. Hence, if we let

$$W_i = [\mathbf{w}_i, \mathbf{u}_1^{(i)}, \ldots, \mathbf{u}_{n_i}^{(i)}] \qquad (1 \leqslant i \leqslant s)$$
$$W_h = [\mathbf{w}_h] \qquad (s+1 \leqslant h \leqslant r),$$

Theorem 2.8.8 implies that $V = W_1 \oplus \cdots \oplus W_s \oplus W_{s+1} \oplus \cdots \oplus W_r$. Moreover, for $1 \leqslant i \leqslant s$ the basis $(\mathbf{w}_i, \mathbf{u}_1^{(i)}, \ldots, \mathbf{u}_{n_i}^{(i)})$ of $W_i$ is $\mathbf{T}$-cyclic by 8.2.11 and 8.2.9, and for $s+1 \leqslant h \leqslant r$ the one-element basis $(\mathbf{w}_h)$ of $W_h$ is $\mathbf{T}$-cyclic since $\mathbf{w}_h\mathbf{T} = \mathbf{0}$. Hence, $\mathbf{T} \restriction W_i$ is nilcyclic for *all* $i$ $(1 \leqslant i \leqslant r)$, and Theorem 8.2.3 is proved. ∎

The preceding proof is constructive in that it shows explicitly how to find the subspaces $W_i$ and their $\mathbf{T}$-cyclic bases, assuming that we have the analogous information about Ran $\mathbf{T}$ in place of $V$. Similarly, the information for Ran $\mathbf{T}$ can be obtained from that for Ran $\mathbf{T}^2$, which in turn can be obtained from that for Ran $\mathbf{T}^3$, and so on. The process is finite in length, since Ran $\mathbf{T}^m = \{\mathbf{0}\}$, where $m$ is the index of $\mathbf{T}$. Thus, a general algorithm for finding the $W_i$ and their $\mathbf{T}$-cyclic bases is as follows:

(8.2.18) (a) Choose any basis $S_0$ of $V$.

(b) Let $S_1$ be a maximal independent subset of $S_0\mathbf{T}$; let $S_2$ be a maximal independent subset of $S_1\mathbf{T}$; and so on, until an integer $m$ is reached such that $S_m$ is empty (equivalently, such that $S_{m-1}\mathbf{T} = \{\mathbf{0}\}$). Then for each $k$ $(0 \leqslant k \leqslant m)$, $S_k$ is a basis of Ran $\mathbf{T}^k$; and $m$ is the index of nilpotence of $\mathbf{T}$. For convenience, let us write

$$S_k = \{\mathbf{x}_1^{(k)}, \ldots, \mathbf{x}_{t_k}^{(k)}\}$$

so that $t_k = \dim \operatorname{Ran} \mathbf{T}^k = \operatorname{rank} \mathbf{T}^k$.

(c) For each value of $k$ from $k = m - 1$ to $k = 0$ (working backwards), we must now express Ran $\mathbf{T}^k$ as a direct sum of subspaces $W_i^{(k)}$ such that each $W_i^{(k)}$ has a **T**-cyclic basis $B_i^{(k)}$. For $k = m - 1$ this is easy: just let

$$B_i^{(m-1)} = \{\mathbf{x}_i^{(m-1)}\} \qquad (1 \leqslant i \leqslant t_{m-1}).$$

For each $k$ in succession ($k = m - 1, m - 2, \ldots, 2, 1$) we then use the method of the proof to obtain the expression for Ran $\mathbf{T}^{k-1}$ from that for Ran $\mathbf{T}^k$. In doing so for fixed $k$, remember that (1) the vectors which correspond to the vectors $\mathbf{w}_1, \ldots, \mathbf{w}_s$ of the proof must be chosen to belong to Ran $\mathbf{T}^{k-1}$, and that (2) the vectors which correspond to the vectors $\mathbf{v}_{s+1}, \ldots, \mathbf{v}_r$ of the proof may be chosen from $S_{k-1}$ and, hence, can be found in finitely many steps.

(d) The process ends when the expression for Ran $\mathbf{T}^0 = V$ is found.

We do not give numerical examples of the above algorithm; however, some of the exercises of this section are related to it.

At various places in the algorithm, the vectors which we obtain are not uniquely determined. In fact, even the subspaces $W_i$ are not uniquely determined by **T**. However, **T** does uniquely determine the *number* of such subspaces, and their *dimensions*, as the next theorem will show. Uniqueness results of this weaker type occur fairly often in algebra; such results usually assert that when a given type of algebraic system is decomposed in a certain way into subsystems which cannot themselves be decomposed further, then these subsystems are unique *up to isomorphism*. When the systems in question are vector spaces, "isomorphic" is, of course, equivalent to "having the same dimension" (*why*?). The interested reader is invited to look up either Ore's theorem on lattices (of which our next theorem is a special case) or the Jordan-Hölder Theorem, for purposes of comparison. See, for example, Hall [5], Chapter 8.

**8.2.19 Theorem.** Let $V$ be finite dimensional, let $\mathbf{T} \in \text{Lin}(V, V)$ be nilpotent, and let $W_1, \ldots, W_r$ be nonzero subspaces of $V$ which satisfy conditions (a), (b), and (c) of Theorem 8.2.3. Then for every positive integer $k$, the number of subspaces $W_i$ having dimension $k$ is equal to

$$\text{rank } \mathbf{T}^{k-1} + \text{rank } \mathbf{T}^{k+1} - 2 \text{ rank } \mathbf{T}^k,$$

and the integer $r$ (the total number of subspaces $W_i$) equals the dimension of Nul **T**. In particular, $r$ and the dimensions of the $W_i$ are uniquely determined by **T**. Moreover, the index of nilpotence of **T** equals the largest of the numbers

$$\dim W_1, \dim W_2, \ldots, \dim W_r.$$

*Proof.* We may assume each $W_i$ has a **T**-cyclic basis $(\mathbf{w}_1^{(i)}, \ldots, \mathbf{w}_{h_i}^{(i)})$; that is,

$$\mathbf{T} : \mathbf{w}_1^{(i)} \to \mathbf{w}_2^{(i)} \to \cdots \to \mathbf{w}_{h_i}^{(i)} \to \mathbf{0} \qquad (i = 1, \ldots, r). \tag{8.2.20}$$

Now apply Theorem 4.6.6 to each of the mappings $\mathbf{T}, \mathbf{T}^2, \ldots$. We deduce that

Ran **T** is spanned by the vectors $\mathbf{w}_j^{(i)}$ such that $j \geqslant 2$;
Ran $\mathbf{T}^2$ is spanned by the vectors $\mathbf{w}_j^{(i)}$ such that $j \geqslant 3$;

and so on; in general, Ran $\mathbf{T}^k$ is spanned by the vectors $\mathbf{w}_j^{(i)}$ such that $j \geqslant k + 1$. Now a subspace $W_i$ will have dimension $\geqslant k$ if and only if the vector $\mathbf{w}_k^{(i)}$ exists. Hence, for each fixed value of $k$,

$$\begin{aligned} &(\text{number of } W_i \text{ with dimension} \geqslant k) \\ &= (\text{number of vectors } \mathbf{w}_k^{(i)},\ 1 \leqslant i \leqslant r) \\ &= (\text{number of } \mathbf{w}_j^{(i)} \text{ such that } j \geqslant k) \\ &\quad - (\text{number of } \mathbf{w}_j^{(i)} \text{ such that } j \geqslant k + 1) \\ &= \dim \operatorname{Ran} \mathbf{T}^{k-1} - \dim \operatorname{Ran} \mathbf{T}^k. \end{aligned} \tag{8.2.21}$$

Hence

$$\begin{aligned} &(\text{number of } W_i \text{ with dimension } k) \\ &= (\text{number of } W_i \text{ with dimension} \geqslant k) \\ &\quad - (\text{number of } W_i \text{ with dimension} \geqslant k + 1) \\ &= (\operatorname{rank} \mathbf{T}^{k-1} - \operatorname{rank} \mathbf{T}^k) - (\operatorname{rank} \mathbf{T}^k - \operatorname{rank} \mathbf{T}^{k+1}) \\ &= \operatorname{rank} \mathbf{T}^{k-1} + \operatorname{rank} \mathbf{T}^{k+1} - 2 \operatorname{rank} \mathbf{T}^k \end{aligned}$$

proving the first assertion of the theorem. Also,

$$\begin{aligned} r &= \text{total number of } W_i \\ &= \text{total number of } W_i \text{ with dimension} \geqslant 1 \\ &= \operatorname{rank} \mathbf{T}^0 - \operatorname{rank} \mathbf{T}^1 \qquad (\text{by } 8.2.21) \\ &= \operatorname{rank} \mathbf{I} - \operatorname{rank} \mathbf{T} \\ &= \dim V - \operatorname{rank} \mathbf{T} \\ &= \dim \operatorname{Nul} \mathbf{T} \qquad (\text{by Theorem } 4.6.7). \end{aligned}$$

Finally, 8.2.20 shows that it takes exactly $h_i$ applications of **T** to map all of the basis vectors of $W_i$ into **0**; hence, if $h = \max(h_i) = \max(\dim W_i)$, it takes exactly $h$ applications of **T** to map all of the basis vectors of $V$ into **0**. Hence, the index of nilpotence of **T** equals $h$, proving the last assertion of the theorem. ∎

Note that since the index of nilpotence equals the dimension of a certain subspace $W_i$ of $V$, this index cannot exceed the dimension of $V$ itself. Thus, we obtain the following corollary:

**8.2.22 Theorem.** (a) If $V$ has finite dimension $n$ and $\mathbf{T} \in \text{Lin}(V, V)$ is nilpotent of index $m$, then $m \leqslant n$. In particular, $\mathbf{T}^n = \mathbf{O}$.

(b) If an $n \times n$ matrix $A$ (over $F$) is nilpotent of index $m$, then $m \leqslant n$ and $A^n = 0$.

(The assertion about $\mathbf{T}^n$ follows from the equations

$$\mathbf{T}^n = \mathbf{T}^m\mathbf{T}^{n-m} = \mathbf{O}\mathbf{T}^{n-m} = \mathbf{O},$$

using the inequality $m \leqslant n$ to show that $\mathbf{T}^{n-m}$ is defined. Theorem 8.2.22(b), of course, follows from 8.2.22(a) via the correspondence between L.T.'s and matrices.)

Theorem 8.2.22(b) shows that we need not compute *all* powers of a matrix $A$ in order to determine whether $A$ is nilpotent; computation of $A^n$, where $n$ is the degree of $A$, will suffice. For example, if

$$A = \begin{bmatrix} 2 & -2 & 3 \\ 2 & -2 & 1 \\ 1 & -1 & 0 \end{bmatrix}$$

then

$$A^3 = \begin{bmatrix} 4 & -4 & 6 \\ 4 & -4 & 2 \\ 2 & -2 & 0 \end{bmatrix};$$

since $A$ is a $3 \times 3$ matrix and $A^3 \neq 0$, it follows that $A$ is not nilpotent. However, if

$$B = \begin{bmatrix} 6 & -3 & 9 \\ 6 & -3 & 9 \\ -2 & 1 & -3 \end{bmatrix}$$

then $B^2 = 0$ and, hence, $B$ is nilpotent.

It is possible to prove Theorem 8.2.22 without referring to Theorem 8.2.19. Some alternate proofs of 8.2.22(a) are suggested in the exercises.

We conclude this section with the following result, which shows that the *characteristic polynomial* of a nilpotent transformation has an especially simple form.

**8.2.23 Theorem.** If $\dim V = n < \infty$ and if $\mathbf{T} \in \text{Lin}(V, V)$ is nilpotent, then $\text{char}(\mathbf{T}; x) = \pm x^n$.

*Proof.* By Theorem 8.2.7, we can choose the matrix of **T** to be upper triangular with zeros on the main diagonal. Hence, by Theorem 5.13.3, we have

$$\text{char}(\mathbf{T}; x) = (0 - x)(0 - x) \cdots (0 - x) = (-x)^n = \pm x^n. \blacksquare$$

In Section 8.5 we shall prove the converse of 8.2.23: if the characteristic polynomial of **T** is $\pm x^n$, then **T** is nilpotent. This provides another way to determine whether a given L.T. (or a given square matrix) is nilpotent.

**Exercises**

1. Use the result of Section 8.1, Exercise 2 to construct an alternate proof of Theorem 8.2.22(a). (SOL)
2. If $\mathbf{T} \in \text{Lin}(V, V)$ is nilpotent of index $m > 0$, then there must exist a vector $\mathbf{u} \in V$ such that $\mathbf{uT}^{m-1} \neq \mathbf{0}$ (why?). If we choose any such $\mathbf{u}$, show that the vectors $\mathbf{u}, \mathbf{uT}, \mathbf{uT}^2, \ldots, \mathbf{uT}^{m-1}$ are linearly independent. Then use this fact to obtain an alternate proof of Theorem 8.2.22(a).
3. In each case, find all values of $a$, $b$, $c$ which make the given matrix nilpotent. (ANS)

   (a) $\begin{bmatrix} 1 & 0 & 2 \\ 2 & 0 & 4 \\ a & 0 & -1 \end{bmatrix}$ (b) $\begin{bmatrix} 2 & 4 \\ a & b \end{bmatrix}$

   (c) $\begin{bmatrix} 0 & 0 & 1 \\ 4 & a & b \\ -2 & 1 & c \end{bmatrix}$ (d) $\begin{bmatrix} 0 & 0 & 1 \\ 0 & 0 & 3 \\ a & b & c \end{bmatrix}$.
4. Let $\dim V = n$ and assume that $\mathbf{T} \in \text{Lin}(V, V)$ is nilpotent. Prove that the following statements are equivalent. (Assume $n \geq 1$.)

   (a) The index of nilpotence of **T** equals $n$ (the dimension of $V$).

   (b) **T** is nilcylic.

   (c) **T** has rank $n - 1$.

   (d) $\dim \text{Nul}\, \mathbf{T} = 1$.

   (e) $V$ cannot be expressed as the direct sum of two proper **T**-stable subspaces.

   (*Suggestion*: Show that (a) $\Rightarrow$ (b) $\Rightarrow$ (c) $\Rightarrow$ (d) $\Rightarrow$ (e) $\Rightarrow$ (a).)
5. Let $\dim V = n$ and assume that $\mathbf{T} \in \text{Lin}(V, V)$ is *nilcyclic*. Prove the following assertions:

   (a) For all integers $k$ from 0 to $n$ (inclusive), the rank of $\mathbf{T}^k$ is $n - k$. (SOL)

(b) For all integers $k$ from 0 to $n$ (inclusive), Ran $\mathbf{T}^k$ = Nul $\mathbf{T}^{n-k}$. (SOL)
(c) If $n \geq 1$, there exists a nonzero vector $\mathbf{u}_n$ in Nul $\mathbf{T}$; for any such vector $\mathbf{u}_n$, if $n \geq 2$, then $\mathbf{u}_n$ has at least one pre-image $\mathbf{u}_{n-1}$ under $\mathbf{T}$; for any such vector $\mathbf{u}_{n-1}$, if $n \geq 3$, then $\mathbf{u}_{n-1}$ has at least one pre-image $\mathbf{u}_{n-2}$ under $\mathbf{T}$; ...; for any such vector $\mathbf{u}_2$, $\mathbf{u}_2$ has at least one pre-image $\mathbf{u}_1$ under $\mathbf{T}$. (SOL)
(d) If $\mathbf{u}_n, \mathbf{u}_{n-1}, ..., \mathbf{u}_1$ are chosen as in part (c), then $(\mathbf{u}_1, \mathbf{u}_2, ..., \mathbf{u}_n)$ is a $\mathbf{T}$-cyclic basis of $V$.

(This gives us one way to find a $\mathbf{T}$-cyclic basis when $\mathbf{T}$ is nilcyclic.)

6. (Generalization of Exercise 5.) Let $\mathbf{T} \in \text{Lin}(V, V)$ be nilpotent of index $m$, and suppose that the integer
$h = \text{rank } \mathbf{T}^{k-1} - \text{rank } \mathbf{T}^k$
is the same for all values of $k$ from 1 to $m$. Prove the following assertions:
(a) For each value of $k$ from 0 to $m$ (inclusive), the rank of $\mathbf{T}^k$ is $h(m - k)$. In particular, dim $V = mh$.
(b) The number $r$ in Theorem 8.2.3 equals $h$, and all of the subspaces $W_1, ..., W_h$ have dimension $m$.
(c) For each value of $k$ from 0 to $m$, Ran $\mathbf{T}^{m-k}$ = Nul $\mathbf{T}^k$ and this space has dimension $hk$.
(d) Let $\{\mathbf{u}_m^{(1)}, ..., \mathbf{u}_m^{(h)}\}$ be any basis of Nul $\mathbf{T}$ (which has dimension $h$ by part (c)). If $m \geq 2$, then each vector $\mathbf{u}_m^{(i)}$ $(1 \leq i \leq h)$ has at least one pre-image $\mathbf{u}_{m-1}^{(i)}$ under $\mathbf{T}$; for any such vector $\mathbf{u}_{m-1}^{(i)}$, if $m \geq 3$, then $\mathbf{u}_{m-1}^{(i)}$ has at least one pre-image $\mathbf{u}_{m-2}^{(i)}$; ...; for any such vector $\mathbf{u}_2^{(i)}$, $\mathbf{u}_2^{(i)}$ has at least one pre-image $\mathbf{u}_1^{(i)}$.
(e) If vectors $\mathbf{u}_k^{(i)}$ $(1 \leq i \leq h, 1 \leq k \leq m)$ are chosen as in part (d) and if we let $W_i = [\mathbf{u}_1^{(i)}, ..., \mathbf{u}_m^{(i)}]$ $(1 \leq i \leq h)$, then for each $i$, the $m$-tuple $(\mathbf{u}_1^{(i)}, ..., \mathbf{u}_m^{(i)})$ is a $\mathbf{T}$-cyclic basis for $W_i$, and these subspaces $W_1, ..., W_h$ satisfy Theorem 8.2.3.
7. In each of the following, $\mathbf{T} \in \text{Lin}(\mathbf{R}_n, \mathbf{R}_n)$ has the given matrix $A$ with respect to the canonical basis. In each case, (a) verify that $\mathbf{T}$ is nilpotent; (b) by computing the ranks of successive powers of $A$, determine the dimensions of the subspaces $W_i$ which satisfy Theorem 8.2.3 (cf. Theorem 8.2.19); (c) find $\mathbf{T}$-cyclic bases for subspaces $W_i$ which satisfy 8.2.3. (Answers to (c) are not unique.) (ANS)

(i) $n = 2$; $A = \begin{bmatrix} 2 & 4 \\ -1 & -2 \end{bmatrix}$.

(ii) $n = 3$; $A = \begin{bmatrix} 1 & 1 & -1 \\ 2 & 2 & -2 \\ 3 & 3 & -3 \end{bmatrix}$.

(iii) $n = 3$; $A = \begin{bmatrix} 1 & 1 & 0 \\ -3 & -2 & 1 \\ -1 & 0 & 1 \end{bmatrix}$.

(iv) $n = 4$; $A = \begin{bmatrix} -1 & 0 & -1 & -1 \\ 1 & 1 & 0 & 1 \\ 1 & 1 & 0 & 1 \\ 0 & -1 & 1 & 0 \end{bmatrix}$.

(v) $n = 4$; $A = \begin{bmatrix} -1 & 1 & -1 & 0 \\ -1 & -3 & 1 & -2 \\ 0 & -6 & 3 & -3 \\ 2 & 0 & 1 & 1 \end{bmatrix}$. (SOL)

(vi) $n = 5$; $A = \begin{bmatrix} 1 & 0 & 0 & -1 & 1 \\ 2 & -1 & 4 & 3 & -2 \\ -1 & 0 & 3 & 4 & -4 \\ 2 & -1 & 0 & -1 & 2 \\ 1 & -1 & 3 & 3 & -2 \end{bmatrix}$.

*8. A square matrix $A$ is called *strictly upper triangular* if it is upper triangular (cf. Sec. 4.5) and its main diagonal consists entirely of zeros. (Equivalently, $A = (a_{ij})$ is strictly upper triangular if $a_{ij} = 0$ whenever $i \geqslant j$.)

(a) If $A$ is strictly upper triangular, what statement (even stronger) can be made about $A^2$? About $A^3$? About $A^k$? (*Suggestion*: Guess the answer after experimenting with some $4 \times 4$ matrices; then prove it formally by induction on $k$.) (ANS)

(b) Deduce that any strictly upper triangular matrix is nilpotent.

9. Interpret Exercise 8 in terms of linear transformations. (SOL)

10. If the dimension of $V$ is finite but nonzero and if $\mathbf{T} \in \operatorname{Lin}(V, V)$ is nilpotent, prove that $\mathbf{T}$ is singular. Show, however,

that the converse is false: **T** may be singular without being nilpotent.

11. (a) If a $3 \times 3$ matrix $A$ is nilpotent of index $m$, prove that the rank of $A$ is exactly $m - 1$.
    (b) Show that a similar statement for $4 \times 4$ matrices is false.
12. Let $S$ and $L$ be (respectively) a plane through $O$ and a line through $O$ in $\mathbf{R}_3$. Suppose also that $\mathbf{T} \in \text{Lin}(\mathbf{R}_3, \mathbf{R}_3)$ is nilpotent. Prove: if $S\mathbf{T} \subseteq L$, then $L\mathbf{T} \subseteq S$.

## 8.3 The (T, λ)-characteristic subspace

Let $\mathbf{T} \in \text{Lin}(V, V)$ and let $\lambda$ be an eigenvalue of $\mathbf{T}$ of multiplicity $m$. (Recall that this means $(x - \lambda)^m$ is a factor of char$(\mathbf{T}; x)$ but $(x - \lambda)^{m+1}$ is not.) The subspace

$$\text{Nul}\,(\mathbf{T} - \lambda\mathbf{I})^m$$

is called the $(\mathbf{T}, \lambda)$-*characteristic subspace* of $V$ and we denote it CH$[\mathbf{T}, \lambda]$. A vector $\mathbf{u} \in V$ belongs to this subspace if and only if $\mathbf{u}(\mathbf{T} - \lambda\mathbf{I})^m = \mathbf{0}$.

**8.3.1** **Example.** Let $\mathbf{T} : \mathbf{R}_3 \to \mathbf{R}_3$ have the matrix

$$A = \begin{bmatrix} 2 & -2 & 3 \\ 10 & -4 & 5 \\ 5 & -4 & 6 \end{bmatrix}$$

with respect to the canonical basis. Since

$$\begin{aligned} \text{char}(\mathbf{T}; x) = \text{char}(A; x) &= -x^3 + 4x^2 - 5x + 2 \\ &= -(x-1)^2(x-2), \end{aligned}$$

1 is an eigenvalue of **T** of multiplicity 2. Hence, the vectors $\mathbf{u} = (x, y, z)$ in the corresponding characteristic subspace can be found by solving the matrix equation $\mathbf{u}(A - I)^2 = 0$. We have

$$(A - I)^2 = \begin{bmatrix} 1 & -2 & 3 \\ 10 & -5 & 5 \\ 5 & -4 & 5 \end{bmatrix}^2 = \begin{bmatrix} -4 & -4 & 8 \\ -15 & -15 & 30 \\ -10 & -10 & 20 \end{bmatrix}$$

The equation $\mathbf{u}(A - I)^2 = 0$ becomes

$$[x \;\; y \;\; z] \begin{bmatrix} -4 & -4 & 8 \\ -15 & -15 & 30 \\ -10 & -10 & 20 \end{bmatrix} = [0 \;\; 0 \;\; 0]$$

which easily reduces to the equation $4x + 15y + 10z = 0$. We recognize the latter as the equation of a plane through $O$ (a 2-dimensional subspace). ∎

The importance of the characteristic subspaces lies primarily in the fact that if the characteristic polynomial of **T** splits then $V$ is actually the direct sum (over all eigenvalues $\lambda$ of **T**) of the $(\mathbf{T}, \lambda)$-characteristic subspaces. In Section 8.4 we shall prove this result and then use it to obtain the Jordan form of a linear transformation. The present section will be devoted to some preliminary results; chief among them is the fact that the dimension of $\mathrm{CH}[\mathbf{T}, \lambda]$ equals the multiplicity of $\lambda$. (In Example 8.3.1 the dimension and the multiplicity were both 2.)

For the remainder of this section, we use the following notations:

$V$ is a vector space of finite dimension over the field $F$.
**T** is a linear transformation of $V \to V$.
$\lambda$ is a fixed eigenvalue of **T**.
$m$ is the multiplicity of $\lambda$.
$\mathbf{S} = \mathbf{T} - \lambda\mathbf{I}$.
$N_k = \mathrm{Nul}\ \mathbf{S}^k = \mathrm{Nul}\ (\mathbf{T} - \lambda\mathbf{I})^k \qquad (\text{all } k \in \mathbf{Z}^+)$.
$d_k = \dim N_k$.

In this notation, the $(\mathbf{T}, \lambda)$-characteristic subspace of $V$ is the subspace $N_m$.

8.3.2 **Lemma.** Each subspace $N_k$ ($k = 1, 2, 3, \ldots$) is both **S**-stable and **T**-stable.

*Proof.* $N_k$ is **S**-stable by Theorem 8.1.4. Also, $N_k$ (being a subspace) is closed under scalar multiplication and, hence, is $\lambda\mathbf{I}$-stable. Since $N_k$ is also closed under addition, it follows that $N_k$ is $(\mathbf{S} + \lambda\mathbf{I})$-stable, that is, **T**-stable. ∎

8.3.3 **Lemma.** For each positive integer $k$, the subspace $\mathrm{Ran}\ \mathbf{S}^k$ is both **S**-stable and **T**-stable.

(*Proof:* similar.)

8.3.4 **Lemma.** $\mathbf{S} \upharpoonright N_k$ is nilpotent of index $\leq d_k$.

*Proof.* First note that $\mathbf{S} \upharpoonright N_k$ belongs to $\mathrm{Lin}(N_k, N_k)$ by Lemma 8.3.2. Since $N_k = \mathrm{Nul}\ \mathbf{S}^k$, $\mathbf{u}\mathbf{S}^k = \mathbf{0}$ for all $\mathbf{u} \in N_k$. Hence, $\mathbf{S}_k \upharpoonright N_k = \mathbf{O}$ so that $\mathbf{S} \upharpoonright N_k$ is nilpotent. The inequality for the index follows from 8.2.22(a). ∎

8.3.5 **Lemma.** For each positive integer $k$, there exists a basis of $N_k$ with respect to which the matrix of $\mathbf{T} \upharpoonright N_k$ has the form

$$\text{(8.3.6)} \qquad \begin{bmatrix} \lambda & \epsilon_1 & & & & \\ & \lambda & \epsilon_2 & & & \\ & & \lambda & & & \\ & & & \ddots & \ddots & \\ & & & & \lambda & \epsilon_{d_k-1} \\ & & & & & \lambda \end{bmatrix}$$

where each $\epsilon_i$ is either 0 or 1 (not necessarily all the same) and all entries not on the main diagonal or superdiagonal are zero. The characteristic polynomial of $\mathbf{T} \restriction N_k$ is given by

$$\text{(8.3.7)} \qquad \operatorname{char}(\mathbf{T} \restriction N_k; x) = \pm(x - \lambda)^{d_k}.$$

*Proof.* By 8.3.4, $\mathbf{S} \restriction N_k$ is nilpotent; hence, by 8.2.7, $\mathbf{S} \restriction N_k$ has a matrix with 1's and 0's on the superdiagonal (the entries $\epsilon_i$ in 8.3.6), zeros elsewhere. Since $\mathbf{T} = \mathbf{S} + \lambda\mathbf{I}$, the matrix of $\mathbf{T} \restriction N_k$ is obtained from that of $\mathbf{S} \restriction N_k$ by adding $\lambda$ to the entries on the main diagonal; this gives us the matrix 8.3.6. Equation 8.3.7 then follows from Theorem 5.13.3. ∎

We now come to the main result of this section. As above, $m$ denotes the multiplicity of $\lambda$.

**8.3.8 Theorem.** (a) For all $k \in \mathbf{Z}^+$, $d_k \leqslant m$. (b) $d_m = m$.

*Proof.* Consider the quotient space $V/N_k$; its elements have the form

$$\mathbf{v}^*_{(k)} = \mathbf{v} + N_k.$$

Since $N_k$ is **T**-stable (Lemma 8.3.2), Theorem 5.13.7 shows that there is an element $\mathbf{T}^*_{(k)}$ of $\operatorname{Lin}(V/N_k, V/N_k)$ defined by

$$\mathbf{v}^*_{(k)}\mathbf{T}^*_{(k)} = (\mathbf{vT})^*_{(k)} \qquad (\text{all } \mathbf{v} \in V)$$

and that

$$\text{(8.3.9)} \qquad \operatorname{char}(\mathbf{T}; x) = \operatorname{char}(\mathbf{T} \restriction N_k; x) \cdot \operatorname{char}(\mathbf{T}^*_{(k)}; x).$$

Substituting 8.3.7 into 8.3.9,

$$\text{(8.3.10)} \qquad \operatorname{char}(\mathbf{T}; x) = \pm(x - \lambda)^{d_k} \operatorname{char}(\mathbf{T}^*_{(k)}; x).$$

Since $m$ is the multiplicity of $\lambda$, the factor $(x - \lambda)$ appears exactly $m$ times in the factorization of $\operatorname{char}(\mathbf{T}; x)$; hence, 8.3.10 implies that $d_k \leqslant m$, proving 8.3.8(a).

To prove 8.3.8(b), suppose $d_m \neq m$. Then $d_m < m$, and 8.3.10 implies that $(x - \lambda)$ must be a factor of char$(\mathbf{T}^*_{(m)}; x)$. Hence, $\lambda$ is an eigenvalue of $\mathbf{T}^*_{(m)}$ so that

$$\mathbf{u}^*_{(m)}\,\mathbf{T}^*_{(m)} = \lambda\mathbf{u}^*_{(m)}$$

for some *nonzero* element $\mathbf{u}^*_{(m)}$ of $V/N_m$. Hence,

$$\begin{aligned}(\mathbf{uT} - \lambda\mathbf{u})^*_{(m)} &= (\mathbf{uT})^*_{(m)} - \lambda\mathbf{u}^*_{(m)} \\ &= \mathbf{u}^*_{(m)}\mathbf{T}^*_{(m)} - \lambda\mathbf{u}^*_{(m)} = \mathbf{0}^*_{(m)} = N_m\end{aligned}$$

which is the same as

(8.3.11) $\mathbf{uT} - \lambda\mathbf{u} \in N_m$.

On the other hand, since $d_m < m$, the mapping $\mathbf{S} \restriction N_m$ is nilpotent of index no greater than $m - 1$ (Lemma 8.3.4). Thus, since the vector $\mathbf{uS} = \mathbf{u}(\mathbf{T} - \lambda\mathbf{I}) = \mathbf{uT} - \lambda\mathbf{u}$ lies in $N_m$ by 8.3.11, it follows that $(\mathbf{uS})\mathbf{S}^{m-1} = \mathbf{0}$. Thus, $\mathbf{uS}^m = \mathbf{0}$, so that $\mathbf{u} \in$ Nul $\mathbf{S}^m = N_m$, from which $\mathbf{u}^*_{(m)} = N_m = \mathbf{0}^*_{(m)}$ But $\mathbf{u}^*_{(m)}$ was chosen to be nonzero, R.A.A. The assumption $d_m \neq m$ having led to a contradiction, it follows that $d_m = m$. ▮

**8.3.12 Corollary.** $N_m = N_{m+1} = N_{m+2} = \cdots$. (That is, $N_m = N_{m+k}$ for all $k \in \mathbf{Z}^+$.)

*Proof.* $N_m \subseteq N_{m+k}$ by 8.1.3 (applied to $\mathbf{S}$ in place of $\mathbf{T}$). But by Theorem 8.3.8 the dimension of $N_{m+k}$ is no larger than that of $N_m$. It follows that $N_m = N_{m+k}$. ▮

**8.3.13 Corollary.** If $\lambda$ is an eigenvalue of $\mathbf{T}$ of multiplicity $m$, then
(a) $V =$ Nul $(\mathbf{T} - \lambda\mathbf{I})^m \oplus$ Ran $(\mathbf{T} - \lambda\mathbf{I})^m$.
(b) The dimension of the $(\mathbf{T}, \lambda)$-eigenspace is $\leq m$.

*Proof.* By 8.3.12, $N_m = N_{2m}$; that is, Nul $(\mathbf{T} - \lambda\mathbf{I})^m =$ Nul $[(\mathbf{T} - \lambda\mathbf{I})^m]^2$. Hence, 8.3.13(a) follows from Theorem 8.1.5. As for 8.3.13(b), the $(\mathbf{T}, \lambda)$-eigenspace is the subspace Nul$(\mathbf{T} - \lambda\mathbf{I}) = N_1$ whose dimension $d_1$ is $\leq m$ by 8.3.8(a). ▮

Using the notation of Corollary 8.3.13, we shall call Ran $(\mathbf{T} - \lambda\mathbf{I})^m$ the $(\mathbf{T}, \lambda)$-*characteristic range* and shall denote it CR$[\mathbf{T}, \lambda]$.

**Exercises**

In each of Exercises 1–5, $\mathbf{T}$ is a linear transformation of $\mathbf{R}_n \to \mathbf{R}_n$ which has the given matrix with respect to the canonical basis.

Find the eigenvalues of **T**; then, for each eigenvalue λ, find the (**T**, λ)-eigenspace, the (**T**, λ)-characteristic subspace, and the (**T**, λ)-characteristic range, and verify Corollary 8.3.13. (ANS)

1. $\begin{bmatrix} 7 & 2 & 6 \\ 14 & 18 & 5 \\ -7 & -6 & 3 \end{bmatrix}.$

2. $\begin{bmatrix} -1 & -6 & 6 & -6 \\ 3 & 6 & -1 & 4 \\ 0 & 3 & -4 & 3 \\ -3 & -1 & -5 & 1 \end{bmatrix}.$

3. $\begin{bmatrix} 0 & 1 & 2 & -1 \\ 0 & -1 & 0 & 0 \\ 1 & 0 & 1 & -1 \\ 1 & 1 & 4 & -2 \end{bmatrix}.$

4. $\begin{bmatrix} -2 & 4 & 2 & -3 \\ 0 & 2 & 0 & 0 \\ -1 & 1 & 1 & -1 \\ 1 & -1 & 0 & 2 \end{bmatrix}.$

5. $\begin{bmatrix} 2 & 1 & 0 \\ -2 & 1 & 2 \\ 1 & 3 & 1 \end{bmatrix}.$

6. Suppose the linear transformation $\mathbf{T} : V \to V$ has distinct eigenvalues $\lambda_1$, $\lambda_2$ of multiplicities $m_1$, $m_2$, respectively. After examining the results of Exercises 1–5 (above), make a conjecture regarding the relation between the ($\mathbf{T}$, $\lambda_1$)-characteristic subspace and the ($\mathbf{T}$, $\lambda_2$)-characteristic range, (a) in the case where dim $V = m_1 + m_2$, and (b) in the case where dim $V > m_1 + m_2$.
7. If $\mathbf{T} \in \text{Lin}(\mathbf{R}_3, \mathbf{R}_3)$ maps $(x, y, z) \to (2x - y - z, 4x + 2y - 4z, 4x + y - 3z)$, then $\text{char}(\mathbf{T}; x) = -(x - 1)(x^2 + 4)$. Find **T**-stable subspaces $W_1$, $W_2$ of $\mathbf{R}_3$ such that $\mathbf{R}_3 = W_1 \oplus W_2$, $\text{char}(\mathbf{T} \restriction W_1; x) = -(x - 1)$, and $\text{char}(\mathbf{T} \restriction W_2; x) = x^2 + 4$.

## 8.4 The Jordan form

Throughout this section $V$ will denote a vector space of finite dimension $n$ over a field $F$, and **T** will denote an element of Lin($V$, $V$). In the preceding section we discussed the characteristic subspace of $V$ corresponding to a fixed eigenvalue λ of **T**. Let us now consider all of the eigenvalues of **T** simultaneously; there is a characteristic subspace corresponding to each eigenvalue. Our first theorem shows that the different characteristic subspaces are *independent* in the sense of Definition 2.8.2.

**8.4.1 Theorem.** Suppose $\lambda_1, \ldots, \lambda_s$ are distinct eigenvalues of **T**, having multiplicities $m_1, \ldots, m_s$, respectively. For each $i$, let $\mathbf{U}_i$ be the linear transformation $(\mathbf{T} - \lambda_i\mathbf{I})^{m_i}$, and let

(8.4.2) $$V^* = \text{Ran}(\mathbf{U}_1\mathbf{U}_2 \cdots \mathbf{U}_s) = V\mathbf{U}_1 \cdots \mathbf{U}_s.$$

Then

(8.4.3) $$V = \text{CH}[\mathbf{T}, \lambda_1] \oplus \cdots \oplus \text{CH}[\mathbf{T}, \lambda_s] \oplus V^*$$

and all $s + 1$ summands in this direct sum are **T**-stable.

*Proof.* The case $s = 1$ is Corollary 8.3.13(a) (both summands being **T**-stable by Lemmas 8.3.2 and 8.3.3). Hence, by induction, we may assume that $s > 1$ and that the theorem is true for $s - 1$ in place of $s$. This implies that

(8.4.4) $$V = \text{CH}[\mathbf{T}, \lambda_1] \oplus \cdots \oplus \text{CH}[\mathbf{T}, \lambda_{s-1}] \oplus V_0$$

where all summands in 8.4.4 are **T**-stable and

(8.4.5) $$V_0 = V\mathbf{U}_1 \cdots \mathbf{U}_{s-1}.$$

By 8.3.7 and 8.3.8(b), the characteristic polynomial of $\mathbf{T} \restriction \text{CH}[\mathbf{T}, \lambda_j]$ is $\pm(x - \lambda_j)^{m_j}$; hence, if we let $\mathbf{T}_0 = \mathbf{T} \restriction V_0$, Theorem 5.13.13 gives

(8.4.6) $$\text{char}(\mathbf{T}; x) = \pm(x - \lambda_1)^{m_1} \cdots (x - \lambda_{s-1})^{m_{s-1}} \cdot \text{char}(\mathbf{T}_0; x).$$

Since $\lambda_s$ is an eigenvalue of **T** of multiplicity $m_s$, equation 8.4.6 implies that $\lambda_s$ is also an eigenvalue of $\mathbf{T}_0$ of multiplicity $m_s$. Thus, we can apply Corollary 8.3.13(a) to $(V_0, \mathbf{T}_0, \lambda_s, m_s)$ in place of $(V, \mathbf{T}, \lambda, m)$. This gives

(8.4.7) $$V_0 = \text{Nul}\,(\mathbf{T}_0 - \lambda_s\mathbf{I})^{m_s} \oplus V^*$$

where both summands are $\mathbf{T}_0$-stable (and, hence, **T**-stable!) and where

$$\begin{aligned} V^* &= \text{Ran}\,(\mathbf{T}_0 - \lambda_s\mathbf{I})^{m_s} = V_0(\mathbf{T}_0 - \lambda_s\mathbf{I})^{m_s} \\ &= V_0(\mathbf{T} - \lambda_s\mathbf{I})^{m_s} \qquad (\text{since } \mathbf{T}, \mathbf{T}_0 \text{ agree on } V_0) \\ &= V_0\mathbf{U}_s = V\mathbf{U}_1 \cdots \mathbf{U}_s \qquad (\text{by 8.4.5}) \end{aligned}$$

which agrees with 8.4.2. Now, by substituting 8.4.7 into 8.4.4, we see that in order to prove 8.4.3 we need only show that

(8.4.8) $$\text{Nul}\,(\mathbf{T}_0 - \lambda_s\mathbf{I})^{m_s} = \text{Nul}\,(\mathbf{T} - \lambda_s\mathbf{I})^{m_s}.$$

But both sides of 8.4.8 have the same dimension $m_s$ (apply 8.3.8(b)

to both $\mathbf{T}_0$ and $\mathbf{T}$), and the left side of 8.4.8 is contained in the right side since $\mathbf{T}_0$ is a restriction of $\mathbf{T}$. Hence, 8.4.8 must hold, and the theorem is proved. ∎

The matrix analog of Theorem 8.4.1 is

**8.4.9 Theorem.** Let dim $V = n$, $\mathbf{T} \in \mathrm{Lin}(V, V)$. Suppose $\lambda_1, \ldots, \lambda_s$ are distinct eigenvalues of $\mathbf{T}$, having multiplicities $m_1, \ldots, m_s$, respectively. Then, with respect to suitable bases, the matrix of $\mathbf{T} \restriction \mathrm{CH}[\mathbf{T}, \lambda_i]$ has the form

$$(8.4.10) \quad J_i = \begin{bmatrix} \lambda_i & \epsilon_1^{(i)} & & & \\ & \lambda_i & \epsilon_2^{(i)} & & \\ & & \lambda_i & \ddots & \\ & & & \ddots & \\ & & & \lambda_i & \epsilon_{m_i-1}^{(i)} \\ & & & & \lambda_i \end{bmatrix}$$

($i = 1, \ldots, s$), where each $\epsilon_j^{(i)}$ is 0 or 1 and all entries not on the diagonal or superdiagonal are zero; and $\mathbf{T}$ itself has the matrix

$$(8.4.11) \quad J = \begin{bmatrix} J_1 & & & & \\ & J_2 & & & \\ & & \ddots & & \\ & & & J_s & \\ & & & & C \end{bmatrix}$$

where $C$ is a square matrix (possibly $0 \times 0$) and all entries outside of the blocks $J_1, \ldots, J_s, C$ are zero.

*Proof.* The form of $J_i$ was obtained in Lemma 8.3.5; by Theorem 8.3.8(b), the degree of $J_i$ is $m_i$. The form of $J$ then follows from Theorems 8.4.1 and 5.13.13. ∎

Of particular importance is the case where the characteristic polynomial of $\mathbf{T}$ splits over $F$ (i.e., equals a product of polynomials of degree 1 over $F$; cf. Sec. 1.16). In this case we can write

$$(8.4.12) \quad \mathrm{char}(\mathbf{T}; x) = (x - \lambda_1)^{m_1} (x - \lambda_2)^{m_2} \cdots (x - \lambda_s)^{m_s}$$

where $\lambda_1, \ldots, \lambda_s$ are distinct elements of $F$. We then have an expression for $V$ of the form 8.4.3. Since dim $\mathrm{CH}[\mathbf{T}, \lambda_i] = m_i$, it follows that dim $V = m_1 + \cdots + m_s +$ dim $V^*$. But 8.4.12 implies that dim $V = m_1 + \cdots + m_s$. Hence, $V^* = \{\mathbf{0}\}$, so that the matrix $C$ in 8.4.11 is the empty matrix. We have, thus, obtained the following result:

**8.4.13 Theorem.** Let $\mathbf{T} \in \mathrm{Lin}(V, V)$ and suppose that the characteristic polynomial of $\mathbf{T}$ satisfies equation 8.4.12 where $\lambda_1, \ldots, \lambda_s$ are distinct elements of $F$. Then $V = \mathrm{CH}[\mathbf{T}, \lambda_1] \oplus \cdots \oplus \mathrm{CH}[\mathbf{T}, \lambda_s]$, and $V$ has a basis with respect to which the matrix of $\mathbf{T}$ is

$$(8.4.14)\quad J = \begin{bmatrix} J_1 & 0 & \ldots & 0 \\ 0 & J_2 & \ldots & 0 \\ \vdots & \vdots & \ddots & \vdots \\ 0 & 0 & \ldots & J_s \end{bmatrix}$$

where the $J_i$ are as in 8.4.10.

A matrix $J$ of the form 8.4.14 (where the $J_i$ are of the form 8.4.10 and the corresponding $\lambda_i$ are distinct) is called a *Jordan matrix*, and the blocks $J_1, \ldots, J_s$ are called the *Jordan blocks* of $J$. Theorem 8.4.13 thus asserts that if $\mathrm{char}(\mathbf{T}; x)$ splits, then the matrix of $\mathbf{T}$ with respect to some basis of $V$ is a Jordan matrix $J$; $J$ is called the *Jordan form* of $\mathbf{T}$. (The corresponding basis of $V$ is called a *Jordan basis*.) As we will show later in the section, the Jordan form of a given linear transformation is essentially unique. A general method for computing it is described below.

We recall from Section 4.8 that if a linear transformation $\mathbf{T} : V \to V$ has the matrix $A$ with respect to one basis of $V$ and the matrix $A'$ with respect to another basis of $V$, then $A$ and $A'$ are *similar*; that is, $A' = P^{-1}AP$ for some nonsingular matrix $P$. Thus, we obtain the following matrix analog of the last part of Theorem 8.4.13:

**8.4.15 Theorem.** If $A$ is a square matrix over $F$ whose characteristic polynomial splits over $F$, then $A$ is similar (over $F$) to a Jordan matrix with entries in $F$.

**8.4.16 Corollary.** If $A$ is *any* square matrix over $F$, then there exists a field $L$ containing $F$, and a Jordan matrix $J$ with entries in $L$, such that $A$ is similar to $J$ over $L$.

The Corollary follows from the fact that any polynomial over $F$ will split over some field $L$ containing $F$ (see [10], Sec. 10.3). If $F = \mathbf{R}$ we can take $L$ to be $\mathbf{C}$ (the complex numbers), by the Fundamental Theorem of Algebra. For example, if

$$A = \begin{bmatrix} 2 & -5 \\ 1 & -2 \end{bmatrix},$$

the polynomial $\mathrm{char}(A; x) = x^2 + 1$ does not split over $\mathbf{R}$, but over $\mathbf{C}$

it equals $(x - i)(x + i)$. Hence, $A$ is similar *over* $\mathbf{C}$ to the Jordan matrix

$$J = \begin{bmatrix} i & 0 \\ 0 & -i \end{bmatrix}$$

(note that the Jordan blocks here are of degree 1). In fact, you should verify by computation that if

$$P = \begin{bmatrix} 2+i & 2-i \\ 1 & 1 \end{bmatrix}$$

then $J = P^{-1}AP$.

**Computation of the Jordan form**

We use the notations of Theorems 8.4.9 and 8.4.13. To compute the Jordan form of $\mathbf{T}$, first factor char($\mathbf{T}$; $x$) to obtain the scalars $\lambda_i$ and the multiplicities $m_i$. Next, the scalars $\epsilon_j^{(i)}$ which appear in the superdiagonal of each Jordan block $J_i$ (see 8.4.10) must be determined. To do so, let us recall where the matrix 8.4.10 came from in the first place. Consulting the proof of Theorem 8.4.9, we find a reference to Lemma 8.3.5, whose proof in turn refers to Theorem 8.2.7, and so on. Ultimately, we find that $J_i$ is obtained by adding $\lambda_i$ to the diagonal entries of a matrix of the form 8.2.5. That is,

$$(8.4.17)\quad J_i = \begin{bmatrix} B_1^{(i)} & & & \\ & B_2^{(i)} & & \\ & & \ddots & \\ & & & B_{r_i}^{(i)} \end{bmatrix} \qquad \text{(an } m_i \times m_i \text{ matrix)}$$

where

$$(8.4.18)\quad B_j^{(i)} = \begin{bmatrix} \lambda_i & 1 & & & & \\ & \lambda_i & 1 & & & \\ & & \lambda_i & & & \\ & & & \ddots & \ddots & \\ & & & & \lambda_i & 1 \\ & & & & & \lambda_i \end{bmatrix}.$$

The $B_j^{(i)}$ are called the *Jordan subblocks* of $J_i$. Referring again to previous proofs, we see that the decomposition of a given Jordan block $J_i$ into subblocks $B_1^{(i)}, \ldots, B_{r_i}^{(i)}$ corresponds to the decomposition of the $(\mathbf{T}, \lambda_i)$-characteristic subspace into a direct sum of subspaces

$$(8.4.19)\quad \mathrm{CH}[\mathbf{T}, \lambda_i] = W_1^{(i)} \oplus \cdots \oplus W_{r_i}^{(i)}$$

such that $(\mathbf{T} - \lambda_i\mathbf{I}) \restriction W_j^{(i)}$ is nilcyclic.

For the time being, let $i$ be fixed. It is obvious (since addition is commutative!) that the order in which we write the subspaces $W_1^{(i)}$,

..., $W_{r_i}^{(i)}$ in equation 8.4.19 does not matter; the $W_j^{(i)}$ (for fixed $i$) may be permuted at will. Hence, the same is true of the subblocks $B_j^{(i)}$ in 8.4.17; if we permute them, the resulting matrix still corresponds to the same linear transformation as before, but the correspondence is with respect to a different ordering of the basis vectors. Except for such permutations, $J_i$ is clearly determined as soon as we know *how many* $B_j^{(i)}$ *of each degree there are*. Now the number of subblocks $B_j^{(i)}$ of degree $k$ equals the number of subspaces $W_j^{(i)}$ of dimension $k$, which by Theorem 8.2.19 is equal to

$$\begin{aligned} &\operatorname{rank}[(\mathbf{T} - \lambda_i\mathbf{I})^{k-1} \upharpoonright CH[\mathbf{T}, \lambda_i]] \\ (8.4.20)\quad &+ \operatorname{rank}[(\mathbf{T} - \lambda_i\mathbf{I})^{k+1} \upharpoonright CH[\mathbf{T}, \lambda_i]] \\ &- 2 \operatorname{rank} [(\mathbf{T} - \lambda_i\mathbf{I})^{k} \upharpoonright CH[\mathbf{T}, \lambda_i]]. \end{aligned}$$

The expression 8.4.20 can be simplified as follows: for every $k$, the space Nul $(\mathbf{T} - \lambda_i\mathbf{I})^k$ is contained in the space $CH[\mathbf{T}, \lambda_i] = \text{Nul}(\mathbf{T} - \lambda_i\mathbf{I})^{m_i}$ (use 8.1.3, if $k \leq m_i$; use 8.3.12, if $k > m_i$). It follows (how?) that

$$(8.4.21)\quad \text{Nul } (\mathbf{T} - \lambda_i\mathbf{I})^k = \text{Nul } [(\mathbf{T} - \lambda_i\mathbf{I})^k \upharpoonright CH[\mathbf{T}, \lambda_i]].$$

By Theorem 4.6.7, the dimensions of the left and right sides of equation 8.4.21 are, respectively, $n - \operatorname{rank} (\mathbf{T} - \lambda_i\mathbf{I})^k$ and $m_i - \operatorname{rank}[(\mathbf{T} - \lambda_i\mathbf{I})^k \upharpoonright CH[\mathbf{T}, \lambda_i]]$. Hence,

$$(8.4.22)\quad \operatorname{rank}[(\mathbf{T} - \lambda_i\mathbf{I})^k \upharpoonright CH[\mathbf{T}, \lambda_i]] = m_i - n + \operatorname{rank} (\mathbf{T} - \lambda_i\mathbf{I})^k.$$

If we substitute 8.4.22 into each of the terms of 8.4.20, the $m_i$'s and $n$'s cancel (since as many of them are subtracted as are added!), and we obtain

**8.4.23 Theorem.** Let $V$, $\mathbf{T}$, $\lambda_i$, $m_i$, $J_i$ be as in Theorem 8.4.9. Then each matrix $J_i$ has the form 8.4.17, where the $B_j^{(i)}$ are as in 8.4.18. For each positive integer $k$, the number of subblocks $B_j^{(i)}$ of degree $k$ appearing in 8.4.17 is equal to

$$(8.4.24)\quad \operatorname{rank} (\mathbf{T} - \lambda_i\mathbf{I})^{k-1} + \operatorname{rank} (\mathbf{T} - \lambda_i\mathbf{I})^{k+1} - 2 \operatorname{rank} (\mathbf{T} - \lambda_i\mathbf{I})^{k}.$$

Computation of the ranks in 8.4.24 is straightforward; indeed, if $\{\mathbf{v}_1, ..., \mathbf{v}_n\}$ is a basis of $V$, then rank $(\mathbf{T} - \lambda_i\mathbf{I})^k$ equals the number of independent vectors among the set $\{\mathbf{v}_1(\mathbf{T} - \lambda_i\mathbf{I})^k, ..., \mathbf{v}_n(\mathbf{T} - \lambda_i\mathbf{I})^k\}$. However, a slightly simpler method of finding the subblocks $B_j^{(i)}$ will suffice if $m_i \leq 3$; it is based on the fact that if $m_i \leq 3$, then the degrees of the subblocks are uniquely determined by the number of 1's on the superdiagonal of $J_i$ (see Exercise 2 below). The latter number, in turn, is given by

8.4.25 **Theorem.** Let hypotheses and notations be as in Theorem 8.4.9. Then the number of 1's on the superdiagonal of the block $J_i$ is equal to $m_i - n + \operatorname{rank}(\mathbf{T} - \lambda_i\mathbf{I})$.

*Proof.* By 8.4.10, the matrix $J_i - \lambda_i I$ consists of the scalars $\epsilon_1^{(i)}, \ldots, \epsilon_{m_i-1}^{(i)}$ on the superdiagonal and *zeros everywhere else*. It easily follows that the number of 1's on the superdiagonal equals the number of independent rows of $J_i - \lambda_i I$, which by definition is the rank of $J_i - \lambda_i I$ and, hence, is the rank of $(\mathbf{T} - \lambda_i\mathbf{I}) \restriction \operatorname{CH}[\mathbf{T}, \lambda_i]$. The theorem now follows by setting $k = 1$ in equation 8.4.22. ∎

The procedures described above will suffice to compute the Jordan form of any transformation $\mathbf{T} \in \operatorname{Lin}(V, V)$ whose characteristic polynomial splits. Even if char $\mathbf{T}$ does not split, the blocks $J_1, \ldots, J_s$ in 8.4.11 can be obtained in this manner but not the additional block $C$. A method for computing the block $C$ is suggested in Exercise 4. (The block $C$, unlike the blocks $J_1, \ldots, J_s$, is *not* uniquely determined by $\mathbf{T}$.)

8.4.26 **Example.** If $\mathbf{T} \in \operatorname{Lin}(\mathbf{R}_4, \mathbf{R}_4)$ has the matrix

$$A = \begin{bmatrix} 0 & 0 & 0 & 1 \\ 1 & 0 & 0 & 1 \\ -3 & 2 & 1 & 1 \\ 3 & -6 & 1 & 4 \end{bmatrix}$$

with respect to the canonical basis, find the Jordan form of $\mathbf{T}$.

*Solution.* By computation,

$$\operatorname{char}(A; x) = x^4 - 5x^3 + 6x^2 + 4x - 8 = (x - 2)^3 (x + 1).$$

This polynomial splits; the eigenvalues are $\lambda_1 = 2$, $\lambda_2 = -1$, with multiplicities $m_1 = 3$ and $m_2 = 1$, respectively. Hence, the Jordan form is

$$\left[\begin{array}{ccc|c} 2 & \epsilon_1^{(1)} & 0 & 0 \\ 0 & 2 & \epsilon_2^{(1)} & 0 \\ 0 & 0 & 2 & 0 \\ \hline 0 & 0 & 0 & -1 \end{array}\right].$$

To compute the numbers $\epsilon_j^{(1)}$, apply Theorem 8.4.25 to the $3 \times 3$ Jordan block corresponding to the eigenvalue $\lambda_1 = 2$. Since the matrix

$$A - \lambda_1 I = A - 2I = \begin{bmatrix} -2 & 0 & 0 & 1 \\ 1 & -2 & 0 & 1 \\ -3 & 2 & -1 & 1 \\ 3 & -6 & 1 & 2 \end{bmatrix}$$

has rank 3, the number of $\epsilon_j^{(1)}$ equal to 1 is

$$m_1 - n + \text{rank}(\mathbf{T} - 2\mathbf{I}) = 3 - 4 + 3 = 2.$$

Hence, $\epsilon_1^{(1)}, \epsilon_2^{(1)}$ are both 1, and the Jordan form of **T** is

$$\begin{bmatrix} 2 & 1 & 0 & 0 \\ 0 & 2 & 1 & 0 \\ 0 & 0 & 2 & 0 \\ 0 & 0 & 0 & -1 \end{bmatrix}. \blacksquare$$

**8.4.27 Example.** If $\mathbf{T} \in \text{Lin}(\mathbf{R}_5, \mathbf{R}_5)$ has characteristic polynomial $-(x-3)^4(x+2)$ and if $\text{rank}(\mathbf{T} - 3\mathbf{I}) = 2$, what is the Jordan form of **T**?

*Solution.* The Jordan form is clearly

$$\left[\begin{array}{cccc|c} 3 & \epsilon_1^{(1)} & 0 & 0 & 0 \\ 0 & 3 & \epsilon_2^{(1)} & 0 & 0 \\ 0 & 0 & 3 & \epsilon_3^{(1)} & 0 \\ 0 & 0 & 0 & 3 & 0 \\ \hline 0 & 0 & 0 & 0 & -2 \end{array}\right]$$

where the number of 1's among $\epsilon_1^{(1)}, \epsilon_2^{(1)}, \epsilon_3^{(1)}$ is equal to

$$m_1 - n + \text{rank}(\mathbf{T} - 3\mathbf{I}) = 4 - 5 + 2 = 1.$$

No matter which one of the three entries $\epsilon_j^{(1)}$ is taken to be 1, the effect will be to decompose the $4 \times 4$ Jordan block $J_1$ into three subblocks of degrees 2, 1, 1, the only difference being the order in which the subblocks occur. Since the subblocks may be permuted at will, it follows that any one of the three possible choices will give the correct Jordan form. ∎

**8.4.28 Example.** Same as Example 8.4.27 except that $\text{rank}(\mathbf{T} - 3\mathbf{I}) = 3$.

*Solution.* Here the number of epsilons equal to 1 is

$$4 - 5 + 3 = 2.$$

If the two 1's occur in adjacent rows, the $4 \times 4$ Jordan block is decomposed into subblocks of degrees 3 and 1. If instead the two 1's are separated by a zero, then the subblocks have degrees 2, 2. Hence, there are two essentially different possibilities for the Jordan form, namely,

$$(8.4.29)\quad (a)\ \left[\begin{array}{ccc:c|c} 3 & 1 & 0 & 0 & 0 \\ 0 & 3 & 1 & 0 & 0 \\ 0 & 0 & 3 & 0 & 0 \\ \hdashline 0 & 0 & 0 & 3 & 0 \\ \hline 0 & 0 & 0 & 0 & -2 \end{array}\right]; \qquad (b)\ \left[\begin{array}{cc:cc|c} 3 & 1 & 0 & 0 & 0 \\ 0 & 3 & 0 & 0 & 0 \\ \hdashline 0 & 0 & 3 & 1 & 0 \\ 0 & 0 & 0 & 3 & 0 \\ \hline 0 & 0 & 0 & 0 & -2 \end{array}\right].$$

In any actual case, we can determine which of the two possibilities occurs by computing the rank $r^*$ of $(\mathbf{T} - 3\mathbf{I})^2$. Indeed, Theorem 8.4.23 (with $k = 1$, $\lambda_1 = 3$, $m_1 = 4$) shows that the $4 \times 4$ Jordan block $J_1$ has exactly $5 + r^* - 6 = r^* - 1$ subblocks of degree 1. Hence, we have case (a) of 8.4.29 if $r^* = 2$, case (b) if $r^* = 1$. ∎

**8.4.30 Example.** The matrix

$$A = \begin{bmatrix} 0 & 3 & -2 & 0 \\ 1 & -2 & 0 & 2 \\ 2 & 0 & 0 & 0 \\ 1 & -4 & 1 & 2 \end{bmatrix}$$

has characteristic polynomial $(x^2 + 1)(x^2 + 4)$. Let $\mathbf{T} \in \mathrm{Lin}(\mathbf{R}_4, \mathbf{R}_4)$ be the L.T. which corresponds to this matrix with respect to the canonical basis of $\mathbf{R}_4$. Find $\mathbf{T}$-stable subspaces $W_1$, $W_2$ of $\mathbf{R}_4$ such that $\mathbf{R}_4 = W_1 \oplus W_2$, $\mathrm{char}(\mathbf{T} \restriction W_1; x) = x^2 + 1$, $\mathrm{char}(\mathbf{T} \restriction W_2; x) = x^2 + 4$. (Cf. remarks at the end of Sec. 5.13.)

*Solution.* The trick is to do most of the work over the larger field $\mathbf{C}$ (over which the given polynomial splits), rather than over $\mathbf{R}$. (This type of approach is useful in many situations; we shall use it in Sec. 8.5 to prove the Cayley-Hamilton Theorem, and in Chapter 9 to classify the isometries of $\mathbf{R}_n$.) We proceed as follows: let $\mathbf{T}^*$ be the element of $\mathrm{Lin}(\mathbf{C}_4, \mathbf{C}_4)$ which has $A$ as its matrix. Then

$$\begin{aligned} \mathrm{char}(\mathbf{T}^*; x) &= \mathrm{char}(A; x) = (x^2 + 1)(x^2 + 4) \\ &= (x - i)(x + i)(x - 2i)(x + 2i). \end{aligned}$$

Note that $\mathbf{T}$ is the restriction of $\mathbf{T}^*$ to $\mathbf{R}_4$ (we may regard $\mathbf{R}_4$ as a subset of $\mathbf{C}_4$ since $\mathbf{R} \subseteq \mathbf{C}$). Now the eigenvalues of $\mathbf{T}^*$ occur in conjugate pairs; consider one such pair, say, the pair $(i, -i)$. Computing the corresponding characteristic subspaces of $\mathbf{C}_4$ in the usual way, we obtain

$$\begin{aligned} &\mathrm{CH}[\mathbf{T}^*, i] = [\mathbf{v}_1], \qquad \text{where } \mathbf{v}_1 = (1, i - 2, 0, 2); \\ &\mathrm{CH}[\mathbf{T}^*, -i] = [\mathbf{v}_2], \qquad \text{where } \mathbf{v}_2 = (1, -i - 2, 0, 2). \end{aligned}$$

Let $W_1^*$ be the sum of these two subspaces:

$$W_1^* = \mathrm{CH}[\mathbf{T}^*, i] \oplus \mathrm{CH}[\mathbf{T}^*, -i] = [\mathbf{v}_1, \mathbf{v}_2]$$

(the sum is direct by Theorem 8.4.1). Then $\mathbf{T}^* \restriction W_1^*$ has characteristic polynomial $(x - i)(x + i) = x^2 + 1$. We now try to find a basis of $W_1^*$ (over **C**) consisting of *real* vectors, that is, vectors in $\mathbf{R}_4$. By inspection, the real vectors

$$\mathbf{w}_1 = (\mathbf{v}_1 + \mathbf{v}_2)/2 = (1, -2, 0, 2)$$
$$\mathbf{w}_2 = (\mathbf{v}_1 - \mathbf{v}_2)/2i = (0, 1, 0, 0)$$

fill the bill: they are independent and belong to $[\mathbf{v}_1, \mathbf{v}_2] = W_1^*$ and, hence, span $W_1^*$ over **C**. Let $W_1$ be the subspace of $\mathbf{R}_4$ spanned by $\mathbf{w}_1$, $\mathbf{w}_2$. Then $W_1 \subseteq W_1^* \cap \mathbf{R}_4$; moreover, $W_1^* \cap \mathbf{R}_4$ has dimension at most 2 over **R** by Corollary 5.9.4. Hence, $W_1 = W_1^* \cap \mathbf{R}_4$. Since $W_1^*$ is $\mathbf{T}^*$-stable, it follows (how?) that $W_1$ is **T**-stable. Clearly, the matrix of $\mathbf{T} \restriction W_1$ with respect to the basis $(\mathbf{w}_1, \mathbf{w}_2)$ of $W_1$ coincides with the matrix of $\mathbf{T}^* \restriction W_1^*$ with respect to the basis $(\mathbf{w}_1, \mathbf{w}_2)$ of $W_1^*$; hence,

$$\text{char}(\mathbf{T} \restriction W_1; x) = \text{char}(\mathbf{T}^* \restriction W_1^*; x) = x^2 + 1.$$

The subspace $W_2$ is obtained similarly: we have

$$W_2^* = \text{CH}[\mathbf{T}^*, 2i] \oplus \text{CH}[\mathbf{T}^*, -2i] = [\mathbf{v}_3, \mathbf{v}_4]$$

where

$$\mathbf{v}_3 = (0, 2i - 2, -i, 2); \qquad \mathbf{v}_4 = (0, -2i - 2, i, 2);$$

and then

$$W_2 = [\mathbf{w}_3, \mathbf{w}_4];$$
$$\mathbf{w}_3 = \frac{\mathbf{v}_3 + \mathbf{v}_4}{2} = (0, -2, 0, 2);$$
$$\mathbf{w}_4 = (\mathbf{v}_3 - \mathbf{v}_4)/2i = (0, 2, -1, 0);$$

and

$$\text{char}(\mathbf{T} \restriction W_2; x) = \text{char}(\mathbf{T}^* \restriction W_2^*; x) = x^2 + 4.$$

Finally, since the sum of all the subspaces $\text{CH}[\mathbf{T}^*, \lambda_i]$ is a direct sum (Theorem 8.4.1), the sum of $W_1^*$ and $W_2^*$ is direct and, hence, so is the sum of $W_1$ and $W_2$ since $W_i \subseteq W_i^*$. It follows that $\dim(W_1 + W_2) = 2 + 2 = 4$, $W_1 \oplus W_2 = \mathbf{R}_4$. ∎

### Uniqueness of the Jordan form

Let $\mathbf{T} \in \text{Lin}(V, V)$ and suppose that the matrix $J$ of **T** with respect to some ordered basis $B$ of $V$ has the form 8.4.14, where the blocks $J_i$ have the form 8.4.10 and $m_i$ is the degree of $J_i$ ($i = 1, \ldots, s$). Then by Theorem 5.13.3, the characteristic polynomial of **T** is given by 8.4.12. Moreover, the matrix $J_i - \lambda_i I$ is nilpotent (cf. Sec. 8.2, Exercise 8, for instance); hence, by Theorem 8.2.22(b), $(J_i - \lambda_i I)^{m_i} =$

0; hence, the $m_i$ basis vectors corresponding to the submatrix $J_i$ of $J$ are mapped into zero by $(\mathbf{T} - \lambda_i\mathbf{I})^{m_i}$ and, hence, belong to CH[**T**, $\lambda_i$]. It follows by a dimension argument that these $m_i$ vectors span CH[**T**, $\lambda_i$] so that $J_i$ is actually the matrix of $\mathbf{T} \restriction$ CH[**T**, $\lambda_i$]. An examination of the proof of Theorem 8.4.23 then shows that the subblocks $B_j^{(i)}$ of $J_i$ are uniquely determined (except for their order) by the subspace CH[**T**, $\lambda_i$], which itself is uniquely determined by **T**. We thus conclude: *the Jordan form of* **T**, *if it exists, is uniquely determined by* **T** *except for* (1) *the order in which the Jordan blocks* $J_1, \ldots, J_s$ *appear, and* (2) *the order in which the Jordan sub-blocks of each block appear within the block.*

**Diagonal Jordan blocks**

We recall that a square matrix is called a *diagonal matrix* if all of its entries are zero except on the main diagonal. If $\mathbf{T} : V \to V$ is an L.T. whose matrix, with respect to some ordered basis $(\mathbf{u}_1, \ldots, \mathbf{u}_n)$ of $V$, is diagonal, say

$$\begin{bmatrix} \lambda_1 & & \\ & \ddots & \\ & & \lambda_n \end{bmatrix}, \tag{8.4.31}$$

then each $\mathbf{u}_i$ is an eigenvector of **T** (in fact, $\mathbf{u}_i\mathbf{T} = \lambda_i\mathbf{u}_i$). Geometrically, we may think of **T** as "stretching" all line segments through the origin by a factor of $\lambda_1$ in the $\mathbf{u}_1$-direction, a factor of $\lambda_2$ in the $\mathbf{u}_2$-direction, ..., a factor of $\lambda_n$ in the $\mathbf{u}_n$-direction. (Try to visualize this!) Conversely, if $V$ has a basis $(\mathbf{u}_1, \ldots, \mathbf{u}_n)$ consisting of eigenvectors of **T**, then $\mathbf{u}_i\mathbf{T} = \lambda_i\mathbf{u}_i$ $(i = 1, \ldots, n)$ for some scalars $\lambda_1, \ldots, \lambda_n$, and the matrix of **T** with respect to this basis is the diagonal matrix 8.4.31.

The following theorem exhibits a sufficient (though not a necessary) condition for **T** to have a diagonal matrix:

**8.4.32 Theorem.** If char **T** splits and each root of char **T** has multiplicity 1, then the Jordan form of **T** exists and is diagonal.

*Proof.* The Jordan form $J$ exists by Theorem 8.4.13. Since each eigenvalue $\lambda_i$ has multiplicity 1, each Jordan block $J_i$ is a $1 \times 1$ matrix (hence, automatically diagonal!), so that $J$ is diagonal. ∎

Even if $J$ itself is not diagonal, some of the Jordan blocks $J_i$ may be diagonal. The following theorem is relevant to this situation.

8.4.33 **Theorem.** Let $V$, $\mathbf{T}$, $\lambda_i$, $m_i$, $n$, $J_i$ be as in Theorem 8.4.9. Then for any fixed $i$ ($1 \leq i \leq s$), the following statements are equivalent:

(a) The Jordan block $J_i$ is a diagonal matrix.
(b) The $(\mathbf{T}, \lambda_i)$-eigenspace has dimension $m_i$.
(c) The $(\mathbf{T}, \lambda_i)$-eigenspace equals the $(\mathbf{T}, \lambda_i)$-characteristic subspace.

*Proof.* (a) holds $\Leftrightarrow$ all of the entries $\epsilon_j^{(i)}$ in 8.4.10 are zero

$\Leftrightarrow m_i - n + \text{rank}(\mathbf{T} - \lambda_i\mathbf{I}) = 0$ (by Theorem 8.4.25)

$\Leftrightarrow \text{rank}(\mathbf{T} - \lambda_i\mathbf{I}) = n - m_i$

$\Leftrightarrow \dim \text{Nul}(\mathbf{T} - \lambda_i\mathbf{I}) = m_i$ (by Theorem 4.6.7)

which is the assertion (b). The equivalence of (b) and (c) follows from 8.1.3 and 8.3.8(b). ∎

**Exercises**

1. Deduce from one of the theorems of this section that if $\lambda_1, \ldots, \lambda_s$ are distinct eigenvalues of $\mathbf{T}$ and $U_i$ is the $(\mathbf{T}, \lambda_i)$-eigenspace ($i = 1, \ldots, s$), then the subspaces $U_1, \ldots, U_s$ are independent. In particular, if $\mathbf{u}_i \in U_i$ for each $i$ and the $\mathbf{u}_i$ are not zero, then the vectors $\mathbf{u}_1, \ldots, \mathbf{u}_s$ are linearly independent. (SOL)
2. Let $J_i$, $B_j^{(i)}$, $m_i$, $r_i$ be as in equations 8.4.17 and 8.4.18 ($m_i$ is the degree of $J_i$; $r_i$ is the number of subblocks $B_j^{(i)}$ of $J_i$). *For fixed i*, show that:
(a) The number of 1's on the superdiagonal of $J_i$ is $m_i - r_i$.
(b) If $m_i \leq 3$, the *degrees* of the subblocks $B_j^{(i)}$ are uniquely determined by the number $r_i$ of such subblocks. (In view of (a), this implies that the degrees are uniquely determined by the number of 1's on the superdiagonal of $J_i$.) Deduce that for fixed $m_i \leq 3$ there are exactly $m_i$ nonequivalent possibilities for $J_i$. Write out these different possibilities for $J_i$ when $m_i = 1$; when $m_i = 2$; when $m_i = 3$.
(c) The result of part (b) is false when $m_i > 3$. In particular, there are five essentially different possibilities for $J_i$ when $m_i = 4$, and seven possibilities when $m_i = 5$. Write them out.
3. Let $\lambda_i$ be an eigenvalue of $\mathbf{T}$ of multiplicity $m_i \leq 6$, let $N^{(i)}$ be the corresponding characteristic subspace, and let $J_i$ be the corresponding Jordan block. Prove: $J_i$ is uniquely determined if the rank of $\mathbf{T} - \lambda_i\mathbf{I}$ and the index of nilpotence of $(\mathbf{T} - \lambda_i\mathbf{I}) \restriction N^{(i)}$ are known. Show, however, that this assertion is false if $m_i = 7$. (SOL)

4. Describe how to find a basis of the subspace $V^*$ in Theorem 8.4.1 if a basis of $V$ is known (and, of course, if $\mathbf{T}$ and the $\lambda_i$ are known). Hence, describe how to find the block $C$ in equation 8.4.11. (ANS)
5. In each of the following, $\mathbf{T} \in \text{Lin}(F_4, F_4)$ and $A$ is the matrix of $\mathbf{T}$ with respect to the canonical basis. Find the Jordan form of $\mathbf{T}$, if it exists. If the Jordan form does not exist, find a matrix of $\mathbf{T}$ in the form 8.4.11.

   (a) $F = \mathbf{R}$; $A = \begin{bmatrix} 2 & 3 & 2 & 3 \\ 2 & 0 & -1 & 1 \\ 1 & 1 & 1 & 1 \\ -3 & -2 & -1 & -3 \end{bmatrix}$.

   (b) $F = \mathbf{C}$; $A = \begin{bmatrix} 0 & 0 & 1 & 1 \\ 0 & 0 & 1 & 0 \\ 0 & -1 & 0 & 0 \\ -1 & 0 & 0 & 0 \end{bmatrix}$. (Note that $F = \mathbf{C}$, not $\mathbf{R}$.) (ANS)

   (c) $F = \mathbf{R}$; $A = \begin{bmatrix} 2 & 1 & 1 & 3 \\ 4 & 0 & 2 & 4 \\ 4 & -3 & 3 & 1 \\ 0 & -1 & -1 & -1 \end{bmatrix}$.

   (d) $F = \mathbf{R}$; $A = \begin{bmatrix} 2 & 1 & 0 & -1 \\ -1 & 2 & 1 & 0 \\ 0 & 1 & 2 & -1 \\ -1 & 0 & 1 & 2 \end{bmatrix}$.

   (e) $F = \mathbf{R}$; $A = \begin{bmatrix} 0 & 2 & 0 & 0 \\ -1 & 2 & 0 & -2 \\ 0 & 1 & 2 & 2 \\ 0 & 1 & 0 & 4 \end{bmatrix}$. (ANS)
6. For each part of Exercise 5 in which the Jordan form exists, find a corresponding Jordan basis of $F_4$. (*Suggestion*: First compute, for each $i$, the subspace $\text{CH}[\mathbf{T}, \lambda_i]$; then, since the restriction of $\mathbf{T} - \lambda_i\mathbf{I}$ to this subspace is nilpotent, the algorithm 8.2.18 can be applied to $(\mathbf{T} - \lambda_i\mathbf{I}, \text{CH}[\mathbf{T}, \lambda_i])$ instead of $(\mathbf{T}, V)$. In this connection, see also Exercises 5 and 6, Sec. 8.2.)
7. Let $\lambda_1, \lambda_2$ be distinct eigenvalues of $\mathbf{T}$. Prove that

   $\text{CH}[\mathbf{T}, \lambda_1] \subseteq \text{CR}[\mathbf{T}, \lambda_2]$.

   Also show that if char $\mathbf{T}$ splits and $\lambda_1, \lambda_2$ are its only roots, then equality holds: $\text{CH}[\mathbf{T}, \lambda_1] = \text{CR}[\mathbf{T}, \lambda_2]$. (Cf. Sec. 8.3, Exercise 6. *Suggestion*: Examine the proof of Theorem 8.4.1.)
8. Let $A$ be an $n \times n$ matrix with *real* entries. By a standard result

on polynomials with real coefficients (cf. Sec. 1.16), if $\lambda$ is a nonreal (complex) root of char $A$, then $\bar{\lambda}$ is also a root of char $A$. Let $J$ be the (complex) Jordan matrix similar to $A$, and let $J_1$, $J_2$ be the Jordan blocks of $J$ which correspond to the eigenvalues $\lambda$, $\bar{\lambda}$, respectively. Prove that the number of Jordan subblocks, and the degrees of the subblocks, are the same for $J_1$ as for $J_2$. Deduce that if we arrange the subblocks in a suitable order, we have $J_2 = \bar{J_1}$.

9. Prove: if **T** is nonsingular and $\lambda$ is an eigenvalue of **T** of multiplicity $m$, then $\lambda^{-1}$ is an eigenvalue of $\mathbf{T}^{-1}$ of multiplicity $m$. (*Suggestion*: See the exercises in Sec. 5.8.)
10. Let $J$ be a Jordan matrix over **R** or **C**. Prove: if $J^k = I$ for some positive integer $k$, then $J$ is a diagonal matrix. Over what wider class of fields is the result valid?

## 8.5 The Cayley-Hamilton Theorem

The main purpose of this section is to prove the following result:

**8.5.1 Theorem.** (Cayley-Hamilton) If $A$ is any $n \times n$ matrix over any field $F$, and if $\operatorname{char}(A; x) = c_0 + c_1x + c_2x^2 + \cdots + (-1)^n x^n$, then

$$c_0I + c_1A + c_2A^2 + \cdots + (-1)^n A^n = 0. \tag{8.5.2}$$

That is, the matrix obtained by substituting $A$ for $x$ in the expression for char$(A; x)$ is the zero matrix.

As a simple example, if

$$A = \begin{bmatrix} 3 & 1 \\ 2 & 0 \end{bmatrix},$$

then $\operatorname{char}(A; x) = x^2 - 3x - 2$. By Theorem 8.5.1, it follows that

$$A^2 - 3A - 2I = 0.$$

This result is easily checked by computation:

$$A^2 - 3A - 2I = \begin{bmatrix} 11 & 3 \\ 6 & 2 \end{bmatrix} - \begin{bmatrix} 9 & 3 \\ 6 & 0 \end{bmatrix} - \begin{bmatrix} 2 & 0 \\ 0 & 2 \end{bmatrix} = \begin{bmatrix} 0 & 0 \\ 0 & 0 \end{bmatrix}.$$

*Proof of 8.5.1.* Let $f(x) = \operatorname{char}(A; x)$, let $f(A)$ be the left side of 8.5.2, and so forth. As noted in the preceding section, there exists a field $L$ containing $F$ such that $f(x)$ splits over $L$:

$$f(x) = c_0 + c_1x + \cdots + (-1)^n x^n \tag{8.5.3}$$
$$= (-1)^n(x - \lambda_1)^{m_1} \cdots (x - \lambda_s)^{m_s}$$

where $\lambda_1, \ldots, \lambda_s$ are distinct elements of $L$. Let $V$ be a vector space of dimension $n$ over $L$ (for instance, we could take $V = L_n$) and let $\mathbf{T}$ be an element of $\text{Lin}(V, V)$ whose matrix is $A$. Then Theorem 8.4.1 is applicable, and by comparing dimensions on both sides of equation 8.4.3 we find that the subspace $V^*$ in that equation must have dimension zero. Thus,

$$\{0\} = V^* = \text{Ran}[(\mathbf{T} - \lambda_1\mathbf{I})^{m_1} \cdots (\mathbf{T} - \lambda_s\mathbf{I})^{m_s}].$$

But a mapping with zero range must be the zero mapping; thus,

$$\mathbf{O} = (\mathbf{T} - \lambda_1\mathbf{I})^{m_1} \cdots (\mathbf{T} - \lambda_s\mathbf{I})^{m_s}. \tag{8.5.4}$$

Since the right side of 8.5.4 is obtained (up to sign) by replacing $x$ by $\mathbf{T}$ in the right side of 8.5.3, we may rewrite 8.5.4 as

$$\mathbf{O} = f(\mathbf{T}) = c_0\mathbf{I} + c_1\mathbf{T} + \cdots + (-1)^n\mathbf{T}^n.$$

(*Question*: Does this last step need further justification?) Since $f(\mathbf{T})$ is the zero mapping, it follows that the corresponding matrix $f(A)$ is the zero matrix, as we wished to prove. ∎

(Note again the device, previously used in Example 8.4.30, of extending the field of scalars.)

By virtue of the correspondence between L.T.'s and matrices, *Theorem 8.5.1 must also hold for L.T.'s*; that is, if $V$ is any finite-dimensional vector space over any field $F$, if $\mathbf{T} \in \text{Lin}(V, V)$, and if

$$\text{char}(\mathbf{T}; x) = c_0 + c_1x + \cdots + (-1)^n x^n,$$

then

$$c_0\mathbf{I} + c_1\mathbf{T} + \cdots + (-1)^n\mathbf{T}^n = \mathbf{O}.$$

One interesting consequence of the Cayley-Hamilton Theorem is a characterization of nilpotent transformations in terms of their characteristic polynomials. As we have already seen (Theorem 8.2.23), the characteristic polynomial of any nilpotent transformation $\mathbf{T}$ is given by

$$\text{char}(\mathbf{T}; x) = \pm x^n.$$

Conversely, if $\mathbf{T}$ is a linear transformation such that $\text{char}(\mathbf{T}; x) = \pm x^n$, then Theorem 8.5.1 implies that $\pm\mathbf{T}^n = \mathbf{O}$ so that $\mathbf{T}$ is nilpotent. Thus, we have proved the following:

8.5.5 **Theorem.** If dim $V = n < \infty$ and $\mathbf{T} \in \text{Lin}(V, V)$, then $\mathbf{T}$ is nilpotent if and only if $\text{char}(\mathbf{T}; x) = (-1)^n x^n$.

**Exercise**

Verify the truth of the Cayley-Hamilton Theorem for each of the matrices of Section 8.3, Exercises 1–5.

# 9 The vector space $\mathbf{C}_n$ and related matters

## 9.0 Introduction

The field **C** of complex numbers has two very nice properties: (a) it contains the field of *real* numbers (the "usual" numbers which we most often deal with), and (b) it satisfies the Fundamental Theorem of Algebra: every polynomial of positive degree over **C** has a root in **C**. (In addition, many of the useful properties of real numbers, e.g., the facts about absolute values of sums and products, carry over to complex numbers.) As a consequence, a problem which on the surface involves only real scalars can sometimes be solved by appropriate use of complex numbers. This idea, which has already arisen in the preceding chapter (cf. Example 8.4.30), has numerous applications in algebra, geometry, and analysis. The present chapter is primarily devoted to three of these applications: (1) the classification of the isometries of $\mathbf{R}_n$, (2) the proof that a real symmetric $n \times n$ matrix has $n$ orthonormal eigenvectors, and (3) a proof (modulo some calculus) of the second-derivative test for extrema of functions of several variables. (1) is of course an extension of the results of Chapter 6. (2) has already been used in our development of the theory of second-degree equations (we "borrowed" this result in Sec. 7.1). (3) is included, not only for its inherent interest, but also to further emphasize the interrelatedness of different branches of mathematics.

The key tool in all of this is the *complex inner product*, whose basic properties are established in Section 9.1. Sections 9.2 and 9.3 contain further preliminary material which helps to lay the groundwork for the major results on isometries in Section 9.4. In Sections 9.5 through 9.7 we develop some elementary properties of Hermitian matrices and Hermitian forms which we need to obtain our results on eigenvectors (Sec. 9.5) and on extrema (Sec. 9.8). More can be said about

Hermitian forms and Hermitian matrices than appears in this chapter; rather than being fully comprehensive, we have tried instead to obtain some specific results of interest as quickly as possible.

## 9.1 Some basic definitions in $\mathbf{C}_n$

As always, $\mathbf{C}$ denotes the field of complex numbers. In Section 1.15 some elementary facts about complex numbers are summarized; you should review them now, before reading further. In particular, recall the definition of $i$ and the definitions of

$$\operatorname{Re} z; \quad \operatorname{Im} z; \quad \bar{z}; \quad |z|$$

when $z$ is a complex number.

9.1.1 **Definition.** If $\mathbf{u} = (a_1, \ldots, a_n)$ is any vector in $\mathbf{C}_n$, we define the vectors $\bar{\mathbf{u}}$, **Re u**, **Im u** as follows:

$$\bar{\mathbf{u}} = (\overline{a_1}, \ldots, \overline{a_n})$$
$$\mathbf{Re\,u} = (\operatorname{Re} a_1, \ldots, \operatorname{Re} a_n)$$
$$\mathbf{Im\,u} = (\operatorname{Im} a_1, \ldots, \operatorname{Im} a_n).$$

9.1.2 **Definition.** A vector $\mathbf{u} = (a_1, \ldots, a_n)$ in $\mathbf{C}_n$ is called a *real vector* if all the $a_i$ are real; equivalently, if $\mathbf{u} \in \mathbf{R}_n$. (Note that $\mathbf{R}_n$ is a subset of $\mathbf{C}_n$ since $\mathbf{R} \subset \mathbf{C}$. However, $\mathbf{R}_n$ is not a *subspace* of $\mathbf{C}_n$ since $\mathbf{R}_n$ is not closed under multiplication by complex scalars.)

The following theorem summarizes some trivial but useful consequences of the above definitions.

9.1.3 **Theorem.** For all vectors $\mathbf{u}, \mathbf{v} \in \mathbf{C}_n$,

(a) **Re u** and **Im u** are real vectors.

(b) $\mathbf{u} = \mathbf{Re\,u} + i\,\mathbf{Im\,u}$.

(c) $\bar{\mathbf{u}} = \mathbf{Re\,u} - i\,\mathbf{Im\,u}$.

(d) $\mathbf{u}$ can be expressed in one and only one way in the form $\mathbf{u} = \mathbf{v} + i\mathbf{w}$, where $\mathbf{v}$ and $\mathbf{w}$ are real vectors. In fact, $\mathbf{v} = \mathbf{Re\,u}$ and $\mathbf{w} = \mathbf{Im\,u}$.

(e) $\mathbf{u}$ is a real vector $\Leftrightarrow \mathbf{u} = \bar{\mathbf{u}}$.

(f) $\overline{\mathbf{u} + \mathbf{v}} = \bar{\mathbf{u}} + \bar{\mathbf{v}}$; $\overline{c\mathbf{u}} = \bar{c}\bar{\mathbf{u}}$ for all $c \in \mathbf{C}$.

(g) $\mathbf{u} + \bar{\mathbf{u}} = 2\,\mathbf{Re\,u}$; $\mathbf{u} - \bar{\mathbf{u}} = 2i\,\mathbf{Im\,u}$.

*Proof.* Exercise. (Part (a) is obvious; all other parts follow easily from the analogous properties of complex numbers, using the definitions of the vector space operations in $\mathbf{C}_n$.)

Let $\mathbf{u} = (a_1, \ldots, a_n)$ and $\mathbf{v} = (b_1, \ldots, b_n)$ be elements of $\mathbf{C}_n$. We define the *distance from* $\mathbf{u}$ *to* $\mathbf{v}$ to be the number

$$(9.1.4)\qquad d(\mathbf{u}, \mathbf{v}) = (|a_1 - b_1|^2 + \cdots + |a_n - b_n|^2)^{1/2}.$$

Note that since the absolute value of a complex number is real, $d(\mathbf{u}, \mathbf{v})$ is *real and nonnegative*. One way of motivating 9.1.4 is as follows: consider the mapping $\mathbf{M}$ of $\mathbf{C}_n \to \mathbf{R}_{2n}$ defined by

$$\mathbf{M}: (z_1, \ldots, z_n) \to (\text{Re } z_1, \text{Im } z_1, \text{Re } z_2, \text{Im } z_2, \ldots, \text{Re } z_n, \text{Im } z_n).$$

Note that when $n = 1$, the mapping $\mathbf{M}$ represents the usual way of identifying complex numbers with points in the real plane (cf. Sec. 1.15, Exercise 1). Now, if $\mathbf{u} = (a_1, \ldots, a_n)$ and $\mathbf{v} = (b_1, \ldots, b_n)$ are elements of $\mathbf{C}_n$, the distance formula in $\mathbf{R}_{2n}$ gives

$$\begin{aligned} d(\mathbf{uM}, \mathbf{vM}) &= \begin{pmatrix} (\text{Re } a_1 - \text{Re } b_1)^2 + (\text{Im } a_1 - \text{Im } b_1)^2 + \cdots \\ + (\text{Re } a_n - \text{Re } b_n)^2 + (\text{Im } a_n - \text{Im } b_n)^2 \end{pmatrix}^{1/2} \\ &= \begin{pmatrix} [\text{Re}(a_1 - b_1)]^2 + [\text{Im}(a_1 - b_1)]^2 + \cdots \\ + [\text{Re}(a_n - b_n)]^2 + [\text{Im}(a_n - b_n)]^2 \end{pmatrix}^{1/2} \\ &= (|a_1 - b_1|^2 + \cdots + |a_n - b_n|^2)^{1/2} \end{aligned}$$

(cf. eq. 1.15.4).

In short, our definition 9.1.4 is such that the mapping $\mathbf{M}$ will preserve distance. A similar argument shows that if we define the *length* of $\mathbf{u} = (a_1, \ldots, a_n)$ to be

$$(9.1.5)\qquad |\mathbf{u}| = (|a_1|^2 + \cdots + |a_n|^2)^{1/2}$$

then the mapping $\mathbf{M}$ preserves length. Again, $|\mathbf{u}|$ is real and nonnegative.

The next theorem shows that several of the basic properties of distance and length in $\mathbf{R}_n$ carry over to $\mathbf{C}_n$ without change.

**9.1.6 Theorem.** If $\mathbf{u}, \mathbf{v} \in \mathbf{C}_n$, then

(a) $d(\mathbf{u}, \mathbf{v}) = |\mathbf{u} - \mathbf{v}|$; $|\mathbf{u}| = d(\mathbf{u}, \mathbf{0})$.
(b) $d(\mathbf{u}, \mathbf{v}) = 0$ if and only if $\mathbf{u} = \mathbf{v}$.
(c) $|\mathbf{u}| = 0$ if and only if $\mathbf{u} = \mathbf{0}$.
(d) $|c\mathbf{u}| = |c|\,|\mathbf{u}|$ for all $c \in \mathbf{C}$.

We leave the proof as an exercise. (In connection with parts (b) and (c), see Sec. 1.12, Exercise 4. Part (d) is an easy consequence of 9.1.5 and 1.15.6.)

If $\mathbf{u} = (a_1, ..., a_n)$ and $\mathbf{v} = (b_1, ..., b_n)$ are elements of $\mathbf{C}_n$, we define

$$\text{(9.1.7)} \quad \langle \mathbf{u}, \mathbf{v} \rangle = a_1\overline{b_1} + a_2\overline{b_2} + \cdots + a_n\overline{b_n} = \sum_{i=1}^{n} a_i\overline{b_i}.$$

The scalar $\langle \mathbf{u}, \mathbf{v} \rangle$ is called the (*complex*) *inner product* of $\mathbf{u}$ and $\mathbf{v}$. The complex inner product is, of course, a generalization of the real dot product (cf. Sec. 3.7); indeed, it is clear that

$$\text{(9.1.8)} \quad \langle \mathbf{u}, \mathbf{v} \rangle = \mathbf{u} \cdot \overline{\mathbf{v}} \qquad (\text{all } \mathbf{u}, \mathbf{v} \in \mathbf{C}_n)$$

and, in particular, $\langle \mathbf{u}, \mathbf{v} \rangle = \mathbf{u} \cdot \mathbf{v}$ whenever $\mathbf{v}$ is a real vector. Thus, any result concerning the complex inner product implies, as a special case, a result concerning the real dot product.

(The bracket notation $\langle\ \rangle$ has been used previously to denote the flat determined by a set of points. However, since points are normally denoted by capital letters and vectors by small letters, no confusion should arise.)

The presence of the complex conjugate in 9.1.7 corresponds to its presence in the definition of the complex absolute value (eq. 1.15.5); it insures that the equation $\langle \mathbf{u}, \mathbf{u} \rangle = |\mathbf{u}|^2$, which already holds in $\mathbf{R}_n$, will remain true in $\mathbf{C}_n$. This property and several others are summarized in the next theorem, which is the complex analog of Theorems 3.7.2 and 3.7.3.

**9.1.9 Theorem.** For all $\mathbf{u}, \mathbf{v}, \mathbf{w} \in \mathbf{C}_n$ and all $c \in \mathbf{C}$,

(a) $\overline{\langle \mathbf{u}, \mathbf{v} \rangle} = \langle \overline{\mathbf{u}}, \overline{\mathbf{v}} \rangle = \langle \mathbf{v}, \mathbf{u} \rangle$.
(b) $\langle \mathbf{u}, \mathbf{v} + \mathbf{w} \rangle = \langle \mathbf{u}, \mathbf{v} \rangle + \langle \mathbf{u}, \mathbf{w} \rangle$; $\langle \mathbf{u} + \mathbf{v}, \mathbf{w} \rangle = \langle \mathbf{u}, \mathbf{w} \rangle + \langle \mathbf{v}, \mathbf{w} \rangle$.
(c) $\langle c\mathbf{u}, \mathbf{v} \rangle = c\langle \mathbf{u}, \mathbf{v} \rangle$; $\langle \mathbf{u}, c\mathbf{v} \rangle = \overline{c}\langle \mathbf{u}, \mathbf{v} \rangle$.
(d) $\langle \mathbf{u}, \mathbf{u} \rangle = |\mathbf{u}|^2$.
(e) $\langle \mathbf{u}, \mathbf{u} \rangle = 0$ if and only if $\mathbf{u} = \mathbf{0}$.

*Proof.* Exercise. (Parts (a) and (d) may be proved directly from the definitions; parts (b) and (c) are most quickly obtained by applying Theorem 3.7.2 and equation 9.1.8.)

We customarily define a *complex inner product space*, or *unitary space*, to be a vector space $V$ over $\mathbf{C}$ on which there is defined a (complex-valued) "inner product" $\langle \mathbf{u}, \mathbf{v} \rangle$ ($\mathbf{u}, \mathbf{v} \in V$) satisfying the conditions of Theorem 9.1.9, with 9.1.9(d) replaced by the statement that $\langle \mathbf{u}, \mathbf{u} \rangle$ is real and nonnegative. Unitary spaces are, of course, analogous to real Euclidean spaces but with **C** as the field of scalars instead of **R**. We give no general treatment of unitary spaces in this book, since $\mathbf{C}_n$ and its subspaces will suffice for our purposes; however, most of what we do in $\mathbf{C}_n$ could easily be generalized to arbitrary finite-dimensional unitary spaces.

It is useful to be able to express complex inner products in terms of lengths. Theorem 3.7.5(c) accomplished this for the real dot product (more generally, for inner products in real Euclidean spaces). The complex inner product $\langle \mathbf{u}, \mathbf{v}\rangle$ can also be expressed in terms of lengths, but not quite so simply since formulas 3.7.5(a) and 3.7.5(b) do not carry over verbatim to $\mathbf{C}_n$. The analogous formulas in $\mathbf{C}_n$ are

**(9.1.10)**
$$\text{(a)}\quad |\mathbf{u}+\mathbf{v}|^2 = |\mathbf{u}|^2 + |\mathbf{v}|^2 + 2\,\mathrm{Re}\langle \mathbf{u}, \mathbf{v}\rangle.$$
$$\text{(b)}\quad |\mathbf{u}-\mathbf{v}|^2 = |\mathbf{u}|^2 + |\mathbf{v}|^2 - 2\,\mathrm{Re}\langle \mathbf{u}, \mathbf{v}\rangle.$$

Indeed, we have

$$\begin{aligned} |\mathbf{u}+\mathbf{v}|^2 &= \langle \mathbf{u}+\mathbf{v}, \mathbf{u}+\mathbf{v}\rangle \qquad \text{(by 9.1.9(d))}\\ &= \langle \mathbf{u}, \mathbf{u}\rangle + \langle \mathbf{v}, \mathbf{v}\rangle + \langle \mathbf{u}, \mathbf{v}\rangle + \langle \mathbf{v}, \mathbf{u}\rangle \qquad \text{(by 9.1.9(b))}\\ &= |\mathbf{u}|^2 + |\mathbf{v}|^2 + \langle \mathbf{u}, \mathbf{v}\rangle + \overline{\langle \mathbf{u}, \mathbf{v}\rangle} \qquad \text{(by 9.1.9(a))}\\ &= |\mathbf{u}|^2 + |\mathbf{v}|^2 + 2\,\mathrm{Re}\langle \mathbf{u}, \mathbf{v}\rangle \qquad \text{(by 1.15.3)}, \end{aligned}$$

proving 9.1.10(a); the proof of 9.1.10(b) is similar. It follows at once from 9.1.10(b) that

**(9.1.11)** $$2\,\mathrm{Re}\langle \mathbf{u}, \mathbf{v}\rangle = |\mathbf{u}|^2 + |\mathbf{v}|^2 - |\mathbf{u}-\mathbf{v}|^2$$

so that at least the real part of $\langle \mathbf{u}, \mathbf{v}\rangle$ is expressible in terms of lengths. But then so is the imaginary part, since

**(9.1.12)** $$\mathrm{Im}\langle \mathbf{u}, \mathbf{v}\rangle = \mathrm{Re}\langle \mathbf{u}, i\mathbf{v}\rangle.$$

We leave the proof of 9.1.12 as an exercise (below). Combining 9.1.11 and 9.1.12, we obtain

**(9.1.13)**
$$\begin{aligned} 2\langle \mathbf{u}, \mathbf{v}\rangle &= 2(\mathrm{Re}\langle \mathbf{u}, \mathbf{v}\rangle + i\,\mathrm{Im}\langle \mathbf{u}, \mathbf{v}\rangle)\\ &= |\mathbf{u}|^2 + |\mathbf{v}|^2 - |\mathbf{u}-\mathbf{v}|^2\\ &\quad + i(|\mathbf{u}|^2 + |i\mathbf{v}|^2 - |\mathbf{u}-i\mathbf{v}|^2) \end{aligned}$$

which is the desired expression. (We could further simplify by noting that $|i\mathbf{v}| = |\mathbf{v}|$.)

Vectors $\mathbf{u}_1, \mathbf{u}_2, \ldots, \mathbf{u}_k$ in $\mathbf{C}_n$ are said to be *orthogonal* if $\langle \mathbf{u}_i, \mathbf{u}_j\rangle = 0$ whenever $i \neq j$; if in addition each $\mathbf{u}_i$ has length 1, the orthogonal vectors $\mathbf{u}_1, \ldots, \mathbf{u}_k$ are *orthonormal*. (These definitions obviously extend those in $\mathbf{R}_n$.) Thus, $\mathbf{u}_1, \ldots, \mathbf{u}_k$ are orthonormal if and only if

**(9.1.14)** $$\langle \mathbf{u}_i, \mathbf{u}_j\rangle = \delta_{ij} \qquad \text{(all } i, j\text{)}.$$

Most of the basic theorems on orthogonal and orthonormal sets in real Euclidean spaces carry over to $\mathbf{C}_n$ with little or no change, as shown by the following theorem.

9.1.15 **Theorem.** (a) The canonical basis in $\mathbf{C}_n$ is an orthonormal set.

(b) Theorems 3.8.2, 3.8.5, 3.8.9, 3.8.15, and 3.8.18 remain true if $E$ is replaced by $\mathbf{C}_n$ (or by any subspace of $\mathbf{C}_n$), inner products in $E$ are replaced by the complex inner product, and "$\mathbf{v} \perp \mathbf{u}$" is replaced by "$\langle \mathbf{v}, \mathbf{u} \rangle = 0$". In particular, any nonzero orthogonal vectors in $\mathbf{C}_n$ are linearly independent, and any subspace of $\mathbf{C}_n$ has an orthonormal basis.

(c) If $\mathbf{u}_1, \ldots, \mathbf{u}_k$ are orthonormal vectors in $\mathbf{C}_n$ and if $a_i, b_i$ ($1 \leqslant i \leqslant k$) are any scalars (complex numbers), then

$$\left\langle \sum_{i=1}^{k} a_i \mathbf{u}_i, \sum_{j=1}^{k} b_j \mathbf{u}_j \right\rangle = \sum_{i=1}^{k} a_i \overline{b_i} = \langle (a_1, \ldots, a_k), (b_1, \ldots, b_k) \rangle.$$

The proofs are the same as in the real case.

**Exercises**

*1. Carry out the details of the proofs of Theorems 9.1.3, 9.1.6, and 9.1.9.

*2. (a) Prove that for all complex numbers $z$, $w$, $\operatorname{Im}(z\overline{w}) = \operatorname{Re}(z\overline{iw})$.

(b) Deduce equation 9.1.12 from the result of part (a).

## 9.2 Real linear transformations of $\mathbf{C}_n$

In studying a linear transformation $\mathbf{T}$ it is helpful to know the Jordan form of $\mathbf{T}$ (see Sec. 8.4). However, $\mathbf{T}$ will not have a Jordan form if its characteristic polynomial fails to split. In particular, if $\mathbf{T} \in \operatorname{Lin}(\mathbf{R}_n, \mathbf{R}_n)$, then char($\mathbf{T}$; $x$) may have irreducible quadratic factors. As we have seen, it is sometimes useful in such a case to extend $\mathbf{T}$ to a linear transformation $\mathbf{T}^*$ of $\mathbf{C}_n \to \mathbf{C}_n$. Since every nonconstant polynomial splits over $\mathbf{C}$, $\mathbf{T}^*$ has a Jordan form, and from it we may be able to obtain other information about $\mathbf{T}^*$ and, hence, also about $\mathbf{T}$. The extension $\mathbf{T}^*$ is called a "real linear transformation" of $\mathbf{C}_n$. Let us state this as a formal definition.

9.2.1 **Definition.** Let $\mathbf{T}^* \in \operatorname{Lin}(\mathbf{C}_n, \mathbf{C}_n)$. $\mathbf{T}^*$ is called a *real* linear transformation (abbreviated "real L.T.") of $\mathbf{C}_n$ if $\mathbf{T}^*$ is an extension of some linear transformation of $\mathbf{R}_n \to \mathbf{R}_n$; equivalently, if $\mathbf{T}^* \restriction \mathbf{R}_n$ is a linear transformation of $\mathbf{R}_n \to \mathbf{R}_n$.

9.2.2 **Theorem.** If $\mathbf{T}^* \in \text{Lin}(\mathbf{C}_n, \mathbf{C}_n)$, the following conditions are equivalent:

(a) $\mathbf{T}^*$ is a real L.T. of $\mathbf{C}_n$.

(b) $\mathbf{T}^*$ maps every real vector into a real vector; that is, whenever $\mathbf{u} \in \mathbf{C}_n$ is a real vector, then so is $\mathbf{uT}^*$.

(c) The matrix $A^*$ of $\mathbf{T}^*$ with respect to the canonical basis of $\mathbf{C}_n$ consists entirely of real entries; that is, $A^*$ is a matrix over $\mathbf{R}$.

*Proof.* If $\mathbf{T}^*$ is real, then, by definition, $\mathbf{T}^*$ maps $\mathbf{R}_n$ into $\mathbf{R}_n$; thus, (a) $\Rightarrow$ (b). Assume (b) holds; since the vectors $\mathbf{e}_i$ of the canonical basis are obviously real vectors, it follows that the vectors $\mathbf{e}_i\mathbf{T}^*$ are real; that is, the row vectors of $A^*$ are real, proving (c). Finally, if all entries in $A^*$ are real, then for any real vector $\mathbf{u}$ the vector $\mathbf{uT}^* = \mathbf{u}A^*$ must be real since its coordinates are obtained by adding and multiplying real numbers. Hence, $\mathbf{T}^*$ maps $\mathbf{R}_n$ into $\mathbf{R}_n$; since $\mathbf{T}^*$ is linear it follows that $\mathbf{T}^* \restriction \mathbf{R}_n$ belongs to $\text{Lin}(\mathbf{R}_n, \mathbf{R}_n)$, proving (a). ▮

9.2.3 **Theorem.** Let $\mathbf{T}^* \in \text{Lin}(\mathbf{C}_n, \mathbf{C}_n)$ and assume that $\mathbf{T}^*$ is a real L.T. Then for all $\mathbf{u} \in \mathbf{C}_n$,

(a) $(\mathbf{Re\,u})\mathbf{T}^* = \mathbf{Re}(\mathbf{uT}^*)$

(b) $(\mathbf{Im\,u})\mathbf{T}^* = \mathbf{Im}(\mathbf{uT}^*)$

(c) $\overline{\mathbf{u}}\mathbf{T}^* = \overline{\mathbf{uT}^*}$.

*Proof.* Exercise. (This exercise is not totally trivial, but it is easy once you see how to begin.)

The remaining results of this section have to do with the eigenvalues and eigenvectors of a real L.T.

9.2.4 **Theorem.** Let $\mathbf{T}^* \in \text{Lin}(\mathbf{C}_n, \mathbf{C}_n)$ be a real L.T., let $\lambda \in \mathbf{C}$ be an eigenvalue of $\mathbf{T}^*$, and let $W$ be the $(\mathbf{T}^*, \lambda)$-eigenspace. Then $\overline{\lambda}$ is an eigenvalue of $\mathbf{T}^*$, and the $(\mathbf{T}^*, \overline{\lambda})$-eigenspace is

$$\overline{W} = \{\overline{\mathbf{w}} : \mathbf{w} \in W\}.$$

*Proof.* $\overline{\mathbf{w}} \in \overline{W} \Leftrightarrow \mathbf{w} \in W \Leftrightarrow \mathbf{w}$ is a $\lambda$-eigenvector $\Leftrightarrow \mathbf{wT}^* = \lambda\mathbf{w}$

$\Leftrightarrow \overline{\mathbf{wT}^*} = \overline{\lambda\mathbf{w}} \Leftrightarrow \overline{\mathbf{w}}\mathbf{T}^* = \overline{\lambda}\,\overline{\mathbf{w}}$ (by 9.2.3 and 9.1.3(f))

$\Leftrightarrow \overline{\mathbf{w}}$ is a $\overline{\lambda}$-eigenvector.

Hence, $\overline{W}$ is the $(\mathbf{T}^*, \overline{\lambda})$-eigenspace. Since $W \neq \{\mathbf{0}\}$, $\overline{W} \neq \{\mathbf{0}\}$ and, hence, $\overline{\lambda}$ is an eigenvalue. ∎

9.2.5 **Theorem.** Let $\mathbf{T}^* \in \text{Lin}(\mathbf{C}_n, \mathbf{C}_n)$ be a real L.T., let $\lambda$ be a real number which is an eigenvalue of $\mathbf{T}^*$, and let $\mathbf{u} \in \mathbf{C}_n$ be a $\lambda$-eigenvector of $\mathbf{T}^*$. Then $\mathbf{Re\ u}$ and $\mathbf{Im\ u}$ are $\lambda$-eigenvectors of $\mathbf{T}^*$.

*Proof.* Exercise. (*Note: The hypothesis that $\lambda$ is real is essential.* If your proof does not use this hypothesis, it's incorrect.)

9.2.6 **Theorem.** Let $\mathbf{T}^* \in \text{Lin}(\mathbf{C}_n, \mathbf{C}_n)$ be a real L.T., let $\lambda$ be a real number which is an eigenvalue of $\mathbf{T}^*$, and let $N$ be the $(\mathbf{T}^*, \lambda)$-eigenspace. Then $N$ has an orthonormal basis consisting entirely of real vectors.

*Proof.* Since $N$ is a subspace of $\mathbf{C}_n$, $N$ certainly has a basis, say, $\{\mathbf{v}_1, ..., \mathbf{v}_k\}$; the $\mathbf{v}_i$ need not be real. However, the real vectors $\mathbf{Re\ v}_i$ and $\mathbf{Im\ v}_i$ lie in $N$ by Theorem 9.2.5, and since $\mathbf{v}_i = \mathbf{Re\ v}_i + i\ \mathbf{Im\ v}_i$ it follows that the set

$$S = \{\mathbf{Re\ v}_1, \mathbf{Im\ v}_1, ..., \mathbf{Re\ v}_k, \mathbf{Im\ v}_k\}$$

spans $N$. By choosing a maximal independent subset of $S$, we obtain a basis $\{\mathbf{w}_1, ..., \mathbf{w}_k\}$ of $N$ such that all of the $\mathbf{w}_i$ are real vectors. Let $N_0$ be the subspace of $\mathbf{R}_n$ spanned by the $\mathbf{w}_i$. By Theorem 3.8.15, $N_0$ has an orthonormal basis $\{\mathbf{u}_1, ..., \mathbf{u}_h\}$ over $\mathbf{R}$ (actually $h = k$, but this point is not essential to the proof). Since the $\mathbf{w}_i$ are then linear combinations of the $\mathbf{u}_j$ over $\mathbf{R}$, they are certainly linear combinations of the $\mathbf{u}_j$ over $\mathbf{C}$. Since $N$ is spanned (over $\mathbf{C}$) by the $\mathbf{w}_i$, it follows that $N$ is spanned by the $\mathbf{u}_j$ (which certainly lie in $N$ since they lie in $N_0$). Moreover, the $\mathbf{u}_j$ are independent over $\mathbf{C}$ since they are orthonormal. (Independence of the $\mathbf{u}_j$ over $\mathbf{C}$ could also be deduced from Theorem 5.9.3.) Hence, the set $\{\mathbf{u}_1, ..., \mathbf{u}_h\}$ is a basis of $N$ having the desired properties. ∎

**Exercises**

*1. Prove Theorem 9.2.3.

*2. Prove Theorem 9.2.5.

## 9.3 Isometries of $\mathbf{C}_n$

A mapping $\mathbf{T} : \mathbf{C}_n \to \mathbf{C}_n$ is called an *isometry* if it preserves distance; that is, if

$$d(\mathbf{u}, \mathbf{v}) = d(\mathbf{uT}, \mathbf{vT}) \qquad (\text{all } \mathbf{u}, \mathbf{v} \in \mathbf{C}_n).$$

This is, of course, the same definition as in the real case. When we classify the isometries of $\mathbf{R}_n$ in Section 9.4, our method will be to extend an isometry of $\mathbf{R}_n$ to one of $\mathbf{C}_n$ and then use properties of isometries in $\mathbf{C}_n$. The needed properties will be established in this section.

9.3.1 **Theorem.** If $\mathbf{T} \in \text{Lin}(\mathbf{C}_n, \mathbf{C}_n)$, then the following three statements are equivalent:

(a) $\mathbf{T}$ is an isometry.
(b) $|\mathbf{u}| = |\mathbf{uT}|$ for all $\mathbf{u} \in \mathbf{C}_n$.
(c) $\langle \mathbf{u}, \mathbf{v} \rangle = \langle \mathbf{uT}, \mathbf{vT} \rangle$ for all $\mathbf{u}, \mathbf{v} \in \mathbf{C}_n$.

The proof is identical with that of Theorem 4.10.9, except that $p(\mathbf{u}, \mathbf{v})$ is replaced by $\langle \mathbf{u}, \mathbf{v} \rangle$ and we must use 9.1.10 and 9.1.13 instead of Theorem 3.7.5. Note that $\mathbf{T}$ is *assumed* to be linear in Theorem 9.3.1, so that we need not use Theorem 4.1.6 in the proof.

9.3.2 **Theorem.** Let $(\mathbf{u}_1, \ldots, \mathbf{u}_n)$ be an ordered orthonormal basis of $\mathbf{C}_n$ and let $\mathbf{T} \in \text{Lin}(\mathbf{C}_n, \mathbf{C}_n)$. Then $\mathbf{T}$ is an isometry if and only if the vectors $\mathbf{u}_1\mathbf{T}, \ldots, \mathbf{u}_n\mathbf{T}$ are orthonormal.

Theorem 9.3.2 is the complex analog of Theorem 4.10.12. Again, the proof is the same as before, except that in two places an expression $\Sigma a_i b_i$ must be replaced by $\Sigma a_i \overline{b_i}$ (using 9.1.15(c) rather than 3.8.17).

When we extend Theorem 4.10.13 to $\mathbf{C}_n$, there is a significant change, arising from the fact that whereas matrix multiplication is described by dot products, orthonormality now relates to the complex inner product. Since the inner product and the dot product are related by equation 9.1.8, orthonormality of the rows of $A$ is equivalent to the equation $A\overline{A}^t = I$ instead of $AA^t = I$. ($\overline{A}$ of course denotes the matrix obtained from $A$ by replacing each entry by its complex conjugate.) Similarly, orthonormality of the columns of $A$ leads to the equation $A^t\overline{A} = I$, which in turn is equivalent to

$$I = I^t = (A^t\overline{A})^t = \overline{A}^t A^{tt} = \overline{A}^t A.$$

Hence, the correct analog of Theorem 4.10.13 is

9.3.3 **Theorem.** Let $\mathbf{T} \in \operatorname{Lin}(\mathbf{C}_n, \mathbf{C}_n)$, and let $A$ be the matrix of $\mathbf{T}$ with respect to some orthonormal basis of $\mathbf{C}_n$. Then the following statements are equivalent:

(a) $\mathbf{T}$ is an isometry.
(b) The rows of $A$ are orthonormal.
(c) The columns of $A$ are orthonormal.
(d) $\overline{A}^t = A^{-1}$.

A matrix $A$ satisfying these conditions is called a *unitary matrix*.

Our next theorem summarizes several facts concerning the eigenvalues and eigenvectors of a linear isometry.

9.3.4 **Theorem.** Let $\mathbf{T} \in \operatorname{Lin}(\mathbf{C}_n, \mathbf{C}_n)$ be an isometry. Let $\lambda_1, \ldots, \lambda_s$ be the distinct eigenvalues of $\mathbf{T}$, and let $m_i$ be the multiplicity of $\lambda_i$ $(1 \leqslant i \leqslant s)$. Then

(a) $\operatorname{char}(\mathbf{T}; x) = (x - \lambda_1)^{m_1} \cdots (x - \lambda_s)^{m_s}$.
(b) For all $i$, $|\lambda_i| = 1$.
(c) The Jordan form of $\mathbf{T}$ is diagonal.
(d) $\mathbf{C}_n$ is the direct sum of the $(\mathbf{T}, \lambda_i)$-eigenspaces; that is,

$$\mathbf{C}_n = \operatorname{Nul}(\mathbf{T} - \lambda_1\mathbf{I}) \oplus \cdots \oplus \operatorname{Nul}(\mathbf{T} - \lambda_s\mathbf{I}).$$

Moreover, each summand is $\mathbf{T}$-stable, and $\dim \operatorname{Nul}(\mathbf{T} - \lambda_i\mathbf{I}) = m_i$ (all $i$).

(e) If $\mathbf{u}$ is a $\lambda_i$-eigenvector of $\mathbf{T}$ and $\mathbf{v}$ is a $\lambda_j$-eigenvector of $\mathbf{T}$ where $i \neq j$, then $\langle \mathbf{u}, \mathbf{v} \rangle = 0$.

*Proof.* (a) is simply the assertion that char $\mathbf{T}$ splits, which follows from the Fundamental Theorem of Algebra. To prove (b), choose a nonzero $\lambda_i$-eigenvector $\mathbf{u}$; then $\mathbf{uT} = \lambda_i\mathbf{u}$, so that by Theorem 9.3.1(b), we have

$$|\mathbf{u}| = |\mathbf{uT}| = |\lambda_i\mathbf{u}| = |\lambda_i|\,|\mathbf{u}|;$$

since $|\mathbf{u}| \neq 0$ it follows that $|\lambda_i| = 1$.

Since char $\mathbf{T}$ splits, there is a basis of $\mathbf{C}_n$ with respect to which the matrix $J$ of $\mathbf{T}$ has the form 8.4.14 (Jordan form). The Jordan blocks $J_i$ of $J$ are given by 8.4.10; letting $N_i = \operatorname{CH}[\mathbf{T}, \lambda_i]$ (the $(\mathbf{T}, \lambda_i)$-characteristic subspace), $J_i$ is the matrix of $\mathbf{T} \restriction N_i$ with respect to some ordered basis $(\mathbf{u}_1^{(i)}, \ldots, \mathbf{u}_{m_i}^{(i)})$ of $N_i$. (Note that $\dim N_i = m_i$ by Theorem 8.3.8.) To prove 9.3.4(c) we must show that for each $i$, all of the entries $\epsilon_j^{(i)}$ on the superdiagonal of $J_i$ are zero. Suppose that for some $i$ this is not true; then there must be a largest integer $j$ $(1 \leqslant j \leqslant m_i - 1)$ such that $\epsilon_j^{(i)} = 1$. Then by definition of the correspondence between L.T.'s and matrices, we have

$$\mathbf{u}_j^{(i)}\mathbf{T} = \lambda_i\mathbf{u}_j^{(i)} + \mathbf{u}_{j+1}^{(i)}$$
$$\mathbf{u}_{j+1}^{(i)}\mathbf{T} = \lambda_i\mathbf{u}_{j+1}^{(i)}.$$

Hence,

$$\begin{aligned} \langle \mathbf{u}_j^{(i)}, \mathbf{u}_{j+1}^{(i)}\rangle &= \langle \mathbf{u}_j^{(i)}\mathbf{T}, \mathbf{u}_{j+1}^{(i)}\mathbf{T}\rangle \qquad \text{(by 9.3.1)} \\ &= \langle \lambda_i\mathbf{u}_j^{(i)} + \mathbf{u}_{j+1}^{(i)}, \lambda_i\mathbf{u}_{j+1}^{(i)}\rangle \\ &= |\lambda_i|^2 \langle \mathbf{u}_j^{(i)}, \mathbf{u}_{j+1}^{(i)}\rangle + \overline{\lambda_i}|\mathbf{u}_{j+1}^{(i)}|^2. \end{aligned} \tag{9.3.5}$$

But $|\lambda_i| = 1$; hence, 9.3.5 reduces to $\overline{\lambda_i}|\mathbf{u}_{j+1}^{(i)}|^2 = 0$. But both $\lambda_i$ and $\mathbf{u}_{j+1}^{(i)}$ are nonzero (why?), so this is impossible. Hence, all of the $\epsilon_j^{(i)}$ are zero after all and 9.3.4(c) is proved. Theorem 9.3.4(d) now follows from Theorems 8.4.33 and 8.4.13 (the summands are **T**-stable by 8.3.2.)

Finally, to prove 9.3.4(e), we have

$$\langle \mathbf{u}, \mathbf{v}\rangle = \langle \mathbf{u}\mathbf{T}, \mathbf{v}\mathbf{T}\rangle = \langle \lambda_i\mathbf{u}, \lambda_j\mathbf{v}\rangle = \lambda_i\overline{\lambda_j}\langle \mathbf{u}, \mathbf{v}\rangle. \tag{9.3.6}$$

Since $i \neq j$ and $\lambda_i\overline{\lambda_i} = |\lambda_i|^2 = 1$, it follows that $\lambda_i\overline{\lambda_j} \neq 1$; hence, 9.3.6 is possible only if $\langle \mathbf{u}, \mathbf{v}\rangle = 0$. ∎

According to 9.3.4(c), the matrix of a linear isometry of $\mathbf{C}_n$ is diagonal with respect to a suitable basis. If **T** is a linear isometry of $\mathbf{R}_n$, we cannot always put the matrix of **T** into diagonal form; however, we can always decompose the matrix into diagonal blocks of degrees 1 and 2, as we will show in the next section. This means that $\mathbf{R}_n$ is the direct sum of **T**-stable subspaces of dimensions $\leq 2$. Since the isometries of 2-dimensional subspaces of $\mathbf{R}_n$ are known (Secs. 6.3 and 6.4), it will then be possible to classify all isometries of $\mathbf{R}_n$.

## 9.4 The isometries of $\mathbf{R}_n$

Let **T** (until further notice) be a linear isometry of $\mathbf{R}_n$. Our object is to determine a simple form for the matrix of **T** and, hence, to determine the geometric character of **T**.

To begin, let $M$ be the matrix of **T** with respect to the canonical basis $(\mathbf{e}_1, ..., \mathbf{e}_n)$. Since $M$ is a matrix over **R**, it is also a matrix over **C**; hence, it corresponds to a linear transformation $\mathbf{T}^*$ of $\mathbf{C}_n \to \mathbf{C}_n$ with respect to the canonical basis. Since $\mathbf{T}^*$ and **T** do the same thing to each $\mathbf{e}_i$, they must do the same thing to each real vector; hence, $\mathbf{T} = \mathbf{T}^* \restriction \mathbf{R}$. $\mathbf{T}^*$ is thus a real linear transformation of $\mathbf{C}_n$. Moreover, since **T** is an isometry, the rows of $M$ are orthonormal by 4.10.13 and, hence, $\mathbf{T}^*$ is an isometry by 9.3.3. It follows by Theorem 9.3.4(b) that all eigenvalues of $\mathbf{T}^*$ have absolute value 1; in particular, the only possible *real* eigenvalues of $\mathbf{T}^*$ are 1 and $-1$. Also, since $\mathbf{T}^*$ is a real L.T., 9.2.4 implies that the nonreal eigenvalues of $\mathbf{T}^*$ may be arranged in pairs of the form $(\lambda_1, \overline{\lambda_1})$, $(\lambda_2, \overline{\lambda_2})$, ..., $(\lambda_r, \overline{\lambda_r})$, where the total number of distinct nonreal eigenvalues is $2r$ (possibly $r = 0$). Let

$U^*$ = the $(\mathbf{T}^*, 1)$-eigenspace, of dimension $d$ (possibly $d = 0$)

$V^*$ = the $(\mathbf{T}^*, -1)$-eigenspace, of dimension $e$ (possibly $e = 0$)

$N_k$ = the $(\mathbf{T}^*, \lambda_k)$-eigenspace, of dimension $m_k$ $(1 \leq k \leq r)$.

It follows from Theorem 9.2.4 that $\overline{N_k}$ (the set of all conjugates of elements of $N_k$) is the $(\mathbf{T}^*, \overline{\lambda_k})$-eigenspace. By Theorem 9.3.4(d), we then have

$$\text{(9.4.1)} \quad \mathbf{C}_n = U^* \oplus V^* \oplus N_1 \oplus \overline{N_1} \oplus \cdots \oplus N_r \oplus \overline{N_r}.$$

By Theorem 9.3.4(e), any two of the summands on the right side of 9.4.1 are orthogonal. (Two subspaces $S_1$, $S_2$ are said to be *orthogonal* if every vector in $S_1$ is orthogonal to every vector in $S_2$.)

By Theorem 9.2.6, the subspace $U^*$ has an orthonormal basis $B_U = \{\mathbf{u}_1, ..., \mathbf{u}_d\}$ consisting of real vectors, and the subspace $V^*$ has an orthonormal basis $B_V = \{\mathbf{v}_1, ..., \mathbf{v}_e\}$ consisting of real vectors.

Let $k$ be fixed and consider the conjugate subspaces $N_k$, $\overline{N_k}$. Let $\{\mathbf{w}_1^{(k)}, ..., \mathbf{w}_{m_k}^{(k)}\}$ be an orthonormal basis of $N_k$ (such a basis exists by 9.1.15(b) ). It easily follows from 9.1.9(a) and 9.1.14 that the set $\{\overline{\mathbf{w}_1^{(k)}}, ..., \overline{\mathbf{w}_{m_k}^{(k)}}\}$ is also orthonormal (and, hence, independent!). Since $\{\mathbf{w}_1^{(k)}, ..., \mathbf{w}_{m_k}^{(k)}\}$ spans $N_k$, $\{\overline{\mathbf{w}_1^{(k)}}, ..., \overline{\mathbf{w}_{m_k}^{(k)}}\}$ must span $\overline{N_k}$ (*why*?) and, hence, is an orthonormal basis of $\overline{N_k}$. Since $N_k$ is orthogonal to $\overline{N_k}$, it follows that the set

$$G_k = \{\mathbf{w}_1^{(k)}, ..., \mathbf{w}_{m_k}^{(k)}, \overline{\mathbf{w}_1^{(k)}}, ..., \overline{\mathbf{w}_{m_k}^{(k)}}\}$$

is an orthonormal basis of $N_k \oplus \overline{N_k}$. If we let

$$M_j^{(k)} = [\mathbf{w}_j^{(k)}, \overline{\mathbf{w}_j^{(k)}}] \qquad (j = 1, ..., m_k),$$

Theorem 2.8.8 implies that

$$\text{(9.4.2)} \quad N_k \oplus \overline{N_k} = M_1^{(k)} \oplus \cdots \oplus M_{m_k}^{(k)}.$$

Moreover, since the set $G_k$ is orthonormal, any two of the subspaces $M_1^{(k)}, ..., M_{m_k}^{(k)}$ must be orthogonal. Since we have already observed that the summands in 9.4.1 are mutually orthogonal, we see by substituting 9.4.2 into 9.4.1 that

(9.4.3) $$\mathbf{C}_n = U^* \oplus V^* \oplus M_1^{(1)} \oplus \cdots \oplus M_{m_1}^{(1)} \oplus \cdots \oplus M_1^{(r)} \oplus \cdots \oplus M_{m_r}^{(r)}$$

and that *any two of the summands in 9.4.3 are orthogonal.*

Now, let

(9.4.4) $$\mathbf{x}_j^{(k)} = (\mathbf{w}_j^{(k)} + \overline{\mathbf{w}_j^{(k)}})/\sqrt{2}; \qquad \mathbf{y}_j^{(k)} = i(\mathbf{w}_j^{(k)} - \overline{\mathbf{w}_j^{(k)}})/\sqrt{2} \qquad (1 \leqslant k \leqslant r,\ 1 \leqslant j \leqslant m_k).$$

For each fixed $k$ and $j$, the two-element set $\{\mathbf{w}_j^{(k)}, \overline{\mathbf{w}_j^{(k)}}\}$ is a subset of $G_k$ and thus orthonormal; hence, a straightforward computation (which we leave as an exercise) shows that the set $\{\mathbf{x}_j^{(k)}, \mathbf{y}_j^{(k)}\}$ is also orthonormal. Since $\mathbf{x}_j^{(k)}$, $\mathbf{y}_j^{(k)}$ clearly lie in $M_j^{(k)}$ and $\dim M_j^{(k)} = 2$, it follows that the set

$$B_j^{(k)} = \{\mathbf{x}_j^{(k)}, \mathbf{y}_j^{(k)}\}$$

is an orthonormal basis of $M_j^{(k)}$. Moreover, $\mathbf{x}_j^{(k)}$ and $\mathbf{y}_j^{(k)}$ are *real* vectors by Theorem 9.1.3(g).

We now let

$$B = B_U \cup B_V \cup \left( \bigcup_{k=1}^{r} \bigcup_{j=1}^{m_k} B_j^{(k)} \right).$$

By 9.4.3 and Theorem 2.8.3, $B$ is a basis of $\mathbf{C}_n$. Since each of the sets $B_U$, $B_V$, $B_j^{(k)}$ is an orthonormal set of real vectors and since the summands in 9.4.3 are mutually orthogonal, it follows that *$B$ is an orthonormal basis of $\mathbf{C}_n$ consisting entirely of real vectors.*

Since the set $B$ is a basis of $\mathbf{C}_n$, it has $n$ elements. Moreover, $B$ is a subset of $\mathbf{R}_n$, and since $B$ is linearly independent over $\mathbf{C}$ it is certainly independent over $\mathbf{R}$. Hence, by Corollary 2.6.14(b), *$B$ is a basis of $\mathbf{R}_n$ over $\mathbf{R}$.* Let us now compute the matrix of our original linear transformation $\mathbf{T}$ with respect to the basis $B$. To do so, we must see what $\mathbf{T}$ does to each element of $B$; equivalently, what $\mathbf{T}^*$ does to these elements, since $\mathbf{T}$ and $\mathbf{T}^*$ coincide on elements of $\mathbf{R}_n$.

For the elements $\mathbf{u}_j$ of $B_U$ this is simple: $U^*$ is the space of eigenvectors corresponding to the eigenvalue 1 of $\mathbf{T}^*$, so that

(9.4.5) $$\mathbf{u}_j\mathbf{T} = \mathbf{u}_j\mathbf{T}^* = 1\mathbf{u}_j = \mathbf{u}_j \qquad (1 \leqslant j \leqslant d).$$

Similarly, the vectors $\mathbf{v}_j$ in $B_V$ are $(-1)$-eigenvectors, so that

(9.4.6) $$\mathbf{v}_j\mathbf{T} = \mathbf{v}_j\mathbf{T}^* = -\mathbf{v}_j \qquad (1 \leqslant j \leqslant e).$$

The vectors $\mathbf{w}_j^{(k)}$ in 9.4.4 belong to $N_k$ and are, thus, $\lambda_k$-eigenvectors, while the vectors $\overline{\mathbf{w}_j^{(k)}}$ are $\overline{\lambda}_k$-eigenvectors. If we write $\lambda_k = a_k + b_k i$ ($a_k, b_k \in \mathbf{R}$), we obtain

$$
\begin{aligned}
\mathbf{x}_j^{(k)}\mathbf{T} &= \mathbf{x}_j^{(k)}\mathbf{T}^* = (\mathbf{w}_j^{(k)}\mathbf{T}^* + \overline{\mathbf{w}_j^{(k)}\mathbf{T}^*})/\sqrt{2} \\
&= (\lambda_k \mathbf{w}_j^{(k)} + \overline{\lambda_k}\ \overline{\mathbf{w}_j^{(k)}})/\sqrt{2} \\
&= \frac{1}{\sqrt{2}}[(a_k + b_k i)\mathbf{w}_j^{(k)} + (a_k - b_k i)\overline{\mathbf{w}_j^{(k)}}] \\
&= a_k[(\mathbf{w}_j^{(k)} + \overline{\mathbf{w}_j^{(k)}})/\sqrt{2}] + b_k[i(\mathbf{w}_j^{(k)} - \overline{\mathbf{w}_j^{(k)}})/\sqrt{2}] \\
&= a_k \mathbf{x}_j^{(k)} + b_k \mathbf{y}_j^{(k)}.
\end{aligned}
$$

A similar computation (which we leave to you) gives

$$\mathbf{y}_j^{(k)}\mathbf{T} = \mathbf{y}_j^{(k)}\mathbf{T}^* = -b_k \mathbf{x}_j^{(k)} + a_k \mathbf{y}_j^{(k)}.$$

Moreover, $a_k^2 + b_k^2 = |\lambda_k|^2 = 1$ so that $a_k = \cos\theta_k$, $b_k = \sin\theta_k$ for some oriented angle $\theta_k$. We thus have

(9.4.7)
$$
\begin{aligned}
\mathbf{x}_j^{(k)}\mathbf{T} &= (\cos\theta_k)\mathbf{x}_j^{(k)} + (\sin\theta_k)\mathbf{y}_j^{(k)} \\
\mathbf{y}_j^{(k)}\mathbf{T} &= (-\sin\theta_k)\mathbf{x}_j^{(k)} + (\cos\theta_k)\mathbf{y}_j^{(k)}.
\end{aligned}
$$

It follows from equations 9.4.5, 9.4.6, and 9.4.7 that the matrix of $\mathbf{T}$ with respect to the basis $B$ has the block form

(9.4.8)
$$A = \begin{bmatrix} I &&&&&&& \\ & -I &&&&&& \\ && A_1^{(1)} &&&&& \\ &&& \ddots &&&& \\ &&&& A_{m_1}^{(1)} &&& \\ &&&&& \ddots && \\ &&&&&& A_1^{(r)} & \\ &&&&&&& \ddots \\ &&&&&&&& A_{m_r}^{(r)} \end{bmatrix}$$

where

(9.4.9)
$$A_j^{(k)} = \begin{bmatrix} \cos\theta_k & \sin\theta_k \\ -\sin\theta_k & \cos\theta_k \end{bmatrix} \qquad (1 \leq k \leq r,\ 1 \leq j \leq m_k).$$

Note that $\theta_k$ is not 0° or 180°, since $\sin\theta_k$ is the imaginary part of the nonreal eigenvalue $\lambda_k$ and, hence, is nonzero. If we let $U$, $V$, $S_j^{(k)}$ be, respectively, the subspaces of $\mathbf{R}_n$ spanned by the sets $B_U$, $B_V$, $B_j^{(k)}$ (respectively), then the blocks $I$, $-I$, $A_j^{(k)}$ in 9.4.8 are the matrices of $\mathbf{T} \restriction U$, $\mathbf{T} \restriction V$, $\mathbf{T} \restriction S_j^{(k)}$, respectively, and

(9.4.10)
$$\mathbf{R}_n = U \oplus V \oplus S_1^{(1)} \oplus \cdots \oplus S_{m_1}^{(1)} \oplus \cdots \oplus S_1^{(r)} \oplus \cdots \oplus S_{m_r}^{(r)}.$$

The summands in 9.4.10 are still mutually orthogonal (since the basis $B$ is an orthonormal set); they are also $\mathbf{T}$-stable since the images of the basis vectors of each subspace lie in the subspace. It is possible that $U$ or $V$ (or both) may be $\{\mathbf{0}\}$; thus, in 9.4.8, the block $I$ or $-I$ (or both) may have degree 0 (i.e., may fail to appear).

The mapping $\mathbf{T}^*$ has now served its purpose; it has enabled us to find a simple form for the matrix of $\mathbf{T}$. *We shall thus put* $\mathbf{T}^*$ *aside and shall work entirely within the space* $\mathbf{R}_n$ *for the remainder of this section.* The block $-I$ in 9.4.8 has degree $e$ (the dimension of $V$). Let $m_0$ be the greatest integer not exceeding $e/2$; then $e = 2m_0$ or $e = 2m_0 + 1$. If we let

$$A_j^{(0)} = \begin{bmatrix} -1 & 0 \\ 0 & -1 \end{bmatrix} = -I_2 \qquad (j = 1, \ldots, m_0)$$

then $A_j^{(0)}$ is of the form 9.4.9 (with $\theta_0 = 180°$), and

$$\textbf{(9.4.11)} \quad -I_e = \begin{bmatrix} -I_t & & & \\ & A_1^{(0)} & & \\ & & \ddots & \\ & & & A_{m_0}^{(0)} \end{bmatrix}$$

where $t = 0$ or 1. Corresponding to the decomposition 9.4.11 of $-I_e$ into blocks, we have a decomposition of the subspace $V$ into smaller subspaces

$$V = V_* \oplus S_1^{(0)} \oplus \cdots \oplus S_{m_0}^{(0)}$$

where $\dim V_* = 0$ or 1, $\dim S_j^{(0)} = 2$, and the summands are $\mathbf{T}$-stable and mutually orthogonal (why?).

It will now be convenient to change notation as follows: if

$$m = \sum_{k=0}^{r} m_k$$

then $m$ is the total number of subspaces $S_j^{(k)}$, and we shall now denote these subspaces $W_1, \ldots, W_m$. Furthermore, if $S_j^{(k)} = W_h$, then we shall write $A_h$ instead of $A_j^{(k)}$, $B_h$ instead of $B_j^{(k)}$, $\theta_h$ instead of $\theta_k$, and so on. In our new notation, we can summarize our results up to this point as follows:

**9.4.12 Theorem.** Let $\mathbf{T}$ be a linear isometry of $\mathbf{R}_n$. Then there exist subspaces $U, V_*, W_1, \ldots, W_m$ of $\mathbf{R}_n$ such that the following statements hold:

(a) $\mathbf{R}_n = U \oplus V_* \oplus W_1 \oplus \cdots \oplus W_m$.

(b) The summands on the right side of (a) are mutually orthogonal, and each summand is $\mathbf{T}$-stable.

(c) $\dim V_* = 0$ or 1.

(d) $\mathbf{T} \restriction U = \mathbf{I}$, and $\mathbf{T} \restriction V_* = -\mathbf{I}$.

(e) For each $j$ $(1 \leqslant j \leqslant m)$, $\dim W_j = 2$ and the matrix of $\mathbf{T} \restriction W_j$ has the form

$$A_j = \begin{bmatrix} \cos\theta_j & \sin\theta_j \\ -\sin\theta_j & \cos\theta_j \end{bmatrix} \qquad (\theta_j \neq 0°)$$

with respect to an orthonormal basis of $W_j$; that is, $\mathbf{T} \restriction W_j$ is a rotation through a nonzero angle. (*Note*: $m$ could be 0, in which case there are no subspaces $W_j$.)

Let us interpret Theorem 9.4.12 geometrically. The interpretation is simplest when $n$ is small; for instance, if $n = 3$, the equations dim $W_i = 2$, dim $V_* \leq 1$ imply that there are only four possibilities for the expression 9.4.12(a), namely:

(a) $\mathbf{R}_3 = U \oplus W_1$; dim $U = 1$.
(b) $\mathbf{R}_3 = V_* \oplus W_1$; dim $V_* = 1$.
(c) $\mathbf{R}_3 = U$.
(d) $\mathbf{R}_3 = U \oplus V_*$; dim $V_* = 1$, dim $U = 2$.

Case (c) is trivial: $\mathbf{T} = \mathbf{I}$. In all other cases, $\mathbf{R}_3$ is the direct sum of a 2-dimensional $\mathbf{T}$-stable subspace and a 1-dimensional space of eigenvectors; this was precisely the assertion of Theorem 6.5.1. You should have no difficulty in showing that cases (a), (d), (b) correspond, respectively, to the three types (a), (b), (c) listed in Theorem 6.5.3.

In the general case ($n$ arbitrary), Theorem 9.4.12 leads to the following result:

**9.4.13 Theorem.** Let $\mathbf{T}$ be any linear isometry of $\mathbf{R}_n$. If $U$, $V_*$, $W_1$, ..., $W_m$ are as in Theorem 9.4.12, then

(9.4.14) $$\mathbf{T} = \mathbf{T}_*\mathbf{T}_1 \cdots \mathbf{T}_m$$

where $\mathbf{T}_*$ is either the identity mapping (if $V_* = \{\mathbf{0}\}$) or the reflection through the $(n-1)$-dimensional subspace $V_*^{\perp}$ (if dim $V_* = 1$); each $\mathbf{T}_i$ $(1 \leq i \leq m)$ is a rotation about $W_i^{\perp}$ through a nonzero angle; and the $m + 1$ factors $\mathbf{T}_*$, $\mathbf{T}_1$, ..., $\mathbf{T}_m$ commute with each other. Moreover, $m \leq n/2$; and if $\mathbf{T}_*$ is a reflection, then $m \leq (n-1)/2$.

*Proof.* For each subspace $W$ appearing on the right side of 9.4.12(a) ($W = U, V_*, W_1, ..., W_m$), the sum of the other $m + 1$ subspaces is orthogonal to $W$ (by 9.4.12(b)) and, hence, must equal $W^{\perp}$ by a dimension argument. Since every element of $\mathbf{R}_n$ can be written uniquely in the form $\mathbf{w} + \mathbf{w}'$ ($\mathbf{w} \in W$, $\mathbf{w}' \in W^{\perp}$), the mapping $\mathbf{T}_W : \mathbf{R}_n \to \mathbf{R}_n$ defined by

(9.4.15) $$(\mathbf{w} + \mathbf{w}')\mathbf{T}_W = \mathbf{w}\mathbf{T} + \mathbf{w}' \qquad (\mathbf{w} \in W, \mathbf{w}' \in W^{\perp})$$

is well defined. It is easy to see that $\mathbf{T}_W$ is linear; $\mathbf{T}_W$ agrees with $\mathbf{T}$ on elements of $W$ but fixes all elements of $W^{\perp}$. To simplify notation, we write $\mathbf{T}_*$ in place of $\mathbf{T}_{V_*}$, $\mathbf{T}_i$ in place of $\mathbf{T}_{W_i}$. Thus, if $\mathbf{v}$ is any vector in $\mathbf{R}_n$ and we write

(9.4.16) $$\mathbf{v} = \mathbf{u} + \mathbf{v}_* + \mathbf{w}_1 + \cdots + \mathbf{w}_m \qquad (\mathbf{u} \in U, \mathbf{v}_* \in V_*, \mathbf{w}_i \in W_i)$$

(cf. 9.4.12(a)), then $\mathbf{vT}_*$ is obtained from $\mathbf{v}$ by replacing $\mathbf{v}_*$ by $\mathbf{v}_*\mathbf{T}$ in 9.4.16, and $\mathbf{vT}_i$ is obtained from $\mathbf{v}$ by replacing $\mathbf{w}_i$ by $\mathbf{w}_i\mathbf{T}$. It easily follows that the effect of applying the mappings $\mathbf{T}_*, \mathbf{T}_1, \ldots, \mathbf{T}_m$ in succession (in any order) is the same as the effect of applying $\mathbf{T}$. Hence, $\mathbf{T} = \mathbf{T}_*\mathbf{T}_1, \cdots \mathbf{T}_m$ and the $m + 1$ factors on the right side commute with each other. By comparing 9.4.12(e) and 9.4.15 with Definition 6.4.2, we see that $\mathbf{T}_i$ is a rotation about $W_i^{\perp}$. The description of $\mathbf{T}_*$ in Theorem 9.4.13 follows from Section 3.8, Exercise 13. Finally, the inequalities for $m$ follow from the fact that the sum of the dimensions of $U, V_*, W_1, \ldots, W_m$ is $n$. ∎

So far we have discussed only isometries which are linear; let us now see what happens when the requirement of linearity is removed. If $\mathbf{M}$ is *any* isometry of $\mathbf{R}_n$, then by Theorem 4.10.6 there exists a linear isometry $\mathbf{T}$ and a translation $\mathscr{T}_{\mathbf{v}}$ such that $\mathbf{M} = \mathbf{T}\mathscr{T}_{\mathbf{v}}$. Using the notations of Theorem 9.4.12, we may write $\mathbf{v}$ in the form 9.4.16 and $\mathbf{T}$ in the form 9.4.14. Then

(9.4.17) $$\mathbf{M} = \mathbf{T}\mathscr{T}_{\mathbf{v}} = \mathbf{T}_*\mathbf{T}_1 \cdots \mathbf{T}_m\mathscr{T}_{\mathbf{u}}\mathscr{T}_{\mathbf{v}_*}\mathscr{T}_{\mathbf{w}_1} \cdots \mathscr{T}_{\mathbf{w}_m}.$$

We know that $\mathbf{T}_*, \mathbf{T}_1, \ldots, \mathbf{T}_m$ commute with each other, and that any two translations commute with each other. In addition, it is easy to verify (from the definitions of the various mappings in question) that

$\mathscr{T}_{\mathbf{u}}$ commutes with all of the mappings $\mathbf{T}_*, \mathbf{T}_1, \ldots, \mathbf{T}_m$;
$\mathscr{T}_{\mathbf{v}_*}$ commutes with all of the mappings $\mathbf{T}_1, \ldots, \mathbf{T}_m$;
$\mathscr{T}_{\mathbf{w}_i}$ commutes with all of the mappings $\mathbf{T}_*, \mathbf{T}_1, \ldots, \mathbf{T}_m$ except possibly $\mathbf{T}_i$.

Hence, we may rewrite 9.4.17 in the form

$$\mathbf{M} = (\mathbf{T}_1\mathscr{T}_{\mathbf{w}_1})(\mathbf{T}_2\mathscr{T}_{\mathbf{w}_2}) \cdots (\mathbf{T}_m\mathscr{T}_{\mathbf{w}_m})(\mathbf{T}_*\mathscr{T}_{\mathbf{v}_*})\mathscr{T}_{\mathbf{u}}.$$

For each $i$ ($1 \leqslant i \leqslant m$), $\mathbf{T}_i\mathscr{T}_{\mathbf{w}_i}$ is a rotation about some $(n-2)$-flat parallel to $W_i^{\perp}$, by Theorem 6.4.4. Also, $\mathbf{T}_*\mathscr{T}_{\mathbf{v}_*}$ is the reflection $\mathscr{M}_S$ through the flat $S = V_*^{\perp} + \frac{1}{2}\mathbf{v}_*$, by Theorem 3.8.30. (If dim $V_* = 0$, then $V_* = \mathbf{R}_n$ so that $\mathscr{M}_S = \mathbf{I}$.) Finally, $\mathbf{u} \in U$ and, hence, $\mathbf{u}$ is orthogonal to every one of the subspaces $W_1, \ldots, W_m, V_*$ by 9.4.12(b). We summarize our results as follows:

**9.4.18 Theorem.** Let $\mathbf{M}$ be any isometry of $\mathbf{R}_n$. Then for some integer $m \geq 0$, there exist mutually orthogonal subspaces $W_1, \ldots, W_m$, $V_*$ of $\mathbf{R}_n$ such that $\dim W_i = 2$ $(1 \leq i \leq m)$, $\dim V_* = 0$ or $1$, and

$$\mathbf{M} = \mathcal{R}_1 \cdots \mathcal{R}_m \mathcal{M}_S \mathcal{T}_{\mathbf{u}} \tag{9.4.19}$$

where (a) each $\mathcal{R}_i$ $(1 \leq i \leq m)$ is a nontrivial rotation about some $(n-2)$-flat $S_i$ parallel to $W_i^{\perp}$; (b) $S$ is a flat parallel to $V_*^{\perp}$ (so that $S$ is a hyperplane if $\dim V_* = 1$, while $\mathcal{M}_S = \mathbf{I}$ if $V_* = \{\mathbf{0}\}$); and (c) $\mathbf{u}$ is a vector orthogonal to all of the subspaces $W_1, \ldots, W_m, V_*$.

Moreover, $m \leq n/2$, and if $\dim V_* = 1$, then $m \leq (n-1)/2$.

For example, when $n = 3$, we must have $m = 0$ or $1$, so that the right side of 9.4.19 reduces to one of the following: $\mathcal{R}_1\mathcal{M}_S$, $\mathcal{R}_1\mathcal{T}_{\mathbf{u}}$, $\mathcal{M}_S\mathcal{T}_{\mathbf{u}}$, $\mathcal{T}_{\mathbf{u}}$. (Note that if $2m + \dim V_* = n$, then $\dim U = 0$, $\mathbf{u} = \mathbf{0}$, and the term $\mathcal{T}_{\mathbf{u}} = \mathbf{I}$ drops out of the equation.) It is not hard to verify that these four possibilities agree with those listed in Section 6.5, Exercise 5. (Both $\mathcal{R}_1\mathcal{T}_{\mathbf{u}}$ and $\mathcal{T}_{\mathbf{u}}$ correspond to part (c) of the exercise, bearing in mind that a "rotation" may be $\mathbf{I}$ if not otherwise specified.)

**9.4.20 Theorem.** Using the same notations as in Theorem 9.4.19, $\mathbf{M}$ has determinant 1 if and only if $\mathcal{M}_S = \mathbf{I}$ (equivalently, $V_* = \{\mathbf{0}\}$); that is, if and only if $\mathbf{M}$ is a product of rotations or a product of rotations times a translation.

*Proof.* By Theorem 6.4.5(d), all rotations have determinant 1, and a translation clearly has determinant 1 since its linear part is $\mathbf{I}$. On the other hand, any reflection through an $(n-1)$-flat has determinant $-1$ by Theorem 5.6.13. Theorem 9.4.20 follows at once. (Recall that the "only if" part of this theorem was "borrowed" at the end of Sec. 6.6.) ∎

**Exercises**

1. Various assertions in this section were left to you to verify. Do so.
2. Prove: every rotation in $\mathbf{R}_n$ is a product of two reflections through hyperplanes. (This generalizes Sec. 6.5, Exercise 6(a).)
3. (a) Prove that every isometry of $\mathbf{R}_n$ is a product of at most $n + 1$ reflections through hyperplanes.

   (b) Prove: if $\mathbf{S}_1, \ldots, \mathbf{S}_h, \mathbf{S}_1', \ldots, \mathbf{S}_k'$ are reflections through hyperplanes such that

$\mathbf{S}_1 \cdots \mathbf{S}_h = \mathbf{S}'_1 \cdots \mathbf{S}'_k$,

then $h \equiv k \pmod 2$.

4. Using the notations of Theorem 9.4.18,

(a) Prove that for all points $P \in \mathbf{R}_n$, we have $P\mathcal{R}_i - P \in W_i$, $P\mathcal{M}_S - P \in V_*$, $P\mathcal{T}_\mathbf{u} - P = \mathbf{u}$.

(b) Deduce from part (a): if $P$ is fixed by $\mathbf{M}$, then $P$ is fixed by each of the mappings $\mathcal{R}_1, \ldots, \mathcal{R}_m, \mathcal{M}_S, \mathcal{T}_\mathbf{u}$.

(c) Deduce from part (b): $\mathbf{M}$ has fixed points if and only if $\mathcal{T}_\mathbf{u} = \mathbf{I}$ (i.e., $\mathbf{u} = \mathbf{0}$), and in this case the set of fixed points of $\mathbf{M}$ is a flat with direction space $U = (W_1 + \cdots + W_m + V_*)^\perp$. (*Suggestion*: Use Theorem 3.4.11 and certain exercises from Sec. 3.8.)

## 9.5 Eigenvectors of Hermitian and real symmetric matrices

In Section 7.1 we asserted that if $\mathbf{T} \in \text{Lin}(\mathbf{R}_n, \mathbf{R}_n)$, and if the matrix of $\mathbf{T}$ (with respect to some orthonormal basis) is symmetric, then $\mathbf{R}_n$ has an orthonormal basis consisting of eigenvectors of $\mathbf{T}$. In this section, we give both a proof and a generalization (Theorem 9.5.7 below). The generalization requires the following definition.

**9.5.1 Definition.** A square matrix $A$ over $\mathbf{C}$ is said to be *Hermitian* if $A = \overline{A}^t$; that is, if $a_{ij} = \overline{a_{ji}}$ for all $i, j$.

In particular, if $A$ has real entries, then $\overline{A} = A$ so that $A$ is Hermitian $\Leftrightarrow$ $A$ is symmetric.

We begin with the following result, which we will need as a lemma but has some interest in itself.

**9.5.2 Theorem.** Let $V$ be a subspace of either $\mathbf{R}_n$ or $\mathbf{C}_n$; let $\mathbf{T} \in \text{Lin}(V, V)$, and let $A$ be the matrix of $\mathbf{T}$ with respect to some orthonormal basis of $V$. Then $A$ is Hermitian if and only if

$$\langle \mathbf{u}\mathbf{T}, \mathbf{v} \rangle = \langle \mathbf{u}, \mathbf{v}\mathbf{T} \rangle \qquad (\text{all } \mathbf{u}, \mathbf{v} \in V). \tag{9.5.3}$$

*Proof.* Let the given orthonormal basis of $V$ be $(\mathbf{u}_1, \ldots, \mathbf{u}_r)$. Then

$$\mathbf{u}_i\mathbf{T} = \sum_{k=1}^{r} a_{ik}\mathbf{u}_k \qquad (\text{all } i);$$

hence, for all $i$ and $j$,

(9.5.4) $$\langle \mathbf{u}_i\mathbf{T}, \mathbf{u}_j\rangle = \sum_{k=1}^{r} a_{ik}\langle \mathbf{u}_k, \mathbf{u}_j\rangle = \sum_{k=1}^{r} a_{ik}\delta_{kj} = a_{ij}.$$

Interchanging the roles of $i$ and $j$,

$$\langle \mathbf{u}_j\mathbf{T}, \mathbf{u}_i\rangle = a_{ji}.$$

Thus, by Theorem 9.1.9(a), we have

(9.5.5) $$\langle \mathbf{u}_i, \mathbf{u}_j\mathbf{T}\rangle = \overline{\langle \mathbf{u}_j\mathbf{T}, \mathbf{u}_i\rangle} = \overline{a_{ji}}.$$

Comparing 9.5.4 with 9.5.5, we see that $A$ is Hermitian if and only if 9.5.3 holds for the vectors $\mathbf{u}_1, \ldots, \mathbf{u}_r$. But the latter easily implies that 9.5.3 holds for *all* vectors $\mathbf{u}, \mathbf{v} \in V$. (*Method*: Express $\mathbf{u}$ and $\mathbf{v}$ as linear combinations of the basis vectors and expand.) The theorem follows. ∎

9.5.6 **Theorem.** If $A$ is a Hermitian matrix, then all complex roots of char($A$; $x$) are real.

*Proof.* Let $\lambda$ be any complex root of char($A$; $x$). Let $n$ be the degree of $A$; then $A$ corresponds to a certain linear transformation $\mathbf{T} : \mathbf{C}_n \to \mathbf{C}_n$ (with respect to the canonical basis, if we want to be specific), and $\lambda$ is an eigenvalue of $\mathbf{T}$. Thus, there exists a nonzero vector $\mathbf{u} \in \mathbf{C}_n$ such that $\mathbf{uT} = \lambda\mathbf{u}$. By 9.1.9(c), we have

$$\lambda\langle \mathbf{u}, \mathbf{u}\rangle = \langle \lambda\mathbf{u}, \mathbf{u}\rangle = \langle \mathbf{uT}, \mathbf{u}\rangle = \langle \mathbf{u}, \mathbf{uT}\rangle \qquad \text{(by Theorem 9.5.2)}$$

$$= (\mathbf{u}, \lambda\mathbf{u}\rangle = \bar{\lambda}\langle \mathbf{u}, \mathbf{u}\rangle$$

and since $\langle \mathbf{u}, \mathbf{u}\rangle \neq 0$ it follows that $\lambda = \bar{\lambda}$, that is, $\lambda$ is real. ∎

9.5.7 **Theorem.** Let $V$ be a subspace of either $\mathbf{R}_n$ or $\mathbf{C}_n$; let $\mathbf{T} \in$ Lin($V$, $V$), and assume that the matrix of $\mathbf{T}$, with respect to some orthonormal basis of $V$, is Hermitian.

Then $V$ has an orthonormal basis $\{\mathbf{u}_1, \ldots, \mathbf{u}_r\}$ such that all of the $\mathbf{u}_i$ ($1 \leq i \leq r$) are eigenvectors of $\mathbf{T}$.

*Proof.* The theorem being trivial if dim $V = 0$, we may assume by induction that $V$ has positive dimension $r$ and that the theorem holds for all subspaces having dimension less than $r$.

By the Fundamental Theorem of Algebra, char($\mathbf{T}$; $x$) has a complex root $\lambda$; by Theorem 9.5.6, $\lambda$ is real. Hence, $\lambda$ is an eigenvalue of $\mathbf{T}$, *regardless of whether V is a subspace of* $\mathbf{R}_n$ *or* $\mathbf{C}_n$, since in either case $\lambda$ lies in the field of scalars. Thus, $\mathbf{u}_1\mathbf{T} = \lambda\mathbf{u}_1$ for some $\mathbf{u}_1 \neq \mathbf{0}$ in $V$. By replacing $\mathbf{u}_1$ by $\mathbf{u}_1/|\mathbf{u}_1|$, we may assume that $|\mathbf{u}_1| = 1$.

We claim that the subspace

$$W = \{\mathbf{v} \in V : \langle \mathbf{v}, \mathbf{u}_1 \rangle = 0\} = V \cap [\mathbf{u}_1]^{\perp}$$

is **T**-stable. Indeed, if $\mathbf{v} \in W$, then $\langle \mathbf{v}, \mathbf{u}_1 \rangle = 0$ so that

$$\begin{aligned} \langle \mathbf{vT}, \mathbf{u}_1 \rangle &= \langle \mathbf{v}, \mathbf{u}_1\mathbf{T} \rangle \qquad \text{(by 9.5.2)} \\ &= \langle \mathbf{v}, \lambda\mathbf{u}_1 \rangle = \bar{\lambda}\langle \mathbf{v}, \mathbf{u}_1 \rangle = \bar{\lambda}0 = 0 \end{aligned}$$

and, hence, $\mathbf{vT} \in W$, proving our claim.

Since $W$ is **T**-stable, $\mathbf{T} \restriction W$ belongs to Lin($W$, $W$); since 9.5.3 holds for all $\mathbf{u}, \mathbf{v} \in V$, it certainly holds for all $\mathbf{u}, \mathbf{v} \in W$ and, hence, the matrix of $\mathbf{T} \restriction W$ is Hermitian. Moreover, $\dim W = \dim V - 1 = r - 1$ by Theorem 9.1.15(b). Hence, by induction, Theorem 9.5.7 holds for ($\mathbf{T} \restriction W$, $W$) in place of (**T**, $V$); that is, $W$ has an orthonormal basis $\{\mathbf{u}_2, ..., \mathbf{u}_r\}$ whose members are eigenvectors of $\mathbf{T} \restriction W$ (hence, automatically eigenvectors of **T**). In addition, $\mathbf{u}_1$ is an eigenvector of **T**, $|\mathbf{u}_1| = 1$, and $\mathbf{u}_1$ is orthogonal to $\mathbf{u}_2, ..., \mathbf{u}_r$ by definition of $W$. It follows that $\{\mathbf{u}_1, ..., \mathbf{u}_r\}$ is an orthonormal basis of $V$ consisting entirely of eigenvectors of **T**. ▮

**9.5.8 Corollary.** If $A$ is a Hermitian $n \times n$ matrix, then there exist $n$ orthonormal vectors $\mathbf{u}_1, ..., \mathbf{u}_n \in \mathbf{C}_n$ such that, if each $\mathbf{u}_i$ is regarded as a $1 \times n$ row matrix, then

$$\mathbf{u}_i A = \lambda_i \mathbf{u}_i \qquad (i = 1, ..., n)$$

for certain real numbers $\lambda_1, ..., \lambda_n$. If $A$ is real-symmetric, the vectors $\mathbf{u}_i$ may be chosen to belong to $\mathbf{R}_n$.

**Exercises**

1. Give an alternate proof of Theorem 9.5.7 along the following lines:
   (a) Use Theorem 9.5.6 to show that the matrix of **T** with respect to some basis of $V$ is a Jordan matrix $J$.
   (b) Prove that $J$ is diagonal. (*Hint*: See the proof of Theorem 9.3.4(c).)
   (c) Prove that eigenvectors corresponding to different eigenvalues of **T** are orthogonal.
   (d) Deduce 9.5.7 from the above results.
2. Deduce from Theorem 9.5.7 that if $A$ is a Hermitian matrix, then there exists a nonsingular matrix $P$ such that $P^{-1} = \bar{P}^t$ and $P^{-1}AP$ is a diagonal matrix with real entries. Moreover, if the entries in $A$ are real, then $P$ may be chosen to have real entries. (SOL)

3. In each case, find a matrix $P$ satisfying the conditions of the preceding exercise.

   (a) $A = \begin{bmatrix} 0 & i & i \\ -i & 0 & -1 \\ -i & -1 & 0 \end{bmatrix}$. (SOL)

   (b) $A = \begin{bmatrix} 19 & -4 & 10 \\ -4 & 7 & 14 \\ 10 & 14 & 10 \end{bmatrix}$. (ANS)

*4. Prove: if $A$, $B$ are $n \times n$ matrices over $\mathbf{C}$ such that $A$ is Hermitian, then $BA\overline{B}^t$ is Hermitian.

5. Prove the following partial converse of Theorems 9.5.6–9.5.7: let $V$ be a subspace of either $\mathbf{R}_n$ or $\mathbf{C}_n$, let $\mathbf{T} \in \text{Lin}(V, V)$, and assume that (a) the eigenvalues of $\mathbf{T}$ are all real, and (b) $V$ has an orthonormal basis consisting of eigenvectors of $\mathbf{T}$. Then the matrix of $\mathbf{T}$, with respect to any given orthonormal basis of $V$, is Hermitian. (SUG)

## 9.6 Symmetric and Hermitian forms

In studying conics and quadrics in Chapter 7, we were led to consider an expression of the form

$$\sum_{i=1}^{n} \sum_{j=1}^{n} a_{ij} x_i x_j \tag{9.6.1}$$

where the matrix $A = (a_{ij})$ was symmetric. More generally, we can consider the expression

$$\sum_{i=1}^{n} \sum_{j=1}^{n} a_{ij} x_i \overline{x_j}$$

where the $x_i$ belong to $\mathbf{C}$ rather than $\mathbf{R}$ and the matrix $A$ is Hermitian rather than real-symmetric. Such expressions arise often enough in mathematics to deserve some study. (We shall see them in Sec. 9.8 in connection with the second-derivative test.) We thus introduce the following definitions:

**9.6.2 Definition.** A *symmetric form on* $\mathbf{R}_n$ (more simply, a *real symmetric form*) is a function $f: \mathbf{R}_n \to \mathbf{R}$ which has the form

$$f(x_1, \ldots, x_n) = \sum_{i=1}^{n} \sum_{j=1}^{n} a_{ij} x_i x_j \qquad (\text{all } (x_1, \ldots, x_n) \in \mathbf{R}_n), \tag{9.6.3}$$

where the matrix $A = (a_{ij})$ is an $n \times n$ symmetric matrix over **R**. We call $f$ the symmetric form associated with the matrix $A$, and $A$ is called the matrix of $f$.

**9.6.4 Definition.** A *Hermitian form on* $\mathbf{C}_n$ is a function $f: \mathbf{C}_n \to \mathbf{C}$ which has the form

$$f(x_1, \ldots, x_n) = \sum_{i=1}^{n} \sum_{j=1}^{n} a_{ij} x_i \overline{x_j} \qquad (\text{all } (x_1, \ldots, x_n) \in \mathbf{C}_n), \tag{9.6.5}$$

where the matrix $A = (a_{ij})$ is an $n \times n$ Hermitian matrix over **C**. We call $f$ the Hermitian form associated with the matrix $A$, and $A$ is called the matrix of $f$.

*Comments.* (1) The matrix of a symmetric or Hermitian form $f$ is uniquely determined by $f$ (see Exercises 1 and 2 below). (2) We could be slightly more general and define Hermitian forms on subspaces of $\mathbf{C}_n$, but this will not be necessary for our purposes. (3) In both 9.6.3 and 9.6.5, we have used standard function notation: $f(X)$ instead of $Xf$ to denote the image of $X$ under $f$. Since $f$ is a mapping of a somewhat different character from the other mappings which we have studied, this change in notation should cause no confusion.

If we think of the vector $X = (x_1, \ldots, x_n)$ as a row matrix, then the right side of 9.6.3 is the sole entry in the $1 \times 1$ matrix $XAX^t$ (cf. eq. 7.1.17). Thus, if we identify any $1 \times 1$ matrix $[c]$ with the scalar $c$ itself, we may rewrite equation 9.6.3 in the form

$$f(X) = XAX^t.$$

A similar argument shows that equation 9.6.5 is the same as

$$f(X) = XA\overline{X}^t.$$

By generalizing the proof of equation 7.1.19 to the complex case (using the complex form of Theorem 9.5.7), we obtain the following theorem:

**9.6.6 Theorem.** Let $f$ be either a symmetric form on $\mathbf{R}_n$ or a Hermitian form on $\mathbf{C}_n$; let $A$ be the matrix of $f$, and let $\mathbf{u}_1, \ldots, \mathbf{u}_n$, $\lambda_1 \ldots, \lambda_n$ be as in Corollary 9.5.8. Then for all vectors $X = (x_1, \ldots, x_n)$ in the domain of $f$,

$$f(X) = \sum_{i=1}^{n} \lambda_i |x_i'|^2,$$

where $x_1', \ldots, x_n'$ are the components of $X$ with respect to the ordered basis $(\mathbf{u}_1, \ldots, \mathbf{u}_n)$.

Since the $\lambda_i$ are all real, we obtain the following corollary of Theorem 9.6.6:

**9.6.7** **Theorem.** If $f$ is a Hermitian form, then $f(X)$ is a real number for all $X$.

**Exercises**

1. Let $f$, $A$ satisfy equation 9.6.3; that is, $A$ is a real symmetric matrix and $f$ is the corresponding symmetric form.
   (a) Prove that $f(\mathbf{e}_i) = a_{ii}$ for all $i$, and $f(\mathbf{e}_i + \mathbf{e}_j) = a_{ii} + 2a_{ij} + a_{jj}$ for all $i$ and $j$, where $\mathbf{e}_1, \ldots, \mathbf{e}_n$ denote (as usual) the vectors in the canonical basis.
   (b) Deduce from part (a) that $A$ is uniquely determined by $f$; that is, if $B = (b_{ij})$ is another real symmetric matrix such that
   $$\sum_{i=1}^{n}\sum_{j=1}^{n} a_{ij}x_ix_j = \sum_{i=1}^{n}\sum_{j=1}^{n} b_{ij}x_ix_j$$
   for all $(x_1, \ldots, x_n) \in \mathbf{R}_n$, then $B = A$.
2. State and prove a result for Hermitian forms analogous to Exercise 1 for real symmetric forms.

## 9.7 Positive definite matrices

**9.7.1** **Definition.** Let $f$ be either a symmetric form on $\mathbf{R}_n$ or a Hermitian form on $\mathbf{C}_n$; let $A$ be the matrix of $f$. We say that $f$ (or $A$) is *positive definite* if $f(X) > 0$ for all vectors $X \neq \mathbf{0}$ in the domain of $f$. We say that $f$ (or $A$) is *positive semidefinite* if $f(X) \geqslant 0$ for all $X$.

The main result of this section (Theorem 9.7.4) shows that positive definiteness of a matrix may be characterized entirely in terms of matrix properties, without reference to the corresponding form $f$. In the next section this characterization will be applied to the problem of finding relative extrema of real-valued functions of $n$ variables. As usual, we need a few preliminary results.

**9.7.2** **Theorem.** A Hermitian (or real symmetric) matrix $A$ is positive definite if and only if all of its eigenvalues are positive.

*Proof.* Let $f$ be the associated form. If all the eigenvalues $\lambda_i$ are positive, then Theorem 9.6.6 implies that $f(X) > 0$ unless $x_i' = 0$ for all $i$, in which case $X = \Sigma x_i'\mathbf{u}_i = \mathbf{0}$. Thus, $f$ is positive definite. Conversely, suppose some eigenvalue $\lambda_i$ is not positive; since $f(\mathbf{u}_i) = \lambda_i$ by Theorem 9.6.6, it follows that $f$ is not positive definite. ∎

*Remark*: If $A$ is a real symmetric matrix corresponding to the real symmetric form $f$, then $A$ is *a priori* Hermitian and, hence, corresponds to a Hermitian form $f^*$. Since the complex eigenvalues of $A$ are the same as the real eigenvalues (Theorem 9.5.6), it follows that positive definiteness of $A$ is unaffected by whether we regard $A$ as a real symmetric matrix or as a complex Hermitian matrix. In other words, *$f$ is positive definite if and only if $f^*$ is*. It follows that in discussing positive definiteness it will suffice to treat the Hermitian case only. For the remainder of the section, we shall do so.

**9.7.3** **Theorem.** Let $A$ be a positive definite (Hermitian) $n \times n$ matrix. Then:

(a) $A_S$ is Hermitian and positive definite, for every subset $S$ of $E_n$. (For the notations $A_S$ and $E_n$, see the beginning of Sec. 5.12.)

(b) $BA\overline{B}^t$ is Hermitian and positive definite, for every nonsingular $n \times n$ matrix $B$ over $\mathbf{C}$.

(c) $\det A > 0$.

*Proof of (a).* Let $S = \{k_1, ..., k_r\} \subseteq \{1, ..., n\}$. It is trivial that $A_S$ is Hermitian. For each vector $\mathbf{v} = (a_1, ..., a_r)$ in $\mathbf{C}_r$, let us define a vector $X_\mathbf{v} = (x_1, ..., x_n)$ in $\mathbf{C}_n$ by the equations

$$x_{k_i} = a_i \qquad (1 \leq i \leq r)$$
$$x_j = 0 \qquad (j \notin S).$$

It easily follows from Definition 9.6.4 that if $f : \mathbf{C}_n \to \mathbf{C}$ and $g : \mathbf{C}_r \to \mathbf{C}$ are the Hermitian forms associated with $A$ and $A_S$ (respectively), then

$$g(\mathbf{v}) = f(X_\mathbf{v}) \qquad (\text{all } \mathbf{v} \in \mathbf{C}_r).$$

Since $A$ is positive definite, $f(X_\mathbf{v}) > 0$ whenever $X_\mathbf{v} \neq \mathbf{0}$; hence, $g(\mathbf{v}) > 0$ whenever $\mathbf{v} \neq \mathbf{0}$; hence, $A_S$ is positive definite.

*Proof of (b).* The matrix $BA\overline{B}^t$ is Hermitian by Section 9.5, Exercise 4. Let $f$ be the Hermitian form associated with $A$, and let $X$ be any nonzero vector. Since $B$ is nonsingular, $XB \neq \mathbf{0}$ (why?); since $A$ is positive definite, it follows that

$$0 < f(XB) = (XB)A(\overline{XB})^t = X(BA\overline{B}^t)\overline{X}^t.$$

Since this holds for all $X \neq \mathbf{0}$, $BA\overline{B}^t$ is positive definite.

*Proof of* (c). By Theorem 9.7.2, the eigenvalues $\lambda_1, \ldots, \lambda_n$ of $A$ are all positive. If $A'$ is the Jordan matrix similar to $A$ (cf. Theorem 8.4.15 or Corollary 8.4.16), then $A'$ is a triangular matrix whose diagonal entries are the $\lambda_i$. Hence,

$$\lambda_1\lambda_2 \cdots \lambda_n = \det A' = \det A$$

and 9.7.3(c) follows. ▮

**9.7.4 Theorem.** If $A$ is a Hermitian $n \times n$ matrix, the following conditions are equivalent:

(a) All of the eigenvalues of $A$ are positive.

(b) $A$ is positive definite.

(c) For every subset $S$ of $E_n$, $\det A_S > 0$.

(d) For every integer $k \in E_n$,

$$\sum_{\# S=k} \det A_S > 0$$

where the sum runs over all subsets of $E_n$ having exactly $k$ elements.

(e) The coefficients of the characteristic polynomial of $A$ have alternating signs; that is, if

$$\operatorname{char}(A; x) = c_0 + c_1x + \cdots + c_nx^n,$$

then $c_k$ is positive whenever $k$ is even, negative whenever $k$ is odd.

(f) For every $k \in E_n$, $\det A_{E_k} > 0$.

Note that $A_{E_k}$ is the $k \times k$ submatrix which appears in the upper left corner of $A$. Thus, for example, the fact that the real-symmetric matrix

$$\begin{bmatrix} 1 & 2 & -1 \\ 2 & 6 & -3 \\ -1 & -3 & 2 \end{bmatrix}$$

is positive definite follows from the three inequalities

$$\det[1] > 0; \quad \det\begin{bmatrix} 1 & 2 \\ 2 & 6 \end{bmatrix} > 0; \quad \det\begin{bmatrix} 1 & 2 & -1 \\ 2 & 6 & -3 \\ -1 & -3 & 2 \end{bmatrix} > 0.$$

*Proof of 9.7.4.* By Theorem 9.7.2, (a) ⇒ (b). We obtain (b) ⇒ (c) by combining the first and third parts of Theorem 9.7.3. It is trivial that (c) ⇒ (d). The implication (d) ⇒ (e) follows from Theorem 5.12.1 (note that the inequality in part (d) holds trivially when $k = 0$). To prove that (e) ⇒ (a), suppose $A$ has an eigenvalue $\lambda$ which is not positive. Since $\lambda$ is real, this means that $\lambda \leqslant 0$. Assuming 9.7.4(e), it follows that each summand in the expression

$$c_0 + c_1\lambda + \cdots + c_n\lambda^n$$

is nonnegative; since $c_0$ is *strictly positive*, it follows that char($A$; $\lambda$) $\neq$ 0, contradicting the fact that $\lambda$ is an eigenvalue. Hence, all eigenvalues must be positive, and we have proved that (e) $\Rightarrow$ (a). Thus, conditions (a) through (e) are all equivalent. Since obviously (c) $\Rightarrow$ (f), the proof can be completed by showing that (f) implies one of the other five conditions.

The proof of the latter is by induction on $n$, the implication being trivial when $n = 1$. Assume then that condition (f) implies conditions (a) through (e) for all Hermitian matrices of degree $k$, and let $A$ be a Hermitian matrix of degree $n = k + 1$ which satisfies condition (f). Clearly, (f) also holds for the submatrix $A^* = A_{E_k}$ instead of $A$; since $A^*$ is of degree $k$, it follows from the induction hypothesis that conditions (a) through (e) hold for $A^*$. In particular, $A^*$ is positive definite and, hence, nonsingular (Theorem 9.7.3(c)). Therefore, Theorem 5.10.12 (Cramer's Rule) implies the existence of a unique solution $(x_1, \ldots, x_k)$ of the system

$$\text{(9.7.5)} \qquad \sum_{j=1}^{k} a_{ij}x_j = -a_{i,k+1} \qquad (i = 1, 2, \ldots, k).$$

Now, let

$$B = \begin{bmatrix} 1 & & & & \\ & 1 & & & \\ & & \ddots & & \\ & & & 1 & \\ \overline{x_1} & \overline{x_2} & \cdots & \overline{x_k} & 1 \end{bmatrix}$$

(a matrix of degree $k + 1$), where the $x_i$ satisfy 9.7.5 and where all entries in $B$ are zero except in the last row and on the main diagonal. Since det $B = 1$, $B$ is nonsingular. Using 9.7.5, a straightforward computation (left to you) shows that

$$\text{(9.7.6)} \qquad BA\overline{B}^t = \left[\begin{array}{ccc|c} & & & 0 \\ & A^* & & \vdots \\ & & & 0 \\ \hline 0 & \cdots & 0 & c \end{array}\right]$$

where $c$ is some scalar. By assumption, $A$ satisfies condition (f) and, hence, both det $A$ and det $A^*$ are positive; moreover, since det $\overline{B}$ = det $B = 1$, we have

$$\det A = \det BA\bar{B}^t = c \det A^*.$$

Hence, $c > 0$. Since $A$ is Hermitian, so is $BA\bar{B}^t$; moreover, we see from 9.7.6 that the eigenvalues of $BA\bar{B}^t$ are precisely $c$ together with the eigenvalues of $A^*$, and the latter are positive since condition (a) holds for $A^*$. Thus, all eigenvalues of $BA\bar{B}^t$ are positive, so that $BA\bar{B}^t$ is positive definite by Theorem 9.7.2. It follows from 9.7.3(b) that $B^{-1}(BA\bar{B}^t)(\overline{B^{-1}})^t$ is positive definite; that is, $A$ is positive definite. Hence, (f) $\Rightarrow$ (b) for Hermitian matrices of degree $k + 1$, as we wished to show. ∎

The theorems of this section concerning positive definite matrices have analogs concerning positive semidefinite matrices. Specifically, Theorems 9.7.2 and 9.7.3 remain true, with only trivial changes in the proofs, if we replace "positive definite" by "positive semidefinite", "$> 0$" by "$\geqslant 0$", and so on. The same goes for Theorem 9.7.4 if part (f) is excluded; see Exercise 2 below.

The literature on Hermitian matrices and Hermitian forms contains some nice results (particularly with reference to positive definiteness) which are not part of our presentation here. The interested reader might consult [4], pages 97–131.

**Exercises**

1. Carry out the verification of equation 9.7.6.
2. If the phrases "positive definite", "positive", "negative", "$> 0$" are replaced by "positive semidefinite", "nonnegative", "nonpositive", "$\geqslant 0$", respectively,
   (a) Show that conditions (a) through (e) of 9.7.4 are still equivalent. (One minor change will be needed in the proof that (e) $\Rightarrow$ (a).)
   (b) Show that condition (f) does *not* always imply the other conditions. (Give a counterexample.) (SOL)
3. Determine which of the following matrices are (a) positive definite, (b) positive semidefinite. For each matrix which is not positive semidefinite, find a vector $X$ such that $f(X) < 0$, where $f$ is the associated Hermitian form. (ANS)

$$\text{(i)} \begin{bmatrix} 2 & 1 & 0 \\ 1 & 0 & 2 \\ 0 & 2 & 1 \end{bmatrix}. \quad \text{(ii)} \begin{bmatrix} 3 & 1 & 0 \\ 1 & 2 & 1 \\ 0 & 1 & 1 \end{bmatrix}. \quad \text{(iii)} \begin{bmatrix} 1 & 2 & 3 \\ 2 & 4 & 6 \\ 3 & 6 & 1 \end{bmatrix}.$$

$$\text{(iv)} \begin{bmatrix} 0 & i & i \\ -i & 0 & i \\ -i & -i & 0 \end{bmatrix}. \quad \text{(v)} \begin{bmatrix} 2 & i & 2-i \\ -i & 2 & 0 \\ 2+i & 0 & 4 \end{bmatrix}. \quad \text{(vi)} \begin{bmatrix} 1 & 2 & 2 \\ 2 & 4 & 4 \\ 2 & 4 & 6 \end{bmatrix}.$$

4. (a) For which values of $x$ is the matrix

$$A = \begin{bmatrix} 2 & 1 & -3 \\ 1 & 1 & 2 \\ -3 & 2 & x \end{bmatrix}$$

positive definite? (ANS)

(b) For which values of $x$ is the matrix

$$A = \begin{bmatrix} 1 & 2 & 2 \\ 2 & 4 & 4 \\ 2 & 4 & x \end{bmatrix}$$

positive semidefinite? (ANS)

5. Deduce from Theorem 9.7.4 that if all eigenvalues of a Hermitian matrix $A$ have the same sign, then all entries on the main diagonal of $A$ have the same sign. (Note that this provides a new solution to Sec. 7.4, Exercise 2.)
6. Let $A$ be a positive definite $n \times n$ Hermitian matrix, and let $B$ be a positive semidefinite $n \times n$ real diagonal matrix. (That is,

$$B = \begin{bmatrix} \lambda_1 & & & \\ & \lambda_2 & & \\ & & \ddots & \\ & & & \lambda_n \end{bmatrix}$$

where $\lambda_i \geqslant 0$ for all $i$.) Prove that $A + B$ is positive definite, and that $\det (A + B) > \det A$ provided $B \neq 0$. (*Hint*: First prove this when only one of the $\lambda_i$ is nonzero.) (SOL)
7. Prove that the results of the preceding exercise remain valid if $B$ is an arbitrary nonzero positive semidefinite Hermitian matrix of degree $n$, not necessarily diagonal. (SUG)

## 9.8 Maxima and minima of real functions of $n$ variables

In this section, $F$ will denote a real-valued function of $n$ real variables $x_1, \ldots, x_n$. (That is, the domain and range of $F$ are subsets of $\mathbf{R}_n$ and $\mathbf{R}$, respectively.) We shall further assume that all second-order partial derivatives of $F$ exist and are continuous throughout some $\epsilon$-neighborhood $\mathcal{N}$ of the point

$$C = (c_1, \ldots, c_n).$$

(The "$\epsilon$-neighborhood" of $C$ consists of all points $X$ such that $d(C, X) < \epsilon$.) Under this assumption, Taylor's Formula with Remainder asserts that for all points $X = (x_1, \ldots, x_n)$ in $\mathcal{N}$,

(9.8.1)
$$F(X) = F(C) + \sum_{i=1}^{n} \frac{\partial F}{\partial x_i}(C) \cdot (x_i - c_i) + \frac{1}{2} \sum_{i=1}^{n} \sum_{j=1}^{n} \frac{\partial^2 F}{\partial x_i\, \partial x_j}(Z) \cdot (x_i - c_i)(x_j - c_j),$$

where $Z = (z_1, \ldots, z_n)$ is some point on the line segment from $C$ to $X$, and where $(\partial F/\partial x_i)(C)$ denotes the partial derivative $\partial F/\partial x_i$ evaluated at the point $C$, and so on. (Proofs may be found in many calculus books; see, e.g., [13], Sec. 17-10.)

It is common to denote by $\nabla F(X)$ the vector

$$\left(\frac{\partial F}{\partial x_1}(X), \ldots, \frac{\partial F}{\partial x_n}(X)\right)$$

whose components are the first-order partial derivatives of $F$ evaluated at the point $X$. Similarly, $\nabla^2 F(X)$ denotes the $n \times n$ matrix whose entry in the $(i, j)$ position is the second-order partial derivative $\partial^2 F/\partial x_i \partial x_j$ evaluated at the point $X$. Since a well-known theorem of calculus asserts that

(9.8.2)
$$\frac{\partial^2 F}{\partial x_i\, \partial x_j} = \frac{\partial^2 F}{\partial x_j\, \partial x_i}$$

(assuming both sides of 9.8.2 are continuous), *the matrix* $\nabla^2 F(X)$ *is symmetric at all points* $X \in \mathcal{N}$. If $X = (x_1, \ldots, x_n)$ and $C = (c_1, \ldots, c_n)$ are regarded as row vectors (row matrices), equation 9.8.1 may be written in the form

(9.8.3)
$$F(X) = F(C) + \nabla F(C) \cdot (X - C) + \tfrac{1}{2}(X - C)\,[\nabla^2 F(Z)]\,(X - C)^t \qquad (\text{all } X \in \mathcal{N}).$$

The middle term on the right side of 9.8.3 is a dot product of two vectors; the last term is a matrix product.

By a standard theorem, if $F$ has a relative minimum at the point $C$, then all first-order partial derivatives of $F$ are zero at $C$; that is, $\nabla F(C) = \mathbf{0}$. Conversely, suppose $\nabla F(C) = \mathbf{0}$. Then 9.8.3 implies that if

(9.8.4)
$$(X - C)\,[\nabla^2 F(Z)]\,(X - C)^t > 0,$$

then $F(X) > F(C)$; hence, if 9.8.4 holds for all $X \neq C$ in $\mathcal{N}$, $F$ has a relative minimum at $C$. But, by definition, a real-symmetric matrix $A$ is positive definite if and only if $XAX^t > 0$ for all $X \neq \mathbf{0}$. Since $X \in \mathcal{N}$ implies $Z \in \mathcal{N}$, we conclude: *if* $\nabla F(C) = \mathbf{0}$ *and if the matrix* $\nabla^2 F$ *is positive definite throughout* $\mathcal{N}$, *then* $F$ *has a relative minimum at* $C$.

The condition that $\nabla^2 F$ be positive definite throughout a neighborhood of $C$ may be weakened: it suffices that $\nabla^2 F$ be positive definite *at C alone*. Indeed, if $\nabla^2 F(C)$ is positive definite, then

$$\det[\nabla^2 F(C)]_{E_k} > 0$$

for $k = 1, 2, \ldots, n$, by Theorem 9.7.4. Since a determinant is a continuous function of its entries and since the second-order partial derivatives of $F$ were assumed continuous throughout $\mathcal{N}$, it follows that

$$\det[\nabla^2 F(X)]_{E_k} > 0 \qquad (k = 1, 2, \ldots, n)$$

for all $X$ in some neighborhood $\mathcal{M}$ of $C$. Applying Theorem 9.7.4 again, it follows that $\nabla^2 F$ is positive definite throughout $\mathcal{M}$, as desired.

**9.8.5 Example.** Show that the function

$$F(x, y, z) = x^2y + 6y^2 - 6yz + xz^2 - 5x - 22y + 10z$$

has a relative minimum at the point $(2, 1, -1)$.

*Solution.* We write $F_x$ instead of $\partial F/\partial x$, etc. By computation,

$$F_x = 2xy + z^2 - 5; \qquad F_y = x^2 + 12y - 6z - 22;$$

$$F_z = -6y + 2xz + 10;$$

$$\nabla^2 F = \begin{bmatrix} 2y & 2x & 2z \\ 2x & 12 & -6 \\ 2z & -6 & 2x \end{bmatrix}.$$

The second-order partial derivatives are obviously continuous everywhere. At the given point $(2, 1, -1)$, $F_x = F_y = F_z = 0$ and

$$\nabla^2 F(2, 1, -1) = \begin{bmatrix} 2 & 4 & -2 \\ 4 & 12 & -6 \\ -2 & -6 & 4 \end{bmatrix}. \tag{9.8.6}$$

We apply condition 9.7.4(f) to the matrix 9.8.6:

$$\det[2] = 2 > 0; \qquad \det\begin{bmatrix} 2 & 4 \\ 4 & 12 \end{bmatrix} = 8 > 0;$$

$$\det\begin{bmatrix} 2 & 4 & -2 \\ 4 & 12 & -6 \\ -2 & -6 & 4 \end{bmatrix} = 8 > 0$$

and, hence, the matrix 9.8.6 is positive definite. Thus, $F$ has a relative minimum at $(2, 1, -1)$. ▮

In the case $n = 2$, we have

$$\nabla^2 F = \begin{bmatrix} F_{xx} & F_{xy} \\ F_{yx} & F_{yy} \end{bmatrix} = \begin{bmatrix} F_{xx} & F_{xy} \\ F_{xy} & F_{yy} \end{bmatrix}$$

and condition 9.7.4(f) for positive definiteness of $\nabla^2 F$ reduces to

$$F_{xx} > 0; \qquad F_{xx}F_{yy} - F_{xy}^2 > 0. \tag{9.8.7}$$

Most calculus books state the second derivative test for relative minima of functions of two real variables in a form like 9.8.7. (See [13], Sec. 17-11.) Theorem 9.7.4 has enabled us to generalize the test to functions of any number of variables.

We have shown that if $\nabla F(C) = \mathbf{0}$ and $\nabla^2 F(C)$ is positive definite, then $F$ has a relative minimum at $C$. Conversely, if $F$ has a relative minimum at $C$, the matrix $\nabla^2 F(C)$, while not necessarily positive definite, must at least be *positive semidefinite*; that is,

$$\det[\nabla^2 F(C)]_S \geq 0$$

for all subsets $S$ of $E_n$. (This condition on the determinants is the analog of 9.7.4(c); see Sec. 9.7, Exercise 2.) The proof is as follows: suppose $\nabla^2 F(C)$ is not positive semidefinite. If $g_X$ is the real-symmetric form associated with the matrix $\nabla^2 F(X)$ (all $X$), this means that $g_C$ takes on a negative value, say, $g_C(\mathbf{u}) < 0$. By continuity, it follows that $g_Y(\mathbf{u}) < 0$ for all points $Y$ in some neighborhood $\mathcal{M}$ of $C$; we lose no generality in assuming that $\mathcal{M} = \mathcal{N}$. If $k$ is any scalar small enough so that the point $X = C + k\mathbf{u}$ lies in $\mathcal{N}$, then 9.8.3 (together with the condition $\nabla F(C) = \mathbf{0}$) gives

$$F(X) = F(C) + \tfrac{1}{2} g_Z(k\mathbf{u}) = F(C) + \tfrac{1}{2} k^2 g_Z(\mathbf{u}) < F(C), \tag{9.8.8}$$

$g_Z(\mathbf{u})$ being negative since $Z \in \mathcal{N}$. Since 9.8.8 holds for *all* sufficiently small values of $k$, $F$ cannot have a relative minimum at $C$.

### Relative maxima

So far we have discussed relative minima only. To take care of relative maxima, we need merely observe that a function $F$ has a maximum at $C$ if and only if the function $-F$ has a minimum there. In the case $n = 2$, for instance, the sufficient conditions for a maximum are $\nabla F(C) = \mathbf{0}$ and

$$-F_{xx} > 0; \qquad (-F_{xx})(-F_{yy}) - (-F_{xy})^2 > 0 \tag{9.8.9}$$

(compare with 9.8.7). We may, of course, rewrite 9.8.9 as

$$F_{xx} < 0; \qquad F_{xx}F_{yy} - F_{xy}^2 > 0.$$

For arbitrary $n$, a similar argument shows that the condition

$$\det[\nabla^2 F(C)]_{E_k} > 0 \qquad \text{(all } k\text{)}$$

must be replaced by

$$(-1)^k \det[\nabla^2 F(C)]_{E_k} > 0 \qquad \text{(all } k\text{)}$$

when considering maxima rather than minima.

**Concluding remark**

This section illustrates the applicability of linear algebra to analysis. Additional applications to analysis are discussed in Appendix E.

**Exercises**

1. Assume that $F$ is a real-valued function of $n$ real variables and that the second-order partial derivatives of $F$ are continuous throughout a neighborhood of the point $C$. Let

   $k_0 + k_1x + \cdots + k_nx^n$

   be the characteristic polynomial of the matrix $\nabla^2F(C)$. Show that if $\nabla F(C) = \mathbf{0}$ and all of the numbers $k_i$ have the same sign (i.e., all are positive or all are negative), then $F$ has a relative maximum at $C$. Conversely, if $F$ has a relative maximum at $C$, then all of the $k_i$ which are *nonzero* have the same sign. (SUG)
2. Let $F : \mathbf{R}_n \to \mathbf{R}$ and assume that (a) the second-order partial derivatives of $F$ are continuous throughout $\mathbf{R}_n$, and (b) $\nabla^2F(X)$ is positive definite at *all* points $X \in \mathbf{R}_n$. Show that at any point $C$ such that $\nabla F(C) = \mathbf{0}$, $F$ has an *absolute* minimum, and that at most one such point $C$ can exist.
3. Find all relative extrema (maxima and minima) of each of the following functions $F$. Also, discuss absolute extrema if possible. (ANS)
   (a) $F(x, y) = x^4 + 2x^2 + 4y^2 - 4xy - 4x + 8y$.
   (b) $F(x, y) = x^3 + 6xy + y^3$.
   (c) $F(x, y) = (x - 2y)/(x^2 + y^2 + 1)$. (SUG)
   (d) $F(x, y, z) = x^2 + y^2 + z^2 - xy + yz + 4z$.
   (e) $F(x, y, z) = \frac{1}{2}x^4 + y^2 + z^2 - xz + 6y$.

# Appendixes

## A Proof of Theorem 4.10.4

Let $E$, $E'$ be real Euclidean spaces with $\dim E \leqslant \dim E' < \infty$; let $S$ be a subset of $E$, and let $\mathbf{M}$ be an isometry of $S$ into $E'$. Our object is to extend $\mathbf{M}$ to an isometry of $E \to E'$. In view of Section 4.10, Exercise 7, we lose no generality in assuming that $\dim E = \dim E'$. We may also assume, without loss of generality, that

(1) $\quad \mathbf{0} \in S; \qquad \mathbf{0M} = \mathbf{0}.$

Indeed, suppose the theorem has been proved in the case where (1) holds. The general case (not assuming (1)) is then treated as follows: choose a fixed element $\mathbf{u} \in S$ (the case $S = \varnothing$ being trivial) and let $\mathbf{v} = \mathbf{uM}$, $S_0 = S - \mathbf{u} = \{\mathbf{s} - \mathbf{u} : \mathbf{s} \in S\}$. Then $\mathbf{0} \in S_0$, and by Theorem 4.10.5 the mapping

$$\mathbf{M}' = (\mathscr{T}_{\mathbf{u}} \upharpoonright S_0)\mathbf{M}\mathscr{T}_{-\mathbf{v}}$$

is an isometry of $S_0 \to E'$ which maps $\mathbf{0} \to \mathbf{0}$. (The symbol $\upharpoonright$ denotes restriction; cf. the end of Sec. 1.2.) Hence, by the case already proved, $\mathbf{M}'$ can be extended to an isometry $\mathbf{T} : E \to E'$. The mapping $\mathscr{T}_{-\mathbf{u}}\mathbf{T}\mathscr{T}_{\mathbf{v}}$ is then an isometry of $E \to E'$ which extends $\mathbf{M}$.

It thus suffices to prove the following modified version of Theorem 4.10.4:

**4.10.4A Theorem.** Let $E$, $E'$ be real Euclidean spaces of the same finite dimension $n$, let $S \subseteq E$, let $\mathbf{M} : S \to E'$ be an isometry, and assume that (1) holds. Then $\mathbf{M}$ can be extended to an isometry $\mathbf{M}^* : E \to E'$.

Our proof of 4.10.4A is by induction on the integer

$$s = \dim\langle S\rangle + \dim\langle S\mathbf{M}\rangle,$$

where $\langle A \rangle$ denotes the flat determined by $A$ (cf. Sec. 3.5). If either dimension is zero, the theorem is trivial (*why*?); thus, we may assume that $\langle S \rangle$ and $\langle S\mathbf{M} \rangle$ are at least one dimensional and that Theorem 4.10.4A holds for all smaller values of $s$.

Let $k = \max(\dim\langle S \rangle, \dim\langle S\mathbf{M} \rangle)$. We define sets $A$ and $B$, Euclidean spaces $V$ and $V'$, and a mapping $\mathbf{N} : A \to B$, as follows:

(a) If $\dim\langle S \rangle \geqslant \dim\langle S\mathbf{M} \rangle$, let $A = S$, $B = S\mathbf{M}$, $\mathbf{N} = \mathbf{M}$, $V = E$, $V' = E'$.

(b) If $\dim\langle S\mathbf{M} \rangle > \dim\langle S \rangle$, let $A = S\mathbf{M}$, $B = S$, $\mathbf{N} = \mathbf{M}^{-1}$, $V = E'$, $V' = E$.

In either case,

$$k = \dim\langle A \rangle \geqslant \dim\langle B \rangle \geqslant 1,$$

$\mathbf{N}$ is an isometry of $A$ onto $B$, and (1) implies that

$$(2) \qquad \mathbf{0} \in A; \qquad \mathbf{0N} = \mathbf{0}.$$

As a notational convenience, the image of any point $P \in A$ under $\mathbf{N}$ will be denoted $P'$. By (2), the identities

$$|\mathbf{u}| = d(\mathbf{u}, \mathbf{0});$$

$$p(\mathbf{u}, \mathbf{v}) = \tfrac{1}{2}\,[d(\mathbf{u}, \mathbf{0})^2 + d(\mathbf{v}, \mathbf{0})^2 - d(\mathbf{u}, \mathbf{v})^2]$$

imply that

$$(3) \qquad |P| = |P'|; \quad p(P, Q) = p(P', Q') \quad (\text{all } P, Q \in A).$$

Since $\dim\langle A \rangle = k$ and $\mathbf{0} \in A$, Exercise 5 of Section 3.5.5 implies that $A$ contains $k$ $b$-independent points $P_0, \ldots, P_{k-1}$ such that $P_0 = O$ (actually, we could choose $k + 1$ such points $P_i$ but this will not be necessary). Let $A_0 = A \cap \langle P_0, \ldots, P_{k-1} \rangle$; then $P_0, \ldots, P_{k-1}$ lie in $A_0$ and

$$\langle A_0 \rangle = \langle P_0, \ldots, P_{k-1} \rangle.$$

Since $\dim\langle A_0 \rangle = k - 1 < \dim\langle A \rangle$, the induction hypothesis implies that $\mathbf{N} \restriction A_0$ can be extended to an isometry $\mathbf{H} : V \to V'$. Moreover, since $P_0 = O$ lies in $A_0$, $\mathbf{0H} = \mathbf{0N} = \mathbf{0}$; hence, $\mathbf{H}$ is linear by Theorem 4.1.6. This implies that

$$\left(\sum_{j=1}^{k-1} b_j P_j\right) \mathbf{H} = \sum_{j=1}^{k-1} b_j P'_j$$

for all scalars $b_j$ and, hence,

$$(4) \qquad \left|\sum_{j=1}^{k-1} b_j P_j\right| = \left|\sum_{j=1}^{k-1} b_j P'_j\right| \quad (\text{all } b_j \in \mathbf{R}).$$

(Here, $P'_j = P_j\mathbf{N} = P_j\mathbf{H}$ since $\mathbf{H}$ extends $\mathbf{N} \restriction A_0$.) Let us also note that by part (b) of 4.10.8 (applied to $\mathbf{H}$), the points $P'_0, \ldots, P'_{k-1}$ are $b$-independent and, hence, $P'_1, \ldots, P'_{k-1}$ are *linearly* independent.

Since $O = P'_0$ lies in $B$, the flat $\langle B \rangle$ is a subspace; by assumption, $\dim\langle B \rangle \leq k$. Hence, $\langle B \rangle \subseteq W$ for some $k$-dimensional subspace $W$ of $V'$. Since $P'_1, \ldots, P'_{k-1}$ are linearly independent elements of $W$, Theorem 3.8.18 implies that the vectors in $W$ which are orthogonal to $P'_1, \ldots, P'_{k-1}$ constitute a one-dimensional subspace $[\mathbf{q}]$; the vectors $P'_1, \ldots, P'_{k-1}, \mathbf{q}$ then form a basis of $W$. Similarly, the $k$-dimensional subspace $\langle A \rangle$ has a basis $\{P_1, \ldots, P_{k-1}, \mathbf{y}\}$ such that $\mathbf{y}$ is orthogonal to $P_1, \ldots, P_{k-1}$. We may clearly assume that $\|\mathbf{y}\| = \|\mathbf{q}\| = 1$, and we do so.

For any point $P \in A$, its image $P' = P\mathbf{N}$ lies in $B$ and thus in $W$; hence, we may write

$$P = \sum_{j=1}^{k-1} a_j P_j + t\mathbf{y}; \quad P' = \sum_{j=1}^{k-1} b_j P'_j + t'\mathbf{q}.$$

We then have, for all $i$ $(1 \leq i \leq k-1)$,

$$\begin{aligned} p\left(\sum_{j=1}^{k-1} a_j P'_j - P', P'_i\right) &= \sum_{j=1}^{k-1} a_j p(P'_j, P'_i) - p(P', P'_i) \\ &= \sum_{j=1}^{k-1} a_j p(P_j, P_i) - p(P, P_i) \qquad \text{(by (3))} \\ &= p(-t\mathbf{y}, P_i) = 0 \qquad (\text{since } \mathbf{y} \perp P_i). \end{aligned}$$

We have thus shown that the vector

$$\sum_{j=1}^{k-1} a_j P'_j - P' = \sum_{j=1}^{k-1} (a_j - b_j) P'_j - t'\mathbf{q}$$

is orthogonal to all $P'_i$ $(1 \leq i \leq k-1)$; hence, this vector lies in $[\mathbf{q}]$. Since $\mathbf{q}$ and the vectors $P'_j$ are linearly independent, it follows that $a_j - b_j = 0$ $(1 \leq j \leq k-1)$, so that $b_j = a_j$ and, hence,

$$P' = \sum_{j=1}^{k-1} a_j P'_j + t'\mathbf{q}.$$

By (3) and (4), we have $\|P'\| = \|P\|$, $\|\Sigma a_j P'_j\| = \|\Sigma a_j P_j\|$; also, $\mathbf{y} \perp P_j$ and $\mathbf{q} \perp P'_j$ for $1 \leq j \leq k-1$. It follows (by the Pythagorean Theorem) that $\|t\mathbf{y}\| = \|t'\mathbf{q}\|$ and, hence, $t' = \pm t$. We have thus shown that for *every* point

$$(5) \qquad P = \sum_{j=1}^{k-1} a_j P_j + t\mathbf{y}$$

in $A$, the image of $P$ under $\mathbf{N}$ is

$$(6) \qquad P' = \sum_{j=1}^{k-1} a_j P'_j + \epsilon t\mathbf{q} \qquad (\epsilon = \pm 1).$$

Moreover, it is not hard to show, using the second half of (3), that the value of $\epsilon$ in equation (6) must be the same for all points $P \in A$.

We now extend **N** to all of $V$. By Theorem 4.10.14(a) (whose proof does not depend on 4.10.4!), there is a linear isometry $\mathbf{T}: V \to V'$ which maps $\langle A \rangle$ onto $W$. By Theorem 3.8.18, any vector in $V$ is uniquely expressible in the form $P + \mathbf{v}$, where $P \in \langle A \rangle$ and $\mathbf{v} \in \langle A \rangle^\perp$. Define $\mathbf{N}^*: V \to V'$ by

(7) $\quad \mathbf{N}^*: P + \mathbf{v} \to P' + \mathbf{v}\mathbf{T} \qquad (P \in \langle A \rangle, \mathbf{v} \in \langle A \rangle^\perp),$

where we write $P$ in the form (5) and then *define* $P'$ by (6). Clearly, $\mathbf{N}^*$ extends **N**, by the result of the preceding paragraph. Using the same notations, the Pythagorean Theorem gives

$$
\begin{aligned}
\|P + \mathbf{v}\|^2 = \|P\|^2 + \|\mathbf{v}\|^2 &= \Big\|\sum_{j=1}^{k-1} a_j P_j\Big\|^2 + t^2 + \|\mathbf{v}\|^2 \\
&= \Big\|\sum_{j=1}^{k-1} a_j P_j'\Big\|^2 + t^2 + \|\mathbf{v}\mathbf{T}\|^2 \\
&\qquad \text{(by (4) and since } \mathbf{T} \text{ is a linear isometry)} \\
&= \|P'\|^2 + \|\mathbf{v}\mathbf{T}\|^2 \\
&= \|P' + \mathbf{v}\mathbf{T}\|^2
\end{aligned}
$$

(the last step follows from the fact that since **T** maps $\langle A \rangle$ onto $W$, $\mathbf{v}\mathbf{T}$ is orthogonal to all elements $P'$ of $W$). We have thus shown that $\mathbf{N}^*$ preserves lengths. Also, it is clear from (5), (6), and (7) that $\mathbf{N}^*$ is linear. Hence, by Theorem 4.10.9, $\mathbf{N}^*$ is a linear isometry. By 4.10.7, $\mathbf{N}^*$ is one-to-one and onto and, hence, has an inverse. Letting

$$
\begin{aligned}
\mathbf{M}^* &= \mathbf{N}^* && \text{if } A = S \\
\mathbf{M}^* &= (\mathbf{N}^*)^{-1} && \text{if } A = S\mathbf{M},
\end{aligned}
$$

it follows that $\mathbf{M}^*$ is a linear isometry of $E$ onto $E'$ which extends **M**.

## B The cycle representation of permutations

As in Section 5.1, let $E_n = \{1, 2, ..., n\}$ and let $\mathbf{S}_n$ be the group of all permutations of 1, 2, ..., $n$. *Cycles* were defined in Section 5.1. Our object here is to give a rigorous proof of the following result:

**5.1.6 Theorem.** Every element $\pi$ of $\mathbf{S}_n$ can be expressed as a product of disjoint cycles; this expression for $\pi$ is unique except for the order of the factors and the inclusion or exclusion of cycles of length 1.

Our method is to define a certain equivalence relation (depending on $\pi$) on the set $E_n$, and then show that the restrictions of $\pi$ to the corresponding (disjoint!) equivalence classes give us the disjoint cycles that we seek. Our treatment is similar, but not identical, to that in [11].

Let $\pi \in \mathbf{S}_n$ be a fixed permutation. We define a relation "$\tilde{\pi}$" on $E_n$ as follows: if $x, y \in E_n$, $x \tilde{\pi} y$ if and only if some integral power of $\pi$ maps $x \to y$; that is, if and only if

$$x\pi^k = y \qquad (\text{some } k \in \mathbf{Z}).$$

To see that "$\tilde{\pi}$" is an equivalence relation, we simply observe that (1) $x\pi^0 = x$; (2) if $x\pi^k = y$, then $y\pi^{-k} = x$; and (3) if $x\pi^k = y$ and $y\pi^h = z$, then $x\pi^{k+h} = z$.

The equivalence classes with respect to the relation "$\tilde{\pi}$" are called *$\pi$-orbits*; thus, by Theorem 1.4.2, every element of $E_n$ lies in one and only one $\pi$-orbit, so that the $\pi$-orbits are mutually disjoint. If $\mathscr{E}_1, \ldots, \mathscr{E}_m$ are the distinct $\pi$-orbits, then

(1) $$E_n = \mathscr{E}_1 \cup \cdots \cup \mathscr{E}_m.$$

Let us look at what $\pi$ does to the elements of a fixed orbit $\mathscr{E}_j$. Choose any element $x$ of $\mathscr{E}_j$; then by definition of our equivalence relation,

(2) $$x, x\pi, x\pi^2, x\pi^3, \ldots$$

all lie in $\mathscr{E}_j$. The elements in the sequence (2) cannot all be different since there are only finitely many elements in $E_n$. Hence, there is a least integer $k \geqslant 1$ such that $x\pi^k$ equals a term which precedes it in the sequence; say, $x\pi^k = x\pi^h$, where $0 \leqslant h < k$. If $h > 0$, then by applying $\pi^{-1}$ to both sides, we get $x\pi^{k-1} = x\pi^{h-1}$, $0 \leqslant h - 1 < k - 1$, contradicting the choice of $k$ as "least". Hence, $h = 0$, $x\pi^k = x$, and the elements $x, \ldots, x\pi^{k-1}$ are all different. Moreover, since $x\pi^k = x$, it follows that for every integer $i$, $x\pi^i$ equals one of the elements $x, \ldots, x\pi^{k-1}$. (Proof: Using the division algorithm to write $i = qk + r$, where $0 \leqslant r < k$, we have $x\pi^i = x\pi^r$. Alternate proof: Use induction on $i$.) Letting $x_1 = x, x_2 = x\pi, \ldots, x_k = x\pi^{k-1}$, we conclude: *the k elements of any fixed $\pi$-orbit $\mathscr{E}_j$* can be arranged in an order $x_1, \ldots, x_k$ such that

$$\pi: x_1 \to x_2 \to \cdots \to x_k \to x_1.$$

It follows at once that if we define a permutation $\pi_j$ by

(3) $$x\pi_j = \begin{Bmatrix} x\pi & \text{if } x \in \mathscr{E}_j \\ x & \text{if } x \notin \mathscr{E}_j \end{Bmatrix},$$

then $\pi_j$ is a cycle (see Def. 5.1.1); in fact, letting $x_1, \ldots, x_k$ be as above, $\pi_j = (x_1\ x_2 \cdots x_k)$. By definition, every element moved by $\pi_j$ lies in the orbit $\mathscr{E}_j$; since the orbits $\mathscr{E}_1, ..., \mathscr{E}_m$ are disjoint, it follows that the cycles $\pi_1, ..., \pi_m$ defined by (3) are disjoint.

Let $x$ be any fixed element of $E_n$. By (1), $x$ must lie in some $\pi$-orbit, say, the orbit $\mathscr{E}_j$. Then $x\pi$ also lies in $\mathscr{E}_j$; hence, all cycles $\pi_i$ with $i \neq j$ leave $x$ and $x\pi$ fixed. Hence, we have

$$x(\pi_1 \cdots \pi_{j-1}) = x$$
$$x(\pi_1 \cdots \pi_{j-1}\pi_j) = [x(\pi_1 \cdots \pi_{j-1})]\pi_j = x\pi_j = x\pi \quad \text{(by (3))}$$
$$x(\pi_1 \cdots \pi_j\pi_{j+1} \cdots \pi_m) = (x\pi)(\pi_{j+1} \cdots \pi_m) = x\pi.$$

Since $x$ was arbitrary, this shows that $\pi$ and $\pi_1\pi_2 \cdots \pi_m$ do the same thing to every $x \in E_n$. Hence,

$$\pi = \pi_1\pi_2 \cdots \pi_m \tag{4}$$

which expresses $\pi$ as a *product of disjoint cycles*.

To prove uniqueness of the expression, suppose also $\pi = \sigma_1\sigma_2 \cdots \sigma_r$, where the $\sigma$'s are disjoint cycles of lengths $> 1$. For any particular cycle $\sigma_j = (x_1\ x_2 \cdots x_k)$, each $x_i$ $(1 \leqslant i \leqslant k)$ has the same image under $\pi$ as under $\sigma_j$ (why?). It easily follows that the images of $x_1$ under all possible *powers* of $\pi$ are precisely the elements $x_1, ..., x_k$. Hence, $\{x_1, ..., x_k\}$ is a $\pi$-orbit. If we label this orbit $\mathscr{E}_j$, it follows at once that the cycle $\pi_j$ defined by (3) is equal to $\sigma_j$. Thus, if the $\pi$-orbits $\mathscr{E}_j$ are numbered in a suitable order, we have

$$\sigma_1 = \pi_1, ..., \sigma_r = \pi_r.$$

Moreover, if $r < m$, then by canceling equal factors from the equation $\pi = \pi_1 \cdots \pi_m = \sigma_1 \cdots \sigma_r$, we obtain

$$\pi_{r+1} \cdots \pi_m = \mathbf{I}. \tag{5}$$

If any *one* of the disjoint cycles $\pi_{r+1}, ..., \pi_m$ moves an element $x$ of $E_n$ (say, it maps $x \to y$), then the product $\pi_{r+1} \cdots \pi_m$ maps $x \to y$ contradicting (5). Hence, each of the cycles $\pi_{r+1}, ..., \pi_m$ moves no elements; that is, each of these cycles has length 1. It follows that the expressions

$$\pi_1 \cdots \pi_m; \qquad \sigma_1 \cdots \sigma_r$$

are identical except for 1-cycles, as we wished to show.

## C Proof of Theorem 5.7.4

Theorem 5.7.4 states that for any $m \times n$ matrix $A$ over the field $F$,

$$\det AA^t = \sum_B (\det B)^2 \tag{1}$$

where the sum is taken over all $m \times m$ submatrices $B$ of $A$. For convenience in the proof, we shall denote the $m \times m$ submatrices of $A$ by $E(m, A)$; thus, (1) may be rewritten in the form

$$\det AA^t = \Sigma\, [\det E(m, A)]^2. \tag{2}$$

**Lemma.** *If $\mathcal{R}$ is a row operation, then* (2) *holds for a given matrix $A$ if and only if it holds for the matrix $\mathcal{R}(A)$ in place of $A$.*

*Proof of lemma.* Let $\mathcal{C}$ be the column operation which does the same thing to columns that $\mathcal{R}$ does to rows. By 5.5.3, $\mathcal{R}$ multiplies all determinants by a scalar $k \neq 0$ which depends only on $\mathcal{R}$; evidently so does $\mathcal{C}$, *with the same k*. Hence, if (2) holds for $\mathcal{R}(A)$ in place of $A$, we have

$$\begin{aligned} k^2 \det AA^t &= \det[\,(\mathcal{R}(AA^t))\mathcal{C}] = \det[\,(\mathcal{R}(A)A^t)\mathcal{C}] = \det[(\mathcal{R}(A)) \cdot (A^t\mathcal{C})] \quad \text{(by 5.5.5)} \\ &= \det[\mathcal{R}(A) \cdot (\mathcal{R}(A))^t] \\ &= \Sigma\, [\det E(m, \mathcal{R}(A))]^2 \qquad \text{(since (2) holds for } \mathcal{R}(A)) \\ &= \Sigma\, [k \det E(m, A)]^2 \qquad \text{(since } \mathcal{R} \text{ does the same thing to } E(m, A) \text{ as to } A) \\ &= k^2\, \Sigma\, [\det E(m, A)]^2 \end{aligned}$$

so that by canceling $k^2 \neq 0$ we see that (2) holds for $A$. Similarly, if (2) holds for $A$, then it holds for $\mathcal{R}(A)$. This proves the lemma.

Our proof of (2) will be by induction on $n$, the number of *columns* of $A$. If $n = m = 1$, then the proof is trivial. If $n = 1 < m$, then the rows of $A$ are dependent (why?) and, hence, so are the rows of $AA^t$ (Sec. 4.4, Exercise 12), so that $\det AA^t = 0$; the right side of (2) is also zero in this case since it is the empty sum (no $m \times m$ submatrices exist). Hence, (2) holds for $n = 1$. Assume (2) holds for $n = k$, and let $A$ be an $m \times (k + 1)$ matrix. If $A = 0$, the proof is trivial; hence, we may assume some column $A^{(h)}$ is nonzero. This column can be reduced to the column (1, 0, ..., 0) by elementary *row* operations on $A$; hence, by the lemma we may assume without loss of generality that

$$A^{(h)} = \begin{bmatrix} 1 \\ 0 \\ \vdots \\ 0 \end{bmatrix}.$$

Let $A'$ be the matrix obtained from $A$ by deleting the column $A^{(h)}$, and let $A^*$ be the matrix obtained from $A'$ by deleting the first row; also, let $B$ be the (1, 1) minor of $A'(A')^t$. A straightforward application of the definition of matrix multiplication gives

$$AA^t = A'(A')^t + \begin{bmatrix} 1 & 0 & \dots & 0 \\ 0 & 0 & \dots & 0 \\ \vdots & \vdots & & \vdots \\ 0 & 0 & \dots & 0 \end{bmatrix}$$

from which it is not hard to show (using, e.g., the expansion by minors of the first row) that

(3) $\quad \det AA^t = \det[A'(A')^t] + \det B.$

Now the $m \times m$ submatrices of $A$ are of two types: (a) those of $A'$, and (b) those consisting of the column $A^{(h)}$ together with an $m \times (m - 1)$ submatrix of $A'$. If we expand the determinant of an $m \times m$ matrix of type (b) using minors of the column $A^{(h)}$, we obtain $\pm 1$ times the determinant of an $(m - 1) \times (m - 1)$ submatrix of $A^*$. Hence,

(4) $$\begin{aligned} \Sigma\,[\det E(m, A)]^2 &= \Sigma\,[\det E(m, A')]^2 + \Sigma\,[\det E(m - 1, A^*)]^2 \\ &= \det[A'(A')^t] + \det\,[A^*(A^*)^t] \end{aligned}$$

since $A'$ and $A^*$ have $k$ columns and we have assumed that the theorem holds for $n = k$. In addition, the definitions of $A^*$ and $B$ easily imply that

$$A^*(A^*)^t = B.$$

Hence, by comparing (3) with (4) we see that (2) holds for $A$, q.e.d.

## D Proof of Theorem 4.9.9

Let $\mathbf{M}$ be a one-to-one mapping of $\mathbf{R}_n$ onto $\mathbf{R}_n$ which maps collinear points into collinear points. Our object is to show that $\mathbf{M}$ is affine, and that if $\mathbf{0M} = \mathbf{0}$ then $\mathbf{M}$ is linear. Throughout the proof, the images under $\mathbf{M}$ of points $P, Q, P_1, \dots$ will be denoted $P', Q', P_1'$, and so forth.

**Lemma 1.** If $P_1, \dots, P_k$ are any points in $\mathbf{R}_n$, then

$$\langle P_1, \dots, P_k\rangle \mathbf{M} \subseteq \langle P_1', \dots, P_k'\rangle.$$

*Proof.* We use induction on $k$. The lemma is trivial if $k = 1$. The case $k = 2$ is covered by the assumptions that $\mathbf{M}$ is one-to-one and preserves collinearity. Assume then that the lemma holds for a given positive value of $k$, and let $P_1, \dots, P_{k+1}$ be $k + 1$ points in $\mathbf{R}_n$. For any point $Q$ in $\langle P_1, \dots, P_{k+1}\rangle$, we must show that $Q' \in \langle P_1', \dots, P_{k+1}'\rangle$. This is true by induction hypothesis if $Q \in \langle P_1, \dots, P_k\rangle$; hence, we may assume that

$$Q = a_1P_1 + \cdots + a_{k+1}P_{k+1},$$

where

$$a_{k+1} \neq 0; \qquad a_1 + \cdots + a_{k+1} = 1.$$

Let

$$R_1 = \frac{1}{2} a_{k+1}P_{k+1} + \left(1 - \frac{1}{2} a_{k+1}\right)P_1$$

$$R_2 = (2 - a_1 - a_{k+1})P_1 - (a_2P_2 + \cdots + a_kP_k).$$

Then $R_1 \neq R_2$ and $Q = 2R_1 - R_2 \in \langle R_1, R_2\rangle$. But $R_1 \in \langle P_1, P_{k+1}\rangle$ and, hence, $R_1' \in \langle P_1', P_{k+1}'\rangle$ (by the case $k = 2$); also, $R_2 \in \langle P_1, ..., P_k\rangle$ and, hence, $R_2' \in \langle P_1', ..., P_k'\rangle$. Hence, $Q' \in \langle R_1', R_2'\rangle \subseteq \langle P_1', ..., P_{k+1}'\rangle$. ∎

**Lemma 2.** If $S$ is a $k$-flat in $\mathbf{R}_n$, then $S\mathbf{M}$ is a $k$-flat.

*Proof.* Write $S = \langle P_1, ..., P_{k+1}\rangle$, where the $P_i$ are $b$-independent. By Lemma 1, $S\mathbf{M}$ is at least a subset of some $k$-flat $S^*$. If $S\mathbf{M}$ is not all of $S^*$, then we may choose a point $Q \in S^*$ which is not in $S\mathbf{M}$. Since $\mathbf{M}$ is onto, $Q = P_{k+2}'$ for some point $P_{k+2}$; since $Q \notin S\mathbf{M}$, $P_{k+2} \notin S$ and, hence, $P_1, ..., P_{k+2}$ are $b$-independent (Sec. 3.5, Exercise 2). By Section 3.5, Exercise 5, we may extend $\{P_1, ..., P_{k+2}\}$ to a $b$-independent set $\{P_1, ..., P_{n+1}\}$ which determines $\mathbf{R}_n$. Since $\mathbf{M}$ is onto, Lemma 1 implies that $\langle P_1', ..., P_{n+1}'\rangle = \mathbf{R}_n$. Hence, the $P_i'$ are $b$-independent; hence, $P_1', ..., P_{k+2}'$ cannot all lie in the $k$-flat $S^*$, R.A.A. Hence, $S\mathbf{M}$ is all of $S^*$. ∎

**Lemma 3.** $\mathbf{M}$ maps noncollinear points into noncollinear points.

*Proof.* Let $P_1, P_2, P_3$ be noncollinear. By Lemma 2, $\langle P_1, P_2\rangle\mathbf{M}$ is a line; since $\mathbf{M}$ is one-to-one, $P_3' = P_3\mathbf{M}$ must lie outside of this line. ∎

**Lemma 4.** If $L_1$, $L_2$ are distinct parallel lines, then $L_1\mathbf{M}$, $L_2\mathbf{M}$ are parallel lines.

*Proof.* $L_1\mathbf{M}$ and $L_2\mathbf{M}$ are certainly lines (Lemma 2). Since $L_1$, $L_2$ are parallel, they lie in a plane $S$ (Theorem 3.4.9); hence, $L_1\mathbf{M}$, $L_2\mathbf{M}$ lie in a plane $S^*$ by Lemma 2. Since $L_1$ and $L_2$ do not meet, neither do their images ($\mathbf{M}$ being one-to-one); hence, $L_1\mathbf{M}$ and $L_2\mathbf{M}$ are parallel by Theorem 3.4.9. ∎

**Lemma 5.** If $\mathbf{0M} = \mathbf{0}$, then $\mathbf{M}$ preserves addition.

*Proof.* If $\mathbf{u}$, $\mathbf{v}$ are linearly independent, then $\mathbf{uM}$, $\mathbf{vM}$ are independent by Lemma 3; hence, we have parallelograms as in Figure D1. Since $\mathbf{u} + \mathbf{v}$ lies on the line through $\mathbf{v}$ parallel to $\langle O, \mathbf{u}\rangle$, its image $(\mathbf{u} + \mathbf{v})\mathbf{M}$ lies on the line $L_1$ through $\mathbf{vM}$ parallel to $\langle O, \mathbf{uM}\rangle$ (Lemma 4); similarly, $(\mathbf{u} + \mathbf{v})\mathbf{M}$ lies on the line $L_2$ through $\mathbf{uM}$ parallel to $\langle O, \mathbf{vM}\rangle$. But $L_1 \cap L_2$ is the point $\mathbf{uM} + \mathbf{vM}$; hence,

(1) $\qquad (\mathbf{u} + \mathbf{v})\mathbf{M} = \mathbf{uM} + \mathbf{vM}$

in this case.

Suppose instead that $\mathbf{u}, \mathbf{v}$ are linearly dependent. If $\mathbf{u} = \mathbf{0}$ or $\mathbf{v} = \mathbf{0}$, equation (1) is trivially true; hence, we may assume $\mathbf{u} = a\mathbf{v} \neq \mathbf{0}$, $[\mathbf{u}] = [\mathbf{v}]$. Since $n \geqslant 2$, we may choose $\mathbf{w} \notin [\mathbf{u}]$. Then by the cases already proved,

$$\begin{aligned}\mathbf{wM} + (\mathbf{u} + \mathbf{v})\mathbf{M} &= (\mathbf{w} + \mathbf{u} + \mathbf{v})\mathbf{M} = (\mathbf{w} + \mathbf{u})\mathbf{M} + \mathbf{vM}\\ &= \mathbf{wM} + \mathbf{uM} + \mathbf{vM}\end{aligned}$$

and (1) follows by cancellation. ∎

**Lemma 6.** If $\mathbf{0M} = \mathbf{0}$, then $\mathbf{M}$ preserves scalar multiplication.

*Proof.* Let $\mathbf{u}$ be any fixed nonzero vector in $\mathbf{R}_n$. Since $n \geqslant 2$, we may choose $\mathbf{v} \in \mathbf{R}_n$ such that the vectors $\mathbf{u}, \mathbf{v}$ are linearly independent. As before, we write $\mathbf{u}' = \mathbf{uM}$, $\mathbf{v}' = \mathbf{vM}$. Since $[\mathbf{u}]$ is the line $\langle \mathbf{0}, \mathbf{u}\rangle$ and $\mathbf{0M} = \mathbf{0}$, Lemma 2 implies $[\mathbf{u}]\mathbf{M} = [\mathbf{u}']$. Hence, for any scalar $a$,

(2) $\qquad (a\mathbf{u})\mathbf{M} = a'\mathbf{u}'$

for some scalar $a'$ depending on $a$. Similarly, we have

$$(a\mathbf{v})\mathbf{M} = a^*\mathbf{v}' \qquad (a^* \text{ depending on } a);$$
$$(a(\mathbf{u} + \mathbf{v}))\mathbf{M} = \hat{a}((\mathbf{u} + \mathbf{v})\mathbf{M}) \qquad (\hat{a} \text{ depending on } a).$$

Hence,

(3) $$\begin{aligned}\hat{a}\mathbf{u}' + \hat{a}\mathbf{v}' &= \hat{a}(\mathbf{u}' + \mathbf{v}') = \hat{a}(\mathbf{uM} + \mathbf{vM})\\ &= \hat{a}(\,(\mathbf{u} + \mathbf{v})\mathbf{M}) \qquad \text{(by Lemma 5)}\\ &= (a(\mathbf{u} + \mathbf{v}))\mathbf{M} = (a\mathbf{u} + a\mathbf{v})\mathbf{M}\end{aligned}$$

Figure D1

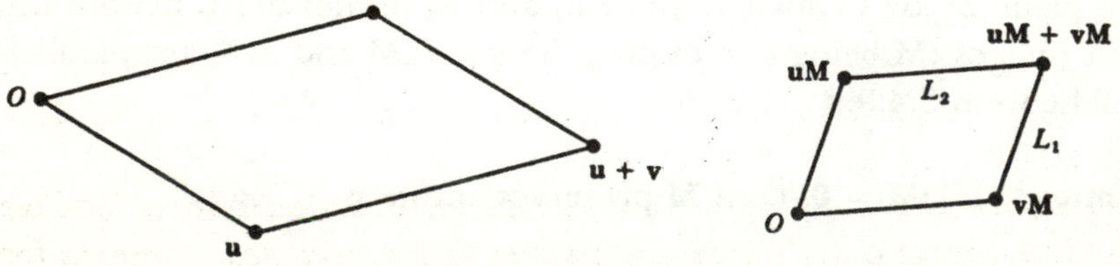


(vectors regarded as points)

$$= (a\mathbf{u})\mathbf{M} + (a\mathbf{v})\mathbf{M} \qquad \text{(by Lemma 5)}$$
$$= a'\mathbf{u}' + a^*\mathbf{v}'.$$

Since $\mathbf{u}$, $\mathbf{v}$ are linearly independent, $\mathbf{u}'$, $\mathbf{v}'$ are independent (Lemma 3) and, hence, equation (3) implies that $\hat{a} = a' = a^*$. This is true *for all scalars a*. Hence, for all $a$, $b$, we have

$$\begin{aligned} a'\mathbf{u}' + (ab)'\mathbf{v}' &= (a\mathbf{u})\mathbf{M} + ((ab)\mathbf{v})\mathbf{M} \\ &= (a\mathbf{u} + ab\mathbf{v})\mathbf{M} \qquad \text{(Lemma 5)} \\ &= (a(\mathbf{u} + b\mathbf{v}))\mathbf{M} \\ &= \bar{a}((\mathbf{u} + b\mathbf{v})\mathbf{M}) \qquad \text{(for some scalar } \bar{a} \text{ depending on } a \text{ and } b) \\ &= \bar{a}(\mathbf{u}\mathbf{M} + (b\mathbf{v})\mathbf{M}) = \bar{a}(\mathbf{u}' + b'\mathbf{v}') \\ &= \bar{a}\mathbf{u}' + \bar{a}b'\mathbf{v}'. \end{aligned}$$

Hence, $\bar{a} = a'$, $(ab)' = \bar{a}b' = a'b'$. Similarly,

$$\begin{aligned} (a + b)'\mathbf{u}' &= ((a + b)\mathbf{u})\mathbf{M} = (a\mathbf{u} + b\mathbf{u})\mathbf{M} = (a\mathbf{u})\mathbf{M} + (b\mathbf{u})\mathbf{M} \\ &= a'\mathbf{u}' + b'\mathbf{u}' = (a' + b')\mathbf{u}' \end{aligned}$$

so that $(a + b)' = a' + b'$. We have thus shown: *the mapping* $a \to a'$ *is a ring homomorphism of* $\mathbf{R} \to \mathbf{R}$. But the only ring homomorphisms of $\mathbf{R} \to \mathbf{R}$ are $\mathbf{O}$ and $\mathbf{I}$ (Sec. 1.15), and the mapping $a \to a'$ is not the zero mapping since $1'\mathbf{u}' = (1\mathbf{u})\mathbf{M} = \mathbf{u}\mathbf{M} = \mathbf{u}'$, $1' = 1$. Hence, $a \to a'$ is the identity mapping, $a' = a$ for all $a$, and (2) becomes

(4) $\qquad (a\mathbf{u})\mathbf{M} = a\mathbf{u}' = a(\mathbf{u}\mathbf{M}).$

In proving (4) we assumed $\mathbf{u} \neq \mathbf{0}$, but (4) holds by hypothesis when $\mathbf{u} = \mathbf{0}$. This proves Lemma 6. ∎

Lemmas 5 and 6 show that *if* $\mathbf{0M} = \mathbf{0}$, *then* $\mathbf{M}$ *is linear*. It follows, by an argument like the proof of Theorem 4.10.6, that even if $\mathbf{0M} \neq \mathbf{0}$, $\mathbf{M}$ is a linear mapping times a translation; that is, $\mathbf{M}$ is affine.

## E Some applications to analysis

The theory of vector spaces and matrices, in addition to its importance for algebra and geometry, has applications in various branches of analysis such as multivariable calculus and differential equations. One such application was exhibited in Section 9.8; some others will be presented here. Since our purpose is primarily illustrative, we omit some of the proofs.

First, some terminology. We shall assume that the reader is familiar with the definition of partial derivatives and knows what it means for a function of $n$ real variables to be continuous. Let $G$ be a mapping of $\mathbf{R}_n \to \mathbf{R}_n$; for convenience, the image under $G$ of any point $P = (x_1, ..., x_n)$ will be denoted $P' = (x'_1, ..., x'_n)$. Since the numbers $x'_j$ depend on the numbers $x_i$, each $x'_j$ may be regarded as a function of the $n$ real variables $x_1, ..., x_n$. Thus, we may write

$$x'_i = g^{(i)}(x_1, ..., x_n).$$

In other words, $g^{(i)}(P)$ is the $i$th coordinate of the point $G(P)$. We call the functions $g^{(i)}$ the *coordinate functions* corresponding to $G$. For example: if $n = 2$ and

$$G(x, y) = (x^2 + y^2, 2xy + 3),$$

then the corresponding coordinate functions $g^{(i)}$ are given by

$$g^{(1)}(x, y) = x^2 + y^2$$
$$g^{(2)}(x, y) = 2xy + 3.$$

As in Section 9.8, we denote by $\nabla g^{(i)}(P)$ the vector

$$\left(\frac{\partial g^{(i)}}{\partial x_1}(P), ..., \frac{\partial g^{(i)}}{\partial x_n}(P)\right)$$

whose coordinates are the first-order partial derivatives of $g^{(i)}$ evaluated at the point $P$ (provided these derivatives exist). The $n \times n$ matrix $J_G(P)$ which is defined by the equations

$$(J_G(P))_{ij} = \frac{\partial g^{(i)}}{\partial x_j}(P) \qquad (1 \leq i \leq n, 1 \leq j \leq n)$$

is called the *Jacobian matrix of G (at P)*; its entry in the $(i, j)$ position is the partial derivative of the $i$th coordinate function with respect to the $j$th variable, evaluated at $P$. It is clear that the row vectors of $J_G(P)$ are $\nabla g^{(1)}(P), ..., \nabla g^{(n)}(P)$.

If $G$ is a linear transformation, the partial derivatives $\partial g^{(i)}/\partial x_j$ are constants (why?) and, hence, $J_G = J_G(P)$ is the same for every point $P$. In this case, we know that $G$ is one-to-one and onto (i.e., has an inverse) if and only if the matrix $J_G$ is nonsingular (Sec. 5.6). This result can be generalized to the nonlinear case if we restrict our considerations to a neighborhood of a given point. (If $P_0$ is a fixed point in $\mathbf{R}_n$, the set of all points $P \in \mathbf{R}_n$ such that $d(P_0, P) < \epsilon$ is called the *$\epsilon$-neighborhood* of $P$. By a *neighborhood* of $P$ we mean any subset of $\mathbf{R}_n$ which includes, for some $\epsilon > 0$, the $\epsilon$-neighborhood of $P$. In the language of topology, $\mathcal{N}$ is a neighborhood of $P$ if $P$ lies in the interior of $\mathcal{N}$.) We state the relevant theorem, as follows:

**Theorem A.** Let $G : \mathbf{R}_n \to \mathbf{R}_n$. Assume that the first-order partial derivatives of the coordinate functions of $G$ exist and are continuous throughout some neighborhood $\mathcal{N}$ of the point $P_0 \in \mathbf{R}_n$, and that the Jacobian matrix $J_G(P_0)$ is nonsingular. Let $G(P_0) = P_0'$.

Then there exists a neighborhood $\mathcal{N}_0$ of $P_0$ such that:

(a) $\mathcal{N}_0 \subseteq \mathcal{N}$.

(b) $G$ maps $\mathcal{N}_0$ one-to-one onto some neighborhood $\mathcal{N}'$ of $P_0'$.

(c) Letting $F: \mathcal{N}' \to \mathcal{N}_0$ be the inverse of $G : \mathcal{N}_0 \to \mathcal{N}'$, the coordinate functions of $F$ have continuous first-order partial derivatives throughout $\mathcal{N}'$.

Theorem A is known as the *Inverse Function Theorem.* For a proof, see [2], Section 5.7, or [3], Section 3.3, Subsections 6 and 8.

We shall use the Inverse Function Theorem to derive a very general form of the *method of Lagrange multipliers*, which relates to the problem of finding relative extrema of a function $f(x_1, ..., x_n)$ subject to "side conditions" of the form

(1) $\qquad H_1 = H_2 = \cdots = H_k = 0,$

where the $H_i$ are functions of $n$ real variables. Formally, we say that *$f$ has a relative maximum at $P_0$ subject to the conditions (1)* if

(a) $H_1(P_0) = \cdots = H_k(P_0) = 0$; and

(b) $f(P_0) \geqslant f(P)$ for all points $P$ in some neighborhood of $P_0$ which satisfy the conditions $H_1(P) = \cdots = H_k(P) = 0$. A relative minimum subject to side conditions is defined similarly. The applicable theorem here is the following:

**Theorem B.** Let $f$, $H_1, ..., H_k$ be real-valued functions of $n$ real variables whose first-order partial derivatives all exist and are continuous throughout a neighborhood of the point $P_0$ in $\mathbf{R}_n$. Assume that $f$ has a relative maximum or minimum at $P_0$ subject to the conditions $H_1 = \cdots = H_k = 0$.

Then the vectors $\nabla f(P_0), \nabla H_1(P_0), ..., \nabla H_k(P_0)$ are linearly dependent.

It is not hard to prove Theorem B if Theorem A is assumed. Indeed, suppose the vectors $\nabla f(P_0)$, $\nabla H_1(P_0)$, ..., $\nabla H_k(P_0)$ are independent; then by Corollary 2.6.14(a) we can extend this set of vectors to a basis of $\mathbf{R}_n$ by adjoining $n - (k + 1)$ further vectors $\mathbf{u}_{k+1}, ..., \mathbf{u}_{n-1}$. If we define functions $H_{k+1}, ..., H_{n-1}$ by

$$H_i(X) = \mathbf{u}_i \cdot X \qquad (i = k + 1, ..., n - 1;\ \text{all } X \in \mathbf{R}_n),$$

then $\nabla H_i(X) = \mathbf{u}_i$. Thus, the $n$ vectors $\nabla f(P_0)$, $\nabla H_1(P_0)$, ..., $\nabla H_{n-1}(P_0)$ are independent, so that the matrix whose rows are these $n$ vectors is nonsingular. Hence, Theorem A is applicable to the mapping $G$ whose coordinate functions are $f, H_1, \ldots, H_{n-1}$. Since the neighborhood $\mathcal{N}$ of Theorem A may be taken to be any sufficiently small neighborhood of $P_0$, it follows that $G$ maps every sufficiently small neighborhood of $P_0$ *onto* a neighborhood of the point

$$G(P_0) = (f(P_0), 0, \ldots, 0, H_{k+1}(P_0), \ldots, H_{n-1}(P_0)).$$

In particular, this implies that every neighborhood of $P_0$ contains a point $P$ such that

$$G(P) = (f(P_0) + h, 0, \ldots, 0, H_{k+1}(P_0), \ldots, H_{n-1}(P_0))$$

where $h > 0$. But then $f(P) = f(P_0) + h > f(P_0)$ and $H_1(P) = \cdots = H_k(P) = 0$, so that $f$ cannot have a relative maximum at $P_0$ subject to the conditions (1). The argument regarding a relative minimum is similar. ▮

If the vectors $\nabla H_1(P_0), \ldots, \nabla H_k(P_0)$ are linearly independent, then linear *dependence* of $\nabla f(P_0), \ldots, \nabla H_k(P_0)$ implies that $\nabla f(P_0)$ is a linear combination of $\nabla H_1(P_0), \ldots, \nabla H_k(P_0)$ (Theorem 2.4.10). This means that at the point $P_0$, we may write

$$\nabla f = \lambda_1 \nabla H_1 + \cdots + \lambda_k \nabla H_k \qquad (\lambda_i \in \mathbf{R}),$$

which is equivalent to the system of equations

$$(2) \qquad \begin{aligned} \partial f/\partial x_1 &= \lambda_1(\partial H_1/\partial x_1) + \cdots + \lambda_k(\partial H_k/\partial x_1) \\ &\cdots \\ \partial f/\partial x_n &= \lambda_1(\partial H_1/\partial x_n) + \cdots + \lambda_k(\partial H_k/\partial x_n) \end{aligned} \qquad \text{(at the point } P_0\text{)}.$$

The scalars $\lambda_i$ are usually called *Lagrange multipliers*. The equations (1) and (2) together constitute a system of $k + n$ equations in $k + n$ unknowns, the "unknowns" being the $k$ scalars $\lambda_i$ and the $n$ coordinates of $P_0$. The set of solutions of the system, together with the set of points (if any) at which $\nabla H_1, \ldots, \nabla H_k$ are dependent, will include all points $P_0$ at which $f$ has an extremum subject to the given side conditions. (Perhaps not *quite* all such points; we must also consider points, if any, at which the continuity hypothesis fails to hold.) In practice, solving the equations is not always a straightforward matter (especially if the equations are nonlinear), but sometimes the equations are nice enough to yield the solution readily.

**Example**

If the temperature at any point $(x, y)$ on the unit circle (the circle of radius 1 about the origin in $\mathbf{R}_2$) is $3x^2 + 3y^2 - 2x + 4y$ degrees centigrade, find the hottest and coldest points on the circle.

*Solution.* The *existence* of the desired points follows from the fact that $f$ is continuous and the unit circle is a closed bounded set. Since the points on the circle satisfy the equation $x^2 + y^2 = 1$, we must find points $P_0$ at which the function

(3) $$f(x, y) = 3x^2 + 3y^2 - 2x + 4y$$

has a maximum or minimum subject to the condition

(4) $$H(x, y) = x^2 + y^2 - 1 = 0.$$

Clearly, the partial derivatives of $f$ and $H$ are continuous everywhere. Also, since $H_x = 2x$ and $H_y = 2y$ cannot be simultaneously zero at any point on the circle (why not?), the vector $\nabla H(P_0)$ is always nonzero and, thus, linearly independent (as a one-element set). Hence, the preceding discussion applies. Equations (2) in this case are

(5) $$6x - 2 = \lambda(2x); \qquad 6y + 4 = \lambda(2y).$$

The solutions of (4)–(5) are

(6) $$x = 1/\sqrt{5}, \qquad y = -2/\sqrt{5}, \qquad \lambda = 3 - \sqrt{5}$$

and

(7) $$x = -1/\sqrt{5}, \qquad y = 2/\sqrt{5}, \qquad \lambda = 3 + \sqrt{5}$$

(we leave the derivation to you!). Substituting (6) into (3), we get $f(x, y) = 3 - 2\sqrt{5} = -1.47$; substituting (7) into (3), we get $f(x, y) = 3 + 2\sqrt{5} = 7.47$. Hence, $(-1/\sqrt{5}, 2/\sqrt{5})$ is the warmest point on the circle and $(1/\sqrt{5}, -2/\sqrt{5})$ is the coldest point. ∎

**Applications to differential equations**

A differential equation (D.E.) of the form

(8) $$p_0(x)y + p_1(x)y' + p_2(x)y'' + \cdots + p_{n-1}(x)y^{(n-1)} + y^{(n)} = 0,$$

where $y = y(x)$ is a function of $x$, is called a *homogeneous linear differential equation*. (The equation is "homogeneous" because the right side is zero, "linear" because the various derivatives of $y$ occur separately and to the first power only; expressions such as $y^2$, $(y'')^3$, $y''y'$, etc. are absent.) It is not hard to verify that the set of all real-valued functions $y = y(x)$ which satisfy a given D.E. of this form is a vector space over **R**. (Functions are added, and multiplied by scalars, pointwise; cf. eq. 2.1.3.) The following theorem tells us the *dimension* of this vector space under rather general conditions.

**Theorem C.** Let $n$ be a positive integer; let $c$, $d$ be real numbers with $c < d$, and let $p_0(x), \ldots, p_{n-1}(x)$ be continuous real-valued functions of $x$ on the interval $c < x < d$.

Then the set of all real-valued functions $y = y(x)$ on the interval $c < x < d$ which satisfy the differential equation (8) is a vector space of dimension $n$ over $\mathbf{R}$.

Once the dimension is known to be $n$, the general solution of (8) can be obtained simply by finding $n$ linearly independent solutions, which must then constitute a basis of the space of all solutions. For example, the D.E.

(9) $\qquad y'' + y = 0$

has the form (8) with $n = 2$; moreover, $y = \sin x$ and $y = \cos x$ are solutions of (9). We show that these two solutions are independent, as follows: suppose $a$, $b$ are scalars such that

(10) $\qquad a \sin x + b \cos x = 0 \qquad (\text{all } x \in \mathbf{R}).$

Differentiating (10), we obtain

(11) $\qquad a \cos x - b \sin x = 0 \qquad (\text{all } x \in \mathbf{R}).$

Equations (10) and (11) may be regarded as a system of two homogeneous equations in the two "unknowns" $a$ and $b$. The matrix of the "system" is

(12) $$\begin{bmatrix} \sin x & \cos x \\ \cos x & -\sin x \end{bmatrix}$$

which has determinant $-1$ and is, thus, nonsingular. (We actually need nonsingularity only for one value of $x$.) It follows that $a = b = 0$ (cf. Sec. 5.10), so that the functions $\sin x$, $\cos x$ are indeed linearly independent. Hence, the general real solution of (9) is

$$y = a \sin x + b \cos x \qquad (a, b \in \mathbf{R}).$$

(At the end of Sec. 4.2 we obtained the general solution of

$$x^2y'' + xy' = y$$

on the interval $0 < x < \infty$ by similar reasoning. This D.E. has the form (8) on any $x$ interval which excludes 0, since if $x \neq 0$ we can divide by $x^2$ to make the coefficient of the highest derivative equal to 1.)

In our proof (above) that the functions $\sin x$, $\cos x$ are independent, we were led to consider the matrix (12). The first row of this matrix consists of the two given functions, while the second row consists of their derivatives. More generally, if $y_1(x), \ldots, y_n(x)$ are $n$ solutions of equation (8) on an interval $(c, d)$, we may differentiate the equation

(13) $\qquad a_1y_1(x) + \cdots + a_ny_n(x) = 0 \qquad (c < x < d)$

$n - 1$ times to obtain a system of $n$ homogeneous linear equations in the $n$ "unknowns" $a_1, \ldots, a_n$. The matrix of this system is

$$(14) \qquad \begin{bmatrix} y_1(x) & y_2(x) & \cdots & y_n(x) \\ y_1'(x) & y_2'(x) & \cdots & y_n'(x) \\ \vdots & \vdots & & \vdots \\ y_1^{(n-1)}(x) & y_2^{(n-1)}(x) & \cdots & y_n^{(n-1)}(x) \end{bmatrix}.$$

The determinant of the matrix (14) is called the *Wronskian*, or *Wronskian determinant*, of the functions $y_1, \ldots, y_n$. Using the same reasoning as before, we see that if the Wronskian is nonzero at even one point of the interval $c < x < d$, then the only solution of (13) is $a_1 = \cdots = a_n = 0$ so that the functions $y_1, \ldots, y_n$ are linearly independent. Conversely, it can be shown that if $y_1, \ldots, y_n$ are $n$ independent solutions of equation (8) on $(c, d)$, then their Wronskian is nonzero *at every point $x$ in* $(c, d)$. (We are still assuming, as in Theorem C, that $p_0(x), \ldots, p_{n-1}(x)$ are continuous.) In other words, the Wronskian of any set of $n$ solutions of (8) on a given interval will be either everywhere zero or everywhere nonzero on the interval. This assertion does *not* hold for fewer than $n$ solutions.

**Example**

Show that there do not exist continuous functions $p(x)$, $q(x)$ such that the differential equation

$$p(x)y + q(x)y' = y''$$

has both $y = x^2$ and $y = x$ as solutions on the interval $(-1, 1)$.

*Solution.* The given D.E. has the form (8) with $n = 2$; hence, the Wronskian of any two solutions must be everywhere zero or everywhere nonzero. But

$$\det \begin{bmatrix} x^2 & x \\ 2x & 1 \end{bmatrix}$$

is zero at $x = 0$, nonzero at $x = 1/2$. ▮

**Exercises**

1. Carry out the derivation of the solution of equations (4)–(5).
2. Find the minimum value of $x^2 + (y - 5)^2$ subject to the condition $x^2 = 16y$, where $x, y \in \mathbf{R}$. (Assume that the minimum exists.) Interpret the result geometrically. (ANS)
3. Is the preceding problem equivalent to the problem, "Find the minimum value of $16y + (y - 5)^2$ where $y \in \mathbf{R}$"? Why or why not?

# F Equations of selected conics and 3-quadrics

### I. The ellipse and hyperbola

We work in the space $\mathbf{R}_2$. An *ellipse* is the locus of points, the sum of whose distances from two fixed points $F_1$, $F_2$ (the "foci") is a positive constant greater than $d(F_1, F_2)$. The definition of *hyperbola* is obtained from that of "ellipse" by replacing "sum" by "absolute difference" and "greater" by "less". Let $2c = d(F_1, F_2)$ and let $2a$ be the "positive constant" (sum or difference). To make things simple, we consider the case where the foci are located at $(c, 0)$ and $(-c, 0)$. The definition of ellipse (hyperbola) is then equivalent to the equation

(1) $$\left| d[(x, y), (c, 0)] \pm d[(x, y), (-c, 0)] \right| = 2a,$$

the $\pm$ sign being plus for the ellipse, minus for the hyperbola. Using the distance formula and squaring both sides of (1), we obtain

$$[(x - c)^2 + y^2] + [(x + c)^2 + y^2] \pm 2\,([\,(x - c)^2 + y^2]\,[(x + c)^2 + y^2])^{1/2} = 4a^2.$$

Now transfer the square root term to the right side and the term $4a^2$ to the left side; then square both sides again, expand, cancel whatever cancels, and divide by 16. We leave it to you to verify that the resulting equation (after rearranging and combining terms) is

$$a^2(a^2 - c^2) = (a^2 - c^2)x^2 + a^2y^2$$

which when divided by $a^2(a^2 - c^2)$ becomes

$$1 = x^2/a^2 + y^2/(a^2 - c^2).$$

Clearly, all steps are reversible. Letting $b = \left| a^2 - c^2 \right|^{1/2}$, and remembering that $a > c$ for the ellipse whereas $a < c$ for the hyperbola, we obtain

(2) $$1 = x^2/a^2 + y^2/b^2 \qquad (a \geqslant b)$$

as the equation of the ellipse, and

(3) $$1 = x^2/a^2 - y^2/b^2$$

as the equation of the hyperbola. For an ellipse or hyperbola with foci along the $y$ axis instead of the $x$ axis, the roles of $x$ and $y$ in equation (2) or (3) are reversed.

If $K$ is any orthogonal coordinate system in $\mathbf{R}_2$, the locus of points whose $K$ coordinates $(x', y')$ satisfy the equation

(4) $$1 = (x')^2/a^2 + (y')^2/b^2$$

(with $a \geqslant b$) is congruent to the ellipse (2) and, hence, is also an ellipse (cf. the remarks following Theorem 4.10.23). Conversely, given any two points $F_1$, $F_2$ in $\mathbf{R}_2$, there exists an orthogonal coordinate system $K$ such that the $K$ coordinates of $F_1$, $F_2$ are, respectively, $(c, 0)$, $(-c, 0)$, where $c = \frac{1}{2}d(F_1, F_2)$ (compare with Example 7.1.44). Since the distance formula remains valid in $K$ coordinates (Sec. 3.10, Exercise 2), the same argument as before shows that an ellipse having $F_1$, $F_2$ as foci will have an equation of the form (4) in $K$ coordinates. Such an ellipse is shown in Figure F1. The $x'$ and $y'$ axes intersect at the point $P_0 = (0, 0)_K$; this point is the midpoint of the foci and is called the *center* of the ellipse. The $x'$ axis meets the ellipse at two points $(\pm a, 0)_K$ (proof: set $y' = 0$ in (4)); these points are called the *vertices* of the ellipse, and the segment joining them is the *major axis*. Likewise, the $y'$ axis meets the curve at the points $(0, \pm b)_K$; the segment joining these two points is the *minor axis*. Since (4) implies that $|x'| \leqslant a$ and $|y'| \leqslant b$, the ellipse is a bounded locus.

The discussion of the hyperbola is similar. The foci $F_1$, $F_2$ of any given hyperbola have coordinates $(c, 0)$, $(-c, 0)$ in some orthogonal coordinate system $K$; the equation of the hyperbola in $K$ coordinates is then

(5) $\quad 1 = (x')^2/a^2 - (y')^2/b^2,$

where $b^2 = c^2 - a^2$ (see Fig. F2). Again, the point $P_0 = (0, 0)_K$ is called the center and the points $V_i = (\pm a, 0)_K$ are the vertices. The line through the foci (the $x'$ axis) is called the *transverse axis*. The $y'$ axis does not meet the curve, and the locus is not bounded. In fact, the two lines $y'/b = \pm x'/a$ are *asymptotes*; the hyperbola "approaches" these

Figure F1

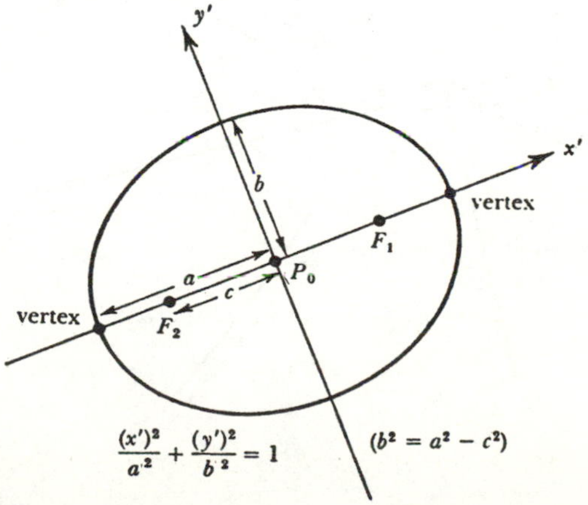

lines as $x'$ approaches $\pm\infty$. (Formally, since (5) is the same as $y'/b = \pm((x')^2/a^2 - 1)^{1/2}$, this means that $((x')^2/a^2 - 1)^{1/2} - x'/a \to 0$ as $x' \to \infty$, a fact whose proof we leave to you.)

An ellipse whose two foci coincide is called a *circle*. (The foci of a hyperbola cannot coincide.) Equivalently, a circle is the locus of points having a fixed constant distance (the *radius*) from a fixed point (the *center*, which coincides with the focus). If $a$, $b$, $c$ are as above, then $a$ is the radius, $c = 0$, $b = a$ so that equation (4) becomes

$$(x')^2 + (y')^2 = a^2.$$

The concepts of vertex, major axis, and minor axis have no meaning in the case of a circle.

## II. The parabola

A *parabola* in $\mathbf{R}_2$ is the locus of points equidistant from a fixed line $D$ (the "directrix") and a fixed point $F$ (the "focus") which does not lie on $D$. The line through $F$ perpendicular to $D$ is called the *axis* of the parabola; the point on the axis which lies halfway from $F$ to $D$ is the *vertex*. The definition of parabola easily implies that the vertex lies on the parabola. If $K$ is the orthogonal coordinate system in which

Figure F2

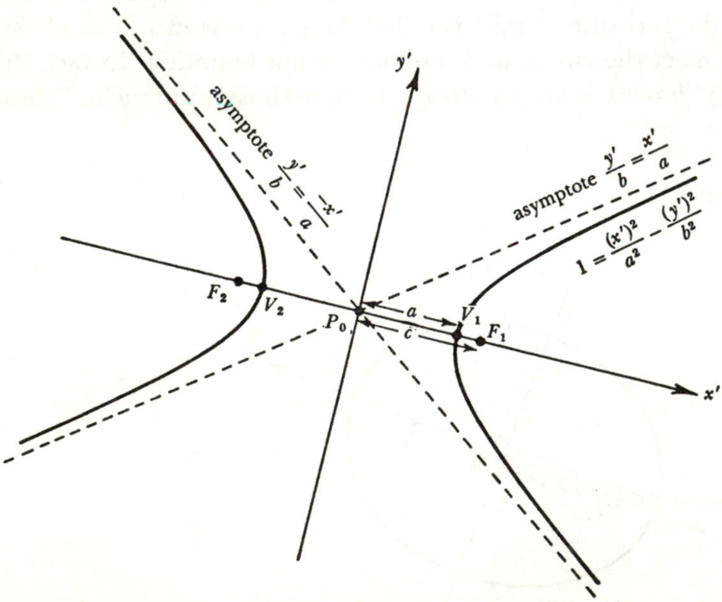

the axis of the parabola is the $y'$ axis and the vertex is the "origin" (see Fig. F3), then the focus has $K$ coordinates $(0, k)$ and the equation of the directrix is $y' = -k$, where $2k$ is the distance from focus to directrix. The distance from any point $(x', y')_K$ to the directrix is $|y' + k|$ (why?); hence, the equation of the parabola in $K$ coordinates is $d[(x', y')_K, (0, k)_K] = |y' + k|$, which when squared becomes

$$(x')^2 + (y' - k)^2 = (y' + k)^2.$$

Expanding and canceling, we obtain

$$(x')^2 = 4ky'.$$

Similarly, the equation of a parabola with vertex $(0, 0)_K$ and focus on the $x'$ axis would have the form $(y')^2 = 4kx'$. The sign of $k$ indicates whether the focus lies on the positive or negative half of the axis.

### III. Surfaces of revolution and near-revolution in $\mathbf{R}_3$

As usual, $x$, $y$, $z$ will denote coordinates in $\mathbf{R}_3$; thus, for example, the plane $[\mathbf{e}_1, \mathbf{e}_2]$ is the "$xy$ plane", consisting of all points whose $z$ coordinate is zero. Consider a curve $C$ in the $xy$ plane whose equation is given by $y = f(x)$. If we rotate the curve about the $x$ axis (think of this as a rigid *motion*, not a rigid *mapping*, cf. Sec. 6.6), the set of all points through which the curve passes during the motion is a surface $S$ which we call a *surface of revolution* (see Fig. F4). The cross sections of $S$ perpendicular to the $x$ axis are circles.

To find the equation of $S$, we argue as follows: given a point $P = (x, y, z)$ in $\mathbf{R}_3$, the point on the $x$ axis which is closest to $P$ is $(x, 0, 0)$

Figure F3

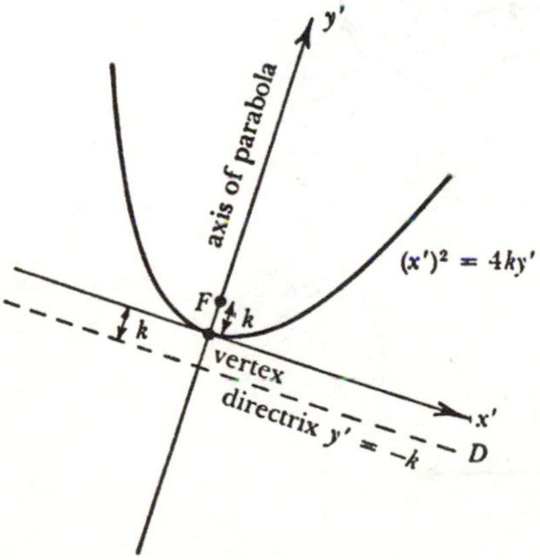

(Fig. F5) and, thus, the distance from $P$ to the $x$ axis is $(y^2 + z^2)^{1/2}$. In particular, any point on the curve $C$ has the form $(x, f(x), 0)$ so that its distance from the $x$ axis is

$$(f(x)^2 + 0^2)^{1/2} = |f(x)|.$$

A rotation about the $x$ axis carries this point into a point $(x, y, z)$ having the same $x$ coordinate and the same distance from the $x$ axis; thus, $(y^2 + z^2)^{1/2} = |f(x)|$. Equivalently,

$$y^2 + z^2 = (f(x))^2 \tag{6}$$

which is the equation of the surface $S$.

Suppose we now "stretch" $S$ by a constant factor $k$ in some fixed direction perpendicular to the $x$ axis, say, the $y$ direction. That is, we replace each point $(a, b, c)$ by the point $(a, kb, c)$. The surface $S'$ thus obtained is called a *surface of near-revolution*. Figure F6 shows the relation between $S$ and $S'$ when $0 < k < 1$: $S'$ lies inside $S$ and is tangent to $S$ at those points of $S$ whose $y$ coordinate is 0. (In this case, $S$ has been "shrunk" rather than "stretched".) If, instead, $k > 1$, $S'$ would lie outside $S$. The equation of $S'$ is

Figure F4

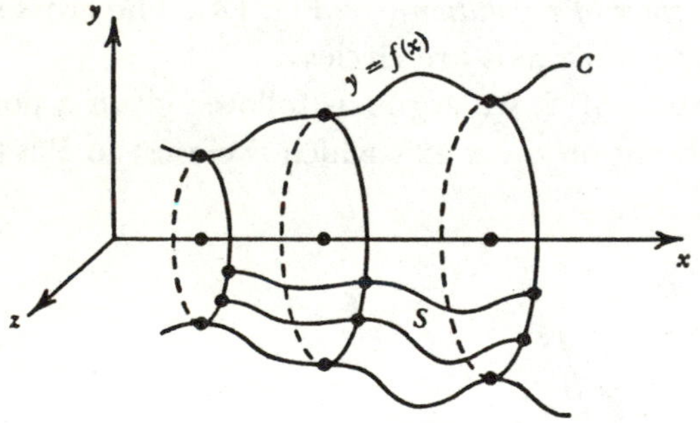


Figure F5

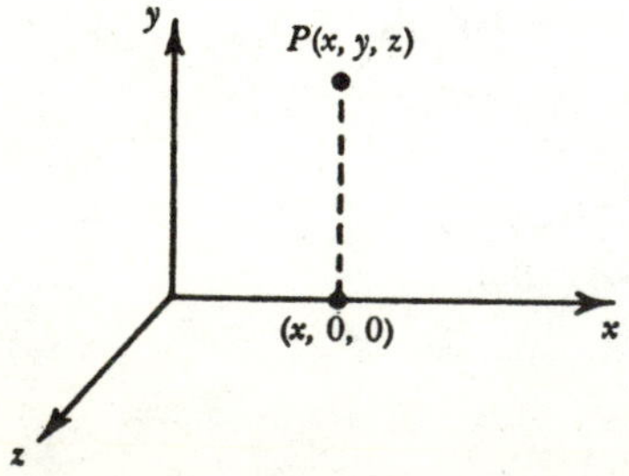

(7) $$\frac{y^2}{k^2} + z^2 = (f(x))^2$$

since the point $(a, b, c)$ satisfies (6) if and only if the point $(a, kb, c)$ satisfies (7). Note that if $x$ is regarded as constant, (7) becomes the equation of an ellipse; thus, the cross sections of $S'$ perpendicular to the $x$ axis are ellipses rather than circles.

It is clear that if the three variables in equation (6) or (7) are permuted, the equation still represents a surface of revolution or near-revolution but with a different axis or a different stretching direction or both. For instance, the surface of revolution

(8) $$x^2 + z^2 = (f(y))^2$$

is obtained by rotating the curve $x = f(y)$ in the $xy$ plane (or the curve $z = f(y)$ in the $yz$ plane) about the $y$ axis. The surface of near-revolution

$$x^2 + z^2/4 = (f(y))^2$$

is obtained by stretching the surface (8) in the $z$ direction by a factor of 2.

Similar considerations apply in other orthogonal coordinate systems. If $K$ is such a system in $\mathbf{R}_3$ and if $K$ coordinates are denoted $(x', y', z')$, then the act of rotating the curve $y' = f(x')$ about the $x'$ axis produces the surface $(y')^2 + (z')^2 = (f(x'))^2$; the act of stretching this surface by a factor of $k$ in, say, the $z'$ direction then produces the surface

$$(y')^2 + (z')^2/k^2 = (f(x'))^2.$$

**IV. The standard 3-quadrics**

Most nonsingular 3-quadrics (and some of the singular ones) are surfaces of revolution or near-revolution. For instance, an *ellipsoid*

Figure F6

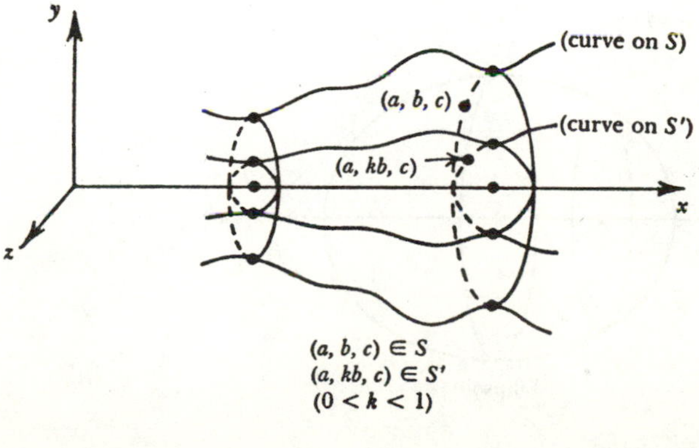

is obtained by rotating an ellipse about its major axis and then stretching or shrinking it (by a constant factor) in a direction perpendicular to that axis. If we choose our coordinates so that (4) is the equation of the ellipse and the direction of "stretch" is the $z'$ direction (Figure F7), then the rotation alone transforms (4) into the equation

$$1 = (x')^2/a^2 + [(y')^2 + (z')^2]/b^2$$

and the subsequent stretch transforms the latter equation into

$$1 = (x')^2/a^2 + [(y')^2 + (z')^2/k^2]/b^2. \tag{9}$$

If we let $c = bk$, (9) becomes the equation of an ellipsoid in standard form:

$$1 = (x')^2/a^2 + (y')^2/b^2 + (z')^2/c^2.$$

Certain other second-degree equations can be analyzed in similar fashion. For example:

(a) The equation

$$1 = (x')^2/a^2 + (y')^2/b^2 - (z')^2/c^2 \tag{10}$$

represents the result of rotating the hyperbola $1 = (x')^2/a^2 - (z')^2/c^2$ about the $z'$ axis and then stretching by a factor of $b/a$ in the $y'$ direction. A surface of this type is called a *hyperboloid of one sheet* (see Fig. F8); the $z'$ axis is its *principal axis*. By regarding $z'$ as constant in equation (10), we see that the cross sections of the hyperboloid perpendicular to the $z'$ axis are ellipses. Similarly, the cross sections perpendicular to the $x'$ axis (or to the $y'$ axis) are hyperbolas.

(b) The equation

$$(x')^2/a^2 + (y')^2/b^2 - (z')^2/c^2 = 0$$

represents the result of rotating the straight line $x'/a = z'/c$ about the $z'$ axis and then stretching in the $y'$ direction. A surface of this type is called an *elliptic cone* (see Fig. F9); the $z'$ axis is its principal axis.

Figure F7

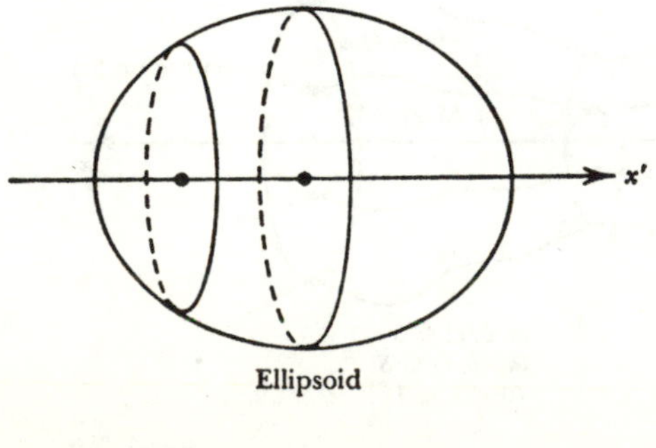


Ellipsoid

Cross sections perpendicular to the $z'$ axis are ellipses. If $b = a$, then the ellipses are circles and the cone is a surface of revolution (a *circular cone*).

(c) The equation

$$1 = (x')^2/a^2 - (y')^2/b^2 - (z')^2/c^2$$

represents the result of rotating the hyperbola $1 = (x')^2/a^2 - (y')^2/b^2$ about the $x'$ axis and then stretching in the $z'$ direction. This surface, a *hyperboloid of 2 sheets*, is sketched in Figure F10.

(d) The equation

$$z' = a^2(x')^2 + b^2(y')^2 \qquad (a \neq 0,\ b \neq 0)$$

represents the result of rotating the parabola $z' = a^2(x')^2$ about the $z'$ axis and then stretching in the $y'$ direction. This surface, an *elliptic paraboloid*, is sketched in Figure F11; its cross sections perpendicular to the $z'$ axis are ellipses.

The only nonsingular 3-quadric which is not a surface of near-revolution is the *hyperbolic paraboloid*, whose equation (in a suitable coordinate system) has the form

$$z' = a^2(x')^2 - b^2(y')^2 \qquad (a \neq 0,\ b \neq 0).$$

Figure F8

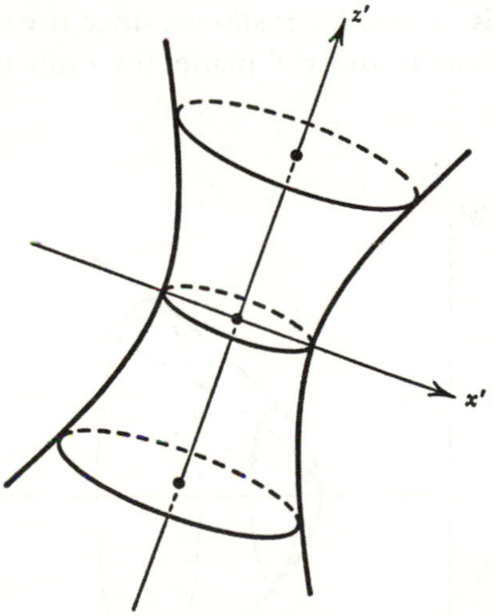


Hyperboloid of one sheet

Figure F9

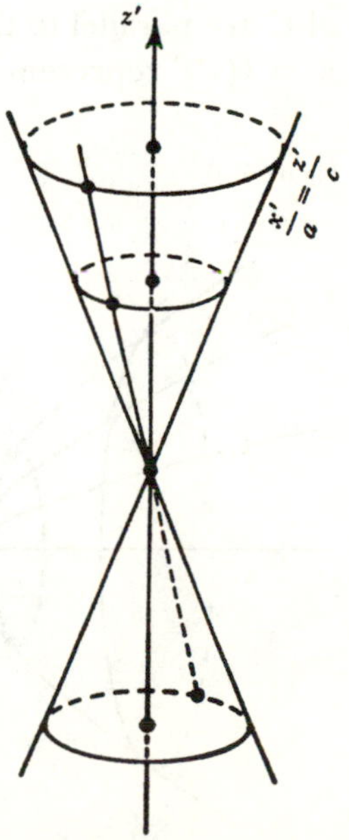


Elliptic cone

Regarding $z'$ as constant, we see that the cross sections perpendicular to the $z'$ axis are hyperbolas; regarding $x'$ or $y'$ as constant, we see that the cross sections perpendicular to the $x'$ and $y'$ axes are parabolas. Near the point $(0, 0, 0)_K$ ($K$ being the coordinate system), the surface is saddle-shaped. If you are ambitious, try a sketch; for the less ambitious (among whom is the author!), sketches may be found in many calculus books, including, for example, [13].

**V. Cylinders**

Let $S$ be a plane in $\mathbf{R}_3$, let $C$ be a curve (more generally, a set of points) in $S$, and let $\mathcal{Q}$ be the surface in $\mathbf{R}_3$ consisting of all points whose perpendicular projection on $S$ lies on $C$. Thus, to every point on $C$, there corresponds a line on $\mathcal{Q}$ which is perpendicular to the plane of $C$. Such a surface $\mathcal{Q}$ is called a *cylinder*. If $C$ is an ellipse, $\mathcal{Q}$ is an *elliptic cylinder*; if $C$ is a parabola, $\mathcal{Q}$ is a *parabolic cylinder*; and so on.

If we choose our coordinate system $K$ in such a way that the plane $S$ is the $x'y'$ plane (cf. Fig. F12), then the equation of the surface $\mathcal{Q}$ does not involve $z'$, since, if a point $P$ lies on $\mathcal{Q}$, then all other points with the same $x'$ and $y'$ coordinates also lie on $\mathcal{Q}$. In fact, it is evident that the equation of $\mathcal{Q}$ in $K$ coordinates coincides with the $x'y'$ equation of the curve $C$, and that the lines on $\mathcal{Q}$ which pass through the points of $C$ are parallel to the $z'$ axis. Thus, for instance, since the equation $y' = k(x')^2$ represents a parabola in the $x'y'$ plane, the same equation

Figure F10

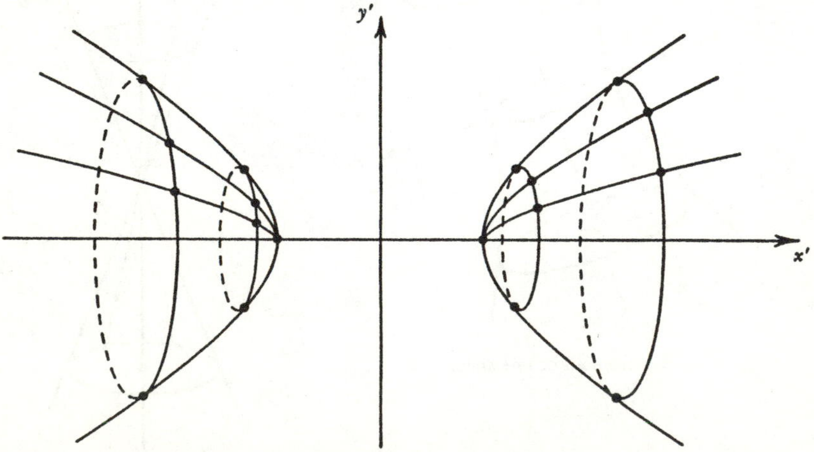


Hyperboloid of two sheets

represents a parabolic cylinder in $\mathbf{R}_3$, on which lie infinitely many lines in the $z'$ direction (cf. Chap. 7, Fig. 7.1.2). Similarly, any equation in $\mathbf{R}_3$ which involves only two coordinates will represent some type of cylinder. In particular, Table 7.4.1 implies that every singular 3-quadric, with the exception of the elliptic cone and the single point, may be regarded as a cylinder.

## G Hints and answers to selected exercises

**Section 1.3**

5. (a) $\mathbf{T}_1 = \mathbf{I}_B\mathbf{T}_1$ (by 1.3.5(a)) $= (\mathbf{T}_2\mathbf{S})\mathbf{T}_1 = \mathbf{T}_2(\mathbf{S}\mathbf{T}_1) = \mathbf{T}_2\mathbf{I}_A = \mathbf{T}_2$.
(b) If $\mathbf{T}_1$ and $\mathbf{T}_1'$ are two right inverses of $\mathbf{S}$, then $\mathbf{T}_1 = \mathbf{T}_2$ and $\mathbf{T}_1' = \mathbf{T}_2$ by part (a); hence, $\mathbf{T}_1 = \mathbf{T}_1'$, showing that $\mathbf{T}_1$ is unique. Similar reasoning shows that $\mathbf{T}_2$ is unique.
9. (a) Since $\mathbf{S}\mathbf{T}_1 = \mathbf{I}_A$, $\mathbf{T}_1$ has a left inverse; hence, $\mathbf{T}_1$ is onto by Theorem 1.3.6(b).

**Section 1.4**

4. Proof of transitive property: assume $a \sim b$, $b \sim c$. Thus, $a - b = 5x$ for some integer $x$, and $b - c = 5y$ for some integer $y$. (*Note*: Since the integers $x$ and $y$ may be different, we must represent them by different letters!) Then
$$a - c = (a - b) + (b - c) = 5x + 5y = 5(x + y).$$

Figure F11

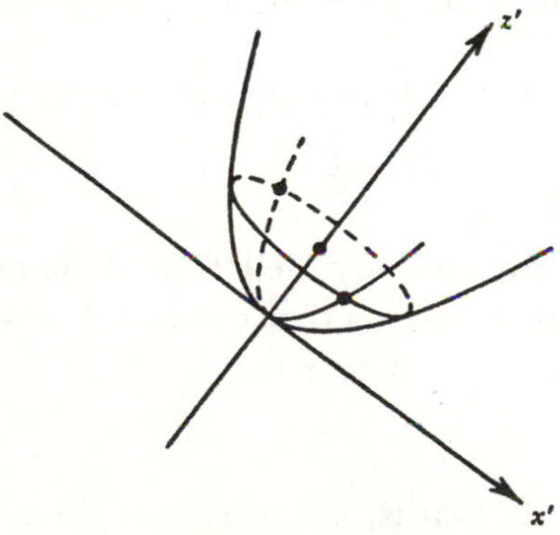


Elliptic paraboloid

Figure F12

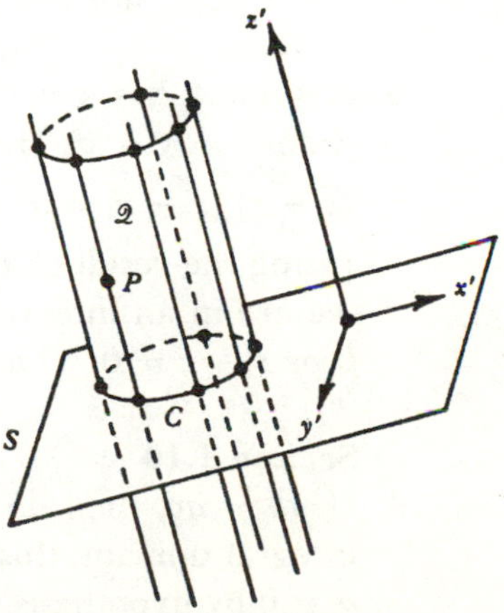


Elliptic cylinder

Thus, $a - c$ is 5 times an integer, that is, $a \sim c$.

**Section 1.8**

2. (a) If the mapping is denoted $\mathbf{T}$, then $x\mathbf{T} = \sqrt{x}$ (all $x \in \mathbf{R}^+$). The operation preserved by $\mathbf{T}$ is multiplication; $\mathbf{R}^+$ is a group under multiplication.
(b) $\sqrt{1} = 1$; $\sqrt{1/x} = 1/\sqrt{x}$.
4. For all $x, y \in S$,
$$\begin{aligned}(x * y)(\mathbf{T}_1\mathbf{T}_2) &= [(x * y)\mathbf{T}_1]\mathbf{T}_2 = [(x\mathbf{T}_1) \circ (y\mathbf{T}_1)]\mathbf{T}_2 \\ &= [(x\mathbf{T}_1)\mathbf{T}_2] \square [(y\mathbf{T}_1)]\mathbf{T}_2 \\ &= [x(\mathbf{T}_1\mathbf{T}_2)] \square [y(\mathbf{T}_1\mathbf{T}_2)].\end{aligned}$$
5. (a) For all $x \in S$, we have
$$(x\mathbf{T}) \circ (e\mathbf{T}) = (x * e)\mathbf{T} = x\mathbf{T}$$
$$(e\mathbf{T}) \circ (x\mathbf{T}) = (e * x)\mathbf{T} = x\mathbf{T}$$
by 1.8.3. Since $\mathbf{T}$ is onto, *every* element of $S'$ has the form $x\mathbf{T}$ where $x \in S$; hence, the above equations imply that $e\mathbf{T}$ is an identity of $S'$.
6. The mapping which takes every element of $S$ into the zero element will do the trick. If a less trivial mapping is desired, consider the mapping $\mathbf{T}$ defined by
$$\mathbf{T} : [n]_{(6)} \to [4n]_{(6)} \qquad \text{(all } n\text{)}.$$
(Notation here is the same as in Sec. 1.13.) In showing that this mapping satisfies equation 1.8.2, you will need the fact that $4 \equiv 16 \pmod 6$.

**Section 1.9**

1. $a(b - c) = a(b + (-c)) = ab + a(-c) = ab + (-ac) = ab - ac$.
4. Assume $y$ has a square root $x$. If $s$ is any square root of $y$, then $s^2 = y = x^2$. Hence,
$$(s + x)(s - x) = s^2 - x^2 = 0$$
(using the result of Exercise 3 to simplify the left side). Since we are in an integral domain, it follows that either $s + x = 0$ or $s - x = 0$. That is, either $s = -x$ or $s = x$.

**Section 1.10**

3. If $ab = ac$, then $ab - ac = 0$, that is, $a(b - c) = 0$. In an integral domain, this implies that $a = 0$ or $b - c = 0$. Since $a \neq 0$ by hypothesis, $b - c = 0$ so that $b = c$.

4. Imitate the proof for subgroups of a group.
5. (a) By assumption, there is some element $a \in F$ such that $a\mathrm{T} \neq 0$. Then $1(a\mathrm{T}) = a\mathrm{T} = (1a)\mathrm{T} = (1\mathrm{T})(a\mathrm{T})$, and Theorem 1.10.2(e) then gives $1\mathrm{T} = 1$.
   (b) Imitate the proof of Theorem 1.8.4(b).
6. (1) $S$ is closed under addition; (2) $S$ is closed under multiplication; (3) $1_F \in S$; (4) for every $a \in S$, $-a \in S$; (5) for every $a \neq 0$ in $S$, $a^{-1} \in S$.

**Section 1.11**

2. Parts (c), (d), (e), (f), (i), (j), (o), (p), and (q) remain true if strong inequalities ($<$, $>$) are replaced by weak ones ($\leqslant$, $\geqslant$). For example, the weak analog to (o) would read: If $a \in D^+$ and $ab \leqslant ac$, then $b \leqslant c$.
4. Since $(\sqrt{a}\,\sqrt{b})^2 = (\sqrt{a})^2\,(\sqrt{b})^2 = ab$, $\sqrt{a}\,\sqrt{b}$ is a square root of $ab$; since $\sqrt{a}$ and $\sqrt{b}$ are nonnegative, so is $\sqrt{a}\,\sqrt{b}$. Hence, $\sqrt{a}\,\sqrt{b}$ is the nonnegative square root of $ab$; that is, $\sqrt{a}\,\sqrt{b} = \sqrt{ab}$.
7. (a) Use parts (l) and (m) of 1.11.2.
   (d) Assume $c/a \leqslant c/b$ and use parts (i) and (q) of 1.11.2 to obtain a contradiction.

**Section 1.12**

1. The argument in question is located at the top of page 67 of [10], immediately following Definition 3.11. If we use function notation and write $f(n)$ instead of $a^n$, then to say that Definition 3.11 is valid is the same as saying that there exists a unique function (mapping) $f$ with domain $\mathbf{Z}^+$ such that $f(1) = a$ and $f(k+1) = f(k) \cdot a$. Likewise, McCoy's statement $S_n$ becomes, "$f(n)$ is defined by 3.11"; equivalently, "$n$ belongs to the domain of $f$". But such a statement is meaningful only if we presuppose the *existence* of the "domain of $f$", which in turn presupposes the existence of $f$ itself, which is what we are trying to show!
2. It must be shown that $(a')^n * a^n = a^n * (a')^n = e$. Use parts (d) and (e) of 1.12.16.
3. (a) Let $\mathscr{S}(n)$ be the statement "$2^n > n$". Since $2^1 = 2 > 1$ (*why?*), $\mathscr{S}(1)$ is true. Suppose now that $k$ is a positive integer such that $\mathscr{S}(k)$ is true, that is, $2^k > k$. Then

$$\begin{aligned} 2^{k+1} &= 2^k \cdot 2 && \text{(by 1.12.11)} \\ &= 2^k + 2^k && \text{(why?)} \end{aligned}$$

$$> k + k \quad \text{(by the truth of } \mathscr{S}(k) \text{ and 1.11.2(f))}$$
$$\geqslant k + 1 \quad (\textit{why?})$$

and, hence, $2^{k+1} > k + 1$, that is, $\mathscr{S}(k+1)$ is true. Therefore, the Principle of Induction implies $\mathscr{S}(n)$ is true for all $n \in \mathbf{Z}^+$.

6. $\sum_{i=m}^{m} a_i = a_m; \qquad \sum_{i=m}^{k+1} a_i = \left(\sum_{i=m}^{k} a_i\right) + a_{k+1} \qquad (k \in \mathbf{Z},\ k \geqslant m).$

8. When $n = 1$, the equation reduces to $ca_1 = ca_1$. Assuming the equation holds for $n = k$, we have

$$\sum_{i=1}^{k+1} (ca_i) = \sum_{i=1}^{k} (ca_i) + ca_{k+1} \quad \text{(by 1.12.13)}$$
$$= c\left(\sum_{i=1}^{k} a_i\right) + ca_{k+1} \quad \text{(equation assumed true for } n = k)$$
$$= c\left[\left(\sum_{i=1}^{k} a_i\right) + a_{k+1}\right] \quad \text{(left distributive law in rings)}$$
$$= c\left(\sum_{i=1}^{k+1} a_i\right) \qquad \text{(by 1.12.13)}$$

so that the desired equation holds for $n = k + 1$. Hence, it is valid for all $n \in \mathbf{Z}^+$ by the Principle of Induction.

**Section 1.14**

1. *Proof of Lemma 1.14.9.* Assume $S'$ is a subset of the $\beta$-finite set $S$, and let $\mathbf{T}' : S' \to S'$ be one-to-one. Define $\mathbf{T} : S \to S$ by $x\mathbf{T} = x\mathbf{T}'$ if $x \in S'$, $x\mathbf{T} = x$ if $x$ is an element of $S$ which is not in $S'$. Since $\mathbf{T}'$ is one-to-one, so is $\mathbf{T}$; hence, $\mathbf{T}$ is onto; hence, $\mathbf{T}'$ is onto.
5. By Lemma 1.14.9, $A'$ is finite; hence, $A'$ has $r$ elements, for some integer $r \geqslant 0$. By Exercise 2, $A \cup A' = S$ has $m + r$ elements; hence, Theorem 1.14.11 implies $m + r = n$, so that $r = n - m$.
6. For $n = 1$, this is easy. Assuming that the statement is true for $n = k$, suppose that $A = \{a_1, ..., a_k, a_{k+1}\}$ and $B = \{b_1, ..., b_k, b_{k+1}\}$. For each fixed $i \in E_{k+1}$, let $B_i$ be the set of all elements of $B$ except $b_i$; then $B_i$ has $k$ elements (Exercise 5). The set of all one-to-one mappings of $\{a_1, ..., a_k\}$ onto $B_i$ can be put into one-to-one correspondence with the set of all one-to-one mappings of $A$ onto $B$ which map $a_{k+1} \to b_i$; since our statement was assumed true for sets of $k$ elements, it follows that there are exactly $k!$ one-to-one mappings of $A$ onto $B$

which map $a_{k+1} \to b_i$. Since there are $k + 1$ choices for the integer $i$, it follows from Exercise 4 that the total number of one-to-one mappings of $A$ onto $B$ is $k!(k + 1) = (k + 1)!$.

8. Let $D$ be an OID. By considering the mapping $x \to x + 1$, show that $D^+$ is infinite; then use Lemma 1.14.9 to deduce that $D$ itself is infinite.

**Section 2.2**

2. (a) *Closure under addition*: Let $\mathbf{u}, \mathbf{v} \in S \cap T$. Then $\mathbf{u}, \mathbf{v} \in S$ and $\mathbf{u}, \mathbf{v} \in T$ (def. of intersection); hence, $\mathbf{u} + \mathbf{v} \in S$ and $\mathbf{u} + \mathbf{v} \in T$ (since $S$ and $T$ are subspaces); hence, $\mathbf{u} + \mathbf{v} \in S \cap T$ (def. of intersection).
4. If $S_i$ denotes the set of all solutions of the equation $a_{i1}x_1 + \cdots + a_{in}x_n = 0$, then $S = S_1 \cap S_2 \cap \cdots \cap S_m$ (why?). By Exercise 3, each $S_i$ is a subspace; hence, $S$ is a subspace by Exercise 2.
5. Only (b) and (g) are subspaces.
7. (a), (c), (e), (f), (g) are subspaces.
12. Since $T$ is a subspace, $\{\mathbf{0}\} \subseteq T$; hence, $S + \{\mathbf{0}\} \subseteq S + T$, that is, $S \subseteq S + T$.
14. (a) Any element of $B$ has the form $(a, -\frac{2}{3}a)$, where $a \in \mathbf{R}$; any element of $C$ has the form $(b, \frac{1}{3} - \frac{2}{3}b)$, where $b \in \mathbf{R}$. The sum of these two elements is $(a + b, \frac{1}{3} - \frac{2}{3}(a + b))$ which has the form of an element of $C$; hence, $B + C \subseteq C$. Conversely, since $B$ is a subspace, we have $C \subseteq B + C$ by Exercise 12.
(c) $D + E$ consists of all points $(x, y)$ such that $4 \leqslant x \leqslant 6$.

**Section 2.3**

2. (a) One such expression is $(8, 5, 2) = 5\mathbf{u} - 6\mathbf{v} + \mathbf{w}$; there are infinitely many others.
(b) A vector $(a, b, c)$ can be so expressed (i.e., lies in $[\mathbf{u}, \mathbf{v}, \mathbf{w}]$) if and only if the numbers $a$, $b$, $c$ form an arithmetic progression; that is, if and only if $2b = a + c$.

**Section 2.4**

2. (a) and (d) are dependent sets; (b) and (c) are independent sets.
3. (b) No.
6. One such subset is $\{(1, 0), (i, 0)\}$; there are many others.
7. No.

**Section 2.6**

2. Only (b) is a basis.
4. The desired subset will have 2 elements in part (a), 3 elements in part (b), one element in part (c).
8. *Exercise 6(b)*: Let $S = \{(a, b, a + b) : a \in \mathbf{R}, b \in \mathbf{R}\}$. We assume that we have already shown $S$ to be a subspace. Since $S$ is not all of $\mathbf{R}_3$, $\dim S < 3$; since the vectors (0, 1, 1), (1, 0, 1) lie in $S$ and are independent, it follows that $\dim S = 2$ and that any two independent vectors in $S$ constitute a basis.
   *Exercise 6(d)*: The given subset is a one-dimensional subspace; any nonzero vector in it will constitute a basis.
   *Exercise 7(c)*: The set $\{1, x, x^2\}$ is a basis.
11. (a) By Theorem 2.6.17, $\dim(S \cap T) + \dim(S + T) = \dim S + \dim T = 2 + 2 = 4$. Also, $S + T$ contains both $S$ and $T$ (Sec. 2.2, Exercise 12). If $T$ is a *proper* subset of $S + T$, then $T \subset S + T \subseteq V$; hence, $2 < \dim(S + T) \leq 3$ (Theorems 2.6.16(a) and 2.6.16(b)); hence, $\dim(S + T) = 3$; hence, $\dim(S \cap T) = 1$. On the other hand, if $T = S + T$, then $S \subseteq S + T = T$; hence, $S = T$ by Theorem 2.6.16(b).
    (b) Two-dimensional subspace = plane through the origin; one-dimensional subspace = line through the origin. Hence: two distinct planes through the origin (in 3-space) intersect in a line. Hence (by translation of origin): if two distinct planes in a 3-dimensional space intersect, their intersection is a line.
13. The components of (0, 2) are −4, 2.
14. $(x, y) = \dfrac{xd - yc}{ad - bc}(a, b) + \dfrac{ay - bx}{ad - bc}(c, d).$

**Section 2.8**

2. Sum is not direct.
3. Sum is direct.

**Section 3.1**

4. If $\mathbf{v} \parallel \mathbf{u}$, then $\mathbf{v} = c\mathbf{u}$ and the pair $(\mathbf{u}, \mathbf{v})$ is dependent by Theorem 2.4.6. Conversely, suppose $(\mathbf{u}, \mathbf{v})$ is dependent; since the set $\{\mathbf{u}\}$ is independent, $\mathbf{v} \in [\mathbf{u}]$ by Theorem 2.4.10, that is, $\mathbf{v} \parallel \mathbf{u}$.
5. 3.1.13(a) is a special case of 2.3.5(b); 3.1.13(b) is a special case of 2.4.10 (with $A = \{\mathbf{u}\}$); 3.1.13(c) is a special case of 2.3.13(e).

**Section 3.2**

3. If $A \neq C$, then $A, B, C$ are collinear $\Leftrightarrow B \in L(A, C) \Leftrightarrow$ **Ar** $AB \mid$ **Ar** $AC$ (by Def. 3.2.1) $\Leftrightarrow$ (**Ar** $AB$, **Ar** $AC$) is dependent (by Sec. 3.1, Exercise 4). If $A = C \neq B$, then all three points lie on $L(A, B)$, and the vectors **Ar** $AB$, **Ar** $AC$ are dependent since one of them is zero. If $A = B = C$, then all three points lie on $L(A, D)$, where $D$ is any point different from $A$, and again the pair (**Ar** $AB$, **Ar** $AC$) = (**0**, **0**) is dependent.
4. It suffices to show that if **Ar** $AB \mid$ **Ar** $AD$, then $A, B, C, D$ all lie on the same line. But if **Ar** $AB \mid$ **Ar** $AD$, then also the vector **Ar** $AC$ = **Ar** $AB$ + **Ar** $BC$ is parallel to **Ar** $AD$; hence, $B$ and $C$ lie on $L(A, D)$.
5. The right side of equation 3.2.16 is symmetric with respect to $A$ and $B$.
11. By 3.2.2(b) (or 3.2.3(a)), $L(A, B) = L(P_1, P_3)$; hence, $P_2 \in L(P_1, P_3)$ and we may write $P_2 = aP_1 + bP_3$, where $a + b = 1$ (eq. 3.2.10). Substituting $P_i = a_iA + b_iB$, this gives

    $$P_2 = (aa_1 + ba_3)A + (ab_1 + bb_3)B$$

    and the sum of the coefficients of $A$ and $B$ in the latter equation is

    $$a(a_1 + b_1) + b(a_3 + b_3) = a + b = 1.$$

    Hence, by uniqueness of the barycentric coordinates,

    $$a_2 = aa_1 + ba_3$$
    $$b_2 = ab_1 + bb_3.$$

    Without loss of generality, we may assume $a_1 < a_3$. Since $a_2 = aa_1 + ba_3$ and $a + b = 1$, we have

    $$a_2 - a_1 = b(a_3 - a_1)$$
    $$a_3 - a_2 = a(a_3 - a_1)$$

    so that $a, b$ are both positive if and only if $a_1 < a_2 < a_3$. This proves that statements (a) and (b) are equivalent. The proof that (a) $\Leftrightarrow$ (c) is similar.
14. This follows easily from Exercise 8.
17. We may write **Ar** $PA_i = a_i$ **Ar** $PB_i$ for each $i$ ($i = 1, 2, 3$), where the $a_i$ are scalars (why?). Since **Ar** $A_1A_2 \mid$ **Ar** $B_1B_2$, the same argument as in Example 3.2.17 shows that $a_1 = a_2$; similarly, from **Ar** $A_2A_3 \mid$ **Ar** $B_2B_3$, we get $a_2 = a_3$. Hence, $a_1 = a_3$. Again arguing as in Example 3.2.17, we deduce that **Ar** $A_1A_3 \mid$ **Ar** $B_1B_3$.
18. *Hint*: See Exercise 4.

**Section 3.4**

1. Let $S_i$ have direction space $W_i$ ($i = 1, 2$). If $S_1$ is parallel to the $h$-flat $S'$, then $S'$ has direction space $W_1$; hence, $S' \subseteq S_2$ implies that $W_1 \subseteq W_2$ (Theorem 2.7.7(d)), so that (a) $\Rightarrow$ (b). Conversely, if $W_1 \subseteq W_2$ and we write $S_2 = W_2 + \mathbf{u}$, then $W_1 + \mathbf{u}$ is an $h$-flat contained in $S_2$ and parallel to $S_1$ (why?), so that (b) $\Rightarrow$ (a).
2. We may write $S_1 = W_1 + \mathbf{v}$, $S_2 = W_2 + \mathbf{u}$, where $W_i$ is the direction space of $S_i$; since $S_1 \parallel S_2$, $W_1 \subseteq W_2$ and, hence, $W_1 + \mathbf{v} \subseteq W_2 + \mathbf{v}$. Since $W_2 + \mathbf{v}$, $W_2 + \mathbf{u}$ are cosets of the same subspace, they are either equal or disjoint, and the desired result follows.
3. Let $S$, $H$ have direction spaces $W_1$, $W_2$, respectively.
(a) If $S \cap H = \varnothing$, then by Theorem 3.4.8 $\dim(W_1 \cap W_2) \geqslant \dim W_1$ so that $W_1 \subseteq W_2$ and, hence, $S \parallel H$.
(b) Let $S \cap H \neq \varnothing$. If $W_1 \subseteq W_2$, then $S \parallel H$ and, hence, $S \subseteq H$ by Exercise 2. If $W_1 \nsubseteq W_2$, then $W_1 + W_2 = V$ (why?); now use Theorems 2.6.17 and 3.4.11.
4. Write $S_i = W_i + \mathbf{u}_i$, $W_i$ = direction space of $S_i$ ($i = 1, 2$). For each $\mathbf{w} \neq \mathbf{0}$ in $W_1$, the line $[\mathbf{w}] + \mathbf{u}_1$ lies in $S_1$, hence, is parallel to some line $L_2$ contained in $S_2$. By 3.2.5(f), $L_2$ is a coset of $[\mathbf{w}]$; since $L_2 \subseteq S_2$, 2.7.7(d) implies that $[\mathbf{w}] \subseteq W_2$ and, hence, $\mathbf{w} \in W_2$. Since $\mathbf{w}$ was arbitrary, $W_1 \subseteq W_2$ and, hence, $S_1 \parallel S_2$.

**Section 3.5**

2. Use a dimension argument.
5. Use the result of Exercise 2.
8. (a) 2. (b) 3. (c) 1.
9. (a) Yes. (b) Yes.
10. Since $E$ lies on $\langle A, B\rangle$ and $\langle C, D\rangle$ we can write $E = a_1A + b_1B$ and $E = c_1C + d_1D$, where $a_1 + b_1 = c_1 + d_1 = 1$. Similarly, $F = a_2A + d_2D = b_2B + c_2C$ and $G = a_3A + c_3C = b_3B + d_3D$, with $a_2 + d_2 = b_2 + c_2 = a_3 + c_3 = b_3 + d_3 = 1$. Now suppose $eE + fF + gG = 0$ with $e + f + g = 0$. Using only the equations for $E, F, G$ in terms of $A$, $B$, and $D$, substitution gives

$$(ea_1 + fa_2)A + (eb_1 + gb_3)B + (fd_2 + gd_3)D = 0,$$

where the sum of the coefficients on the left side is zero. Since $A$, $B$, $D$ are noncollinear, Theorem 3.5.6 implies that $ea_1 + fa_2 = 0$. By similar reasoning (using different equations for

$E$, $F$, and $G$) we obtain $ea_1 + ga_3 = 0$ and $fa_2 + ga_3 = 0$. Hence, the result of Exercise 7 implies $ea_1 = fa_2 = ga_3 = 0$, from which $e = f = g = 0$ (note that $a_1 \neq 0$ since $E \neq B$, etc.). It follows from Theorem 3.5.6 that $E$, $F$, $G$ are noncollinear.

**Section 3.6**

1. Imitate the proof of Section 3.2, Exercise 11.
3. We can take $\alpha = b_2/(b_2 - a_2)$, $\beta = -a_2/(b_2 - a_2)$. The conclusion that $\overline{PQ}$ meets $L_2$ follows from the fact that in the expression for $\alpha P + \beta Q$ as a barycentric combination of $P_0$, $P_1$, $P_2$, the coefficient of $P_2$ is 0.
8. (2, 0, 6, 2).
9. (b) Solving $(1, 5/2, -1) = aP_0 + bP_1 + cP_2 + dP_3$ with $a + b + c + d = 1$, we get $a = 1$, $b = \frac{1}{2}$, $c = -\frac{1}{2}$, $d = 0$. Since at least one coefficient is negative, the point lies outside the tetrahedron. (It would lie inside the tetrahedron if $a$, $b$, $c$, $d$ were all positive, on the tetrahedron if at least one coefficient were zero and all nonzero coefficients were positive.)
11. (a) Use a dimension argument. (Cf. Sec. 3.4, Exercise 8.)
    (d) Show that the set in question is equal to $S + (1 - k)P_2$.
12. Show that the set in question is equal to $\langle B, C\rangle + k(A - C)$.
14. (a) $P = -\frac{1}{3}A + \frac{2}{3}B + \frac{2}{3}C$; (b) 6; (c) 6.
19. By Section 3.2, Exercise 22, there exists a line $L'$ through $B_3$ which intersects $\langle P, A_2\rangle$ and $\langle P, A_4\rangle$ in points $B_2'$, $B_4'$, respectively, such that $B_3 \in \overline{B_2'B_4'}$ and $|\overline{B_2'B_3}|/|\overline{B_3B_4'}| = 3$. To show $L' = L$ (which implies the desired result), it suffices to show that $L'$ is parallel to $\langle P, A_1\rangle$. But if $L'$ is not parallel to $\langle P, A_1\rangle$, then $L'$ intersects $\langle P, A_1\rangle$ at $B_1'$ (say), and by Example 3.6.9, we have

$$\frac{1}{4} = \frac{|\overline{A_1A_2}|/|\overline{A_2A_4}|}{|\overline{A_1A_3}|/|\overline{A_3A_4}|} = \frac{|\overline{B_1'B_2'}|/|\overline{B_2'B_4'}|}{|\overline{B_1'B_3}|/|\overline{B_3B_4'}|} = \frac{1}{4}\cdot\frac{|\overline{B_1'B_2'}|}{|\overline{B_1'B_3}|}$$

so that $|\overline{B_1'B_2'}| = |\overline{B_1'B_3}|$. This is possible only if $B_1'$ lies between $B_2'$ and $B_3$; but the latter contradicts Example 3.6.17.

**Section 3.7**

7. (a) The desired segment intersects the given line at the point $(-3, 4, 0)$.
   (b) The desired segment intersects the given line at the point $(5, 1, 0, 3)$.

**Section 3.8**

1. (a) One orthogonal basis is $\{(3, 0, 1, 1), (1, 2, -1, -2), (-1, 2, 3, 0)\}$; there are many others.
(b) *First solution*: Using the notation of the proof of Theorem 3.8.9, take $E = \mathbf{R}_4$, $k = 1$, $\mathbf{w}_1 = (2, 1, -1, 1)$, $\mathbf{u}_i^* = \mathbf{e}_i$. We obtain

$$\{ (2, 1, -1, 1), (3, -2, 2, -2), (0, 2, 1, -1), (0, 0, 1, 1) \}$$

as an orthogonal basis of $\mathbf{R}_4$. The last three of these vectors constitute an orthogonal basis of $W^\perp$.
*Second solution*: By inspection, the vectors $\mathbf{w}_2 = (1, 0, 0, -2)$ and $\mathbf{w}_3 = (0, 1, 1, 0)$ are orthogonal to each other and to the vector $\mathbf{w}_1 = (2, 1, -1, 1)$. Applying the Gram-Schmidt process just once, we obtain $\mathbf{w}_4 = (4, -5, 5, 2)$. The set $\{\mathbf{w}_2, \mathbf{w}_3, \mathbf{w}_4\}$ is an orthogonal basis of $W^\perp$.
(There are many other solutions.)
(c) *First solution*: $S$ is the orthogonal complement of the subspace $W = [(3, 0, 6, 5), (1, 1, -2, -1)]$. One orthogonal basis of $W$ is $\{(3, 0, 6, 5), (8, 5, -4, 0)\}$; extend this set to an orthogonal basis of $\mathbf{R}_4$. The two additional vectors constitute an orthogonal basis of $S$.
*Second solution*: Since $\dim S = 4 - 2 = 2$, any two linearly independent solutions of the given system of equations will constitute a basis of $S$, from which an *orthogonal* basis is obtained in the usual way.
4. *Suggestion*: Apply 3.8.18(b) to $W^\perp$ in place of $W$.
5. This question admittedly requires knowledge external to algebra. *Suggestion*: Use the Weierstrass Approximation Theorem.
9. (a) $(-2, -5, 5)$. (b) $(6, -6, 1)$. (c) $(3, -1, 5, 7)$.
10. The reflection point is $(1, -4, 4, 4)$.
13. For the first part, use 3.7.10 with $A = Q = \mathbf{w}$; for the second part, use 3.7.10 with $A = O$, $Q = \mathbf{v}$.
14. If $\mathbf{v} \in S \cap W^\perp$, then $\mathbf{v} + \mathbf{x} \in (S + \mathbf{x}) \cap W^\perp$ (why?). Hence, by 3.8.30(b), we have

$$\mathcal{M}_S = \mathcal{M}_W \mathcal{T}_{2\mathbf{v}}$$
$$\mathcal{M}_{S+\mathbf{x}} = \mathcal{M}_{W+\mathbf{v}+\mathbf{x}} = \mathcal{M}_W \mathcal{T}_{2(\mathbf{v}+\mathbf{x})} = (\mathcal{M}_W \mathcal{T}_{2\mathbf{v}})\mathcal{T}_{2\mathbf{x}} = \mathcal{M}_S \mathcal{T}_{2\mathbf{x}}.$$

19. Use a dimension argument (cf. Theorem 3.4.8 and Ex. 18).
21. Let $\mathbf{Ar}\, AB = \mathbf{u}$, $\mathbf{Ar}\, AC = \mathbf{w}$, $\mathbf{Ar}\, AD = \mathbf{v}$; then $\mathbf{Ar}\, DC = \mathbf{w} - \mathbf{v}$, $\mathbf{Ar}\, BC = \mathbf{w} - \mathbf{u}$, $\mathbf{Ar}\, BD = \mathbf{v} - \mathbf{u}$, $\mathbf{Ar}\, AD - \mathbf{Ar}\, BC = \mathbf{v} + \mathbf{u} - \mathbf{w}$. We are given

$$|\mathbf{u}| = |\mathbf{w} - \mathbf{v}|$$
$$|\mathbf{v}| = |\mathbf{w} - \mathbf{u}|.$$

Squaring both sides,

(1) $|\mathbf{u}|^2 = |\mathbf{w}|^2 + |\mathbf{v}|^2 - 2\mathbf{w} \cdot \mathbf{v}$

(2) $|\mathbf{v}|^2 = |\mathbf{w}|^2 + |\mathbf{u}|^2 - 2\mathbf{w} \cdot \mathbf{u}.$

If we add (1) and (2), cancel terms appearing on both sides, and divide by $-2$, we get

$$0 = -|\mathbf{w}|^2 + \mathbf{w} \cdot \mathbf{v} + \mathbf{w} \cdot \mathbf{u} = -\mathbf{w} \cdot \mathbf{w} + \mathbf{w} \cdot (\mathbf{v} + \mathbf{u})$$
$$= \mathbf{w} \cdot (\mathbf{v} + \mathbf{u} - \mathbf{w})$$
$$= \mathbf{Ar}\, AC \cdot (\mathbf{Ar}\, AD - \mathbf{Ar}\, BC).$$

If, instead, we subtract (1) from (2), transpose terms and divide by 2, we get $|\mathbf{v}|^2 - |\mathbf{u}|^2 = \mathbf{w} \cdot \mathbf{v} - \mathbf{w} \cdot \mathbf{u}$, that is, $(\mathbf{v} + \mathbf{u}) \cdot (\mathbf{v} - \mathbf{u}) = \mathbf{w} \cdot (\mathbf{v} - \mathbf{u})$, so that

$$0 = (\mathbf{v} + \mathbf{u} - \mathbf{w}) \cdot (\mathbf{v} - \mathbf{u}) = (\mathbf{Ar}\, AD - \mathbf{Ar}\, BC) \cdot \mathbf{Ar}\, BD.$$

Assertion (a) follows. If $\mathbf{Ar}\, AD \neq \mathbf{Ar}\, BC$, then $\dim[\mathbf{Ar}\, AD - \mathbf{Ar}\, BC] = 1$ and, hence, $\dim[\mathbf{Ar}\, AD - \mathbf{Ar}\, BC]^\perp = 1$, so that by part (a) the vectors $\mathbf{Ar}\, AC$, $\mathbf{Ar}\, BD$ both lie in the same one-dimensional subspace and thus are parallel, proving (b). Figure G1 shows the two possibilities.

**Section 3.10**

1. (a) $P_0 = (5, 4)$; $\mathbf{u}_1 = (2/\sqrt{5}, 1/\sqrt{5})$; $\mathbf{u}_2 = (-1/\sqrt{5}, 2/\sqrt{5})$.
   (b) $(-2\sqrt{5}, -\sqrt{5})$.

**Section 4.1**

8. (a) The line (one-dimensional subspace) $[(1, 2, 3)]$.
   (b) The line consisting of all points of the form $(2, 1, z)$ $(z \in \mathbf{R})$.
11. (a) The image of $c\mathbf{v}$ is $c^*(\mathbf{v}T)$, not $c(\mathbf{v}T)$.
    (b) The image of any point $aA + bB$ $(a + b = 1)$ on the line $\langle A, B\rangle$ is $a^*(AT) + b^*(BT)$; since $a^* + b^*$ still equals 1

Figure G1

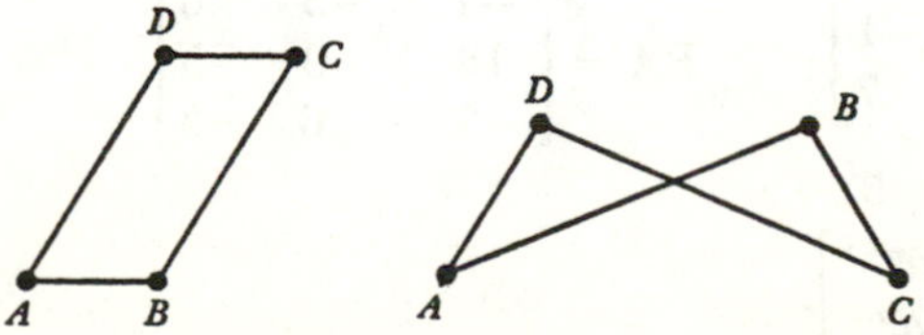

(why?), it follows that all points on $\langle A, B\rangle$ have their images on the line $\langle A\mathbf{T}, B\mathbf{T}\rangle$.

**Section 4.2**

8. $x' = x + y/m$; $y' = y$; matrix is $\begin{bmatrix} 1 & 0 \\ 1/m & 1 \end{bmatrix}$.
9. Matrix is $\begin{bmatrix} k & mk \\ mk & m^2k \end{bmatrix}$, where $k = 1/(1 + m^2)$.
10. If the orthogonal (ordered) basis $(\mathbf{w}_1, \ldots, \mathbf{w}_k)$ of $W$ is extended to an orthogonal basis $(\mathbf{w}_1, \ldots, \mathbf{w}_k, \mathbf{w}_{k+1}, \ldots, \mathbf{w}_n)$ of $E$, then the perpendicular projection of $c_1\mathbf{w}_1 + \cdots + c_n\mathbf{w}_n$ on $W$ is $c_1\mathbf{w}_1 + \cdots + c_k\mathbf{w}_k$ (why?). It follows that $\mathbf{T}$ satisfies equations of the form 4.2.14.
11. $\begin{bmatrix} 2/3 & 1/3 & -2/3 \\ 1/3 & 2/3 & 2/3 \\ -2/3 & 2/3 & -1/3 \end{bmatrix}$.
12. Since $(1, 2) \to (7, 1)$, equations 4.2.14 give $a_{11} + 2a_{21} = 7$, $a_{12} + 2a_{22} = 1$. Similarly, since $(3, -2) \to (5, -5)$, we have $3a_{11} - 2a_{21} = 5$, $3a_{12} - 2a_{22} = -5$. Solving the first and third of these four equations for $a_{11}$ and $a_{21}$, and the second and fourth equations for $a_{12}$ and $a_{22}$, we find that the matrix of $\mathbf{T}$ is $A = \begin{bmatrix} 3 & -1 \\ 2 & 1 \end{bmatrix}$.
13. (a) $\begin{bmatrix} 1 & 1 \\ 2 & -1 \end{bmatrix}$. (b) $\begin{bmatrix} 1 & 1 \\ 2 & -1 \end{bmatrix}$.

**Section 4.3**

4. $\mathbf{ST}$ is the reflection through the $y$ axis; $\mathbf{TS}$ is the reflection through the $x$ axis.
5. $\mathbf{ST}$ is the translation $X \to X + (8, 0)$; $\mathbf{TS}$ is the translation $X \to X - (8, 0)$.

**Section 4.4**

2. $A + B = \begin{bmatrix} 5 & 3 \\ 2 & 1 \end{bmatrix}$; $AB = \begin{bmatrix} 2 & 3 \\ 12 & 3 \end{bmatrix}$; $BA = \begin{bmatrix} 7 & 8 \\ 2 & -2 \end{bmatrix}$.
3. $AB = \begin{bmatrix} -2 & -1 \\ 5 & 7 \end{bmatrix}$; $BA = \begin{bmatrix} -1 & -3 & 0 \\ 18 & 9 & 5 \\ -7 & 6 & -3 \end{bmatrix}$.
5. $(3, 4)\mathbf{T} = (11, 5)$.
8. $A^n = \begin{bmatrix} a^n & na^{n-1} \\ 0 & a^n \end{bmatrix}$.

10. The linear transformation $\mathbf{M} : F_m \to F_n$ which maps $\mathbf{e}_i \to \mathbf{u}_i$ has matrix $B = R(\mathbf{u}_1, \ldots, \mathbf{u}_m)$ (why?). Similarly, the linear transformation $\mathbf{M}' : F_m \to F_p$ which maps $\mathbf{e}_i \to \mathbf{u}_i\mathbf{T}$ has matrix $B' = R(\mathbf{u}_1\mathbf{T}, \ldots, \mathbf{u}_m\mathbf{T})$. Since $\mathbf{MT}$ and $\mathbf{M}'$ both map $\mathbf{e}_i \to \mathbf{u}_i\mathbf{T}$, $\mathbf{MT} = \mathbf{M}'$ (why?) and, hence, $BA = B'$.
11. Both matrices are equal to $\begin{bmatrix} 5 & -3 & 2 \\ 24 & 9 & 7 \end{bmatrix}$.
12. *Suggestion*: Use 4.4.11 together with properties of the dot product (cf. Theorem 3.7.2).

**Section 4.5**

4. (a) $(2^n - 1)^2$. (b) $\binom{2n}{n}$.
5. Assume that $A$ has $n$ columns, $B$ has $h$ columns, and $i_1 < i_2 < \cdots < i_h$ are the indices of those columns of $A$ which appear in $B$. Under the linear transformation

   $\mathbf{T} : (c_1, \ldots, c_n) \to (c_{i_1}, \ldots, c_{i_h})$

   of $F_n \to F_h$, each row of $B$ is the image of the corresponding row of $A$. Now apply the contrapositive of Theorem 4.1.4(d).
8. (b) If $A$ is diagonal and $B = (b_{ij})$ commutes with $A$, then $a_{ii}b_{ij} = b_{ij}a_{jj}$ for all $i$ and $j$ (why?). If $i \neq j$, then $a_{ii} \neq a_{jj}$ implies $b_{ij} = 0$.
9. (b) Let $i, j$ be fixed, and let $M$ be the matrix having the entry 1 in the $(i, j)$ position and zeros everywhere else. By computing the $(i, h)$ entry in the product $AM = MA$, we obtain $a_{ii}\,\delta_{jh} = a_{jh}$ for all $h$. The result easily follows (*how*?).

**Section 4.6**

3. Ran $\mathbf{T}$ = Nul $\mathbf{T}$ = $[(1, 1)]$ (= the line $x = y$).
4. Both Ran $\mathbf{T}$ and Nul $\mathbf{T}$ have dimension 2.
5. (b) Let $\mathbf{T}$ map $\mathbf{e}_1 \to \mathbf{e}_1$, $\mathbf{e}_2 \to \mathbf{0}$, $\mathbf{e}_3 \to \mathbf{e}_2$, $\mathbf{e}_4 \to \mathbf{0}$.
9. If $\mathbf{T}$ is the perpendicular projection mapping of $E$ onto $W$, then Ran $\mathbf{T} = W$ and Nul $\mathbf{T} = W^{\perp}$.

**Section 4.7**

2. The most general right inverse of $A$ is

$$\begin{bmatrix} 1 - 5x & -2 - 5y \\ x & y + 1 \\ x & y \end{bmatrix},$$

where $x, y$ are arbitrary.

3. (a) No. (b) Yes. (c) No.
4. (a) Yes. (b) No.

**Section 4.8**

1. (a) $A = \begin{bmatrix} 1 & 2 \\ -1 & 0 \end{bmatrix}$; $A' = \begin{bmatrix} 11 & -4 \\ 28 & -10 \end{bmatrix}$;

$P = \begin{bmatrix} -5 & 2 \\ 3 & -1 \end{bmatrix}$; $Q = \begin{bmatrix} 1 & 2 \\ 3 & 5 \end{bmatrix}$.

(c) Since $(1, -1) = \mathbf{v}_1 - \mathbf{v}_2 = -8\mathbf{v}_1' + 3\mathbf{v}_2'$, we have $X = [1 \quad -1]$, $X' = [-8 \; 3]$; etc.

4. $P$ does not have an inverse.
5. (a) Yes; $P$ may be chosen to be any matrix of the form

$\begin{bmatrix} b - c & b \\ c & b + c \end{bmatrix}$

such that $b^2 - c^2 - bc \neq 0$. For instance, we may take

$P = \begin{bmatrix} 1 & 1 \\ 0 & 1 \end{bmatrix}$.

(b) No.

(c) Yes; one choice for $P$ is $\begin{bmatrix} 1 & 1 \\ 0 & -1 \end{bmatrix}$.

(d) No.

(e) Yes.

**Section 4.9**

1. $x' = 3x - y + 1$; $y' = 2x + 3y - 3$.
2. $x' = -x + 3y + 4$; $y' = y - 1$.
9. (a) Yes, if the points $AM$, $BM$ are distinct.
   (b) Yes, if the points $AM$, $BM$, $CM$ are noncollinear.
15. $x' = \dfrac{x + my - mb}{1 + m^2}$; $y' = \dfrac{mx + m^2 y + b}{1 + m^2}$.
16. Method: Let $W$ be the direction space of $S$, extend an orthonormal basis of $W$ to one of $E$, and then show that the equations of **M** with respect to this basis have the form 4.9.2.

**Section 4.10**

3. Suggestions for proofs: For (a), use Theorem 4.10.16. For (b), first make some "simplifying assumptions" about the location of the vertices.

5. "Only if" follows from 4.10.3; the converse is obtainable using 4.10.16 or 4.10.4.
6. (a) **M** exists by the preceding exercise and must map

$$\left(\frac{1}{2}, \frac{5}{2}\right) \to \left(\frac{9}{10}, \frac{13}{10}\right)$$

(why?). Also, by Section 4.9, Exercise 4, the linear part **S** of **M** maps the subspace [(1, 1)] onto [(1, 7)] and, hence, maps the vector (1, 1) onto $\pm(1/5, 7/5)$ since **S** preserves length. Now continue as in Example 4.9.7. The two solutions are

(i) $x' = (4x - 3y + 10)/5$
$y' = (3x + 4y - 5)/5$

(ii) $x' = (3x - 4y + 13)/5$
$y' = (-4x - 3y + 16)/5.$

(b) No solution.

(c) Show that the linear part of **M** maps $(7, 1) \to \pm(5, 5)$ and $(1, -7) \to \pm(5, -5)$; etc.

(d) $x' = (-5x + 12y + 37)/13$; $y' = (12x + 5y - 16)/13$.

(e) Two possible solutions. One solution is

$$x' = (4x + 3y + 4)/5; \qquad y' = (-3x + 4y + 7)/5.$$

7. *First proof*: By Theorem 4.10.14(a) there is a linear isometry **M** which maps $E$ onto the subspace $W = [\mathbf{e}_1, ..., \mathbf{e}_m]$ of $\mathbf{R}_n$. Since **M** preserves addition, scalar multiplication, and inner product, we may "identify" $E$ with $W$, and the assertion follows. *Second proof*: Choose an orthonormal basis $(\mathbf{u}_1, ..., \mathbf{u}_m)$ of $E$; define $E^*$ to be the set of all *formal expressions*

$$a_1\mathbf{u}_1 + \cdots + a_m\mathbf{u}_m + a_{m+1}\mathbf{u}_{m+1} + \cdots + a_n\mathbf{u}_n \quad (a_1, ..., a_n \in \mathbf{R})$$

(where for the time being we regard $\mathbf{u}_{m+1}, ..., \mathbf{u}_n$ as mere symbols); define addition, scalar multiplication, and inner product on $E^*$ in the obvious ways; and so forth. (This approach is like that used to construct polynomial rings in [10], Chap. 9.)

**Section 5.1**

1. (a) (2 8 5) (4 9) (6 7); index is 4.
(b) (2 4 6 5); index is 3.
(c) Index is 7.
5. At one point in the proof you will need to use the fact that $-1 \equiv 1 \pmod 2$.

7. *Suggestion*: First show that the effect of multiplying a permutation $\pi$ by a transposition is to change $N(\pi)$ from odd to even or from even to odd.

**Section 5.2**

3. $-79$.

**Section 5.3**

3. 24.

**Section 5.4**

3. (a) 11. (b) $-39$.
7. Use 5.3.3 and 5.4.3(b).

**Section 5.5**

1. $\begin{bmatrix} 3 & 1 \\ -2 & 5 \end{bmatrix} \xrightarrow{\mathscr{R}_1} \begin{bmatrix} 3 & 1 \\ 4 & 7 \end{bmatrix} \xrightarrow{\mathscr{R}_2} \begin{bmatrix} -1 & -6 \\ 4 & 7 \end{bmatrix}$; $\begin{bmatrix} 3 & 1 \\ -2 & 5 \end{bmatrix} \xrightarrow{\mathscr{R}_2}$ $\begin{bmatrix} 5 & -4 \\ -2 & 5 \end{bmatrix} \xrightarrow{\mathscr{R}_1} \begin{bmatrix} 5 & -4 \\ 8 & -3 \end{bmatrix}$.

4. Both $\mathscr{R}(AB)$ and $[\mathscr{R}(A)]B$ equal $\begin{bmatrix} -2 & 2 \\ 4 & 4 \\ 8 & 3 \end{bmatrix}$.

7. Let $S = \{i_1, i_2, \ldots, i_k\}$, where $i_1 < i_2 < \cdots < i_k$. The mapping $\mathbf{T} : F_n \to F_k$ defined by

   $\mathbf{T} : (c_1, c_2, \ldots, c_n) \to (c_{i_1}, \ldots, c_{i_k})$

   is linear and maps $A_{(i)} \to A'_{(i)}, B_{(i)} \to B'_{(i)}$. Hence, by Theorem 4.6.6 $\mathbf{T}$ maps the row space of $A$ onto that of $A'$, and the row space of $B$ onto that of $B'$. The result follows. (Compare with solution to Sec. 4.5, Exercise 5.)

**Section 5.6**

2. (b) $-14$.

   (c) $A^* = \begin{bmatrix} 1 & 0 & -1 & 0 & 1/2 \\ 0 & 1 & 1/3 & 0 & 0 \\ 0 & 0 & 0 & 1 & -1/2 \\ 0 & 0 & 0 & 0 & 0 \end{bmatrix}$; $B^* = \begin{bmatrix} 1 & 0 & 0 & 0 \\ 0 & 1 & 0 & 0 \\ 0 & 0 & 1 & 0 \\ 0 & 0 & 0 & 1 \end{bmatrix}$;

   $C^* = \begin{bmatrix} 1 & 0 & -7/4 \\ 0 & 1 & 3/2 \\ 0 & 0 & 0 \\ 0 & 0 & 0 \end{bmatrix}$.

3. (a) Yes. (b) No.
5. Since any nonsingular matrix can be reduced to $I$ by operations of the types described in Theorems 5.4.1 and 5.4.5, det $A$ is uniquely determined when $A$ is nonsingular. When $A$ is singular, the proof of Theorem 5.4.8 still works since it uses only 5.4.5 and 5.4.2, and 5.4.2 can be obtained by letting $c = 0$ in 5.4.1.
6. (a) Let $r$ be the dimension of the row space (of $A$ or $B$); then both $A$ and $B$ have $r$ nonzero rows. Say the first nonzero entry in the $i$th row of $A$ $(1 \leq i \leq r)$ appears in the $(s_i)$th column, while the first nonzero entry in the $i$th row of $B$ appears in the $(t_i)$th column. If for any $i$, we had $s_i < t_i$, then the submatrix of $A$ consisting of the first $s_i$ columns would have more nonzero rows than the corresponding submatrix of $B$, contradicting Section 5.5, Exercise 7. Similarly, $t_i < s_i$ is impossible; hence, $t_i = s_i$ for all $i$. Now for any $h$ $(1 \leq h \leq r)$,

$$A_{(h)} \in (\text{row space of } A) = (\text{row space of } B)$$

and, thus,

$$A_{(h)} = d_1 B_{(1)} + \cdots + d_r B_{(r)}$$

for some scalars $d_1, \ldots, d_r$. Hence, for each $i$ $(1 \leq i \leq r)$, we have

$$\begin{aligned} \delta_{hi} &= a_{hs_i} \qquad \text{(by 5.6.1(c))} \\ &= \sum_{j=1}^{r} d_j b_{js_i} = \sum_{j=1}^{r} d_j b_{jt_i} = \sum_{j=1}^{r} d_j \delta_{ji} = d_i \end{aligned}$$

so that $A_{(h)} = \sum_i d_i B_{(i)} = \sum_i \delta_{hi} B_{(i)} = B_{(h)}$. The result follows.

8. *Suggestion*: **T** is singular $\Leftrightarrow$ det **T** $= 0$.

**Section 5.7**

4. Area $= \pi ab$. Method: show that the linear transformation having matrix

$$\begin{bmatrix} a & 0 \\ 0 & b \end{bmatrix}$$

maps the given circle onto the given ellipse.

5. It suffices to prove this for a circle. (Why?)
6. Area $= 12$.

**Section 5.8**

2. $A^{-1} = \begin{bmatrix} -7 & 5 & -9 \\ 2 & -1 & 2 \\ 1 & -1 & 2 \end{bmatrix}$.

**Section 5.9**

1. $\dim W = 2$.
3. If $A$, $B$ correspond to linear transformations **T**, **S**, then by Section 4.6, Exercise 7, we have Ran **TS** ⊆ Ran **S**, which implies (a). To prove (b), apply (a) to $B^t$, $A^t$ in place of $A$, $B$ (respectively) and use Theorem 4.5.3(c).
4. Let $A$ correspond to $\mathbf{T} \in \operatorname{Lin}(U, V)$ and let $B$ correspond to $\mathbf{S} \in \operatorname{Lin}(V, W)$, where $U$, $V$, $W$ have dimensions $m$, $n$, $p$, respectively. Let $\mathbf{S}_0 = \mathbf{S} \restriction (U\mathbf{T})$ (the restriction of **S** to the domain $U\mathbf{T}$). Then $U(\mathbf{TS}) = (U\mathbf{T})\mathbf{S} = (U\mathbf{T})\mathbf{S}_0$ and, hence,
$$\begin{aligned} \text{rank } \mathbf{TS} &= \dim U(\mathbf{TS}) = \dim (U\mathbf{T})\mathbf{S}_0 \\ &= \dim U\mathbf{T} - \dim \operatorname{Nul} \mathbf{S}_0 \qquad \text{(by 4.6.7)} \\ &\geqslant \dim U\mathbf{T} - \dim \operatorname{Nul} \mathbf{S} \qquad \text{(since Nul } \mathbf{S}_0 \subseteq \text{Nul } \mathbf{S}) \\ &= \dim U\mathbf{T} - (n - \dim V\mathbf{S}) \qquad \text{(by 4.6.7)} \\ &= \dim U\mathbf{T} + \dim V\mathbf{S} - n \\ &= \text{rank } \mathbf{T} + \text{rank } \mathbf{S} - n. \end{aligned}$$
5. $AB = 0$ means that the row space $V_1$ of $A$ is orthogonal to the column space $V_2$ of $B$; hence, $V_1 \subseteq V_2^\perp$. Now use Theorem 3.8.18(b).
6. Use Exercises 3 and 4.
10. If a matrix $A$ has finitely many rows but infinitely many columns, the proof of Theorem 5.6.4 still works; hence, so does the proof of 5.6.6; hence, so does the proof that $r = d$ in Theorem 5.9.1; hence, so does the proof of 5.9.3.

**Section 5.10**

1. (a) $x_1 = x_3 - 2x_4 + 1$; $x_2 = 3x_3 + x_4 + 2$; $x_3$ = any value; $x_4$ = any value. Set of solutions is the plane $[(1, 3, 1, 0), (-2, 1, 0, 1)] + (1, 2, 0, 0)$.
   (b) $x_1 = 2$; $x_2 = 2x_3 - 1$; $x_4 = 3x_3$; $x_3$ = any value. Set of solutions is the line $[(0, 2, 1, 3)] + (2, -1, 0, 0)$.
   (c) No solution. (d) Unique solution is the point $(-1, 2, 4)$.
2. (b) $x = -1$, $y = 4$, $z = 2$.
3. $W_1 \cap W_2 = [(1, 3, 2)]$.
6. Any basis of $S^\perp$ may be taken to be the rows of $A$.

8. By Section 5.4, Exercise 7, $\mathbf{u} \times \mathbf{v}$ is perpendicular to $\mathbf{u}$ and $\mathbf{v}$ and, hence, is a solution; by Section 5.9, Exercise 8, this solution is nonzero; and by Theorem 5.10.14, the set of all solutions is a one-dimensional subspace. The result follows.
11. Use the result of the preceding exercise.

**Section 5.11**

1. No.
2. (a) $\text{char}(\mathbf{T}; x) = -x(x-1)^2$; eigenvalues are 0, 1; $(\mathbf{T}, 0)$-eigenspace is the line $[(1, 1, 1)]$; $(\mathbf{T}, 1)$-eigenspace is the plane $[(-3, 1, 1)]^\perp$.
   (d) $\text{char}(\mathbf{T}; x) = -(x+1)^2(x+2)$; $(\mathbf{T}, -1)$-eigenspace is the line $[(0, 1, 1)]$; $(\mathbf{T}, -2)$-eigenspace is the line $[(1, -1, 1)]$.
5. If $\mathbf{T} : F_3 \to F_3$ has matrix $A$ with respect to the canonical basis, if $\lambda$ is an eigenvalue of $\mathbf{T}$, and if some two of the three columns of $A - \lambda I$ are linearly independent, then the cross product of these two column vectors will span the $(\mathbf{T}, \lambda)$-eigenspace. Compare with Section 5.10, Exercise 8.
8. By Section 4.5, Exercise 3 (or by Chap. 9, eq. 9.5.3), $(\mathbf{uT}) \cdot \mathbf{v} = \mathbf{u} \cdot (\mathbf{vT})$ for all $\mathbf{u}$, $\mathbf{v}$. If $\mathbf{uT} = \lambda_1\mathbf{u}$ and $\mathbf{vT} = \lambda_2\mathbf{v}$, where $\lambda_1 \neq \lambda_2$, then $\lambda_1(\mathbf{u} \cdot \mathbf{v}) = (\lambda_1\mathbf{u}) \cdot \mathbf{v} = \mathbf{uT} \cdot \mathbf{v} = \mathbf{u} \cdot \mathbf{vT} = \mathbf{u} \cdot (\lambda_2\mathbf{v}) = \lambda_2(\mathbf{u} \cdot \mathbf{v})$, so that $(\lambda_1 - \lambda_2)(\mathbf{u} \cdot \mathbf{v}) = 0$ and, hence, $\mathbf{u} \cdot \mathbf{v} = 0$.
10. $\text{char}(\mathbf{T}; x)$ has $x$ as a factor; equivalently, the constant term of the polynomial is zero.

**Section 5.14**

1. $(-7, -4, 1, 6)$.
2. Use Theorem 5.4.4.
4. (a) All the $\mathbf{v}$'s except $\mathbf{v}_j$ are unchanged; $\mathbf{v}_j$ is replaced by $\mathbf{v}_j + (-1)^{i+j+1}k\mathbf{v}_i$.
   (b) $\mathbf{v}_i$ is unchanged; all other $\mathbf{v}$'s are multiplied by $c$.

**Section 6.1**

1. $\begin{bmatrix} \frac{4}{5} & -\frac{3}{5} \\ \frac{3}{5} & \frac{4}{5} \end{bmatrix}$.
2. $(3, -1)$.

**Section 6.2**

1. Use 6.2.18 for the first assertion, Theorem 6.2.12 for the rest.
2. Use Theorem 6.2.14.
3. Letting $x = \cos\theta$, $y = \sin\theta$, we will have $2\theta = \psi$ if and only if

$$(1) \qquad x^2 - y^2 = \cos\psi$$
$$(2) \qquad 2xy = \sin\psi.$$

Squaring both (1) and (2) and then adding, we get $(x^2 - y^2)^2 + (2xy)^2 = 1$, that is, $(x^2 + y^2)^2 = 1$, that is,

$$(3) \qquad x^2 + y^2 = 1$$

(showing that any solution $(x, y)$ of (1) and (2) does correspond to an oriented angle). Now, solve for $x^2$ by adding (1) and (3), for $y^2$ by subtracting (1) from (3). The two solutions are

$$(x, y) = \left(\left(\frac{1 + \cos\psi}{2}\right)^{1/2}, \epsilon\left(\frac{1 - \cos\psi}{2}\right)^{1/2}\right)$$

$$(x, y) = \left(-\left(\frac{1 + \cos\psi}{2}\right)^{1/2}, -\epsilon\left(\frac{1 - \cos\psi}{2}\right)^{1/2}\right),$$

where $\epsilon = 1$ or $-1$ according to whether $\sin\psi \geqslant 0$ or $\sin\psi < 0$, so as to be consistent with (2). Since one solution $(x, y)$ is minus the other, the sines of the corresponding oriented angles have opposite signs (except when $y = 0$, i.e., when $\cos\psi = 1$, $\psi = 0°$) and, hence, the orientations are opposite by Definition 6.2.11.

4. Apply the last part of Theorem 6.2.12 to both $\theta$ and $\psi$, and use Theorem 6.2.9.
6. $x' = (5/13)\,x - (12/13)y + 28/13$; $y' = (12/13)\,x + (5/13)y - 16/13$.
9. Letting $\theta = 180°$, the rotation $\mathcal{R}_{O,\theta}$ with matrix $-I$ clearly maps $\mathbf{u} \to -\mathbf{u}$ for all $\mathbf{u} \in \mathbf{R}_2$; hence, $\mathcal{R}_{A,\theta} = \mathcal{T}_{-A}\mathcal{R}_{O,\theta}\mathcal{T}_A$ maps $A + \mathbf{u} \to A - \mathbf{u}$ and thus maps $H_1 \to H_2$, from which Or $\angle\,(H_1, H_2) = \theta = 180°$. The other assertions follow from Theorems 6.2.12 and 6.2.14.
14. Show that the image of $\mathbf{u}$ under the matrix 6.2.13 is orthogonal to $\mathbf{u}$ if and only if $\cos\theta = 0$.

**Section 6.3**

3. For (a), use matrix multiplication; for (b), first show that $[\mathbf{e}_1] = L(0°)$.

5. $\mathcal{R}_{A,2\theta} = \mathcal{T}_{-A}\mathcal{R}_{O,2\theta}\mathcal{T}_A = \mathcal{T}_{-A}\mathcal{M}_{L(\psi)}\mathcal{M}_{L(\psi+\theta)}\mathcal{T}_A$
$= (\mathcal{T}_{-A}\mathcal{M}_{L(\psi)}\mathcal{T}_A)\ (\mathcal{T}_{-A}\mathcal{M}_{L(\psi+\theta)}\mathcal{T}_A)$
$= \mathcal{M}_{L(\psi)+A}\mathcal{M}_{L(\psi+\theta)+A}$ (by Theorem 3.8.32).
The result follows.
6. If $\mathcal{M}_L\mathcal{T}_{\mathbf{u}}$ fixes the point $Q$, show that $Q - \mathbf{u}/2$ is the perpendicular projection of $Q$ on $L$, and hence that $\mathbf{u}$ is perpendicular to the direction space of $L$. Deduce that $\mathcal{M}_L\mathcal{T}_{\mathbf{u}}$ is of Type (d).
7. For (b), use the fact that reflections reverse oriented angles.
9. Using 6.3.9,

$$\mathbf{u} = (2, 2) - [2 \quad 2]\begin{bmatrix} \frac{3}{5} & \frac{4}{5} \\ -\frac{4}{5} & \frac{3}{5} \end{bmatrix} = (12/5, -4/5);$$

$(0, 3)\mathbf{T} = (0, 1)$.
11. $\mathbf{u} = B - A - (B - A)\mathcal{R}_{O,-\theta}$.
12. $(-2\sqrt{3}, \sqrt{3})$.
13. (a) From the given equations, $\mathbf{T} = \mathcal{M}_{L(\theta)}\mathcal{T}_{\mathbf{u}}$, where $\mathbf{u} = (3, 4)$ and $(\cos 2\theta, \sin 2\theta) = (3/5, 4/5)$. Using the half angle formulas (cf. solution to Sec. 6.2, Exercise 3), we obtain
$(\cos\theta, \sin\theta) = \pm\,(2/\sqrt{5}, 1/\sqrt{5})$;
hence, in Theorem 6.3.17 we have $W = [(2, 1)]$,
$\mathbf{w} = (\text{projection of } \mathbf{u} \text{ on } W) = (4, 2)$,
$S_1 = L_1 = L(\theta) + \frac{1}{2}(\mathbf{u} - \mathbf{w}) = [(2, 1)] + (-\frac{1}{2}, 1)$.
$\mathbf{T}$ is the reflection through $L_1$ followed by the translation $\mathcal{T}_{(4,2)}$.
(b) Rotation through $\theta$ about the point (5, 2), where $\cos\theta = 5/13$, $\sin\theta = 12/13$.
(c) Reflection through the line $[(2, 1)] + (-3/2, 3)$.
(d) Reflection through the line $[(\sqrt{3}, 1)] + (-\sqrt{3}, 3)$, followed by the translation $\mathcal{T}_{(2\sqrt{3}, 2)}$.
(e) Rotation of $-120°$ about $(\sqrt{3}, 1)$.
14. Write $\mathcal{R}_{P,\theta} = \mathcal{R}_{O,\theta}\mathcal{T}_{\mathbf{u}}$. Since $\mathcal{R}_{O,\theta}$ is the linear part of $\mathcal{R}_{P,\theta}$, 4.9.6 gives
$\mathcal{R}_{O,\theta}$: $(4, 3) \to (4, -3)$.
Hence, the matrix of $\mathcal{R}_{O,\theta}$ is

$$\frac{1}{5}\cdot\frac{1}{5}\begin{bmatrix} 4 & -3 \\ 3 & 4 \end{bmatrix}\begin{bmatrix} 4 & -3 \\ 3 & 4 \end{bmatrix} = \begin{bmatrix} \frac{7}{25} & -\frac{24}{25} \\ \frac{24}{25} & \frac{7}{25} \end{bmatrix}$$

(see the end of Sec. 6.1), and $\theta$ is the oriented angle whose cosine and sine are 7/25, −24/25, respectively. Since

$\mathcal{R}_{P,\theta} : (0, -2) \to (-6, -2)$

$$\mathcal{R}_{O,\theta}: (0, -2) \to [0\ -2]\begin{bmatrix} \frac{7}{25} & -\frac{24}{25} \\ \frac{24}{25} & \frac{7}{25} \end{bmatrix} = \left(-\frac{48}{25}, -\frac{14}{25}\right),$$

we must have $\mathbf{u} = (-6, -2) - (-48/25, -14/25) = (-102/25, -36/25)$. Finally, we find $P$ by solving the system 6.3.11 with

$$(a, b) = \left(\frac{7}{25}, -\frac{24}{25}\right) \quad \text{and} \quad (c, d) = \left(-\frac{102}{25}, -\frac{36}{25}\right);$$

the answer is $P = (-3, 2)$.

15. $\theta = 90°$; $P = (3, 3)$.
16. $L$ is parallel to the line [ (2, 2) ] which equals $L(\theta)$, where $\cos\theta = \sin\theta = 1/\sqrt{2}$ (why?). Hence, $\mathcal{M}_L = \mathcal{M}_{L(\theta)}\mathcal{T}_{\mathbf{w}}$ for some $\mathbf{w}$. Since $\mathcal{M}_L$ fixes (0, 1) while $M_{L(\theta)}$ maps

$$(0, 1) \to [0\quad 1]\begin{bmatrix} \cos 2\theta & \sin 2\theta \\ \sin 2\theta & -\cos 2\theta \end{bmatrix} = [0\quad 1]\begin{bmatrix} 0 & 1 \\ 1 & 0 \end{bmatrix} = (1, 0),$$

we must have $\mathbf{w} = (0, 1) - (1, 0) = (-1, 1)$. Hence, $\mathcal{M}_L = \mathcal{M}_{L(\theta)}\mathcal{T}_{(-1,1)}$, and the equations of $\mathcal{M}_L$ are

$$x' = y - 1$$
$$y' = x + 1.$$

17. $x' = -\frac{3}{5}x - \frac{4}{5}y + 4; \qquad y' = -\frac{4}{5}x + \frac{3}{5}y + 2.$
18. (a) By applying the result of Section 3.8, Exercise 16(b) to the result of Exercise 13(a), we see that $\mathbf{T}^{-1}$ is the reflection through the line [(2, 1)] + (−1/2, 1) followed by the translation $\mathcal{T}_{(-4,\,-2)}$. The equations of $\mathbf{T}^{-1}$ are

$$x' = \frac{3}{5}x + \frac{4}{5}y - 5; \qquad y' = \frac{4}{5}x - \frac{3}{5}y.$$

These equations can be obtained by the method of Exercise 16. An alternate method is to observe that $\mathbf{T} = \mathcal{M}_L\mathcal{T}_{(3,4)}$, where $\mathcal{M}_L$ has matrix

$$\begin{bmatrix} \frac{3}{5} & \frac{4}{5} \\ \frac{4}{5} & -\frac{3}{5} \end{bmatrix}; \text{ hence,}$$

$$\mathbf{T}^{-1} = \mathcal{T}^{-1}_{(3,4)}\mathcal{M}_L^{-1} = \mathcal{T}_{(-3,-4)}\mathcal{M}_L$$

and the effect of $\mathbf{T}^{-1}$ on any point $(x, y)$ is given by

$$(x, y) \xrightarrow{\mathcal{T}_{(-3,-4)}} (x-3,\ y-4) \xrightarrow{\mathcal{M}_L} [x-3 \quad y-4] \begin{bmatrix} \frac{3}{5} & \frac{4}{5} \\ \frac{4}{5} & -\frac{3}{5} \end{bmatrix}$$

$$= \left(\frac{3}{5}x + \frac{4}{5}y - 5, \frac{4}{5}x - \frac{3}{5}y\right)$$

from which the equations follow.

(b) $x' = (5/13)x + (12/13)y + 16/13$; $y' = -(12/13)x + (5/13)y + 76/13$.

(c) Since **T** is a reflection, $\mathbf{T}^{-1} = \mathbf{T}$.

(d) $x' = \frac{1}{2}x + \frac{1}{2}\sqrt{3}y - 4\sqrt{3}$; $y' = \frac{1}{2}\sqrt{3}x - \frac{1}{2}y + 4$.

$\mathbf{T}^{-1}$ is the reflection through the line $[(\sqrt{3}, 1)] + (-\sqrt{3}, 3)$ followed by the translation $\mathcal{T}_{(-2\sqrt{3},-2)}$.

(e) $x' = -\frac{1}{2}x - \frac{1}{2}\sqrt{3}y + 2\sqrt{3}$; $y' = \frac{1}{2}\sqrt{3}x - \frac{1}{2}y$.

$\mathbf{T}^{-1}$ is the (counterclockwise) rotation through 120° about $(\sqrt{3}, 1)$.

19. Since **T** has a fixed point, it must be of type 6.3.18(a) or (c) (rotation or reflection). If **T** is a rotation, we can find **T** by the method of Example 6.2.20 (or by the method of Exercise 14 above); the equations of **T** are

$$x' = \frac{4}{5}x - \frac{3}{5}y + \frac{7}{5}; \qquad y' = \frac{3}{5}x + \frac{4}{5}y - \frac{1}{5}.$$

If instead **T** is a reflection $\mathcal{M}_L$, then $L$ must be the line through the points (1, 2) and $P$, where

$$P = \text{midpoint}\ ((4, 6), (1, 7)) = (5/2, 13/2)$$

(*why?*). The method of Exercise 16 is applicable, and we find that the equations of **T** are

$$x' = -\frac{4}{5}x + \frac{3}{5}y + \frac{3}{5}; \qquad y' = \frac{3}{5}x + \frac{4}{5}y - \frac{1}{5}.$$

20. *Solution when $\theta$, $\psi$ are both clockwise angles*: Draw rays $H_1$, $H_2$ with initial points $A$, $B$, respectively, such that the oriented angle from $H_1$ to $H(A, B)$ is $\theta/2$ and the oriented angle from $H(B, A)$ to $H_2$ is $\psi/2$. The intersection point $H_1 \cap H_2$ is fixed by the product $\mathcal{R}_{A,\theta}\mathcal{R}_{B,\psi}$ and, hence, must be the desired point $C$. (In Fig. G2,

$C \xrightarrow{\mathcal{R}_{A,\theta}} D \xrightarrow{\mathcal{R}_{B,\psi}} C$.)

25. $\mathbf{T} = \mathcal{R}_{O,\theta}$ where $\theta = \pm 60°$.

**Section 6.4**

3. Use 6.3.6 (applied to $W$ instead of $\mathbf{R}_2$, where $S$ has direction space $U = W^{\perp}$) together with 6.4.1.
4. By 6.4.4(a), the linear part of $\mathbf{T}$ is a rotation $\mathbf{T}^*$ about the direction space $U$ of $S$. If $\mathbf{T}^* = \mathbf{I}$, then $\mathbf{T}$ is a translation; in this case, since $\mathbf{T}$ has fixed points it follows that $\mathbf{T} = \mathbf{I}$, from which $\mathbf{M} = \mathbf{I}$ and the result is trivial. Hence, we may assume $\mathbf{T}^* \neq \mathbf{I}$. By 4.9.3, the linear part of $\mathbf{M} = \mathcal{T}_{-\mathbf{v}}\mathbf{T}\mathcal{T}_{\mathbf{v}}$ is $\mathbf{IT^*I} = \mathbf{T}^*$. Hence, by 6.4.4(b), $\mathbf{M}$ is a rotation about some flat $S'$ with direction space $U$. Moreover, since $\mathbf{T}$ fixes every point of $S$, $\mathbf{M} = \mathcal{T}_{-\mathbf{v}}\mathbf{T}\mathcal{T}_{\mathbf{v}}$ fixes every point of $S + \mathbf{v}$. It follows from Exercise 3 that $S + \mathbf{v} = S'$, which is the desired result.

**Section 6.5**

2. $x' = \frac{1}{8}[(2 + 3\sqrt{2})x + (2 + \sqrt{2})y + (-2 + 4\sqrt{2})z]$
   $y' = \frac{1}{8}[(-6 + \sqrt{2})x + (2 + 3\sqrt{2})y + 2z]$
   $z' = \frac{1}{8}[-2x + (2 - 4\sqrt{2})y + (4 + 2\sqrt{2})z]$.
3. One orthonormal basis for the given plane is $\{\mathbf{w}_1, \mathbf{w}_2\}$, where $\mathbf{w}_1 = (1/\sqrt{2}, -1/\sqrt{2}, 0)$, $\mathbf{w}_2 = (2/3, 2/3, 1/3)$; the vectors $\mathbf{w}_1$, $\mathbf{w}_2$ are fixed by $\mathcal{M}_W$, whereas $\mathbf{u} = \mathbf{w}_1 \times \mathbf{w}_2 = (-1/3\sqrt{2}, -1/3\sqrt{2}, 4/3\sqrt{2})$ is multiplied by $-1$. Hence, the matrix of $\mathcal{M}_W$ is $Q^tA'Q$, where the rows of $Q$ are $\mathbf{u}$, $\mathbf{w}_1$, $\mathbf{w}_2$ and $A'$ is a diagonal matrix with diagonal entries $-1, 1, 1$. Carrying out the computation, we find that $\mathcal{M}_W$ has equations

   $x' = \frac{1}{9}(8x - y + 4z)$; $y' = \frac{1}{9}(-x + 8y + 4z)$;
   $z' = \frac{1}{9}(4x + 4y - 7z)$.

Figure G2

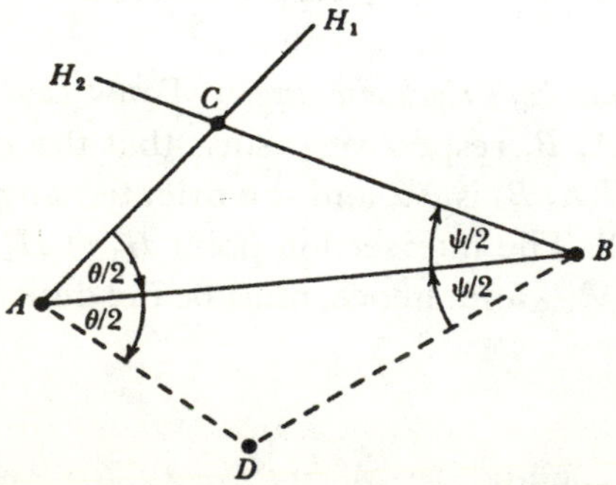

5. (a) By assumption, $\mathbf{M} = \mathbf{T}\mathcal{T}_{\mathbf{z}}$, where $\mathbf{T}$ is a rotation about some one-dimensional subspace $[\mathbf{u}]$. If $\mathbf{T} = \mathbf{I}$, the conclusion is immediate. Assume $\mathbf{T} \neq \mathbf{I}$. Writing $\mathbf{z} = k\mathbf{u} + \mathbf{w}$ ($k \in \mathbf{R}$, $\mathbf{w} \in [\mathbf{u}]^{\perp}$), Exercise 4 tells us that $\mathbf{T}\mathcal{T}_{\mathbf{w}}$ is a rotation $\mathcal{R}$ about some line $L$ having direction space $[\mathbf{u}]$. Hence,
$$\mathbf{M} = \mathbf{T}\mathcal{T}_{\mathbf{z}} = \mathbf{T}\mathcal{T}_{\mathbf{w}}\mathcal{T}_{k\mathbf{u}} = \mathcal{R}\mathcal{T}_{k\mathbf{u}}$$
and since $k\mathbf{u} \in [\mathbf{u}]$, $\mathcal{T}_{k\mathbf{u}}$ is in the direction of $L$.
(b) Here $\mathbf{M} = \mathcal{M}_W\mathcal{T}_{\mathbf{z}}$, where $W$ is a 2-dimensional subspace. Write $\mathbf{z} = \mathbf{w} + \mathbf{u}$, where $\mathbf{w} \in W$, $\mathbf{u} \in W^{\perp}$. If $\mathbf{v} = \mathbf{u}/2$ and $S = W + \mathbf{v}$, then Theorem 3.8.30 implies that $\mathcal{M}_W\mathcal{T}_{\mathbf{u}} = \mathcal{M}_S$. Hence,
$$\mathbf{M} = \mathcal{M}_W\mathcal{T}_{\mathbf{u}+\mathbf{w}} = \mathcal{M}_S\mathcal{T}_{\mathbf{w}}$$
and $\mathbf{w} \in W$ so that $\mathbf{w}$ is in the direction of $S$.
(c) Here $\mathbf{M} = \mathbf{T}\mathcal{T}_{\mathbf{z}} = \mathcal{R}\mathcal{M}_W\mathcal{T}_{\mathbf{z}}$, where $\mathcal{R}$ is a rotation about $[\mathbf{u}]$ and $W = [\mathbf{u}]^{\perp}$. By part (b), $\mathcal{M}_W\mathcal{T}_{\mathbf{z}} = \mathcal{M}_S\mathcal{T}_{\mathbf{w}}$ for some $\mathbf{w} \in W$ and some flat $S$ with direction space $W$. By Section 3.8, Exercise 17, $\mathcal{M}_S$ commutes with $\mathcal{T}_{\mathbf{w}}$; hence, $\mathbf{M} = \mathcal{R}\mathcal{M}_S\mathcal{T}_{\mathbf{w}} = (\mathcal{R}\mathcal{T}_{\mathbf{w}})\mathcal{M}_S$. Since $\mathcal{R}\mathcal{T}_{\mathbf{w}}$ is a rotation about a line $L$ parallel to $[\mathbf{u}]$ (Exercise 4), the result follows. (Note that $L$ and $S$ are indeed perpendicular; cf. Sec. 3.8, Exercise 19.)
6. (a) Let $\mathbf{T} \neq \mathbf{I}$ be a rotation about $L$; let $[\mathbf{u}]$ be the direction space of $L$, $W = [\mathbf{u}]^{\perp}$, $P = L \cap W$. By Section 6.3, Exercise 5, the rotation $\mathbf{T} \restriction W = \mathcal{R}$ is the product of two reflections of $W$ through lines $[\mathbf{w}_1] + P$, $[\mathbf{w}_2] + P$ ($\mathbf{w}_1, \mathbf{w}_2 \in W$). Now show that for any point $Q \in W$, the perpendicular segment from $Q$ to the line $[\mathbf{w}_i] + P$ is also the perpendicular segment from $Q$ to the plane $S_i = [\mathbf{u}, \mathbf{w}_i] + P$; deduce that the reflection of $W$ through $[\mathbf{w}_i] + P$ is the restriction to $W$ of the reflection of $\mathbf{R}_3$ through $S_i$ and, hence, that $\mathbf{T}$ and $\mathcal{M}_{S_1}\mathcal{M}_{S_2}$ agree on $W$. Also, $\mathbf{T}$ and $\mathcal{M}_{S_1}\mathcal{M}_{S_2}$ agree on $L$ since they both fix every point of $L$. It follows from Theorem 4.9.4 that $\mathbf{T} = \mathcal{M}_{S_1}\mathcal{M}_{S_2}$ (consider, e.g., the $b$-independent points $P$, $\mathbf{u} + P$, $\mathbf{w}_1, + P$, $\mathbf{w}_2 + P$).
*Alternate proof*: In the linear case, the matrix of $\mathbf{T}$ with respect to a suitable orthonormal basis $(\mathbf{u}, \mathbf{w}_1, \mathbf{w}_2)$ is the matrix 6.5.5, which can be written in the form
$$\begin{bmatrix} 1 & 0 & 0 \\ 0 & \cos 2\theta & \sin 2\theta \\ 0 & -\sin 2\theta & \cos 2\theta \end{bmatrix} = \begin{bmatrix} 1 & 0 & 0 \\ 0 & 1 & 0 \\ 0 & 0 & -1 \end{bmatrix} \begin{bmatrix} 1 & 0 & 0 \\ 0 & \cos 2\theta & \sin 2\theta \\ 0 & \sin 2\theta & -\cos 2\theta \end{bmatrix}$$
showing that $\mathbf{T}$ is the reflection through $[\mathbf{u}, \mathbf{w}_1]$ times the reflection through $[\mathbf{u}, (\cos\theta)\mathbf{w}_1 + (\sin\theta)\mathbf{w}_2]$. The nonlinear

case can be reduced to the linear case by using Section 6.4, Exercise 5, and Theorem 3.8.32.

(b) This assertion follows from Exercises 5 and 6(a) together with Section 3.8, Exercise 15.

9. *Suggestion*: Let $[\mathbf{u}]$ be the direction space of the given parallel axes, let $W = [\mathbf{u}]^{\perp}$, and apply the results of Section 6.3 (Theorem 6.3.16 and Exercise 11) to the rotations $\mathbf{T}_i \restriction W$. (Cf. 6.4.1.)

10. (a) Let $P$ be fixed by $\mathbf{T} = \mathbf{T}_1\mathbf{T}_2$. If $P \in L_1$, then $P = (P\mathbf{T}_1)\mathbf{T}_2 = P\mathbf{T}_2$; hence, $P$ is fixed by $\mathbf{T}_2$; hence, $P \in L_2$, R.A.A. Hence, $P \notin L_1$. Since $P$ and $P\mathbf{T}_1$ are equidistant from any point on $L_1$ (why?), it follows that the perpendicular bisector of the segment from $P$ to $P\mathbf{T}_1$ is a plane containing $L_1$ (see Sec. 3.7, Exercises 5(a) and 6(d)). Similarly, since $P\mathbf{T}_1$ is fixed by $\mathbf{T}_2\mathbf{T}_1$ (proof: $(P\mathbf{T}_1)(\mathbf{T}_2\mathbf{T}_1) = (P\mathbf{T}_1\mathbf{T}_2)\mathbf{T}_1 = P\mathbf{T}_1$), the same argument shows that the perpendicular bisector of the segment from $P\mathbf{T}_1$ to $(P\mathbf{T}_1)\mathbf{T}_2$ (i.e., from $P\mathbf{T}_1$ to $P$) is a plane containing $L_2$. But then the same plane contains two skew lines, R.A.A.

(b) By Theorem 6.4.4(a), the linear part of $\mathbf{T}_i$ is a rotation $\mathbf{T}_i^*$ about the direction space of $L_i$. If $\mathbf{T}_1\mathbf{T}_2$ is a translation, then $\mathbf{T}_1^*\mathbf{T}_2^* = \mathbf{I}$ (Theorem 4.9.3); hence, $\mathbf{T}_1^*$ and $\mathbf{T}_2^*$ are rotations about the same axis; hence, $L_1$ and $L_2$ have the same direction space and are parallel, R.A.A.

(c) Since $\mathbf{T}_1$, $\mathbf{T}_2$ each have determinant 1, so does $\mathbf{T}_1\mathbf{T}_2$. Now see Exercise 5.

**Section 7.1**

1. (a) Parabola; vertex is $\bar{P} = (8.76, -5.08)$; equation in the coordinate system $(\bar{P}; (1, 2)/\sqrt{5}, (2, -1)/\sqrt{5})$ is $5\bar{x}^2 - (2/\sqrt{5})\bar{y} = 0$; focus is $(8.8, -5.1)$.

(b) Hyperbola, with center at $P_0 = (1, 1/2)$; equation in the coordinate system $(P_0; (-1, 4)/\sqrt{17}, (4, 1)/\sqrt{17})$ is $16(y'')^2 - (x'')^2 = 4$. Foci are $(3, 1)$ and $(-1, 0)$; asymptotes are the lines $x = 1$, $15x + 8y = 19$.

(c) Ellipse; equation in coordinate system

$((-3/\sqrt{5}, 4/\sqrt{5}); (1, 2)/\sqrt{5}, (2, -1)/\sqrt{5})$

is $(x'')^2/4 + (y'')^2/9 = 1$.

(d) Equation in coordinate system $((-7/5, 1/5); (4/5, 3/5), (-3/5, 4/5))$ is $20(x'')^2 - 5(y'')^2 = 0$ or $y'' = \pm 2x''$. In $(x, y)$ coordinates, the graph consists of the two lines $11x + 2y + 15 = 0$, $x + 2y + 1 = 0$.

(e) Parabola with vertex $\bar{P} = (0, 1)$; equation in coordinate system $(\bar{P}; (1, -1)/\sqrt{2}, (1, 1)/\sqrt{2})$ is $\bar{x}^2 - 4\sqrt{2}\bar{y} = 0$; focus is $(1, 2)$.

(f) Ellipse with center at the origin, vertices (on major axis) at $\pm(\sqrt{3}, -\sqrt{3})$, foci at $\pm(\sqrt{2}, -\sqrt{2})$.

2. (a) $9x^2 + 16y^2 - 24xy - 370x - 340y + 1025 = 0$.
   (b) $91x^2 + 84y^2 - 24xy - 814x - 552y - 4361 = 0$.
3. (a) Hyperboloid of 2 sheets; center is $P_0 = (1/2, -1/2, -1/2)$; principal axis is the line $P_0 + [(1, 1, 1)]$.
   (b) Elliptic paraboloid; vertex is $P_0 = (-2, 1, 0)$; principal axis is the line $P_0 + [(1, -1, 0)]$.
   (c) Circular cone with vertex $(1, 1, 1)$, axis $(1, 1, 1) + [(2, -2, 1)]$.

**Section 7.2**

1. (a) Rank = 3. (b) Rank = 3; determinant = 27.
   (c) Determinant = $36^2$.
   (d) Rank = 2. (e) Rank = 4; determinant = −1. (f) Rank = 3.
2. *Hint*: It follows from 4.10.19(b) that $\mathscr{H}$ is actually the graph of equation 7.1.3 in some orthogonal coordinate system.

**Section 7.3**

1. (a) Ellipse. (b) Hyperbola. (c) Two intersecting lines.
   (d) Point. (e) Line. (f) Empty.
2. (a)

$$x^2 + y^2 + 4xy + 2x + 6y + (2/3)$$
$$= [x + (2 + \sqrt{3})y + (1 - 1/\sqrt{3})]\,[x + (2 - \sqrt{3})y + (1 + 1/\sqrt{3})];$$
$$4x^2 + 9y^2 - 12xy + 4x - 6y + 1 = (2x - 3y + 1)^2.$$

**Section 7.4**

1. (a) Hyperboloid of one sheet. (b) Hyperboloid of two sheets.
   (c) Elliptic cone. (d) Hyperboloid of two sheets.
   (e) Ellipsoid. (f) Hyperboloid of one sheet.
   (g) Elliptic cylinder. (h) Parabolic cylinder.
   (i) Two intersecting planes. (j) Empty. (k) Hyperbolic paraboloid.
2. Show that this is implied by the conditions on $\sigma_1, \sigma_2, \sigma_3$ (cf. Table 7.4.1).

**Section 7.5**

1. (a) Hyperbola. (b) Parabola.
2. One solution is $\sqrt{3}y - 3z - 8 = 0$.

**Section 7.6**

2. See Example 7.7.3 in Section 7.7.
4. (a) Use 4.9.8(d).
   (c) By 4.9.10(b), Ran $\mathbf{T} = H_2$; by 4.9.10(a), $\mathbf{T}^{-1} : H_2 \to H_1$ is affine and, hence, we obtain a mapping $(\mathbf{T}^{-1})^* : H_2^* \to H_1^*$. Now show that $(\mathbf{T}^{-1})^*\mathbf{T}^* = \mathbf{I}$ and $\mathbf{T}^*(\mathbf{T}^{-1})^* = \mathbf{I}$, and deduce the result.

**Section 7.7**

1. Proof that the left side is contained in the right side: let $P \in \bigcap_\alpha (W_\alpha\phi)$. By 7.7.2(a), we can write $\{P\} = [\mathbf{u}]\phi$, where $[\mathbf{u}]$ is a one-dimensional subspace of $V'$. Then $[\mathbf{u}]\phi = \{P\} \subseteq W_\alpha\phi$ for all $\alpha$; hence, $[\mathbf{u}] \subseteq W_\alpha$ for all $\alpha$; hence, $[\mathbf{u}] \subseteq \bigcap_\alpha W_\alpha$; hence, $\{P\} = [\mathbf{u}]\phi \subseteq \left(\bigcap_\alpha W_\alpha\right)\phi$.
5. Let $S_i^* = U_i\phi$ as in 7.7.2, where the $U_i$ are subspaces of $V'$; let $d = n - \Sigma\, e_i$. If Theorem 3.4.5 is applied to $(U_i, V')$ in place of $(W_i, V)$, the numbers $e_i$ remain the same since $(n + 1) - \dim U_i = n - \dim W_i$. Thus, 3.4.6 gives

   $$\dim(U_1 \cap \cdots \cap U_k) \geqslant n + 1 - \sum_{i=1}^{k} e_i = d + 1.$$

   By 7.7.5, $S_1^* \cap \cdots \cap S_k^* = (U_1 \cap \cdots \cap U_k)\phi$; hence, $\dim(S_1^* \cap \cdots \cap S_k^*) \geqslant d$.

   Moreover, $S_1^* \cap \cdots \cap S_k^* = (S_1 \cap \cdots \cap S_k) \cup (\mathcal{P}_{S_1} \cap \cdots \cap \mathcal{P}_{S_k})$; since $S_1 \cap \cdots \cap S_k = \phi$, it follows that $\mathcal{P}_{S_1} \cap \cdots \cap \mathcal{P}_{S_k}$ is a flat at infinity of dimension $\geqslant d$. Hence, $\mathcal{P}_{S_1} \cap \cdots \cap \mathcal{P}_{S_k} = \mathcal{P}_S$ for some flat $S \subseteq V$ of dimension $\geqslant d + 1$. By 7.6.6, $S$ is parallel to each $S_i$; hence, the direction space of $S$ is contained in each $W_i$ and, hence, in $W_1 \cap \cdots \cap W_k$. It follows that $\dim(W_1 \cap \cdots \cap W_k) \geqslant d + 1$, q.e.d.

**Section 8.1**

1. *Hint*: See Section 4.7, Exercise 6.

**Section 8.2**

1. Since $\mathbf{T}$ is nilpotent, Ran $\mathbf{T}^j = \{\mathbf{0}\}$ for all sufficiently large $j$. If for any $k$, we have Ran $\mathbf{T}^k =$ Ran $\mathbf{T}^{k+1}$, then by Section 8.1, Exercise 2, we have Ran $\mathbf{T}^k =$ Ran $\mathbf{T}^j$ for all $j > k$, and hence Ran $\mathbf{T}^k = \{\mathbf{0}\}$, $\mathbf{T}^k = \mathbf{O}$, $k \geqslant m$. It follows that all of the inclusions in the sequence 8.1.2 are *proper* until we come to Ran $\mathbf{T}^m = \{\mathbf{0}\}$. In particular, if $n = \dim V$, then

$$n > \text{rank } \mathbf{T} > \text{rank } \mathbf{T}^2 > \cdots > \text{rank } \mathbf{T}^m = 0$$

from which we must have $m \leqslant n$.

3. (a) $a = -1/2$. (b) $a = -1$, $b = -2$. (c) $a = b = -2$, $c = 2$. (d) $c = 0$, $a = -3b$, $b$ is arbitrary.

5. (a) If $(\mathbf{v}_1, \ldots, \mathbf{v}_n)$ is a $\mathbf{T}$-cyclic basis, then

$$\begin{aligned} V\mathbf{T}^k &= [\mathbf{v}_1\mathbf{T}^k, \ldots, \mathbf{v}_{n-k}\mathbf{T}^k, \mathbf{v}_{n-k+1}\mathbf{T}^k, \ldots, \mathbf{v}_n\mathbf{T}^k] \\ &= [\mathbf{v}_{k+1}, \ldots, \mathbf{v}_n, \mathbf{0}, \ldots, \mathbf{0}] = [\mathbf{v}_{k+1}, \ldots, \mathbf{v}_n] \end{aligned}$$

and the assertion follows.

(b) Since $(V\mathbf{T}^k)\mathbf{T}^{n-k} = V\mathbf{T}^n = \{\mathbf{0}\}$, it follows that $V\mathbf{T}^k \subseteq$ Nul $\mathbf{T}^{n-k}$. But by part (a), we have

$$\dim V\mathbf{T}^k = n - k = n - \text{rank } \mathbf{T}^{n-k} = \dim \text{Nul } \mathbf{T}^{n-k}.$$

Hence, $V\mathbf{T}^k =$ Nul $\mathbf{T}^{n-k}$, by Theorem 2.6.16(b).

(c) By part (d) of the preceding exercise, Nul $\mathbf{T} \neq \{\mathbf{0}\}$ and, hence, $\mathbf{u}_n$ exists. Suppose that $\mathbf{u}_n, \ldots, \mathbf{u}_{k+1}$ exist where $k + 1 \geqslant 2$. To show that $\mathbf{u}_k$ exists (thus completing the proof by backward induction), we observe that $\mathbf{u}_{k+1} \in$ Nul $\mathbf{T}^{n-k}$ by the definition of $\mathbf{u}_n, \ldots, \mathbf{u}_{k+1}$. Since Nul $\mathbf{T}^{n-k} =$ Ran $\mathbf{T}^k$ by part (b), it follows that $\mathbf{u}_{k+1} \in$ Ran $\mathbf{T}^k \subseteq$ Ran $\mathbf{T}$ (by 8.1.2); hence, $\mathbf{u}_{k+1}$ has a pre-image under $\mathbf{T}$, that is, $\mathbf{u}_k$ exists.

7. (i) $r = 1$, $W_1 = \mathbf{R}_2$. (ii) $r = 2$, dimensions are 1, 2. (iii) $r = 1$, $W_1 = \mathbf{R}_3$. (iv) $r = 2$, dimensions are 2, 2.
(v) (Solution) under $\mathbf{T}$, we have

$$\begin{array}{|c|cl|c|cl|c|cl}
\hline
\mathbf{e}_1 & \longrightarrow & \fbox{(-1, 1, -1, 0)} & \longrightarrow & \fbox{(0, 2, -1, 1)} & \longrightarrow & \mathbf{0} \\
\mathbf{e}_2 & \longrightarrow & \fbox{(-1, -3, 1, -2)} & \longrightarrow & (0, 2, -1, 1) & \longrightarrow & \mathbf{0} \\
\mathbf{e}_3 & \longrightarrow & (0, -6, 3, -3) & \longrightarrow & (0, 0, 0, 0) & \longrightarrow & \mathbf{0} \\
\mathbf{e}_4 & \longrightarrow & (2, 0, 1, 1) & \longrightarrow & (0, -4, 2, -2) & \longrightarrow & \mathbf{0}. \\
\hline
\end{array}$$

The successive sets $S_k$ described in steps (a) and (b) of the algorithm 8.2.18 are enclosed in boxes. Hence,

Ran $\mathbf{T}^2 = [\,(0, 2, -1, 1)]$

Ran $\mathbf{T} = [\,(-1, 1, -1, 0), (0, 2, -1, 1)]$

$V = \mathbf{R}_4 = [\mathbf{e}_1, (-1, 1, -1, 0), (0, 2, -1, 1)] \oplus [\mathbf{w}]$

where in each case we use a dimension argument to obtain the number of additional direct summands (none in going from Ran $\mathbf{T}^2$ to Ran $\mathbf{T}$, one in going from Ran $\mathbf{T}$ to $\mathbf{R}_4$). To find the vector $\mathbf{w}$, we apply the method of the proof of Theorem 8.2.3: first extend $[\mathbf{e}_1, (-1, 1, -1, 0), (0, 2, -1, 1)]$ to a basis of $\mathbf{R}_4$ by adjoining one more vector (the vector $\mathbf{v} = \mathbf{e}_2$ is a possible choice here); then write $\mathbf{vT}$ in terms of the basis already found for Ran $\mathbf{T}$, in this case,

$$\mathbf{vT} = \mathbf{e}_2\mathbf{T} = (-1, -3, 1, -2) = (-1, 1, -1, 0) - 2(0, 2, -1, 1)$$
$$= \mathbf{e}_1\mathbf{T} - 2(-1, 1, -1, 0)\mathbf{T}.$$

Hence, by 8.2.17, we can choose

$$\mathbf{w} = \mathbf{v} - \mathbf{e}_1 + 2(-1, 1, -1, 0) = (-3, 3, -2, 0).$$

Our final expression is, thus,

$$\mathbf{R}_4 = [\,(1, 0, 0, 0), (-1, 1, -1, 0), (0, 2, -1, 1)\,] \oplus [\,(-3, 3, -2, 0)\,].$$

(vi) $r = 2$; dimensions are 2, 3.

8. (a) $(A^k)_{ij} = 0$, whenever $i + k > j$.
9. Let $A$ be the matrix of a linear transformation $\mathbf{T} : V \rightarrow V$ with respect to an ordered basis $(\mathbf{u}_1, ..., \mathbf{u}_n)$ of $V$. The condition that $a_{ij} = 0$ whenever $j \leqslant i$ means that $\mathbf{u}_i\mathbf{T} \in [\mathbf{u}_{i+1}, ..., \mathbf{u}_n]$. Hence, $\mathbf{u}_i\mathbf{T}^2 \in [\mathbf{u}_{i+1}\mathbf{T}, ..., \mathbf{u}_n\mathbf{T}] \subseteq [\mathbf{u}_{i+2}, ..., \mathbf{u}_n]$, etc.; in general, $\mathbf{u}_i\mathbf{T}^k \in [\mathbf{u}_{i+k}, ..., \mathbf{u}_n]$, which means that $(A^k)_{ij} = 0$ whenever $j < i + k$.

**Section 8.3**

1. $\lambda = 7$: multiplicity = 2; eigenspace = $[(1, 2, 4)]$; CH$[\mathbf{T}, 7]$ is the plane $-2x + 17y - 8z = 0$; Ran$(\mathbf{T} - 7\mathbf{I})^2 = [(1, 1, 1)]$.
   $\lambda = 14$: multiplicity = 1; eigenspace = CH$[\mathbf{T}, 14] = [\,(1, 1, 1)\,]$; Ran$(\mathbf{T} - 14\mathbf{I})$ is the plane $-2x + 17y - 8z = 0$.
2. $\lambda = 2$: multiplicity = 2; eigenspace = $[(0, 1, -1, 1)]$; CH$[\mathbf{T}, 2] = \{\,(a, b, 2a - b, b - a): a, b \in \mathbf{R}\}$; Ran$(\mathbf{T} - 2\mathbf{I})^2 = [(1, 2, -2, 2), (1, 1, 0, 1)]$.
   $\lambda = -1$: multiplicity = 2; eigenspace = CH$[\mathbf{T}, -1] = [(1, 2, -2, 2), (1, 1, 0, 1)]$; Ran$(\mathbf{T} + \mathbf{I})^2 = [(0, -1, 1, -1), (3, 7, -1, 4)] = \{\,(a, b, 2a - b, b - a): a, b \in \mathbf{R}\}$.
3. $\lambda = -1$ is the only eigenvalue; multiplicity = 2; eigenspace = $[\mathbf{e}_2]$; CH$[\mathbf{T}, -1] = \{\,(a, b, -a, 0) : a, b \in \mathbf{R}\}$; Ran$(\mathbf{T} + \mathbf{I})^2 = [(1, 0, 1, -1), \mathbf{e}_3]$.
4. $\lambda = 2$: multiplicity = 1; eigenspace = CH$[\mathbf{T}, 2] = [\mathbf{e}_2]$; Ran(**T**

$-$ **2I**) is the hyperplane $\{(x_1, x_2, x_3, x_4) \in \mathbf{R}_4 : x_1 + x_2 = 0\}$.
$\lambda = 1$: multiplicity $= 1$; eigenspace $=$ CH[**T**, 1] $= [(0, 0, 1, 1)]$; Ran(**T** $-$ **I**) is the hyperplane $\{(x_1, x_2, x_3, x_4) \in \mathbf{R}_4 : x_1 = x_4\}$.

5. $\lambda = 2$: multiplicity $= 1$; eigenspace $=$ CH[**T**, 2] $= [(-5, 1, 2)]$; Ran(**T** $-$ 2**I**) is the plane $x + z = 0$.
$\lambda = -1$: multiplicity $= 1$; eigenspace $=$ CH[**T**, $-$ 1] $= [(1, 1, -1)]$; Ran(**T** + **I**) is the plane $x - 3y + 4z = 0$.
$\lambda = 3$: multiplicity $= 1$; eigenspace $=$ CH[**T**, 3] $= [(-1, 1, 1)]$; Ran(**T** $-$ 3**I**) is the plane $x + y + 2z = 0$.

**Section 8.4**

1. By 8.1.3, $U_i \subseteq$ CH[**T**, $\lambda_i$]; by 8.4.3, the subspaces CH[**T**, $\lambda_i$] are independent and, hence, so are the $U_i$.
3. The index of nilpotence of $(\mathbf{T} - \lambda_i\mathbf{I}) \restriction N^{(i)}$ is the degree of the largest subblock of $J_i$ (cf. Theorem 8.2.19); the rank of $\mathbf{T} - \lambda_i\mathbf{I}$ determines the number of subblocks (how?). Since an integer $m_i \leq 6$ cannot be expressed in two different ways as a (nonordered) sum of positive integers having a given number of summands and a given largest summand (verify this!), the result follows. For $m_i = 7$, a counterexample is
$7 = 3 + 3 + 1 = 3 + 2 + 2.$
4. If $\{\mathbf{u}_1, ..., \mathbf{u}_n\}$ is a basis of $V$, any maximal independent subset of the set $\{\mathbf{u}_1\mathbf{U}_1 \cdots \mathbf{U}_s, ..., \mathbf{u}_n\mathbf{U}_1 \cdots \mathbf{U}_s\}$ is a basis of $V^*$. Once such a basis is known, $C$ can be computed since $C$ is the matrix of $\mathbf{T} \restriction V^*$.
5. (b) char(**T**; $x$) $= x^4 + 2x^2 + 1 = (x - i)^2 (x + i)^2$;

$$J = \left[\begin{array}{cc|cc} i & 1 & 0 & 0 \\ 0 & i & 0 & 0 \\ \hline 0 & 0 & -i & 1 \\ 0 & 0 & 0 & -i \end{array}\right].$$

(e) char(**T**; $x$) $= (x - 2)^4$;

$$J = \left[\begin{array}{ccc|c} 2 & 1 & 0 & 0 \\ 0 & 2 & 1 & 0 \\ 0 & 0 & 2 & 0 \\ \hline 0 & 0 & 0 & 2 \end{array}\right].$$

**Section 9.5**

2. Let $n$ be the degree of $A$ and let $\mathbf{T} \in \text{Lin}(\mathbf{C}_n, \mathbf{C}_n)$ have matrix $A$ with respect to the canonical basis. By 9.5.7, **T** has ortho-

normal eigenvectors $\mathbf{u}_1, \ldots, \mathbf{u}_n$; say, $\mathbf{u}_i\mathbf{T} = \lambda_i\mathbf{u}_i$. By 9.5.6, all of the $\lambda_i$ are real. With respect to the basis $(\mathbf{u}_1, \ldots, \mathbf{u}_n)$, the matrix $A'$ of $\mathbf{T}$ is diagonal, with (real) diagonal entries $\lambda_1, \ldots, \lambda_n$. If $Q$ is the matrix whose row vectors are $\mathbf{u}_1, \ldots, \mathbf{u}_n$, the results of Section 4.8 imply that $A' = P^{-1}AP$, where $P = Q^{-1}$. Moreover, since the rows of $Q$ are orthonormal, $Q^{-1} = \overline{Q}^t$ (Theorem 9.3.3); since $P = Q^{-1}$, it follows that $P^{-1} = \overline{P}^t$ (why?).

If $A$ has real entries, choose $\mathbf{T}$ to be a linear transformation of $\mathbf{R}_n$ rather than $\mathbf{C}_n$. The vectors $\mathbf{u}_i$ then belong to $\mathbf{R}_n$ so that $Q$, hence also $P$, has real entries.

3. (a) One such matrix is

$$P = \begin{bmatrix} 0 & 2i/\sqrt{6} & -i/\sqrt{3} \\ 1/\sqrt{2} & 1/\sqrt{6} & 1/\sqrt{3} \\ -1/\sqrt{2} & 1/\sqrt{6} & 1/\sqrt{3} \end{bmatrix}.$$

(The conjugates of the columns are orthonormal eigenvectors of $A$; cf. the solution to the preceding exercise.) The first two columns of $P$ could be replaced by any other two orthonormal columns which span the same subspace of $\mathbf{C}_3$.

(b) One such matrix is $P = \dfrac{1}{3}\begin{bmatrix} 1 & -2 & 2 \\ 2 & 2 & 1 \\ -2 & 1 & 2 \end{bmatrix}$.

5. *Hint*: Use the result of Exercise 4.

**Section 9.7**

2. (a) The proof of (e) $\Rightarrow$ (a) goes as follows: if $\lambda < 0$, then each summand of $c_0\lambda^0 + \cdots + c_n\lambda^n$ is nonnegative and $c_n\lambda^n \neq 0$, so that char$(A; \lambda) \neq 0$, R.A.A.

(b) See part (iii) of the next exercise.

3. (i) Not positive semidefinite since $\lambda = -\sqrt{3}$ is a negative eigenvalue. Since $X = (1, -2-\sqrt{3}, 1+\sqrt{3})$ is a $\lambda$-eigenvector, Theorem 9.6.6 implies that $f(X) = \lambda|X|^2 < 0$; this may also be verified by direct computation of $XA\overline{X}^t = -18 - 12\sqrt{3}$ using matrix multiplication.

(ii) Positive definite.

(iii) Not positive semidefinite. (Even though the analog to condition (f) of Theorem 9.7.4 holds, the other conditions do not.) One counterexample is $f(1, 2, -3) = -56$.

(iv) Not positive semidefinite; $f(-1 + i\sqrt{3}, 1 + i\sqrt{3}, 2) = -12\sqrt{3}$.

(v) Positive definite.

(vi) Positive semidefinite.

4. (a) Condition 9.7.4(f) holds $\Leftrightarrow \det A > 0 \Leftrightarrow x > 29$.

(b) $x \geq 4$.

6. Suppose only $\lambda_r$ (say) is nonzero. For each $k$, let $F_k = E_k - \{r\} = \{m \in \mathbf{Z} : 1 \leq m \leq k, m \neq r\}$. It is easy to see (using, say, the expansion by minors) that

$$\det(A + B)_{E_k} = \begin{Bmatrix} \det A_{E_k} & (k < r) \\ \det A_{E_k} + \lambda_r \det A_{F_k} & (k \geq r) \end{Bmatrix}.$$

Since the right side is positive (Theorem 9.7.4(c)), so is the left side and, hence, $A + B$ is positive definite by 9.7.4(f). In addition, the case $k = n$ gives

$$\det(A + B) = \det A + \lambda_r \det A_{F_n} > \det A.$$

The general case (any number of nonzero $\lambda$'s) may be derived from the special case by induction on the number of nonzero $\lambda$'s.

7. Use Theorem 9.7.3(b) together with Section 9.5, Exercise 2.

**Section 9.8**

1. Compare with Theorem 9.7.4(e).

3. (a) Absolute minimum at $(0, -1)$.

(b) Relative maximum at $(-2, -2)$.

(c) Relative minimum at $(-1/\sqrt{5}, 2/\sqrt{5})$; relative maximum at $(1/\sqrt{5}, -2/\sqrt{5})$. It is possible to show that these extrema are in fact absolute extrema. (Use the fact that $F(x, y) \to 0$ as $x^2 + y^2 \to \infty$ and that a continuous function on a closed disk takes on an absolute maximum value and an absolute minimum value.)

(d) Absolute minimum at $(1, 2, -3)$.

(e) Relative minima at $(1/2, -3, 1/4)$ and $(-1/2, -3, -1/4)$. Again, it is possible to show that these minima are absolute, although this cannot be deduced from the matrices alone.

**Appendix E**

2. Answer: 25. The answer may be interpreted as the square of the shortest distance from the point $(0, 5)$ to the curve $x^2 = 16y$.

# References


1. Birkhoff, G., and MacLane, S., *A Survey of Modern Algebra*. New York: Macmillan, 1947.
2. Buck, R. C., *Advanced Calculus*, Second edition. New York: McGraw-Hill, 1965.
3. Courant, R., *Differential and Integral Calculus*, Volume 2. London: Blackie and Son, Ltd., 1936.
4. Gel'fand, I. M., *Lectures on Linear Algebra*, Interscience Tracts in Pure and Applied Mathematics, No. 9. New York: Wiley, 1961.
5. Hall, M., *The Theory of Groups*. New York: Macmillan, 1959.
6. Hamilton, N. T., and Landin, I., *Set Theory and the Structure of Arithmetic*. Boston: Allyn and Bacon, 1961.
7. Hausner, M., *A Vector Space Approach to Geometry*. New York: Prentice-Hall, 1965.
8. Johnson, R. E., and Kiokemeister, F. L., *Calculus with Analytic Geometry*, Second edition. Boston: Allyn and Bacon, 1960.
9. LeVeque, W. J., *Topics in Number Theory*, Volume II. Reading, Mass.: Addison-Wesley, 1956.
10. McCoy, N., *Introduction to Modern Algebra*, Third edition. Boston: Allyn and Bacon, 1975.
11. Mostow, G. D., Sampson, J. H., and Meyer, J.-P., *Fundamental Structures of Algebra*. New York: McGraw-Hill, 1963.
12. Pedoe, D., *An Introduction to Projective Geometry*. New York: Macmillan, 1963.
13. Protter, M. H., and Morrey, C. B., *College Calculus with Analytic Geometry*, Third edition. Reading, Mass.: Addison-Wesley, 1977.
14. Stoll, R., *Set Theory and Logic*. San Francisco: Freeman, 1961.
15. Strang, G., *Linear Algebra and Its Applications*. New York: Academic Press, 1976.

# Index of notation

| I | Alphabetic (specific) (English and Greek) | *page* |
|---|---|---|
| ANS | Answer appears in Appendix G | xii |
| **Ar** | Arrow (vector) | 119 |
| **C** | Complex numbers | 56 |
| CH | Characteristic subspace | 502 |
| char | Characteristic (of a field) | 151 |
| | Characteristic polynomial | 350 |
| cos | Cosine | 199,200,202,382 |
| CR | Characteristic range | 505 |
| $d(...)$ | Distance | 121,174,524 |
| deg | Degree (of a polynomial) | 61 |
| det | Determinant | 295 |
| dim | Dimension | 99 |
| $\mathbf{e}_1, ..., \mathbf{e}_n$ | Canonical basis (of $F_n$) | 97 |
| $E_n$ | Set of integers from 1 to $n$ | 52,288,354 |
| $H(P, Q)$ | Half line from $P$ through $Q$ | 201 |
| Ha | Set of half lines (with given initial point) | 370 |
| $I, I_n$ | Identity matrix | 240 |
| $\mathbf{I}, \mathbf{I}_A$ | Identity mapping | 8 |
| Im | Imaginary part (of a complex number) | 57 |
| **Im** | Imaginary part (of a vector) | 523 |
| ind | Index (of a permutation) | 291 |
| $J_G$ | Jacobian matrix of (the function) $G$ | 566 |
| $L(A, B)$ | Line through points $A$, $B$ | 125 |
| Lin $(V, W)$ | Set of linear transformations from $V$ into $W$ | 228 |
| $\mathcal{M}_S$ | Reflection through $S$ | 193 |
| Mat $(m, n; F)$ | Set of $m \times n$ matrices over $F$ | 235 |
| Nul | Nullspace | 245 |
| $O$ | Origin | 119 |

## II Alphabetic (nonspecific)

## III Numerical



## IV Symbols

# Index